The Chemistry Maths Book

Second Edition

Erich Steiner

University of Exeter

without mathematics the sciences cannot be understood, nor made clear, nor taught, nor learned.

(Roger Bacon, 1214 – 1292)

OXFORD
UNIVERSITY PRESS

OXFORD
UNIVERSITY PRESS

Great Clarendon Street, Oxford OX2 6DP

Oxford University Press is a department of the University of Oxford.
It furthers the University's objective of excellence in research, scholarship,
and education by publishing worldwide in

Oxford New York

Auckland Cape Town Dar es Salaam Hong Kong Karachi
Kuala Lumpur Madrid Melbourne Mexico City Nairobi
New Delhi Shanghai Taipei Toronto

With offices in

Argentina Austria Brazil Chile Czech Republic France Greece
Guatemala Hungary Italy Japan Poland Portugal Singapore
South Korea Switzerland Thailand Turkey Ukraine Vietnam

Oxford is a registered trade mark of Oxford University Press
in the UK and in certain other countries

Published in the United States
by Oxford University Press Inc., New York

British Library Cataloguing in Publication Data

Data available

Library of Congress Cataloging in Publication Data

Data available

Typeset by Graphicraft Limited, Hong Kong
Printed in Great Britain on acid-free paper by
Ashford Colour Press Ltd, Gosport, Hampshire

ISBN 978–0–19–920535–6

7 9 10 8

Preface

This book describes the mathematics required for the full range of topics that make up a university degree course in chemistry. It has been designed as a textbook for courses in 'mathematics for chemists'.

Structure of the book

The subject is developed in a logical and consistent way with few assumptions made of prior knowledge of mathematics. The material is organized in three largely independent parts: Chapters 1 to 15 on algebra, the calculus, differential equations, and expansions in series; Chapters 16 to 19 on vectors, determinants, and matrices; Chapters 20 and 21 are introductions to the big topics of numerical analysis and statistics.

A feature of the book is the extensive use of examples to illustrate every important concept and method in the text. Some of these examples have also been used to demonstrate applications of the mathematics in chemistry and several basic concepts in physics. The exercises at the end of each chapter are an essential element of the development of the subject, and have been designed to give the student a working understanding of the material in the text. The text is accompanied by a 'footnote history' of mathematics.

Several topics in chemistry are given extended treatments. These include the concept of pressure–volume work in thermodynamics in Chapter 5, periodic systems in Chapter 8, the differential equations of chemical kinetics in Chapter 11, and several applications of the Schrödinger equation in Chapters 12 and 14. In addition, the contents of several chapters are largely dictated by their applications in the physical sciences: Chapter 9, the mathematics of thermodynamics; Chapters 10 and 16, the description of systems and processes in three dimensions; Chapter 13 (advanced), some important differential equations and special functions in mathematical chemistry and physics; Chapter 15 (advanced), intermolecular forces, wave analysis, and Fourier transform spectroscopy; Chapters 18 and 19, molecular symmetry and symmetry operations, molecular orbital theory, molecular dynamics, and advanced quantum mechanics.

Global changes in this edition

1. An overall reorganization has been carried out to link the text and examples more closely to the exercises at the end of each chapter. The symbol ➤ has been placed at appropriate places within the body of the text as a pointer to the relevant exercises. New examples and exercises have been inserted to give a more complete coverage of the development of the mathematics.
2. In addition to the solutions to the numerical exercises given at the back of the book, a full set of worked solutions of the end-of-chapter exercises has been placed on the book's companion website at www.oxfordtextbooks.co.uk/orc/steiner2e
3. The opportunity has been taken to consolidate the several major and many minor corrections and improvements that have been made over the years since publication of the first edition in 1996. A small number of new historical footnotes have been

added to accompany new material. The material within some chapters has been reordered to make the development of the subject more logical.

Other principal changes

Chapter 1. A new section, *Factorization, factors, and factorials*, fills a gap in the coverage of elementary topics. The *rules of precedence for arithmetic operations* has been brought forward from chapter 2 and extended with examples and exercises, providing further revision and practice of the arithmetic that is so important for the understanding of the material in subsequent chapters. The biggest change in the chapter, reflected in the change of title to *Numbers, variables, and units*, is a rewritten and much enlarged section on units to make it a more authoritative and useful account of this important but often neglected topic. It includes new examples of the type met in the physical sciences, a brief subsection on dimensional analysis, and a new example and exercise on the structure of atomic units.

Chapter 2. Parts of the chapter have been rewritten to accommodate more discussion of the factorization and manipulation of algebraic expressions.

Chapter 7. Numerous small changes have been made, including an introduction to the multinomial expansion, and revision of the discussion of the Taylor series.

Chapter 9. Section 9.8 has been rewritten to clarify the relevance of line integrals to change of state in thermodynamics.

Chapter 13. The section on the Frobenius method has been revised, with new and more demanding examples and exercises.

Chapter 19. Sections 19.2 and 19.3 on eigenvalues and eigenvectors have been rewritten, with new examples and exercises, to improve the flow and clarity of the discussion.

Acknowledgements

I wish to express my gratitude to colleagues and students at Exeter University and other institutions for their often helpful comments on the previous edition of the book, for pointing out errors and obscurities, and for their suggestions for improvements. I particularly want to thank Anthony Legon for his valuable comments on the revised chapters 1 and 9.

I also wish to thank the reviewers of this book for their generous response to the proposal of a second edition, and the staff of Oxford University Press for their patience and help.

Above all, I want to thank my wife Mary, without whom nothing could have been done.

Erich Steiner, Exeter, June 2007

Contents

1	**Numbers, variables, and units**	**1**
	1.1 Concepts	1
	1.2 Real numbers	3
	1.3 Factorization, factors, and factorials	7
	1.4 Decimal representation of numbers	9
	1.5 Variables	13
	1.6 The algebra of real numbers	14
	1.7 Complex numbers	19
	1.8 Units	19
	1.9 Exercises	29
2	**Algebraic functions**	**31**
	2.1 Concepts	31
	2.2 Graphical representation of functions	32
	2.3 Factorization and simplification of expressions	34
	2.4 Inverse functions	37
	2.5 Polynomials	40
	2.6 Rational functions	50
	2.7 Partial fractions	52
	2.8 Solution of simultaneous equations	55
	2.9 Exercises	58
3	**Transcendental functions**	**62**
	3.1 Concepts	62
	3.2 Trigonometric functions	63
	3.3 Inverse trigonometric functions	72
	3.4 Trigonometric relations	73
	3.5 Polar coordinates	77
	3.6 The exponential function	80
	3.7 The logarithmic function	83
	3.8 Values of exponential and logarithmic functions	86
	3.9 Hyperbolic functions	87
	3.10 Exercises	89
4	**Differentiation**	**93**
	4.1 Concepts	93
	4.2 The process of differentiation	94
	4.3 Continuity	97
	4.4 Limits	98
	4.5 Differentiation from first principles	100
	4.6 Differentiation by rule	102
	4.7 Implicit functions	110

4.8	Logarithmic differentiation	111
4.9	Successive differentiation	113
4.10	Stationary points	114
4.11	Linear and angular motion	118
4.12	The differential	119
4.13	Exercises	122

5	**Integration**	**126**
5.1	Concepts	126
5.2	The indefinite integral	127
5.3	The definite integral	132
5.4	The integral calculus	142
5.5	Uses of the integral calculus	147
5.6	Static properties of matter	148
5.7	Dynamics	152
5.8	Pressure–volume work	157
5.9	Exercises	160

6	**Methods of integration**	**163**
6.1	Concepts	163
6.2	The use of trigonometric relations	163
6.3	The method of substitution	165
6.4	Integration by parts	173
6.5	Reduction formulas	176
6.6	Rational integrands. The method of partial fractions	179
6.7	Parametric differentiation of integrals	184
6.8	Exercises	187

7	**Sequences and series**	**191**
7.1	Concepts	191
7.2	Sequences	191
7.3	Finite series	196
7.4	Infinite series	203
7.5	Tests of convergence	204
7.6	MacLaurin and Taylor series	208
7.7	Approximate values and limits	214
7.8	Operations with power series	219
7.9	Exercises	221

8	**Complex numbers**	**225**
8.1	Concepts	225
8.2	Algebra of complex numbers	226
8.3	Graphical representation	228
8.4	Complex functions	235
8.5	Euler's formula	236
8.6	Periodicity	240

| | 8.7 | Evaluation of integrals | 244 |
| | 8.8 | Exercises | 245 |

9	**Functions of several variables**		**247**
	9.1	Concepts	247
	9.2	Graphical representation	248
	9.3	Partial differentiation	249
	9.4	Stationary points	253
	9.5	The total differential	258
	9.6	Some differential properties	262
	9.7	Exact differentials	272
	9.8	Line integrals	275
	9.9	Multiple integrals	281
	9.10	The double integral	283
	9.11	Change of variables	285
	9.12	Exercises	289

10	**Functions in 3 dimensions**		**294**
	10.1	Concepts	294
	10.2	Spherical polar coordinates	294
	10.3	Functions of position	296
	10.4	Volume integrals	299
	10.5	The Laplacian operator	304
	10.6	Other coordinate systems	307
	10.7	Exercises	312

11	**First-order differential equations**		**314**
	11.1	Concepts	314
	11.2	Solution of a differential equation	315
	11.3	Separable equations	318
	11.4	Separable equations in chemical kinetics	322
	11.5	First-order linear equations	328
	11.6	An example of linear equations in chemical kinetics	330
	11.7	Electric circuits	332
	11.8	Exercises	334

12	**Second-order differential equations. Constant coefficients**		**337**
	12.1	Concepts	337
	12.2	Homogeneous linear equations	337
	12.3	The general solution	340
	12.4	Particular solutions	344
	12.5	The harmonic oscillator	348
	12.6	The particle in a one-dimensional box	352
	12.7	The particle in a ring	356
	12.8	Inhomogeneous linear equations	359
	12.9	Forced oscillations	363
	12.10	Exercises	365

13   Second-order differential equations. Some special functions   **368**

   13.1   Concepts   368
   13.2   The power-series method   369
   13.3   The Frobenius method   371
   13.4   The Legendre equation   375
   13.5   The Hermite equation   381
   13.6   The Laguerre equation   384
   13.7   Bessel functions   385
   13.8   Exercises   389

14   Partial differential equations   **391**

   14.1   Concepts   391
   14.2   General solutions   392
   14.3   Separation of variables   393
   14.4   The particle in a rectangular box   395
   14.5   The particle in a circular box   398
   14.6   The hydrogen atom   401
   14.7   The vibrating string   410
   14.8   Exercises   413

15   Orthogonal expansions. Fourier analysis   **416**

   15.1   Concepts   416
   15.2   Orthogonal expansions   416
   15.3   Two expansions in Legendre polynomials   421
   15.4   Fourier series   425
   15.5   The vibrating string   432
   15.6   Fourier transforms   433
   15.7   Exercises   441

16   Vectors   **444**

   16.1   Concepts   444
   16.2   Vector algebra   445
   16.3   Components of vectors   448
   16.4   Scalar differentiation of a vector   453
   16.5   The scalar (dot) product   456
   16.6   The vector (cross) product   462
   16.7   Scalar and vector fields   466
   16.8   The gradient of a scalar field   467
   16.9   Divergence and curl of a vector field   469
   16.10 Vector spaces   471
   16.11 Exercises   471

17   Determinants   **474**

   17.1   Concepts   474
   17.2   Determinants of order 3   476
   17.3   The general case   481

17.4	The solution of linear equations	483
17.5	Properties of determinants	488
17.6	Reduction to triangular form	493
17.7	Alternating functions	494
17.8	Exercises	496

18   Matrices and linear transformations   499

18.1	Concepts	499
18.2	Some special matrices	502
18.3	Matrix algebra	505
18.4	The inverse matrix	513
18.5	Linear transformations	516
18.6	Orthogonal matrices and orthogonal transformations	521
18.7	Symmetry operations	524
18.8	Exercises	529

19   The matrix eigenvalue problem   532

19.1	Concepts	532
19.2	The eigenvalue problem	534
19.3	Properties of the eigenvectors	537
19.4	Matrix diagonalization	543
19.5	Quadratic forms	546
19.6	Complex matrices	551
19.7	Exercises	555

20   Numerical methods   558

20.1	Concepts	558
20.2	Errors	558
20.3	Solution of ordinary equations	562
20.4	Interpolation	566
20.5	Numerical integration	573
20.6	Methods in linear algebra	581
20.7	Gauss elimination for the solution of linear equations	581
20.8	Gauss–Jordan elimination for the inverse of a matrix	584
20.9	First-order differential equations	585
20.10	Systems of differential equations	590
20.11	Exercises	592

21   Probability and statistics   595

21.1	Concepts	595
21.2	Descriptive statistics	595
21.3	Frequency and probability	601
21.4	Combinations of probabilities	603
21.5	The binomial distribution	604
21.6	Permutations and combinations	607
21.7	Continuous distributions	613
21.8	The Gaussian distribution	615

21.9   More than one variable   618
21.10   Least squares   619
21.11   Sample statistics   623
21.12   Exercises   624

Appendix. Standard integrals   **627**

Solutions to exercises   **631**

Index   **653**

1 Numbers, variables, and units

1.1 Concepts

Chemistry, in common with the other physical sciences, comprises

(i) **experiment**: the observation of physical phenomena and the measurement of physical quantities, and

(ii) **theory**: the interpretation of the results of experiment, the correlation of one set of measurements with other sets of measurements, the discovery and application of rules to rationalize and interpret these correlations.

Both experiment and theory involve the manipulation of numbers and of the symbols that are used to represent numbers and physical quantities. Equations containing these symbols provide relations amongst physical quantities. Examples of such equations are

1. the equation of state of the ideal gas

$$pV = nRT \tag{1.1}$$

2. Bragg's Law in the theory of crystal structure

$$n\lambda = 2d \sin \theta \tag{1.2}$$

3. the Arrhenius equation for the temperature dependence of rate of reaction

$$k = Ae^{-E_a/RT} \tag{1.3}$$

4. the Nernst equation for the emf of an electrochemical cell

$$E = E^{\ominus} - \frac{RT}{nF} \ln Q \tag{1.4}$$

When an equation involves physical quantities, the expressions on the two sides of the equal sign[1] must be of the same kind as well as the same magnitude.

---

[1] The sign for equality was introduced by Robert Recorde (*c.* 1510–1558) in his *The whetstone of witte* (London, 1557); 'I will sette as I doe often in woorke use, a paire of parallels, or Gemowe (twin) lines of one lengthe, thus: =, bicause noe.2. thynges can be moare equalle.'

EXAMPLE 1.1 The equation of state of the ideal gas, (1.1), can be written as an equation for the volume,

$$V = \frac{nRT}{p}$$

in which the physical quantities on the right of the equal sign are the pressure p of the gas, the temperature T, the amount of substance n, and the molar gas constant $R = 8.31447 \, \text{J K}^{-1} \, \text{mol}^{-1}$.

We suppose that we have one tenth of a mole of gas, $n = 0.1$ mol, at temperature $T = 298$ K and pressure $p = 10^5$ Pa. Then

$$V = \frac{nRT}{p} = \frac{0.1 \, \text{mol} \times 8.31447 \, \text{J K}^{-1} \, \text{mol}^{-1} \times 298 \, \text{K}}{10^5 \, \text{Pa}}$$

$$= \left(\frac{0.1 \times 8.31447 \times 298}{10^5} \right) \times \left(\frac{\text{mol J K}^{-1} \, \text{mol}^{-1} \, \text{K}}{\text{Pa}} \right)$$

$$= 2.478 \times 10^{-3} \, \text{m}^3$$

The quantities on the right side of the equation have been expressed in terms of SI units (see Section 1.8), and the combination of these units is the SI unit of volume, m^3 (see Example 1.17).

Example 1.1 demonstrates a number of concepts:

(i) Function. Given any particular set of values of the pressure p, temperature T, and amount of substance n, equation (1.1) allows us to calculate the corresponding volume V. The value of V is determined by the values of p, T, and n; we say

$$V \text{ is a } \textbf{function} \text{ of } p, T, \text{ and } n.$$

This statement is usually expressed in mathematics as

$$V = f(p, T, n)$$

and means that, for given values of p, T and n, the value of V is given by the value of a function $f(p, T, n)$. In the present case, the function is $f(p, T, n) = nRT/p$. A slightly different form, often used in the sciences, is

$$V = V(p, T, n)$$

which means that V is *some* function of p, T and n, which may or may not be known.

Algebraic functions are discussed in Chapter 2. Transcendental functions, including the trigonometric, exponential and logarithmic functions in equations (1.2) to (1.4), are discussed in Chapter 3.

(ii) Constant and variable. Equation (1.1) contains two types of quantity:

Constant: a quantity whose value is fixed for the present purposes. The quantity $R = 8.31447$ J K^{-1} mol^{-1} is a constant physical quantity.[2] A constant number is any particular number; for example, $a = 0.1$ and $\pi = 3.14159\ldots$

Variable: a quantity that can have any value of a given set of allowed values. The quantities p, T, and n are the variables of the function $f(p, T, n) = nRT/p$.

Two types of variable can be distinguished. An **independent variable** is one whose value does not depend on the value of any other variable. When equation (1.1) is written in the form $V = nRT/p$, it is implied that the independent variables are p, T, and n. The quantity V is then the **dependent variable** because its value depends on the values of the independent variables. We could have chosen the dependent variable to be T and the independent variables as p, V, and n; that is, $T = pV/nR$. In practice, the choice of independent variables is often one of mathematical convenience, but it may also be determined by the conditions of an experiment; it is sometimes easier to measure pressure p, temperature T, and amount of substance n, and to calculate V from them.

Numbers are discussed in Sections 1.2 to 1.4, and variables in Section 1.5. The algebra of numbers (arithmetic) is discussed in Section 1.6.

(iii) A physical quantity is always the product of two quantities, a number and a **unit**; for example $T = 298.15$ K or $R = 8.31447$ J K^{-1} mol^{-1}. In applications of mathematics in the sciences, numbers by themselves have no meaning unless the units of the physical quantities are specified. It is important to know what these units are, but the mathematics does not depend on them. Units are discussed in Section 1.8.

1.2   Real numbers

The concept of number, and of counting, is learnt very early in life, and nearly every measurement in the physical world involves numbers and counting in some way. The simplest numbers are the **natural numbers**, the 'whole numbers' or signless integers 1, 2, 3, … It is easily verified that the addition or multiplication of two natural numbers always gives a natural number, whereas subtraction and division may not. For example $5 - 3 = 2$, but $5 - 6$ is not a natural number. A set of numbers for which the operation of *subtraction* is always valid is the set of **integers**, consisting of all positive and negative whole numbers, and zero:

$$\cdots \quad -3 \quad -2 \quad -1 \quad 0 \quad +1 \quad +2 \quad +3 \quad \cdots$$

The operations of addition and subtraction of both positive and negative integers are made possible by the rules

$$m + (-n) = m - n$$
$$m - (-n) = m + n \tag{1.5}$$

[2] The values of the fundamental physical constants are under continual review. For the latest recommended values, see the NIST (National Institute of Standards and Technology) website at www.physics.nist.gov

so that, for example, the subtraction of a negative number is equivalent to the addition of the corresponding positive number. The operation of multiplication is made possible by the rules

$$(-m) \times (-n) = +(m \times n)$$
$$(-m) \times (+n) = -(m \times n)$$

(1.6)

Similarly for division. Note that $-m = (-1) \times m$.

EXAMPLES 1.2 Addition and multiplication of negative numbers

$$2 + (-3) = 2 - 3 = -1 \qquad 2 - (-3) = 2 + 3 = 5$$
$$(-2) \times (-3) = 2 \times 3 = 6 \qquad (2) \times (-3) = -2 \times 3 = -6$$
$$(-6) \div (-3) = 6 \div 3 = 2 \qquad 6 \div (-3) = -6 \div 3 = -2$$

> Exercises 1–7

In equations (1.5) and (1.6) the letters m and n are symbols used to represent any pair of integers; they are **integer variables**, whose values belong to the (infinite) set of integers.

Division of one integer by another does not always give an integer; for example $6 \div 3 = 2$, but $6 \div 4$ is not an integer. A set of numbers for which the operation of *division* is always valid is the set of **rational numbers**, consisting of all the numbers $m/n = m \div n$ where m and n are integers (m/n, read as 'm over n', is the more commonly used notation for 'm divided by n'). The definition excludes the case $n = 0$ because division by zero is not defined (see Section 1.6), but integers are included because an integer m can be written as $m/1$. The rules for the combination of rational numbers (and of fractions in general) are

$$\frac{m}{n} + \frac{p}{q} = \frac{mq + np}{nq}$$

(1.7)

$$\frac{m}{n} \times \frac{p}{q} = \frac{mp}{nq}$$

(1.8)

$$\frac{m}{n} \div \frac{p}{q} = \frac{m}{n} \times \frac{q}{p} = \frac{mq}{np}$$

(1.9)

where, for example, mq means $m \times q$.

EXAMPLES 1.3 Addition of fractions

(1) Add $\dfrac{1}{2}$ and $\dfrac{1}{4}$.

The number **one half** is equal to **two quarters** and can be added to **one quarter** to give **three quarters**:

$$\frac{1}{2}+\frac{1}{4}=\frac{2}{4}+\frac{1}{4}=\frac{3}{4}$$

The value of a fraction like $1/2$ is unchanged if the numerator and the denominator are both multiplied by the same number:

$$\frac{1}{2}=\frac{1\times 2}{2\times 2}=\frac{2}{4}$$

and the general method of adding fractions is (a) find a common denominator for the fractions to be added, (b) express all the fractions in terms of this common denominator, (c) add.

(2) Add $\dfrac{2}{3}$ and $\dfrac{4}{5}$.

A common denominator is $3\times 5=15$. Then

$$\frac{2}{3}+\frac{4}{5}=\frac{2\times 5}{3\times 5}+\frac{3\times 4}{3\times 5}=\frac{10}{15}+\frac{12}{15}=\frac{10+12}{15}=\frac{22}{15}$$

(3) Add $\dfrac{1}{4}$ and $\dfrac{5}{6}$.

A common denominator is $4\times 6=24$, but the lowest (smallest) common denominator is 12:

$$\frac{1}{4}+\frac{5}{6}=\frac{3}{12}+\frac{10}{12}=\frac{13}{12}$$

▸ Exercises 8–13

EXAMPLE 1.4 Multiplication of fractions

$$\frac{2}{3}\times\frac{4}{5}=\frac{2\times 4}{3\times 5}=\frac{8}{15}$$

This can be interpreted as taking two thirds of $4/5$ (or four fifths of $2/3$).

▸ Exercises 14–17

EXAMPLE 1.5 Division of fractions

$$\frac{2}{3} \div \frac{4}{5} = \frac{2}{3} \times \frac{5}{4} = \frac{10}{12}$$

The number $10/12$ can be simplified by 'dividing top and bottom' by the common factor 2: $10/12 = 5/6$ (see Section 1.3).

> Exercises 18–21

Every rational number is the solution of a linear equation

$$mx = n \tag{1.10}$$

where m and n are integers; for example, $3x = 2$ has solution $x = 2/3$. Not all numbers are rational however. One solution of the quadratic equation

$$x^2 = 2$$

is $x = \sqrt{2}$, the positive square root of 2 (the other solution is $-\sqrt{2}$), and this number cannot be written as a rational number m/n; it is called an **irrational number**. Other irrational numbers are obtained as solutions of the more general quadratic equation

$$ax^2 + bx + c = 0$$

where a, b, and c are arbitrary integers, and of higher-order algebraic equations; for example, a solution of the cubic equation

$$x^3 = 2$$

is $x = \sqrt[3]{2}$, the cube root of 2. Irrational numbers like $\sqrt{2}$ and $\sqrt[3]{2}$ are called **surds**.

The rational and irrational numbers obtained as solutions of **algebraic equations** of type

$$a_0 + a_1 x + a_2 x^2 + a_3 x^3 + \cdots + a_n x^n = 0 \tag{1.11}$$

where a_0, a_1, ..., a_n are integers, are called **algebraic numbers**; these numbers can be expressed exactly in terms of a finite number of rational numbers and surds. There exist also other numbers that are not algebraic; they are not obtained as solutions of any finite algebraic equation. These numbers are irrational numbers called **transcendental numbers**; 'they transcend the power of algebraic methods'

(Euler).[3] The best known and most important of these are the Euler number e and the Archimedean number π.[4] These are discussed in Section 1.4.

The rational and irrational numbers form the **continuum of numbers**; together they are called the **real numbers**.

1.3 Factorization, factors, and factorials

Factorization is the decomposition of a number (or other quantity) into a product of other numbers (quantities), or **factors**; for example

$$30 = 2 \times 3 \times 5$$

shows the decomposition of the natural number 30 into a product of **prime numbers**; that is, natural numbers that cannot be factorized further (the number 1 is not counted as a prime number). The **fundamental theorem of arithmetic** is that every natural number can be factorized as a product of prime numbers in only one way.[5]

EXAMPLES 1.6   Prime number factorization

    (1) $4 = 2 \times 2 = 2^2$

    (2) $12 = 2 \times 2 \times 3 = 2^2 \times 3$

    (3) $315 = 3 \times 3 \times 5 \times 7 = 3^2 \times 5 \times 7$

    (4) $5120 = 2 \times 2 \times 2 \times 2 \times 2 \times 2 \times 2 \times 2 \times 2 \times 2 \times 5 = 2^{10} \times 5$

▸ Exercises 22–25

Factorization and cancellation of **common factors** can be used for the simplification of fractions. For example, in

$$\frac{12}{42} = \frac{\cancel{6} \times 2}{\cancel{6} \times 7} = \frac{2}{7}$$

---

[3] Leonhard Euler (1707–1783). Born in Switzerland, he worked most of his life in St Petersburg and in Berlin. One of the world's most prolific mathematicians, he wrote 'voluminous papers and huge textbooks'. He contributed to nearly all branches of mathematics and its application to physical problems, including the calculus, differential equations, infinite series, complex functions, mechanics, and hydrodynamics, and his name is associated with many theorems and formulas. One of his important, if unspectacular, contributions was to mathematical notation. He introduced the symbol e, gave the trigonometric functions their modern definition, and by his use of the symbols sin, cos, i, and π made them universally accepted.

[4] The symbol π was first used by William Jones (1675–1749) in a textbook on mathematics, *Synopsis palmariorum matheseos* (A new introduction to the mathematics) in 1706. Euler's adoption of the symbol ensured its acceptance.

[5] A version of the fundamental theorem of arithmetic is given by Propositions 31 and 32 in Book VII of Euclid's *Stoichia* (Elements). Euclid was one of the first teachers at the Museum and Library of Alexandria founded by Ptolemy I in about 300 BC after he had gained control of Egypt when Alexander's empire broke up in 323 BC.

cancellation of the common factor 6 is equivalent to dividing both numerator and denominator by 6, and such an operation does not change the value of the fraction.

EXAMPLES 1.7 Simplification of fractions

(1) $\dfrac{4}{24} = \dfrac{2^2}{2^3 \times 3} = \dfrac{\cancel{2} \times \cancel{2}}{\cancel{2} \times \cancel{2} \times 2 \times 3} = \dfrac{1}{6}$

(2) $\dfrac{15}{25} = \dfrac{3 \times \cancel{5}}{5^{\cancel{2}}} = \dfrac{3}{5}$

(3) $\dfrac{105}{1470} = \dfrac{\cancel{3} \times \cancel{5} \times \cancel{7}}{2 \times \cancel{3} \times \cancel{5} \times 7^{\cancel{2}}} = \dfrac{1}{2 \times 7} = \dfrac{1}{14}$

➤ Exercises 26–29

In general, the purpose of factorization is to express a quantity in terms of simpler quantities (see Section 2.3 for factorization of algebraic expressions).

Factorials

The **factorial** of n is the number whose factors are the first n natural numbers:

$$\begin{aligned} n! &= 1 \times 2 \times 3 \times \cdots \times n \\ &= n \times (n-1) \times (n-2) \times \cdots \times 2 \times 1 \end{aligned} \tag{1.12}$$

(read as 'n factorial'). Consecutive factorials are related by the recurrence relation

$$(n+1)! = (n+1) \times n!$$

for example, $3! = 3 \times 2 \times 1 = 6$ and $4! = 4 \times 3 \times 2 \times 1 = 4 \times 3! = 24$. In addition, the factorial of zero is defined as $0! = 1$.

EXAMPLES 1.8 Factorials

(1) $1! = 1 \times 0! = 1 \times 1 = 1$

(2) $5! = 5 \times 4! = 5 \times 4 \times 3! = 5 \times 4 \times 3 \times 2! = 5 \times 4 \times 3 \times 2 = 120$

(3) $\dfrac{5!}{3!} = \dfrac{5 \times 4 \times \cancel{3!}}{\cancel{3!}} = 5 \times 4 = 20$

(4) $\dfrac{7!}{3!4!} = \dfrac{7 \times \cancel{6} \times 5 \times \cancel{4!}}{\cancel{3} \times \cancel{2} \times \cancel{4!}} = 7 \times 5 = 35$

➤ Exercises 30–36

1.4   Decimal representation of numbers

These are the nine figures of the Indians

$$9 \quad 8 \quad 7 \quad 6 \quad 5 \quad 4 \quad 3 \quad 2 \quad 1$$

With these nine figures, and with this sign 0 which in Arabic is called zephirum, any number can be written, as will below be demonstrated.

(Fibonacci)[6]

In the decimal system of numbers, the ten digit symbols 0 to 9 (Hindu-Arabic numerals)[7] are used for zero and the first nine positive integers; the tenth positive integer is denoted by 10. A larger integer, such as 'three hundred and seventy-two' is expressed in the form

$$300 + 70 + 2 = 3 \times 10^2 + 7 \times 10 + 2$$

and is denoted by the symbol 372, in which the value of each digit is dependent on its position in the symbol for the number. The decimal system has **base** 10, and is the only system in common use.

Although rational numbers can always be expressed exactly as ratios of integers, this is not so for irrational numbers. For computational purposes, a number that is not an integer is conveniently expressed as a **decimal fraction**;[8] for example, $5/4 = 1.25$. The general form of the decimal fraction

(integral part).(fractional part)

consists of an integer to the left of the decimal point, the integral part of the number, and one or more digits to the right of the decimal point, the decimal or fractional part of the number. The value of each digit is determined by its position; for example

$$234.567 = 200 + 30 + 4 + \frac{5}{10} + \frac{6}{100} + \frac{7}{1000}$$

$$= 2 \times 10^2 + 3 \times 10^1 + 4 \times 10^0 + 5 \times 10^{-1} + 6 \times 10^{-2} + 7 \times 10^{-3}$$

---

[6] Leonardo of Pisa, also called Fibonacci (c. 1170–after 1240). The outstanding mathematician of the Latin Middle Ages. In his travels in Egypt, Syria, Greece, and Sicily, Fibonacci studied Greek and Arabic (Muslim) mathematical writings, and became familiar with the Arabic positional number system developed by the Hindu mathematicians of the Indus valley of NW India. Fibonacci's first book, the *Liber abaci*, or Book of the Abacus, (1202, revised 1228) circulated widely in manuscript, but was published only in 1857 in *Scritti di Leonardo Pisano*. The first chapter opens with the quotation given above in the text.

[7] One of the principal sources by which the Hindu-Arabic decimal position system was introduced into (Latin) Europe was Al-Khwarizmi's *Arithmetic*. Muhammad ibn Musa Al-Khwarizmi (Mohammed the son of Moses from Khorezm, modern Khiva in Uzbekistan) was active in the time of the Baghdad Caliph Al-Mamun (813–833), and was probably a member of his 'House of Wisdom' (Academy) at a time when Baghdad was the largest city in the world. Al-Khwarizmi's *Algebra* was widely used in Arabic and in Latin translation as a source on linear and quadratic equations. The word algorithm is derived from his name, and the word algebra comes from the title, *Liber algebrae et almucabala*, of Robert of Chester's Latin translation (c. 1140) of his work on equations.

[8] The use of decimal fractions was introduced into European mathematics by the Flemish mathematician and engineer Simon Stevin (1548–1620) in his *De Thiende* (The art of tenths) in 1585. Although decimal fractions were used by the Chinese several centuries earlier, and the Persian astronomer Al-Kashi used decimal and sexagesimal fractions in his *Key to Arithmetic* early in the fifteenth century, the common use of decimal fractions in European mathematics can be traced directly to Stevin, especially after John Napier modified the notation into the present one with the decimal point (or decimal comma as is used in much of continental Europe). It greatly simplified the operations of multiplication and division.

where $10^{-n} = 1/10^n$ and $10^0 = 1$ (see Section 1.6).

➤ Exercises 37–42

A number with a finite number of digits after (to the right of) the decimal point can always be written in the rational form m/n; for example $1.234 = 1234/1000$. The converse is not always true however. The number $1/3$ cannot be expressed exactly as a finite decimal fraction:

$$\frac{1}{3} = 0.333\ldots$$

the dots indicating that the fraction is to be extended indefinitely. If quoted to four decimal places, the number has lower and upper bounds 0.3333 and 0.3334:

$$0.3333 < \frac{1}{3} < 0.3334$$

where the symbol < means 'is less than'; other symbols of the same kind are ≤ for 'is less than or equal to', > for 'is greater than', and ≥ for 'is greater than or equal to'. Further examples of nonterminating decimal fractions are

$$\frac{1}{7} = 0.142857\,142857\ldots, \qquad \frac{1}{12} = 0.083333\,333333\ldots$$

In both cases a finite sequence of digits after the decimal point repeats itself indefinitely, either immediately after the decimal point, as the sequence 142857 in $1/7$, or after a finite number of leading digits, as 3 in $1/12$. This is a characteristic property of rational numbers.

EXAMPLE 1.9 Express $1/13$ as a decimal fraction. By long division,

$$
\begin{array}{r}
0.07692307\ldots \\
13\overline{)1.00} \\
\underline{91} \\
90 \\
\underline{78} \\
120 \\
\underline{117} \\
30 \\
\underline{26} \\
40 \\
\underline{39} \\
100 \\
\vdots
\end{array}
$$

The rational number $1/13 = 0.076923\,076923\ldots$ is therefore a nonterminating decimal fraction with repeating sequence 076923 after the decimal point.

➤ Exercises 43–46

An irrational number cannot be represented exactly in terms of a finite number of digits, and the digits after the decimal point do not show a repeating sequence. The number $\sqrt{2}$ has approximate value to 16 significant figures,

$$\sqrt{2} = 1.414213\ 562373\ 095\ldots$$

and can, in principle, be computed to any desired accuracy by a numerical method such as the Newton–Raphson method discussed in Chapter 20.[9]

The Archimedean number π

The number π is defined as the ratio of the circumference of a circle to its diameter. It is a transcendental number,[10] and has been computed to many significant figures; it was quoted to 127 decimal places by Euler in 1748. Its value to 16 significant figures is

$$\pi = 3.14159\ 26535\ 89793\ldots$$

The value of π has been of practical importance for thousands of years. For example, an Egyptian manuscript dated about 1650 BC (the Rhind papyrus in the British Museum) contains a prescription for the calculation of the volume of a cylindrical granary from which the approximate value $256/81 \approx 3.160$ can be deduced. A method for generating accurate approximations was first used by Archimedes[11] who determined the bounds

$$\frac{223}{71} < \pi < \frac{22}{7}$$

and the upper bound has an error of only 2 parts in a thousand.

---

[9] A clay tablet (YBC 7289, Yale Babylonian Collection) dating from the Old Babylonian Period (c. 1800–1600 BC) has inscribed on it a square with its two diagonals and numbers that give $\sqrt{2}$ to three sexagesimal places: $\sqrt{2} = 1, 24, 51, 10 = 1 + 24/60 + 51/60^2 + 10/60^3 \approx 1.41421296$, correct to 6 significant decimal figures.

[10] The proof of the irrationality of π was first given in 1761 by Johann Heinrich Lambert (1728–1777), German physicist and mathematician. He is also known for his introduction of hyperbolic functions into trigonometry. The number π was proved to be transcendental by Carl Louis Ferdinand von Lindemann (1852–1939) in 1882 by a method similar to that used by Hermite for e.

[11] Archimedes (287–212 BC) was born in Syracuse in Sicily. He made contributions to mathematics, mechanics, and astronomy, and was a great mechanical inventor. His main contributions to mathematics and the mathematical sciences are his invention of methods for determining areas and volumes that anticipated the integral calculus and his discoveries of the first law of hydrostatics and of the law of levers.

The Euler number *e*

The number *e* is defined by the 'infinite series' (see Chapter 7)

$$e = 1 + \frac{1}{1!} + \frac{1}{2!} + \frac{1}{3!} + \frac{1}{4!} + \cdots$$

$$= 2.71828\ 18284\ 59045\ldots$$

The value of *e* can be computed from the series to any desired accuracy. The number was shown to be a transcendental number by Hermite in 1873.[12]

EXAMPLE 1.10 Show that the sum of the first 10 terms of the series gives an approximate value of *e* that is correct to at least 6 significant figures.

$$e = 1 + 1 + \frac{1}{2} + \frac{1}{6} + \frac{1}{24} + \frac{1}{120} + \frac{1}{720} + \frac{1}{5040} + \frac{1}{40320} + \frac{1}{362880} + \frac{1}{3628800} + \cdots$$

$$\approx 1 + 1 + 0.5 + 0.166667 + 0.041667 + 0.008333 + 0.001389 + 0.000198$$
$$+ 0.000025 + 0.000003 + 0.0000003$$

$$\approx 2.71828$$

The value is correct to the 6 figures quoted because every additional term in the series is at least ten times smaller than the preceding one.

Significant figures and rounding

In practice, arithmetic involving only integers gives exact answers (unless the numbers are too large to be written). More generally, a number in the decimal system is approximated either with some given number of decimal places or with a given number of significant figures, and the result of an arithmetic operation is also approximate. In the **fixed-point** representation, all numbers are given with a fixed number of decimal places; for example,

$$3.142, \quad 62.358, \quad 0.013, \quad 1.000$$

have 3 decimal places. In the **floating-point** representation, used more widely in the sciences, the numbers are given with a fixed number of 'significant figures', with zeros on the left of a number not counted. For example,

$$3210 = 0.3210 \times 10^4, \quad 003.210 = 0.3210 \times 10^1, \quad 0.003210 = 0.3210 \times 10^{-2}$$

all have 4 significant figures.

---

[12] Charles Hermite (1822–1901). French mathematician, professor at the Sorbonne, is known for his work in algebra and number theory. His work on the algebra of complex numbers ('Hermitian forms') became important in the formulation of quantum theory. The Hermite differential equation and the Hermite polynomials are important in the solution of the Schrödinger equation for the harmonic oscillator.

A number whose exact (decimal) representation involves more than a given number of digits is reduced most simply by **truncation**; that is, by removing or replacing by zeros all superfluous digits on the right. For example, to 4 decimal places or 5 significant figures, 3.14159 is truncated to 3.1415. Truncation is not recommended because it can lead to serious computational errors. A more sensible (accurate) approximation of π to five figures is 3.1416, obtained by **rounding up**. The most widely accepted rules for rounding are:

(i) If the first digit dropped is *greater than or equal to* 5, the preceding digit is increased by 1; the number is *rounded up*.

(ii) If the first digit dropped is *less than* 5, the preceding digit is left unchanged; the number is *rounded down*. For example, for 4, 3, 2, and 1 decimal places,

$$7.36284 \quad \text{is} \quad 7.3628, \quad 7.363, \quad 7.36, \quad 7.4$$

Errors arising from truncation and rounding are discussed in Section 20.2.

➤ Exercises 47–49

1.5 Variables

In the foregoing sections, symbols (letters) have been used to represent arbitrary numbers. A quantity that can take as its value any value chosen from a given set of values is called a **variable**. If $\{x_1, x_2, x_3, \ldots, x_n\}$ is a set of objects, not necessarily numbers, then a variable x can be defined in terms of this set such that x can have as its value any member of the set; the set forms the **domain** of the variable. In (real) number theory, the objects of the set are real numbers, and a **real variable** can have as its domain either the whole continuum of real numbers or a subset thereof. If the domain of the variable x is an interval a to b,

$$a \leq x \leq b$$

then x is a **continuous variable** in the interval, and can have any value in the continuous range of values a to b (including a and b). If the domain consists of a discrete set of values, for example the n numbers $x_1, x_2, x_3, \ldots, x_n$, then x is called a **discrete variable**. If the domain consists of integers, x is an **integer variable**. If the set consists of only one value then the variable is called a **constant variable**, or simply a **constant**.

In the physical sciences, variables are used to represent both numbers and physical quantities. In the ideal-gas example discussed in Section 1.1, the physical quantities p, V, n, and T are continuous variables whose numerical values can in principle be any positive real numbers. Discrete variables are normally involved whenever objects are counted as opposed to measured. Typically, an integer variable is used for the counting and the counted objects form a sample of some discrete set. In some cases however a physical quantity can have values, some of which belong to a discrete set and others to a continuous set. This is the case for the energy levels and the observed spectral frequencies of an atom or molecule.

EXAMPLE 1.11 The spectrum of the hydrogen atom

The energy levels of the hydrogen atom are of two types:

(i) Discrete (quantized) energy levels with (negative) energies given by the formula (in atomic units, see Section 1.8)

$$E_n = -\frac{1}{2n^2}, \quad n = 1, 2, 3, \ldots$$

The corresponding states of the atom are the 'bound states', in which the motion of the electron is confined to the vicinity of the nucleus. Transitions between the energy levels give rise to discrete lines in the spectrum of the atom.

(ii) Continuous energy levels, with all positive energies, $E > 0$. The corresponding states of the atom are those of a free (unbound) electron moving in the presence of the electrostatic field of the nuclear charge. Transitions between these energy levels and those of the bound states give rise to continuous ranges of spectral frequencies.

1.6 The algebra of real numbers

The importance of the concept of variable is that variables can be used to make statements about the properties of whole sets of numbers (or other objects), and it allows the formulation of a set of rules for the manipulation of numbers. The set of rules is called the **algebra**.

Let a, b, and c be variables whose values can be any real numbers. The basic rules for the combination of real numbers, the algebra of real numbers or the arithmetic, are

1. $a + b = b + a$ (commutative law of addition)
2. $ab = ba$ (commutative law of multiplication)
3. $a + (b + c) = (a + b) + c$ (associative law of addition)
4. $a(bc) = (ab)c$ (associative law of multiplication)
5. $a(b + c) = ab + ac$ (distributive law)

The operations of addition and multiplication and their inverses, subtraction and division, are called **arithmetic operations**. The symbols $+$, $-$, $\times$ and $\div$ (or $/$) are called **arithmetic operators**. The result of adding two numbers, $a + b$, is called the **sum** of a and b; the result of multiplying two numbers, $ab = a \times b = a \cdot b$, is called the **product** of a and b.[13]

---

[13] In 1698 Leibniz wrote in a letter to Johann Bernoulli: 'I do not like $\times$ as a symbol for multiplication, as it easily confounded with x ... often I simply relate two quantities by an interposed dot'. It is becoming accepted practice to place the 'dot' in the 'high position' to denote multiplication ($2 \cdot 5 = 2 \times 5$) and in the 'low position', on the line, for the decimal point ($2.5 = 5/2$). An alternative convention, still widely used, is to place the dot on the line for multiplication ($2.5 = 2 \times 5$) and high for the decimal point ($2 \cdot 5 = 5/2$).

EXAMPLES 1.12 Examples of the rules of arithmetic

rule	examples
1. $a + b = b + a$	$2 + 3 = 3 + 2 = 5$
2. $ab = ba$	$2 \times 3 = 3 \times 2 = 6$
3. $a + (b + c) = (a + b) + c$	$\begin{cases} 2 + (3 + 4) = 2 + 7 = 9, \text{ and} \\ (2 + 3) + 4 = 5 + 4 = 9 \end{cases}$
4. $a(bc) = (ab)c$	$\begin{cases} 2 \times (3 \times 4) = 2 \times 12 = 24, \text{ and} \\ (2 \times 3) \times 4 = 6 \times 4 = 24 \end{cases}$
5. $a(b + c) = ab + ac$	$\begin{cases} 2 \times (3 + 4) = 2 \times 7 = 14, \text{ and} \\ 2 \times (3 + 4) = (2 \times 3) + (2 \times 4) = 6 + 8 = 14 \end{cases}$
	$-2(3 + 4) = (-2 \times 3) + (-2 \times 4) = -6 - 8 = -14$
	$-2(3 - 4) = -2 \times 3 - 2 \times (-4) = -6 + 8 = 2$

A corollary to rule **5** is

$(a + b)(c + d) = a(c + d) + b(c + d)$ $\quad$ $(2 + 3)(4 + 5) = 2(4 + 5) + 3(4 + 5) = 18 + 27 = 45$

Three rules define the properties of zero and unity:

6. $a + 0 = 0 + a = a$ $\quad$ (addition of zero)

7. $a \times 0 = 0 \times a = 0$ $\quad$ (multiplication by zero)

8. $a \times 1 = 1 \times a = a$ $\quad$ (multiplication by unity)

We have already seen that subtraction of a number is the same as addition of its negative, and that division by a number is the same as multiplication by its inverse. However, division by zero is not defined; there is no number whose inverse is zero. For example, the number $1/a$, for positive values of a, becomes arbitrarily large as the value of a approaches zero; we say that $1/a$ **tends to infinity** as a tends to zero:

$$\frac{1}{a} \to \infty \quad \text{as} \quad a \to 0$$

Although 'infinity' is represented by the symbol ∞, it is not a number. If it were a number then, by the laws of algebra, the equations $1/0 = \infty$ and $2/0 = \infty$ would imply $1 = 2$.

The **modulus** of a real number a is defined as the positive square root of a^2; $|a| = +\sqrt{a^2}$ (read as 'mod a'). It is the 'magnitude' of the number, equal to $+a$ if a is positive, and equal to $-a$ if a is negative:

$$|a| = \begin{cases} +a & \text{if } a > 0 \\ -a & \text{if } a < 0 \end{cases} \tag{1.13}$$

For example, $|3| = 3$ $\,and\,$ $|-3| = 3$.

The index rule

Numbers are often written in the form a^m, where a is called the **base** and m is the **index** or **exponent**; for example, $100 = 10^2$ with base 10 and exponent 2, and $16 = 2^4$ with base 2 and exponent 4. When m is a positive integer, a^m is the mth power of a; for $m = 3$,

$$a^3 = a \times a \times a, \quad (-a)^3 = (-a) \times (-a) \times (-a) = (-1)^3 \times a^3 = -a^3$$

Numbers are also defined with negative and non-integral exponent. In practice, the number a^m is read 'a to the power m' or 'a to the m', even when m is not a positive integer. The rule for the product of numbers in base–index form is

9. $a^m a^n = a^{m+n}$ (index rule)

For example,

$$a^3 a^2 = (a \times a \times a) \times (a \times a) = a \times a \times a \times a \times a = a^5 = a^{3+2}$$

Three auxiliary rules are

10. $a^m / a^n = a^{m-n}$ **11.** $(a^m)^n = (a^n)^m = a^{m \times n}$ **12.** $(ab)^m = a^m b^m$

Rule **10** defines numbers with zero and negative exponents. Thus, setting $m = n$,

$$a^n / a^n = a^{n-n} = a^0 = 1$$

and any number raised to power zero is unity; for example, $2^3 / 2^3 = 2^{3-3} = 2^0 = 1$ because $2^3 / 2^3 = 1$. Also, setting $m = 0$ in rule **10**,

$$a^0 / a^n = 1 / a^n = a^{-n}$$

so that the inverse of a^n is a^{-n}. In particular, $1/a = a^{-1}$.

EXAMPLES 1.13 The index rule

rule	examples
9. $a^m a^n = a^{m+n}$	(a) $2^3 \times 2^2 = 2^{3+2} = 2^5$
	(b) $3^6 \times 3^{-3} = 3^{6-3} = 3^3$
	(c) $2^{1/2} \times 2^{1/4} = 2^{1/2+1/4} = 2^{3/4}$
10. $a^m / a^n = a^{m-n}$	(d) $2^{3/4} / 2^{1/4} = 2^{3/4-1/4} = 2^{1/2}$
	(e) $2^4 / 2^{-2} = 2^{4-(-2)} = 2^{4+2} = 2^6$
	(f) $3^4 / 3^4 = 3^{4-4} = 3^0 = 1$

11. $(a^m)^n = (a^n)^m = a^{m \times n}$

(g) $(2^2)^3 = (2^2) \times (2^2) \times (2^2) = 2^{2 \times 3} = 2^6$

(h) $(2^{1/2})^2 = 2^{(1/2) \times 2} = 2^1 = 2$

(i) $(2^3)^{4/3} = (2^{4/3})^3 = 2^{3 \times 4/3} = 2^4$

(j) $(2^{\sqrt{2}})^{\sqrt{2}} = 2^{\sqrt{2} \times \sqrt{2}} = 2^2 = 4$

12. $(ab)^m = a^m b^m$

(k) $(2 \times 3)^2 = 2^2 \times 3^2$

(l) $(-8)^{1/3} = (-1)^{1/3} \times 8^{1/3} = (-1) \times 2 = -2$

▶ Exercises 50–65

Example 1.13(h) shows that $2^{1/2} \times 2^{1/2} = 2$. It follows that $2^{1/2} = \sqrt{2}$, the square root of 2. In general, for positive integer m, $a^{1/m}$ is the mth root of a:

$$a^{1/m} = \sqrt[m]{a}$$

Thus, $2^{1/3}$ is a cube root of 2 because $(2^{1/3})^3 = 2^{(1/3) \times 3} = 2^1 = 2$. More generally, for rational exponent m/n, rule **11** gives

$$a^{m/n} = (a^m)^{1/n} = (a^{1/n})^m$$

or, equivalently,

$$a^{m/n} = \sqrt[n]{a^m} = (\sqrt[n]{a})^m$$

so that $a^{m/n}$ is both the nth root of the mth power of a and the mth power of the nth root. For example,

$$4^{3/2} = (4^3)^{1/2} = (4^{1/2})^3 = 8$$

Although the index rules have been demonstrated only for integral and rational indices, they apply to all numbers written in the base–index form. When the exponent m is a variable, a^m is called an exponential function (see Section 3.6 for real exponents and Chapter 8 for complex exponents). If $x = a^m$ then $m = \log_a x$ is the logarithm of x to base a (see Section 3.7).

Rules of precedence for arithmetic operations

An arithmetic expression such as

$$2 + 3 \times 4$$

is ambiguous because its value depends on the order in which the arithmetic operations are applied. The expression can be interpreted in two ways:

$$(2 + 3) \times 4 = 5 \times 4 = 20$$

with the parentheses indicating that the addition is to performed first, and

$$2 + (3 \times 4) = 2 + 12 = 14$$

in which the multiplication is performed first. Ambiguities if this kind can always be resolved by the proper use of parentheses or other brackets. In case of more complicated expressions, containing nested brackets, the convention is to use parentheses as the innermost brackets, then square brackets, then braces (curly brackets). Evaluation then proceeds from the innermost bracketed expressions outwards; for example

$$\left\{\left[(2+3)\times 4\right]+5\right\}\times 6 = \left\{\left[5\times 4\right]+5\right\}\times 6 = \left\{20+5\right\}\times 6 = 25\times 6 = 150$$

As shown, increasing sizes of brackets can help to clarify the structure of the expression.

If in doubt use brackets.

Arithmetic expressions are generally evaluated by following the **rules of precedence** for arithmetic operations:

1. Brackets take precedence over arithmetic operators.
2. Exponentiation (taking powers) takes precedence over multiplication/division and addition/subtraction.
3. Multiplication and division take precedence over addition and subtraction.
4. Addition and subtraction are performed last.

EXAMPLES 1.14 Rules of precedence for arithmetic operations

(1) $\begin{cases} 2+3\times 4 = 2+(3\times 4) = 2+12 = 14 & \text{(rule 3)} \\ \text{but} \quad (2+3)\times 4 = 5\times 4 = 20 & \text{(rule 1)} \end{cases}$

(2) $\begin{cases} 2+3\times 4\times 5+6 = 2+(3\times 4\times 5)+6 = 2+60+6 = 68 & \text{(rule 3)} \\ \text{but} \quad (2+3)\times 4\times(5+6) = 5\times 4\times 30 = 600 & \text{(rule 1)} \end{cases}$

(3) $\begin{cases} 2+3^2 = 2+9 = 11 & \text{(rule 2)} \\ \text{but} \quad (2+3)^2 = 5^2 = 25 & \text{(rule 1)} \end{cases}$

(4) $\begin{cases} 9+16^{1/2} = 9+4 = 11 & \text{(rule 2)} \\ \text{but} \quad (9+16)^{1/2} = (25)^{1/2} = 5 & \text{(rule 1)} \end{cases}$

(5) $\begin{cases} 3\times 4^2 = 3\times(4^2) = 3\times 16 = 48 & \text{(rule 2)} \\ \text{but} \quad (3\times 4)^2 = (12)^2 = 144 & \text{(rule 1)} \end{cases}$

(6) $\begin{cases} 2\times 6\div 3 = (2\times 6)\div 3 = 12\div 3 = 4 \\ \text{and} \quad 2\times 6\div 3 = 2\times(6\div 3) = 2\times 2 = 4 \end{cases}$

What not to do: $(a+b)^n \neq a^n + b^n$, where $\neq$ means 'is not equal to'. Thus,

in case (3): $(2+3)^2 \neq 2^2 + 3^2$,   in case (4): $(9+16)^{1/2} \neq 9^{1/2} + 16^{1/2}$

➤ Exercises 66–77

1.7 Complex numbers

The solutions of algebraic equations are not always real numbers. For example, the solutions of the equation

$$x^2 = -1 \text{ are } x = \pm\sqrt{-1}$$

and these are not any of the numbers described in Section 1.2. They are incorporated into the system of numbers by defining the square root of -1 as a new number which is usually represented by the symbol i (sometimes j) with the property

$$i^2 = -1$$

The two square roots of an arbitrary negative real number $-x^2$ are then ix and $-ix$. For example,

$$\sqrt{-16} = \sqrt{(16)\times(-1)} = \sqrt{16} \times \sqrt{-1} = \pm 4i$$

Such numbers are called **imaginary** to distinguish them from real numbers. More generally, the number

$$z = x + iy$$

where x and y are real is called a **complex number**.

Complex numbers obey the same rules of algebra as real numbers; it is only necessary to remember to replace i^2 by -1 whenever it occurs. They are discussed in greater detail in Chapter 8.

EXAMPLE 1.15 Find the sum and product of the complex numbers $z_1 = 2 + 3i$ and $z_2 = 4 - 2i$.

Addition: $z_1 + z_2 = (2+3i) + (4-2i) = (2+4) + (3i-2i) = 6 + i$

Multiplication: $z_1 z_2 = (2+3i)(4-2i) = 2(4-2i) + 3i(4-2i)$
$$= 8 - 4i + 12i - 6i^2 = 8 + 8i + 6 = 14 + 8i$$

➤ Exercises 78, 79

1.8 Units

A physical quantity has two essential attributes, **magnitude** and **dimensions**. For example, the quantity '2 metres' has the dimensions of length and has magnitude equal to twice the magnitude of the metre. The metre is a constant physical quantity that defines the dimensions of the quantity and provides a scale for the specification of the magnitude of an arbitrary length; it is a **unit** of length. In general, a physical quantity is the product of a number and a unit. All physical quantities can be expressed in terms of the seven 'base' quantities whose names and symbols are listed in the first two columns of Table 1.1.

Table 1.1 Base physical quantities and SI units

Physical quantity	Symbol	Dimension	Name of SI unit	Symbol for SI unit
length	l	L	metre	m
mass	m	M	kilogram	kg
time	t	T	second	s
electric current	I	I	ampere	A
temperature	T	θ	kelvin	K
amount of substance	n	N	mole	mol
luminous intensity	I_v	J	candela	cd

The symbols in column 3 define the dimensions of the base physical quantities, and the dimensions of all other quantities (derived quantities) can be expressed in terms of them. For example, velocity (or more precisely, speed) is distance travelled in unit time, $l/t = lt^{-1}$, and has dimensions of length divided by time, LT^{-1}. The dimensions of a physical quantity are independent of the system of units used to describe its value. Every system of units must, however, conform with the dimensions.

A variety of systems of units are in use, many tailored to the needs of particular disciplines in the sciences. The recommended system for the physical sciences, and for chemistry in particular, is the International System of Units (SI)[14] which is based on the seven base units whose names and symbols are listed in columns 4 and 5 in Table 1.1. Every physical quantity has an SI unit determined by its dimensionality. The SI units of length and time are the metre, m, and the second, s; the corresponding SI unit of velocity is metre per second, $m/s = m\ s^{-1}$ (see Example 1.16(i)). In addition to the base units, a number of quantities that are particularly important in the physical sciences have been given SI names and symbols. Some of these are listed in Table 1.2.

We note that some physical quantities have no dimensions. This is the case for a quantity that is the ratio of two others with the same dimensions; examples are relative density, relative molar mass, and mole fraction. A less obvious example is (plane) angle which is defined as the ratio of two lengths (see Section 3.2).

---

[14] SI (Système International d'Unités) is the international standard for the construction and use of units (see the NIST website at www.physics.nist.gov). In addition, IUPAC (International Union of Pure and Applied Chemistry) provides the standard on chemical nomenclature and terminology, and on the measurement and evaluation of data (see www.iupac.org).

Table 1.2 SI derived units with special names and symbols

Physical quantity	Name	Symbol	Description	SI unit
frequency	hertz	Hz	events per unit time	s^{-1}
force	newton	N	mass $\times$ acceleration	$kg\,m\,s^{-2}$
pressure	pascal	Pa	force per unit area	$N\,m^{-2}$
energy, work, heat	joule	J	force $\times$ distance	$N\,m$
power	watt	W	work per unit time	$J\,s^{-1}$
electric charge	coulomb	C	current $\times$ time	$A\,s$
electric potential	volt	V	work per unit charge	$J\,C^{-1}$
electric capacitance	farad	F	charge per unit potential	$C\,V^{-1}$
electric resistance	ohm	Ω	potential per unit current	$V\,A^{-1}$
electric conductance	siemens	S	current per unit potential	Ω^{-1}
magnetic flux	weber	Wb	work per unit current	$J\,A^{-1}$
magnetic flux density	tesla	T	magnetic flux per unit area	$Wb\,m^{-2}$
inductance	henry	H	magnetic flux per unit current	$Wb\,A^{-1}$
plane angle	radian	rad	angle subtended by unit arc at centre of unit circle	1
solid angle	steradian	sr	solid angle subtended by unit surface at centre of unit sphere	1

> ‣ Exercises 80–90

EXAMPLES 1.16 Dimensions and units

(i) **Velocity** is rate of change of position with time, and has dimensions of length/time: LT^{-1}.

In general, the unit of a derived quantity is obtained by replacing each base quantity by its corresponding unit. In SI, the unit of velocity is meters per second, $m\,s^{-1}$. In a system in which, for example, the unit of length is the yard (yd) and the unit of time is the minute (min), the unit of velocity is yards per minute, $yd\,min^{-1}$. This 'non-SI' unit is expressed in terms of the SI unit by means of **conversion factors** defined within SI. Thus $1\,yd = 0.9144\,m$ (exactly), $1\,min = 60\,s$, and

$$1\,yd\,min^{-1} = (0.9144\,m) \times (60\,s)^{-1} = (0.9144/60)\,m\,s^{-1} = 0.01524\,m\,s^{-1}$$

(ii) **Acceleration** is rate of change of velocity with time, and has dimensions of velocity/time:

$$[LT^{-1}] \times [T^{-1}] = LT^{-2}, \text{ with SI unit } m\,s^{-2}.$$

The standard acceleration of gravity is $g = 9.80665 \text{ m s}^{-2} = 980.665 \text{ Gal}$, where

$$\text{Gal} = 10^{-2} \text{ m s}^{-2} \text{ (cm s}^{-2}\text{) is called the } \textbf{galileo}.$$

(iii) **Force** has dimensions of mass × acceleration:

$$[M] \times [LT^{-2}] = MLT^{-2}, \text{ with SI unit the } \textbf{newton}, \ N = \text{kg m s}^{-2}.$$

(iv) **Pressure** has dimensions of force per unit area:

$$[MLT^{-2}]/[L^2] = ML^{-1}T^{-2}, \text{ with SI unit the } \textbf{pascal}, \ Pa = N \text{ m}^{-2} = \text{kg m}^{-1} \text{ s}^{-2}$$

Widely used alternative non-SI units for pressure are:

'standard pressure': $\text{bar} = 10^5 \text{ Pa}$
atmosphere: $\text{atm} = 101325 \text{ Pa}$
torr: $\text{Torr} = (101325/760) \text{ Pa} \approx 133.322 \text{ Pa}$

(v) **Work, energy and heat** are quantities of the same kind, with the same dimensions and unit. Thus, work has dimensions of force × distance:

$$[MLT^{-2}] \times [L] = ML^2T^{-2}, \text{ with SI unit the } \textbf{joule}, \ J = N \text{ m} = \text{kg m}^2 \text{ s}^{-2}$$

and kinetic energy $\frac{1}{2}mv^2$ has dimensions of mass × (velocity)^2:

$$[M] \times [LT^{-1}]^2 = ML^2T^{-2}.$$

▸ Exercises 91–94

Dimensional analysis

The terms on both sides of an equation that contains physical quantities must have the same dimensions. Dimensional analysis is the name given to the checking of equations for dimensional consistency.

EXAMPLE 1.17 For the ideal-gas equation $pV = nRT$, equation (1.1), the dimensions of pV (using Tables 1.1 and 1.2) are those of work (or energy): $[ML^{-1}T^{-2}] \times [L^3] = ML^2T^{-2}$. The corresponding expression in terms of SI units is

$$Pa \times m^3 = N \text{ m}^{-2} \times m^3 = N \text{ m} = J.$$

For nRT,

$$(\text{mol})(J \text{ K}^{-1} \text{ mol}^{-1})(K) = J$$

as required. It follows that when equation (1.1) is written in the form $V = nRT/p$, as in Example 1.1, the dimensions of V are energy/pressure, with SI unit

$$\frac{\text{J}}{\text{Pa}} = \frac{\cancel{\text{N}}\,\text{m}}{\cancel{\text{N}}\,\text{m}^{-2}} = \text{m}^3 \text{ for the volume.}$$

➤ Exercise 95

Large and small units

Decimal multiples of SI units have names formed from the names of the units and the prefixes listed in Table 1.3. For example, a picometre is $\text{pm} = 10^{-12}$ m, a decimetre is $\text{dm} = 10^{-1}$ m. These units of length are frequently used in chemistry; molecular bond lengths in picometres, and concentrations in moles per decimetre cube, $\text{mol dm}^{-3} = 10^3 \text{ mol m}^{-3}$.

Table 1.3 SI prefixes

Multiple	Prefix	Symbol	Multiple	Prefix	Symbol
10	deca	da	10^{-1}	deci	d
10^2	hecto	h	10^{-2}	centi	c
10^3	kilo	k	10^{-3}	milli	m
10^6	mega	M	10^{-6}	micro	μ
10^9	giga	G	10^{-9}	nano	n
10^{12}	tera	T	10^{-12}	pico	p
10^{15}	peta	P	10^{-15}	femto	f
10^{18}	exa	E	10^{-18}	atto	a
10^{21}	zetta	Z	10^{-21}	zepto	z
10^{24}	yotta	Y	10^{-24}	yocto	y

➤ Exercises 96–103

The quantities that are of interest in chemistry often have very different magnitudes from those of the SI units themselves, particularly when the properties of individual atoms and molecules are considered. For example, the mole is defined as the amount of substance that contains as many elementary entities (atoms or molecules) as there are atoms in 12 g (0.012 kg) of ^{12}C. This number is given by Avogadro's constant, $N_A \approx 6.02214 \times 10^{23} \text{ mol}^{-1}$. The mass of an atom of ^{12}C is therefore

$$m(^{12}\text{C}) = 12/(6.02214 \times 10^{23}) \text{ g} \approx 2 \times 10^{-26} \text{ kg}$$

or $m(^{12}\text{C}) = 12$ u, where

$$\text{u} = 1/(6.02214 \times 10^{23}) \text{ g} \approx 1.66054 \times 10^{-27} \text{ kg}$$

is called the **unified atomic mass unit** (sometimes called a **Dalton**, with symbol **Da**).

EXAMPLES 1.18 Molecular properties: mass, length and moment of inertia

(i) **mass.** Atomic and molecular masses are often given as relative masses: A_r for an atom and M_r for a molecule, on a scale on which $A_r(^{12}C) = 12$. On this scale, $A_r(^1H) = 1.0078$ and $A_r(^{16}O) = 15.9948$. The corresponding relative molar mass of water is

$$M_r(^1H_2{}^{16}O) = 2 \times A_r(^1H) + A_r(^{16}O) = 18.0105,$$

the molar mass is

$$M(^1H_2{}^{16}O) = 18.0105 \text{ g mol}^{-1} = 0.01801 \text{ kg mol}^{-1},$$

and the mass of the individual molecule is

$$m(^1H_2{}^{16}O) = M_r(^1H_2{}^{16}O) \times u = 2.9907 \times 10^{-26} \text{ kg}$$

(ii) **length.** The bond length of the oxygen molecule is $R_e = 1.2075 \times 10^{-10}$ m, and molecular dimensions are usually quoted in appropriate units such as the picometre $pm = 10^{-12}$ m or the nanometre $nm = 10^{-9}$ m in spectroscopy, and the Ångström $Å = 10^{-10}$ m or the Bohr radius $a_0 = 0.529177 \times 10^{-10}$ m $= 0.529177$ Å in theoretical chemistry. Thus, for O_2, $R_e = 1.2075$ Å $= 120.75$ pm.

(iii) **reduced mass and moment of inertia.** The moment of inertia of a system of two masses, m_A and m_B, separated by distance R is $I = \mu R^2$, where μ is the reduced mass, given by

$$\frac{1}{\mu} = \frac{1}{m_A} + \frac{1}{m_B}, \qquad \mu = \frac{m_A m_B}{m_A + m_B}$$

Relative atomic masses can be used to calculate the reduced mass of a diatomic molecule. Thus for CO, $A_r(^{12}C) = 12$ and $A_r(^{16}O) = 15.9948$, and these are the atomic masses in units of the unified atomic mass unit u. Then

$$\mu(^{12}C^{16}O) = \left(\frac{12 \times 15.9948}{27.9948} \right) \left(\frac{u^2}{u} \right) = 6.8562 \text{ u}$$

$$= 6.8562 \times 1.66054 \times 10^{-27} \text{ kg} = 1.1385 \times 10^{-26} \text{ kg}$$

The bond length of CO is 112.81 pm $= 1.1281 \times 10^{-10}$ m, so that the moment of inertia of the molecule is

$$I = \mu R^2 = (1.1385 \times 10^{-26} \text{ kg}) \times (1.1281 \times 10^{-10} \text{ m})^2$$
$$= 1.4489 \times 10^{-46} \text{ kg m}^2$$

The reduced mass and moment of inertia are of importance in discussions of vibrational and rotational properties of molecules.

▸ Exercises 104, 105

EXAMPLES 1.19 Molecular properties: wavelength, frequency, and energy

The wavelength λ and frequency v of electromagnetic radiation are related to the speed of light by

$$c = \lambda v \tag{1.14}$$

(see Example 3.7), where $c = 2.99792 \times 10^8 \text{ m s}^{-1} \approx 3 \times 10^8 \text{ m s}^{-1}$. The energy of a photon is related to the frequency of its associated wave via Planck's constant $h = 6.62608 \times 10^{-34} \text{ J s}$:

$$E = hv \tag{1.15}$$

In a spectroscopic observation of the transition between two states of an atom or molecule, the frequency of the radiation emitted or absorbed is given by $hv = |\Delta E|$, where $\Delta E = E_2 - E_1$ is the energy of transition between states with energies E_1 and E_2. Different spectroscopic techniques are used to study the properties of atoms and molecules in different regions of the electromagnetic spectrum, and different units are used to report the characteristics of the radiation in the different regions. The values of frequency and wavelength are usually recorded in multiples of the SI units of hertz ($\text{Hz} = \text{s}^{-1}$) and metre (m), respectively, but a variety of units is used for energy. For example, the wavelength of one of the pair of yellow D lines in the electronic spectrum of the sodium atom is $\lambda = 589.76 \text{ nm} = 5.8976 \times 10^{-7} \text{ m}$. By equation (1.14), this corresponds to frequency

$$v = \frac{c}{\lambda} = \frac{2.99792 \times 10^8 \text{ m s}^{-1}}{5.8975 \times 10^{-7} \text{ m}} = 5.0833 \times 10^{14} \text{ s}^{-1}$$

and by equation (1.15), the corresponding energy of transition is

$$\Delta E = hv = (6.62608 \times 10^{-34} \text{ J s}) \times (5.0833 \times 10^{14} \text{ s}^{-1}) = 3.368 \times 10^{-19} \text{ J}$$

Energies are often quoted in units of the electron volt, eV, or as molar energies in units of kJ mol^{-1}. The value of eV is the product of the protonic charge e (see Table 1.4) and the SI unit of electric potential $\text{V} = \text{J C}^{-1}$ (Table 1.2): $\text{eV} = 1.60218 \times 10^{-19} \text{ J}$. The corresponding molar energy is

$$\begin{aligned} \text{eV} \times N_A &= (1.60218 \times 10^{-19} \text{ J}) \times (6.02214 \times 10^{23} \text{ mol}^{-1}) \\ &= 96.486 \text{ kJ mol}^{-1} \end{aligned} \tag{1.16}$$

where N_A is Avogadro's constant. For the sodium example,

$$\Delta E = 3.368 \times 10^{-19} \text{ J} = 2.102 \text{ eV} \equiv 202.8 \text{ kJ mol}^{-1}$$

Very often, the characteristics of the radiation are given in terms of the **wavenumber**

$$\tilde{v} = \frac{1}{\lambda} = \frac{v}{c} = \frac{\Delta E}{hc} \tag{1.17}$$

This has dimensions of inverse length and is normally reported in units of the reciprocal centimetre, cm^{-1}. For the sodium example, $\lambda = 5.8976 \times 10^{-5}$ cm and

$$\tilde{\nu} = \frac{1}{5.8976 \times 10^{-5} \ cm} = 16956 \ cm^{-1}$$

The second line of the sodium doublet lies at $16973 \ cm^{-1}$, and the fine structure splitting due to spin–orbit coupling in the atom is $17 \ cm^{-1}$.

In summary, the characteristics of the radiation observed in spectroscopy can be reported in terms of frequency ν in Hz (s^{-1}), wavelength λ in (multiples of) m, energy ΔE in units of eV, molar energy in units of kJ mol^{-1}, and wavenumber $\tilde{\nu}$ in units of cm^{-1}. These quantities are related by equations (1.14) to (1.17). Conversion factors for energy are

$$1 \ eV = 1.60218 \times 10^{-19} \ J \equiv 96.486 \ kJ \ mol^{-1} \equiv 8065.5 \ cm^{-1}$$

➤ Exercises 106

Approximate calculations

Powers of 10 are often used as a description of **order of magnitude**; for example, if a length A is two orders of magnitude larger than length B then it is about $10^2 = 100$ times larger. In some calculations that involve a wide range of orders of magnitude it can be helpful, as an aid to avoiding errors, to calculate the order of magnitude of the answer before embarking on the full detailed calculation. The simplest way of performing such an 'order of magnitude calculation' is to convert all physical quantities to base SI units and to approximate the magnitude of each by an appropriate power of ten, possibly multiplied by an integer. Such calculations are often surprisingly accurate.

EXAMPLE 1.20 Order of magnitude calculations

(i) For the calculation of volume in Example 1.1 (ignoring units),

$$V = \frac{nRT}{p} = \frac{0.1 \times 8.31447 \times 298}{10^5} = 2.478 \times 10^{-3}$$

Two estimates of the answer are

(a) $V \approx \dfrac{10^{-1} \times 10 \times 10^2}{10^5} = 10^{-3}$ (b) $V \approx \dfrac{10^{-1} \times 8 \times 300}{10^5} = 2.4 \times 10^{-3}$

(ii) For the calculation of the moment of inertia of CO in Example 1.18 (ignoring units), $\mu = 1.1385 \times 10^{-26} \approx 10^{-26}$ and $R = 1.1281 \times 10^{-10} \approx 10^{-10}$, and an order of magnitude estimate of the moment of inertia is $I = \mu R^2 \approx (10^{-26}) \times (10^{-10})^2 = 10^{-46}$ (accurate value 1.4489×10^{-46}).

➤ Exercise 107

Atomic units

The equations of motion in quantum mechanics are complicated by the presence of the physical quantities m_e, the rest mass of the electron, e, the charge on the proton, h, Planck's constant, and ε_0, the permittivity of a vacuum. For example, the Schrödinger equation for the motion of the electron about the stationary nucleus in the hydrogen atom is

$$-\frac{h^2}{8\pi^2 m_e}\nabla^2\psi - \frac{e^2}{4\pi\varepsilon_0 r}\psi = E\psi \qquad (1.18)$$

The four experimentally determined quantities can be used as base units for the construction of **atomic units** for all physical quantities whose dimensions involve length, mass, time, and electric current (the first four entries in Table 1.1). Some of the atomic units are listed in Table 1.4. The atomic units of length and energy have been given names: the unit of length, a_0, is called the **bohr**, and is the most probable distance of the electron from the nucleus in the ground state of the hydrogen atom (the radius of the ground-state orbit in the 'old quantum theory' of Bohr). The unit of energy, E_h, is called the **hartree**, and is equal to twice the ionization energy of the hydrogen atom. Atomic units are widely used in quantum chemistry. The convention is to express each physical quantity in an expression in atomic units, and then to delete the unit from the expression; for example, for a distance r, the dimensionless quantity r/a_0 is replaced by r. If this is done to equation (1.18) the resulting dimensionless equation is

$$-\frac{1}{2}\nabla^2\psi - \frac{1}{r}\psi = E\psi \qquad (1.19)$$

Table 1.4 Atomic units

Physical quantity	Atomic unit	Value in SI units
mass	m_e	9.10938×10^{-31} kg
charge	e	1.60218×10^{-19} C
angular momentum	$\hbar = h/2\pi$	1.05457×10^{-34} J s
length	$a_0 = 4\pi\varepsilon_0\hbar^2/m_e e^2$	5.29177×10^{-11} m
energy	$E_h = m_e e^4/16\pi^2\varepsilon_0^2\hbar^2$	4.35974×10^{-18} J
time	$\hbar/E_h$	2.41888×10^{-17} s
electric current	$eE_h/\hbar$	6.62362×10^{-3} A
electric potential	E_h/e	2.72114×10^{1} V
electric dipole moment	ea_0	8.47835×10^{-30} C m
electric field strength	E_h/ea_0	5.14221×10^{11} V m^{-1}
electric polarizability	$4\pi\varepsilon_0 a_0^3$	1.64878×10^{-41} F m^2
magnetic dipole moment	$e\hbar/m_e$	1.85480×10^{-23} J T^{-1}
magnetic flux density	$\hbar/ea_0^2$	2.35052×10^{5} T
magnetizability	$e^2 a_0^2/m_e$	7.89104×10^{-29} J T^{-2}

In this form the equation is often referred to as the 'Schrödinger equation in atomic units'. The results of computations are then numbers that must be reinterpreted as physical quantities. For example, the quantity E in equation (1.18) is an energy. Solution of equation (1.19) gives the numbers $E = -1/2n^2$, for all positive integers n, and these numbers are then interpreted as the energies $E = -1/2n^2\, E_{\rm h}$.

EXAMPLE 1.21 The atomic unit of energy

By Coulomb's law, the potential energy of interaction of charges q_1 and q_2 separated by distance r is

$$V = \frac{q_1 q_2}{4\pi\varepsilon_0 r}$$

where $\varepsilon_0 = 8.85419 \times 10^{-12}\,\mathrm{F\,m^{-1}}$ is the permittivity of a vacuum. For charges $q_1 = Z_1 e$ and $q_2 = Z_2 e$ separated by distance $r = R a_0$,

$$V = \left(\frac{Z_1 Z_2}{R}\right)\left(\frac{e^2}{4\pi\varepsilon_0 a_0}\right)$$

(i) To show that the unit is the hartree unit $E_{\rm h}$ in Table 1.4, use $a_0 = 4\pi\varepsilon_0 \hbar^2 / m_e e^2$:

$$\frac{e^2}{4\pi\varepsilon_0 a_0} = \left(\frac{e^2}{4\pi\varepsilon_0}\right) \div \left(\frac{4\pi\varepsilon_0 \hbar^2}{m_e e^2}\right) = \left(\frac{e^2}{4\pi\varepsilon_0}\right) \times \left(\frac{m_e e^2}{4\pi\varepsilon_0 \hbar^2}\right) = \frac{m_e e^4}{16\pi^2 \varepsilon_0^2 \hbar^2} = E_{\rm h}$$

(ii) To calculate the value of $E_{\rm h}$ in SI units, use the values of e and a_0 given in Table 1.4. Then

$$\frac{e^2}{4\pi\varepsilon_0 a_0} = \left(\frac{1.60218^2}{4 \times 3.14159 \times 8.85419 \times 5.29177}\right) \times \left(\frac{10^{-19} \times 10^{-19}}{10^{-12} \times 10^{-11}}\right) \times \left(\frac{\mathrm{C}^2}{\mathrm{F\,m^{-1}\,m}}\right)$$

$$= (4.35975 \times 10^{-3}) \times (10^{-15}) \times (\mathrm{C^2\,F^{-1}})$$

From the definitions of the coulomb C and farad F in Table 1.2, $\mathrm{F = C^2\,J^{-1}}$ so that $\mathrm{C^2\,F^{-1} = J}$. Therefore

$$\frac{e^2}{4\pi\varepsilon_0 a_0} = 4.35975 \times 10^{-18}\,\mathrm{J} = E_{\rm h}$$

▸ Exercise 108

1.9 Exercises

Section 1.2

Calculate and express each result in its simplest form:

1. $3 + (-4)$ **2.** $3 - (-4)$ **3.** $(-3) - (-4)$ **4.** $(-3) \times (-4)$ **5.** $3 \times (-4)$

6. $8 \div (-4)$ **7.** $(-8) \div (-4)$ **8.** $\dfrac{1}{4} + \dfrac{1}{8}$ **9.** $\dfrac{3}{4} - \dfrac{5}{7}$ **10.** $\dfrac{2}{9} - \dfrac{5}{6}$

11. $\dfrac{1}{14} + \dfrac{2}{21}$ **12.** $\dfrac{1}{18} - \dfrac{2}{27}$ **13.** $\dfrac{11}{12} + \dfrac{3}{16}$ **14.** $\dfrac{1}{2} \times \dfrac{3}{4}$ **15.** $2 \times \dfrac{3}{4}$

16. $\dfrac{2}{3} \times \dfrac{5}{6}$ **17.** $\left(-\dfrac{2}{3}\right) \times \left(-\dfrac{3}{4}\right)$ **18.** $\dfrac{3}{4} \div \dfrac{4}{5}$ **19.** $\dfrac{2}{3} \div \dfrac{5}{3}$ **20.** $\dfrac{2}{15} \div \dfrac{4}{5}$

21. $\dfrac{1}{3} \div \dfrac{1}{9}$

Section 1.3

Factorize in prime numbers:

22. 6   **23.** 80   **24.** 256   **25.** 810

Simplify by factorization and cancellation:

26. $\dfrac{3}{18}$ **27.** $\dfrac{21}{49}$ **28.** $\dfrac{63}{294}$ **29.** $\dfrac{768}{5120}$

Find the value of:

30. $2!$ **32.** $7!$ **33.** $10!$

Evaluate by cancellation:

33. $\dfrac{3!}{2!}$ **34.** $\dfrac{6!}{3!}$ **35.** $\dfrac{5!}{3!2!}$ **36.** $\dfrac{10!}{7!3!}$

Section 1.4

Express as decimal fractions:

37. 10^{-2} **38.** 2×10^{-3} **39.** $2 + 3 \times 10^{-4} + 5 \times 10^{-6}$ **40.** $\dfrac{3}{8}$ **41.** $\dfrac{1}{25}$ **42.** $\dfrac{5}{32}$

Find the repeating sequence of digits in the nonterminating decimal fraction representation of:

43. $\dfrac{1}{9}$ **44.** $\dfrac{1}{11}$ **45.** $\dfrac{1}{21}$ **46.** $\dfrac{1}{17}$

Use the rules of rounding to give each of the following to 8, 7, 6, 5, 4, 3, 2 and 1 significant figures:

47. $1/13 = 0.07692\,3076923$ **48.** $\sqrt{2} = 1.414213\,562373$ **49.** $\pi = 3.141592\,653589$

Section 1.6

Simplify if possible:

50. $a^2 a^3$ **51.** $a^3 a^{-3}$ **52.** $a^3 a^{-4}$ **53.** a^3/a^2 **54.** a^5/a^{-4} **55.** $(a^3)^4$

56. $(a^2)^{-3}$ **57.** $(1/a^2)^{-4}$ **58.** $a^{1/2} a^{1/3}$ **59.** $(a^2)^{3/2}$ **60.** $(a^3 b^6)^{2/3}$

61. $(a^3 + b^3)^{1/3}$ **62.** $9^{1/2}$ **63.** $8^{2/3}$ **64.** $32^{3/5}$ **65.** $27^{-4/3}$

Evaluate:

66. $7 - 3 \times 2$ **67.** $7 - (3 \times 2)$ **68.** $(7 - 3) \times 2$ **69.** $7 + 3 \times 4 - 5$

70. $(7 + 3) \times 4 - 5$ **71.** $4 \div 2 \times 7 - 2$ **72.** $4 \div 2 + 7 \times 2$ **73.** $8 \times 2 \div 4 \div 2$

74. $3 + 4^2$ **75.** $3 + 4 \times 5^2$ **76.** $25 + 144^{1/2}$ **77.** $(5^2 + 12^2)^{1/2}$

Section 1.7

Find the sum and product of the pairs of complex numbers:

78. $z_1 = 3 + 5i, z_2 = 4 - 7i$ **79.** $z_1 = 1 - 6i, z_2 = -5 - 4i$

Section 1.8

For each of the following dimensions give its SI unit in terms of base units (column 5 of Table 1.1) and, where possible, in terms of the derived units in Table 1.2; identify a physical quantity for each:

80. L^3 **81.** ML^{-3} **82.** NL^{-3} **83.** MLT^{-1} **84.** MLT^{-2} **85.** ML^2T^{-2}

86. $ML^{-1}T^{-2}$ **87.** IT **88.** $ML^2I^{-1}T^{-3}$ **89.** $ML^2T^{-2}N^{-1}$ **90.** $ML^2T^{-2}N^{-1}\theta^{-1}$

91. Given that 1 mile (mi) is 1760 yd and 1 hour (h) is 60 min, express a speed of 60 miles per hour in **(i)** $m\,s^{-1}$, **(ii)** $km\,h^{-1}$.

92. **(i)** What is the unit of velocity in a system in which the unit of length is the inch (in = 2.54×10^{-2} m) and the unit of time is the hour (h)? **(ii)** Express this in terms of base SI units. **(iii)** A snail travels at speed 1.2 in min^{-1}. Express this in units $yd\,h^{-1}$, $m\,s^{-1}$, and $km\,h^{-1}$.

93. The non-SI unit of mass called the (international avoirdupois) pound has value 1 lb = 0.45359237 kg. The 'weight' of the mass in the presence of gravity is called the pound-force, lbf. Assuming that the acceleration of gravity is $g = 9.80665\,m\,s^{-2}$, **(i)** express 1 lbf in SI units, **(ii)** express, in SI units, the pressure that is denoted (in some parts of the world) by psi = 1 lbf in^{-2}, **(iii)** calculate the work done (in SI units) in moving a body of mass 200 lb through distance 5 yd against the force of gravity.

94. The vapour pressure of water at 20°C is recorded as $p(H_2O, 20°C) = 17.5$ Torr. Express this in terms of **(i)** the base SI unit of pressure, **(ii)** bar, **(iii)** atm.

95. The root mean square speed of the particles of an ideal gas at temperature T is $c = (3RT/M)^{1/2}$, where $R = 8.31447\,J\,K^{-1}\,mol^{-1}$ and M is the molar mass. Confirm that c has dimensions of velocity.

Express in base SI units

96. dm^{-3} **97.** $cm\,ms^{-2}$ **98.** $g\,dm^{-3}$ **99.** $mg\,pm\,\mu s^{-2}$ **100.** $dg\,mm^{-1}\,ns^{-2}$

101. $GHz\,\mu m$ **102.** $kN\,dm$ **103.** $mmol\,dm^{-3}$

104. Given relative atomic masses $A_r(^{14}N) = 14.0031$ and $A_r(^1H) = 1.0078$, calculate **(i)** the relative molar mass of ammonia, $M_r(^{14}N^1H_3)$, **(ii)** the molecular mass and **(iii)** the molar mass.

105. The bond length of HCl is $R_e = 1.2745 \times 10^{-10}$ m and the relative atomic masses are $A_r(^{35}Cl) = 34.9688$ and $A_r(^1H) = 1.0078$. **(i)** Express the bond length in (a) pm, (b) Å and (c) a_0. Calculate **(ii)** the reduced mass of the molecule and **(iii)** its moment of inertia.

106. The origin of the fundamental aborption band in the vibration–rotation spectrum of $^1H^{35}Cl$ lies at wavenumber $\tilde{v} = 2886\,cm^{-1}$. Calculate the corresponding **(i)** frequency, **(ii)** wavelength, and **(iii)** energy in units of eV and $kJ\,mol^{-1}$.

107. In the kinetic theory of gases, the mean speed of the particles of gas at temperature T is $\bar{c} = (8RT/\pi M)^{1/2}$, where M is the molar mass. **(i)** Perform an order-of-magnitude calculation of $\bar{c}$ for N_2 at 298.15 K ($M = 28.01\,g\,mol^{-1}$). **(ii)** Calculate $\bar{c}$ to 3 significant figures.

108. In the Bohr model of the ground state of the hydrogen atom, the electron moves round the nucleus in a circular orbit of radius $a_0 = 4\pi\varepsilon_0\hbar^2/m_e e^2$, now called the Bohr (radius). Given $\varepsilon_0 = 8.85419 \times 10^{-12}\,F\,m^{-1}$, use the units and values of m_e, e and $\hbar$ given in Table 1.4 to confirm **(i)** that a_0 is a length, and **(ii)** the value of a_0 in Table 1.4.

2 Algebraic functions

2.1 Concepts

When the equation of state (1.1) of the ideal gas is written in the form

$$V = \frac{nRT}{p}$$

it is implied that the value of the volume V (the dependent variable) is determined by the values of the pressure p, temperature T, and amount of substance n (the independent variables). In general, a dependent variable is said to be a function of the variable or variables on which it depends.[1] In this example, V is a function of the three variables p, T, and n. In the present chapter we are concerned with functions of only one variable; the case of more than one independent variable is discussed in Chapter 9.

Let the variable y be a function of the variable x. For example, in equation

$$y = 2x^2 - 3x + 1 \tag{2.1}$$

the expression on the right of the equal sign defines the function f,

$$f(x) = 2x^2 - 3x + 1 \tag{2.2}$$

whose value for any given value of x is to be assigned to the variable y (read $f(x)$ as 'f of x'). The function f is the **rule** for calculating y from x.

A function has a numerical value when numerical values are assigned to the variables.

EXAMPLE 2.1 The values of the function (2.2) when $x = 2$, $x = 1$, and $x = 0$ are

$$f(2) = 2 \times 2^2 - 3 \times 2 + 1 = 3$$
$$f(1) = 2 \times 1^2 - 3 \times 1 + 1 = 0$$
$$f(0) = 2 \times 0^2 - 3 \times 0 + 1 = 1$$

▸ Exercises 1–3

The concept of function is more general than this however, because the variable x can be replaced by another variable, by a function, or by a more complicated quantity such as a differential operator or a matrix.

---

[1] The word 'function' was first used in this context by the German mathematician Gottfried Wilhelm Leibniz (1646–1716).

EXAMPLE 2.2 Replace the variable x in (2.2) by the variable a.

$$f(a) = 2a^2 - 3a + 1$$

➤ Exercise 4

EXAMPLE 2.3 Replace the variable x in (2.2) by the function $h + 2$.

$$\begin{aligned}
f(h+2) &= 2(h+2)^2 - 3(h+2) + 1 \\
&= 2(h^2 + 4h + 4) - 3(h+2) + 1 \\
&= 2h^2 + 8h + 8 - 3h - 6 + 1 \\
&= 2h^2 + 5h + 3 \\
&= g(h)
\end{aligned}$$

where

$$g(x) = 2x^2 + 5x + 3$$

is a new function of x that is related to $f(x)$ by $g(x) = f(x+2)$.

➤ Exercises 5, 6

EXAMPLE 2.4 Replace the variable x in (2.2) by the differential operator $\dfrac{d}{dx}$ (see Chapter 4).

$$f\left(\frac{d}{dx}\right) = 2\left(\frac{d}{dx}\right)^2 - 3\left(\frac{d}{dx}\right) + 1 = 2\frac{d^2}{dx^2} - 3\frac{d}{dx} + 1$$

is a new differential operator.

EXAMPLE 2.5 By the equation of state of the ideal gas, the volume is a function of pressure, temperature, and amount of substance,

$$V = f(p, T, n) = nRT/p$$

and by the calculation performed in Example 1.1

$$f(10^5 \text{ Pa}, 298 \text{ K}, 0.1 \text{ mol}) = 2.478 \times 10^{-3} \text{ m}^3$$

2.2 Graphical representation of functions

A real function may be visualized either by tabulation or graphically by plotting. Consider the function

$$y = f(x) = x^2 - 2x - 3 \tag{2.3}$$

For each value of x there exists a value of y. A table can be drawn, such as Table 2.1, giving values of y corresponding to a set of values of x. In addition, each pair of numbers (x, y) in the table may be regarded as defining the position of a point in a plane, and can be plotted in a graph as in Figure 2.1.

Table 2.1

x	y
−3	12
−2	5
−1	0
0	−3
1	−4
2	−3
3	0
4	5
5	12

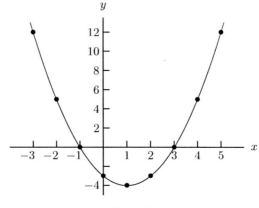

Figure 2.1

The function given by equation (2.3) is called a **quadratic** function because the highest power of x is the square (in plane geometry, quadrature is the act of squaring; that is, finding a square whose area is equal to that of a given figure). It is an example of a general class of functions called polynomials; polynomials and other algebraic functions are discussed in the following sections.

➤ Exercises 7, 8

The cartesian coordinate system[2]

The position of a point in a plane is specified uniquely by its **coordinates** in a given **coordinate system**. The most generally useful system is the cartesian (rectangular) coordinate system shown in Figure 2.2.

[2] René Descartes (1596–1650), or Renatus Cartesius, French philosopher and mathematician. He attributed his search for a universal mathematics to a mystical experience in 1619 in which 'full of enthusiasm, I discovered the foundation of a wonderful science'. He developed the relation between algebra and geometry in his *Géométrie*, published as an appendix to the *Discours de la méthode pour bien conduire sa raison et chercher la vérité dans les sciences* (Discourse on the method of good reasoning and seeking truth in the sciences), 1637. Before Descartes, the quantities x, x^2, and x^3 were always associated with the geometric concepts of line, area, and volume. Descartes discarded this restriction, 'the root (x), the square, the cube, *etc.* are merely magnitudes in continuous proportion'. His work marks the beginning of modern algebra. Descartes introduced the convention of using letters at the beginning of the alphabet $(a, b, ...)$ for constants and parameters, and letters at the end (x, y, z) for unknowns and variables. The *Géométrie* contains a formulation of the fundamental theorem of algebra, and the first use of the words 'real' and 'imaginary' in the context of complex numbers.

Coordinate geometry was also developed by Fermat at about the same time as Descartes, but his work was not published until 1679, after his death. Pierre de Fermat (1601–1665), lawyer at the provincial parliament of Toulouse, made important contributions to the theory of numbers and to the calculus.

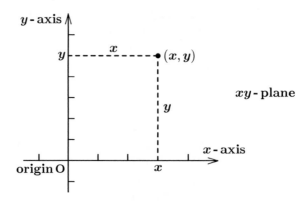

Figure 2.2

A **frame of reference** is defined in the plane, consisting of a fixed point called the **origin** of the coordinate system, and two perpendicular axes (directed lines), called the **coordinate axes**, which intersect at the origin. In Figure 2.2 the origin is labelled O, the coordinate axes are the x and y axes, and the plane is called the xy-plane. The position of a point in the plane is specified by the ordered pair (x, y), where x is the x-**coordinate** or **abscissa** and y is the y-**coordinate** or **ordinate**. A point with coordinates (x, y) lies at perpendicular distance $|x|$ from the y-axis and $|y|$ from the x-axis. It lies to the right of the (vertical) y-axis if $x > 0$ and to the left if $x < 0$; it lies above the x-axis if $y > 0$ and below if $y < 0$. The origin has coordinates $(0, 0)$.

In an actual example, suitable scales are marked on the coordinate axes, and each pair of numbers in a table such as Table 2.1 is plotted as a point on the graph. If the function is known to vary smoothly between the plotted points (as it is in this example) then the points may be joined by a smooth curve. The curve is the graphical representation of the function.

2.3   Factorization and simplification of expressions

The structure of an algebraic expression can often be simplified and clarified by the process of factorization (see Section 1.3 for factorization of numbers). For example, in the expression

$$3xy + 6x^2$$

each term of the sum can be written as the product of the common factor $(3x)$ and another term:

$$3xy + 6x^2 = (3x) \times (y) + (3x) \times (2x)$$

Therefore

$$3xy + 6x^2 = (3x) \times (y + 2x) = 3x(y + 2x)$$

and the algebraic expression has been written as the product of the two factors $(3x)$ and $(y + 2x)$.

The inverse operation of factorization is usually called **expansion** or **multiplying out**.

EXAMPLE 2.6 Factorize:

(i) $2xy^2 - 4x^2y + 6xy$

The expression $2xy$ is a common factor. Therefore

$$2xy^2 - 4x^2y + 6xy = (2xy) \times (y) - (2xy) \times (2x) + (2xy) \times (3) = 2xy(y - 2x + 3)$$

(ii) $x^2 - 5x - 6$

The aim is to express the quadratic function as the product of two linear functions; that is, to find numbers a and b such that

$$x^2 - 5x - 6 = (x + a)(x + b)$$

Expansion of the product gives

$$(x + a)(x + b) = x(x + b) + a(x + b) = x^2 + bx + ax + ab$$

and, therefore,

$$x^2 - 5x - 6 = x^2 + (a + b)x + ab$$

For this equation to be true for all values of x it is necessary that the coefficient of each power of x be the same on both sides of the equal sign: $a + b = -5$ and $ab = -6$. The two numbers whose sum is -5 and whose product is 6 are $a = -6$ and $b = 1$. Therefore

$$x^2 - 5x - 6 = (x - 6)(x + 1)$$

(iii) $x^2 - 9$

Let $x^2 - 9 = (x + a)(x + b) = x^2 + (a + b)x + ab$. In this case there is no term linear in x: $a + b = 0$, so that $b = -a$ and $ab = -a^2 = -9$. Therefore $a = \sqrt{9} = \pm 3$ and the factorization is

$$x^2 - 9 = (x + 3)(x - 3)$$

This is an example of the general form $x^2 - a^2 = (x + a)(x - a)$.

(iv) $x^4 - 5x^2 + 4$

The quartic in x is a quadratic in disguise. Replacement of x^2 by y, followed by factorization gives

$$y^2 - 5y + 4 = (y - 1)(y - 4)$$

Therefore

$$x^4 - 5x^2 + 4 = (x^2 - 1)(x^2 - 4)$$

Both the quadratic factors have the form $x^2 - a^2 = (x + a)(x - a)$ discussed in case (iii) above:

$$x^2 - 1 = (x + 1)(x - 1) \quad \text{and} \quad x^2 - 4 = (x + 2)(x - 2)$$

Therefore,

$$x^4 - 5x^2 + 4 = (x + 1)(x - 1)(x + 2)(x - 2)$$

▸ Exercises 9–16

The expansion

$$(x + a)(x + b) = x^2 + (a + b)x + ab \tag{2.4}$$

used in Examples 2.6, has geometric interpretation as the area of a rectangle of sides $(x + a)$ and $(x + b)$, as illustrated in Figure 2.3.[3]
Other important general forms are

$(a + b)^2 = a^2 + 2ab + b^2$ square of side $(a + b)$

$(a - b)^2 = a^2 - 2ab + b^2$ square of side $|a - b|$ (2.5)

$(a + b)(a - b) = a^2 - b^2$ difference of squares

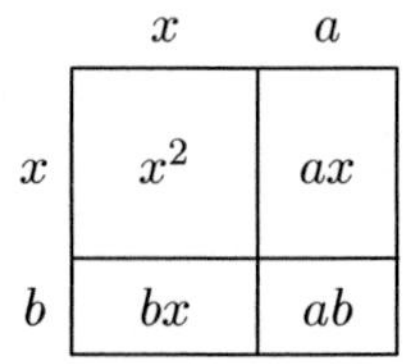

Figure 2.3

The first two equations of (2.5) can be combined by using the symbol $\pm$, meaning 'plus or minus':

$$(a \pm b)^2 = a^2 \pm 2ab + b^2 \tag{2.6}$$

in which either the upper symbol is used on *both* sides of the equation or the lower symbol is used on both sides. Sometimes the symbol $\mp$ is used in a similar way; for example, $a \mp b = \pm c$ represents the pair of equations $a - b = +c$ and $a + b = -c$.
Factorization can be used to simplify algebraic fractions. For example, in

$$\frac{xy + 2x^2}{4x + 6xy}$$

---

[3] Euclid, 'The Elements', Book II, Propositions 4 and 7 are the geometric equivalents of the first two equations (2.5) for the squares of $(a + b)$ and $(a - b)$.

both numerator and denominator have the common factor x, and can be divided by this factor (when $x \neq 0$) without changing the value of the fraction:

$$\frac{xy + 2x^2}{4x + 6xy} = \frac{\cancel{x}(y + 2x)}{2\cancel{x}(2 + 3y)} = \frac{y + 2x}{2(2 + 3y)}$$

EXAMPLE 2.7 Simplification of fractions

(1) $\dfrac{4x}{2y} = \dfrac{\cancel{2} \times (2x)}{\cancel{2} \times (y)} = \dfrac{2x}{y}$

(2) $\dfrac{3 + 6y}{9 + 18x} = \dfrac{\cancel{3}(1 + 2y)}{3^{\cancel{2}}(1 + 2x)} = \dfrac{1 + 2y}{3(1 + 2x)}$

(3) $\dfrac{a^2 - b^2}{a^2 + 2ab + b^2} = \dfrac{\cancel{(a + b)}(a - b)}{\cancel{(a + b)}(a + b)} = \dfrac{(a - b)}{(a + b)}$

➤ Exercises 17–22

2.4 Inverse functions

Given some function f and the equation $y = f(x)$, it is usually possible to define, at least for some values of x and y, a function g such that $x = g(y)$. This new function is the **inverse function** of f and is denoted by the symbol f^{-1} (not to be confused with the reciprocal $1/f$):

$$\text{if } y = f(x) \text{ then } x = f^{-1}(y) \qquad (2.7)$$

EXAMPLE 2.8 If $y = f(x) = 2x + 3$, find $x = f^{-1}(y)$.

To find x in terms of y,

(i) subtract 3 from both sides of the equation: $y = 2x + 3 \rightarrow y - 3 = 2x$

(ii) divide both sides by 2: $\rightarrow \dfrac{y - 3}{2} = x$

Therefore $x = (y - 3)/2 = f^{-1}(y)$.

In this example, y is a **single-valued** function of x; that is, for each value of x there exists just one value of y. Similarly, x is a single-valued function of y.

➤ Exercises 23–25

EXAMPLE 2.9 If $y = \dfrac{ax + b}{cx + d}$, express x in terms of y.

To solve for x,

(i) multiply both sides of the equation by $(cx+d)$: $(cx+d)y = ax+b$

(ii) expand the l.h.s.: $cxy+dy = ax+b$

(iii) subtract $(ax+b)$ from both sides: $cxy+dy-ax-b = 0$

(iv) collect the terms in x^1 and x^0: $(cy-a)x+(dy-b) = 0$

(v) subtract $(dy-b)$ from both sides: $(cy-a)x = -(dy-b)$

(vi) divide both sides by $(cy-a)$: $x = -\dfrac{(dy-b)}{(cy-a)} = f^{-1}(y)$

We note that step (vi) is not valid if $(cy-a)=0$ because division by zero has no meaning. Such complications can normally be ignored.

This example demonstrates the type of algebraic manipulation routinely used in the solution of real problems.

▸ Exercises 26–29

───

EXAMPLE 2.10 If $y=f(x)=x^2+1$, express x in terms of y.

We have

$$y=x^2+1, \quad x^2=y-1, \quad x=\pm\sqrt{y-1}=f^{-1}(y)$$

y is a single-valued function of x, but x is a **double-valued** function of y (except for $y=1$); that is, for each real value of $y>1$ there exist two real values of x (if $y<1$ then x is complex).

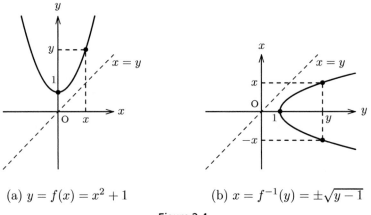

(a) $y = f(x) = x^2 + 1$ (b) $x = f^{-1}(y) = \pm\sqrt{y-1}$

Figure 2.4

Figure 2.4 shows how the graphs of the function and its inverse are related; graph (b) is obtained from (a) by interchanging the x and y axes, or by rotation around the line $x=y$. Graph (b) also shows the double-valued nature of the inverse function.

In physical applications it is usually obvious from the context which value is to be chosen. It is also seen, when x and y are real numbers, that whereas y is defined for all values of x, $-\infty < x < +\infty$, x is only defined for $y \geq 1$.

▸ Exercises 30, 31

The finding of the inverse function is not always so straightforward.

EXAMPLE 2.11 $y = f(x) = x^5 - 2x$

In this case the inverse function f^{-1} exists for all values of x, but it cannot be written in simple algebraic form, although it can be tabulated and plotted as in Figure 2.5. As in Example 2.10, graph (b) has been obtained from graph (a) by rotation around the line $x = y$.

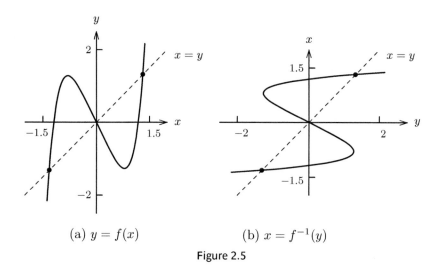

(a) $y = f(x)$ (b) $x = f^{-1}(y)$

Figure 2.5

In Example 2.11 the functional dependence of y on x is given explicitly by the right side of the equation; y is an **explicit function** of x. On the other hand, x cannot be written as an explicit function of y, and the equation defines x as an **implicit function** of y. In general, an equation of the form

$$f(x, y) = 0$$

where f is a function of both x and y, gives either variable as a function of the other. In many equations in the physical sciences it is either not possible to express one variable as an explicit function of the others, or it may not be convenient to do so.

EXAMPLE 2.12 The van der Waals equation

The equation of state for a 'slightly imperfect gas' is

$$\left(p + \frac{n^2 a}{V^2}\right)(V - nb) - nRT = 0 \tag{2.8}$$

In this case, both T and p are easily expressed as explicit functions of the other variables:

$$T = \frac{1}{nR}\left(p + \frac{n^2 a}{V^2}\right)(V - nb), \qquad p = \frac{nRT}{V - nb} - \frac{n^2 a}{V^2}$$

For V, equation (2.8) can be rearranged into

$$V^3 - n\left(b + \frac{RT}{p}\right)V^2 + \frac{n^2 a}{p}V - \frac{n^3 ab}{p} = 0$$

which is a cubic equation in V. It is possible to write down explicit solutions of a cubic equation, but these are complicated and seldom used. In this case, it is most convenient to regard equation (2.8) as defining V as an implicit function of p, T, and n. For any set of values of the independent variables and of the constants, equation (2.8) can be solved numerically by an iterative method such as the Newton–Raphson method described in Chapter 20.

➤ Exercises 32–35

2.5 Polynomials

The general polynomial of **degree** n has the form

$$f(x) = a_0 + a_1 x + a_2 x^2 + \cdots + a_n x^n \tag{2.9}$$

where the **coefficients** $a_0, a_1, \ldots, a_n$ are constants, and n is a positive integer. If $n = 0$ the function is the constant a_0. The polynomial is often written in short-hand notation as

$$f(x) = \sum_{i=0}^{n} a_i x^i \tag{2.10}$$

where the symbol Σ represents summation. The notation tells us to add together the terms $a_i x^i$ in which the integer variable i takes in turn the values $0, 1, 2, \ldots, n$:

$$\sum_{i=0}^{n} a_i x^i = (a_0 x^0) + (a_1 x^1) + (a_2 x^2) + \cdots + (a_n x^n)$$
$$= a_0 + a_1 x + a_2 x^2 + \cdots + a_n x^n$$

(remembering that $x^0 = 1$ and $x^1 = x$).

Only real coefficients are discussed here; the case of complex coefficients is shown in Section 8.4 to involve no new principles.

EXAMPLE 2.13 Write out in full:

(1) $\displaystyle\sum_{i=0}^{3} ix^i = 0 \times x^0 + 1 \times x^1 + 2 \times x^2 + 3 \times x^3 = x + 2x^2 + 3x^3$

(2) $\displaystyle\sum_{n=0}^{3} \frac{x^{2n-1}}{n+1} = \frac{x^{-1}}{1} + \frac{x^1}{2} + \frac{x^3}{3} + \frac{x^5}{4} = \frac{1}{x} + \frac{x}{2} + \frac{x^3}{3} + \frac{x^5}{4}$

(3) $\displaystyle\sum_{i=2}^{4} (-x)^i = (-x)^2 + (-x)^3 + (-x)^4 = x^2 - x^3 + x^4$

▸ Exercises 36–39

Degree $n = 1$: linear function

$$f(x) = a_0 + a_1 x \tag{2.11}$$

This is the simplest type of function, and is better known in the form

$$y = mx + c \tag{2.12}$$

The graph of the function is a straight line. It has slope m, and intercepts the vertical y-axis (when $x = 0$) at the point $y = c$, as shown in Figure 2.6.

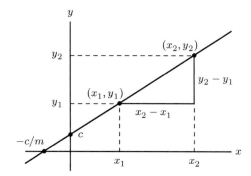

Figure 2.6

If we take any two points on the line, with coordinates (x_1, y_1) and (x_2, y_2), then

$$y_1 = mx_1 + c$$
$$y_2 = mx_2 + c$$

and

$$m = \frac{y_2 - y_1}{x_2 - x_1} \tag{2.13}$$

defines the constant slope. The line crosses the horizontal x-axis at one point:

$$y = 0 \quad \text{when} \quad x = -\frac{c}{m} \tag{2.14}$$

This value of x is called the **root** of the linear function. In general, the roots of a polynomial function are those values of the variable for which the value of the function is zero; that is, the roots are the solutions of the **polynomial equation**

$$f(x) = 0 \tag{2.15}$$

EXAMPLE 2.14 Find the equation of the straight line that passes through the points $(-1, -6)$ and $(3, 2)$.

Let the line be $y = mx + c$. Then:

at point $(x_1, y_1) = (-1, -6)$, $-6 = -m + c$

at point $(x_2, y_2) = (3, 2)$, $2 = 3m + c$

Solution of the pair of simultaneous equations (see Section 2.8) gives $m = 2$ and $c = -4$. Therefore

$$y = 2x - 4$$

The graph of the line is shown in Figure 2.7. The line has slope $m = 2$, which means that the value of y increases twice as fast as that of x. The line crosses the y-axis at $y = c = -4$, and crosses the x-axis at $x = 2$.

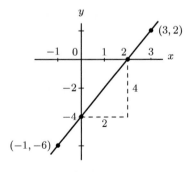

Figure 2.7

▸ Exercises 40–43

Degree $n = 2$: quadratic function

$$f(x) = a_0 + a_1 x + a_2 x^2 \qquad (2.16)$$

The quadratic function is usually written as

$$y = ax^2 + bx + c \qquad (2.17)$$

A typical graph is shown in Figure 2.1 in Section 2.2. The shape of the curve is that of a parabola. When the constant a is positive, the function has a single minimum value (turning point), and is symmetrical about a vertical line that passes through the point of minimum value. For the function in Figure 2.1,

$$f(x) = x^2 - 2x - 3 \qquad (2.18)$$

this minimum point has coordinates $(x, y) = (1, -4)$. The graph crosses the x-axis, when $f(x) = 0$, at the two points $x = -1$ and $x = 3$. These are the roots of the quadratic function, and they are the solutions of the **quadratic equation**

$$x^2 - 2x - 3 = 0 \qquad (2.19)$$

In this example the roots are easily obtained by factorization:

$$x^2 - 2x - 3 = (x + 1)(x - 3)$$

and the function is zero when either of the linear factors is zero:

$$x^2 - 2x - 3 = 0 \quad \text{when} \quad \begin{cases} \text{either} & x + 1 = 0 \implies x = -1 \\ \text{or} & x - 3 = 0 \implies x = 3 \end{cases}$$

(the symbol $\implies$ means 'implies')

   Whilst it is possible to factorize a variety of quadratic functions by trial and error, as in Examples 2.6, the roots can *always* be found by formula:[4]

$$ax^2 + bx + c = 0$$

when

$$x = \frac{-b \pm \sqrt{b^2 - 4ac}}{2a} \qquad (2.20)$$

[4] A clay tablet (YBC 6967, Yale Babylonian Collection) of the Old Babylonian Period (c. 1800–1600 BC) has inscribed on it in the Sumerian cuneiform script the following problem (in modern notation): given that $xy = 60$ and $x - y = 7$, find x and y. The prescription given for the (positive) solution corresponds to $x = \sqrt{(7/2)^2 + 60} + (7/2)$ and $y = \sqrt{(7/2)^2 + 60} - (7/2)$. The method and prescriptive approach is almost identical to that used by Al-Khwarizmi two and a half millennia later. Modern algebra became possible with the development of a general abstract notation in the 15th to 17th centuries. One important step was taken by François Viète (1540–1603). French lawyer, politician, cryptoanalyst, and amateur mathematician, he made contributions to trigonometry and algebra. He is best remembered as the man who, in his *In artem analyticem isagoge* (Introduction to the analytical art) of 1591, introduced the systematic use of symbols (letters) into the theory of equations, distinguishing between constants and variables.

The two roots are

$$x_1 = \frac{-b + \sqrt{b^2 - 4ac}}{2a}, \qquad x_2 = \frac{-b - \sqrt{b^2 - 4ac}}{2a} \qquad (2.21)$$

and the quadratic has factorized form

$$ax^2 + bx + c = a(x - x_1)(x - x_2) \qquad (2.22)$$

EXAMPLE 2.15 The roots of the quadratic function $f(x) = x^2 - 2x - 3$.

We have $a = 1$, $b = -2$, and $c = -3$ in formula (2.20). The roots are therefore

$$x = \frac{+2 \pm \sqrt{4 + 12}}{2} = 1 \pm 2 = -1 \text{ or } 3$$

and the factorized form of the function is $x^2 - 2x - 3 = (x + 1)(x - 3)$.

EXAMPLE 2.16 Find the roots of the quadratic function $f(x) = 2x^2 + 6x + 3$.

We have $a = 2$, $b = 6$, and $c = 3$ in formula (2.20), and the roots are

$$x = \frac{-6 \pm \sqrt{36 - 24}}{4} = \frac{1}{2}\left(-3 \pm \sqrt{3}\right)$$

▸ Exercises 44–46

The quantity

$$b^2 - 4ac \qquad (2.23)$$

in (2.20) is called the **discriminant** of the quadratic function. Its value in Examples 2.15 and 2.16 is positive, and the function has two real roots, but in other examples it can have zero or negative value. A graphical explanation of the three possible types of discriminant is shown in Figure 2.8.

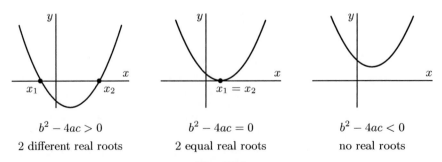

$b^2 - 4ac > 0$ $b^2 - 4ac = 0$ $b^2 - 4ac < 0$

2 different real roots   2 equal real roots   no real roots

Figure 2.8

EXAMPLE 2.17 Case $\sqrt{b^2 - 4ac} = 0$.

The quadratic

$$2x^2 - 8x + 8 = 2(x - 2)^2$$

has zero discriminant and the **double root** (two equal roots) $x = 2$.

➤ Exercises 47, 48

When the discriminant is negative, formula (2.20) requires the taking of the square root of a negative number, and the result is not a real number. In this case the roots of the quadratic are a pair of complex numbers involving the square root of -1: $i = \sqrt{-1}$ (see Section 8.2)

EXAMPLE 2.18 Case $\sqrt{b^2 - 4ac} < 0$.

The quadratic

$$x^2 - 4x + 13$$

has a pair of complex roots x_1 and x_2 given by

$$x = \frac{4 \pm \sqrt{-36}}{2} = 2 \pm 3i$$

The roots are $x_1 = 2 + 3i$ and $x_2 = 2 - 3i$ and the factorized form of the quadratic is $(x - x_1)(x - x_2)$.

➤ Exercises 49, 50

➤ Exercises 51, 52

For very large ('large enough') values of $|x|$ the term in x^2 in the quadratic $f(x) = ax^2 + bx + c$ is very much larger in magnitude than the other two terms. Thus, dividing the function by x^2,

$$\frac{f(x)}{x^2} = a + \frac{b}{x} + \frac{c}{x^2} \to a \quad \text{as } x \to \pm\infty$$

This means that for large enough positive and negative values of x, the function behaves like the simpler function ax^2, and can sometimes be replaced by it. In general for the polynomial of degree n, equation (2.9),

$$\frac{f(x)}{x^n} = a_n + \frac{a_{n-1}}{x} + \frac{a_{n-2}}{x^2} + \cdots + \frac{a_0}{x^n} \rightarrow a_n \quad \text{as } x \rightarrow \pm\infty \qquad (2.24)$$

EXAMPLE 2.19 **Behaviour of a quadratic for large values of the variable**

For values of $|x|$ larger than about 100 the function $f(x) = x^2 + x - 1$ differs from x^2 by less than 1%. The difference decreases like $1/x$ as x increases: 0.1% for $|x| = 10^3$, 0.001%, for $|x| = 10^5$, and 10^{-8}% for $|x| = 10^{10}$.

Quadratic functions are important in the physical sciences because they are used to model vibrational motions of many kinds. The simplest kind of vibrational motion is simple harmonic motion and, for example, a ball rolling forwards and backwards in a parabolic container (a 'parabolic potential well') performs simple harmonic motion.

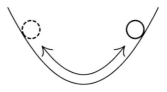

Figure 2.9

Other examples are the swings of a pendulum, the vibrations of atoms in molecules and solids, the oscillating electric and magnetic fields in electromagnetic radiation.

EXAMPLE 2.20 **The classical simple harmonic oscillator**

Figure 2.10

The simple (linear) harmonic oscillator consists of a body moving in a straight line under the influence of a force

$$F = -kx$$

whose magnitude is proportional to the displacement x of the body from the fixed point O, the point of equilibrium, and whose direction is towards this point. The (positive) quantity k is called the force constant and the negative sign ensures that the force acts in the direction opposite to that of the displacement. For a body of mass m, the energy of the system is

$$E = \frac{1}{2}mv^2 + \frac{1}{2}kx^2$$

where v is the velocity of the body. The expression for the energy is a quadratic function of the variables v and x; $\frac{1}{2}mv^2$ is the kinetic energy and $\frac{1}{2}kx^2$ is the potential energy. In the absence of external forces the total energy is constant (see Section 12.5 for a more complete discussion of the harmonic oscillator).

EXAMPLE 2.21 The simple harmonic oscillator in quantum mechanics

The stationary states of the simple harmonic oscillator in quantum mechanics are given by the solutions of the time-independent Schrödinger equation

$$-\frac{\hbar^2}{2m}\frac{d^2\psi}{dx^2} + \frac{1}{2}kx^2\psi = E\psi$$

where $\psi = \psi(x)$ is the wave function (see Example 13.11). The equation can be written in the form

$$\mathcal{H}\psi = E\psi$$

where the Hamiltonian operator for harmonic motion

$$\mathcal{H} = -\frac{\hbar^2}{2m}\frac{d^2}{dx^2} + \frac{1}{2}kx^2$$

is a quadratic function of x and of $\dfrac{d}{dx}$, since $\left(\dfrac{d}{dx}\right)^2 = \dfrac{d^2}{dx^2}$.

The general polynomial

A polynomial of degree n can always be factorized as the product of n linear factors

$$\begin{aligned} f(x) &= a_0 + a_1 x + a_2 x^2 + \cdots + a_n x^n \\ &= a_n(x - x_1)(x - x_2) \cdots (x - x_n) \end{aligned} \tag{2.25}$$

This is called the **fundamental theorem of algebra**, and was first proved by the great mathematician Gauss.[5] The function is zero when any of the linear factors is zero, and the numbers x_1, x_2, ..., x_n are the n roots of the polynomial; that is, they are the solutions of the polynomial equation $f(x) = 0$. Some of the roots may be equal (multiple roots) and some may be complex. A polynomial of odd degree ($n = 1, 3, 5, ...$) always has at least one real root because its graph must cross the x-axis at least

---

[5] Carl Friedrich Gauss (1777–1855), child prodigy and professor at Göttingen, he made substantial contributions to every important branch of pure and applied mathematics; the theory of numbers, geometry, algebra, statistics, perturbation theory, electromagnetic theory. He invented or initiated new branches of mathematics, including the theory of functions of a complex variable and the differential geometry, that formed the basis for much of 19th century mathematics. He gave his first proof (there were four) of the fundamental theorem of algebra in his doctoral thesis, *Proof of the theorem that every rational integral function in one variable can be resolved into real factors of first or second degree*, 1799.

once. In general it has an odd number of real roots. A polynomial of even degree $(n = 2, 4, 6, \ldots)$ has an even number of real roots, or no real roots if the curve does not cross the x-axis.

EXAMPLE 2.22 Factorization of a cubic

A polynomial of degree 3 can have all three roots real or it can have one real root and two complex roots; for example, in Figure 2.11,

(a) three real roots: $x^3 - 6x^2 + 11x - 6 = (x-1)(x-2)(x-3)$

(b) three real roots, one double: $x^3 - 5x^2 + 7x - 3 = (x-1)^2(x-3)$

(c) one real root and two complex roots: $x^3 - 3x^2 + 4x - 2 = (x-1)(x^2 - 2x + 2)$

The roots of the quadratic factor are $1 \pm i$, where $i = \sqrt{-1}$,

$$x^2 - 2x + 2 = [x - (1 + i)][x - (1 - i)]$$

and the fully factorized form of the cubic is

$$x^3 - 3x^2 + 4x - 2 = (x-1)(x-1-i)(x-1+i)$$

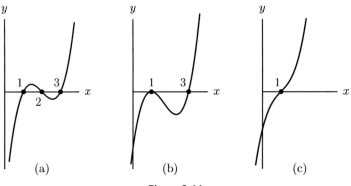

(a)   (b)   (c)

Figure 2.11

EXAMPLE 2.23 Given that $x-1$ is a factor, find the roots of the cubic $x^3 - 7x^2 + 16x - 10$.

If $x-1$ is a factor then the cubic function can be written as

$$x^3 - 7x^2 + 16x - 10 = (x-1)(ax^2 + bx + c)$$
$$= ax^3 + (b - a)x^2 + (c - b)x - c$$

For this equation to true for all values of x it is necessary that the coefficient of each power of x be the same on both sides of the equal sign:

$$1 = a, \quad -7 = b - a, \quad 16 = c - b, \quad -10 = -c$$

Then $a = 1$, $b = -6$, $c = 10$, and the quadratic factor is

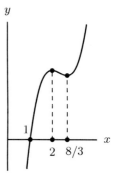

$$x^2 - 6x + 10 \quad \text{with roots} \quad x = \frac{6 \pm \sqrt{36 - 40}}{2} = 3 \pm i$$

(see also Examples 2.25 and 2.26). The cubic is therefore of type (c) in Example 2.22, with one real root, $x_1 = 2$ and the two complex roots $x_2 = 3 + i$ and $x_3 = 3 - i$. The graph of the function, Figure 2.12, shows however that, like types (a) and (b), the function has local maximum and minimum values (or turning points; see Section 4.10) at $x = 2$ and $x = 8/3$, respectively.

Figure 2.12

> Exercises 53–55

EXAMPLE 2.24 Factorization of a quartic

Three cases may be considered.

(i) 4 real roots; for example

$$x^4 - x^3 - 7x^2 + x + 6 = (x - 1)(x + 1)(x + 2)(x - 3)$$

(ii) 2 real roots and 2 complex roots; for example

$$x^4 - 2x^3 + x^2 + 2x - 2 = (x - 1)(x + 1)(x^2 - 2x + 2)$$
$$= (x - 1)(x + 1)(x - 1 - i)(x - 1 + i)$$

(iii) 4 complex roots; for example

$$x^4 - 2x^3 + 3x^2 - 2x + 2 = (x^2 + 1)(x^2 - 2x + 2)$$
$$= (x - i)(x + i)(x - 1 - i)(x - 1 + i)$$

> Exercise 56

Examples 2.22 to 2.24 demonstrate that, *if complex numbers are disallowed*, a polynomial can always be factorized as the product of some linear factors, one for each real root, and, at most, quadratic factors, all real.[6] The theorem is used in Section 2.7 for the construction of partial fractions.

---

[6] This is the statement of the fundamental theorem of algebra given in Gauss' first proof of 1799.

Algebraic functions

Polynomials are the simplest examples of **algebraic functions**. More generally, an equation of the kind

$$P(x)y^n + Q(x)y^{n-1} + \cdots + U(x)y + V(x) = 0 \qquad (2.26)$$

where $P(x)$, $Q(x)$, $\ldots$, $V(x)$ are polynomials of any (finite) degree in x, defines the variable y as an algebraic function of x. For example, the equation

$$y^3 + (x+1)y^2 + (x^2 + 3x + 2)y + (x^3 + 2x^2 - x - 1) = 0$$

is a cubic equation in y, and can be solved for each value of x.[7] Functions that cannot be defined in this way in terms of a finite number of polynomials are called **transcendental functions**. Examples are the trigonometric functions, the exponential function, and the logarithmic function; these functions are discussed in Chapter 3.

2.6   Rational functions

Let $P(x)$ and $Q(x)$ be two polynomials

$$
\begin{aligned}
P(x) &= a_0 + a_1 x + a_2 x^2 + \cdots + a_n x^n \\
Q(x) &= b_0 + b_1 x + b_2 x^2 + \cdots + b_m x^m
\end{aligned}
\qquad (2.27)
$$

A **rational function**, or **algebraic fraction**, is an algebraic function that has the general form

$$y = f(x) = \frac{P(x)}{Q(x)} = \frac{a_0 + a_1 x + a_2 x^2 + \cdots + a_n x^n}{b_0 + b_1 x + b_2 x^2 + \cdots + b_m x^m} \qquad (2.28)$$

Examples of rational functions are

$$\text{(i) } \frac{1}{x} \qquad \text{(ii) } \frac{x+2}{x+1} \qquad \text{(iii) } \frac{3x^2 + 2x - 1}{x+2} \qquad \text{(iv) } \frac{x-1}{3x^2 + 2x - 1} \qquad (2.29)$$

In each case the function is defined for all values of x for which the denominator is not zero, since division by zero is not permitted. For example, the function (i) in (2.29) is not defined at $x = 0$, and (iii) is not defined at $x = -2$. In general, the rational function

---

[7] The formula for the general solution of the cubic equation was discovered in Bologna in the early 16th century by Scipio del Ferro and Nicolo Tartaglia. The method of solution (Cardano's method) was described by Girolamo Cardano (1501–1576) in his *Ars magna* of 1545. Cardano showed that some solutions are complex. The book also contains a description of a method of solving quartic equations due to Ludovico Ferrari (1522–1565). The Norwegian mathematician Niels Henrik Abel (1802–1829) proved in his *On the algebraic resolution of equations* (1824) that there does not exist an algebraic solution of the general quintic equation, or of any polynomial equation of degree greater than 4. 'Abel's short life was filled with poverty and tragedy'; he died of consumption at the age of 27. He gave the first rigorous proof of the binomial theorem, made early contributions to group theory, and did important and innovative work on the theory of elliptic and other higher transcendental functions. The general equation of the fifth degree was solved in terms of elliptic functions by Hermite in 1858.

(2.28) is defined for all values of x with the exception of the roots of the polynomial $Q(x)$ in the denominator, for which $Q(x) = 0$.

The graph of the function $y = 1/x$ in Figure 2.13 demonstrates some typical properties of rational functions. As x approaches zero from the right $(x > 0)$ the value of $1/x$ becomes arbitrarily large; we say that $y = 1/x$ tends to infinity as x tends to zero. Similarly, y tends to minus infinity as x tends to zero from the negative side. The point $x = 0$ is called a **point of singularity**, and all rational functions have at least one such point, one for each root of $Q(x)$. The graph also shows that as $x \to 0$ from either side, the curve approaches the y-axis arbitrarily closely but does not cross it. The y-axis is the line $x = 0$ and is called an **asymptote** to the curve; we say that the curve approaches the line $x = 0$ **asymptotically**. The line $y = 0$ (the x-axis) is also an asymptote.

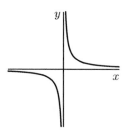

Figure 2.13

Division of one polynomial by another

The function (2.28) is called a **proper** rational function if the degree n of the numerator $P(x)$ is smaller than the degree m of the denominator $Q(x)$, as in examples (i) and (iv) of (2.29). Otherwise, as in examples (ii) and (iii), it is called **improper**. In ordinary number theory, an improper fraction is one whose value is greater than or equal to 1; for example $5/2$ or $2/2$. An improper fraction can always be reduced to a combination of proper fractions by division. An improper rational function is reduced to a combination of proper functions by **algebraic division**.

EXAMPLE 2.25 Divide $x^3 - 7x^2 + 16x - 11$ by $x - 1$.

By adapting the method of ordinary long division, we write

$$
\begin{array}{r}
x^2 - 6x + 10 \\
x - 1 \overline{\smash{)}\, x^3 - 7x^2 + 16x - 11} \\
\underline{x^3 - x^2} \\
-6x^2 + 16x - 11 \\
\underline{-6x^2 + 6x} \\
10x - 11 \\
\underline{10x - 10} \\
-1
\end{array}
$$

x into x^3 goes x^2 times

subtract

x into $-6x^2$ goes $-6x$ times

subtract

x into $10x$ goes 10 times

subtract

remainder

It follows that

$$\frac{x^3 - 7x^2 + 16x - 11}{x - 1} = x^2 - 6x + 10 - \frac{1}{x - 1}$$

EXAMPLE 2.26 Divide $x^3 - 7x^2 + 16x - 10$ by $x - 1$.

The cubic in this example is (number) 1 larger than that in Example 2.25, and there is no remainder of the division. It follows that the cubic can be factorized:

$$\frac{x^3 - 7x^2 + 16x - 10}{x - 1} = \frac{(x-1)(x^2 - 6x + 10)}{x - 1} = x^2 - 6x + 10$$

(see Example 2.23). In this case the fact that the rational function is not defined at $x = 1$ either before or after cancellation of the factor $(x - 1)$ has no practical consequences and can be ignored.

EXAMPLE 2.27 Divide $x + 2$ by $x + 1$.

In this case it is not necessary to resort to long division:

$$\frac{x+2}{x+1} = \frac{(x+1)+1}{x+1} = \frac{x+1}{x+1} + \frac{1}{x+1} = 1 + \frac{1}{x+1}$$

➤ Exercises 57–60

2.7 Partial fractions

Consider

$$\frac{1}{x^2 + 3x + 2} = \frac{1}{(x+1)(x+2)} = \frac{1}{x+1} - \frac{1}{x+2} \tag{2.30}$$

The quadratic denominator has been expressed as the product of two linear factors, and the fraction has been decomposed into two simpler **partial fractions**. We will see in Chapters 6 and 11 that the decomposition into partial fractions is an important tool in the solution of some differential equations in the theory of reaction rates, and in integration in general. A proper rational function $P(x)/Q(x)$, whose denominator can be factorized, can always be decomposed into simpler partial fractions. The following examples demonstrate three of the important simple cases. All others can be treated in the same way.

EXAMPLE 2.28 Two linear factors in the denominator

$$\frac{x+2}{(x-3)(x+4)} = \frac{5}{7(x-3)} + \frac{2}{7(x+4)}$$

To derive this result, write

$$\frac{x+2}{(x-3)(x+4)} = \frac{A}{x-3} + \frac{B}{x+4} = \frac{A(x+4)+B(x-3)}{(x-3)(x+4)}$$

It is required therefore that

$$x+2 = A(x+4) + B(x-3)$$

for *all* values of x. The values of A and B can be obtained from this equation of the numerators by the 'method of equating coefficients' described in Example 2.23. Alternatively, they are obtained directly by making suitable choices of the variable x. Thus

when $x = 3$: $5 = 7A$ and $A = 5/7$

when $x = -4$: $2 = 7B$ and $B = 2/7$

▸ Exercises 61–63

EXAMPLE 2.29 Three linear factors in the denominator

$$\frac{3x^2 + 4x - 2}{(x-1)(x-2)(x+3)} = \frac{A}{x-1} + \frac{B}{x-2} + \frac{C}{x+3}$$

$$= \frac{A(x-2)(x+3) + B(x-1)(x+3) + C(x-1)(x-2)}{(x-1)(x-2)(x+3)}$$

Then,

$$3x^2 + 4x - 2 = A(x-2)(x+3) + B(x-1)(x+3) + C(x-1)(x-2)$$

so that

when $x = 1$: $5 = -4A$ and $A = -5/4$

when $x = 2$: $18 = 5B$ and $B = 18/5$

when $x = -3$: $13 = 20C$ and $C = 13/20$

Therefore

$$\frac{3x^2 + 4x - 2}{(x-1)(x-2)(x+3)} = \frac{1}{20}\left[-\frac{25}{x-1} + \frac{72}{x-2} + \frac{13}{x+3} \right]$$

▸ Exercise 64

EXAMPLE 2.30 A repeated linear factor in the denominator

$$\frac{3x+1}{(x+3)^2} = \frac{A}{x+3} + \frac{B}{(x+3)^2} = \frac{A(x+3)+B}{(x+3)^2} = \frac{Ax+(3A+B)}{(x+3)^2}$$

and it follows that $A = 3$ and $B = -8$.

▶ Exercise 65

In the general case, the decomposition of a proper rational function $P(x)/Q(x)$ in partial fractions depends on the nature of the roots of the denominator $Q(x)$ (see Section 2.5 on the roots of the general polynomial).

(i) All the roots are real.
In this case $Q(x)$ can be factorized as the product of real linear factors; if Q has degree n then

$$Q(x) = a(x - x_1)(x - x_2) \cdots (x - x_n) \qquad (2.31)$$

where $x_1, x_2, \ldots, x_n$ are the roots. If all the roots are different then $P(x)/Q(x)$ can be decomposed into the sum of n simple fractions, as in Examples 2.28 and 2.29:

$$\frac{P(x)}{Q(x)} = \frac{c_1}{x - x_1} + \frac{c_2}{x - x_2} + \cdots + \frac{c_n}{x - x_n} \qquad (2.32)$$

If some of the roots are equal then there are additional terms, as in Example 2.30, with powers of the linear factor in the denominator. For example, if $x_1 = x_2 = x_3$ in $Q(x)$ then

$$\frac{P(x)}{Q(x)} = \frac{c_1}{x - x_1} + \frac{c_2}{(x - x_1)^2} + \frac{c_3}{(x - x_1)^3} + \frac{c_4}{x - x_4} + \cdots + \frac{c_n}{x - x_n}$$

$$= \frac{ax^2 + bx + c}{(x - x_1)^3} + \frac{c_4}{x - x_4} + \cdots + \frac{c_n}{x - x_n} \qquad (2.33)$$

(ii) Some of the roots are complex.
If complex numbers are allowed then the discussion of (i) above is applicable. If complex numbers are *disallowed* then the denominator $Q(x)$ can be factorized as the product of linear factors, one for each real root, and one or more real quadratic factors, one for each pair of complex conjugate roots (Section 8.2). The decomposition of the rational function in partial fractions then contains, in addition to the terms discussed in (i) above, one fraction for each quadratic factor, of the form

$$\frac{ax+b}{x^2 + px + q} \qquad (2.34)$$

or if the same quadratic factor occurs m times,

$$\frac{a_1x+b_1}{x^2+px+q}+\frac{a_2x+b_2}{(x^2+px+q)^2}+\cdots+\frac{a_mx+b_m}{(x^2+px+q)^m} \tag{2.35}$$

For example (see Example 2.22c),

$$\frac{1}{x^3-3x^2+4x-2}=\frac{1}{(x-1)(x^2-2x+2)}=\frac{1}{x-1}+\frac{1-x}{x^2-2x+2}$$

2.8 Solution of simultaneous equations

Consider the pair of linear equations

(1) $x+y=3$

(2) $x-y=1$

Equation (1) defines y as one function of x

$$y=3-x$$

whereas equation (2) defines y as a second function of x

$$y=x-1$$

Figure 2.14 shows that the graphs of these linear functions cross at the point $(x, y) = (2, 1)$; the two equations have the **simultaneous solution** $x=2$, $y=1$.

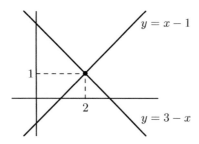

Figure 2.14

In general, an algebraic equation in two variables x and y defines either variable as an algebraic function of the other. For example, the equation

$$p(x)y^n+q(x)y^{n-1}+\cdots+u(x)y+v(x)=0 \tag{2.36}$$

defines one function $y=f(x)$. A second algebraic equation,

$$p'(x)y^m+q'(x)y^{m-1}+\cdots+u'(x)y+v'(x)=0 \tag{2.37}$$

defines a second function $y = g(x)$. The two equations have *simultaneous* solutions for those values of x for which $f(x)$ and $g(x)$ are equal. Graphically, the simultaneous real solutions are those points, if any, at which the graphs of $y = f(x)$ and $y = g(x)$ cross. For example, the two linear equations

$$a_0 x + b_0 y = c_0$$
$$a_1 x + b_1 y = c_1$$

$$(2.38)$$

can be solved to give the solution

$$x = \frac{c_0 b_1 - c_1 b_0}{a_0 b_1 - a_1 b_0}, \qquad y = \frac{a_0 c_1 - a_1 c_0}{a_0 b_1 - a_1 b_0}$$

$$(2.39)$$

We note that this solution exists only if the denominator $(a_0 b_1 - a_1 b_0)$ is not zero. Graphically, the equations (2.38) then represent two straight lines, and the solution is the point at which the lines cross.

EXAMPLE 2.31 Solve

(1) $x + y = 3$
(2) $2x + 3y = 4$

To solve for y, multiply equation (1) by 2:

(1′) $2x + 2y = 6$
(2) $2x + 3y = 4$

and subtract (1′) from (2) to give $y = -2$. Substitution for y in (1) then gives $x = 5$. The lines therefore cross at point $(x, y) = (5, -2)$

▸ Exercises 66, 67

EXAMPLE 2.32 Solve

(1) $x + y = 3$
(2) $2x + 2y = 4$

To solve, subtract twice (1) from (2):

(1) $x + y = 3$
(2′) $0 = -2$

The second equation is not possible. The equations are said to be **inconsistent** and there is no solution. This is an example for which the denominator $(a_0 b_1 - a_1 b_0)$ in equation (2.39) is zero and, graphically, it corresponds to parallel lines.

▸ Exercise 68

EXAMPLE 2.33 Solve

$$(1) \quad x+ \ y=3$$
$$(2) \quad 2x+2y=6$$

In this case, doubling equation (1) gives equation (2) and there is effectively only one independent equation; both equations represent the same line. The equations are said to be **linearly dependent** and it is only possible to obtain a partial solution; both equations give $x=3-y$ for all values of y. We will return to the general problem of solving systems of linear equations in Chapters 17 and 20.

▸ Exercise 69

EXAMPLE 2.34 Three linear equations

$$(1) \quad x+ \ y+ \ z= \ 3$$
$$(2) \quad 2x+3y+4z=12$$
$$(3) \quad x- \ y-2z=-5$$

To solve, first eliminate x from equations (2) and (3) by subtracting $2 \times (1)$ from (2) and (1) from (3):

$$(1) \quad x+ \ y+ \ z= \ 3$$
$$(2') \qquad \ y+2z= \ 6$$
$$(3') \quad -2y-3z=-8$$

Now eliminate y from (3') by adding $2 \times (2')$ to (3'):

$$(1) \quad x+y+ \ z=3$$
$$(2') \qquad y+2z=6$$
$$(3'') \qquad \quad z=4$$

The equations can now be solved in reverse order: (3'') is $z=4$, then (2') is $y+8=6$ so that $y=-2$, and (1) is $x-2+4=3$ so that $x=1$.

The method used in this example is a general systematic method for solving any number of simultaneous linear equations. It is discussed further in Chapter 20.

▸ Exercise 70

EXAMPLE 2.35 A line and an ellipse

$$(1) \qquad \quad 3x+4y \ = \ 4$$
$$(2) \quad 2x^2+3xy+2y^2=16$$

Equation (1) can be solved for y in terms of x and the result substituted in equation (2) of the ellipse (see Section 19.5 on quadratic forms). Thus, from (1), $y=1-3x/4$, and (2) becomes

$$x^2-16=(x+4)(x-4)=0$$

This has roots $x_1 = -4$ and $x_2 = 4$, with corresponding value of y, $y_1 = 4$ and $y_2 = -2$. In this case the two solutions are the points at which the straight line crosses the ellipse; $(x_1, y_1) = (-4, 4)$ and $(x_2, y_2) = (4, -2)$, as in Figure 2.15.

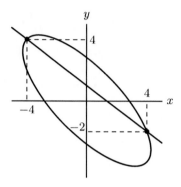

Figure 2.15

> Exercises 71–73

2.9 Exercises

Section 2.1

1. Find the values of $y = 2 - 3x$ for:
   (i) $x = 0$ (ii) $x = 2$ (iii) $x = -3$ (iv) $x = 2/3$
2. Find the values of $y = 2x^2 + 3x - 1$ for:
   (i) $x = 0$ (ii) $x = 1$ (iii) $x = -1$ (iv) $x = -2/3$
3. Given $f(x) = x^3 - 3x^2 + 4x - 3$, find:
   (i) $f(5)$ (ii) $f(0)$ (iii) $f(-2)$ (iv) $f(-2/3)$
4. If $f(x) = 2x^2 + 4x + 3$, what is:
   (i) $f(a)$ (ii) $f(y^2)$
5. If $f(x) = x^2 - 3x - 4$, what is:
   (i) $f(a + 3)$ (ii) $f(a^2 + 1)$ (iii) $f(x + 1)$ (iv) $f(x^2 - 3x - 4)$
6. If $f(x) = 2x - 1$ and $g(x) = 3x + 1$, express $f(g)$ as a function of x.

Section 2.2

Make a table of (x, y) values and sketch a fully labelled graph of the quadratic:
7. $y = x^2 - 4x$ 8. $y = -x^2 - x + 2$

Section 2.3

Factorize:
9. $6x^2y^2 - 2xy^3 - 4y^2$ 10. $x^2 + 6x + 5$ 11. $x^2 + x - 6$
12. $x^2 - 8x + 15$ 13. $x^2 - 4$ 14. $4x^2 - 9$
15. $2x^2 + x - 6$ 16. $x^4 - 10x^2 + 9$

Simplify if possible:

17. $\dfrac{x}{3x^2 + 2x}$ **18.** $\dfrac{x+2}{x+4}$ **19.** $\dfrac{x^2 - 4}{x - 2}$ **20.** $\dfrac{x^2 + 3x + 2}{x + 2}$ **21.** $\dfrac{x^2 - 9}{x^2 + 5x + 6}$

22. $\dfrac{2x^2 - 3x + 1}{x^2 - 3x + 2}$

Section 2.4

Find x as a function of y:

23. $y = x - 2$ **24.** $y = \dfrac{1}{2}(3x + 1)$ **25.** $y = \dfrac{1}{3}(2 - x)$ **26.** $y = \dfrac{x}{1 - x}$

27. $y = \dfrac{2x + 3}{3x - 2}$ **28.** $y = \dfrac{x - 1}{2x + 1}$ **29.** $y = \dfrac{x^2 - 1}{x^2 + 1}$

For $y = f(x)$, **(i)** find x as a function of y, **(ii)** sketch graphs of $y = f(x)$ and $x = f^{-1}(y)$:

30. $y^2 = x^2 + 1$ **31.** $y = (x^2 - 1)^2$

32. The virial equation of state of a gas can be approximated at low pressure as

$$pV_{\mathrm{m}} = RT\left(1 + \frac{B}{V_{\mathrm{m}}}\right)$$

where p is the pressure, V_{m} is the molar volume, T is the temperature, R is the gas constant, and B is the second virial coefficient. Express B as an explicit function of the other variables.

33. Kohlrausch's law for the molar conductivity Λ_{m} of a strong electrolyte at low concentration c is

$$\Lambda_{\mathrm{m}} = \Lambda_{\mathrm{m}}^0 - K\sqrt{c}$$

where Λ_{m}^0 is the molar conductivity at infinite dilution and K is a constant. Express c as an explicit function of Λ_{m}.

34. The Langmuir adsorption isotherm

$$\theta = \frac{Kp}{1 + Kp}$$

gives the fractional coverage θ of a surface by adsorbed gas at pressure p, where K is a constant. Express p in terms of θ.

35. In Example 2.12 on the van der Waals equation, verify the explicit expressions given for T and p, and the cubic equation in V.

Section 2.5

Expand (write out in full):

36. $\displaystyle\sum_{n=0}^{2}(n + 1)x^n$ **37.** $\displaystyle\sum_{i=0}^{3} i x^{i-1}$ **38.** $\displaystyle\sum_{k=1}^{3} k(k + 1)x^{-k}$ **39.** $\displaystyle\sum_{n=0}^{3} n! x^{n^2}$

Find the equation and sketch the graph of the straight line that passes through the points:

40. $(-2, -5)$ and $(1, 4)$ **41.** $(-1, 6)$ and $(3, -2)$

42. Explain how $\mathcal{K}$ and Λ_m^0 in Kohlrausch's law (Exercise 33),

$$\Lambda_m = \Lambda_m^0 - \mathcal{K}\sqrt{c}$$

can be obtained graphically from the results of measurements of Λ_m over a range of concentration c.

43. The Debye equation

$$\frac{\varepsilon_r - 1}{\varepsilon_r + 2} = \frac{\rho}{M}\frac{N_A}{3\varepsilon_0}\left(\alpha + \frac{\mu^2}{3kT}\right)$$

relates the relative permittivity (dielectric constant) ε_r of a pure substance to the dipole moment μ and polarizability α of the constituent molecules, where ρ is the density at temperature T, and M, N_A, k, and ε_0 are constants. Explain how μ and α can be obtained graphically from the results of measurements of ε_r and ρ over a range of temperatures.

Find the roots and sketch the graphs of the quadratic functions:

44. $x^2 - 3x + 2$ **45.** $-2x^2 - 3x + 2$ **46.** $3x^2 - 3x - 1$ **47.** $-x^2 + 6x - 9$

48. $4x^2 + 4x + 1$ **49.** $x^2 + x + 2$ **50.** $-3x^2 + 3x - 1$

51. If $y = \dfrac{2x^2 + x + 1}{2x^2 + x - 1}$ find x as a function of y.

52. The acidity constant K_a of a weak acid at concentration c is

$$K_a = \frac{\alpha^2 c}{1 - \alpha}$$

where α is the degree of ionization. Express α in terms of K_a and c (remember that α, K_a, and c are positive quantities).

Given that $x - 1$ is a factor of the cubic, **(i)** find the roots, **(ii)** sketch the graph:

53. $x^3 + 4x^2 + x - 6$ **54.** $x^3 - 6x^2 + 9x - 4$ **55.** $x^3 - 3x^2 + 3x - 1$

56. Given that $x^2 - 1$ is a factor of the quartic $x^4 - 5x^3 + 5x^2 + 5x - 6$, **(i)** find the roots, **(ii)** sketch the graph.

Section 2.6

Use algebraic division to reduce the rational function to proper form:

57. $\dfrac{2x - 1}{x + 3}$ **58.** $\dfrac{3x^3 - 2x^2 - x + 4}{x + 2}$ **59.** $\dfrac{x^3 + 2x^2 - 5x - 6}{x + 1}$

60. $\dfrac{2x^4 - 3x^3 + 4x^2 - 5x + 6}{x^2 - 2x - 2}$

Section 2.7

Express in terms of partial fractions:

61. $\dfrac{1}{(x - 1)(x + 2)}$ **62.** $\dfrac{x + 2}{x(x + 3)}$ **63.** $\dfrac{x - 2}{x^2 + 3x + 2}$ **64.** $\dfrac{2x^2 - 5x + 7}{x(x - 1)(x + 2)}$

65. $\dfrac{x^2 + 2x - 1}{(x - 1)^2(x + 2)}$

Section 2.8

Solve the simultaneous equations:

66. $x + y = 3, x - y = 1$

67. $3x - 2y = 1, 2x + 3y = 2$

68. $3x - 2y = 1, 6x - 4y = 3$

69. $3x - 2y = 1, 6x - 4y = 2$

70. $x - 2y + 3z = 3, 2x - y - 2z = 8, 3x + 3y - z = 1$

71. $2x - y = 1, x^2 - xy + y^2 = 1$

72. $2x - y = 2, x^2 - xy + y^2 = 1$

73. $2x - y = 3, x^2 - xy + y^2 = 1$

3 Transcendental functions

3.1 Concepts

The mathematical description of physical phenomena often involves functions other than the algebraic functions discussed in Chapter 2. The most important of these transcendental functions are the trigonometric functions, discussed in Sections 3.2 to 3.5, and the exponential function and its inverse function, the logarithmic function, discussed in Sections 3.6 to 3.8. Other functions are defined in terms of these 'elementary functions', and a brief description of the hyperbolic functions is given in Section 3.9.

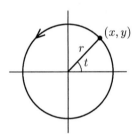

(i) circular motion:

$$x(t) = r \cos t, \quad y(t) = r \sin t$$

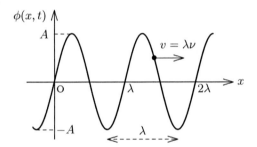

(ii) wave motion:

$$\phi(x, t) = A \sin 2\pi (x/\lambda - \nu t)$$

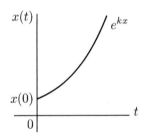

(iii) exponential growth:

$$x(t) = x(0) \, e^{kx}$$

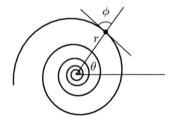

(iv) logarithmic spiral:

$$\ln r = \theta \cot \phi; \quad r = e^{\theta \cot \phi}$$

Figure 3.1

Some applications are illustrated in Figure 3.1. Three of these involve trigonometric functions. Trigonometry as a distinct branch of mathematics has its origins as a tool for the construction of astronomical tables. It is concerned with the use of trigonometric functions for the solution of geometric problems involving triangles,[1] and is important in structural and architectural design, astronomy, and navigation. The trigonometric functions are important in the physical sciences for the description of periodic motion, including circular motion, as in figure 3.1(i), and wave motion, Figure 3.1(ii). They are also essential for the description of systems with periodic structure, as in the use of Bragg's law, equation (1.2) in Chapter 1, for the interpretation of the diffraction of X-rays from the surfaces of crystal lattices.

The exponential function e^x has the unique and definitive property that the slope of its graph at any point (its derivative, see Chapter 4) is equal to the value of the function at that point. This makes it particularly useful for the modeling of all types of first-order rate processes, in which the rate of change (growth or decay) of a property is proportional to the value of the property. Figure 3.1(iii) show exponential growth (with k positive) which provides a model of, for example, population explosion, bacterial growth in a culture, and a nuclear chain reaction. When k is negative, exponential decay provides a model of first-order chemical reactions, radioactive decomposition of nuclei and, more trivially, the popularity of fashions. In quantum theory, the function e^{-r} is the 1s orbital of the hydrogen atom if r is the distance of the electron from the nucleus.

The logarithmic function $\ln x$ is the inverse function of e^x. It is often used as an alternative to the exponential, and is involved in the definition and representation of fundamental physical concepts in thermodynamics, such as entropy and chemical potential. The Arrhenius and Nernst equations, (1.3) and (1.4), provide an example of the dual relation of the inverse functions. Another example is given in Figure 3.1(iv) by alternative equations for the logarithmic spiral observed in natural phenomena, as in the growth of shells of molluscs, flight behaviour of some birds and, on a larger scale, the shape of hurricanes and some galaxies.

3.2   Trigonometric functions

Geometric definitions

The principal trigonometric functions of the angle θ, the internal angle at A of the right-angle triangle in Figure 3.2, are the **sine** (sin), the **cosine** (cos), and the **tangent** (tan):

$$\sin\theta = \frac{\text{opposite}}{\text{hypotenuse}} = \frac{BC}{AC} \qquad (3.1)$$

$$\cos\theta = \frac{\text{adjacent}}{\text{hypotenuse}} = \frac{AB}{AC} \qquad (3.2)$$

$$\tan\theta = \frac{\text{opposite}}{\text{adjacent}} = \frac{BC}{AB} = \frac{\sin\theta}{\cos\theta} \qquad (3.3)$$

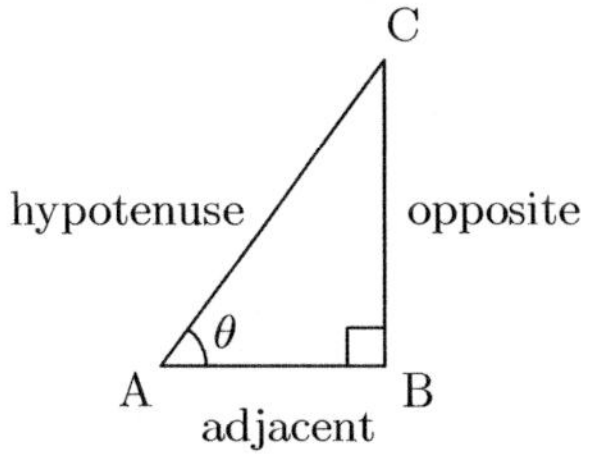

Figure 3.2

---

[1] The earliest 'trigonometric tables' were constructed in about 150 BC by the astronomer Hipparchus of Nicaea (modern Iznik in Turkey), to whom we owe the 360° circle, and by Claudius Ptolemy of Alexandria (*c.* 100–178 AD) whose *Syntaxis mathematica* (Mathematical Synthesis), known as the *Almagest*, was called by the Arabs *al-magisti* (the greatest) and whose tables were used by astronomers for over a thousand years. The tables in the Hindu *Siddhantas* (about 400 AD) are essentially tabulations of the sine function. Arab mathematicians (about 950 AD) added new tabulations and theorems. European trigonometry was developed by Johann Müller (Regiomontanus) of Königsberg (1436–1476), and by Georg Joachim Rheticus (1514–1576) of Wittenburg, a student of Copernicus, whose *Opus palatinum de triangulis* (The palatine work on triangles), 1596, focused for the first time on the properties of the right-angled triangle. François Viète (1540–1603) built on this work with extensive new tables for all six common functions, new formulas, and the use of trigonometric functions for the solution of algebraic equations. The word 'trigonometry' came into use in about 1600.

Several derived functions are also defined, the most important being the **secant** (sec), **cosecant** (cosec), and **cotangent** (cot):

$$\sec\theta = \frac{1}{\cos\theta}, \quad \operatorname{cosec}\theta = \frac{1}{\sin\theta}, \quad \cot\theta = \frac{1}{\tan\theta} = \frac{\cos\theta}{\sin\theta} \tag{3.4}$$

EXAMPLE 3.1 For the angles in Figure 3.3,

$$\sin\theta = \frac{4}{5} \qquad \cos\theta = \frac{3}{5} \qquad \tan\theta = \frac{4}{3}$$

$$\operatorname{cosec}\theta = \frac{5}{4} \qquad \sec\theta = \frac{5}{3} \qquad \cot\theta = \frac{3}{4}$$

$$\sin\phi = \frac{3}{5} \qquad \cos\phi = \frac{4}{5} \qquad \tan\phi = \frac{3}{4}$$

$$\operatorname{cosec}\phi = \frac{5}{3} \qquad \sec\phi = \frac{5}{4} \qquad \cot\phi = \frac{4}{3}$$

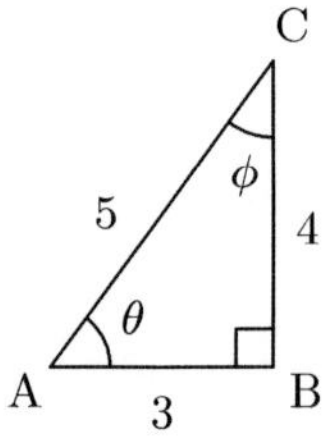

Figure 3.3

▸ Exercise 1

One of the best known properties of the right-angled triangle is the theorem of Pythagoras.[2]

$$AB^2 + BC^2 = AC^2 \tag{3.5}$$

This can be written as a trigonometric equation by dividing both sides by AC^2:

$$\frac{AB^2}{AC^2} + \frac{BC^2}{AC^2} = 1$$

or

$$\sin^2\theta + \cos^2\theta = 1 \tag{3.6}$$

(a quantity like $(\sin\theta)^2$, the square of the sine of θ, is usually written as $\sin^2\theta$).

---

[2] Pythagoras (*c.* 580–*c.* 500 BC). Born on the island of Samos, he travelled widely and was one of the principal importers of Egyptian and Babylonian mathematics and astronomy into the Greek world. He settled in Croton, in Southern Italy, where he founded a religious and philosophical society with a strong mathematical basis; motto 'all is number'. He is reputed to have coined the word 'mathematics'; that which is learned. Attributions to him of mathematical discoveries are traditional; Pythagoras' theorem was known in the old Babylonian period, and the existence of irrational numbers was possibly a discovery of the later Pythagoreans in about 400 BC. The Pythagorean school introduced the systematic study of the principles of mathematics; number theory and geometry.

The general form of the theorem of Pythagoras is: 'In a right-angled triangle, the area of the figure on the hypotenuse is equal to the sum of the areas of the similar figures on the other two sides' (Euclid, 'The elements', Book VI, Proposition 31 in the Heath translation).

EXAMPLE 3.2 For the triangle in Figure 3.3, $3^2 + 4^2 = 5^2$ and

$$\sin^2\theta + \cos^2\theta = \left(\frac{4}{5}\right)^2 + \left(\frac{3}{5}\right)^2 = \frac{4^2 + 3^2}{5^2} = \frac{25}{25} = 1$$

➤ Exercises 2

Units of angle

The ordinary unit of angle is the degree. Figure 3.4 shows the right angle (90°), the angle on a line (180°) and the angle round a point (360°).

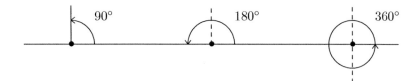

$90°$ $180°$ $360°$

Figure 3.4

The unit of angle that is always used in mathematical and scientific applications is the radian (SI symbol rad), defined in terms of the properties of the circle. Figure 3.5 shows a circle of radius r, and an arc of length s subtending the angle θ at the centre. The length of the arc is proportional to the size of the angle; for example, doubling the angle θ doubles the length s. It follows that the ratio of s to the circumference of the circle $(2\pi r)$ is equal to the ratio of θ to the complete angle around the centre (360°):

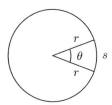

Figure 3.5

$$\frac{s}{2\pi r} = \frac{\theta}{360°} \tag{3.7}$$

so that

$$\theta = \frac{s}{r} \times \frac{360°}{2\pi} \tag{3.8}$$

The unit of angle, the radian, is defined by

$$1 \text{ rad} = \frac{360°}{2\pi} \approx 57°18' \tag{3.9}$$

(rad is the SI unit of angle; sometimes a superscript c is used, so that $1^c = 1$ rad). In practice, the symbol for the unit is omitted, and an angle given without unit is assumed to be in radians; for example

$$90° = \frac{\pi}{2}, \qquad 180° = \pi, \qquad 360° = 2\pi, \qquad \sin\frac{\pi}{6} = \sin 30°$$

(angles are often quoted as multiples of π). The length of an arc of a circle is $s = r\theta$ when θ is in radians.

EXAMPLE 3.3 Radians and degrees

(1) The angle 40° in units of the radian is

$$40° = 40° \times \frac{2\pi}{360°} = \frac{2\pi}{9} \approx 0.7 \text{ rad}$$

(2) The angle 0.5 rad in degrees is

$$0.5 = 0.5 \times \frac{360°}{2\pi} = \frac{90°}{\pi} \approx 28.6°$$

(3) The length of the arc that subtends an angle of $\theta = 2$ rad at the centre of a circle of radius $r = 3$ is

$$s = r\theta = 3 \times 2 = 6$$

➤ Exercises 3–5

Trigonometric functions for all angles

The geometric definition of the trigonometric functions in terms of ratios of the sides of a right-angled triangle restricts the functions to values of angles in the range 0° to 90°, or

$$0 \le \theta \le \frac{\pi}{2}$$

The definitions can be extended to all values of the angle by consideration of the coordinates of a point on a circle, as shown in Figure 3.6.

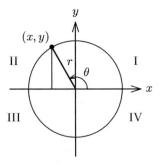

Figure 3.6

The trigonometric functions are defined for all points on the circle as

$$\sin\theta = \frac{y}{r} \qquad \cos\theta = \frac{x}{r} \qquad \tan\theta = \frac{y}{x} \qquad\qquad (3.10)$$

In the first quadrant, I, with angles $0 < \theta < \pi/2$, the values of x and y are both positive and

$$\sin\theta > 0 \qquad \cos\theta > 0 \qquad \tan\theta > 0$$

In the second quadrant however, with $\pi/2 < \theta < \pi$, x is negative and y is positive. The signs of the trigonometric functions in all four quadrants are given in Table 3.1.

Table 3.1 Signs of the trigonometric functions

Quadrant	I	II	III	IV
Angles	$0 < \theta < \pi/2$	$\pi/2 < \theta < \pi$	$\pi < \theta < 3\pi/2$	$3\pi/2 < \theta < 2\pi$
	$0 < \theta < 90°$	$90° < \theta < 180°$	$180° < \theta < 270°$	$270° < \theta < 360°$
$\sin\theta$	$+$	$+$	$-$	$-$
$\cos\theta$	$+$	$-$	$-$	$+$
$\tan\theta$	$+$	$-$	$+$	$-$

EXAMPLE 3.4 By Table 3.2, $\sin\pi/3 = \sqrt{3}/2$, $\cos\pi/3 = 1/2$, and $\tan\pi/3 = \sqrt{3}$. What are the values of the sine, cosine, and tangent of $2\pi/3$, $4\pi/3$, and $5\pi/3$?

The angle $2\pi/3$ is in the second quadrant, $4\pi/3$ in the third, $5\pi/3$ in the fourth. Therefore, by Figure 3.7,

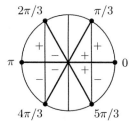

Figure 3.7

$$\sin 2\pi/3 = +\sqrt{3}/2 \qquad \cos 2\pi/3 = -1/2 \qquad \tan 2\pi/3 = -\sqrt{3}$$

$$\sin 4\pi/3 = -\sqrt{3}/2 \qquad \cos 4\pi/3 = -1/2 \qquad \tan 4\pi/3 = +\sqrt{3}$$

$$\sin 5\pi/3 = -\sqrt{3}/2 \qquad \cos 5\pi/3 = +1/2 \qquad \tan 5\pi/3 = -\sqrt{3}$$

▸ Exercises 6

Special values

The values for the angles on the boundaries of the four quadrants, and for a few other angles, are listed in Table 3.2.

Table 3.2 Some special values

θ	0°	90°	180°	270°	360°	30°	60°	45°
θ/rad	0	$\pi/2$	π	$3\pi/2$	2π	$\pi/6$	$\pi/3$	$\pi/4$
$\sin\theta$	0	1	0	-1	0	$1/2$	$\sqrt{3}/2$	$1/\sqrt{2}$
$\cos\theta$	1	0	-1	0	1	$\sqrt{3}/2$	$1/2$	$1/\sqrt{2}$
$\tan\theta$	0	∞	0	$-\infty$	0	$1/\sqrt{3}$	$\sqrt{3}$	1

EXAMPLE 3.5 Demonstrate that $\sin \pi/2 = 1$ and $\cos \pi/2 = 0$.

By Figure 3.8,

$$\sin\theta = a/c, \qquad \cos\theta = b/c$$

As the angle θ increases to 90°, the magnitude of side b approaches zero whilst that of side a aproach the value of c. Therefore

$$\sin\theta \to c/c = 1 \quad \text{and} \quad \cos\theta \to 0/c = 0 \quad \text{as} \quad \theta \to \pi/2$$

Figure 3.8

▸ Exercises 7

EXAMPLE 3.6 Verify the values of the trigonometric functions for $\theta = \pi/6$ in Table 3.2.

Draw an equilateral triangle with sides of length 2 and bisect the triangle as in Figure 3.9. Then, by Pythagoras, $h = \sqrt{2^2 - 1^2} = \sqrt{3}$, and $\sin \pi/6 = 1/2$, $\cos \pi/6 = \sqrt{3}/2$, $\tan \pi/6 = 1/\sqrt{3}$

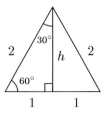

Figure 3.9

▸ Exercise 8

Negative angles

Each point on the circle can be reached by either anti-clockwise rotation or by clockwise rotation. An angle is defined to have positive value for an anti-clockwise rotation and negative for a clockwise rotation.

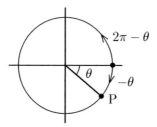

Figure 3.10

The point P in Figure 3.10, corresponding to the negative angle $-\theta$, can be reached by anti-clockwise rotation through angle $(2\pi - \theta)$, and the two angles have the same trigonometric values:

$$\sin(-\theta) = \sin(2\pi - \theta) = -\sin\theta$$
$$\cos(-\theta) = \cos(2\pi - \theta) = +\cos\theta \qquad (3.11)$$
$$\tan(-\theta) = \tan(2\pi - \theta) = -\tan\theta$$

Further angles

The range of allowed values of the angle can be extended further by allowing one or more complete rotations around the centre. Each complete rotation adds or subtracts 2π, and the angles $\theta \pm 2\pi n$, for all values of the integer $n = 0, 1, 2, 3 \ldots$, have the same trigonometric values:

$$\sin(\theta \pm 2\pi n) = \sin\theta, \qquad \cos(\theta \pm 2\pi n) = \cos\theta \qquad (3.12)$$

In addition, the tangent repeats every half rotation,

$$\tan(\theta \pm \pi n) = \tan\theta \qquad (3.13)$$

We see that, whereas every angle corresponds to a point on the circle, each point corresponds to an infinite number of angles. The graphs of sine, cosine, and tangent are shown in Figure 3.11.[3]

---

[3] The graph of the sine function was first drawn in 1635 by Gilles Personne de Roberval (1602–1675).

(a) $y = \sin x$

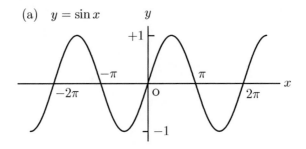

(b) $y = \cos x$

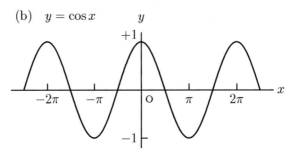

(c) $y = \tan x$

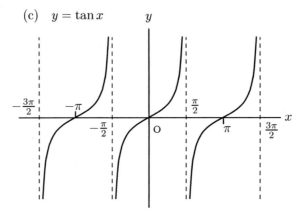

Figure 3.11

➤ Exercises 9, 10

Periodic functions

A function $f(x)$ with the property

$$f(x \pm a) = f(x) \tag{3.14}$$

is called a **periodic function** of x with **period** a; the value of the function is unchanged when x is replaced by $x + a$ or $x - a$. The sine and cosine functions are periodic functions with period 2π, the tangent is periodic with period π. The sine and cosine curves in Figure 3.11 are called **harmonic waves**, and the functions form the basis for the description of all forms of waves and other oscillatory motions.

EXAMPLE 3.7 A harmonic wave travelling in the positive x direction (a plane wave) is described by the wave function

$$\phi(x, t) = A \sin 2\pi \left(\frac{x}{\lambda} - vt \right)$$

(or by the equivalent cosine function), and is shown in Figure 3.12 at time $t = 0$.

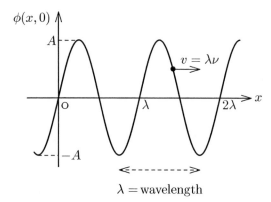

Figure 3.12

The shortest distance between equivalent points on the curve (the period with respect to changes in x) is the wavelength λ. The speed of propagation of the wave is $v = \lambda v$, where v is the frequency, or number of oscillations per unit time, and $\tau = 1/v$ is the (time) period, the time of a complete oscillation. The number A is the amplitude of the wave.

> Exercises 11

EXAMPLE 3.8 In classical mechanics, Newton's second law of motion states that the force acting on a body is equal to the mass of the body times its acceleration:

$$f = ma$$

For the simple harmonic oscillator (see Example 2.20 and Section 12.5), the force is $f = -kx$ and the acceleration is the second derivative of distance x with respect to time (see Chapter 4 on differentiation):

$$-kx = m \frac{d^2 x}{dt^2}$$

This is one of the simplest second-order differential equations (Chapter 12) and a solution is

$$x(t) = A \cos \omega t$$

where A, the amplitude, is the maximum displacement from equilibrium and $\omega = \sqrt{k/m}$ is called the angular frequency. The displacement $x(t)$ is periodic with respect to time, with period

$$\tau = \frac{2\pi}{\omega} = \frac{1}{\nu}$$

where $\nu = \frac{1}{2\pi}\sqrt{\frac{k}{m}}$ is the frequency of vibration. A plot of displacement against time is very similar to that in Figure 3.12 (see Figure 12.3).

3.3 Inverse trigonometric functions

If $y = \sin x$ then x is the angle whose sine is y, and is given by the inverse sine function $\sin^{-1} y$:

$$\text{if } y = \sin x \text{ then } x = \sin^{-1} y \qquad (3.15)$$

Because of the possible confusion between the notation for the inverse sine, $\sin^{-1} y$, and the inverse of the sine, $(\sin y)^{-1}$, an alternative notation for the inverse trigonometric functions is often used:

$$\arcsin y = \sin^{-1} y, \qquad \arccos y = \cos^{-1} y, \qquad \arctan y = \tan^{-1} y \qquad (3.16)$$

The inverse functions are multi-valued functions; for example, as indicated in Figure 3.13, many angles have the same sine:

$$\sin x = \sin(\pi - x) = \sin(x \pm 2n\pi) \qquad (3.17)$$

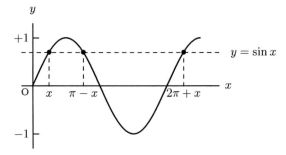

Figure 3.13

To overcome this ambiguity, a **principal value** has been defined for each inverse function:

$$x = \sin^{-1} y \qquad -\frac{\pi}{2} \leq x \leq \frac{\pi}{2} \qquad \text{(quadrants I and IV)}$$

$$x = \cos^{-1} y \qquad 0 \leq x \leq \pi \qquad \text{(quadrants I and II)} \qquad (3.18)$$

$$x = \tan^{-1} y \qquad -\frac{\pi}{2} \leq x \leq \frac{\pi}{2} \qquad \text{(quadrants I and IV)}$$

They are the values computed, for example, on a pocket calculator or other computer.

EXAMPLE 3.9

$$\cos\frac{\pi}{6} = \cos\left(-\frac{\pi}{6}\right) = \cos\left(\frac{13\pi}{6}\right) = \cdots = \frac{\sqrt{3}}{2}$$

Angle $\pi/6$ lies in the first quadrant and is the principal value of the inverse function:

$$\cos^{-1}\left(\frac{\sqrt{3}}{2}\right) = \frac{\pi}{6}$$

▸ Exercises 12, 13

3.4 Trigonometric relations

The sine and cosine rules

The angles and sides of a triangle (Figure 3.14) are related by two rules:

sine rule:

$$\frac{\sin A}{a} = \frac{\sin B}{b} = \frac{\sin C}{c} \qquad (3.19)$$

cosine rule:

$$a^2 = b^2 + c^2 - 2bc \cos A \qquad (3.20)$$

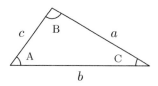

Figure 3.14

Proofs

For the sine rule the proof is, using Figure 3.15,

$$\sin A = \frac{h}{c}, \quad \sin C = \frac{h}{a}$$

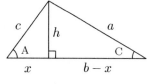

Figure 3.15

so that $h = c \sin A = a \sin C$. Then

$$\frac{\sin A}{a} = \frac{\sin C}{c}$$

Similarly for the third angle and side.

For the cosine rule, from Figure 3.15 and Pythagoras' theorem,

$$a^2 = h^2 + (b-x)^2 = h^2 + b^2 + x^2 - 2bx, \quad c^2 = h^2 + x^2$$

and the rule follows since $x = c \cos A$.

EXAMPLE 3.10 Given the lengths $a = 2$, $c = 3$, and the angle $B = \pi/3$ of the triangle in Figure 3.14, find the third side and the other angles.

Given two sides and the included angle, we use the cosine rule to find b:

$$b^2 = a^2 + c^2 - 2ac \cos B = 4 + 9 - 12 \cos \pi/3$$

and $\cos \pi/3 = 1/2$. Therefore $b^2 = 4 + 9 - 6 = 7$ and $b = \sqrt{7}$.
  To find the other two angles we use the cosine rule for one; for example,

$$a^2 = b^2 + c^2 - 2bc \cos A$$

so that

$$\cos A = (b^2 + c^2 - a^2)/2bc = 2/\sqrt{7} \quad \text{and} \quad A = \cos^{-1}\left(\frac{2}{\sqrt{7}}\right) \approx 40.89°$$

Then $C \approx 180° - 60° - 40.89° = 79.11°$.

▶ Exercises 14–17

Compound-angle identities

The sines and cosines of the sum and difference of two angles are

$$\sin(x+y) = \sin x \cos y + \cos x \sin y$$
$$\sin(x-y) = \sin x \cos y - \cos x \sin y$$

$$\tag{3.21}$$

$$\cos(x+y) = \cos x \cos y - \sin x \sin y$$
$$\cos(x-y) = \cos x \cos y + \sin x \sin y$$

$$\tag{3.22}$$

Also, putting $x = y$,

$$\sin 2x = 2 \sin x \cos x \tag{3.23}$$

$$\cos 2x = \cos^2 x - \sin^2 x$$
$$= 1 - 2 \sin^2 x = 2 \cos^2 x - 1 \tag{3.24}$$

where the alternative expressions for $\cos 2x$ are obtained from $\sin^2 x + \cos^2 x = 1$.

EXAMPLE 3.11 Express $\sin 5\theta$ and $\sin \theta$ in terms of the sines and cosines of 2θ and 3θ.

From equations (3.21),

$$\sin 5\theta = \sin(3\theta + 2\theta) = \sin 3\theta \cos 2\theta + \cos 3\theta \sin 2\theta$$
$$\sin \theta = \sin(3\theta - 2\theta) = \sin 3\theta \cos 2\theta - \cos 3\theta \sin 2\theta$$

‣ Exercises 18–21

EXAMPLE 3.12 Express $\sin 3\theta \cos 2\theta$ in terms of $\sin \theta$ and $\sin 5\theta$.

From equations (3.21) it follows that

$$\sin x \cos y = \frac{1}{2} \Big[\sin(x + y) + \sin(x - y) \Big] \tag{3.25}$$

and, therefore,

$$\sin 3\theta \cos 2\theta = \frac{1}{2} \Big[\sin 5\theta + \sin \theta \Big]$$

‣ Exercises 22

EXAMPLE 3.13 Express $\sin 3\theta \sin 2\theta$ and $\cos 3\theta \cos 2\theta$ in terms of $\cos \theta$ and $\cos 5\theta$.

From equations (3.22) it follows that

$$\sin x \sin y = \frac{1}{2} \Big[\cos(x - y) - \cos(x + y) \Big]$$
$$\cos x \cos y = \frac{1}{2} \Big[\cos(x - y) + \cos(x + y) \Big] \tag{3.26}$$

and, therefore,

$$\sin 3\theta \sin 2\theta = \frac{1}{2} \Big[\cos \theta - \cos 5\theta \Big]$$

$$\cos 3\theta \cos 2\theta = \frac{1}{2} \Big[\cos \theta + \cos 5\theta \Big]$$

‣ Exercises 23

EXAMPLE 3.14 Express $\sin\left(\dfrac{\pi}{2}\pm\theta\right)$ and $\cos\left(\dfrac{\pi}{2}\pm\theta\right)$ in terms of $\sin\theta$ and $\cos\theta$.

From equations (3.21),

$$\sin\left(\frac{\pi}{2}\pm\theta\right)=\sin\frac{\pi}{2}\cos\theta\pm\cos\frac{\pi}{2}\sin\theta$$

Figure 3.11 shows that $\sin\dfrac{\pi}{2}=1$ and $\cos\dfrac{\pi}{2}=0$ (see also Example 3.5). Therefore

$$\sin\left(\frac{\pi}{2}\pm\theta\right)=\cos\theta$$

Similarly, using equations (3.22),

$$\cos\left(\frac{\pi}{2}\pm\theta\right)=\cos\frac{\pi}{2}\cos\theta\mp\sin\frac{\pi}{2}\sin\theta=\mp\sin\theta$$

‣ Exercises 24

The expressions for the sum and difference of angles are important for the calculation of integrals (see Chapter 6) and for the description of the combination (interference) of waves.

EXAMPLE 3.15 The harmonic wave travelling in the x-direction described in Example 3.7 and shown in Figure 3.12 has wave function

$$\phi_+ = A\sin 2\pi\left(\frac{x}{\lambda}-vt\right)$$

The same wave travelling in the opposite direction is (replacing t by $-t$)

$$\phi_- = A\sin 2\pi\left(\frac{x}{\lambda}+vt\right)$$

As the waves overlap they interfere to give a new wave whose wave function is a linear combination of the form

$$\psi = a\phi_+ + b\phi_- = a\,A\sin 2\pi\left(\frac{x}{\lambda}-vt\right)+b\,A\sin 2\pi\left(\frac{x}{\lambda}+vt\right)$$

Making use of the expressions (3.21) for the sines of the sum and difference of angles

$$\psi = a\left[A\sin\frac{2\pi x}{\lambda}\cos 2\pi vt - A\cos\frac{2\pi x}{\lambda}\sin 2\pi vt\right]$$

$$+ b\left[A\sin\frac{2\pi x}{\lambda}\cos 2\pi vt + A\cos\frac{2\pi x}{\lambda}\sin 2\pi vt\right]$$

$$= (a+b)A\sin\frac{2\pi x}{\lambda}\cos 2\pi vt - (a-b)A\cos\frac{2\pi x}{\lambda}\sin 2\pi vt$$

An important special case is obtained for $a = b = 1$:

$$\psi = 2A\sin\frac{2\pi x}{\lambda}\cos 2\pi vt$$

This is called a **standing wave**. Its shape is given by the x-dependent factor $2A\sin\frac{2\pi x}{\lambda}$, as shown in Figure 3.16.

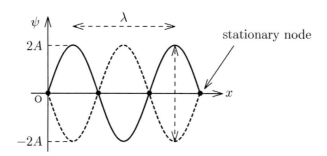

Figure 3.16

There is **constructive interference**; the amplitude has doubled but there has been no change in wavelength. The motion in time is give by the periodic function $\cos 2\pi vt$. The wave oscillates with frequency v, but the positions of the nodes (zeros) of the wave do not move; that is, the wave is stationary in space.

▸ Exercises 25, 26

3.5   Polar coordinates

The position of a point in a plane can be specified by its coordinates with respect to a given frame of reference (see Section 2.2), as shown in Figure 3.17.

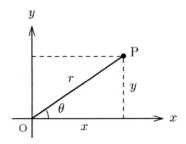

Figure 3.17

The figure shows that the position of the point P can be specified not only by its cartesian coordinates (x, y), but also by the pair of numbers (r, θ), the distance r of the point from the origin and the angle θ which gives the orientation of the line OP with respect to the positive x-direction (with the usual convention about sign). The numbers r and θ are called the **polar coordinates** of the point in the plane.[4] Whereas the cartesian coordinates can take any real positive and negative values, r is necessarily positive, with values $r = 0$ to ∞, and the angle θ, has values $\theta = 0$ to 2π.

The two sets of coordinates are related by $\cos\theta = x/r$ and $\sin\theta = y/r$, so that

$$x = r\cos\theta \qquad y = r\sin\theta \tag{3.27}$$

and the conversion from polar to cartesian coordinates is straightforward.

EXAMPLE 3.16 Find the cartesian coordinates of a point whose polar coordinates are $r = 2$ and $\theta = \pi/6$.

Making use of the values displayed in Table 3.2,

$$x = r\cos\theta = 2\cos\frac{\pi}{6} = \sqrt{3}$$

$$y = r\sin\theta = 2\sin\frac{\pi}{6} = 1$$

> Exercises 27, 28

The polar coordinates can be obtained from the pair of equations

$$r^2 = x^2 + y^2 \qquad\qquad \tan\theta = \frac{y}{x}$$

However, because $\tan\theta$ is a periodic function, period π, the angle θ is not uniquely determined by the inverse tangent $\tan^{-1}(y/x)$, and some care needs to be taken in

---

[4] In his *Methodus fluxionum* (Method of fluxions), written in about 1671, Newton suggested eight new types of coordinate system. The polar coordinates were his 'seventh manner; for spirals'.

converting from cartesian coordinates to polar coordinates. The correct value of θ is determined by the quadrant in which the point (x, y) lies. When $x > 0$, the point lies in the first or fourth quadrant (see Figure 3.6), and the angle is the principal value, $\theta = \tan^{-1}(y/x)$. When $x < 0$, the point lies in the second or third quadrant, and the angle is $\theta = [\text{principal value} + \pi]$. Therefore,

$$r = +\sqrt{x^2 + y^2}, \quad \theta = \tan^{-1}\left(\frac{y}{x}\right) \qquad \text{when } x > 0$$

$$\theta = \tan^{-1}\left(\frac{y}{x}\right) + \pi \qquad \text{when } x < 0$$

(3.28)

EXAMPLE 3.17 Find the polar coordinates (r, θ) of the point whose cartesian coordinates are $(x, y) = (3, 4)$.

We have

$$r^2 = x^2 + y^2 = 25, \qquad r = 5$$
$$\tan\theta = y/x = 4/3, \qquad \theta = \tan^{-1}(4/3) \approx 53°$$

The point $(3, 4)$ lies in the first quadrant, and the angle is the principal value of the inverse tangent.

➤ Exercises 29

EXAMPLE 3.18 Find the polar coordinates (r, θ) of the point whose cartesian coordinates are $(x, y) = (-1, 2)$.

We have

$$r^2 = x^2 + y^2 = 5, \qquad r = +\sqrt{5}$$
$$\tan\theta = \frac{y}{x} = -2, \qquad \theta = \tan^{-1}(-2) + \pi$$

Use of the inverse tangent facility of a pocket calculator gives the principal value, $\tan^{-1}(-2) \approx -63°$. But the point $(-1, 2)$ lies in the second quadrant, where $90° < \theta < 180°$, and the correct angle is

$$\tan^{-1}(-2) + \pi \approx -63° + 180° = 117°$$

with the same tangent value.

➤ Exercises 30

➤ Exercise 31

3.6   The exponential function

An exponential function has the form

$$f(x) = a^x \tag{3.29}$$

in which the number a is the **base** of the function and the variable x is the **exponent**. The best known example is the function 10^x which, when x is an integer, is used for the representation of numbers in the decimal system. The exponential function that occurs in statistics and in the physical sciences, *the* exponential function, is

$$\exp(x) = e^x \tag{3.30}$$

in which the base e is the Euler number (see Section 1.4).

➤ Exercise 32

The function is defined by the infinite series (see Chapter 7)[5]

$$e^x = \sum_{n=0}^{\infty} \frac{x^n}{n!} = 1 + \frac{x}{1!} + \frac{x^2}{2!} + \frac{x^3}{3!} + \frac{x^4}{4!} + \cdots \tag{3.31}$$

The value of the exponential function can be computed from the series to any desired accuracy, although the number of terms required increases rapidly as $|x|$ increases (see Example 1.10 for $x = 1$).

EXAMPLE 3.19 Calculate $\exp(-2x^2)$ to 8 significant figures for $x = 0.1$.

The value of the exponential can be obtained directly by substituting (-0.02) for x in (3.31). Alternatively,

$$\exp(-2x^2) = 1 + \frac{(-2x^2)}{1!} + \frac{(-2x^2)^2}{2!} + \frac{(-2x^2)^3}{3!} + \frac{(-2x^2)^4}{4!} + \frac{(-2x^2)^5}{5!} + \cdots$$

$$= 1 - 2x^2 + 2x^4 - \frac{4}{3}x^6 + \frac{2}{3}x^8 - \frac{4}{15}x^{10} + \cdots$$

Substituting for $x = 0.1$, and holding 2 significant figures more than the required 8 for intermediate values,

$$\exp(-0.02) \approx 1 - 0.02 + 0.000\,2 - 0.000\,001\,333\,3 + 0.000\,000\,006\,7$$
$$- 0.3 \times 10^{-10} + \cdots$$

---

[5] In his influential textbook *Introductio in analysin infinitorum* of 1748, Euler defined the exponential and logarithmic functions in terms of infinite series, and showed them to be inverse functions. In this book Euler laid the foundation for the branch of mathematics called analysis, the study of infinite processes, as in infinite series and in the calculus.

The terms are decreasing rapidly in magnitude, and the first five are sufficient (compare Example 1.10):

$$\exp(-0.02) \approx 0.980\ 198\ 67$$

▸ Exercises 33, 34

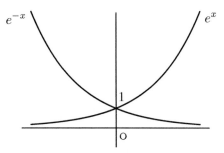

Figure 3.18

The graphs of e^x and its reciprocal e^{-x} are shown in Figure 3.18. The graphs of all the exponential functions a^x are very similar, with properties (for all $a > 0$)

$$a^0 = 1, \quad a^x \to \infty \text{ as } x \to \infty, \quad a^x \to 0 \text{ as } x \to -\infty \tag{3.32}$$

It will be seen in Chapter 4 that the unique property of e^x is that the slope of its graph at any point is equal to the value of the function at that point:

$$\frac{de^x}{dx} = e^x \tag{3.33}$$

The exponential function occurs in nearly all branches of applied mathematics, including statistics, kinetics, electromagnetic theory, quantum mechanics, and statistical mechanics.

▸ Exercises 35

EXAMPLE 3.20 Exponential growth and decay

Exponential growth arises when the rate of growth of a system at any time is proportional to the size of the system at that time. If $x(t)$ is the size at time t then (see Chapter 4) the rate of change of x is

$$\frac{dx}{dt} = \pm kx(t)$$

with the $+$ sign for growth and the $-$ sign for decay. The proportionality factor k is called the **rate constant**. The solution of the differential equation is

$$x(t) = x_0 e^{\pm kt}$$

where x_0 is the size at time $t=0$. As an example, consider a system whose size x is doubled after every time interval τ. Starting with size $x = x_0$, the size after time τ is $2x_0$, after time 2τ it is $4x_0$, after time 3τ it is $8x_0$, and so on. After time t,

$$x = 2^{t/\tau} x_0$$

Equating this with the solution of the differential equation shows that the rate constant k is inversely proportional to the time interval τ: $k = (\ln 2)/\tau$, where $\ln 2$ is the natural logarithm of the number 2 (see Section 3.7).

EXAMPLE 3.21 Atomic orbitals

The $1s$ orbital for an electron in the ground state of the hydrogen atom is

$$\psi = e^{-r}$$

where r is the distance of the electron from the nucleus. All the orbitals for the hydrogen atom have the form

$$\psi = f(x, y, z) e^{-ar}$$

where (x, y, z) are the cartesian coordinates of the electron relative to the nucleus at the origin, and a is a constant. The function $f(x, y, z)$ is a polynomial in x, y, and z, and determines the shape of the orbital; for example, $f = z$ gives a p_z orbital.

EXAMPLE 3.22 The normal distribution

The normal or Gaussian distribution in statistics is described by the probability density function

$$p(x) = \frac{1}{\sigma\sqrt{2\pi}} \exp\left[-\frac{1}{2}\left(\frac{x-\mu}{\sigma} \right)^2 \right]$$

where μ is the mean and σ is the standard deviation of the distribution (see Section 21.8). The probability function forms the basis for the statistical analysis of a wide range of phenomena; for example, error analysis of the results of experiments in the physical sciences, sample analysis in population studies, sample analysis for quality control in the manufacturing industry.

➤ Exercise 36

3.7   The logarithmic function

The logarithmic function[6] is the inverse function of the exponential:

$$\text{if} \quad y = a^x \quad \text{then} \quad x = \log_a y \tag{3.34}$$

and $\log_a y$ is called the logarithm to base a of y. The most important logarithmic functions are the **ordinary logarithm**, to base 10,

$$y = 10^x, \quad x = \log_{10} y = \lg y \tag{3.35}$$

and the **natural logarithm** (sometimes called the Napierian logarithm), to base e,

$$y = e^x, \quad x = \log_e y = \ln y \tag{3.36}$$

The ordinary logarithm $\log_{10}$ is sometimes given the symbol lg. The natural logarithm $\log_e$ is nearly always given the symbol ln.

It follows from equations (3.34) to (3.36) that

$$y = \log_a a^y = \log_{10} 10^y = \ln e^y \tag{3.37}$$

EXAMPLE 3.23

$$\log_2 2^3 = 3, \qquad \log_{10} 10^3 = 3, \qquad \lg 10^{-2} = -2,$$

$$\log_e e^3 = 3, \qquad \ln e^{-1/2} = -1/2, \qquad \log_a a^0 = \log_a 1 = 0$$

➤ Exercises 37

We note that the logarithm of 1 to any base is zero.

The graph of $\ln x$ and of its inverse function e^x are shown in Figure 3.19.[7]

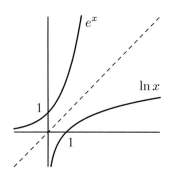

Figure 3.19

---

[6] John Napier (1550–1617), Scottish baron and amateur mathematician, published his invention of what he called logarithms in the *Mirifici logarithmorum canonis descriptio* (A description of the wonderful canon of logarithms) in 1614. Napier's logarithms were based on a logarithm of $10^7 = 0$. The first table of common logarithms, with log 1 = 0 and log 10 = 1, was published, after a famous consultation with Napier, by Henry Briggs (1561–1630), professor of geometry at Oxford, in the *Arithmetica logarithmica* in 1624. Logarithms greatly simplified computations involving multiplication and division.

[7] The graph of a log function was first drawn in 1646 by Evangelista Torricelli (1608–1647).

The graphs of all the logarithmic functions are similar, with properties

$$\log 1 = 0, \qquad \log x \to +\infty \text{ as } x \to \infty, \qquad \log x \to -\infty \text{ as } x \to 0 \qquad (3.38)$$

We note that (real) $\log x$ is not defined for negative values of x.

The combination properties of logarithms are the properties of the indices (exponents) of the corresponding exponentials; for the natural logarithm:

$$\ln x + \ln y = \ln xy \qquad (3.39)$$

$$\ln x - \ln y = \ln \frac{x}{y} \qquad (3.40)$$

$$\ln x^n = n \ln x \qquad (3.41)$$

EXAMPLE 3.24 Addition of logarithms. To prove equation (3.39),

$$\text{let} \quad x = e^a \quad \text{and} \quad y = e^b, \quad \text{so that} \quad xy = e^{a+b}$$

Then, by definition (3.36)

$$\ln x = a, \qquad \ln y = b, \qquad \ln xy = a + b$$

Therefore $\ln x + \ln y = \ln xy$.

EXAMPLE 3.25 Combinations of logarithms.

$$\ln 2 + \ln 4 = \ln(2 \times 4) = \ln 8 \qquad\qquad \ln 30 = \ln(2 \times 3 \times 5) = \ln 2 + \ln 3 + \ln 5$$

$$\ln 6 - \ln 3 = \ln \frac{6}{3} = \ln 2 \qquad\qquad\qquad \ln 2^3 = 3 \ln 2 = \ln 2 + \ln 2 + \ln 2$$

$$\ln \frac{1}{2} = \ln 2^{-1} = -\ln 2 \qquad\qquad\qquad \ln \frac{1}{2} = \ln 1 - \ln 2 = 0 - \ln 2 = -\ln 2$$

$$\frac{1}{3} \ln 27 = \ln(27)^{1/3} = \ln 3 \qquad\qquad -2\ln 5 = \ln 5^{-2} = \ln \frac{1}{25}$$

▸ Exercises 38

EXAMPLE 3.26 Simplify the expression $\ln(1 - x^2) + \ln(1 + x)^{-1} - \ln(1 - x)$.

From the rule $\ln x^n = n \ln x$, it follows that $\ln(1 + x)^{-1} = -\ln(1 + x)$. Then

$$\ln(1-x^2) + \ln(1+x)^{-1} - \ln(1-x) = \ln(1-x^2) - \ln(1+x) - \ln(1-x)$$

$$= \ln\frac{(1-x^2)}{(1+x)(1-x)} = \ln\left(\frac{1-x^2}{1-x^2}\right)$$

$$= \ln 1 = 0$$

> Exercises 39

EXAMPLE 3.27 What *not* to do.

A surprisingly common error is to put

$$\ln(x+y) = \ln x + \ln y$$

This is not in general true. For example,

$$\ln(1+2) = \ln 3 \qquad \text{but} \qquad \ln 1 + \ln 2 = \ln 2 \qquad (\ln 1 = 0)$$

The only case for which $\ln(x+y) = \ln x + \ln y$ is when $x+y = xy$; that is, when $x = y/(y-1)$.

Before the invention of the microchip and of the pocket calculator in the early 1970's, the ordinary logarithm was used mainly as an aid to long multiplication and division; for example, the multiplication of numbers can be replaced by the addition of their logarithms. There are now only a few uses of $\log_{10}$ in the physical sciences; for example in the definitions of pH as a measure of hydrogen-ion concentration, and of pK where K is an equilibrium constant.

EXAMPLE 3.28 pH as a measure of hydrogen-ion concentration

The pH of an aqueous solution is defined as

$$pH = -\log_{10}[H^+]$$

where $[H^+]$ is the 'hydrogen-ion concentration' in units of $mol\ dm^{-3}$ (moles per litre). Then

$$[H^+] = 10^{-pH}\ mol\ dm^{-3}$$

For example, a pH of 7 (neutral) corresponds to $[H^+] = 10^{-7}\ mol\ dm^{-3}$.

The pH is an example of the use of the logarithm as a scale of measure. Other examples of logarithmic scales are the Richter scale for the 'strength of earthquake', the bel scale of loudness, and the scale of star magnitudes.

> Exercises 40–42

Conversion factors

The logarithm occurs in a number of expressions for physical properties and processes, and the ordinary logarithm is still sometimes found, instead of the natural logarithm, in the scientific literature and in textbooks. In general, the conversion factor from one base, a, to another, b, is given by

$$\log_b x = \log_b a \times \log_a x \qquad (3.42)$$

Thus, if $x = a^y$, then

$$\log_a x = y$$

$$\log_b x = \log_b a^y = y \log_b a = (\log_a x) \times (\log_b a)$$

The conversion factors between ordinary and natural logarithms are therefore

$$\lg x = \log_{10} x = \log_{10} e \times \log_e x \approx 0.43429448 \times \log_e x$$
$$\ln x = \log_e x = \log_e 10 \times \log_{10} x \approx 2.30258093 \times \log_{10} x$$

3.8   Values of exponential and logarithmic functions

Table 3.3 shows values of $\ln x$, $x \ln x$, e^x, and e^{-x} for a wide range of values of x.

Table 3.3

x	$\ln x$	$x \ln x$	e^x	e^{-x}
0	$-\infty$	0	1	1
$\vdots$	$\vdots$	$\vdots$	$\vdots$	$\vdots$
	slowly			
$\vdots$	$\vdots$	$\vdots$	$\vdots$	$\vdots$
10^{-6}	-13.8	-0.00001	1.000001	0.9999990
10^{-3}	-6.9	-0.007	1.001	0.9990
1	0	0	2.7	0.37
10	2.3	23	2×10^4	5×10^{-5}
10^2	4.6	460	3×10^{43}	4×10^{-44}
10^3	6.9	6908	$\vdots$	$\vdots$
$\vdots$	$\vdots$	$\vdots$	$\vdots$	$\vdots$
	slowly		fast	fast
$\vdots$	$\vdots$	$\vdots$	$\vdots$	$\vdots$
∞	∞	∞	∞	0

The following conclusions can be drawn from the table.

(a) $\ln x$ varies slowly compared to x. In fact it varies more slowly than any power of x:

$$\text{as } x \to 0, \qquad \ln x \to -\infty \qquad \text{but} \qquad x^a \ln x \to 0 \tag{3.43}$$

$$\text{as } x \to \infty, \qquad \ln x \to \infty \qquad \text{but} \qquad x^{-a} \ln x \to 0 \tag{3.44}$$

for any positive value of a, however small.

(b) e^x varies rapidly compared with x. In fact it varies more rapidly than any power of x:

$$\text{as } x \to \infty, \qquad e^x \to \infty \qquad \text{and} \qquad x^{-a} e^x \to \infty \tag{3.45}$$

$$\text{as } x \to \infty, \qquad e^{-x} \to 0 \qquad \text{and} \qquad x^a e^{-x} \to 0 \tag{3.46}$$

for any positive value of a, however large.

EXAMPLE 3.29 Plots of xe^{-x} and $x^{-1}e^x$.

Figure 3.20 shows that whereas e^{-x} decreases monotonically with increasing value of x, the function xe^{-x} first increases and passes through a maximum before the exponential decay takes over. This is a characteristic behaviour of atomic orbitals. A $1s$ orbital has the form e^{-r}, where r is the distance from the nucleus, but all other orbitals behave with distance like $r^l e^{-r}$, where l is a positive integer.

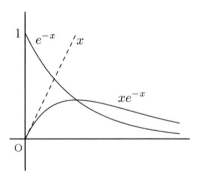

Figure 3.20

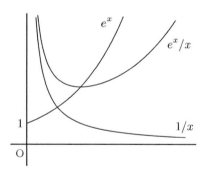

Figure 3.21

▸ Exercises 43

3.9 Hyperbolic functions

Hyperbolic functions have their origin in geometry in the description of the properties of the hyperbola. The properties of the functions are readily derived from the properties of the other transcendental functions described in this chapter, and only a brief discussion is presented here.

The hyperbolic cosine and sine are defined in terms of the exponential function as

$$\cosh x = \frac{1}{2}\left(e^x + e^{-x}\right), \quad \sinh x = \frac{1}{2}\left(e^x - e^{-x}\right) \tag{3.47}$$

(read as 'cosh' and 'shine'). Their graphs are shown in Figure 3.22.

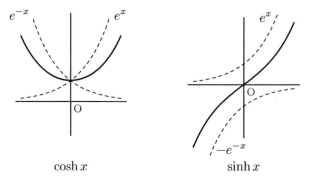

$$\cosh x \qquad\qquad \sinh x$$

Figure 3.22

Both the notation and the properties of the hyperbolic functions parallel those of the trigonometric (circular) functions; for example,

$$\cosh^2 x - \sinh^2 x = 1$$

$$\sinh(x \pm y) = \sinh x \cosh y \pm \cosh x \sinh y \tag{3.48}$$

$$\cosh(x \pm y) = \cosh x \cosh y \pm \sinh x \sinh y$$

The functions are called 'hyperbolic' because, if $x = a \cosh t$ and $y = a \sinh t$ are the coordinates of a point, where t is a parameter, then by the first equation (3.48),

$$x^2 - y^2 = a^2$$

and this is the equation of a hyperbola.

The inverse hyperbolic functions

The inverse hyperbolic functions are defined in the same way as the inverse trigonometric functions; for example, if $x = \cosh y$ then $y = \cosh^{-1} x$. These functions can be expressed in terms of the logarithmic function:

$$\cosh^{-1} x = \ln\left[x \pm \sqrt{x^2 - 1}\right] \qquad\qquad (x \geq 1)$$

$$\sinh^{-1} x = \ln\left[x + \sqrt{x^2 + 1}\right] \qquad\qquad (3.49)$$

$$\tanh^{-1} x = \frac{1}{2}\ln\left[\frac{1+x}{1-x}\right] \qquad\qquad (|x| < 1)$$

The relation for $\cosh^{-1} x$ shows that there exist two values of the function for each value of $x\ (> 1)$. These two values differ only in sign, and the positive value is defined as the principal value.

EXAMPLE 3.30 To show that $\cosh^{-1} x = \ln\left[x \pm \sqrt{x^2 - 1}\right]$.

If $x = \cosh y$ then, because $\cosh^2 y - \sinh^2 y = 1$, it follows that $\sinh y = \pm\sqrt{x^2 - 1}$ and

$$\ln\left[x \pm \sqrt{x^2 - 1}\right] = \ln\left[\cosh y + \sinh y\right] = \ln e^y$$

from the definitions (3.47). The result follows since $\ln e^y = y = \cosh^{-1} x$.

3.10 Exercises

Section 3.2

1. In Figure 3.23, the right-angled triangle ABC has sides $a = 12$ and $b = 5$. Find c and the sin, cos, tan, cosec, sec and cot of the internal angles A and B.
2. For the triangle in Exercise 1, find (i) $\sin^2 A + \cos^2 A$, (ii) $\sin^2 B + \cos^2 B$.
3. Express the following angles in radians:
   (i) $5°$ (ii) $87°$ (iii) $120°$ (iv) $260°$ (v) $540°$
   (vi) $720°$
4. Express the following angles in degrees:
   (i) $\pi/10$ (ii) $\pi/4$ (iii) $\pi/6$ (iv) $\pi/3$ (v) $3\pi/8$ (vi) $7\pi/8$
5. For a circle of radius $r = 4$, find
   (i) the angle subtended at the centre of the circle by arc of length 6,
   (ii) the length of arc that subtends angle $\pi/10$ at the centre of the circle,
   (iii) the length of arc that subtends angle $\pi/2$ at the centre of the circle,
   (iv) the circumference of the circle.
6. Use Table 3.2 to find the sine, cosine and tangent of (i) $3\pi/4$, (ii) $5\pi/4$, (iii) $7\pi/4$.
7. By considering the limit $\theta \to 0$ of an internal angle of a right-angled triangle, show that
   (i) $\sin 0 = 0$ (ii) $\cos 0 = 1$
8. Use the properties of the right-angled isosceles triangle to verify the values of the trigonometric functions for $\theta = \pi/4$ in Table 3.2.
9. Sketch diagrams to show that
   (i) $\sin(\pi - \theta) = \sin \theta$ (ii) $\cos(\pi - \theta) = -\cos \theta$ (iii) $\sin(\pi + \theta) = -\sin \theta$
   (iv) $\cos(\pi + \theta) = -\cos \theta$ (v) $\sin(\pi/2 - \theta) = \cos \theta$ (vi) $\cos(\pi/2 - \theta) = \sin \theta$
10. Find the period and sketch the graph $(-\pi \leq x \leq 2\pi)$ of (i) $\sin 2x$, (ii) $\cos 3x$.
11. Sketch the graph of the harmonic wave $\phi(x, t) = \sin 2\pi(x - t)$ as a function of $x\ (-1 \leq x \leq 2)$ for values of time t, (i) $t = 0$, (ii) $t = 1/4$, (iii) $t = 1/2$.

Figure 3.23

Section 3.3

12. Find the principal values of

    (i) $\sin^{-1}\left(\frac{1}{2}\right)$ (ii) $\sin^{-1}(1)$ (iii) $\cos^{-1}\left(\frac{1}{2}\right)$ (iv) $\cos^{-1}(-1)$

13. The Bragg equation for the reflection of radiation of wavelength λ from the planes of a crystal is $n\lambda = 2d\sin\theta$ where d is the separation of the planes, θ is the angle of incidence of the radiation, and n is an integer. Calculate the angles θ at which X-rays of wavelength 1.5×10^{-10} m are reflected by planes separated by 3.0×10^{-10} m.

Section 3.4

14. Given the side $a = 1$ and angles $A = \pi/4$ and $B = \pi/3$ of a triangle ABC, Figure 3.24, find the third angle and the other two sides.

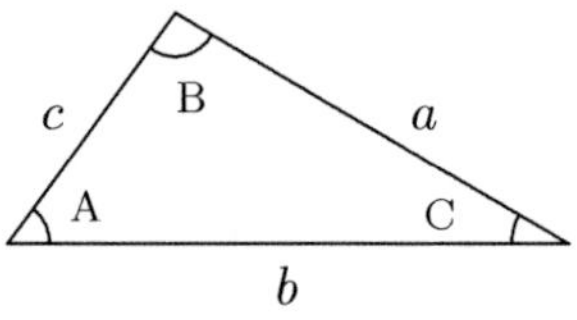

15. Given the sides $a = 2$, $b = 2.5$ and $c = 3$ of a triangle ABC, find the angles.

Figure 3.24

16. Given the sides $a = 3$, $b = 4$, and included angle $C = \pi/4$ of triangle ABC, find the third side and the other two angles.

17. Given the sides $a = \sqrt{2}$, $b = 3$, and included angle $C = \pi/4$ of the triangle ABC, find the third side and the other two angles.

18. Express in terms of in terms of the sines and cosines of 2θ and 5θ:

    (i) $\sin 7\theta$,   (ii) $\sin 3\theta$,   (iii) $\cos 7\theta$,   (iv) $\cos 3\theta$.

19. Express (i) $\sin 3\theta$ in terms of $\sin\theta$, (ii) $\cos 3\theta$ in terms of $\cos\theta$.

20. Express $\cos 4x$ in terms of

    (i) $\sin 2x$ and $\cos 2x$,   (ii) $\sin 2x$ only,   (iii) $\cos 2x$ only,   (iv) $\sin x$ only,
    (v) $\cos x$ only.

21. Given $\sin 10° = 0.1736$, $\sin 30° = 1/2$, $\sin 50° = 0.7660$, find $\cos 20°$ (without using a calculator).

22. Express in terms of the sines of $8x$ and $2x$: (i) $\sin 5x \cos 3x$, (ii) $\cos 5x \sin 3x$.

23. Express in terms of the cosines of $8x$ and $2x$: (i) $\sin 5x \sin 3x$, (ii) $\cos 5x \cos 3x$.

24. Express (i) $\sin(\pi \pm \theta)$ and (ii) $\cos(\pi \pm \theta)$ in terms of $\sin\theta$ and $\cos\theta$.

25. The function $\psi(x, t) = \sin \pi x \cos 2\pi t$ represents a standing wave. Find the values of time t for which ψ has (i) maximum amplitude, (ii) zero amplitude. (iii) Sketch the wave function between $x = 0$ and $x = 3$ at (a) $t = 0$, (b) $t = 1/8$.

26. The function

$$\phi(x) = a\sin\frac{2\pi x}{\lambda} + b\cos\frac{2\pi x}{\lambda}$$

represents the superposition of two harmonic waves with the same wavelength λ. Show that ϕ is (i) also harmonic with the same wavelength, and (ii) can be written as

$$\phi(x) = A\sin\left(\frac{2\pi x}{\lambda} + \alpha\right)$$

where $A = \sqrt{a^2 + b^2}$ and $\tan\alpha = b/a$.

Section 3.5

27. Find the cartesian coordinates of the points whose polar coordinates are

    (i) $r = 3$, $\theta = \pi/3$,   (ii) $r = 3$, $\theta = 5\pi/3$.

28. Find the cartesian coordinates of the points whose polar coordinates are
    (i) $r = 3$, $\theta = 2\pi/3$,   (ii) $r = 3$, $\theta = 4\pi/3$.
29. Find the polar coordinates of the points whose cartesian coordinates are
    (i) $(3, 2)$,   (ii) $(3, -2)$.
30. Find the polar coordinates of the points whose cartesian coordinates are
    (i) $(-3, 2)$,   (ii) $(-3, -2)$.
31. A solution of the equation of motion for the harmonic oscillator is given in Example 3.8
    as $x(t) = A \cos \omega t$. Show that $x(t)$ can be interpreted as the x-coordinate of a point
    moving with constant angular speed ω in a circle in the xy-plane, with centre at the origin
    and radius A.

Section 3.6

32. Simplify
    (i) $e^2 e^3$ (ii) $e^3 e^{-3}$ (iii) $e^3 e^{-4}$ (iv) e^3/e^2 (v) e^5/e^{-4}
33. (i) Write down the expansion of $e^{-x/3}$ in powers of x to terms in x^5.
    (ii) Use the expansion to calculate an approximate value of $e^{-1/3}$. Determine how many
    significant figures of this value are correct, and quote your answer to this number of figures.
34. (i) Write down the expansion of e^{-x^3} in powers of x to terms in x^{15}.
    Use the expansion to calculate an approximate value of e^{-x^3} that is correct to 12
    significant figures for the following values of x, in each case giving the smallest number of
    terms required: (ii) 10^{-1}, (iii) 10^{-2}, (iv) 10^{-3}, (v) 10^{-4}, (vi) 10^{-5}.
35. Sketch the graphs of e^{2x} and e^{-2x} for values $-1.5 \leq x \leq 1.5$.
36. For a system composed of N identical molecules, the Boltzmann distribution

    $$\frac{n_i}{N} = e^{-\varepsilon_i/kT}$$

    gives the average fraction of molecules in the molecular state i with energy ε_i.
    (i) Show that the ratio n_i/n_j of the populations of states i and j depends only on the
    difference in energy of the two states. (ii) What is the ratio for two states with the same
    energy (degenerate states)?

Section 3.7

37. Simplify:
    (i) $\log_{10} 100$ (ii) $\log_2 16$ (iii) $\ln e^{-5}$ (iv) $\ln e^{x^2}$ (v) $\ln e^{-(ax^2+bx+c)}$
    (vi) $\ln e^{-kt}$
38. Express the following as the log of a single number:
    (i) $\ln 2 + \ln 3$ (ii) $\ln 2 - \ln 3$ (iii) $5 \ln 2$ (iv) $\ln 3 + \ln 4 - \ln 6$
39. Simplify:
    (i) $\ln x^3 - \ln x$ (ii) $\ln (2x^3 - 3x^2) + \ln x^{-2}$ (iii) $\ln (x^5 - 3x^2) + 2 \ln x^{-1} - \ln (x^3 - 3)$
    (iv) $\ln e^x$ (v) $\ln e^{x^2+3} - \ln e^3$
40. The barometric formula

    $$p = p_0 e^{-Mgh/RT}$$

    gives the pressure of a gas of molar mass M at altitude h, when p_0 is the pressure at sea
    level. Express h in terms of the other variables.
41. The chemical potential of a gas at pressure p and temperature T is

    $$\mu = \mu^{\ominus} + RT \ln \frac{f}{p^{\ominus}}$$

where $f = \gamma p$ is the fugacity and γ is the fugacity coefficient. Express p as an explicit function of the other variables.

42. In a first-order decomposition reaction, A $\rightarrow$ products, the amount of substance A at time t is

$$x(t) = x(0)e^{-kt}$$

where $x(0)$ is the initial amount of A, and k is the rate constant. The time taken for the amount of A to fall to half of its initial value is called the half-life, $\tau_{1/2}$, of the reaction. Find the half-life for rate constants: (i) $k = 3 \text{ s}^{-1}$, (ii) $k = 10^{-5} \text{ s}^{-1}$.

Section 3.8

43. As in Example 3.30, sketch (i) x^2, e^{-x}, $x^2 e^{-x}$, (ii) x^{-2}, e^x, $x^{-2}e^x$.

4 Differentiation

4.1 Concepts

In the physical sciences we are interested in the value of a physical quantity and how it is related to other physical quantities. In addition, we are interested in how the value of the physical quantity changes on going from one state of the system to another, and in the rate of change with respect to time or with respect to some other physical quantity.

Consider the equation of state of the ideal gas

$$pV = nRT$$

Because any one of the four variables p, V, T, and n can be expressed as a function of the other three, the state of the system is determined by three of the four quantities. If the temperature T is changed by an amount ΔT (read as 'delta t'), keeping the pressure p and the amount of substance n fixed, the volume changes from

$$V = \frac{nRT}{p} \quad \text{to} \quad V + \Delta V = \frac{nR(T + \Delta T)}{p}$$

and the change in volume is

$$\Delta V = \frac{nR}{p}\Delta T$$

Figure 4.1 shows that the graph of V against T (at constant p and n) is a straight line with gradient, or slope,

$$\frac{\Delta V}{\Delta T} = \frac{nR}{p} = \text{constant}$$

The gradient is the change in V per unit change in T, or the **rate of change** of V with respect to T.

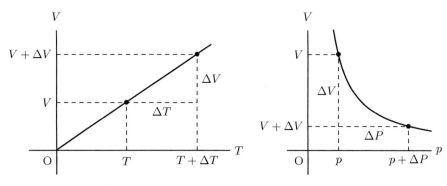

Figure 4.1   Figure 4.2

If the pressure p of the gas is changed by an amount Δp at constant T and n, the change in the volume is

$$\Delta V = \frac{nRT}{p+\Delta p} - \frac{nRT}{p} = -nRT\frac{\Delta p}{p(p+\Delta p)}$$

Figure 4.2 shows that the graph of V against p (at constant T and n) is not a straight line. The gradient at any point on the curve is defined as the gradient of the tangent to the curve at that point. The gradient changes from point to point, and is not given by $\Delta V/\Delta p$.

The branch of mathematics concerned with the determination of gradients and, therefore, with rates of change is the **differential calculus**.

4.2 The process of differentiation

Let the value of a variable x change continuously from p to q. The difference $(q-p)$ is called the **change** or **increment** in x. In the differential calculus this change is denoted by*

$$\Delta x = q - p \tag{4.1}$$

We note that $\Delta x > 0$ if $q > p$, and $\Delta x < 0$ if $q < p$.

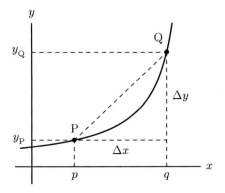

Figure 4.3

Let $y = f(x)$ be a function of x that changes continuously and smoothly from point P to point Q (Figure 4.3). The values of y at P and Q are $y_P = f(p)$ and $y_Q = f(q)$. The change Δy in y corresponding to change Δx in x is therefore

$$y_Q - y_P = \Delta y = f(q) - f(p)$$

The quantity

$$\frac{y_Q - y_P}{q - p} = \frac{\Delta y}{\Delta x} \tag{4.2}$$

is the gradient of the line PQ, and can be interpreted as the *average* rate of change of y with respect to x between P and Q.

---

* Sometimes, if the change in x is supposed to be 'small', δx is used instead of Δx.

More generally, if the variable changes by Δx, from $p=x$ to $q=x+\Delta x$, the corresponding change in the function is

$$\Delta y = f(x+\Delta x)-f(x) \tag{4.3}$$

and the corresponding average rate of change is

$$\frac{\Delta y}{\Delta x} = \frac{f(x+\Delta x)-f(x)}{\Delta x} \tag{4.4}$$

EXAMPLE 4.1 The general quadratic function, $y=ax^2+bx+c$:

At point P $y_\mathrm{P}=f(x)=ax^2+bx+c$

At point Q $y_\mathrm{Q}=f(x+\Delta x)=a(x+\Delta x)^2+b(x+\Delta x)+c$

The change in the function on going from P to Q is therefore

$$\Delta y = y_\mathrm{Q} - y_\mathrm{P} = \left[a(x+\Delta x)^2 + b(x+\Delta x)+c\right]-\left[ax^2+bx+c\right]$$

$$= (2ax+b)\Delta x + a(\Delta x)^2$$

and the gradient of the line PQ is

$$\frac{\Delta y}{\Delta x} = (2ax+b)+a\,\Delta x \tag{4.5}$$

▸ Exercise 1

Figure 4.4 shows how the quantity $\Delta y/\Delta x$ changes as the point Q is moved along the curve towards P (as Δx is decreased in magnitude).

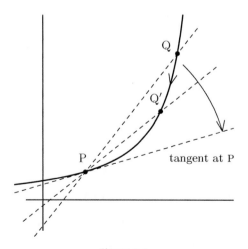

Figure 4.4

As Q moves through Q′ towards P, the gradient of the line PQ approaches the gradient of the tangent at P. We express this as

$$\text{gradient at P} = \lim_{Q \to P} \left(\frac{\Delta y}{\Delta x} \right) \tag{4.6}$$

(read as 'the limit as Q goes to P'). At the same time, the magnitude of Δx goes to zero, $\Delta x \to 0$ as $Q \to P$, and the limit can be expressed as[1]

$$\text{gradient at } x = \lim_{\Delta x \to 0} \left(\frac{\Delta y}{\Delta x} \right) \tag{4.7}$$

EXAMPLE 4.2 For the quadratic function $y = ax^2 + bx + c$ in Example 4.1,

$$\frac{\Delta y}{\Delta x} = (2ax + b) + a \Delta x \quad \text{so that} \quad \lim_{\Delta x \to 0} \left(\frac{\Delta y}{\Delta x} \right) = 2ax + b$$

The limit is a function of x, and it gives the gradient or slope of the curve at each value of x (each point on the curve).

▸ Exercises 2, 3

The process of taking the limit in (4.7) is called **differentiation**. In the differential calculus, the limit is denoted by the symbol $\dfrac{dy}{dx}$:

$$\frac{dy}{dx} = \lim_{\Delta x \to 0} \left(\frac{\Delta y}{\Delta x} \right) = \lim_{\Delta x \to 0} \left\{ \frac{f(x + \Delta x) - f(x)}{\Delta x} \right\} \tag{4.8}$$

(read as 'dy by dx').[2] It is called the **differential coefficient** of the function or, simply, the **derivative** of the function. We note that the symbol $\dfrac{dy}{dx}$ does *not* mean the quantity '*dy*' divided by the quantity '*dx*'; the symbol represents the limit, and the taking of the limit as given by the right side of (4.8).

---

[1] This method of finding the tangent at a point on a curve is essentially that given by Fermat in his *Method of finding maxima and minima* in about 1630. This work marks the beginning of the differential calculus. A method similar to Fermat's but involving quantities equivalent to Δx and Δy was described by Barrow in *Lectiones geometriae*, published in 1670. In these lectures Isaac Barrow (1630–1677), theologian and professor of geometry at Cambridge, gave a 'state of the art' account of infinitesimal methods. The formulation of the method of tangents was included 'on the advice of a friend', Newton, who succeeded him in his chair when Barrow became chaplain to Charles II in 1669.

[2] The notation is derived from Leibniz's formulation of the calculus. Gottfried Wilhelm Leibniz (1646–1716), philosopher, diplomat and mathematician, discovered his form of the calculus in the years 1672–1676 whilst on diplomatic service in Paris, where he came under the influence of the physicist and mathematician Christiaan Huygens, inventor of the pendulum clock. His first account of the differential calculus was the *Nova methodus pro maximis et minimis, itemque tangentibus* (A new method for maxima and minima, and also for tangents), published in 1684.

An alternative symbol for the derivative of the function $f(x)$ is $f'(x)$:

$$f'(x) = \frac{df(x)}{dx} \tag{4.9}$$

Another symbol is Df, by which is implied that the derivative of f is obtained by acting (operating) on f with the **differential operator** D,

$$D = \frac{d}{dx}, \qquad Df = \frac{d}{dx}f = \frac{df}{dx} \tag{4.10}$$

The concept of differential and other operators is widely used in the physical sciences, and will be discussed in later chapters.

4.3   Continuity

In the discussion of taking the limit in Section 4.2 it was assumed that the function $y = f(x)$ is a **continuous function** of x, and that the limit defined by equation (4.8) exists and is unique. Generally speaking, a function is continuous if its graph is an uninterrupted curve. A given function $y = f(x)$ may be tested for continuity at a point, x_1 say, by letting the independent variable x move continuously from the right side and from the left side towards the specified point x_1 as shown in Figure 4.5.

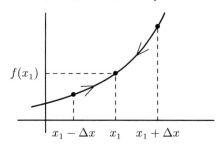

Figure 4.5

If $f(x_1 + \Delta x)$ and $f(x_1 - \Delta x)$ approach the *same* value $f(x_1)$ as $\Delta x \to 0$ then the function is said to be continuous at x_1, and we write

$$\lim_{x \to x_1} f(x) = f(x_1) \tag{4.11}$$

If this holds for every value of x_1 in a certain interval $a \le x_1 \le b$, then the function is continuous in the interval. Three types of **discontinuity** are illustrated in Figure 4.6.

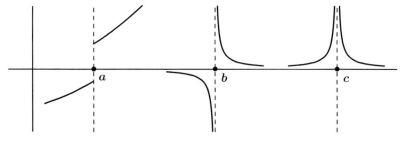

Figure 4.6

Point a. The function has a **finite discontinuity** at $x = a$. For example, the function

$$f(x) = \begin{cases} x+1 & \text{if } x > 0 \\ x-1 & \text{if } x \le 0 \end{cases}$$

is discontinuous at $x = 0$ where its value is -1. However, if we approach the point $x = 0$ from the right, $f(x)$ approaches the value $+1$.

Point b. The function has an **infinite discontinuity** at $x = b$. If we approach this point from the left, the value of the function tends to $-\infty$; if we approach from the right the function tends to $+\infty$. Infinity is not a number, and the function is not defined at $x = b$. An example is $1/(x-1)$, discontinuous at $x = 1$.

Point c. The function tends to infinity if we approach the point $x = c$ from either side; for example, $1/x^2$ at $x = 0$.

In these three cases, the nature of the discontinuities is obvious from the graphs; they are said to be **essential discontinuities**. In some cases however, the discontinuity is not obvious from the graph. For example, the function $f(x) = x/x$ has constant value equal to 1 for all values of $x \ne 0$, but it is not defined at $x = 0$ because $0/0$ is indeterminate and has no meaning. Such a discontinuity is said to be **removable**. If we redefine the function such that $f = x/x$ when $x \ne 0$ and $f(x) = 1$ when $x = 0$, then the function becomes continuous for every value of x; the discontinuity has been removed with no change to the function except at an isolated point.

▸ Exercises 4–6

4.4 Limits

Consider the rational function

$$y = \frac{x^2 - 4}{x - 2}$$

which is continuous for all values of x except $x = 2$. Table 4.1 shows that, whereas both the numerator and the denominator go to zero at $x = 2$, their *ratio* approaches the value $y = 4$ as $x \to 2$ from both sides:

$$\lim_{x \to 2}\left(\frac{x^2 - 4}{x - 2}\right) = 4$$

Table 4.1 Values of $y = (x^2 - 4)/(x - 2)$

x	2.1	2.01	2.001	2.0001	...	1.9999	1.999	1.99	1.9
y	4.1	4.01	4.001	4.0001	...	3.9999	3.999	3.99	3.9

This is an example of a removable discontinuity; we have

$$\frac{x^2 - 4}{x - 2} = \frac{\cancel{(x-2)}(x+2)}{\cancel{(x-2)}} = x + 2 \qquad \text{if} \quad x \neq 2$$

and the discontinuity can be removed by redefining the function to have value $y = 4$ when $x = 2$. This example is important because taking the limit in differentiation always involves letting the denominator go to zero.

➤ Exercises 7–10

The finding of limits is necessary whenever a quantity becomes indeterminate. In addition to the case $0/0$, the indeterminate forms most commonly met in the physical sciences are ∞/∞ and $\infty - \infty$.

EXAMPLE 4.3

$$\lim_{x \to \infty} \left(\frac{2x^2 + 5}{x^2 + 3x} \right) = 2$$

Both numerator and denominator tend to infinity as $x \to \infty$, but the ratio remains finite. Thus, dividing both numerator and denominator by x^2, which is allowed if $x \to \infty$, we obtain

$$\frac{2x^2 + 5}{x^2 + 3x} = \frac{2 + 5/x^2}{1 + 3/x} \to \frac{2}{1} = 2 \quad \text{as} \quad x \to \infty$$

➤ Exercises 11–13

EXAMPLE 4.4

$$\lim_{x \to 0} \left\{ \left(2x + \frac{1}{x} \right)^2 - \left(3x - \frac{1}{x} \right)^2 \right\} = 10$$

Both squared terms tend to infinity as $x \to 0$, but the difference remains finite. By expanding the squared terms and simplifying, we get

$$\left(2x + \frac{1}{x} \right)^2 - \left(3x - \frac{1}{x} \right)^2 = \left(4x^2 + 4 + \frac{1}{x^2} \right) - \left(9x^2 - 6 + \frac{1}{x^2} \right)$$

$$= -5x^2 + 10 \to 10 \quad \text{as} \quad x \to 0$$

➤ Exercises 14, 15

EXAMPLE 4.5

$$\lim_{x \to \infty} \left[\ln(2x+3) - \ln(x-2) \right] = \ln 2$$

From the properties of the logarithm,

$$\ln(2x+3) - \ln(x-2) = \ln\left(\frac{2x+3}{x-2} \right) \; \to \; \ln 2 \quad \text{as} \quad x \to \infty$$

➤ Exercises 16, 17

4.5   Differentiation from first principles

A function is said to be **differentiable** at a point if the limit in equation (4.8),

$$\frac{dy}{dx} = \lim_{\Delta x \to 0} \left(\frac{\Delta y}{\Delta x} \right) = \lim_{\Delta x \to 0} \left\{ \frac{f(x+\Delta x) - f(x)}{\Delta x} \right\} \tag{4.8}$$

exists and is unique. A necessary condition for this to be true is that the function be continuous at the point, but not all continuous functions are everywhere differentiable. For example, the function

$$f(x) = |x| = \begin{cases} x & \text{if } x \geq 0 \\ -x & \text{if } x < 0 \end{cases}$$

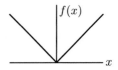

is continuous at all values of x but Figure 4.7 shows that its slope changes abruptly at $x=0$, from value -1 when $x<0$ to value $+1$ for $x \geq 0$. The function has a **cusp** at $x=0$ and the derivative is not defined by equation (4.8).

Figure 4.7

In general, a function is differentiable if it is continuous and 'smooth', with no essential discontinuities or cusps. For any such function, the taking of the limit in equation (4.8) is called **differentiation from first principles**, and was demonstrated in Examples 4.1 and 4.2 for the general quadratic. All functions can be differentiated in this way. Let $y=f(x)$ be a function of x, and let the function change from y to $y+\Delta y = f(x+\Delta x)$ when the variable changes from x to $x+\Delta x$. The steps for differentiating from first principles are:

(1) subtract y from $y+\Delta y$ to obtain Δy as a function of x and Δx, and simplify as much as possible,

(2) divide both sides of the equation by Δx,

(3) find the limit of $\dfrac{\Delta y}{\Delta x}$ as $\Delta x \to 0$; this gives the required derivative $\dfrac{dy}{dx}$.

➤ Exercises 18, 19

EXAMPLE 4.6 Find $\dfrac{dy}{dx}$ from first principles for $y = \dfrac{1}{x}$.

(1) $y = f(x) = \dfrac{1}{x}, \quad y + \Delta y = f(x + \Delta x) = \dfrac{1}{x + \Delta x}$

$$\Delta y = f(x + \Delta x) - f(x) = \frac{1}{x + \Delta x} - \frac{1}{x} = \frac{-\Delta x}{x(x + \Delta x)}$$

(2) $\dfrac{\Delta y}{\Delta x} = \dfrac{-1}{x(x + \Delta x)}$ (3) $\dfrac{dy}{dx} = \lim_{\Delta x \to 0} \left(\dfrac{\Delta y}{\Delta x} \right) = -\dfrac{1}{x^2}$

➤ Exercise 20

EXAMPLE 4.7 Differentiate $\sqrt{x}$.

(1) $y = \sqrt{x}, \quad y + \Delta y = \sqrt{x + \Delta x}$

$$\Delta y = \sqrt{x + \Delta x} - \sqrt{x} = \frac{(\sqrt{x + \Delta} - \sqrt{x}) \times (\sqrt{x + \Delta x} + \sqrt{x})}{\sqrt{x + \Delta x} + \sqrt{x}}$$

$$= \frac{(x + \Delta x) - x}{\sqrt{x + \Delta x} + \sqrt{x}} = \frac{\Delta x}{\sqrt{x + \Delta x} + \sqrt{x}}$$

(making use of the relation $(a - b)(a + b) = a^2 - b^2$).

(2) $\dfrac{\Delta y}{\Delta x} = \dfrac{1}{\sqrt{x + \Delta x} + \sqrt{x}}$ (3) $\dfrac{dy}{dx} = \lim_{\Delta x \to 0} \left(\dfrac{\Delta y}{\Delta x} \right) = \dfrac{1}{2\sqrt{x}}$

➤ Exercise 21

EXAMPLE 4.8 Differentiate e^x.

(1) $y = e^x, \quad y + \Delta y = e^{x + \Delta x} = e^x \times e^{\Delta x}$

From the definition of the exponential function as an infinite series, equation (3.31),

$$e^{\Delta x} = 1 + \Delta x + \frac{(\Delta x)^2}{2!} + \frac{(\Delta x)^3}{3!} + \cdots$$

Therefore,

$$y + \Delta y = e^x \times (1 + \Delta x + \frac{(\Delta x)^2}{2} + \frac{(\Delta x)^3}{6} + \cdots)$$

$$= e^x + e^x (\Delta x + \frac{(\Delta x)^2}{2} + \frac{(\Delta x)^3}{6} + \cdots)$$

and

$$\Delta y = e^x \left(\Delta x + \frac{(\Delta x)^2}{2} + \frac{(\Delta x)^3}{6} + \cdots \right)$$

(2) $\dfrac{\Delta y}{\Delta x} = e^x \left(1 + \dfrac{\Delta x}{2} + \dfrac{(\Delta x)^2}{6} + \cdots \right)$ (3) $\dfrac{dy}{dx} = e^x$

➤ Exercise 22

4.6 Differentiation by rule

Whilst all functions can be differentiated from first principles, differentiation is performed in practice by following a set of rules.[3] These rules can be proved from first principles, as in Examples 4.6 to 4.8. The derivatives of the more important elementary functions are listed in Table 4.2.

Table 4.2 Differentiation of elementary functions

Type	Function	Derivative
$c =$ constant	c	0
power of x	x^a	ax^{a-1}
trigonometric	$\sin x$	$\cos x$
	$\cos x$	$-\sin x$
	$\tan x$	$\sec^2 x$
exponential	e^x	e^x
hyperbolic	$\cosh x$	$\sinh x$
	$\sinh x$	$\cosh x$
logarithmic	$\ln x$	$1/x$

The first example in Table 4.2 states that the rate of change of a constant is zero. In general, if a function is independent of a variable x then its derivative with respect to x is zero. Examples of the derivative of a 'power' of x, x^a, have been given in Examples 4.6 for $a = -1$ and 4.7 for $a = 1/2$. Other examples, using a variety of notations, are given in Examples 4.9.

---

[3] Many of the rules of differentiation appeared in Leibniz's 1684 paper on the differential calculus, with the d–notation and the name *calculus differentialis* for the finding of tangents.

Isaac Newton (1642–1727) developed his ideas on the calculus in the year 1665–1666 (Trinity College, Cambridge, was closed because of the plague); he later maintained that during this time he discovered the binomial theorem, the calculus, the law of gravitation and the nature of colours. He wrote the first of three accounts of the calculus in 1669 in *De analysi per aequationes numero terminorum infinitas*, but published only in 1711. The first account to be published appeared in 1687 in *Philosophiae naturalis principia mathematica*, probably the most influential scientific treatise of all time. In the first edition of the *Principia*, Newton acknowledged that Leibniz also had a similar method. By the third edition of 1726, the reference to Leibniz had been deleted when questions of priority had led to a bitter quarrel between supporters of the two men.

EXAMPLES 4.9 Differentiating powers

(i) $y = x^5$ $\qquad\qquad \dfrac{dy}{dx} = 5x^4$

(ii) $f(x) = x^{-1/2}$ $\qquad f'(x) = -\dfrac{1}{2}x^{-3/2}$

(iii) $f(x) = x^{0.3}$ $\qquad \dfrac{d}{dx}f(x) = 0.3x^{-0.7}$

▸ Exercises 23–26

EXAMPLE 4.10 For the ideal gas example discussed in Section 4.1,

$$V = \frac{nRT}{p}, \qquad \frac{dV}{dp} = nRT\,\frac{d}{dp}\left(\frac{1}{p}\right) = nRT\left(-\frac{1}{p^2}\right) = \frac{-nRT}{p^2}$$

since nRT is constant at constant T and n. We note that whereas V is inversely proportional to p, it is directly proportional to $1/p$; that is, V is a linear function of $1/p$, and the graph of V against $1/p$ is a straight line with slope

$$\frac{dV}{d(1/p)} = nRT.$$

The rules for differentiating combinations of elementary functions are summarized in Table 4.3; in these rules, x is the independent variable, y, u and v are functions of x, and a is a constant.

Table 4.3 Differentiation of combinations of functions

Type	Rule
1. multiple of a function	$\dfrac{d}{dx}(a\,u) = a\,\dfrac{du}{dx}$
2. sum of functions	$\dfrac{d}{dx}(u + v) = \dfrac{du}{dx} + \dfrac{dv}{dx}$
3. product rule	$\dfrac{d}{dx}(uv) = u\,\dfrac{dv}{dx} + v\,\dfrac{du}{dx}$
4. quotient rule	$\dfrac{d}{dx}\left(\dfrac{u}{v}\right) = \left(v\,\dfrac{du}{dx} - u\,\dfrac{dv}{dx}\right)\Big/ v^2$
5. chain rule	$\dfrac{d}{dx}f(u) = \dfrac{df}{du} \times \dfrac{du}{dx}$
6. inverse rule	$\dfrac{dx}{dy} = 1\Big/\left(\dfrac{dy}{dx}\right)$ or $\dfrac{dx}{dy} \times \dfrac{dy}{dx} = 1$

Linear combination of functions

A linear combination of the functions u, v, and w of x has the form

$$y = au(x) + bv(x) + cw(x) \qquad (4.12)$$

where a, b, and c are constants. Such a function can be differentiated term by term; by Rules 1 and 2 in Table 4.3,

$$\frac{dy}{dx} = a\frac{du}{dx} + b\frac{dv}{dx} + c\frac{dw}{dx} \qquad (4.13)$$

EXAMPLE 4.11 Differentiate $y = 2x^3 + 3e^x - \frac{1}{2}\ln x$.

By Equation (4.13),

$$\frac{dy}{dx} = 2\frac{d}{dx}x^3 + 3\frac{d}{dx}e^x - \frac{1}{2}\frac{d}{dx}\ln x = 2 \times 3x^2 + 3 \times e^x - \frac{1}{2} \times \frac{1}{x}$$

$$= 6x^2 + 3e^x - \frac{1}{2x}$$

> Exercises 27, 28

The product rule

The function

$$y = (2x + 3x^2)(5 + 7x^3)$$

can be differentiated by treating it as the product $y = uv$ where $u = (2x + 3x^2)$ and $v = (5 + 7x^3)$. Then, by Rule 3 in Table 4.3,

$$\frac{dy}{dx} = u\frac{dv}{dx} + v\frac{du}{dx}$$

$$= (2x + 3x^2)\frac{d}{dx}(5 + 7x^3) + (5 + 7x^3)\frac{d}{dx}(2x + 3x^2)$$

$$= (2x + 3x^2)(21x^2) + (5 + 7x^3)(2 + 6x)$$

This may now be simplified. In this example it is equally simple to multiply out the original product and differentiate term by term, but in many cases the brute force approach is more difficult than use of the product rule.

EXAMPLE 4.12 Product rule

The function

$$y = (2x + 3x^2)\sin x$$

is easily differentiated only by means of the product rule. Let $y = uv$ where $u = (2x + 3x^2)$ and $v = \sin x$. Then

$$\frac{dy}{dx} = u\frac{dv}{dx} + v\frac{du}{dx}$$

$$= (2x + 3x^2)\frac{d}{dx}\sin x + \sin x\frac{d}{dx}(2x + 3x^2)$$

$$= (2x + 3x^2)\cos x + (2 + 6x)\sin x$$

‣ Exercises 29–32

The quotient rule

By Rule 4 in Table 4.3,

$$\frac{d}{dx}\left(\frac{u}{v}\right) = \left(v\frac{du}{dx} - u\frac{dv}{dx}\right)\Big/v^2$$

EXAMPLE 4.13 Differentiate $y = \dfrac{2x + 3x^2}{5 + 7x^3}$

Let $y = u/v$ where $u = (2x + 3x^2)$ and $v = (5 + 7x^3)$. Then

$$\frac{dy}{dx} = \left((5 + 7x^3)\frac{d}{dx}(2x + 3x^2) - (2x + 3x^2)\frac{d}{dx}(5 + 7x^3)\right)\Big/(5 + 7x^3)^2$$

$$= \frac{(5 + 7x^3)(2 + 6x) - (2x + 3x^2)(21x^2)}{(5 + 7x^3)^2}$$

‣ Exercises 33–36

The chain rule (function of a function)

The polynomial

$$y = f(x) = (2x^2 - 1)^3$$

can be differentiated by first expanding the cube and then differentiating term by term:

$$y = (2x^2 - 1)^3 = 8x^6 - 12x^4 + 6x^2 - 1$$

$$\frac{dy}{dx} = 48x^5 - 48x^3 + 12x = 12x(2x^2 - 1)^2$$

A simpler way is to use the chain rule. Let $u = 2x^2 - 1$ and rewrite y as a function of u:

$$y = g(u) = u^3$$

Then by Rule 5 in Table 4.3,

$$\frac{dy}{dx} = \frac{dy}{du} \times \frac{du}{dx} = (3u^2) \times (4x)$$

and, substituting for u,

$$\frac{dy}{dx} = 12x(2x^2 - 1)^2$$

In this example, y has been considered in *two ways*:

(i) as a function of x: $f(x) = (2x^2 - 1)^3$;

(ii) as a function of u: $g(u) = u^3$ where u is the function $u = 2x^2 - 1$. The substitution $u = 2x^2 - 1$ highlights the structure of the function, that of a cube, and makes the chain rule the natural method of differentiation.

EXAMPLE 4.14 Differentiate $y = (2x^2 - 1)^{5/2}$.

Put $y = u^{5/2}$, where $u = 2x^2 - 1$. Then

$$\frac{dy}{dx} = \frac{dy}{du} \times \frac{du}{dx} = \left(\frac{5}{2} u^{3/2} \right)(4x) = 10x(2x^2 - 1)^{3/2}$$

▶ Exercises 37–40

Table 4.4 shows the generalization of Table 4.2 for elementary functions of a function, $f(u)$ where $u = u(x)$.

Table 4.4 The chain rule

Type	Function	Derivative
power of u	u^a	$a\,u^{a-1}\dfrac{du}{dx}$
trigonometric	$\sin u$	$\cos u\,\dfrac{du}{dx}$
	$\cos u$	$-\sin u\,\dfrac{du}{dx}$
	$\tan u$	$\sec^2 u\,\dfrac{du}{dx}$
exponential	e^u	$e^u\,\dfrac{du}{dx}$
logarithmic	$\ln u$	$\dfrac{1}{u}\dfrac{du}{dx}$

EXAMPLES 4.15 The chain rule

(i) $y = \sin 2x = \sin u$, where $u = 2x$

$$\frac{dy}{dx} = \frac{dy}{du} \times \frac{du}{dx} = (\cos u) \times (2) = 2\cos 2x$$

(ii) $y = \cos(2x^2 - 1) = \cos u$, where $u = 2x^2 - 1$

$$\frac{dy}{dx} = \frac{dy}{du} \times \frac{du}{dx} = (-\sin u) \times (4x) = -4x\sin(2x^2 - 1)$$

(iii) $y = e^{2x^2 - 1} = e^u$, where $u = 2x^2 - 1$

$$\frac{dy}{dx} = \frac{dy}{du} \times \frac{du}{dx} = (e^u) \times (4x) = 4xe^{2x^2 - 1}$$

(iv) $y = \ln(2x^2 - 1) = \ln u$, where $u = 2x^2 - 1$

$$\frac{dy}{dx} = \frac{dy}{du} \times \frac{du}{dx} = \left(\frac{1}{u}\right) \times (4x) = \frac{4x}{2x^2 - 1}$$

(v) $y = \ln(\sin x) = \ln u$, where $u = \sin x$

$$\frac{dy}{dx} = \frac{dy}{du} \times \frac{du}{dx} = \left(\frac{1}{u}\right)(\cos x) = \frac{\cos x}{\sin x} = \cot x$$

> Exercises 41–55

Example 4.15(i) demonstrates the important special case of $u(x) = ax$, for which $\dfrac{du}{dx} = a$. Then

$$\frac{dy}{dx} = \frac{dy}{du} \times \frac{du}{dx} = a\frac{dy}{du}$$

For example,

$$\frac{d}{dx}\cos 3x = -3\sin 3x, \qquad \frac{d}{dx}e^{2x} = 2e^{2x}$$

Inverse functions

If $y = f(x)$, the inverse function of f is defined by $x = f^{-1}(y)$. By Rule 6 in Table 4.3, the derivatives of function and inverse function are related by

$$\frac{dy}{dx} \times \frac{dx}{dy} = \left[\frac{d}{dx}f(x)\right] \times \left[\frac{d}{dy}f^{-1}(y)\right] = 1$$

The inverse rule

$$\frac{dx}{dy} = \frac{1}{\left(\dfrac{dy}{dx}\right)} \tag{4.14}$$

is used when it is more difficult to differentiate the function than its inverse.

EXAMPLE 4.16 Use of the inverse rule

If y is defined implicitly by

$$x = y^5 - 2y$$

(see Example 2.11), then $\dfrac{dy}{dx}$ can be found as the inverse of $\dfrac{dx}{dy}$:

$$\frac{dx}{dy} = 5y^4 - 2, \qquad \frac{dy}{dx} = \frac{1}{5y^4 - 2}$$

▸ Exercises 56–59

Particularly important examples of the differentiation of inverse functions are given in Table 4.5, where the inverse trigonometric functions have their principal values.

Table 4.5

$$\frac{d}{dx}\left(\sin^{-1}\frac{x}{a}\right) = \frac{1}{\sqrt{a^2 - x^2}}$$

$$\frac{d}{dx}\left(\cos^{-1}\frac{x}{a}\right) = \frac{-1}{\sqrt{a^2 - x^2}}$$

$$\frac{d}{dx}\left(\tan^{-1}\frac{x}{a}\right) = \frac{a}{a^2 + x^2}$$

EXAMPLE 4.17 Inverse trig functions

If $y = \sin^{-1}\left(\dfrac{x}{a}\right)$ then $x = a\sin y$ and

$$\frac{dx}{dy} = a\cos y, \qquad \frac{dy}{dx} = \frac{1}{a\cos y}$$

If y has its principal value then it lies between $-\frac{\pi}{2}$ and $\frac{\pi}{2}$, and $\cos y$ is positive.
Then

$$a\cos y = a\sqrt{1 - \sin^2 y} = \sqrt{a^2 - a^2\sin^2 y} = \sqrt{a^2 - x^2}$$

and

$$\frac{dy}{dx} = \frac{1}{\sqrt{a^2 - x^2}}$$

▶ Exercises 60 – 62

The derivatives of the inverse hyperbolic functions (Section 3.9) are given in Table 4.6.

Table 4.6

$$\frac{d}{dx}\left(\sinh^{-1}\frac{x}{a}\right) = \frac{1}{\sqrt{x^2 + a^2}}$$

$$\frac{d}{dx}\left(\cosh^{-1}\frac{x}{a}\right) = \frac{1}{\sqrt{x^2 - a^2}}$$

$$\frac{d}{dx}\left(\tanh^{-1}\frac{x}{a}\right) = \frac{a}{a^2 - x^2}$$

EXAMPLE 4.18 Inverse hyperbolic functions

If $y = \sinh^{-1}\left(\dfrac{x}{a}\right)$ then $x = a\,\sinh y$ and

$$\frac{dx}{dy} = a\cosh y, \qquad \frac{dy}{dx} = \frac{1}{a\cosh y}$$

Because $\cosh y > 0$ (see Figure 3.22), $a\cosh y = a\sqrt{1 + \sinh^2 y} = \sqrt{a^2 + x^2}$, and

$$\frac{dy}{dx} = \frac{1}{\sqrt{a^2 + x^2}}$$

‣ Exercises 63, 64

4.7 Implicit functions

Every functional relationship between two variables x and y can be expressed in the implicit form (see Section 2.4)

$$f(x, y) = 0 \tag{4.15}$$

In all the examples of the applications of the rules of differentiation discussed in Section 4.6 it has been possible to express at least one of the variables as an explicit function of the other. In some cases however neither variable can be so expressed and the rules must be applied to the implicit function itself. Let equation (4.15) define y as a function of x. Then the change in y that accompanies a change in x is such that equation (4.15) is always true. It follows that the rate of change of the implicit function $f(x, y)$ is zero for all allowed changes:

$$\frac{d}{dx} f(x, y) = 0 \tag{4.16}$$

EXAMPLE 4.19 Find dy/dx for $f(x, y) = y^5 - 2y - x = 0$.

We have

$$\frac{df}{dx} = \frac{d}{dx}(y^5) - 2\frac{d}{dx}(y) - \frac{d}{dx}(x) = 5y^4\frac{dy}{dx} - 2\frac{dy}{dx} - 1$$

$$= (5y^4 - 2)\frac{dy}{dx} - 1 = 0$$

Solving for $\dfrac{dy}{dx}$ gives $\dfrac{dy}{dx} = \dfrac{1}{5y^4 - 2}$

This is the result obtained in Example 4.16 by differentiation of the inverse function.

▸ Exercises 65–68

4.8 Logarithmic differentiation

For some functions it is easier to differentiate the natural logarithm than the function itself. For example, if

$$y = u^a v^b w^c \cdots \tag{4.17}$$

where u, v, w, … are functions of x, and a, b, c, … are constants, then

$$\ln y = a \ln u + b \ln v + c \ln w + \cdots \tag{4.18}$$

and

$$\frac{d}{dx} \ln y = a \frac{d}{dx} \ln u + b \frac{d}{dx} \ln v + c \frac{d}{dx} \ln w +$$

Then, because $\dfrac{d \ln y}{dx} = \dfrac{1}{y} \dfrac{dy}{dx}$,

$$\frac{1}{y} \frac{dy}{dx} = \frac{a}{u} \frac{du}{dx} + \frac{b}{v} \frac{dv}{dx} + \frac{c}{w} \frac{dw}{dx} + \tag{4.19}$$

This method of differentiating is called **logarithmic differentiation**. When, $y = uv$ the method reproduces the product rule; when $y = u/v$, it reproduces the quotient rule.

EXAMPLES 4.20 Logarithmic differentiation

(i) $y = \left(\dfrac{1+x}{1-x}\right)^{1/2}$, $\ln y = \frac{1}{2}\left[\ln(1+x) - \ln(1-x)\right]$

Then

$$\frac{d}{dx} \ln y = \frac{1}{y} \frac{dy}{dx} = \frac{1}{2}\left(\frac{1}{1+x} + \frac{1}{1-x}\right) = \frac{1}{1-x^2}$$

and, multiplying by y,

$$\frac{dy}{dx} = \frac{1}{1-x^2}\left(\frac{1+x}{1-x}\right)^{1/2} = \frac{1}{(1+x)^{1/2}(1-x)^{3/2}}$$

(ii) $y = a^x$, $\ln y = x \ln a$.

Then

$$\frac{d}{dx}\ln y = \frac{1}{y}\frac{dy}{dx} = \ln a \quad \text{and, therefore,} \quad \frac{dy}{dx} = a^x \ln a$$

(iii) $y = x^x$, $\ln y = x \ln x$.

Applying the product rule,

$$\frac{d}{dx}\ln y = x\frac{d}{dx}\ln x + \ln x\frac{d}{dx}x = 1 + \ln x$$

and, therefore,

$$\frac{1}{y}\frac{dy}{dx} = 1 + \ln x \quad \text{and} \quad \frac{dy}{dx} = x^x(1 + \ln x)$$

> Exercises 69–72

EXAMPLE 4.21 Logarithmic plots

For a system undergoing exponential decay (first-order kinetics), the size of the system is given by (see Example 3.20)

$$x(t) = x_0 e^{-kt}$$

and, taking the logarithm of both sides,

$$\ln x = \ln x_0 - kt$$

A plot of $\ln x$ against t, Figure 4.8, gives a straight line with slope $d(\ln x)/dt = -k$, and intercept $\ln x_0$ with the axis $t = 0$. This example is important because it demonstrates the standard way of calculating the rate constant k from a linear plot of experimental values of x and t.

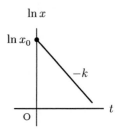

Figure 4.8

> Exercise 73

4.9 Successive differentiation

The derivative of a function can itself be differentiated if it satisfies the continuity and smoothness conditions discussed in Sections 4.3 to 4.5. For example, the cubic

$$y = x^3 + x^2 + x + 1$$

has (first) derivative

$$\frac{dy}{dx} = 3x^2 + 2x + 1$$

and this may be differentiated to give the **second derivative**, or **second differential coefficient**,

$$\frac{d}{dx}\left(\frac{dy}{dx}\right) = \frac{d^2 y}{dx^2} = 6x + 2$$

Successive differentiation gives the third and fourth derivatives

$$\frac{d^3 y}{dx^3} = 6, \qquad \frac{d^4 y}{dx^4} = 0$$

and all higher derivatives are zero. Alternative notations (see Section 4.2) are $f'(x)$, $f''(x)$, $f'''(x)$, ... for derivatives of $f(x)$, or Df, $D^2 f$, $D^3 f$, ... , where D is the differential operator $\dfrac{d}{dx}$.

> Exercise 74

A polynomial of degree n can have only up to the first n derivatives nonzero, but other simple functions can be differentiated indefinitely. In particular, the exponential function e^x has all its derivatives equal to e^x.

EXAMPLE 4.22 Derivatives of sin ax.

$$f(x) = \sin ax$$
$$f'(x) = a \cos ax$$
$$f''(x) = -a^2 \sin ax = -a^2 f(x)$$

In general, for the nth derivative,

$$f^{(n)}(x) = \begin{cases} (-1)^{(n-1)/2} a^n \cos ax & \text{if } n \text{ is odd} \\ (-1)^{n/2} a^n \sin ax & \text{if } n \text{ is even} \end{cases}$$

For example,

$$n = 3: \qquad f^{(3)}(x) = f'''(x) = -a^3 \cos ax$$
$$n = 4: \qquad f^{(4)}(x) = (-1)^2 a^4 \sin ax = a^4 \sin ax$$

> Exercises 75–77

The first derivative $f'(x)$ of a function $f(x)$ is the rate of change of the function, or the slope of its graph at point x. The second derivative $f''(x)$ is the rate of change of slope, and is related to the curvature at x.

4.10   Stationary points

Consider the cubic (Figure 4.9)

$$y = x(x-3)^2$$

The function goes to $\pm\infty$ as $x \to \pm\infty$, but the graph shows that the function has a **local maximum** at point A, at $x = 1$, where its value is greater than at all neighbouring points. The function also has a (local) **minimum** at point B, at $x = 3$, where its value is smaller than at all neighbouring points. Points of maximum and minimum value are called **turning points**.

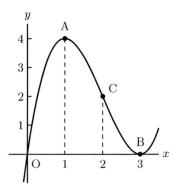

Figure 4.9

The determination of the maximum and minimum values of a function is of importance in the physical sciences because, for example, (i) equations of motion are often formulated as 'variation principles', whereby solutions are obtained as maxima or minima of some variational function (see Example 4.25), (ii) the fitting of a theoretical curve to a set of experimental points can be expressed in terms of a 'minimum deviation' principle, as in the method of least squares discussed in Chapter 21.

Consider the curve in the neighbourhood of the maximum at point A in Figure 4.9. To the left of A the gradient is positive and the value of the function is increasing. To the right of A the gradient is negative and y is decreasing. At the point A itself the function has zero gradient (the tangent to the curve is horizontal), and the rate of change of y with respect to x is zero. The point is called a **stationary point**, and the value of the function at the point is called a **stationary value**. Similar considerations apply for the minimum at B, and the general condition for a stationary point is that the first derivative of the function be zero:

$$\frac{dy}{dx} = 0 \qquad \text{at a stationary point} \qquad (4.20)$$

To distinguish between maximum and minimum values, it is necessary to consider the second derivative. On moving through the maximum at A from left to right, the gradient decreases from positive values, through zero at A, to negative values. It follows that the rate of change of the gradient is negative at A, and this is a sufficient condition for the function to have maximum value at this point:

$$\text{for a maximum:} \qquad \frac{dy}{dx} = 0 \quad \text{and} \quad \frac{d^2 y}{dx^2} < 0 \qquad\qquad (4.21)$$

Similar considerations applied to the minimum at B show that

$$\text{for a minimum:} \qquad \frac{dy}{dx} = 0 \quad \text{and} \quad \frac{d^2 y}{dx^2} > 0 \qquad\qquad (4.22)$$

For the cubic shown in Figure 4.9,

$$y = x(x-3)^2 = x^3 - 6x^2 + 9x$$

$$\frac{dy}{dx} = 3x^2 - 12x + 9 = 3(x-1)(x-3) = 0 \quad \text{when} \quad x = 1 \quad \text{or} \quad x = 3$$

$$\frac{d^2 y}{dx^2} = 6x - 12 \quad \begin{cases} < 0 & \text{when } x = 1, \quad \text{a maximum} \\ > 0 & \text{when } x = 3, \quad \text{a minimum} \\ = 0 & \text{when } x = 2, \quad \text{a point of inflection} \end{cases}$$

The point C, at $x = 2$ in Figure 4.9, is an example of a **point of inflection**, at which the gradient is a maximum or minimum, with $\dfrac{d^2 y}{dx^2} = 0$. The slope of the curve decreases (becomes more negative) between A and C and increases between C and B, with minimum value at C. This is an example of a simple point of inflection with

$$\frac{dy}{dx} \neq 0 \quad \text{and} \quad \frac{d^2 y}{dx^2} = 0$$

▸ Exercises 78 – 82

When $\dfrac{dy}{dx} = 0$ and $\dfrac{d^2 y}{dx^2} = 0$ at a point then the nature of the point is determined by the first nonzero higher derivative. Two examples are

$$\text{(i)} \quad \frac{dy}{dx} = 0, \qquad \frac{d^2 y}{dx^2} = 0, \qquad \frac{d^3 y}{dx^3} \neq 0 \qquad\qquad (4.23)$$

This is a point of inflection which is also a stationary point (but not a turning point), and is the case discussed in Example 4.23.

$$\text{(ii)} \quad \frac{dy}{dx} = 0, \qquad \frac{d^2 y}{dx^2} = 0, \qquad \frac{d^3 y}{dx^3} = 0, \qquad \frac{d^4 y}{dx^4} \neq 0 \qquad\qquad (4.24)$$

This is a maximum or minimum, depending on the sign of the fourth derivative. For example, the function $y = (x - 1)^4$ satisfies (4.24), with a minimum at $x = 1$.

EXAMPLE 4.23 Find the stationary points of the quartic $y = 3x^4 - 4x^3 + 1$.

For the stationary values,

$$\frac{dy}{dx} = 12x^3 - 12x^2 = 12x^2(x - 1)$$

$$= 0 \text{ when } x = 0 \text{ or } x = 1.$$

To determine the kinds of stationary points,

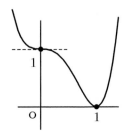

Figure 4.10

$$\frac{d^2y}{dx^2} = 36x^2 - 24x \quad \begin{cases} = 0 & \text{when } x = 0 \\ > 0 & \text{when } x = 1, \text{ a minimum} \end{cases}$$

To determine the kind of stationary point at $x = 0$,

$$\frac{d^3y}{dx^3} = 72x - 24 \neq 0 \text{ when } x = 0, \text{ a point of inflection}$$

The function therefore has a single turning point, a minimum, at $x = 1$, when $y = 0$, and a point of inflection at $x = 0$, when $y = 1$. In fact, the function can be factorized,

$$y = (x - 1)^2(3x^2 + 2x + 1)$$

and has a double root $x = 1$, and two complex roots.

➤ Exercise 83

EXAMPLE 4.24 In Hückel molecular orbital theory, the possible values of the orbital energies of the π electrons of ethene (C_2H_4) are given by the stationary values of the quantity

$$\varepsilon = \alpha + 2c(1 - c^2)^{1/2}\beta$$

where α and β are constant 'Hückel parameters', and c is a variable. For the stationary values of ε,

$$\frac{d\varepsilon}{dc} = 2(1 - c^2)^{1/2}\beta - 2c^2(1 - c^2)^{-1/2}\beta = 0$$

Division by 2β and multiplication by $(1-c^2)^{1/2}$ gives

$$(1-c^2)-c^2 = 0, \qquad \text{or} \qquad c = \pm\frac{1}{\sqrt{2}}$$

Then $\varepsilon = \alpha \pm \beta$.

EXAMPLE 4.25 Snell's law of refraction in geometric optics.[4]

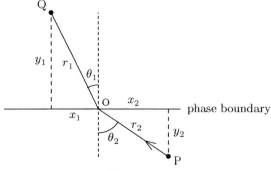

Figure 4.11

A ray of light travels between points P and Q across a phase boundary at O. In the upper region, the speed of light is $v_1 = c/\eta_1$, where c is the speed of light in vacuum and η_1 is the refractive index of the phase. In the lower region the speed of light is $v_2 = c/\eta_2$. Snell's law of refraction is

$$\frac{\sin\theta_1}{\sin\theta_2} = \frac{v_1}{v_2} = \frac{\eta_2}{\eta_1}$$

The law can be derived from a 'principle of least time', that the path followed is that of least time.[5] The total time travelled from P to Q through point O is (distance/speed in each phase),

$$t = \frac{r_1}{v_1} + \frac{r_2}{v_2}$$

The problem is to find point O such that t is a minimum. Choosing x_1 as the independent variable, we have

$$r_1 = (x_1^2 + y_1^2)^{1/2}, \qquad r_2 = (x_2^2 + y_2^2)^{1/2} = \left[(X-x_1)^2 + y_2^2\right]^{1/2}$$

---

[4] Willebrord van Roijen Snell (1591–1626), Dutch mathematician and physicist, formulated the law of refraction in 1621.

[5] This use of the principle of least time, proposed by Fermat, was one of the examples used by Leibniz in his 1684 paper to demonstrate his method of finding maxima and minima.

in which y_1, y_2, and $X = x_1 + x_2$ are constant. Then

$$\frac{dt}{dx_1} = \frac{1}{v_1}\frac{dr_1}{dx_1} + \frac{1}{v_2}\frac{dr_2}{dx_1} = \frac{1}{v_1}\frac{x_1}{r_1} - \frac{1}{v_2}\frac{x_2}{r_2} = \frac{\sin\theta_1}{v_1} - \frac{\sin\theta_2}{v_2}$$

$$= 0 \;\text{ for a minimum}$$

Hence Snell's law.

▸ Exercises 84 – 86

4.11 Linear and angular motion

The description of the motion of bodies in space is an important application of the differential calculus. We consider here only the simplest kinds of motion; motion in a straight line and motion in a circle. More general kinds of motion are discussed in Chapter 16.

Linear motion

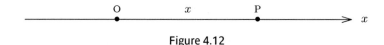

Figure 4.12

Consider a body moving in a straight line, along the x-direction say. Let O be a fixed point and let P be the position of the body at time t. The distance $OP = x$ is then a function of time; $x = f(t)$.

If the body moves from point x to point $x + \Delta x$ in time interval Δt, then the average rate of change of x in the interval is

$$\frac{\Delta x}{\Delta t} = \text{average velocity in interval } \Delta t$$

The limit of this as $\Delta t \to 0$ is the instantaneous rate of change of x with respect to t. It is the **linear velocity**, or simply the velocity, at time t,

$$\text{velocity} = v = \frac{dx}{dt} \qquad\qquad (4.25)$$

When v is positive, x is increasing and the body is moving to the right. When v is negative, x is decreasing and the body is moving to the left. Velocity is in fact a vector quantity, having both magnitude and direction; vectors are discussed in Chapter 16. The magnitude of the velocity is the **speed**. The derivative of the velocity is the **acceleration**,

$$\text{acceleration} = \frac{dv}{dt} = \frac{d^2 x}{dt^2} \qquad\qquad (4.26)$$

Derivatives with respect to time are sometimes written in a 'dot notation':[6]

$$v = \dot{x} = \frac{dx}{dt}, \qquad \dot{v} = \ddot{x} = \frac{d^2x}{dt^2} \qquad (4.27)$$

> Exercise 87

Angular motion

Consider a body moving in a circle of radius r. Let O be a fixed point on the circle and let the position of the body at time t be given by the angle θ (Figure 4.13). The rate of change of θ with respect to time is called the **angular velocity**:

$$\text{angular velocity} = \omega = \lim_{\Delta t \to 0} \frac{\Delta \theta}{\Delta t} = \frac{d\theta}{dt} = \dot{\theta} \qquad (4.28)$$

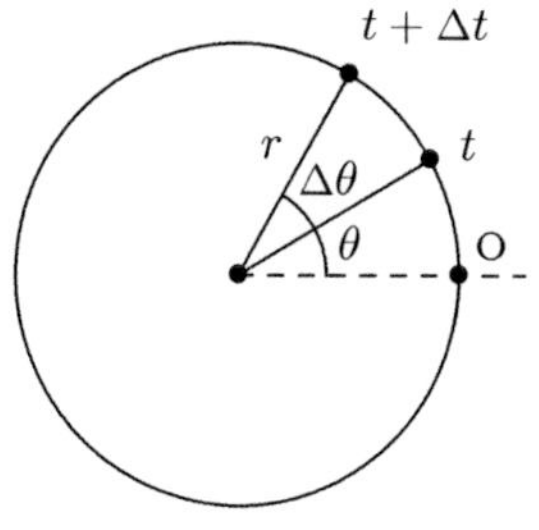

Figure 4.13

In addition, because arc length s is related to subtended angle θ by $s = r\theta$, we have $\dot{s} = r\dot{\theta}$ or

$$v = r\omega \qquad (4.29)$$

for the relation between linear and angular velocities for motion in a circle (see Chapter 16 for a more general discussion of velocity in terms of vectors). The angular acceleration is $\dot{\omega} = \ddot{\theta}$.

> Exercise 88

4.12   The differential

The first derivative of a function $y = f(x)$ is defined by equation (4.8),

$$\frac{dy}{dx} = f'(x) = \lim_{\Delta x \to 0} \left(\frac{\Delta y}{\Delta x} \right) = \lim_{\Delta x \to 0} \left\{ \frac{f(x + \Delta x) - f(x)}{\Delta x} \right\}$$

and, as has been emphasized before, the symbol dy/dx does not mean the quantity dy divided by dx, but represents the value of the limit; in this sense $f'(x)$, or y', is a better symbol for the derivative. It is nevertheless tempting to write

$$dy = f'(x)\, dx$$

and, when properly interpreted, this is a useful way of describing changes.

---

[6] In his 1671 paper on the calculus, *Methodus fluxionum et serierum infinitorum*, Newton considered variables like x and y as flowing quantities, or *fluents*, and wrote $\dot{x}$ and $\dot{y}$ for their rates of change, or *fluxions*. Dotted fluxions were still being used by English mathematicians when the Cambridge undergraduates George Peacock (1791–1858), John Herschel (1792–1871), and Charles Babbage (1792–1871) founded the Analytical Society in 1813. One of the aims of the Society was to promote 'the principles of pure *d*-ism as opposed to the *dot*-age of the university'.

Consider the cubic

$$y = f(x) = x^3, \qquad \frac{dy}{dx} = f'(x) = 3x^2$$

Let Δy be the change in y accompanying the change Δx in x:

$$\Delta y = f(x + \Delta x) - f(x) = (x + \Delta x)^3 - x^3$$
$$= 3x^2 \, \Delta x + 3x(\Delta x)^2 + (\Delta x)^3$$

The derivative dy/dx is obtained by dividing this expression by Δx and letting $\Delta x \to 0$. Another way of looking at the limit is to consider Δx as a 'very small' change. If Δx is made small enough then the term in $(\Delta x)^3$ becomes much smaller than the term in $(\Delta x)^2$ which in turn becomes much smaller than the term in Δx,

$$(\Delta x)^3 \ll (\Delta x)^2 \ll \Delta x$$

For example, $\Delta x = 10^{-3}$, $(\Delta x)^2 = 10^{-6}$, and $(\Delta x)^3 = 10^{-9}$. An approximate expression for the change in y is then

$$\Delta y \approx 3x^2 \, \Delta x = f'(x) \, \Delta x$$

and this is often a useful way of approximating small changes. The quantity $f'(x)\Delta x$ would be the change in y if Δx were small enough. It is *useful* to consider an arbitrary small change dx, an '**infinitesimal change**', such that terms in $(dx)^2$ and higher can be set to zero. The corresponding change in y

$$dy = f'(x) \, dx \qquad (4.30)$$

is called the **differential** of y.[7]

The use of the differential will become clear in later chapters. It is important in the physical sciences because fundamental theorems are sometimes expressed in differential form; in particular, the laws of thermodynamics are nearly always expressed in terms of differentials.

> Exercises 89–91

EXAMPLE 4.26 The differential area of a circle

The area of a circle as a function of the radius is

$$A(r) = \pi r^2$$

---

[7] Leibniz's formulation of the calculus was in terms of differentials. His 1684 paper contains the formulas $dx^n = nx^{n-1} \, dx$, for the infinitesimal change or differential of x^n, and $dxy = xdy + ydx$ for the product rule (see Example 4.27).

If the radius is increased by amount Δr, the corresponding change in the area is

$$\Delta A = \pi(r + \Delta r)^2 - \pi r^2 = 2\pi r\,\Delta r + \pi(\Delta r)^2$$

and this is the area of a circular ring of radius r and width Δr. When Δr is small enough,

$$\Delta A \approx 2\pi r\,\Delta r = \frac{dA}{dr}\,\Delta r$$

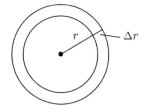

Figure 4.14

The corresponding 'differential area' is

$$dA = 2\pi r\,dr = \text{circumference} \times \text{width}$$

› Exercise 92

EXAMPLE 4.27 Differential form of the product rule

If $y = uv$, then the change in y accompanying changes in u and v is

$$\Delta y = (u + \Delta u)(v + \Delta v) - uv = u\,\Delta v + v\,\Delta u + (\Delta u)(\Delta v)$$

and the differential form is

$$dy = d(uv) = u\,dv + v\,du$$

If y, u, and v are functions of x, this expression is equivalent to the normal form of the product rule: 'division by dx' gives

$$\frac{dy}{dx} = u\frac{dv}{dx} + v\frac{du}{dx}$$

An important application of the differential is as a formal procedure for changing the independent variable. Consider $y = f(x)$ and its differential

$$dy = f'(x)\,dx \tag{4.31}$$

Let x be a function of some other variable, t, such that $x = g(t)$ with differential

$$dx = g'(t)\,dt \tag{4.32}$$

Substitution of this in (4.31) then gives

$$dy = f'(x)\,dx = f'(x)g'(t)\,dt$$

or

$$dy = \left(\frac{dy}{dx} \times \frac{dx}{dt} \right) dt \tag{4.33}$$

This is the differential of y with respect to the variable t. Division by dt then gives the Chain Rule:

$$\frac{dy}{dt} = \frac{dy}{dx} \times \frac{dx}{dt} \tag{4.34}$$

Although there have been conceptual difficulties with this type of manipulation of differentials (infinitesimals or 'infinitely small changes'), operations with differentials can always be shown to duplicate methods involving finite changes (Δ's instead of d's) and limits.[8]

4.13 Exercises

Section 4.2

1. For $y = x^3$, find **(i)** the change Δy in y that corresponds to change Δx in x, **(ii)** $\Delta y / \Delta x$.

2. For $y = x^3$, find $\displaystyle\lim_{\Delta x \to 0} \left(\frac{\Delta y}{\Delta x} \right)$

3. For the Langmuir isotherm $\theta = \dfrac{Kp}{1 + Kp}$ find **(i)** the change $\Delta\theta$ in θ that corresponds to change Δp in p, **(ii)** $\displaystyle\lim_{\Delta p \to 0} \left(\frac{\Delta\theta}{\Delta p} \right)$

Section 4.3

Find the discontinuities of the following functions and state which are essential and which removable. Sketch graphs to demonstrate your answers.

4. $\dfrac{1}{x+1}$ **5.** $\dfrac{x^2}{x}$ **6.** $\dfrac{2x}{x^2 - 3x}$

Section 4.4

Find the limits:

7. $\displaystyle\lim_{x \to 0} \left(\frac{x^2}{x} \right)$ **8.** $\displaystyle\lim_{x \to 0} \left(\frac{x}{x^2} \right)$ **9.** $\displaystyle\lim_{x \to 0} \left(\frac{x+1}{x+3} \right)$

---

[8] In fact, developments in mathematical logic between 1920 and 1960 have led to the development of a 'non-standard analysis' which involves an extension to the number system to include infinitesimals (Abraham Robinson, *Non-standard analysis*, Princeton, 1996).

10. $\lim_{x \to 1} \left(\dfrac{x-1}{x^2-1} \right)$ **11.** $\lim_{x \to \infty} \left(\dfrac{x+1}{x+3} \right)$ **12.** $\lim_{x \to \infty} \left(\dfrac{x-1}{x^2-1} \right)$

13. $\lim_{x \to \infty} \left(\dfrac{x^2-1}{x+1} \right)$ **14.** $\lim_{x \to 0} \left[\left(4x^2 - \dfrac{1}{x^2} \right) + \left(2x - \dfrac{1}{x} \right)^2 \right]$ **15.** $\lim_{x \to 0} \left(\dfrac{e^{2x}-1}{x} \right)$

16. $\lim_{x \to 0} (\ln x - \ln 2x)$ **17.** $\lim_{x \to \infty} \left[\ln(x-4) - \ln(3x+2) \right]$

Section 4.5

Differentiate from first principles:

18. $2x^2 + 3x + 4$ **19.** x^4 **20.** $2/x^2$ **21.** $x^{3/2}$ **22.** e^{-x}

Section 4.6

Differentiate by rule:

23. x^3 **24.** $x^{5/4}$ **25.** $x^{1/3}$ **26.** $1/x^3$

27. $1 - 2x + 3x^2 - 4x^3 + 5 \sin x - 6 \cos x + 7e^x - 8 \ln x$

28. The virial equation of state of a gas at low pressure is $pV = nRT \left(1 - \dfrac{nB}{V} \right)$. Find $\dfrac{dp}{dV}$

at constant T and n (assume B is also constant).

Products and quotients

Differentiate

29. $(1 - 4x^2) \cos x$ **30.** $(2 + 3x)e^x$ **31.** $e^x \cos x$ **32.** $x \ln x$

33. $(1 + 2x + 3x^2)/(3 + x^3)$ **34.** $(1 - 4x^2)/\sin x$ **35.** $\cos x / \sin x$ **36.** $(\ln x)/x$

Chain rule

Differentiate

37. $(1 + x)^5$ **38.** $\sqrt{2 + x^2}$ **39.** $\dfrac{1}{3 - x^2}$ **40.** $\dfrac{3}{(2x^2 - 3x - 1)^{1/2}}$

41. $\sin 4x$ **42.** e^{-2x} **43.** $e^{2x^2 - 3x + 1}$ **44.** $\ln(2x^2 - 3x + 1)$

45. $\cos(2x^2 - 3x + 1)$ **46.** $e^{\sin x}$ **47.** $\ln(\cos x)$ **48.** $e^{-\cos(3x^2 + 2)}$

49. $\ln \left(\dfrac{2+x}{3-x} \right)$ **50.** $\ln(\sin 2x + \sin^2 x)$ **51.** $3x^2(2 + x)^{1/2}$ **52.** $\sin x \cos 2x$

53. $\tan 4x \cos^2 2x$ **54.** $x^2 e^{2x^2 + 3}$ **55.** $\dfrac{3x^2}{(2 + x^2)^{1/2}}$

Inverse functions

56. If $x = 2y^2 - 3y + 1$, find $\dfrac{dy}{dx}$.

Find $\dfrac{dV}{dp}$ at constant T and n for the following equations of state (assume that B, a and b are constants).

57. $pV = nRT\left(1 + \dfrac{nB}{V}\right)$ **58.** $p(V - nb) - nRT = 0$ **59.** $\left(p + \dfrac{n^2 a}{V^2}\right)(V - nb) = nRT$

Differentiate

60. $\sin^{-1} 2x$ **61.** $\tan^{-1} x^2$ **62.** $\cos^{-1}\left(\dfrac{1-x}{1+x}\right)$ **63.** $\sinh^{-1} 2x$ **64.** $\tanh^{-1} x^2$

Section 4.7

Find $\dfrac{dy}{dx}$:

65. $x^2 + y^2 = 4$ **66.** $y^3 + 3x + x^2 - 1 = 0$ **67.** $x = y \ln xy$

68. $y^2 + \dfrac{2}{y} - x^2 y^2 + 3x + 2 = 0$

Section 4.8

Differentiate:

69. $\left(\dfrac{3-x}{4+x}\right)^{1/3}$ **70.** $\dfrac{(1+x^2)(x-1)^{1/2}}{(2x+1)(3x^2+2x-1)^{1/3}}$ **71.** $\sin^{1/2} x \cos^3(x^2+1)\tan^{1/3} 2x$

72. Show that the equations

$$\frac{d \ln p}{dT} = \frac{\Delta H_{\text{vap}}}{RT^2} \quad \text{and} \quad \frac{dp}{dT} = \frac{p\Delta H_{\text{vap}}}{RT^2}$$

are equivalent expressions of the Clausius–Clapeyron equation.

73. The decomposition of dinitrogen pentoxide in tetrachloromethane at $T = 45°C$ has stoichiometry:

$$N_2O_5 \rightarrow 2NO_2 + \tfrac{1}{2}O_2$$

and obeys first-order kinetics. From the volumes of oxygen liberated after various times t, the following concentrations of N_2O_5 were obtained:

$x = [N_2O_5]/\text{mol dm}^{-3}$	2.33	1.91	1.36	1.11	0.72	0.55
t/s	0.0	319	867	1196	1877	2315

Plot a graph of $\ln x$ against t/s and determine the rate constant.

Section 4.9

74. Find all the nonzero derivatives of the function $y = 3x^5 + 4x^4 - 3x^3 + x^2 - 2x + 1$.

75. Find $\dfrac{dy}{dx}, \dfrac{d^2 y}{dx^2}, \dfrac{d^3 y}{dx^3}, \dfrac{d^4 y}{dx^4}$ for the function $y = \ln x$.

76. Find a general formula for the nth derivative of e^{3x}.

77. Find a general formula for the nth derivative of $\cos 2x$.

Section 4.10

Find the maximum and minimum values and the points of inflection of the following functions. In each case, sketch the graph and show the positions of these points.

78. $y = x^2 - 3x + 2$ **79.** $y = x^3 - 7x^2 + 15x - 9$ **80.** $y = 4x^3 + 6x^2 + 3$

81. $y = xe^{-x}$ (see Figure 3.20)

82. Confirm that the cubic $y = x^3 - 7x^2 + 16x - 10$, discussed in Example 2.23, has local maximum and minimum values at $x = 2$ and $x = 8/3$.

83. Find the maximum and minimum values and the points of inflection of $y = 2x^5 - 5x^4 + 3$. Sketch a graph to show the positions of these points.

84. The Lennard-Jones potential for the interaction of two molecules separated by distance R is

$$U(R) = \frac{A}{R^{12}} - \frac{B}{R^6}$$

where A and B are constants. The equilibrium separation R_e is that value of R at which $U(R)$ is a minimum and the binding energy is $D_e = -U(R_e)$. Express **(i)** A and B in terms of R_e and D_e, **(ii)** $U(R)$ in terms of R, R_e and D_e.

85. The probability that a molecule of mass m in a gas at temperature T has speed v is given by the Maxwell–Boltzmann distribution

$$f(v) = 4\pi \left(\frac{m}{2\pi kT} \right)^{3/2} v^2 e^{-mv^2/2kT}$$

where k is Boltzmann's constant. Find the most probable speed (for which $f(v)$ is a maximum).

86. The concentration of species B in the rate process $A \xrightarrow{k_1} B \xrightarrow{k_2} C$, consisting of two consecutive irreversible first-order reactions, is given by (when $k_1 \neq k_2$)

$$[B] = [A]_0 \frac{k_1}{k_2 - k_1} (e^{-k_1 t} - e^{-k_2 t})$$

(i) Find the time t, in terms of the rate constants k_1 and k_2, at which B has its maximum concentration, and **(ii)** show that the maximum concentration is

$$[B]_{max} = [A]_0 \left(\frac{k_1}{k_2} \right)^{k_2/(k_2 - k_1)}$$

Section 4.11

87. A particle moving along a straight line travels the distance $s = 2t^2 - 3t$ in time t. **(i)** Find the velocity v and acceleration a at time t. **(ii)** Sketch graphs of s and v as functions of t in the interval $t = 0 \to 2$, **(iii)** find the stationary values, and describe the motion of the particle.

88. A particle moving on the circumference of a circle of radius $r = 2$ travels distance $s = t^3 - 2t^2 - 4t$ in time t. **(i)** Express the distance travelled in terms of the angle θ subtended at the centre of the circle, **(ii)** find the angular velocity ω and acceleration $\dot{\omega}$ around the centre of the circle, **(iii)** Sketch graphs of θ, ω and $\dot{\omega}$ as functions of t in the interval $t = 0 \to 4$, **(iv)** find the stationary values, and describe the motion of the particle.

Section 4.12

Find the differential dy:

89. $y = 2x$ **90.** $y = 3x^2 + 2x + 1$ **91.** $y = \sin x$

92. The volume of a sphere of radius r is $V = 4\pi r^3/3$. Derive the differential dV from first principles. Give a geometric interpretation of the result.

5 Integration

5.1 Concepts

Consider a body moving along a curve from point A at time $t = t_A$ to point B at time $t = t_B$. Let the distance from A along the curve at some intermediate time t be $s(t)$, as illustrated in Figure 5.1.

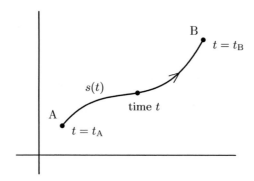

Figure 5.1

If $v(t)$ is the velocity along the curve at time t then, by the discussion of Section 4.11, $v(t) = ds/dt$ is the gradient of the graph of the function $s(t)$. Therefore, if $s(t)$ is a known function, $v(t)$ is obtained by differentiation. Conversely, if $v(t)$ is a known function then $s(t)$ is that function whose derivative is $v(t)$; that is, to find $s(t)$ we need to *reverse* the differentiation $ds/dt = v(t)$.

The distance travelled between two points is equal to the average velocity multiplied by the time taken. For the motion shown in Figure 5.1, the distance AB along the curve is therefore

$$d = \bar{v} \times (t_B - t_A) \tag{5.1}$$

where $\bar{v}$ is the average (or mean) value of $v(t)$ between A and B. For example, if the body undergoes constant acceleration from $v = v_A$ at A to $v = v_B$ at B, as illustrated in Figure 5.2,[1] the average velocity is $\bar{v} = (v_A + v_B)/2$ and the total distance travelled is

$$d = \frac{1}{2}(v_A + v_B) \times (t_B - t_A)$$

---

[1] This graph of velocity as a function of time for a body moving with uniform acceleration appeared in *Quaestiones super geometriam Euclides* by Nicole Oresme (c. 1323–1382), Dean of Rouen Cathedral and Bishop of Lisieux. This was possibly the first graph of a variable physical quantity. Oresme also considered the extension of his 'latitude of forms' to the representation of the 'quality' of a surface by a body in three dimensions and 'the quality of a body will no doubt be represented by something having four dimensions in a different kind of quantity'.

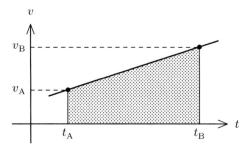

Figure 5.2

It is readily verified that this distance is equal to the area, shaded in the figure, bounded by the line $v(t)$ and the t-axis between t_A and t_B. We shall see that this last result is valid for any velocity function $v(t)$. We shall also see that the solution of a physical problem is often equivalent to finding the area enclosed by an appropriate curve.

This example demonstrates the two central problems of seventeenth-century European mathematics; the 'problem of tangents' and the 'problem of quadrature'. The first of these, to find the tangent lines to an arbitrary curve, led to the invention of the differential calculus, the subject of Chapter 4. The second, to find the area enclosed by a given curve, led to the invention of the integral calculus.[2] The demonstration by Leibniz and by Newton that differentiation and integration are essentially inverse operations is one of the landmarks of the history of mathematics.

The concept of integration as the inverse operation to differentiation leads to the definition of the **indefinite integral**. The concept of the integral as an area leads to the definition of the **definite integral**.

5.2   The indefinite integral

Let $y = F(x)$ be a function of x whose derivative is $F'(x) = \dfrac{dy}{dx}$. The indefinite integral of the derivative is defined by

---

[2] Integration has its origins in the Greek 'method of exhaustion' for finding areas and volumes. Archimedes in his *Quadrature of the parabola* attributed the method to Eudoxus of Cnidus (*c.* 408–355 BC). In the *Method*, discovered in Constantinople in 1906 after being 'lost' for over a thousand years, Archimedes describes how a plane area can be regarded as a sum of line segments. In 1586, Stevin described how the centroid of a triangle can be obtained by considering the area as made up of a large number of parallelograms. Johann Kepler (1571–1630), best known for his *Astronomia nova* of 1609, computed areas and volumes by considering them to be composed of infinitely many infinitesimal elements. His work on volumes appeared in 1615 in *Nova stereometria doliorum vinariorum* (New solid geometry of wine barrels). Galileo Galilei (1564–1642) made use of the infinitely small in his work on dynamics. Bonaventura Cavalieri (1598–1647), a follower of Galileo, described in his influential *Geometria indivisibilibus continuorum*, 1635, how an area can be thought of as made up of lines or 'indivisibles' and a volume of areas, and developed a geometric method for finding the integral of x^n for positive integers n. At about the same time, Fermat solved the same problem for positive and negative integers (except $n = -1$) and for fractions by dividing his areas into suitable rectangular strips. The case of $n = -1$ was treated by Gregoire de Saint Vincent (1584–1667). Roberval integrated the sine function in 1635, and Torricelli the log function in 1646. Other contributors include Pascal, whose *Traité des sinus du quart de cercle* of 1658 Leibniz said inspired his discovery of the fundamental theorem, John Wallis (1616–1703), whose work on infinite processes influenced Newton and to whom we owe the symbol ∞, the Scot James Gregory (1638–1675) whose work on infinite series and the calculus anticipated that of Newton, and Barrow, whose lectures Newton attended and a copy of whose *Lectiones* was bought by Leibniz when on a visit to London in 1673. The final step in the synthesis of the differential and integral calculus was taken by Newton (Footnote 3, Chapter 4) and by Leibniz, who published the first account of his integral calculus, *Analysi indivisibilium atque infinitorum* (Analysis of indivisibles and infinities) in 1686.

$$\int F'(x)\,dx = F(x) + C \qquad\qquad (5.2)$$

where C is an arbitrary constant. For example, if $y = x^2$ then $\dfrac{dy}{dx} = 2x$, and the indefinite integral of the function $2x$ is

$$\int 2x\,dx = x^2 + C \qquad because \qquad \frac{d}{dx}(x^2 + C) = 2x$$

The symbol $\int$ is called the **integral sign**; it is an elongated 'S' (for summation) and has its origins in Leibniz's formulation of the integral calculus; its significance will become clearer in Section 5.4 when we discuss the integral as the limit of a sum. The function to be integrated, $F'(x)$ in (5.2), is called the **integrand**, x is the **variable of integration** and dx is called the **element** of x. C is an arbitrary constant called the **integration constant**. It is included as part of the value of the indefinite integral because, given $y = F(x)$ with derivative $F'(x)$, the function $(y + C)$ also has derivative $F'(x)$

$$\frac{d}{dx}(y + C) = \frac{dy}{dx} + \frac{dC}{dx} = \frac{dy}{dx} \qquad\qquad (5.3)$$

Table 5.1 is a short list of 'standard integrals' involving some of the more important elementary functions (compare Table 4.2). Each entry in the list can be checked by differentiation of the right side of the equation; for example,

$$\frac{d}{dx}\Big[\ln(ax + b) + C\Big] = \frac{a}{ax + b} \qquad so\ that \qquad \int \frac{1}{ax + b}\,dx = \frac{1}{a}\ln(ax + b) + C$$

General methods of integration and further standard integrals are discussed in Chapter 6. A more comprehensive list of standard integrals is given in the Appendix.

Table 5.1 Elementary integrals

1. $\displaystyle\int x^a\,dx \quad = \dfrac{x^{a+1}}{a+1} + C \qquad a \neq -1$

2. $\displaystyle\int e^{ax}\,dx \quad = \dfrac{1}{a}e^{ax} + C$

3. $\displaystyle\int \sin ax\,dx \; = -\dfrac{1}{a}\cos ax + C$

4. $\displaystyle\int \cos ax\,dx = \dfrac{1}{a}\sin ax + C$

5. $\displaystyle\int \dfrac{1}{ax + b}\,dx = \dfrac{1}{a}\ln(ax + b) + C$

EXAMPLES 5.1 Indefinite integrals

(i) $\displaystyle \int \frac{dx}{x^2} = \int x^{-2}\, dx = \frac{x^{-2+1}}{-2+1} + C = \frac{x^{-1}}{-1} + C = -\frac{1}{x} + C$

(ii) $\displaystyle \int \frac{dx}{\sqrt{x}} = \int x^{-1/2}\, dx = \frac{x^{(-1/2)+1}}{(-1/2)+1} + C = \frac{x^{1/2}}{1/2} + C = 2x^{1/2} + C = 2\sqrt{x} + C$

(iii) $\displaystyle \int dx = \int 1\, dx = x + C$

(iv) $\displaystyle \int e^{2t}\, dt = \frac{1}{2}e^{2t} + C$

(v) $\displaystyle \int \cos 2\theta\, d\theta = \frac{1}{2}\sin 2\theta + C$

(vi) $\displaystyle \int \frac{1}{x+3}\, dx = \ln(x+3) + C$

We note that if we put $C = \ln A$ in (vi) then $\displaystyle \int \frac{1}{x+3}\, dx = \ln(x+3) + \ln A = \ln A(x+3)$

▸ Exercises 1–10

In every case, the effect of the operation $\int \ldots\, dx$ is to reverse the effect of differentiation; the *integral of the derivative* of a function retrieves the function. Also, differentiating both sides of equation (5.2) gives

$$\frac{d}{dx}\int F'(x)\, dx = \frac{d}{dx}\Big[F(x) + C \Big] = F'(x) \tag{5.4}$$

so that the *derivative of the integral* of a function retrieves that function.

Differential and integral operators

An alternative way of describing the operations of differentiation and integration, that does not involve Leibniz's symbolism, makes use of the differential operator $D = d/dx$ introduced in Section 4.2, with property

$$DF(x) = F'(x)$$

The effect of D on $F(x)$ is to transform it into its derivative $F'(x)$, and a corresponding inverse operator D^{-1}, an **integral operator**, can be defined whose effect is to reverse that of D. Thus

$$D^{-1}F'(x) = F(x) + C$$

is the operator form of equation (5.2). The operators D and D^{-1} have the property

$$\mathrm{D}^{-1}\mathrm{D} = \mathrm{D}\,\mathrm{D}^{-1} = 1$$

true of any pair of inverse operators. Then

$$\mathrm{D}^{-1}\,\mathrm{D}\left[F(x) + C\right] = \mathrm{D}\,\mathrm{D}^{-1}\left[F(x) + C\right] = F(x) + C$$

is equivalent to

$$\int \frac{d}{dx}\left[F(x) + C\right] dx = \frac{d}{dx}\int\left[F(x) + C\right] dx = F(x) + C$$

Value of the integration constant

In an actual problem the value of the integration constant C in (5.2) is determined by some auxiliary condition. Consider, for example, a curve whose gradient at every point is given by the function $2x$. The equation of such a curve is

$$y = \int 2x\,dx = x^2 + C \tag{5.5}$$

This equation represents a **family of curves**, one curve for each value of C (Figure 5.3).

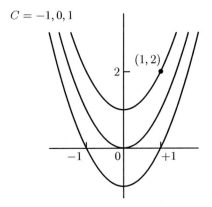

$C = -1, 0, 1$

Figure 5.3

If it is required that the curve pass through a *particular* point, the point $(1, 2)$ say, the auxiliary condition on (5.5) is that $y = 2$ when $x = 1$; that is, $y = 2 = 1 + C$. Therefore, $C = 1$, and $y = x^2 + 1$ is the equation of the particular curve whose gradient is $dy/dx = 2x$ and which passes through the point $(1, 2)$.

It is shown in later chapters how, in a physical problem, the value of the integration constant is determined by the nature and state of the system.

EXAMPLE 5.2 Find $y = \int 2x\, dx$ subject to conditions (i) $(x, y) = (2, 0)$, (ii) $y = 10$ when $x = 3$.

We have $y = \int 2x\, dx = x^2 + C$. Then

(i) $0 = 2^2 + C$, $C = -4$, $y = x^2 - 4$ (ii) $10 = 3^2 + C$, $C = 1$, $y = x^2 + 1$

▸ Exercises 11, 12

Two rules

1. For a multiple of a function:

$$\int a\,u(x)\, dx = a \int u(x)\, dx \tag{5.6}$$

2. For a sum of functions:

$$\int \left[u(x) + v(x) \right] dx = \int u(x)\, dx + \int v(x)\, dx \tag{5.7}$$

It follows from these rules that the integral of a linear combination of functions,

$$f(x) = a\,u(x) + b\,v(x) + c\,w(x) + \cdots \tag{5.8}$$

can be written as the sum of integrals

$$\int f(x)\, dx = a \int u(x)\, dx + b \int v(x)\, dx + c \int w(x)\, dx + \tag{5.9}$$

EXAMPLE 5.3 Integral of a linear combination of functions

$$\int \left(3x^3 + 2 \sin 3x - e^{-x} \right) dx = 3 \int x^3\, dx + 2 \int \sin 3x\, dx - \int e^{-x} dx$$

$$= 3 \times \frac{x^4}{4} + 2 \times \left(-\frac{1}{3} \cos 3x \right) - \left(-e^{-x} \right) + C = \frac{3}{4}x^4 - \frac{2}{3} \cos 3x + e^{-x} + C$$

▸ Exercises 13–15

5.3   The definite integral

The integral calculus was invented to solve the problem of finding the area enclosed by a given curve. In this section we introduce, without proof, the definite integral as a measure of area, and show how it is related to the indefinite integral. For many problems in the physical sciences, this brief introduction is sufficient. Other problems however require a more intimate understanding of the integral calculus, and we will return to this in Section 5.4.

Let $y=f(x)$ be a function of x, continuous in the interval $a \le x \le b$. The shaded area in Figure 5.4 is known as the **area under the curve**; this is the area enclosed by the graph of $y=f(x)$, the x-axis, and the verticals at $x=a$ and $x=b$. In general, this area is equal to the width, $(b-a)$, multiplied by the average height, $\bar{y}$, of the curve above the axis (the average value of the function between a and b),

$$A = (b-a)\bar{y} \qquad\qquad (5.10)$$

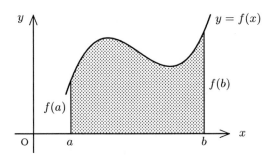

Figure 5.4

In the special case of a linear function, as in Figure 5.5, the average height is

$$\bar{y} = \frac{1}{2}\Big[f(a) + f(b) \Big]$$

and the area under the curve (straight line) is that of a rectangle of width $(b-a)$ and height $\bar{y}$.

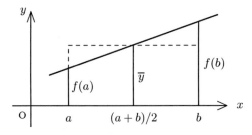

Figure 5.5

In the general case, when $y=f(x)$ is not necessarily a linear function, the integral calculus (see Section 5.4) tells us that the area is given by the **definite integral**

$$A = \int_a^b f(x)\, dx = F(b) - F(a) \tag{5.11}$$

where $F(x)$ is the function whose derivative is $f(x) = F'(x) = \dfrac{dF}{dx}$. The numbers $F(a)$ and $F(b)$ are the values of $F(x)$ at the **limits of integration** a and b; a is called the lower limit, b the upper limit, and the interval a to b is called the **range of integration**.

The difference $F(b) - F(a)$ in equation (5.11) is often denoted by $\left[F(x)\right]_a^b$, so that

$$\int_a^b f(x)\, dx = \left[F(x)\right]_a^b = F(b) - F(a) \tag{5.12}$$

It follows that in order to calculate the value of the definite integral it is normally necessary first to evaluate the corresponding indefinite integral. For example, let $y=f(x)=2x+3$. The indefinite integral is

$$\int f(x)\, dx = \int (2x+3)\, dx = x^2 + 3x + C = F(x)$$

The definite integral of $f(x)$ in the range $x=a$ to $x=b$ ('the integral from a to b') is then

$$\int_a^b (2x+3)\, dx = \left[x^2 + 3x + C\right]_a^b = (b^2 + 3b + C) - (a^2 + 3a + C)$$

$$= (b^2 + 3b) - (a^2 + 3a)$$

We note that the constant of integration C cancels for a definite integral, and can always be omitted.

EXAMPLES 5.4 Definite integrals

(i) $\displaystyle\int_1^4 (2x+3)\, dx = \left[x^2 + 3x\right]_1^4 = (4^2 + 3\times 4) - (1^2 + 3\times 1) = 28 - 4 = 24$

(ii) $\displaystyle\int_2^3 x^2\, dx = \left[\dfrac{x^3}{3}\right]_2^3 = \dfrac{3^3}{3} - \dfrac{2^3}{3} = \dfrac{19}{3}$

(iii) $\displaystyle\int_2^4 \dfrac{dx}{x^2} = \int_2^4 x^{-2}\, dx = \left[-\dfrac{1}{x}\right]_2^4 = \left(-\dfrac{1}{4}\right) - \left(-\dfrac{1}{2}\right) = -\dfrac{1}{4} + \dfrac{1}{2} = \dfrac{1}{4}$

(iv) $\displaystyle\int_a^b dx = \int_a^b 1\,dx = \left[x\right]_a^b = b - a$

(v) $\displaystyle\int_0^{\pi/2} \sin\theta\,d\theta = \left[-\cos\theta\right]_0^{\pi/2} = \left(-\cos\frac{\pi}{2}\right) - \left(-\cos 0\right) = (0) - (-1) = 1$

(vi) $\displaystyle\int_{-1}^{+1} e^{-2t}\,dt = \left[-\frac{1}{2}e^{-2t}\right]_{-1}^{+1} = \left(-\frac{1}{2}e^{-2}\right) - \left(-\frac{1}{2}e^2\right) = \frac{1}{2}\left(e^2 - e^{-2}\right)$

(vii) $\displaystyle\int_2^3 \frac{dx}{x} = \left[\ln x\right]_2^3 = \ln 3 - \ln 2 = \ln\frac{3}{2}$

▸ Exercises 16–25

Average value of a function

Because the definite integral is identified as the area under the curve it follows from equation (5.10) that the average value of the function $y=f(x)$ in the interval $a\le x\le b$ is

$$\bar{y} = \frac{A}{b-a} = \frac{\displaystyle\int_a^b f(x)\,dx}{\displaystyle\int_a^b dx} \qquad (5.13)$$

EXAMPLE 5.5 Find the average value of $\sin\theta$ in the interval $0\le\theta\le\pi/2$.

By Example 5.4(v),

$$\overline{\sin\theta} = \frac{1}{\pi/2}\int_0^{\pi/2} \sin\theta\,d\theta = \frac{2}{\pi}$$

▸ Exercises 26–28

Three properties of the definite integral

Let $f(x) = F'(x)$ so that, by equation (5.11),

$$\int_a^b f(x)\,dx = F(b) - F(a)$$

1. The value of the integral does not depend on the symbol used for the variable of integration:

$$\int_a^b f(x)\,dx = \int_a^b f(t)\,dt = \int_a^b f(u)\,du = \cdots \qquad (5.14)$$

In each case, the value of the integral is $F(b) - F(a)$.

2. If c is a third limit of integration, not necessarily between a and b:

$$\int_a^b f(x)\, dx = \int_a^c f(x)\, dx + \int_c^b f(x)\, dx \qquad (5.15)$$

This is true because the value of the integral can be written as

$$F(b) - F(a) = \left[F(c) - F(a) \right] + \left[F(b) - F(c) \right]$$

If c lies between a and b then the area represented by the integral on the left side of equation (5.15) is equal to the sum of the areas represented by the integrals on the right side.

3. When the limits are interchanged, the value of the integral changes sign:

$$\int_b^a f(x)\, dx = -\int_a^b f(x)\, dx \qquad (5.16)$$

This follows because $F(a) - F(b) = -\left[F(b) - F(a) \right]$.

EXAMPLE 5.6 Properties

(i) $\displaystyle A = \int_2^3 x\, dx + \int_3^5 x\, dx = \frac{1}{2}(3^2 - 2^2) + \frac{1}{2}(5^2 - 3^2)$

$$= \frac{1}{2}(5^2 - 2^2) = \int_2^5 x\, dx.$$

(ii) $\displaystyle \int_2^5 x^2\, dx = \frac{1}{3}(5^3 - 2^3) = \frac{117}{3},$

$$\int_5^2 x^2\, dx = \frac{1}{3}(2^3 - 5^3) = -\frac{117}{3} = -\int_2^5 x^2\, dx.$$

> Exercises 29, 30

Negative areas

Consider the integral

$$\int_0^{2\pi} \sin x\, dx = \left[-\cos x \right]_0^{2\pi} = (-\cos 2\pi) - (-\cos 0) = (-1) - (-1) = 0$$

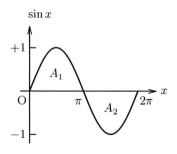

Figure 5.6

Figure 5.6 shows that the integrand, $\sin x$, has positive values when $0 < x < \pi$, and negative values when $\pi < x < 2\pi$. The total 'area under the curve' is the sum of the two areas labelled A_1 and A_2 in the figure, and the integral can be written as

$$\int_0^{2\pi} \sin x \, dx = A_1 + A_2$$

where

$$A_1 = \int_0^{\pi} \sin x \, dx = +2 \qquad \text{and} \qquad A_2 = \int_{\pi}^{2\pi} \sin x \, dx = -2$$

This example shows that areas corresponding to negative values of the integrand make negative contributions to the total. In this case the positive and negative contributions cancel, and the average value of $\sin x$ in the interval is zero.

➤ Exercise 31

Integration of discontinuous functions

The function

$$f(x) = \begin{cases} 2x & \text{if } x < 2 \\ 2x+1 & \text{if } x \geq 2 \end{cases}$$

is discontinuous at $x=2$, but Figure 5.7 shows that the function can be integrated across the discontinuity if the range of integration is split at the point of discontinuity. Thus,

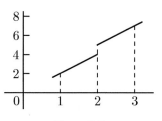

Figure 5.7

$$\int_1^3 f(x) \, dx = \int_1^2 f(x) \, dx + \int_2^3 f(x) \, dx$$

$$= \int_1^2 2x \, dx + \int_2^3 (2x+1) \, dx = 3 + 6 = 9$$

This can always be done when the integrand has only a finite number of finite discontinuities within the range of integration. A similar technique is used when the function is continuous but has a discontinuous derivative.

EXAMPLE 5.7 The function

$$f(x) = e^{-|x|} = \begin{cases} e^{-x} & \text{if } x \ge 0 \\ e^{x} & \text{if } x \le 0 \end{cases}$$

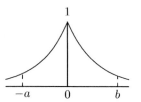

Figure 5.8

is continuous at $x=0$, but not smooth, the slope of its graph changing discontinuously from $+1$ to -1 on passing through $x=0$ from left to right (Figure 5.8). The function can be integrated if the range of integration is split at the point of gradient discontinuity:

$$\int_{-a}^{b} e^{-|x|} dx = \int_{-a}^{0} e^{x} dx + \int_{0}^{b} e^{-x} dx = \left[e^{x} \right]_{-a}^{0} + \left[-e^{-x} \right]_{0}^{b}$$

$$= \left(1 - e^{-a}\right) + \left(1 - e^{-b}\right) = 2 - e^{-a} - e^{-b}$$

▸ Exercises 32–34

Improper integrals

A definite integral is called **improper** when the integrand has an infinite discontinuity at a point within the range of integration. If the discontinuity is at the point $x=c$, where $a \le c \le b$, then the integral is defined as the limit, for $\varepsilon > 0$,

$$\int_{a}^{b} f(x)\,dx = \lim_{\varepsilon \to 0} \left\{ \int_{a}^{c-\varepsilon} f(x)\,dx + \int_{c+\varepsilon}^{b} f(x)\,dx \right\} \qquad (5.17)$$

As shown in Figure 5.9, the point c is excluded because the integrand is not defined there. When the limit in (5.17) is finite and unique then the value of the integral is the 'area under the curve'.

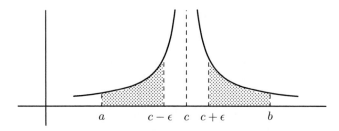

Figure 5.9

If the discontinuity lies at one end of the range of integration, at $x=a$ say, the integral is defined as

$$\int_{a}^{b} f(x)\,dx = \lim_{\varepsilon \to 0} \int_{a+\varepsilon}^{b} f(x)\,dx \qquad (5.18)$$

EXAMPLES 5.8 Improper integrals

(i) The function $1/\sqrt{x}$ is not defined at $x=0$ and the definite integral between limits $x=0$ and $x=1$ is defined as

$$\int_0^1 \frac{dx}{\sqrt{x}} = \lim_{\varepsilon \to 0} \int_\varepsilon^1 \frac{dx}{\sqrt{x}}$$

Then

$$\int_\varepsilon^1 \frac{dx}{\sqrt{x}} = \int_\varepsilon^1 x^{-1/2}\, dx = \left[2x^{1/2} \right]_\varepsilon^1 = 2 - 2\varepsilon^{1/2}$$

and, letting $\varepsilon \to 0$,

$$\int_0^1 \frac{dx}{\sqrt{x}} = 2$$

(ii) $\displaystyle\int_0^1 \frac{dx}{x} = \lim_{\varepsilon \to 0} \int_\varepsilon^1 \frac{dx}{x} = \lim_{\varepsilon \to 0} \left[\ln x \right]_\varepsilon^1 = \lim_{\varepsilon \to 0} (-\ln \varepsilon)$

The limit does not exist because $\ln \varepsilon \to -\infty$ as $\varepsilon \to 0$.

(iii) In the general case of the integral of an inverse power, $a \neq 1$

$$\int_0^1 \frac{dx}{x^a} = \lim_{\varepsilon \to 0} \left[\left(\frac{1}{1-a} \right) \frac{1}{x^{a-1}} \right]_\varepsilon^1 = \left(\frac{1}{1-a} \right) \lim_{\varepsilon \to 0} \left(1 - \frac{1}{\varepsilon^{a-1}} \right)$$

When $a < 1$ the limit has value $1/(1-a)$, but when $a > 1$ the limit is infinite and the integral is not defined.

▸ Exercise 35

Infinite integrals

It often happens in applications in the physical sciences that one or both of the limits of integration are infinite. Integrals with infinite ranges of integration are called **infinite integrals**. If the upper limit is infinite then the definite integral is defined by

$$\int_a^\infty f(x)\, dx = \lim_{b \to \infty} \int_a^b f(x)\, dx \tag{5.19}$$

For example,

$$\int_0^\infty e^{-r}\, dr = \lim_{b\to\infty} \int_0^b e^{-r}\, dr = \lim_{b\to\infty} \left[-e^{-r}\right]_0^b$$

$$= \lim_{b\to\infty} \left[-e^{-b} + 1\right] = 1$$

using the property of the exponential that $e^{-b} \to 0$ as $b\to\infty$. In practice, infinite integrals are treated in exactly the same way as ordinary integrals, but some care must be taken when assigning the infinite value to the variable.

EXAMPLE 5.9 Find the value of the infinite integral $I = \displaystyle\int_4^\infty \frac{dx}{(x-1)(x-2)}$.

The integrand can be written in terms of partial fractions (see Section 2.7) as

$$\frac{1}{(x-1)(x-2)} = \frac{1}{x-2} - \frac{1}{x-1}$$

Then

$$I = \lim_{b\to\infty} \int_4^b \left(\frac{1}{x-2} - \frac{1}{x-1}\right) dx = \lim_{b\to\infty} \left[\ln(x-2) - \ln(x-1)\right]_4^b$$

$$= \lim_{b\to\infty} \left[\ln\left(\frac{x-2}{x-1}\right)\right]_4^b = \lim_{b\to\infty} \ln\left(\frac{b-2}{b-1}\right) - \ln\frac{2}{3} = \ln\frac{3}{2}$$

because $(b-2)/(b-1) \to 1$ as $b\to\infty$ and $\ln 1 = 0$

▸ Exercises 36–39

When the limit in the definition (5.19) is finite and unique, the integral is said to be **convergent**. An integral is **divergent** when the limit is indeterminate. For example,

$$\int_1^\infty \frac{dx}{x} = \lim_{b\to\infty} \left[\ln x\right]_1^b = \lim_{b\to\infty} \ln b$$

is divergent because $\ln b \to \infty$ as $b\to\infty$.
    A more subtle example is

$$\int_0^\infty \cos x\, dx = \lim_{b\to\infty} \left[\sin x\right]_0^b = \lim_{b\to\infty} \sin b$$

In this case the value of $\sin b$ oscillates between $+1$ and -1 as b increases, and no unique value can be assigned to the integral as defined here. On the other hand, it is shown in Example 6.13 that, for $a > 0$,

$$\int_0^\infty e^{-ax} \cos x \, dx = \frac{a}{1+a^2}$$

This result is valid for all positive values of the parameter a, however small, and it follows that

$$\lim_{a \to 0} \int_0^\infty e^{-ax} \cos x \, dx = \lim_{a \to 0} \left(\frac{a}{1+a^2} \right) = 0$$

We note that the limit must be taken *after* integration; taking the limit before integration leads to a different, and divergent, integral.

Even and odd functions

When a function has the property

$$f(-x) = f(x) \tag{5.20}$$

it is called an **even function** of x; it has **even parity** (or parity $+1$) and is **symmetric** with respect to the axis $x = 0$ (the y-axis). Examples of even functions are x^2, $e^{-|x|}$ (figure 5.8) and $\cos x$ (shown in Figure 5.10a); in each case the value of the function is unchanged when x is replaced by $-x$. On the other hand, a function with the property

$$f(-x) = -f(x) \tag{5.21}$$

is said to be an **odd function**; it has **odd parity** (or parity -1) and is **antisymmetric** with respect to the axis $x = 0$. Examples of odd functions are x^3, $\sin x$ (shown in Figure 5.10b), and $x \cos x$; in each case the value of the function changes sign when x is replaced by $-x$. The product of two even functions or of two odd functions is even; the product of an even function and an odd function is odd. Thus $x \cos x$ is an odd function, with x odd and $\cos x$ even.

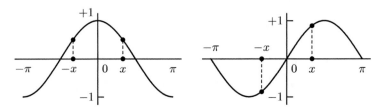

(a) $y = \cos x$; even function   (b) $y = \sin x$; odd function

Figure 5.10

In general, an arbitrary function is neither even nor odd; $f(-x) \neq \pm f(x)$. It is always possible however to express a function as the sum of an even component and an odd component; we can write

$$f(x) = \frac{1}{2}\left[f(x) + f(-x)\right] + \frac{1}{2}\left[f(x) - f(-x)\right]$$

$$= f_+(x) + f_-(x)$$

(5.22)

where $f_+(x)$ is the even component of $f(x)$ and $f_-(x)$ is the odd component. For example, an arbitrary polynomial

$$f(x) = a_0 + a_1 x + a_2 x^2 + \cdots + a_n x^n$$

can be written as

$$f(x) = \left[a_0 + a_2 x^2 + a_4 x^4 + \cdots\right] + \left[a_1 x + a_3 x^3 + a_5 x^5 + \cdots\right]$$

and, because every even power of x is an even function whilst every odd power of x is an odd function, the first set of terms forms the even component of the polynomial and the second set forms the odd component.

The integral properties of functions of well-defined symmetry are of great importance in the physical sciences. The area represented by the integral

$$A = \int_{-a}^{+a} f(x)\, dx$$

(5.23)

is the sum of the areas to the left and right of the $x = 0$ axis:

$$A = \int_{-a}^{0} f(x)\, dx + \int_{0}^{+a} f(x)\, dx = A_< + A_>$$

(5.24)

If $f(x)$ is an even function of x, the two areas $A_<$ and $A_>$ are equal in magnitude and sign, and the total area is twice each of them:

$$\int_{-a}^{+a} f(x)\, dx = 2 \int_{0}^{+a} f(x)\, dx, \qquad \text{if } f(x) \text{ even}$$

(5.25)

On the other hand, if $f(x)$ is an odd function of x, then $A_<$ and $A_>$ are equal in magnitude but have *opposite* signs: $A_< = -A_>$, and the value of the integral is zero:

$$\int_{-a}^{+a} f(x)\, dx = 0 \qquad \text{if } f(x) \text{ odd}$$

(5.26)

In the general case, when $f(x)$ has no particular symmetry, the integral is equal to the integral of the even component:

$$\int_{-a}^{+a} f(x)\, dx = \int_{-a}^{+a} f_+(x)\, dx + \int_{-a}^{+a} f_-(x)\, dx = 2 \int_{0}^{+a} f_+(x)\, dx$$

(5.27)

EXAMPLE 5.10

(i) Find the even and odd components of $f(x) = e^x$, (ii) evaluate the definite integrals of $f(x)$ and of its even and odd components over the range $x = -1$ to $x = +1$.

(i) even: $f_+(x) = \dfrac{1}{2}\left[f(x) + f(-x)\right] = \dfrac{1}{2}\left[e^x + e^{-x}\right]$

odd: $f_-(x) = \dfrac{1}{2}\left[f(x) - f(-x)\right] = \dfrac{1}{2}\left[e^x - e^{-x}\right]$

These are the hyperbolic functions $\cosh x$ and $\sinh x$ (Equation (3.47) and Figure 3.22).

(ii) $\displaystyle\int_{-1}^{+1} f(x)\,dx = \int_{-1}^{+1} e^x\,dx = \left[e^x\right]_{-1}^{+1} = e - e^{-1}$

$\displaystyle\int_{-1}^{+1} f_+(x)\,dx = \frac{1}{2}\int_{-1}^{+1}\left[e^x + e^{-x}\right]dx = \frac{1}{2}\left[e^x - e^{-x}\right]_{-1}^{+1} = e - e^{-1}$

$\displaystyle\int_{-1}^{+1} f_-(x)\,dx = \frac{1}{2}\int_{-1}^{+1}\left[e^x - e^{-x}\right]dx = \frac{1}{2}\left[e^x + e^{-x}\right]_{-1}^{+1} = 0$

Only the symmetric component makes a nonzero contribution to the integral of $f(x)$.

▸ Exercises 40–47

The concept of the parity or symmetry of functions is widely used in the physical sciences; the relevant branch of mathematics is called **group theory**. In molecular chemistry, the point groups are used to describe, for example, the symmetry properties of molecular wave functions and of the normal modes of vibration of molecules, and are used to explain and predict the allowed transitions between energy levels that are observed in molecular spectra. In solid-state chemistry, the space groups describe the symmetry properties of lattices and the structure of X-ray diffraction spectra.

5.4 The integral calculus

Let $y = f(x)$ be a function of x, continuous in the interval $a \le x \le b$, as in Figure 5.4. We postulated in Section 5.3 that the 'area under the curve', the shaded region in Figure 5.4, is given by the value of the definite integral of $f(x)$ from a to b. We now look at how this result is derived.

To obtain an estimate of the area, we divide the interval a to b into n subintervals by choosing $n - 1$ arbitrary points on the x-axis, with

$$a = x_0 < x_1 < x_2 < \cdots < x_{n-1} < x_n = b \tag{5.28}$$

and divide the area into n strips by vertical lines at these points as shown in Figure 5.11.

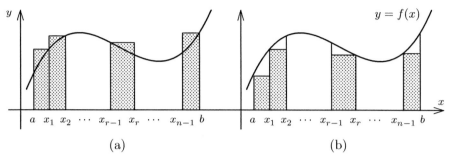

Figure 5.11

An estimate of the area is then obtained by replacing each strip by a rectangle. The width of the rth strip is $\Delta x_r = x_r - x_{r-1}$ and we can choose the height y_r of the rectangle to be any value of the function in the strip. The area of the rth rectangle is $\Delta A_r = y_r \Delta x_r$, and the total area is

$$A \approx \sum_{r=1}^{n} \Delta A_r = \sum_{r=1}^{n} y_r \Delta x_r \qquad (5.29)$$

Two ways of choosing the heights of the rectangles are shown in Figure 5.11. In Figure (a) the height of each rectangle has been chosen to be the *largest* value of $y = f(x)$ in the subinterval; we call this quantity $y_r(\max)$ for the rth strip. In Figure (b) the height has been chosen to be the *smallest* value; we call this quantity $y_r(\min)$ for the rth strip. It is clear that the first choice gives a value that *overestimates* the area under the curve, whereas the second choice gives a value that *underestimates* the area:

$$\sum_{r=1}^{n} y_r(\min) \Delta x_r < A < \sum_{r=1}^{n} y_r(\max) \Delta x_r \qquad (5.30)$$

If we decrease the widths of all the strips by increasing the number of subdivisions, the values of $y_r(\min)$ and $y_r(\max)$ in each strip approach each other (if the function is continuous, as has been assumed) and the two sums in (5.30) converge to the same limit as the number of strips is increased indefinitely; as $n \to \infty$ each sum converges to the limit A, the area under the curve. Therefore, irrespective of the particular choice of heights of the rectangles and of how the divisions of the interval are made, we have

$$A = \lim_{n \to \infty} \sum_{r=1}^{n} y_r \Delta x_r \qquad (5.31)$$

or, because we can choose $y_r = f(x_r)$,

$$A = \lim_{n \to \infty} \sum_{r=1}^{n} f(x_r) \Delta x_r \qquad (\text{all } \Delta x_r \to 0) \qquad (5.32)$$

Following Leibniz, this limit is written as

$$\int_a^b f(x)\,dx = \lim_{n\to\infty} \sum_{r=1}^n f(x_r)\Delta x_r \qquad (5.33)$$

and is called the definite integral of the function $f(x)$ from $x=a$ to $x=b$.

This taking of the limit of the sum of small quantities is the characteristic feature of the integral calculus, just as taking the limit of the ratio of two small quantities is characteristic of the differential calculus. The essential discovery, made independently by Leibniz and by Newton, was that if $F(x)$ is a function whose derivative is $f(x)=F'(x)$ then

$$\int_a^b f(x)\,dx = \int_a^b F'(x)\,dx = F(b) - F(a) \qquad (5.34)$$

This synthesis of the integral and differential calculus is called the **fundamental theorem of the calculus**. The definite integral as defined in this section is known as the **Riemann integral**.[3]

In calculating the area enclosed by a curve it is neither essential nor always convenient that the area be divided into linear strips in the way described above. We illustrate this point in Example 5.11 with two different ways of calculating the area of a circle.

EXAMPLE 5.11 The area of a circle

The equation of the circle of radius a and centre at the origin of coordinates is $x^2 + y^2 = a^2$.

Method 1. Let A be the area of that quarter of the circle that lies in the first quadrant, in which both x and y are positive (Figure 5.12). Then

$$y = \sqrt{a^2 - x^2} \qquad \text{for } 0 \le x \le a.$$

Divide the area into n vertical strips as described in the derivation of equation (5.33) above. An approximate value of the area of the strip between x and $x+\Delta x$ is $\Delta A \approx y\Delta x = \sqrt{a^2 - x^2}\,\Delta x$ and, by equation (5.33), the total area is

$$A = \int_0^a \sqrt{a^2 - x^2}\,dx$$

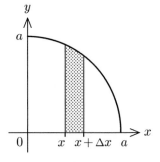

Figure 5.12

[3] Georg Friedrich Bernhard Riemann (1826–1866), professor of mathematics at Göttingen, made contributions to the theory of numbers, functions of a complex variable, and differential equations. In his 1854 lecture *Über die hypothesen welche zu grunde liegen* (On the hypotheses that lie at the foundations of geometry) Riemann developed a system of non-Euclidean geometry and initiated the study of curved metric spaces that ultimately formed the basis for the mathematics of the general theory of relativity.

The integral is evaluated in Example 6.10; $A = \pi a^2/4$, and the area of the circle is four times this.

Method 2. Divide the area of the circle into concentric circular strips as in Figure 5.13. Given that the circumference of a circle of radius r is $2\pi r$ (see Example 5.12), the area of the strip between r and $r + \Delta r$ lies between $2\pi r \Delta r$ and $2\pi(r + \Delta r)\Delta r$:

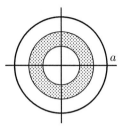

$$2\pi r \Delta r < \Delta A < 2\pi(r + \Delta r)\Delta r \qquad (5.35)$$

and, when Δr is small enough, $\Delta A \approx 2\pi r \Delta r$. The total area is then

Figure 5.13

$$A = \int_0^a 2\pi r \, dr = \pi a^2$$

➤ Exercise 48

We note that the *indefinite* integral can be viewed as a special case of the definite integral, equation (5.34), in which the upper limit of integration, b, has been replaced by the variable x, whilst the lower limit of integration, a, is arbitrary. The quantity $F(a)$ is therefore an arbitrary constant, and can be replaced by the symbol C:

$$\int f(x) \, dx = \int_a^x f(x) \, dx = F(x) - F(a) = F(x) + C \qquad (5.36)$$

The use of differentials

A convenient alternative approach to the definite integral as an area makes use of the concept of the differential introduced in Section 4.12. This approach is widely used in the application of the calculus to the formulation of physical problems.

Consider the expression (5.35) in Example 5.11. Division by Δr gives

$$2\pi r < \frac{\Delta A}{\Delta r} < 2\pi(r + \Delta r)$$

and, in the limit $\Delta r \to 0$,

$$\frac{\Delta A}{\Delta r} \to \frac{dA}{dr} = 2\pi r$$

The quantity dA/dr is the rate of change of the area $A(r)$ of the circle with respect to the radius r. The corresponding differential $dA = 2\pi r \, dr$ is, in the language of differentials, the area of a circular strip of radius r and 'infinitesimal' width dr; it is an **element of area** or **differential area**. The 'sum' of these elements is the integral

$$A = \int_0^A dA = \int_0^a 2\pi r \, dr$$

The integral as a length

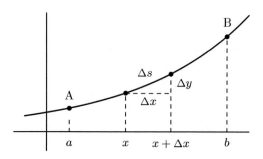

Figure 5.14

Let $y = f(x)$ be continuous in the range $a \leq x \leq b$, and let s be the length of its graph from point A, at $x = a$, to point B, at $x = b$ (Figure 5.14). To calculate this length, we divide the arc AB into segments of length Δs. Then, by Pythagoras' theorem, an approximate value of the length of the segment between x and $x + \Delta x$ is

$$\Delta s \approx \left[(\Delta x)^2 + (\Delta y)^2 \right]^{1/2} = \left[1 + \left(\frac{\Delta y}{\Delta x} \right)^2 \right]^{1/2} \Delta x$$

The corresponding element of length is

$$ds = \left[1 + \left(\frac{dy}{dx} \right)^2 \right]^{1/2} dx \tag{5.37}$$

and the total length of arc AB is

$$s = \int_0^s ds = \int_a^b \left[1 + \left(\frac{dy}{dx} \right)^2 \right]^{1/2} dx \tag{5.38}$$

EXAMPLE 5.12 The circumference of a circle

The equation of the circle of radius a and centre at the origin is $x^2 + y^2 = a^2$ so that $y = \pm(a^2 - x^2)^{1/2}$ and $dy/dx = \mp x/(a^2 - x^2)^{1/2}$. Therefore,

$$1 + \left(\frac{dy}{dx} \right)^2 = 1 + \frac{x^2}{a^2 - x^2} = \frac{a^2}{a^2 - x^2}$$

and the circumference of the circle is

$$s = 4 \int_0^a \left[1 + \left(\frac{dy}{dx} \right)^2 \right]^{1/2} dx = 4a \int_0^a \frac{dx}{\sqrt{a^2 - x^2}}$$

(four times the length of that quarter in the first quadrant, see Example 5.11). The integral is one of the standard integrals listed in Table 6.3:

$$ s = 4a \left[\sin^{-1}\left(\frac{x}{a}\right) \right]_0^a = 4a\left(\sin^{-1} 1 - \sin^{-1} 0 \right) $$

The principal values of the inverse sine of 1 and of 0 are $\pi/2$ and 0 respectively. Therefore $s = 2\pi a$.

> Exercise 49

5.5 Uses of the integral calculus

The discussion of Section 5.4 was concerned with the use of the integral calculus for the determination of geometric properties; in particular, the area of a plane figure and the length of a curve in a plane. The generalization to curves in three dimensions, to nonplanar surfaces, and to volumes is discussed in Chapters 9 and 10.

We saw in Section 5.3, equation (5.13), that the definite integral provides the average value of a function. When the function represents, for example, the distribution of mass in a physical body, the integral calculus can be used to determine **static properties** of the body such as the total mass, the centre of mass, and the moment of inertia in terms of definite integrals involving the mass density. Similarly for a distribution of charge or of any physical property of a system that is distributed in space. This use of the integral calculus for the determination of static properties is introduced in Section 5.6 for properties distributed along a line (the one-dimensional case); the three-dimensional case is discussed in Chapter 10. The same methods are used in Chapter 21 for the analysis of probability distributions in statistics.

The concept of the integral was introduced in Section 5.1 by considering the motion of a body along a line. By Newton's first law of motion, a body moving with a given velocity at any time continues to move along a straight line with the same velocity if no external forces act on the body; $v = $ constant when $F = 0$. On the other hand, by Newton's second law, the acceleration, or rate of change of velocity $a = dv/dt$, experienced by a body of mass m in the presence of an external force F is given by $F = ma$. The use of the integral calculus for the description of the dynamics of physical systems is introduced in Section 5.7 for motion along a straight line; the more general case of motion in two and three dimensions is discussed in Chapter 16. The application to pressure–volume work in thermodynamics is discussed in Section 5.8.

The widest use of the integral calculus is for the solution (integration) of differential equations; in particular, the equations of chemical kinetics and of other rate processes, the equations of motion in classical mechanics (derived from Newton's second law for example), and the equations of motion in quantum mechanics (such as the Schrödinger equation). This use of the integral calculus is discussed in Chapters 11 to 14.

5.6 Static properties of matter

Consider a set of N discrete masses $m_1, m_2, \ldots, m_N$ distributed along a (horizontal) straight line, with mass m_i at position x_i with respect to a fixed point O as in Figure 5.15.

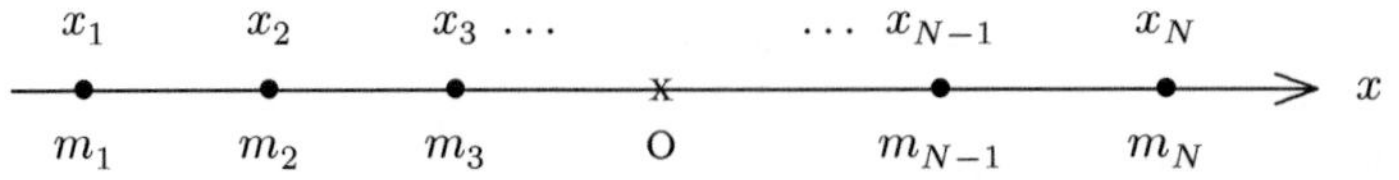

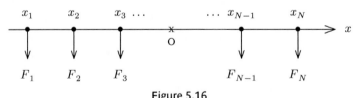

x_1 x_2 x_3 ... ... x_{N-1} x_N

m_1 m_2 m_3 O m_{N-1} m_N

Figure 5.15

The system of masses has the following properties:

(i) total mass: $M = \sum\limits_{i=1}^{N} m_i$

(ii) first moment of mass about O: $\sum\limits_{i=1}^{N} m_i x_i$

(iii) second moment of mass about O: $\sum\limits_{i=1}^{N} m_i x_i^2$ (5.39)

and there are similar definitions for all higher moments.

The first moment of mass, (ii), defines the position X of the **centre of mass** of the system of masses:

$$\sum_{i=1}^{N} m_i x_i = MX \qquad (5.40)$$

The first moment of the system of masses is therefore equal to the moment of the total mass M concentrated at the centre of mass.

x_1 x_2 x_3 ... ... x_{N-1} x_N

F_1 F_2 F_3 F_{N-1} F_N

Figure 5.16

In the presence of gravity, each mass experiences a force $F_i = m_i g$ directed downwards (Figure 5.16), where g is the standard acceleration of gravity ('gravitational constant'). This force F_i is what is commonly called the weight of the body. Multiplication of equation (5.40) by g gives

$$T = \sum_{i=1}^{N} F_i x_i = MgX = FX \qquad (5.41)$$

where $F = \sum_i F_i$ is the total force acting on the system of masses. The position X of the centre of mass is then also called the **centre of gravity**, and the quantity $T = \sum_i F_i x_i$ is called the **moment of force** or **torque** of the system of forces about the point O. If the masses are attached to a uniform rigid rod (itself of negligible mass) then the torque is a measure of the tendency of the forces to rotate the system around O as pivot and, by equation (5.41), it is equal to the torque produced by the total force F concentrated at the centre of mass. If the point O is *at* the centre of mass (gravity) then $X = 0$ and

$$T = \sum_{i=1}^{N} F_i x_i = 0 \tag{5.42}$$

so that the total torque about the centre of mass is zero.

The second moment of mass, (iii), is the **moment of inertia** of the system of masses with respect to the point O. It is the property of the mass distribution that is most important in the description of the dynamics of rotating bodies.

EXAMPLE 5.13 A system of two masses

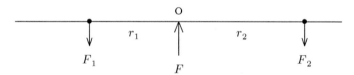

Figure 5.17

Figure 5.17 shows two bodies, masses m_1 and m_2, joined by a rigid rod (of negligible mass). The forces F_1 and F_2 are the weights of the bodies, and F is a counter force acting at the pivot point O. If the pivot is at the centre of mass then, putting $x_1 = -r_1$ and $x_2 = r_2$ in equation (5.42),

$$F_1 r_1 = F_2 r_2 \qquad \text{(law of levers)}$$

and the body is at equilibrium with respect to rotation about the pivot O. The total vertical force acting on the body is $F_1 + F_2 - F$, so that the body is at equilibrium with respect to vertical motion if $F = F_1 + F_2$.

The moment of inertia of the two masses is $I = m_1 r_1^2 + m_2 r_2^2$. When O is at the centre of mass then $X = 0$ in (5.40) so that $m_1 r_1 = m_2 r_2$, and the distances r_1 and r_2 can be written as

$$r_1 = \frac{m_2}{m_1 + m_2} R, \qquad r_2 = \frac{m_1}{m_1 + m_2} R$$

where $R = r_1 + r_2$ is the distance between the masses. The moment of inertia is then

$$I = m_1 r_1^2 + m_2 r_2^2 = \frac{m_1 m_2}{m_1 + m_2} R^2 = \mu R^2$$

where the quantity $\mu = m_1 m_2 / (m_1 + m_2)$ is the **reduced mass** of the system of two masses (see also Example 1.18). The moment of inertia about the centre of mass of a system of two masses is therefore the same as the moment of inertia of a single mass μ at distance R from the centre of mass.

› Exercise 50

The continuous case

Consider a continuous mass distribution such as a straight rod of matter of length l (Figure 5.18).

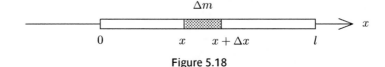

Figure 5.18

Let the mass in the segment x to $x + \Delta x$ be Δm, so that $\Delta m / \Delta x$ is the mass per unit length in the segment. If the mass is distributed evenly over the length, then $\Delta m / \Delta x$ is independent of the choice of segment. It is then the **density** (or, more correctly, the linear mass density) ρ of the body, a constant throughout the length. In this case, the total mass is $M = \rho l$. If the mass of the rod is *not* evenly distributed over its length, then $\Delta m / \Delta x$ is the *average density* in the segment x to $x + \Delta x$, and its value depends on the position of the segment and on its length. As this length is reduced to zero, the ratio approaches the limit

$$\rho(x) = \lim_{\Delta x \to 0} \left(\frac{\Delta m}{\Delta x} \right) = \frac{dm}{dx} \tag{5.43}$$

The value of the function $\rho(x)$ at each point is the density at that point, and the differential quantity $dm = \rho(x) dx$ is the mass of a segment dx at x. The total mass of the body is then

$$M = \int_0^M dm = \int_0^l \rho(x) \, dx \tag{5.44}$$

Therefore, when a discrete mass distribution is replaced by a continuous distribution, the sums of the discrete case are replaced by integrals. Thus, the average density is

$$\bar{\rho} = \frac{M}{l} = \frac{\int_0^l \rho(x) \, dx}{\int_0^l dx} \tag{5.45}$$

The position of the centre of mass is given by (see equation (5.40))

$$MX = \int_0^l \rho(x)x \, dx, \quad \text{or} \quad X = \frac{\int_0^l \rho(x)x \, dx}{\int_0^l \rho(x) \, dx} \tag{5.46}$$

and the moment of inertia with respect to an arbitrary point x_0 on the line is

$$I = \int_0^l \rho(x)(x - x_0)^2 \, dx \tag{5.47}$$

We note that, for example, the total mass can be interpreted as the 'area under the curve' of the graph of the mass density function $\rho(x)$.

EXAMPLE 5.14 Find the total mass, centre of mass, and moment of inertia of the linear distribution of mass of length a and density $\rho(x) = b(a - x)$, $0 \leq x \leq a$.

The total mass is equal to the area $a^2 b/2$ of the triangle shaded in Figure 5.19. Thus

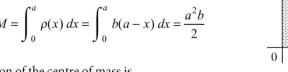

$$M = \int_0^a \rho(x) \, dx = \int_0^a b(a - x) \, dx = \frac{a^2 b}{2}$$

The position of the centre of mass is

Figure 5.19

$$X = \frac{1}{M} \int_0^a \rho(x)x \, dx = \frac{2}{a^2} \int_0^a (a - x)x \, dx = \frac{a}{3}$$

The moment of inertia with respect to an arbitrary point x_0 on the line is

$$I = \int_0^a \rho(x)(x - x_0)^2 \, dx = b \int_0^a (a - x)(x - x_0)^2 \, dx = \frac{ba^2}{12}(a^2 - 4ax_0 + 6x_0^2)$$

The value of x_0 for which I has its minimum value is given by

$$\frac{dI}{dx_0} = 0 = \frac{ba^2}{12}(-4a + 12x_0)$$

so that $x_0 = a/3$. The moment of inertia therefore has its smallest value when computed with respect to the centre of mass; *this is a general result.*

▸ Exercise 51

5.7 Dynamics

Velocity and distance

As in Section 5.1, we consider a body moving along a curve from point A at time $t = t_A$ to point B at time $t = t_B$, and let the distance from A along the curve be $s(t)$ at some intermediate time t. By the discussion of Section 4.11, the quantity $v(t) = ds/dt$ is the velocity of the body at time t. The differential length $ds = v\, dt$ is the distance travelled in the infinitesimal time interval dt, and the total distance is then

$$s = \int_0^s ds = \int_{t_A}^{t_B} v(t)\, dt \qquad (5.48)$$

EXAMPLE 5.15 A body falling under the influence of gravity

A body of mass m falling freely under the influence of gravity experiences the constant acceleration $dv/dt = g$ (air resistance and other frictional forces are neglected). Then $v(t) = gt + v(0)$, where $v(0)$ is the velocity at time $t = 0$. If the body falls from rest at $t = 0$, then $v(t) = gt$, and the distance travelled in the time interval $t = t_A$ to $t = t_B$ is

$$s = \int_{t_A}^{t_B} v(t)\, dt = \int_{t_A}^{t_B} gt\, dt = \frac{1}{2} g \left(t_B^2 - t_A^2 \right)$$

Therefore, given that $g \approx 9.8$ m s^{-2}, the distance travelled in the first second of fall is 4.9 m, and in the following second it is $4.9 \times (2^2 - 1^2)$ m $= 14.7$ m.

> Exercise 52

Force and work

Consider a body moving along the x direction between points A and B with velocity $v = dx/dt$. If a force F acts on the body then, by Newton's second law, the acceleration a experienced by the body is given by

$$F = ma = m \frac{dv}{dt} \qquad (5.49)$$

where m is the mass of the body. Work is done on the body by the application of the force. If the force is constant then the work done is (work $=$ force $\times$ distance) $W = F(x_B - x_A)$.

If the force is *not* constant between A and B, but is a function of position, $F = F(x)$, then the work done is obtained by means of the integral calculus. The work done on the body between positions x and $x + \Delta x$ is $\Delta W \approx F(x)\, \Delta x$. In the limit $\Delta x \to 0$, the corresponding element of work is $dW = F(x)\, dx$ and the total work done is the integral

$$W = \int_0^W dW = \int_{x_A}^{x_B} F(x)\, dx \qquad (5.50)$$

EXAMPLE 5.16 A body falling under the influence of gravity

A body of mass m falling freely under the influence of gravity experiences the constant force $F = mg$ directed downwards. This is the total force acting on the body in the absence of air resistance and other frictional forces. The work done by gravity on the body as it falls through a height h is therefore

$$W = \int_0^h F\,dx = \int_0^h mg\,dx = mgh$$

This is also the work that must be done *against* the force of gravity to raise the body through the distance h.

EXAMPLE 5.17 Electrostatic work

The force acting between two electric charges q_1 and q_2 separated by distance x in a vacuum is given by Coulomb's inverse-square law

$$F(x) = \frac{q_1 q_2}{4\pi\varepsilon_0 x^2}$$

where ε_0 is the permittivity of a vacuum.* Like charges (charges with the same sign, such as two nuclei or two electrons) repel, so that, in Figure 5.20, the force acting on q_2 due to the presence of q_1 acts in the positive x-direction, away from q_1. Unlike charges (of opposite signs, such as the proton and electron in the hydrogen atom) attract, and the force on q_2 is directed towards q_1 (F is negative in Figure 5.20).

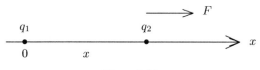

Figure 5.20

Consider two like charges, initially infinitely far apart. Because the charges repel, work must be done on the system to bring q_2 from infinity to the distance x from q_1. The force $-F$ must be applied to overcome the repulsion, and the work done is

$$W = -\int_\infty^x F(x)\,dx = -\frac{q_1 q_2}{4\pi\varepsilon_0}\int_\infty^x \frac{dx}{x^2} = -\frac{q_1 q_2}{4\pi\varepsilon_0}\left[-\frac{1}{x}\right]_\infty^x = \frac{q_1 q_2}{4\pi\varepsilon_0 x}$$

---

* The presence of the factor $4\pi\varepsilon_0$ ensures that the force acting between two charges of one coulomb separated by one metre is one newton; $N = C^2/(4\pi\varepsilon_0\ m^2)$ in terms of SI units.

This work is positive for like charges. The same formula applies to the case of unlike charges, but the work is then negative.

‣ Exercise 53

Work and energy

When work is done on a system by an external force, the energy of the system is increased by the amount of the work done. Conversely, when a system does work against an external force, the energy of the system is decreased by the amount of the work done. The energy of a system is usually expressed as the sum of two parts. For a simple system, with no internal structure, these parts are (i) the kinetic energy, or translational energy, arising from the motion of the system in space, and (ii) the potential energy, arising from the position of the system in space and from the forces acting on the system at that position. In the case of a system with internal structure, the kinetic energy is the sum of the kinetic energies of its constituent parts, and the potential energy is the sum of the potential energies of its parts.

(i) Kinetic energy

The work done on a body by an external force $F(x)$ as the body travels from point A to point B is

$$W_{AB} = \int_A^B F(x)\, dx = m \int_A^B \frac{dv}{dt}\, dx \tag{5.51}$$

Because $v = dx/dt$, the element of length dx can be replaced by the differential $v\, dt$, so that

$$W_{AB} = m \int_A^B \frac{dv}{dt} v\, dt = \frac{1}{2} m \int_A^B \frac{d}{dt}(v^2)\, dt = \frac{1}{2} m \Big[v^2 \Big]_A^B \tag{5.52}$$

in which use has been made of $d(v^2)/dt = 2v\, dv/dt$, and the integration limits now refer to the times at A and B. It follows that the work done between A and B is

$$W_{AB} = \frac{1}{2} m \left(v_B^2 - v_A^2 \right) \tag{5.53}$$

where v_A and v_B are the velocities of the body at A and B respectively. The quantity $\frac{1}{2} mv^2$ is called the **kinetic energy** of the body and is usually denoted by the symbol T (or K). The work done on the body is therefore equal to the change in kinetic energy:

$$W_{AB} = T_B - T_A \tag{5.54}$$

We note that the kinetic energy of a body at rest is defined to be zero.

(ii) Potential energy and total energy

Let the force acting on a body depend only on the position of the body, so that $F = F(x)$. This condition excludes time-dependent forces and, more importantly, dissipative forces such as those due to friction. Then, by the fundamental theorem of the calculus (equation (5.34)), there is a function $f(x)$ such that $F(x) = f'(x)$ and

$$W_{AB} = \int_A^B F(x)\, dx = f(B) - f(A) \tag{5.55}$$

The work done from A to B can therefore be expressed as the change in a quantity which depends only on the end points A and B, and not on the path A to B. This quantity is normally designated by $-V$, and V is called the **potential energy** of the body. The work from A to B is then

$$W_{AB} = \int_A^B F(x)\, dx = V_A - V_B \tag{5.56}$$

where V_A and V_B are the values of the potential energy at A and B. When the work done by a force is independent of the path, the force is called a **conservative force** and is $(-)$ the derivative of the potential energy (function):

$$F(x) = -\frac{dV}{dx}, \qquad V(x) = -\int F(x)\, dx + C \tag{5.57}$$

Three simple but important types of conservative force, and corresponding potential energy, are

(a) $F = \text{constant},$ $V(x) = -Fx + C$ (5.58)

(b) $F = -kx,$ $V(x) = \frac{1}{2}kx^2 + C$ (5.59)

(c) $F = -\dfrac{1}{x^2},$ $V(x) = -\dfrac{1}{x} + C$ (5.60)

where C is an arbitrary constant. The graphs of $V(x)$, for $C = 0$, are illustrated in Figure 5.21. In each case, the gradient of the graph is $dV/dx = -F$ so that the force acts in the direction of decreasing potential energy.

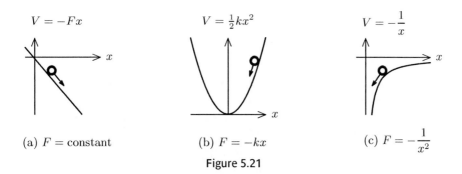

$V = -Fx$	$V = \frac{1}{2}kx^2$	$V = -\dfrac{1}{x}$
(a) $F = \text{constant}$	(b) $F = -kx$	(c) $F = -\dfrac{1}{x^2}$

Figure 5.21

We note that, whilst the kinetic energy has a well defined absolute value, this is not true of the potential energy, for which only relative values are defined by equations (5.56) and (5.57). Examples 5.18 and 5.19 below show how the zero of potential energy is chosen in two different physical situations.

A system in which all the forces are conservative is called a **conservative system**. In such a system, the work done in moving a body round a closed loop A → B → A is zero:

$$W_{ABA} = W_{AB} + W_{BA} = (V_A - V_B) + (V_B - V_A) = 0 \qquad (5.61)$$

Dissipative forces such as those due to friction are not conservative forces because the work done against friction is always positive.

Combining the expressions (5.54) and (5.56) we have the result

$$T_A + V_A = T_B + V_B \qquad (5.62)$$

and it follows that, in a conservative system, the quantity $T + V$ is constant. This quantity is called the **total energy** of the system, $E = T + V$, and (5.62) is an expression of the **principle of the conservation of energy**: if the forces acting on a body are conservative, then the total energy of the body, $T + V$, is conserved.

EXAMPLE 5.18 A body moving under the influence of gravity

Consider a body of mass m at height h above a horizontal surface, as in Figure 5.22. The force of gravity acting on the body is $F = -mg$ (negative because the force acts in the negative x-direction) and the work done on the body as it falls freely from height $x = h$ onto the surface at $x = 0$ is

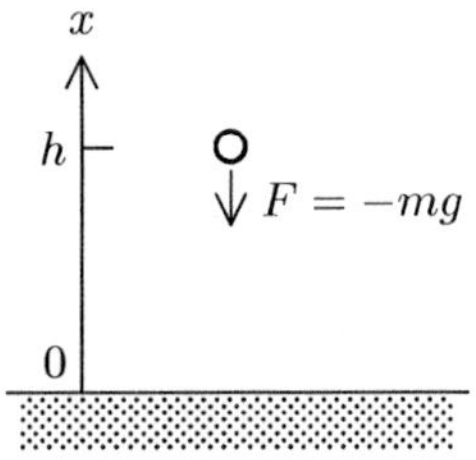

Figure 5.22

$$W = \int_h^0 F \, dx = -mg \int_h^0 dx = mgh$$

This work is the change of potential energy,

$$W = mgh = V(h) - V(0)$$

where $V(x)$ is the potential energy of the body at height x. The natural choice of zero of potential energy in this example is $V(0) = 0$, zero at the surface. Then $V(x) = mgx$ is the potential energy of the body at height x, and the force is related to it by $F = -dV/dx = -mg$.

Let the body fall from rest at $x = h$ and let the kinetic energy at height x be $T(x)$. Then $T(h) = 0$ and, by equation (5.54), the kinetic energy of the body when it reaches the surface is $T(0) = mgh$. In addition, because the force (a constant) is conservative, the total energy of the body is conserved and is equal to $E = mgh$, which is the potential energy at $x = h$ (where the kinetic energy is zero) and is the kinetic energy at $x = 0$

(where the potential energy is zero). At an intermediate height, $E = mgh = T(x) + V(x)$ and the kinetic energy is

$$T(x) = \frac{1}{2}mv^2 = mg(h - x) \tag{5.63}$$

If both the body and the surface are perfectly elastic, the velocity of the body is reversed on contact with the surface, and the body returns to its original height at $x = h$ in an exact reversal of the falling motion. Thus, solving equation (5.63) for the velocity,

$$v = \pm\sqrt{2g(h - x)}$$

and the velocity of the body is negative as it falls and positive as it rises. In the absence of dissipative forces the bouncing motion is repeated indefinitely.

EXAMPLE 5.19 Electrostatic potential energy

By Example 5.17, the work that must be done *against* the internal force to bring two charges from infinite separation to separation x is $W = q_1 q_2/4\pi\varepsilon_0 x$. This is the same as the work done *by* the internal force in separating the charges:

$$W = \int_x^\infty F(x)\, dx = \frac{q_1 q_2}{4\pi\varepsilon_0 x}$$

The force depends only on the relative positions of the charges and is conservative, so that a potential energy function $V(x)$ exists such that $F(x) = -dV/dx$ and $W = V(x) - V(\infty)$. It is conventional, and convenient, to choose the zero of potential energy of interacting charges to be zero for infinite separation: $V(\infty) = 0$. Then

$$V(x) = \frac{q_1 q_2}{4\pi\varepsilon_0 x}$$

is the electrostatic potential energy of the system of two charges.

> Exercises 54, 55

5.8 Pressure–volume work

Consider a fluid (gas or liquid) enclosed in a uniform cylindrical container, closed at one end and fitted with a piston as shown in Figure 5.23. Let A be the internal cross-sectional area of the cylinder. A fluid with internal pressure p exerts a force of magnitude $|F| = pA$ on the surface of the piston, and the piston moves in or out

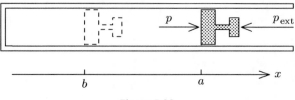

Figure 5.23

according as the external pressure p_{ext} is greater than or less than p. If $p_{ext} > p$ then the fluid is compressed, and the work done by the external force F_{ext} in moving the piston from a to b is

$$W_{ab} = \int_a^b F_{ext} \, dx \qquad (5.64)$$

Now, the external force has magnitude $|F_{ext}| = p_{ext} A$, and a length of cylinder $|dx|$ contains a volume $|dV| = A|dx|$. The work can therefore be written in 'pressure–volume' form as

$$W_{ab} = -\int_a^b p_{ext} \, dV \qquad (5.65)$$

in which the limits of integration now refer to the volume, and the minus sign is included to make the work positive for compression. It can be shown that this expression for the mechanical work done on a thermodynamic system is independent of the shape of the container.

To compress the fluid, it is necessary that the external pressure be greater than the internal pressure of the fluid. Let $p_{ext} = (p + \Delta p)$ where Δp is a positive excess pressure that, for simplicity, can be assumed to be constant throughout the compression. Then, with $V_a > V_b$,

$$W_{ab} = -\int_a^b p \, dV - \Delta p \int_a^b dV = -\int_a^b p \, dV + \Delta p (V_a - V_b)$$
$$> -\int_a^b p \, dV \qquad (5.66)$$

Conversely, to allow the fluid to expand from V_b to V_a it is necessary that the external pressure be smaller than the internal pressure. If $p_{ext} = (p - \Delta p)$ then

$$W_{ba} = -\int_b^a p \, dV + \Delta p (V_a - V_b)$$
$$> -\int_b^a p \, dV \qquad (5.67)$$

The total work done on the fluid in the cycle $a \to b \to a$ is therefore

$$W_{aba} = W_{ab} + W_{ba} = 2\Delta p(V_a - V_b) > 0 \tag{5.68}$$

and the process is said to be **irreversible**. This is analogous to the case in ordinary dynamics when nonconservative dissipative forces are present. The process can be made reversible only by letting the excess pressure, Δp in our example, approach zero. In this (ideal) limit the work is **reversible** and

$$W_{ab} = -W_{ba} = -\int_a^b p \, dV \tag{5.69}$$

Expansion of a gas

The equation of state of a gas is a relation of the form $f(p, V, T) = 0$ amongst the three thermodynamic quantities p, V, and T (for a given amount of gas). For example, the equation of state for the ideal gas can be written as $pV - nRT = 0$. There is therefore some freedom in the choice of conditions under which the expansion of the gas can occur.

isobaric expansion

Expansion of a gas can occur at constant pressure; for example, the gas can be heated to expand against a constant external pressure such as atmospheric pressure. The reversible work done by the gas against the external pressure is then

$$W = \int_a^b p \, dV = p \int_a^b dV = p(V_b - V_a) \tag{5.70}$$

For the ideal gas, $pV = nRT$ and

$$W = nR(T_b - T_a) \tag{5.71}$$

isothermal expansion

Expansion of a gas can occur at constant temperature; for example, if the expansion occurs with the container immersed in a heat bath at a given temperature. To calculate the work it is necessary to know the equation of state. For the ideal gas, $p = nRT/V$ and the (reversible) work is

$$W = \int_a^b p \, dV = nRT \int_a^b \frac{dV}{V} = nRT \ln\frac{V_b}{V_a} = -nRT \ln\frac{p_b}{p_a} \tag{5.72}$$

➤ Exercise 56

5.9 Exercises

Section 5.2

Evaluate the indefinite integrals:

1. $\displaystyle\int 2\,dx$ **2.** $\displaystyle\int x^3\,dx$ **3.** $\displaystyle\int x^{2/3}\,dx$ **4.** $\displaystyle\int \frac{dx}{x^3}$ **5.** $\displaystyle\int x^{-1/3}\,dx$

6. $\displaystyle\int \sin 4x\,dx$ **7.** $\displaystyle\int e^{3x}\,dx$ **8.** $\displaystyle\int e^{-2x}\,dx$ **9.** $\displaystyle\int \frac{dx}{x-1}$ **10.** $\displaystyle\int \frac{dx}{3-x}$

Evaluate the indefinite integrals subject to the given conditions:

11. $y = \displaystyle\int x^2\,dx$; $y = 0$ when $x = 3$ **12.** $y = \displaystyle\int \cos 4x\,dx$; $y = 0$ when $x = \pi/4$

13. $I = \displaystyle\int (5x^4 + 2x + 3)\,dx$; $I = 4$ when $x = 2$ **14.** $I = \displaystyle\int \frac{3x^2 + 2x + 1}{x^2}\,dx$; $I = 3$ when $x = 1$

15. $I = \displaystyle\int \left(-4 + 4\cos 2x - \frac{1}{2}e^{2x}\right)dx$; $I = 0$ when $x = 0$

Section 5.3

Evaluate the definite integrals:

16. $\displaystyle\int_{-1}^{+1} (2x^2 + 3x + 4)\,dx$ **17.** $\displaystyle\int_{3}^{5} dx$ **18.** $\displaystyle\int_{1}^{2} \frac{du}{u^3}$ **19.** $\displaystyle\int_{2}^{3} \frac{dv}{v+2}$

20. $\displaystyle\int_{1}^{5} e^{-3t}\,dt$ **21.** $\displaystyle\int_{0}^{\pi/2} \cos\theta\,d\theta$ **22.** $\displaystyle\int_{0}^{\pi} \cos 3\theta\,d\theta$ **23.** $\displaystyle\int_{-\pi/4}^{0} \sin 2x\,dx$

24. **(i)** Show that the rate equation of a first-order decomposition reaction

$$\frac{dx}{dt} = -kx$$

can be written in the logarithmic form

$$\frac{d\ln x}{dt} = -k$$

(ii) Integrate this equation with respect to t over the range 0 to t, and show that

$$\ln\left[\frac{x(t)}{x(0)}\right] = -kt \quad\text{and}\quad x(t) = x(0)e^{-kt}$$

25. The Clausius–Clapeyron equation for liquid–vapour equilibrium is

$$\frac{d\ln p}{dT} = \frac{\Delta H_{vap}}{RT^2}$$

If the enthalpy of vaporization, ΔH_{vap}, is constant in the temperature range T_1 to T_2 show, by integrating both sides of the equation with respect to T, that

$$\ln\left(\frac{p_2}{p_1}\right) = \frac{\Delta H_{vap}}{R}\left(\frac{1}{T_1} - \frac{1}{T_2}\right)$$

where $p_1 = p(T_1)$ and $p_2 = p(T_2)$.

Find the average values in the given intervals:

26. $2x^2 + 3x + 4;$ $-1 \le x \le +1$ **27.** $\cos 3\theta;$ $0 \le \theta \le \pi/2$ **28.** $1;$ $3 \le x \le 5$

Demonstrate and sketch a graph to interpret:

29. $\displaystyle\int_0^3 e^{-x}\, dx = \int_0^1 e^{-x}\, dx + \int_1^2 e^{-x}\, dx + \int_2^3 e^{-x}\, dx$ **30.** $\displaystyle\int_2^6 e^x\, dx = \int_2^6 e^x\, dx - \int_3^6 e^x\, dx$

31. (i) Show that $\displaystyle\int_0^{\pi/2} \cos x\, dx = -\int_{\pi/2}^{\pi} \cos x\, dx.$ **(ii)** Calculate $\displaystyle\int_{-\pi}^0 \cos x\, dx, \int_{-\pi}^{\pi/2} \cos x\, dx,$

$\displaystyle\int_{-\pi}^{\pi} \cos x\, dx.$ **(iii)** Sketch a graph to interpret these results.

Evaluate and sketch a graph to interpret:

32. $\displaystyle\int_{-1}^{+3} f(x)\, dx$ where $f(x) = \begin{cases} x^2 + 2 & \text{if } x < 1 \\ x^2 & \text{if } x \ge 1 \end{cases}$

33. $\displaystyle\int_{-1}^{+1} f(x)\, dx$ where $f(x) = \begin{cases} x & \text{if } x > 0 \\ -x & \text{if } x \le 0 \end{cases}$

34. $\displaystyle\int_{-a}^{+a} f(x)\, dx$ where $f(x) = \begin{cases} e^{-x} & \text{if } x > 0 \\ -e^{+x} & \text{if } x \le 0 \end{cases}$

35. (i) Show that $\dfrac{d}{dx}(x\ln x - x) = \ln x,$ **(ii)** evaluate $\displaystyle\int_0^1 \ln x\, dx.$

Evaluate:

36. $\displaystyle\int_0^{\infty} e^{-3t}\, dt$ **37.** $\displaystyle\int_0^{\infty} e^{-x/2}\, dx$ **38.** $\displaystyle\int_2^{\infty} \dfrac{dx}{x(x-1)}$ **39.** $\displaystyle\int_2^{\infty} \dfrac{dx}{x(x-1)^2}$

For each function, state if it is an even function of x, an odd function, or neither. If neither, give the even and odd components.

40. $\sin 2x$ **41.** $\cos 3x$ **42.** $\sin x \cos x$ **43.** x **44.** x^4 **45.** $3x^2 + 2x + 1$
46. e^{-x} **47.** $(3x^2 + 2x + 1)e^{-x}$

Section 5.4

48. The equation of an ellipse with centre at the origin is $\dfrac{x^2}{a^2} + \dfrac{y^2}{b^2} = 1,$ where, if $a > b$, a is the major axis and b the minor axis (if $a = b$, we have a circle). Use Method 1 in Example 5.11 to find the area of the ellipse.

49. Find the length of the curve $y = \frac{1}{2}x^{3/2}$ between $x = 0$ and $x = 1$.

Section 5.6

50. Three masses, $m_1 = 1$, $m_2 = 2$ and $m_3 = 3$, lie on a straight line with m_1 at $x_1 = -4$, m_2 at $x_2 = -1$ and m_3 at $x_3 = +4$ with respect to a point O on the line. Calculate **(i)** the position of the centre of mass, **(ii)** the moment of inertia with respect to O, and **(iii)** the moment of inertia with respect to the centre of mass.

51. The distribution of mass in a straight rod of length l is given by the density function $\rho(x) = x^2$; $0 \le x \le l$. Find **(i)** the total mass, **(ii)** the mean density, **(iii)** the centre of mass,

(iv) the moment of inertia with respect to an arbitrary point x_0 on the line, **(v)** the moment of inertia with respect to the centre of mass. **(vi)** Show that the moment of inertia has its smallest value when computed with respect to the centre of mass.

Section 5.7

52. A body moves in a straight line with velocity $v = 3t^2$ at time t. Calculate the distance travelled in time interval **(i)** $t = 0 \rightarrow 1$, **(ii)** $t = 1 \rightarrow 2$, **(iii)** $t = 3 \rightarrow 4$.

53. A body of mass m moves in a straight line (the x-direction) under the influence of a force $F = kx$. What is the work done on the body between $x = x_A$ and $x = x_B$?

54. A body of mass m moves in a straight line (the x-direction) under the influence of a force $F = kx$, where k is positive (see Exercise 53). **(i)** Find the potential energy $V(x)$ (choose $V(0) = 0$).

   The body is released from rest at $x = 1$. **(ii)** Find (a) the total energy E and (b) the kinetic energy $T(x)$ as functions of x. **(iii)** Sketch a graph showing the dependence of $V(x)$, $T(x)$, and E on x. **(iv)** Use the graph to describe the motion of the body. **(v)** What would be the motion if the body were released from rest at (a) $x = -1$, (b) $x = 0$?

55. Repeat Exercise 54 with $F = -kx$.

Section 5.8

56. A slightly imperfect gas obeys the van der Waals equation of state

$$\left(p + \frac{n^2 a}{V^2} \right)(V - nb) = nRT$$

Find expressions for the work done by the gas in expanding reversibly from volume V_1 to volume V_2 at **(i)** constant pressure, and **(ii)** constant temperature (assume a and b are constant).

6 Methods of integration

6.1 Concepts

To calculate the value of a definite integral it is normally necessary to evaluate the corresponding indefinite integral; that is, given

$$\int f(x)\, dx = F(x) + C$$

we need to find the function $F(x)$ whose derivative is $f(x) = F'(x)$. We saw in Section 4.6 that every continuous function can be differentiated by the application of a small number of rules. Each such rule, on inversion, provides in principle a rule of integration. In particular, the chain rule and the product rule provide, on inversion, the two principal general methods for calculating integrals; the **method of substitution** (Section 6.3) and **integration by parts** (Section 6.4).

The aim of these general methods of integration, and of the other particular methods discussed in the following sections, is to reduce a given integral to **standard form**; that is, to transform it into an integral whose value is given in a table of **standard integrals**. A standard integral is simply an integral whose value is known and whose form can be used for the evaluation of other 'non-standard' integrals. The most elementary standard integrals are those given in Table 5.1, and a more comprehensive list is given in the Appendix. Extensive tabulations of standard integrals have been published, and it is our purpose in this chapter to show how (some of) these standard integrals are obtained and, most important, how others can be reduced to standard form. It must be remembered, however, that there are many functions whose integrals cannot be expressed in terms of a finite number of elementary functions. In such cases, approximate values are obtained by numerical methods; with the development of computing machines, numerical methods of integration (numerical quadratures) have become routine and accurate, and some of the simpler numerical methods are discussed in Chapter 20.

We consider first the use of trigonometric relations (Section 3.4) for the integration of some trigonometric functions.

6.2 The use of trigonometric relations

Table 6.1 contains a number of integrals that can be evaluated by making use of the trigonometric relations (3.21) to (3.26) discussed in Section 3.4. The useful forms of these relations are

$$\cos^2 x = \frac{1}{2}\left[1 + \cos 2x\right] \qquad\qquad (6.1)$$

$$\sin^2 x = \frac{1}{2}\left[1 - \cos 2x\right] \qquad\qquad (6.2)$$

$$\sin x \cos x = \frac{1}{2}\sin 2x \tag{6.3}$$

$$\sin x \sin y = \frac{1}{2}\Big[\cos(x-y) - \cos(x+y)\Big] \tag{6.4}$$

$$\cos x \cos y = \frac{1}{2}\Big[\cos(x-y) + \cos(x+y)\Big] \tag{6.5}$$

$$\sin x \cos y = \frac{1}{2}\Big[\sin(x-y) + \sin(x+y)\Big] \tag{6.6}$$

More generally, the relations can be used to express a function $\sin^m x \cos^n x$, where m and n are positive integers, in terms of simple sines and cosines, but alternative methods of integration are often simpler to use when m or n is greater than 2.

Table 6.1

1. $\displaystyle\int \cos^2 ax\, dx = \frac{1}{2a}\Big[ax + \sin ax \cos ax\Big] + C$

2. $\displaystyle\int \sin^2 ax\, dx = \frac{1}{2a}\Big[ax - \sin ax \cos ax\Big] + C$

3. $\displaystyle\int \sin ax \cos ax\, dx = \frac{1}{2a}\sin^2 ax + C$

For $a \neq b$:

4. $\displaystyle\int \sin ax \sin bx\, dx = \frac{1}{2}\left[\frac{\sin(a-b)x}{a-b} - \frac{\sin(a+b)x}{a+b}\right] + C$

5. $\displaystyle\int \cos ax \cos bx\, dx = \frac{1}{2}\left[\frac{\sin(a-b)x}{a-b} + \frac{\sin(a+b)x}{a+b}\right] + C$

6. $\displaystyle\int \sin ax \cos bx\, dx = -\frac{1}{2}\left[\frac{\cos(a-b)x}{a-b} + \frac{\cos(a+b)x}{a+b}\right] + C$

EXAMPLES 6.1 The integrals in Table 6.1

Integral 1: $\displaystyle\int \cos^2 2x\, dx$

By equation (6.1), $\cos^2 2x = \frac{1}{2}\Big[1 + \cos 4x\Big]$. Therefore

$$\int \cos^2 2x \; dx = \frac{1}{2}\int \left[1 + \cos 4x\right] dx = \frac{1}{2}\left[x + \frac{1}{4}\sin 4x\right] + C$$

$$= \frac{1}{4}\left[2x + \sin 2x \cos 2x\right] + C$$

Integral 3: $\int \sin 2x \cos 2x \; dx$

By equation (6.3), $\sin 2x \cos 2x = \frac{1}{2}\sin 4x$. Therefore

$$\int \sin 2x \cos 2x \; dx = \frac{1}{2}\int \sin 4x \; dx = -\frac{1}{8}\cos 4x + C$$

$$= -\frac{1}{8}\left[1 - 2\sin^2 2x\right] + C = \frac{1}{4}\sin^2 2x + C - \frac{1}{8}$$

$$= \frac{1}{4}\sin^2 2x + C'$$

where C' is a new arbitrary constant. Most tabulations of indefinite integrals omit the arbitrary constant, and this example shows why different tabulations sometimes give apparently different values for indefinite integrals.

Integral 4: $\int \sin 2x \sin 4x \; dx$

By equation (6.4), $\sin 2x \sin 4x = \frac{1}{2}\left[\cos(-2x) - \cos 6x\right] = \frac{1}{2}\left[\cos 2x - \cos 6x\right]$.

Therefore

$$\int \sin 2x \sin 4x \; dx = \frac{1}{2}\int \left[\cos 2x - \cos 6x\right] dx = \frac{1}{2}\left[\frac{\sin 2x}{2} - \frac{\sin 6x}{6}\right] + C$$

▸ Exercise 1–10

6.3 The method of substitution

The polynomial

$$f(x) = (2x - 1)^3$$

can be integrated by first expanding the cube and then integrating term by term (see the corresponding discussion of the chain rule in Section 4.6):

$$f(x) = (2x - 1)^3 = 8x^3 - 12x^2 + 6x - 1$$

$$\int (2x-1)^3 \, dx = 8 \int x^3 \, dx - 12 \int x^2 \, dx + 6 \int x \, dx - \int dx$$

$$= 2x^4 - 4x^3 + 3x^2 - x + C'$$

$$= \frac{1}{8}(2x - 1)^4 + C$$

$(C = C' - 1/8$ is an arbitrary constant).

A simpler way of integrating the function is to make the substitution

$$u = 2x - 1, \qquad du = \frac{du}{dx} \, dx = 2 \, dx$$

where du is the differential of $u(x)$. Then, $dx = \frac{1}{2} du$ and

$$\int (2x - 1)^3 \, dx = \frac{1}{2} \int u^3 \, du = \frac{1}{8} u^4 + C = \frac{1}{8}(2x - 1)^4 + C$$

The integral has been transformed into a 'standard integral' by changing the variable of integration from x to u. The method of substitution is also called **integration by change of variable**.

In the general case, given the integral of a function $f(x)$ whose form is non-standard, the method of substitution is to find a new variable $u(x)$ such that

$$\int f(x) \, dx = \int g(u) \, du \tag{6.7}$$

where the integral on the right is a standard integral; that is, $g(u)$ *is easier to integrate than* $f(x)$. Differentiating both sides of (6.7) with respect to x gives:

on the left side, by definition of the indefinite integral,

$$\frac{d}{dx}\left(\int f(x) \, dx \right) = f(x)$$

on the right side, by application of the chain rule,

$$\frac{d}{dx}\left(\int g(u) \, du \right) = \frac{d}{du}\left(\int g(u) \, du \right) \times \frac{du}{dx} = g(u)\frac{du}{dx}$$

Therefore

$$f(x) = g(u)\frac{du}{dx} \tag{6.8}$$

and, substituting in (6.7),

$$\int f(x)\,dx = \int g(u)\frac{du}{dx}\,dx = \int g(u)\,du \tag{6.9}$$

The essential skill in applying the method of substitution is the ability to recognize when an integrand can be written in the form (6.8). The transformation of the integral is then achieved by the substitution

$$u = u(x), \qquad du = \frac{du}{dx}\,dx \tag{6.10}$$

or, alternatively, $x = x(u)$, $dx = \dfrac{dx}{du}\,du$.

EXAMPLE 6.2 Show that

$$\int (ax+b)^n\,dx = \frac{1}{a(n+1)}(ax+b)^{n+1} + C$$

where the numbers a, b, and n are arbitrary, except that $n \neq -1$.
   Let $u = ax + b$. Then $du = a\,dx$, and

$$\int (ax+b)^n\,dx = \frac{1}{a}\int u^n\,du = \frac{1}{a}\frac{u^{n+1}}{n+1} + C = \frac{(ax+b)^{n+1}}{a(n+1)} + C$$

➤ Exercises 11, 12

EXAMPLE 6.3 Integrate $\displaystyle\int f(x)\,dx = \int (x+x^2)^{-1/2}(1+2x)\,dx$.

Because $(1+2x) = \dfrac{d}{dx}(x+x^2)$, we make the substitution $u = x + x^2$. Then $du = (1+2x)\,dx$ and

$$f(x) = (x+x^2)^{-1/2}(1+2x) = u^{-1/2}\frac{du}{dx}$$

Therefore

$$\int (x+x^2)^{-1/2}(1+2x)\,dx = \int u^{-1/2}\,du = 2u^{1/2} + C = 2(x+x^2)^{1/2} + C$$

The result is confirmed by differentiation:

$$\frac{d}{dx}\left[2(x+x^2)^{1/2} + C\right] = (x+x^2)^{-1/2}(1+2x)$$

➤ Exercises 13, 14

EXAMPLE 6.4 Integrate $\int e^{-(2x^2+3x+1)}(4x+3)\,dx$.

Let $u = 2x^2 + 3x + 1$. Then $du = (4x+3)\,dx$, and

$$\int e^{-(2x^2+3x+1)}(4x+3)\,dx = \int e^{-u}\,du = -e^{-u} + C = -e^{-(2x^2+3x+1)} + C$$

➤ Exercises 15, 16

There are no all-embracing rules for finding the correct change of variable that will transform an integral to standard form; proficiency in the art of integration is the result of a lot of practice. Some of the simpler types of substitution are summarized in Table 6.2.

Table 6.2

Type	Substitution	Result
1. $\int f(x)f'(x)\,dx$	$u = f(x),\ du = f'(x)\,dx$	$\int u\,du = \frac{1}{2}u^2 + C$
2. $\int \dfrac{f'(x)}{f(x)}\,dx$	$u = f(x),\ du = f'(x)\,dx$	$\int \dfrac{du}{u} = \ln u + C$
3. $\int f(\sin x)\cos x\,dx$	$u = \sin x,\ du = \cos x\,dx$	$\int f(u)\,du$
4. $\int f(\cos x)\sin x\,dx$	$u = \cos x,\ du = -\sin x\,dx$	$-\int f(u)\,du$

EXAMPLES 6.5 Integrals of type 1: $\int f(x)f'(x)\,dx$.

(i) $I = \int \sin ax \cos ax\,dx$.

In this case, $\cos ax$ is proportional to the derivative of $\sin ax$ (and *vice versa*);

$$f(x) = \sin ax, \qquad f'(x) = a\cos ax$$

Therefore, putting $u = \sin ax$, $du = a\cos ax\,dx$,

$$I = \frac{1}{a}\int u\,du = \frac{1}{2a}u^2 + C = \frac{1}{2a}\sin^2 ax + C$$

We note that this integral is identical to case 3 in Table 6.1, and this example demonstrates that there are often several ways of evaluating a particular integral.

(ii) $I = \displaystyle\int \frac{\ln x}{x} \, dx.$

Because $\dfrac{1}{x} = \dfrac{d}{dx} \ln x,$ we let $u = \ln x$ and $du = \dfrac{1}{x} dx.$ Then

$$I = \int u \, du = \frac{1}{2}u^2 + C = \frac{1}{2}(\ln x)^2 + C$$

▸ Exercises 17–20

EXAMPLES 6.6 Integrals of type 2: $\displaystyle\int \frac{f'(x)}{f(x)} \, dx.$

(i) $I = \displaystyle\int \frac{x}{2x^2 + 3} \, dx.$

In this case, x is proportional to the derivative of $2x^2 + 3$: $f(x) = 2x^2 + 3,$ $f'(x) = 4x.$ Therefore, putting $u = 2x^2 + 3$ and $du = 4x \, dx,$

$$I = \frac{1}{4}\int \frac{du}{u} = \frac{1}{4}\ln u + C = \frac{1}{4}\ln(2x^2 + 3) + C$$

(ii) $I = \displaystyle\int \cot x \, dx = \int \frac{\cos x}{\sin x} \, dx.$

Let $u = \sin x.$ Then $du = \cos x \, dx,$ and

$$I = \int \frac{du}{u} = \ln u + C = \ln(\sin x) + C$$

(iii) $I = \displaystyle\int \frac{1}{x \ln x} \, dx.$

Let $u = \ln x.$ Then $du = \dfrac{1}{x} dx,$ and

$$I = \int \frac{du}{u} = \ln u + C = \ln(\ln x) + C$$

▸ Exercises 21–26

EXAMPLES 6.7 Integrals of type 3: $\displaystyle\int f(\sin x)\cos x\, dx$.

(i) $\displaystyle I = \int \sin^a x \cos x\, dx$, where a is an arbitrary number (but $a \neq -1$).

Let $u = \sin x$. Then $du = \cos x\, dx$, and

$$I = \int u^a\, du = \frac{1}{a+1} u^{a+1} + C = \frac{1}{a+1} \sin^{a+1} x + C$$

(ii) $\displaystyle I = \int e^{\sin x} \cos x\, dx$.

Let $u = \sin x$. Then $du = \cos x\, dx$, and

$$I = \int e^u\, du = e^u + C = e^{\sin x} + C$$

> Exercises 27, 28

Trigonometric and hyperbolic substitutions

The standard integrals listed in Table 6.3 can be evaluated by substituting appropriate trigonometric and hyperbolic functions for the variable x.

Table 6.3

1. $\displaystyle\int \frac{dx}{\sqrt{a^2 - x^2}} = \sin^{-1}\left(\frac{x}{a}\right) + C,$		$a^2 > x^2$
2. $\displaystyle\int \frac{dx}{\sqrt{x^2 - a^2}} = \cosh^{-1}\left(\frac{x}{a}\right) + C = \ln\left[x + \sqrt{x^2 - a^2}\right] + C,$		$x^2 > a^2$
3. $\displaystyle\int \frac{dx}{\sqrt{x^2 + a^2}} = \sinh^{-1}\left(\frac{x}{a}\right) + C = \ln\left[x + \sqrt{x^2 + a^2}\right] + C$		
4. $\displaystyle\int \frac{dx}{a^2 + x^2} = \frac{1}{a}\tan^{-1}\left(\frac{x}{a}\right) + C$		
5. $\displaystyle\int \frac{dx}{a^2 - x^2} = \frac{1}{a}\tanh^{-1}\left(\frac{x}{a}\right) + C = \frac{1}{2a}\ln\left[\frac{a+x}{a-x}\right] + C,$		$a^2 > x^2$

Integral **1** in the table is evaluated by means of the substitution $x = a \sin \theta$. Then

$$dx = a \cos \theta \, d\theta \text{ and } \sqrt{a^2 - x^2} = \sqrt{a^2 - a^2 \sin^2 \theta} = a\sqrt{1 - \sin^2 \theta} = a \cos \theta.$$

Therefore

$$\int \frac{dx}{\sqrt{a^2 - x^2}} = \int \frac{a \cos \theta \, d\theta}{a \cos \theta} = \int d\theta = \theta + C = \sin^{-1}\left(\frac{x}{a}\right) + C$$

Similarly, integral **2** is evaluated by means of the substitution $x = a \cosh u$. Integral **4**, evaluated by means of the substitution $x = a \tan \theta$, is used in Section 6.6 for the integration of rational functions. Integral **5** can be evaluated either by means of the substitution $x = a \tanh u$ or by expressing the integrand in terms of partial fractions to give the logarithmic form (see Section 6.6). Alternatively, all the integrals in Table 6.3 are readily obtained by integrating the standard derivatives listed in Tables 4.5 and 4.6.

Such substitutions are useful when the integrand contains the square root of a quadratic function.

EXAMPLE 6.8 Evaluate $\int \sqrt{a^2 - x^2} \, dx$.

Let $x = a \sin \theta$. Then $dx = a \cos \theta \, d\theta$, $\sqrt{a^2 - x^2} = a \cos \theta$, and

$$\int \sqrt{a^2 - x^2} \, dx = a^2 \int \cos^2 \theta \, d\theta$$

This is integral **1** in Table 6.1. Therefore

$$\int \sqrt{a^2 - x^2} \, dx = \frac{a^2}{2} \int (1 + \cos 2\theta) \, d\theta = \frac{a^2}{2} (\theta + \sin \theta \cos \theta) + C$$

Now $\sin \theta = x/a$, $\cos \theta = \sqrt{a^2 - x^2}/a$, and $\theta = \sin^{-1}(x/a)$. Therefore,

$$\int \sqrt{a^2 - x^2} \, dx = \frac{1}{2} a^2 \sin^{-1}\left(\frac{x}{a}\right) + \frac{1}{2} x \sqrt{a^2 - x^2} + C$$

▸ Exercises 29–32

Definite integrals

When the variable of integration of a definite integral is changed, from x to $u(x)$ say, the limits of integration must also be changed. If the range of integration over x is from a to b then the range of integration over $u(x)$ is from $u(a)$ to $u(b)$:

$$\int_a^b f(x)\, dx = \int_{u(a)}^{u(b)} g(u)\, du \qquad (6.11)$$

EXAMPLE 6.9 Integrate $I = \displaystyle\int_0^\pi \cos^2 x \sin x\, dx$.

Substitute $u = \cos x$ and $du = -\sin x\, dx$. When $x = 0$, $u = \cos 0 = +1$; when $x = \pi$, $u = \cos \pi = -1$. Therefore

$$I = -\int_{+1}^{-1} u^2\, du = +\int_{-1}^{+1} u^2\, du$$

since interchanging the limits changes the sign of the integral. Then

$$\int_0^\pi \cos^2 x \sin x\, dx = \int_{-1}^{+1} u^2\, du = \left[\frac{u^3}{3}\right]_{-1}^{+1} = \frac{2}{3}$$

EXAMPLE 6.10 Find the area of the circle whose equation is $x^2 + y^2 = a^2$.

As in Example 5.11, let A be the area of that quarter of the circle that lies in the first quadrant, in which both x and y are positive (Figure 5.12). Then $y = \sqrt{a^2 - x^2}$ and

$$A = \int_0^a \sqrt{a^2 - x^2}\, dx$$

As in Example 6.8, the integral is evaluated by means of the substitution $x = a \sin \theta$, and the new integration limits are $\theta = 0$ when $x = 0$ and $\theta = \pi/2$ when $x = a$. Therefore,

$$A = a^2 \int_0^{\pi/2} \cos^2 \theta\, d\theta = \frac{a^2}{2}\left[\theta + \sin\theta \cos\theta\right]_0^{\pi/2} = \frac{\pi a^2}{4}$$

The area of the circle is four times this.

▸ Exercises 33–39

6.4   Integration by parts

In the integral

$$y = \int x \cos x \, dx$$

the integrand is the product of two quite different types of function; the polynomial x and the trigonometric function $\cos x$. The value of the integral is

$$y = x \sin x + \cos x + C$$

as can be verified by differentiation:

$$\frac{dy}{dx} = \frac{d}{dx}(x \sin x) + \frac{d}{dx}(\cos x)$$

$$= (\sin x + x \cos x) + (-\sin x) = x \cos x$$

The product rule has been used to *differentiate* the product $x \sin x$, and the method of integration by parts is used to *integrate* products of this type. In general, let $y = uv$, where u and v are functions of x. Then

$$\frac{dy}{dx} = u\frac{dv}{dx} + v\frac{du}{dx} \tag{6.12}$$

and, integrating both sides of this equation with respect to x,

$$\int \frac{dy}{dx} \, dx = \int u\frac{dv}{dx} \, dx + \int v\frac{du}{dx} \, dx \tag{6.13}$$

The left-hand side is equal to $y = uv$ by definition, and the equation can be rearranged as

$$\int u\frac{dv}{dx} \, dx = uv - \int v\frac{du}{dx} \, dx \tag{6.14}$$

This equation, the inverse of the product rule, is the rule of integration by parts. Given an integrand like $x \cos x$, one of the factors is identified with u in (6.14), the other with dv/dx. For example, let

$$u = x, \qquad \frac{dv}{dx} = \cos x$$

Then

$$\frac{du}{dx} = 1, \qquad v = \sin x$$

and (6.14) becomes

$$\int x \frac{d}{dx}(\sin x)\, dx = x \sin x - \int \sin x \frac{d}{dx}(x)\, dx$$

$$= x \sin x - \int \sin x\, dx$$

$$= x \sin x + \cos x + C$$

The art of integration by parts is the ability to make the correct choice of u and v. Thus, the choice of $u = \cos x$ and $dv/dx = x$ in our example gives

$$\int x \cos x\, dx = \frac{1}{2}x^2 \cos x - \int \left(\frac{1}{2}x^2\right) \times (-\sin x)\, dx$$

$$= \frac{1}{2}x^2 \cos x + \frac{1}{2}\int x^2 \sin x\, dx$$

and the problem has become *more* difficult. This demonstrates the rule that if one of the factors is a polynomial then, with only one important exception, the polynomial must be chosen as the function u in equation (6.14).

EXAMPLE 6.11 Integrate by parts $\int x^2 \cos x\, dx$.

Let $u = x^2$ and $\dfrac{dv}{dx} = \cos x$. Then $\dfrac{du}{dx} = 2x$ and $v = \sin x$, and

$$\int x^2 \cos x\, dx = x^2 \sin x - 2\int x \sin x\, dx$$

The degree of the polynomial under the integral sign has been decreased by one; x^2 has been replaced by x. Integrating the new integral by parts, with $u = x$ and $\dfrac{dv}{dx} = \sin x$, gives

$$\int x \sin x\, dx = -x \cos x + \int \cos x\, dx = -x \cos x + \sin x$$

Therefore

$$\int x^2 \cos x\, dx = x^2 \sin x - 2\left[-x \cos x + \sin x\right] + C$$

$$= x^2 \sin x + 2x \cos x - 2\sin x + C$$

The results may be verified by differentiation.

▸ Exercises 40–45

In general, a polynomial of degree n can be removed by n successive integrations by parts. The exception to the rule is when the other factor is a logarithmic function.

EXAMPLE 6.12 Integrate by parts $\int x^n \ln x \, dx, \, (n \neq -1)$.

In this case, choosing $u = x^n$ leads to a more complicated integral. The correct choice is $u = \ln x$ and $\dfrac{dv}{dx} = x^n$. Then $\dfrac{du}{dx} = \dfrac{1}{x}$ and $v = x^{n+1}/(n+1)$, and

$$\int x^n \ln x \, dx = \frac{1}{n+1} x^{n+1} \ln x - \frac{1}{n+1} \int x^{n+1} \times \frac{1}{x} \, dx$$

$$= \frac{1}{n+1} x^{n+1} \ln x - \frac{1}{n+1} \int x^n \, dx$$

$$= \frac{1}{n+1} x^{n+1} \ln x - \frac{1}{(n+1)^2} x^{n+1} + C$$

$$= \frac{1}{(n+1)^2} x^{n+1} \left[(n+1) \ln x - 1 \right] + C$$

A special case of this integral is $\int \ln x \, dx = x \ln x - x + C$

▸ Exercises 46–48

Integration by parts is straightforward only if one of the factors is a polynomial.

EXAMPLE 6.13 Integrate $\displaystyle\int_0^\infty e^{-ax} \cos x \, dx, \quad (a > 0)$.

In this case either factor can be chosen as u in (6.14); for example, if $u = e^{-ax}$ and $\dfrac{dv}{dx} = \cos x$ then, for the indefinite integral,

$$I = \int e^{-ax} \cos x \, dx = e^{-ax} \sin x + a \int e^{-ax} \sin x \, dx$$

$$= e^{-ax} \sin x + a \left[-e^{-ax} \cos x - a \int e^{-ax} \cos x \, dx \right]$$

$$= e^{-ax} \sin x - a e^{-ax} \cos x - a^2 \int e^{-ax} \cos x \, dx$$

$$= e^{-ax} \sin x - a e^{-ax} \cos x - a^2 I$$

Then, solving for I,

$$I = \int e^{-ax} \cos x \, dx = \frac{1}{1+a^2} e^{-ax} (\sin x - a \cos x) + C$$

and

$$\int_0^\infty e^{-ax} \cos x \, dx = \frac{a}{1+a^2}$$

> Exercises 49–51

6.5 Reduction formulas

The method of integration by parts can be used to derive formulas for families of related integrals. Consider the integral

$$I_n = \int x^n e^{ax} \, dx$$

where n is a positive integer. Choosing $u = x^n$ and $\dfrac{dv}{dx} = e^{ax}$ in the formula (6.14),

$$I_n = \frac{1}{a} x^n e^{ax} - \frac{n}{a} \int x^{n-1} e^{ax} \, dx$$

or

$$I_n = \frac{1}{a} x^n e^{ax} - \frac{n}{a} I_{n-1} \qquad (6.15)$$

This result is called a **reduction formula** for I_n, or a **recurrence relation** between I_n and I_{n-1}. For example

$$I_3 = \frac{1}{a} x^3 e^{ax} - \frac{3}{a} I_2, \qquad I_2 = \frac{1}{a} x^2 e^{ax} - \frac{2}{a} I_1, \qquad I_1 = \frac{1}{a} x e^{ax} - \frac{1}{a} I_0$$

where

$$I_0 = \int e^{ax} \, dx = \frac{1}{a} e^{ax} + C$$

Then

$$I_3 = \frac{1}{a} e^{ax} \left[x^3 - \frac{3}{a} x^2 + \frac{6}{a^2} x - \frac{6}{a^3} \right] + C$$

More important still, the recurrence relation is ideally suited for the computation of one or several members of a family of integrals.

Reduction formulas are particularly simple for some important definite integrals in the physical sciences. For example, the integral

$$I_n = \int_0^\infty e^{-ar} r^n \, dr \qquad (a > 0) \tag{6.16}$$

where n is a positive integer or zero, occurs in the quantum-mechanical description of the properties of the hydrogen atom. Integration by parts gives

$$I_n = \left[-\frac{1}{a} e^{-ar} r^n \right]_0^\infty + \frac{n}{a} I_{n-1}$$

When $n \neq 0$, the quantity $e^{-ar} r^n$ is zero at both integration limits, $r = 0$ and $r \to \infty$. Therefore

$$I_n = \frac{n}{a} I_{n-1}, \qquad n \geq 1$$

and it follows that

$$I_n = \int_0^\infty e^{-ar} r^n \, dr = \frac{n!}{a^{n+1}} \tag{6.17}$$

where $n! = n(n-1)(n-2) \cdots 1$ is the factorial of n.

EXAMPLE 6.14 Determine a reduction formula for $I_n = \int \cos^n x \, dx$ where n is a positive integer.

Write the integrand as $\cos^{n-1} x \cos x$, and let $u = \cos^{n-1} x$ and $\dfrac{dv}{dx} = \cos x$. Then $v = \sin x$, and

$$I_n = \cos^{n-1} x \sin x + (n-1) \int \cos^{n-2} x \sin^2 x \, dx$$

and, because $\sin^2 x = 1 - \cos^2 x$,

$$I_n = \cos^{n-1} x \sin x + (n-1) \int \cos^{n-2} x \, dx - (n-1) \int \cos^n x \, dx$$

$$= \cos^{n-1} x \sin x + (n-1) I_{n-2} - (n-1) I_n$$

Solving for I_n then gives

$$I_n = \frac{1}{n}\cos^{n-1} x \sin x + \frac{n-1}{n}I_{n-2}$$

‣ Exercises 52–54

EXAMPLE 6.15 Determine a reduction formula for the definite integral $I_n = \int_0^{\pi/2} \cos^n x\, dx.$

From the result of Example 6.14,

$$I_n = \left[\frac{1}{n}\cos^{n-1} x \sin x\right]_0^{\pi/2} + \frac{n-1}{n}I_{n-2}$$

Because $\sin 0 = 0$ and $\cos^{n-1}(\pi/2) = 0$ if $n > 1$, it follows that

$$I_n = \frac{n-1}{n}I_{n-2}, \qquad n \geq 2$$

with

$$I_1 = \int_0^{\pi/2} \cos x\, dx = 1, \qquad I_0 = \int_0^{\pi/2} dx = \frac{\pi}{2}$$

For example,

$$I_4 = \frac{3}{4}I_2 = \frac{3}{4}\times\frac{1}{2}I_0 = \frac{3\pi}{16}, \qquad I_5 = \frac{4}{5}I_3 = \frac{4}{5}\times\frac{2}{3}I_1 = \frac{8}{15}$$

‣ Exercises 55, 56

EXAMPLE 6.16 Determine a reduction formula for $I_n = \int_0^\infty e^{-ar^2} r^n\, dr.$

Because $\dfrac{d}{dr}e^{-ar^2} = -2ar\, e^{-ar^2}$, the integrand is written as $-\dfrac{1}{2a}\left[-2ar\, e^{-ar^2}\right]r^{n-1}.$

Then, by parts,

$$\int_0^\infty e^{-ar^2} r^n\, dr = \left[-\frac{1}{2a}e^{-ar^2} r^{n-1}\right]_0^\infty + \frac{(n-1)}{2a}\int_0^\infty e^{-ar^2} r^{n-2}\, dr$$

Now $e^{-ar^2} \to 0$ as $r \to \infty$ and, if $n > 1$, $r^{n-1} = 0$ when $r = 0$. Then, for $n \geq 2$,

$$\int_0^\infty e^{-ar^2} r^n \, dr = \frac{(n-1)}{2a} \int_0^\infty e^{-ar^2} r^{n-2} \, dr, \qquad I_n = \frac{(n-1)}{2a} I_{n-2}$$

When $n = 1$,

$$I_1 = \int_0^\infty e^{-ar^2} r \, dr = \left[-\frac{1}{2a} e^{-ar^2} \right]_0^\infty = \frac{1}{2a}$$

When $n = 0$,

$$I_0 = \int_0^\infty e^{-ar^2} \, dr = \frac{1}{2} \sqrt{\frac{\pi}{a}} \tag{6.18}$$

This last integral is a standard integral that cannot be evaluated by the methods described in this chapter (see Section 9.11).

The integrals I_n are important in several branches of chemistry. For example, in modern computational methods for the calculation of molecular wave functions, the molecular orbitals are expressed in terms of 'gaussian basis functions'. Such a function is essentially the exponential e^{-ar^2} multiplied by a polynomial, and the use of these functions leads to integrals of the type discussed in this example.

‣ Exercises 57–62

6.6 Rational integrands. The method of partial fractions

A rational algebraic function has the general form $P(x)/Q(x)$ where $P(x)$ and $Q(x)$ are polynomials:

$$\frac{P(x)}{Q(x)} = \frac{a_0 + a_1 x + a_2 x^2 + \cdots + a_n x^n}{b_0 + b_1 x + b_2 x^2 + \cdots + b_m x^m}$$

It was shown in Sections 2.6 and 2.7 that every such function can be expressed as the sum of a polynomial and one or more partial fractions of types (if complex numbers are excluded)

$$\text{(i)} \quad \frac{1}{(x+a)^n}, \qquad \text{(ii)} \quad \frac{ax+b}{(x^2 + px + q)^n} \tag{6.19}$$

where n is a positive integer and the quadratic $x^2 + px + q$ has no real roots; that is, its discriminant $p^2 - 4q$ is negative.

Integrals of type (i)

These are the integrals that occur in the theory of elementary kinetic processes:

$$\int \frac{dx}{(x+a)^n} = \begin{cases} \ln(x+a) + C & \text{if } n = 1 \\[2ex] \dfrac{-1}{(n-1)(x+a)^{n-1}} + C & \text{if } n > 1 \end{cases} \tag{6.20}$$

EXAMPLE 6.17 Integrate $\displaystyle\int \frac{dx}{(2-x)(4-x)}$.

The integrand can be expressed in terms of partial fractions as

$$\frac{1}{(2-x)(4-x)} = \frac{1}{2}\left(\frac{1}{2-x} - \frac{1}{4-x}\right)$$

Therefore

$$\int \frac{dx}{(2-x)(4-x)} = \frac{1}{2}\int\left(\frac{1}{2-x} - \frac{1}{4-x}\right)dx$$

$$= \frac{1}{2}\left[-\ln(2-x) + \ln(4-x)\right] + C = \frac{1}{2}\ln\left(\frac{4-x}{2-x}\right) + C$$

EXAMPLE 6.18 Integrate $\displaystyle\int \frac{5x+1}{x^3 - 3x + 2}\,dx$.

The cubic in the denominator can be factorized as $x^3 - 3x + 2 = (x-1)^2(x+2)$ so that the integrand can be expressed in terms of partial fractions as

$$\frac{5x+1}{x^3 - 3x + 2} = \frac{1}{x-1} + \frac{2}{(x-1)^2} - \frac{1}{x+2}$$

Then

$$\int \frac{5x+1}{x^3 - 3x + 2}\,dx = \int\frac{dx}{x-1} + 2\int\frac{dx}{(x-1)^2} - \int\frac{dx}{x+2}$$

$$= \ln(x-1) - \frac{2}{x-1} - \ln(x+2) + C$$

$$= \ln\left(\frac{x-1}{x+2}\right) - \frac{2}{x-1} + C$$

➤ Exercises 63–65

Integrals of type (ii)

We first consider two special forms.

The numerator is the derivative of the quadratic

In this case, the integral is either of type 2 in Table 6.2 or it is a simple generalization thereof:

$$\int \frac{2x+p}{(x^2 + px + q)^n}\,dx = \int \frac{f'(x)}{[f(x)]^n}\,dx \qquad (6.21)$$

If $u = f(x) = x^2 + px + q$ then $du = f'(x)\,dx = (2x + p)\,dx$, and

$$\int \frac{2x + p}{(x^2 + px + q)^n}\,dx = \int \frac{du}{u^n} = \begin{cases} \ln u + C & \text{if} \quad n = 1 \\ \dfrac{-1}{(n-1)u^{n-1}} + C & \text{if} \quad n > 1 \end{cases} \tag{6.22}$$

Therefore,

$$\int \frac{2x + p}{(x^2 + px + q)^n}\,dx = \begin{cases} \ln(x^2 + px + q) + C & \text{if} \quad n = 1 \\ \dfrac{-1}{(n-1)(x^2 + px + q)^{n-1}} + C & \text{if} \quad n > 1 \end{cases} \tag{6.23}$$

▸ Exercises 66, 67

The numerator is unity: $\displaystyle \int \frac{dx}{(x^2 + px + q)^n}$

The integral is evaluated by first transforming the quadratic in the denominator into the form $u^2 + a^2$. We write

$$x^2 + px + q = \left(x + \frac{p}{2} \right)^2 - \left(\frac{p^2 - 4q}{4} \right) \tag{6.24}$$

where $p^2 - 4q$ is the discriminant of the quadratic. Because it has been assumed that the quadratic has no real roots, the discriminant is negative, but $4q - p^2$ is positive:

$$x^2 + px + q = \left(x + \frac{p}{2} \right)^2 + \left(\frac{4q - p^2}{4} \right) = u^2 + a^2 \tag{6.25}$$

where $u = x + p/2$ and $a^2 = (4q - p^2)/4 > 0$. Then, because $dx = du$,

$$\int \frac{dx}{(x^2 + px + q)^n} = \int \frac{du}{(u^2 + a^2)^n}$$

We now use the trigonometric substitution $u = a \tan \theta$. Then $du = a \sec^2 \theta\,d\theta = (a/\cos^2 \theta)\,d\theta$, and $u^2 + a^2 = a^2(\tan^2 \theta + 1) = a^2/\cos^2 \theta$. Therefore

$$\int \frac{du}{(u^2 + a^2)^n} = \frac{1}{a^{2n-1}} \int \cos^{2n-2} \theta\,d\theta \tag{6.26}$$

When $n = 1$,

$$\int \frac{du}{u^2 + a^2} = \frac{1}{a} \int d\theta = \frac{\theta}{a} + C = \frac{1}{a} \tan^{-1}\left(\frac{u}{a} \right) + C \tag{6.27}$$

which is one of the standard integrals in Table 6.3. When $n > 1$, the integral (6.26) can be evaluated, for example, by reduction as in Example 6.14.

EXAMPLE 6.19 Integrate $\displaystyle\int \frac{dx}{x^2 + 2x + 5}$.

The quadratic function $x^2 + 2x + 5$ can be written as $(x+1)^2 + 2^2$. Then, by equation (6.27), with $a = 2$ and $u = x + 1$,

$$\int \frac{dx}{x^2 + 2x + 5} = \int \frac{dx}{(x+1)^2 + 2^2} = \frac{1}{2} \tan^{-1}\left(\frac{x+1}{2}\right) + C$$

➤ Exercise 68

EXAMPLE 6.20 Integrate $\displaystyle\int \frac{dx}{(x^2 + 2x + 5)^3}$.

By equation (6.26), with $n = 3$, $a = 2$, and $u = x + 1$,

$$\int \frac{dx}{(x^2 + 2x + 5)^3} = \frac{1}{32} \int \cos^4 \theta \, d\theta$$

where $\tan \theta = u/a = (x+1)/2$. Then, by reduction as in Example 6.14,

$$\int \cos^4 \theta \, d\theta = \frac{1}{4} \sin \theta \cos^3 \theta + \frac{3}{4} \int \cos^2 \theta \, d\theta$$

$$= \frac{1}{4} \sin \theta \cos^3 \theta + \frac{3}{8} \sin \theta \cos \theta + \frac{3}{8} \theta + C$$

We need to change variable from θ back to u (and then to x). If $\tan \theta = u/a$ then $\theta = \tan^{-1}(u/a)$ and it is readily verified that

$$\sin \theta = \frac{u}{\sqrt{u^2 + a^2}}, \qquad \cos \theta = \frac{a}{\sqrt{u^2 + a^2}}$$

Then

$$\int \cos^4 \theta \, d\theta = \frac{1}{4} \frac{ua^3}{(u^2 + a^2)^2} + \frac{3}{8} \frac{ua}{u^2 + a^2} + \frac{3}{8} \tan^{-1}\left(\frac{u}{a}\right) + C$$

and

$$\int \frac{dx}{(x^2 + 2x + 5)^3} = \frac{1}{32}\left[\frac{2(x+1)}{(x^2 + 2x + 5)^2} + \frac{3(x+1)}{4(x^2 + 2x + 5)} + \frac{3}{8} \tan^{-1}\left(\frac{x+1}{2}\right)\right] + C$$

➤ Exercise 69

The general form: $\displaystyle\int \frac{ax+b}{(x^2+px+q)^n}\,dx$

The numerator can be written as

$$ax+b=\frac{a}{2}\left(2x+p\right)+\left(b-\frac{ap}{2}\right) \qquad (6.28)$$

so that the integral is expressed in terms of the special cases discussed above:

$$\int \frac{ax+b}{(x^2+px+q)^n}\,dx=\frac{a}{2}\int\frac{2x+p}{(x^2+px+q)^n}\,dx$$

$$+\left(b-\frac{ap}{2}\right)\int\frac{dx}{(x^2+px+q)^n} \qquad (6.29)$$

▶ Exercises 70, 71

Rational trigonometric integrands

By trigonometry,

$$\sin\theta=2\sin\frac{\theta}{2}\cos\frac{\theta}{2}=\frac{2\sin\frac{\theta}{2}\cos\frac{\theta}{2}}{\cos^2\frac{\theta}{2}+\sin^2\frac{\theta}{2}}$$

$$\qquad (6.30)$$

$$\cos\theta=\cos^2\frac{\theta}{2}-\sin^2\frac{\theta}{2}=\frac{\cos^2\frac{\theta}{2}-\sin^2\frac{\theta}{2}}{\cos^2\frac{\theta}{2}+\sin^2\frac{\theta}{2}}$$

Then, dividing the numerators and denominators by $\cos^2\theta/2$ and putting $t=\tan\theta/2$, we obtain

$$\sin\theta=\frac{2t}{1+t^2}, \qquad \cos\theta=\frac{1-t^2}{1+t^2}, \qquad t=\tan\frac{\theta}{2} \qquad (6.31)$$

The trigonometric functions $\sin\theta$ and $\cos\theta$ are rational functions of t. A trigonometric function of θ that becomes a rational (algebraic) function of t when we make the substitution $t=\tan\theta/2$ is called **a rational trigonometric function** of θ. Every such function can be integrated by the methods described earlier in this section. Thus, if the integrand in the integral $\int f(\theta)\,d\theta$ is a rational trigonometric function of θ, the substitution

$$t=\tan\frac{\theta}{2}, \qquad d\theta=\frac{2}{1+t^2}\,dt \qquad (6.32)$$

gives

$$\int f(\theta)\, d\theta = \int \frac{2f(\theta)}{1+t^2}\, dt \qquad\qquad (6.33)$$

where $f(\theta)$ in terms of t is a rational function of t.

▸ Exercise 72

EXAMPLE 6.21 Examples of the substitution $t = \tan\theta/2$.

(i) $\displaystyle\int \frac{d\theta}{\sin\theta} = \int \left(\frac{2}{1+t^2}\right)\Big/\left(\frac{2t}{1+t^2}\right) dt = \int \frac{1}{t}\, dt = \ln t + C = \ln\tan\theta/2 + C$

(ii) $\displaystyle\int \frac{d\theta}{3+5\cos\theta} = \int \left(\frac{2}{1+t^2}\right)\Big/\left(3+5\frac{1-t^2}{1+t^2}\right) dt = \int \frac{2}{8-2t^2}\, dt$

$$= \int \frac{dt}{4-t^2} = \frac{1}{4}\ln\left[\frac{2+t}{2-t}\right] + C$$

$$= \frac{1}{4}\ln\left[\frac{2+\tan\theta/2}{2-\tan\theta/2}\right] + C$$

making use of standard integral 5 in Table 6.3.

▸ Exercises 73–75

This method can be applied in all cases but is not always the simplest in practice. For example, the application of the method to the integration of the elementary trigonometric functions $\sin\theta$ and $\cos\theta$ is considerable more complicated than the use of the standard integrals.

6.7 Parametric differentiation of integrals

Consider the indefinite integral

$$\int e^{-\alpha x}\, dx = -\frac{1}{\alpha} e^{-\alpha x} + C, \qquad (\alpha \neq 0)$$

The integral can be treated as a function of the parameter α; differentiation then gives

$$\frac{d}{d\alpha}\int e^{-\alpha x}\, dx = \frac{d}{d\alpha}\left(-\frac{1}{\alpha} e^{-\alpha x} + C\right) = \left(\frac{1}{\alpha^2} + \frac{x}{\alpha}\right) e^{-\alpha x}$$

In addition, by integration by parts,

$$\int \left(\frac{d}{d\alpha} e^{-\alpha x} \right) dx = -\int xe^{-\alpha x} \ dx = \left(\frac{1}{\alpha^2} + \frac{x}{\alpha} \right) e^{-\alpha x}$$

It follows that

$$\frac{d}{d\alpha} \int e^{-\alpha x} \ dx = \int \left(\frac{d}{d\alpha} e^{-\alpha x} \right) dx$$

so that the order of integration with respect to x and differentiation with respect to α can be interchanged in this case. This result is true in the general case

$$\int f(x, \alpha) \ dx = F(x, \alpha) + C \tag{6.34}$$

if $f(x, \alpha)$ and $\dfrac{d}{d\alpha} f(x, \alpha)$ are continuous functions of x and α:

$$\frac{d}{d\alpha} \int f(x, \alpha) dx = \int \left(\frac{d}{d\alpha} f(x, \alpha) \right) dx = \frac{d}{d\alpha} F(x, \alpha) \tag{6.35}$$

For the corresponding definite integral, if the limits are independent of the parameter α,

$$\frac{d}{d\alpha} \int_a^b f(x, \alpha) \ dx = \int_a^b \left(\frac{d}{d\alpha} f(x, \alpha) \right) dx = \frac{d}{d\alpha} F(b, \alpha) - \frac{d}{d\alpha} F(a, \alpha) \tag{6.36}$$

When one or both of the limits of integration are infinite, it is necessary to ensure that the integral of the function and that of its derivative are both convergent.

EXAMPLE 6.22 Integrate $\displaystyle\int_0^\infty e^{-\alpha x} x^n \ dx.$

The integral was evaluated in Section 6.5 from a reduction formula derived by successive integrations by parts. An alternative method is to differentiate the simple standard integral

$$\int_0^\infty e^{-\alpha x} \ dx = \frac{1}{a}, \quad (a \neq 0)$$

The nth derivative of $e^{-\alpha x}$ with respect to a is $(-1)^n x^n e^{-\alpha x}$ so that

$$\int_0^\infty x^n e^{-\alpha x} \ dx = (-1)^n \frac{d^n}{da^n} \int_0^\infty e^{-\alpha x} dx = (-1)^n \frac{d^n}{da^n} \left(\frac{1}{a} \right) = \frac{n!}{a^{n+1}}$$

EXAMPLE 6.23 Integrate $\displaystyle\int_0^\infty \frac{\sin x}{x}\,dx.$

The integral cannot be evaluated by the standard methods discussed in this chapter; the method of integration by parts does not work. We consider instead the related integral

$$I(\alpha) = \int_0^\infty \frac{e^{-\alpha x}\sin x}{x}\,dx$$

Then

$$\frac{d}{d\alpha} I(\alpha) = -\int_0^\infty e^{-\alpha x}\sin x\,dx$$

and the new integral can be integrated by parts as in Example 6.13. Then

$$\frac{d}{d\alpha} I(\alpha) = -\frac{1}{1+\alpha^2} \tag{6.37}$$

and, integrating with respect to α,

$$I(\alpha) = -\int \frac{d\alpha}{1+\alpha^2} = -\tan^{-1}\alpha + C$$

To obtain the value of the constant of integration, we note that $I(\alpha) \to 0$ as $\alpha \to \infty$, so that

$$C = \lim_{\alpha \to \infty} \tan^{-1}\alpha = \frac{\pi}{2}$$

To retrieve the original integral we now set $\alpha = 0$:

$$I(0) = \int_0^\infty \frac{\sin x}{x}\,dx = -\tan^{-1}0 + \frac{\pi}{2} = \frac{\pi}{2} \tag{6.38}$$

➤ Exercise 76

When the limits of integration also depend on the parameter, the result of differentiating the integral is given by Leibniz's theorem: if $a(\alpha)$ and $b(\alpha)$ are continuous functions of α,

$$\frac{d}{d\alpha}\int_{a(\alpha)}^{b(\alpha)} f(x,\alpha)\,dx = \int_{a(\alpha)}^{b(\alpha)}\left(\frac{d}{d\alpha}f(x,\alpha)\right)dx + f(b,\alpha)\frac{db}{d\alpha} - f(a,\alpha)\frac{da}{d\alpha} \tag{6.39}$$

6.8 Exercises

Section 6.2

Evaluate the indefinite integrals:

1. $\int \sin^2 3x \, dx$

2. $\int \sin 3x \cos 3x \, dx$

3. $\int \sin 3x \cos 2x \, dx$

4. $\int \sin x \cos 3x \, dx$

5. $\int \sin 3x \sin x \, dx$

6. $\int \cos 5x \cos 2x \, dx$

Evaluate the definite integrals:

7. $\int_0^{\pi/2} \cos^2 3x \, dx$

8. $\int_0^{\pi/2} \sin 2x \cos 2x \, dx$

9. $\int_0^{\pi} \sin x \cos 2x \, dx$

10. The wave functions for a particle in a box of length l are

$$\psi_n(x) = \sqrt{\frac{2}{l}} \sin\left(\frac{n\pi x}{l}\right), \qquad n = 1, 2, 3,$$

Show that the functions satisfy the orthonormality conditions

$$\int_0^l \psi_n \psi_m \, dx = \begin{cases} 1 & \text{if } m = n \\ 0 & \text{if } m \neq n \end{cases}$$

Section 6.3

Evaluate the indefinite integrals (use the substitutions in parentheses, when given):

11. $\int (3x + 1)^5 \, dx \quad (u = 3x + 1)$

12. $\int (2x - 1)^{1/2} \, dx$

13. $\int (3x^2 + 2x + 5)^3 (3x + 1) \, dx \quad (u = 3x^2 + 2x + 5)$

14. $\int (2x^3 + 3x - 1)^{1/3}(2x^2 + 1) \, dx$

15. $\int (3x^2 + 2)e^{-(x^3 + 2x)} \, dx \quad (u = x^3 + 2x)$

16. $\int (1 - x)e^{4x - 2x^2} \, dx$

17. $\int x\sqrt{4 - x^2} \, dx \quad (u = 4 - x^2)$

18. $\int \cos x \, e^{2 \sin x} \, dx$

19. $\int e^x (1 + e^x)^{1/2} \, dx \quad (u = 1 + e^x)$

20. $\int x \cos(3x^2 - 1) \, dx$

21. $\int \frac{2x + 1}{x^2 + x + 2} \, dx \quad (u = x^2 + x + 2)$

22. $\int \frac{3x^2 - x}{2x^3 - x^2 + 3} \, dx$

23. $\int \frac{\cos x}{1 - \sin x} \, dx \quad (u = 1 - \sin x)$

24. $\int \tan x \, dx$

25. $\int \frac{x}{\sqrt{4 - x^2}} \, dx$

26. $\int \frac{\tan x}{\ln(\cos x)} \, dx$

27. $\int \sin^3 x \cos x \, dx \quad (u = \sin x)$

28. $\int \ln(\cos x) \sin x \, dx$

29. $\int \dfrac{dx}{4 + x^2}$

30. $\int \dfrac{x^2 dx}{\sqrt{1 - x^2}} \quad (x = \sin \theta)$

31 $\int \dfrac{\sqrt{x}}{1 + x} \, dx \quad (u = \sqrt{x})$

32. **(i)** Use the substitution $x = a \sinh u$ to show that $\int \dfrac{dx}{\sqrt{x^2 + a^2}} = \sinh^{-1}\left(\dfrac{x}{a}\right) + C$.

(ii) Use the substitution $u = x + \sqrt{x^2 + a^2}$ to show that $\int \dfrac{dx}{\sqrt{x^2 + a^2}} = \ln\left[x + \sqrt{x^2 + a^2}\right] + C$.

Evaluate the definite integrals:

33. $\displaystyle\int_1^2 \dfrac{x \, dx}{3x^2 - 2}$

34. $\displaystyle\int_0^{\pi^2} \dfrac{\sin(\sqrt{x} + \pi)}{\sqrt{x}} \, dx$

35. $\displaystyle\int_0^{\pi/2} \sqrt{\sin \theta} \cos \theta \, d\theta$

36. $\displaystyle\int_0^1 \dfrac{dx}{\sqrt{2 - x^2}}$

37. $\displaystyle\int_0^\infty x e^{-x^2} \, dx$

38. Line shapes in magnetic resonance spectroscopy are often described by the Lorentz function

$$g(\omega) = \dfrac{1}{\pi} \dfrac{T}{1 + T^2(\omega - \omega_0)^2}.$$

Find $\displaystyle\int_{\omega_0}^\infty g(\omega) \, d\omega.$

39. An approximate expression for the rotational partition function of a linear rotor is

$$q_r = \int_0^\infty (2J + 1)e^{-J(J+1)\theta_R/T} \, dJ$$

where $\theta_R = \hbar^2/2Ik$ is the rotational temperature, I is the moment of inertia, and k is Boltzmann's constant. Evaluate the integral.

Section 6.4

Evaluate the integrals:

40. $\int x \sin x \, dx$

41. $\int x^3 \sin x \, dx$

42. $\int (x + 1)^2 \cos 2x \, dx$

43. $\int x^2 e^{2x} \, dx$

44. $\int_0^1 x e^x \, dx$

45. $\int_0^\infty x^2 e^{-2x} \, dx$

46. $\int x \ln x \, dx$

47. $\int \dfrac{\ln x}{x^2} \, dx$

48. $\int_0^1 x^2 \ln x \, dx$

49. $\int e^{-x} \sin 2x \, dx$

50. $\int e^{ax} \cos bx \, dx$

51. $\int_0^{\pi/2} e^{-2x} \cos 3x \, dx$

Section 6.5

52. Determine a reduction formula for $\int \sin^n x\, dx$, where n is a positive integer.

53. Show that, for integers $m \geq 0$ and $n \geq 1$,

$$\int \sin^m \theta \cos^n \theta\, d\theta = \frac{\sin^{m+1}\theta \cos^{n-1}\theta}{m+n} + \frac{n-1}{m+n}\int \sin^m \theta \cos^{n-2}\theta\, d\theta$$

54. Use the results of Exercises **52** and **53** to evaluate $\int \sin^5 x \cos^4 x\, dx$.

55. Show that, for integers $m \geq 0$ and $n > 1$,

$$\int_0^{\pi/2} \sin^m \theta \cos^n \theta\, d\theta = \frac{n-1}{m+n}\int_0^{\pi/2} \sin^m \theta \cos^{n-2}\theta\, d\theta$$

Evaluate

56. $\displaystyle\int_0^{\pi/2} \sin^5 x \cos^5 x\, dx.$

57. $\displaystyle\int_0^{\infty} r\, e^{-2r^2}\, dr$

58. $\displaystyle\int_0^{\infty} r^2\, e^{-2r^2}\, dr$

59. $\displaystyle\int_0^{\infty} r^3\, e^{-2r^2}\, dr$

60. The probability that a molecule of mass m in a gas at temperature T has speed v is given by the Maxwell–Boltzmann distribution

$$f(v) = 4\pi\left(\frac{m}{2\pi kT}\right)^{3/2} v^2 e^{-mv^2/2kT}$$

where k is Boltzmann's constant. Find the average speed $\bar{v} = \displaystyle\int_0^{\infty} v f(v)\, dv.$

61. For the Maxwell–Boltzmann distribution in Exercise **60**, find the root mean square speed $\sqrt{\overline{v^2}}$, where $\overline{v^2} = \displaystyle\int_0^{\infty} v^2 f(v)\, dv.$

62. Line shapes in spectroscopy are sometimes analysed in terms of second moments. The second moment of a signal centred at angular frequency ω_0 is

$$\int_{\omega_0}^{\infty} (\omega - \omega_0)^2 g(\omega)\, d\omega$$

where $g(\omega)$ is a shape function for the signal. Evaluate the integral for the gaussian curve

$$g(\omega) = \sqrt{\frac{2}{\pi}}\, T \exp\left[-\frac{1}{2}T^2(\omega - \omega_0)^2\right]$$

Section 6.6

Evaluate the indefinite integrals:

63. $\displaystyle\int \frac{dx}{(2x-1)(x+3)}$

64. $\displaystyle\int \frac{(x+2)}{(x+3)(x+4)}\, dx$

65. $\displaystyle\int \frac{(x^2 - 3x + 3)}{(x+1)(x+2)(x+3)}\, dx$

66. $\displaystyle\int \frac{x+2}{x^2+4x+5}\, dx$ **67.** $\displaystyle\int \frac{x}{(x^2+3)(x^2+4)}\, dx$ **68.** $\displaystyle\int \frac{dx}{x^2+4x+5}$

69. $\displaystyle\int \frac{dx}{(x^2+4x+5)^2}$ **70.** $\displaystyle\int \frac{x}{x^2+4x+5}\, dx$ **71.** $\displaystyle\int \frac{4x+3}{(x^2+4x+5)^2}\, dx$

72. If $t = \tan\dfrac{\theta}{2}$, show that $d\theta = \dfrac{2}{1+t^2}\, dt$ (Equation (6.33))

Evaluate by means of the substitution $t = \tan\theta/2$:

73. $\displaystyle\int \frac{d\theta}{\cos\theta}$ **74.** $\displaystyle\int \frac{d\theta}{5-3\cos\theta}$ **75.** $\displaystyle\int \frac{d\theta}{1+\sin\theta+\cos\theta}$

Section 6.7

76. By differentiation of the integral

$$\int_0^\infty e^{-ax^2}\, dx = \frac{1}{2}\sqrt{\frac{\pi}{a}}$$

with respect to a, show that

$$\int_0^\infty x^{2n} e^{-ax^2}\, dx = \frac{1\cdot 3\cdot 5 \quad (2n-1)}{2^{n+1} a^n}\sqrt{\frac{\pi}{a}}$$

7 Sequences and series

7.1 Concepts

A series is a set of terms that is to be summed. The terms can be numbers, variables, functions, or more complex quantities. A series can be **finite**, containing a finite number of terms,

$$u_1 + u_2 + u_3 + \cdots + u_n$$

or it can be **infinite**,

$$u_1 + u_2 + u_3 + u_4 + \cdots$$

where the dots mean that the sum is to be extended indefinitely (*ad infinitum*). The terms themselves form a **sequence**. Sequences are discussed in Section 7.2, finite series in 7.3, infinite series and tests of convergence in 7.4 and 7.5. In Section 7.6 we discuss how the MacLaurin and Taylor series can be used to represent certain types of function as power series ('infinite polynomials'), and in 7.7 how they are used to obtain approximate values of functions. Some properties of power series are described in Section 7.8.

Series occur in all branches of the physical sciences, and the representation of functions as series is an essential tool for the solution of many physical problems. Some functions, such as the exponential function and other transcendental functions, are defined as series, as are some important physical quantities; for example, the partition function in statistical thermodynamics. We will see in Chapters 12–14 that the differential equations that are important in the physical sciences often have solutions that can only be represented as series. Approximate and numerical methods of solution of problems are often based on series. For example, solutions of the Schrödinger equation are often represented as series, both in formal theory and in approximate methods such as the method of 'linear combination of atomic orbitals' in molecular-orbital theory (LCAO-MO). An important application of series is in the analysis of wave forms in terms of Fourier series and Fourier transforms; Fourier analysis is discussed in Chapter 15.

7.2 Sequences

A sequence is an ordered set of terms

$$u_1, \ u_2, \ u_3, \ \ldots$$

with a rule that specifies each term. For example, the numbers

$$1, \ 3, \ 5, \ 7, \ \ldots$$

form a sequence defined by the general term

$$u_r = 1 + 2(r - 1), \qquad r = 1, 2, 3, \ldots$$

Alternatively, the rule can be expressed as a recurrence relation plus an initial term:

$$u_{r+1} = u_r + 2, \qquad u_1 = 1$$

so that, for example, $u_5 = u_4 + 2 = 7 + 2 = 9$. This sequence is an example of the **arithmetic progression**

$$a, \ a + d, \ a + 2d, \ a + 3d, \ \ldots \tag{7.1}$$

with rule

$$u_{r+1} = u_r + d, \quad u_1 = a \qquad \text{or} \qquad u_r = a + (r - 1)d, \quad r = 1, 2, 3, \ldots \tag{7.2}$$

Another simple, but important, sequence is the **geometric progression**

$$a, \ ax, \ ax^2, \ ax^3, \ \ldots \tag{7.3}$$

with rule

$$u_{r+1} = xu_r, \quad u_1 = a \qquad \text{or} \qquad u_r = ax^{r-1}, \quad r = 1, 2, 3 \ldots \tag{7.4}$$

A sequence of terms u_r is denoted by $\{u_r\}$.

EXAMPLES 7.1 Sequences

(i) Arithmetic progression:

    $0, \ 5, \ 10, \ 15, \ \ldots$ $u_{r+1} = u_r + 5, \quad u_1 = 0$

    $1, \ -1, \ -3, \ -5, \ \ldots$ $u_r = 1 - 2(r - 1), \quad r = 1, 2, 3, \ldots$

(ii) Geometric progression:[1]

    $1, \ 7, \ 49, \ 343, \ \ldots$ $u_{r+1} = 7u_r, \quad u_1 = 1$

    $1, \ \dfrac{1}{2}, \ \dfrac{1}{4}, \ \dfrac{1}{8}, \ \ldots$ $u_r = \dfrac{1}{2^r}, \quad r = 0, 1, 2, 3, \ldots$

    $1, \ -\dfrac{1}{3}, \ \dfrac{1}{9}, \ -\dfrac{1}{27}, \ \ldots$ $u_{r+1} = -\dfrac{1}{3}u_r, \quad u_1 = 1$

---

[1] The Rhind papyrus (*c.* 1650 BC) contains a problem concerning '7 houses, 49 cats, 343 mice, 2401 ears of grain, 16807 hekats'. The version of the 'St. Ives problem' in Fibonacci's *Liber abaci* (1202 AD) is '7 old women went to Rome; each woman had 7 mules and each mule carried 7 sacks; each sack contained 7 loaves; with each loaf were 7 knives, and each knife was put up in 7 sheaths'.

(iii) Harmonic sequence:

$$1, \ \frac{1}{2}, \ \frac{1}{3}, \ \frac{1}{4}, \ \ldots \qquad\qquad u_r = \frac{1}{r}, \quad r = 1, 2, 3, \ldots$$

(iv) Fibonacci sequence ('series'):[2]

$$1, \ 1, \ 2, \ 3, \ 5, \ 8, \ 13, \ \ldots \qquad u_{r+2} = u_{r+1} + u_r, \quad u_0 = u_1 = 1$$

▸ Exercises 1–10

Limits of sequences

The terms of the harmonic sequence $\left\{\dfrac{1}{r}\right\}$,

$$1, \ \frac{1}{2}, \ \frac{1}{3}, \ \frac{1}{4}, \ \ldots$$

decrease in magnitude as r increases, and approach the value zero as r tends to infinity. The quantity

$$\lim_{r \to \infty}\left(\frac{1}{r}\right) = 0$$

is called the **limit** of the sequence, and in this case the limit is finite and unique. Similarly, the sequence

$$\frac{1}{2}, \ \frac{2}{3}, \ \frac{3}{4}, \ \frac{4}{5}, \ \ldots$$

has general term $u_r = (r-1)/r$ and limit $\lim_{r \to \infty}\left[(r-1)/r\right] = 1$. Thus,

$$u_{10} = 0.9, \qquad u_{100} = 0.99, \qquad u_{1000} = 0.999, \qquad \ldots$$

and when $r = 10^n$, the term u_r has n 9's after the decimal point. When the limit u of a sequence is finite and unique the sequence is said to **converge** to the limit u. When the

---

[2] This is the solution to the *paria coniculorum*, or rabbit problem, given in Fibonacci's *Liber abaci*: 'how many pairs of rabbits can be bred from one pair in one year if each pair breeds one other pair every month, and they begin to breed in the second month after birth?' The Fibonacci sequence has been linked with various patterns of growth and behaviour in nature.

limit is not finite or not unique the sequence is **divergent**. For example, the arithmetic progression (7.1) is necessarily divergent for all values of d:

$$\lim_{r \to \infty} \left[a + d(r-1) \right] = \pm\infty$$

The geometric progression (7.3) is interesting because it is convergent for some values of x and divergent for others. It shows the six possible types of behaviour, characteristic of many sequences, illustrated in Figure 7.1.

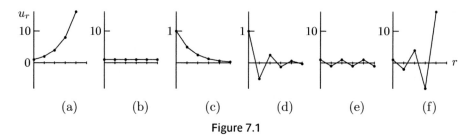

(a)   (b)   (c)   (d)   (e)   (f)

Figure 7.1

The examples shown are (for $a = 1$):

(a) $x > 1$ 1, 2, 4, 8, ...

(b) $x = 1$ 1, 1, 1, 1, ...

(c) $0 < x < 1$ $1, \dfrac{1}{2}, \dfrac{1}{4}, \dfrac{1}{8}, \ldots$

(d) $-1 < x < 0$ $1, -\dfrac{1}{2}, \dfrac{1}{4}, -\dfrac{1}{8}, \ldots$

(e) $x = -1$ 1, −1, 1, −1, ...

(f) $x < -1$ 1, −2, 4, −8, ...

The sequence is convergent only for $x = +1$ and $|x| < 1$, types (b), (c), and (d).

The behaviour of a sequence in the limit does not necessarily depend on the behaviour of a finite part of the sequence. An important example is the sequence of the terms $\{x^r/r!\}$ in the expansion of the exponential function (Section 3.6). This sequence has limit zero for *all* values of x; thus, the ratio of consecutive terms is

$$\left(\frac{x^{r+1}}{(r+1)!} \right) \bigg/ \left(\frac{x^r}{r!} \right) = \frac{x}{r+1} \to 0 \ \text{ as } r \to \infty \text{ for all } x$$

but the terms increase in magnitude when $r + 1 < |x|$, decrease when $r + 1 > |x|$. For example,

$$e^3 = 1 + \frac{3}{1!} + \frac{3^2}{2!} + \frac{3^3}{3!} + \frac{3^4}{4!} + \frac{3^5}{5!} + \cdots = 1 + 3 + 4.5 + 4.5 + 3.375 + 2.025 + \cdots$$

EXAMPLES 7.2 Limits $\lim_{r\to\infty} (u_r)$.

(i) $u_r = \dfrac{r-2}{2r}$

We have $u_r = \dfrac{1}{2} - \dfrac{1}{r}$ so that $u_r \to \dfrac{1}{2}$ as $r \to \infty$.

(ii) $u_r = \dfrac{2r^2 + 2r + 1}{r^2 - r + 1}$

Dividing top and bottom by r^2, $u_r = \dfrac{2 + 2/r + 1/r^2}{1 - 1/r + 1/r^2} \to 2$ as $r \to \infty$

(iii) $u_r = \sin\dfrac{1}{r} \to \sin 0 = 0$ as $r \to \infty$

(iv) The Fibonacci sequence (Example 7.1(iv)) is divergent, but a convergent sequence is obtained from the ratios of consecutive terms, u_{r+1}/u_r. The first 10 terms are 1, 2, 1.5, 1.6666, 1.6, 1.6350, 1.6153, 1.6190. 1.6176, 1.6181, and the limit of the sequence is

$$\lim_{r\to\infty} \left(u_{r+1}/u_r\right) = \phi = (1 + \sqrt{5})/2 \approx 1.618034.$$

Thus, dividing the recurrence relation $u_{r+2} = u_{r+1} + u_r$ by u_{r+1},

$$u_{r+2}/u_{r+1} = 1 + u_r/u_{r+1}$$

and taking the limit $r \to \infty$ gives $\phi = 1 + 1/\phi$. Then $\phi^2 - \phi - 1 = 0$, with positive solution $\phi = (1 + \sqrt{5})/2$. This is identical to the quantity known in geometry as the 'golden section' (ratio).[3]

➤ Exercises 11–17

---

[3] The golden section was known to the Greeks as 'the division of a straight line in extreme and mean ratio' (Euclid, 'The elements', Book II, Proposition 11, Book VI, Proposition 30). Its present name originated in 15th Century Italy when it was taken up by artists as a 'divine proportion' and used in painting and architecture. Luca Pacioli (1445–1517) wrote *De divina proportione* (1509), with illustrations thought to be by Leonardo da Vinci. Pacioli's *Summa de arithmetica, geometrica, proportioni et proportionalita* (Venice, 1494) was one of the first comprehensive mathematics book to be printed; it contained the first published description of double-entry bookkeeping. In his 'Lives of the artists', Giorgio Vasari (1511–1574) accuses Pacioli of plagiarising the mathematical works of Piero della Francesca (*c.* 1420–1492).

7.3   Finite series

Given a sequence $u_1, u_2, u_3, \ldots$ the partial sums

$$S_1 = u_1$$
$$S_2 = u_1 + u_2$$
$$S_3 = u_1 + u_2 + u_3$$
$$\cdots$$

with general term

$$S_n = u_1 + u_2 + u_3 + \cdots + u_n = \sum_{r=1}^{n} u_r \tag{7.5}$$

also form a sequence, $S_1, S_2, S_3, \ldots$ A sequence obtained in this way is called a **series** and, when the sequence converges, the limit

$$S = \lim_{n \to \infty} S_n = \lim_{n \to \infty} \left(\sum_{r=1}^{n} u_r \right) \tag{7.6}$$

is called the **sum of the series**.

The word series is commonly also used for the sums of terms themselves. **A finite series** of n terms is then

$$\sum_{r=1}^{n} u_r = u_1 + u_2 + u_3 + \cdots + u_n$$

with sum S_n, and the **infinite series** is

$$\sum_{r=1}^{\infty} u_r = u_1 + u_2 + u_3 + \cdots$$

with sum (if it exists) given by (7.6).

The arithmetic series

The sum of the first n terms of the arithmetic progression (7.1) is

$$S_n = \sum_{r=1}^{n} [a + (r-1)d] = a + [a+d] + [a+2d] + \cdots + [a + (n-1)d]$$

The value of the finite series is obtained by considering the sum in reverse order,

$$S_n = [a + (n-1)d] + [a + (n-2)d] + \cdots + a$$

Then, addition of the two forms, term by term, gives

$$2S_n = [2a + (n-1)d] + [2a + (n-1)d] + \cdots + [2a + (n-1)d]$$
$$= n[2a + (n-1)d]$$

so that

$$S_n = \frac{n}{2}[2a + (n-1)d] \tag{7.7}$$

In particular, the sum of the first n natural numbers is $(a = d = 1)$

$$S_n = 1 + 2 + 3 + \cdots + n = \frac{1}{2}n(n+1) \tag{7.8}$$

▸ Exercises 18, 19

The geometric series

The sum of the first n terms of the geometric progression (7.3) is

$$S_n = \sum_{r=0}^{n-1} ax^r = a + ax + ax^2 + ax^3 + \cdots + ax^{n-1}$$

To obtain the value of the series, multiply by x,

$$xS_n = ax + ax^2 + ax^3 + ax^4 + \cdots + ax^n$$

and subtract the two series term by term:

$$S_n - xS_n = a - ax^n = a(1 - x^n)$$

Therefore[4]

$$S_n = a\left(\frac{1 - x^n}{1 - x}\right), \qquad (x \neq 1) \tag{7.9}$$

Then, for $a = 1$,

$$\frac{1 - x^n}{1 - x} = 1 + x + x^2 + x^3 + \cdots + x^{n-1} \tag{7.10}$$

---

[4] A discussion of the sum of the geometric series is given in Euclid's 'Elements', Book IX, Proposition 35.

This equation can be regarded in two ways:

(i) the value of the sum $(1+x+x^2+\cdots+x^{n-1})$ is $(1-x^n)/(1-x)$,
(ii) the series $(1+x+x^2+\cdots+x^{n-1})$ is the **expansion** of the function $(1-x^n)/(1-x)$ in powers of x.

This concept of the expansion of one function in terms of a set of (other) functions provides an important tool for the representation of complicated (or unknown) functions in the physical sciences.

▸ Exercises 20–23

The binomial expansion

The binomial expansion is the expansion of the function $(1+x)^n$ in powers of x when n is a positive integer. Examples of such expansions are

$$(1+x)^2 = 1+2x+x^2$$

$$(1+x)^3 = 1+3x+3x^2+x^3$$

$$(1+x)^4 = 1+4x+6x^2+4x^3+x^4$$

In the general case,

$$(1+x)^n = 1+nx+\frac{n(n-1)}{2!}x^2+\frac{n(n-1)(n-2)}{3!}x^3+\cdots+x^n \qquad (7.11)$$

with general term

$$\frac{n(n-1)(n-2)...(n-r+1)}{r!}x^r$$

EXAMPLE 7.3 Expand $(1+x)^6$ in powers of x.

By equation (7.11), with $n=6$,

$$(1+x)^6 = 1+6x+\frac{6\times5}{2\times1}x^2+\frac{6\times5\times4}{3\times2\times1}x^3+\frac{6\times5\times4\times3}{4\times3\times2\times1}x^4$$

$$+\frac{6\times5\times4\times3\times2}{5\times4\times3\times2\times1}x^5+\frac{6\times5\times4\times3\times2\times1}{6\times5\times4\times3\times2\times1}x^6$$

$$= 1+6x+15x^2+20x^3+15x^4+6x^5+x^6$$

▸ Exercises 24, 25

The coefficient of x^r in the expansion (7.11) is

$$\binom{n}{r} = \frac{n(n-1)(n-2)...(n-r+1)}{r!} = \frac{n!}{r!(n-r)!} \tag{7.12}$$

and is called a **binomial coefficient** (sometimes read as 'n choose r'). The binomial coefficients are important in probability theory, where they are often given the symbol ^nC_r (or C_r^n); we shall see in Chapter 21 that $\binom{n}{r}$ is the number of combinations of n objects taken r at a time. In terms of binomial coefficients, the expansion is

$$(1+x)^n = \sum_{r=0}^{n} \binom{n}{r} x^r \tag{7.13}$$

A more general form of the binomial expansion is

$$(x+y)^n = \sum_{r=0}^{n} \binom{n}{r} x^r y^{n-r} \tag{7.14}$$

EXAMPLE 7.4 Calculate the binomial coefficients $\binom{5}{r}$, and use them to expand (i) $(1+x)^5$ and (ii) $(x+3)^5$ in powers of x.

By the definition of the binomial coefficients,

$$\binom{5}{r} = \binom{5}{5-r} = \frac{5!}{r!(5-r)!}, \qquad r = 0, 1, 2, ..., 5$$

Therefore, remembering that $0! = 1$,

$$\binom{5}{0} = \binom{5}{5} = \frac{5!}{0!\,5!} = 1, \qquad \binom{5}{1} = \binom{5}{4} = \frac{5!}{1!\,4!} = 5, \qquad \binom{5}{2} = \binom{5}{3} = \frac{5!}{2!\,3!} = 10$$

(i) By equation (7.13),

$$(1+x)^5 = \binom{5}{0} x^0 + \binom{5}{1} x^1 + \binom{5}{2} x^2 + \binom{5}{3} x^3 + \binom{5}{4} x^4 + \binom{5}{5} x^5$$

$$= 1 + 5x + 10x^2 + 10x^3 + 5x^4 + x^5$$

(ii) By equation (7.14),

$$(x+3)^5 = \binom{5}{0}x^0 3^5 + \binom{5}{1}x^1 3^4 + \binom{5}{2}x^2 3^3 + \binom{5}{3}x^3 3^2 + \binom{5}{4}x^4 3^1 + \binom{5}{5}x^5 3^0$$

$$= 243 + 405x + 270x^2 + 90x^3 + 15x^4 + x^5$$

▸ Exercises 26–33

The binomial coefficients form a pattern of numbers called the Pascal triangle[5]

```
                1
             1     1
          1     2     1
       1     3     3     1
    1     4     6     4     1
 1     5    10    10     5     1
1   ...
```

Each row begins and ends with the number 1 for the coefficients of x^n and y^n in the expansion of $(x+y)^n$, and each interior number is the sum of the two numbers diagonally above it.

The multinomial expansion

This is the generalization of the binomial expansion (7.14),

$$(x_1 + x_2 + \cdots + x_k)^n = \sum_{n_1} \sum_{n_2} \cdots \sum_{n_k} \frac{n!}{n_1! n_2! \cdots n_k!} x_1^{n_1} x_2^{n_2} \cdots x_k^{n_k} \qquad (7.15)$$

in which the k-fold sum is over all positive integer and zero values of $n_1, n_2, \ldots, n_k$, subject to the constraint $n_1 + n_2 + \cdots + n_k = n$. The **multinomial coefficients**

$$\binom{n}{n_1\, n_2 \cdots n_k} = \frac{n!}{n_1! n_2! \cdots n_k!}$$

---

[5] Blaise Pascal (1623–1662). French philosopher and mathematician who made contributions to geometry, the calculus and, with Fermat, developed the mathematical theory of probability (at the instigation of the gambler Antoine Gombard, Chevalier de Méré). The Pascal triangle appears in the *Traité du triangle arithmétique, avec quelques autres petits traités sur la même manière*, published posthumously in 1665. The work contains a discussion of the properties of the binomial coefficients, with applications in games of chance. The triangle was known long before; it appeared in a book by the Chinese mathematician Yang Hui in 1261, and the properties of the binomial coefficients were discussed by the Persian Jamshid Al-Kashi in his *Key to arithmetic*, (c. 1425).

are important in combinatorial theory (Section 21.6) and are used in a popular derivation of the Boltzmann distribution in statistical thermodynamics (Example 21.12).

EXAMPLE 7.5 Expand the trinomial $(a+b+c)^3$.

The possible values of (n_1, n_2, n_3) are

$$(3, 0, 0), \ (0, 3, 0), \ (0, 0, 3), \ (2, 1, 0), \ (2, 0, 1), \ (1, 2, 0), \ (1, 0, 2),$$
$$(0, 2, 1), \ (0, 1, 2), \ (1, 1, 1)$$

with distinct multinomial coefficients

$$\frac{3!}{3!0!0!} = 1, \quad \frac{3!}{2!1!0!} = 3, \quad \frac{3!}{1!1!1!} = 6$$

Therefore

$$(a+b+c)^3 = (a^3 + b^3 + c^3) + 3(a^2b + a^2c + ab^2 + ac^2 + b^2c + bc^2) + 6abc$$

‣ Exercises 34, 35

The method of differences

Many simple finite series can be summed if the general term u_r can be written as the
difference

$$u_r = v_r - v_{r-1}$$

Then

$$\sum_{r=1}^{n} u_r = \sum_{r=1}^{n} v_r - \sum_{r=1}^{n} v_{r-1}$$

$$= (v_1 + v_2 + \cdots + v_{n-1} + v_n) - (v_0 + v_1 + \cdots + v_{n-1})$$

$$= v_n - v_0$$

EXAMPLE 7.6 Find the sum of the series $\displaystyle\sum_{r=1}^{n} \frac{1}{r(r+1)} = \frac{1}{1 \cdot 2} + \frac{1}{2 \cdot 3} + \cdots + \frac{1}{n(n+1)}$.

The general term can be written as

$$\frac{1}{r(r+1)} = \frac{1}{r} - \frac{1}{r+1}$$

and, therefore,

$$\sum_{r=1}^{n}\frac{1}{r(r+1)}=\sum_{r=1}^{n}\frac{1}{r}-\sum_{r=1}^{n}\frac{1}{r+1}$$

$$=\left(\frac{1}{1}+\frac{1}{2}+\frac{1}{3}+\cdots+\frac{1}{n}\right)-\left(\frac{1}{2}+\frac{1}{3}+\cdots+\frac{1}{n}+\frac{1}{n+1}\right)$$

$$=1-\frac{1}{n+1}=\frac{n}{n+1}$$

▶ Exercises 36–39

Some finite series

The arithmetic series (7.8) is the simplest example of the family of series $\sum r^{m}$ where m is a positive integer; that is, the sum of powers of the natural numbers;[6] some of these sums and other finite series are given in Table 7.1.

Table 7.1 Some finite series

$$\sum_{r=1}^{n}r=1+2+3+\cdots+n\quad=\frac{1}{2}n^{2}+\frac{1}{2}n=\frac{1}{2}n(n+1)$$

$$\sum_{r=1}^{n}r^{2}=1+2^{2}+3^{2}+\cdots+n^{2}=\frac{1}{3}n^{3}+\frac{1}{2}n^{2}+\frac{1}{6}n=\frac{1}{6}n(n+1)(2n+1)$$

$$\sum_{r=1}^{n}r^{3}=1+2^{3}+3^{3}+\cdots+n^{3}=\frac{1}{4}n^{4}+\frac{1}{2}n^{3}+\frac{1}{4}n^{2}=\frac{1}{4}n^{2}(n+1)^{2}$$

$$\sum_{r=1}^{n}r^{4}=1+2^{4}+3^{4}+\cdots+n^{4}=\frac{1}{5}n^{5}+\frac{1}{2}n^{4}+\frac{1}{3}n^{3}-\frac{1}{30}n$$

$$\sum_{r=1}^{n}r(r+1)=\frac{1}{3}n(n+1)(n+2)$$

$$\sum_{r=1}^{n}r(r+1)(r+2)=\frac{1}{4}n(n+1)(n+2)(n+3)$$

$$\sum_{r=1}^{n}\frac{1}{r(r+1)}=\frac{n}{n+1}$$

$$\sum_{r=1}^{n}\frac{1}{r(r+1)(r+2)}=\frac{1}{4}-\frac{1}{2(n+1)(n+2)}$$

The sums of many other series can be derived from these.

[6] The general method for generating these sums is due to the Swiss mathematician Jakob Bernoulli (1654–1705), Professor of mathematics at Basel. With his brother Johann (1667–1748), professor of mathematics in Groningen and at Basel, and in collaboration with Leibniz, he also made contributions to the calculus, theory of differential equations, series, and the calculus of variations. It was this collaboration that led to the success of Leibniz's formulation of the calculus; by 1700 most of the elementary calculus (described in this book) had been developed. Newton's method of fluxions was never well known outside England. It was Jakob Bernoulli who first used the word 'integral'.

EXAMPLE 7.7 Find the sum of the first n terms of the series $2 \cdot 5 + 3 \cdot 7 + 4 \cdot 9 + \cdots$

The general term is $u_r = (r+1)(2r+3)$ for $r = 1, 2, 3, \ldots$ Then

$$u_r = 2r(r+1) + 3r + 3$$

and

$$\sum_{r=1}^{n} u_r = 2\sum_{r=1}^{n} r(r+1) + 3\sum_{r=1}^{n} r + 3\sum_{r=1}^{n} 1$$

$$= 2 \times \frac{1}{3}n(n+1)(n+2) + 3 \times \frac{1}{2}n(n+1) + 3 \times n$$

$$= \frac{1}{6}n(4n^2 + 21n + 35)$$

➤ Exercise 40

7.4 Infinite series

The limit of a sequence of partial sums is the (sum of the) infinite series

$$S = \lim_{n \to \infty} \left(\sum_{r=1}^{n} u_n \right) = u_1 + u_2 + u_3 + \cdots \tag{7.16}$$

where the dots mean that the sum is to be extended indefinitely. The series has a sum only if the limit is finite and unique; that is, when the sequence of partial sums converges.

The geometric series

The geometric series is the limit of the sequence of partial sums

$$S_n = 1 + x + x^2 + \cdots + x^{n-1} = \frac{1 - x^n}{1 - x}, \qquad x \neq 1$$

(when $x = 1$ the sum is n and the series diverges). The sequence of sums shows all six types of behaviour illustrated in Figure 7.1 (type (b) only for the trivial case $x = 0$), and converges only when $|x| < 1$. Thus, when $|x| < 1$, $x^n \to 0$ as $n \to \infty$, and

$$\lim_{n \to \infty} \left(\frac{1 - x^n}{1 - x} \right) = \frac{1}{1 - x}$$

so that the geometric series is the expansion of the function $\dfrac{1}{1-x}$ in powers of x:

$$\frac{1}{1-x} = 1 + x + x^2 + x^3 + \cdots, \qquad |x| < 1 \tag{7.17}$$

The series diverges for all other values of x ($|x| \geq 1$).

▸ Exercises 41–45

___

The harmonic series

$$S = 1 + \frac{1}{2} + \frac{1}{3} + \frac{1}{4} + \cdots$$

Despite appearances, this series *diverges*. It can be written as a sum of partial sums,

$$S = 1 + \frac{1}{2} + \left(\frac{1}{3} + \frac{1}{4} \right) + \left(\frac{1}{5} + \frac{1}{6} + \frac{1}{7} + \frac{1}{8} \right)$$

$$+ \left(\frac{1}{9} + \frac{1}{10} + \frac{1}{11} + \frac{1}{12} + \frac{1}{13} + \frac{1}{14} + \frac{1}{15} + \frac{1}{16} \right) + \cdots$$

$$= 1 + \frac{1}{2} + s_1 + s_2 + s_3 + \cdots$$

in which s_n contains 2^n terms of which the last, and smallest, has value $1/2^{n+1}$. Each of these sums is therefore larger than $2^n \times (1/2^{n+1}) = 1/2$; for example,

$$s_1 = \frac{1}{3} + \frac{1}{4} > \frac{1}{4} + \frac{1}{4}, \qquad s_2 = \frac{1}{5} + \frac{1}{6} + \frac{1}{7} + \frac{1}{8} > \frac{1}{8} + \frac{1}{8} + \frac{1}{8} + \frac{1}{8}$$

It follows that

$$S > 1 + \frac{1}{2} + \frac{1}{2} + \frac{1}{2} + \frac{1}{2} + \cdots$$

and the series diverges.[7]

7.5   Tests of convergence

The example of the geometric series demonstrates that it is straightforward to show that a series converges, or otherwise, if a closed formula is known for the partial sums.

___

[7] This demonstration of the divergence of the harmonic series and discussions of other infinite series are found in Oresme's *Quaestiones super geometriam Euclidis* (c. 1350).

The sum of the infinite series is then the limit of the sequence of partial sums. On the other hand, the example of the harmonic series shows that when such a formula is not known less direct procedures must be followed.

Given a series

$$\sum_{r=1}^{\infty} a_r = a_1 + a_2 + a_3 + \cdots$$

a necessary first condition for convergence is that the limit of the sequence $\{a_r\}$ be zero:

$$a_r \to 0 \quad \text{as} \quad r \to \infty$$

If this condition is satisfied, the series can be tested for convergence in a number of ways.

The comparison test

Let

$$A = a_1 + a_2 + \cdots + a_r + \cdots$$
$$B = b_1 + b_2 + \cdots + b_r + \cdots$$

be two series of *positive* terms. Then:

(i) If series B *converges*, then series A converges if $a_r \le b_r$.
(ii) If series B *diverges*, then series A diverges if $a_r \ge b_r$.

EXAMPLE 7.8 The series

$$S = 1 + \frac{1}{2^p} + \frac{1}{3^p} + \frac{1}{4^p} + \cdots$$

converges if $p > 1$ and diverges if $p \le 1$.

(i) $p = 1$: S is the harmonic series, and diverges.
(ii) $p < 1$: each term of S (after the first) is larger than the corresponding terms of the harmonic series, and S diverges.
(iii) $p > 1$: write the series as

$$S = 1 + \left(\frac{1}{2^p} + + \frac{1}{3^p} \right) + \left(\frac{1}{4^p} + \frac{1}{5^p} + \frac{1}{6^p} + \frac{1}{7^p} \right) + \cdots$$

$$= 1 + s_1 + s_2 + \cdots$$

in which s_n contains 2^n terms of which the first, and largest, is $1/2^{np}$. Each sum s_n is then less than $\dfrac{2^n}{2^{np}} = \left(\dfrac{1}{2^{p-1}}\right)^n$ and

$$S < 1 + \left(\frac{1}{2^{p-1}}\right) + \left(\frac{1}{2^{p-1}}\right)^2 + \left(\frac{1}{2^{p-1}}\right)^3 + \cdots$$

The series on the right is a convergent geometric series $\sum x^r$ with $x = 1/2^{p-1} < 1$.

▸ Exercises 46, 47

---

d'Alembert's ratio test[8]

The series $a_1 + a_2 + \cdots + a_r + a_{r+1} + \cdots$

(i) converges if $\displaystyle\lim_{r\to\infty} \left|\frac{a_{r+1}}{a_r}\right| < 1$

(ii) diverges if $\displaystyle\lim_{r\to\infty} \left|\frac{a_{r+1}}{a_r}\right| > 1$

(iii) may do either if the limit equals 1, and further tests are necessary.

EXAMPLES 7.9 The ratio test

(i) The geometric series $1 + x + x^2 + x^3 + \cdots$

The general term is $a_r = x^r$ so that $a_{r+1} = x^{r+1}$ and $\dfrac{a_{r+1}}{a_r} = x$, independent of r. Then

$$\lim_{r\to\infty}\left|\frac{a_{r+1}}{a_r}\right| = |x|$$

and the series converges if $|x| < 1$, diverges if $|x| > 1$, and the test fails for $x = \pm 1$. In this case it is readily shown by inspection that the series diverges when $x = \pm 1$.

---

[8] Jean LeRond d'Alembert (1717–1783). Secretary of the French Academy, he is best known for his contribution to the post-Newtonian development of mechanics with his *principle of virtual work* (d'Alembert's principle) published in his *Traité de Dynamique* (1743). He contributed to the theory of partial differential equations, the calculus, and infinite series. In his *Différentiel* (1754) in the *Encyclopédie des sciences, des arts et des métiers* (1751–1772) he first gave the derivative as the limit of a quotient of increments but, because of the conceptual difficulties associated with the limit as a process consisting of an infinite number of steps, the definition was not accepted until the work of Cauchy (1821). He gave the complete solution of the precession of the equinoxes. Although the ratio test is usually ascribed to him, and sometimes to Cauchy, it was probably first given in 1776 by the Cambridge mathematician Edward Waring (1734–1793).

(ii) The exponential series $1 + \dfrac{x}{1!} + \dfrac{x^2}{2!} + \dfrac{x^3}{3!} + \cdots$

The general term is $a_r = \dfrac{x^r}{r!}$ so that $a_{r+1} = \dfrac{x^{r+1}}{(r+1)!}$ and $\dfrac{a_{r+1}}{a_r} = \dfrac{x}{r+1}$. Then

$$\lim_{r \to \infty} \left| \frac{a_{r+1}}{a_r} \right| = \lim_{r \to \infty} \left| \frac{x}{r+1} \right| = 0$$

and the series converges for all values of x.

➤ Exercises 48, 49

Cauchy's integral test[9]

Let $a_1 + a_2 + \cdots + a_r + \cdots$ be a series of decreasing positive terms, $a_{r+1} < a_r$. Let $a(x)$ be a function of the continuous variable x such that $a(x)$ decreases as x increases and $a(r) = a_r$. The series then

converges if $\displaystyle\int_1^\infty a(x)\,dx$ converges (is finite and unique)

diverges if $\displaystyle\int_1^\infty a(x)\,dx$ diverges

EXAMPLE 7.10 The harmonic series $1 + \dfrac{1}{2} + \dfrac{1}{3} + \cdots$.

In this case, $a_r = \dfrac{1}{r}$ and $a(x) = \dfrac{1}{x}$. Then

$$\int_1^\infty \frac{1}{x}\,dx = \Big[\ln x \Big]_1^\infty$$

and $\ln x \to \infty$ as $x \to \infty$. The harmonic series therefore diverges.

➤ Exercises 50, 51

---

[9] Augustin-Louis Cauchy (1789–1857). The leading French mathematician of the first half of the nineteenth century, he is best known for his work on the theory of functions of a complex variable, with the Cauchy integral theorem and the calculus of residues. He made contributions to, amongst others, partial differential equations, the theory of elasticity, infinite series, and limits (see d'Alembert). He invented the word **determinant** for his class of alternating symmetric functions (1812). The integral test is sometimes named for MacLaurin.

Alternating series

If the terms of the series $a_1 + a_2 + a_3 + \cdots$ become progressively smaller and alternate in sign then the series converges. For example, the alternating harmonic series

$$1 - \frac{1}{2} + \frac{1}{3} - \frac{1}{4} + \cdots$$

converges. We will see in Section 7.6 that the sum of this series is $\ln 2$.

7.6 MacLaurin and Taylor series

Power series

A power series in the variable x has the form of an 'infinite polynomial'

$$c_0 + c_1 x + c_2 x^2 + c_3 x^3 + \cdots$$

where $c_0, c_1, c_2, \ldots$ are constants. The convergence properties of such series can be investigated by the methods described in the previous section. Thus, applying the ratio test, a power series converges when

$$\lim_{n \to \infty} \left| \frac{c_{n+1} x^{n+1}}{c_n x^n} \right| = |x| \lim_{n \to \infty} \left| \frac{c_{n+1}}{c_n} \right| < 1$$

or, equivalently, when

$$|x| < \lim_{n \to \infty} \left| \frac{c_n}{c_{n+1}} \right| = R \tag{7.18}$$

where R is called the **radius of convergence** of the series. The series is therefore convergent when $|x| < R$; it diverges when $|x| > R$, and the case $x = \pm R$ has to be tested by other methods.

The geometric and exponential series are examples of power series. The geometric series has radius of convergence $R = 1$, the exponential series has $R = \infty$ (see Examples 7.9).

EXAMPLES 7.11 Radius of convergence

(i) The coefficient of x^n in the series

$$\sum_{n=1}^{\infty} \frac{x^n}{n}$$

is $c_n = \frac{1}{n}$. By the ratio test, $|c_n/c_{n+1}| = (n+1)/n \to 1$ as $n \to \infty$ and the radius of convergence is $R = 1$. The series therefore converges when $|x| < 1$ and diverges when $|x| > 1$. It also diverges when $x = 1$, when it is the harmonic series, but converges to $\ln 2$ when $x = -1$ (see the MacLaurin series for the logarithmic function).

(ii) The coefficient of x^n in the series

$$\sum_{n=1}^{\infty} \frac{(-3x)^n}{n^2}$$

is $c_n = (-3)^n/n^2$. By the ratio test, $|c_n/c_{n+1}| = (1/3)(n/n+1)^2 \to (1/3)$ as $n \to \infty$ and the radius of convergence is $R = 1/3$. The series therefore converges when $|x| < 1/3$. It also converges when $|x| = 1/3$ (see Example 7.8).

> Exercises 52–57

The MacLaurin series[10]

We saw in Section 7.4 that the geometric series can be regarded as the expansion of the function $1/(1-x)$ in powers of x. Similarly, the exponential series can be regarded as the expansion of that function $f(x) = e^x$ whose derivative is equal to itself, $f'(x) = f(x)$. Many other functions can be expanded in this way.

Let $f(x)$ be a function of x that can be represented as a power series

$$f(x) = c_0 + c_1 x + c_2 x^2 + c_3 x^3 + c_4 x^4 + \cdots \tag{7.19}$$

The coefficients $c_0, c_1, c_2, \ldots$ can be obtained in the following way. The derivatives of the function are

$$f'(x) = \frac{df}{dx} = c_1 + 2\,c_2 x + 3\,c_3 x^2 + 4\,c_4 x^3 +$$

$$f''(x) = \frac{d^2 f}{dx^2} = 2\,c_2 + 6\,c_3 x + 12\,c_4 x^2 +$$

$$f'''(x) = \frac{d^3 f}{dx^3} = 6\,c_3 + 24\,c_4 x +$$

$$\cdots$$

Then, letting $x = 0$,

$$f(0) = c_0, \quad f'(0) = c_1, \quad f''(0) = 2!c_2, \quad f'''(0) = 3!c_3, \quad \cdots$$

and, in general, for the nth derivative,

$$f^{(n)}(0) = \left(\frac{d^n f(x)}{dx^n} \right)_{x=0} = n!\,c_n, \qquad c_n = \frac{1}{n!} f^{(n)}(0)$$

[10] Colin MacLaurin (1698–1746), professor of mathematics at Edinburgh. The series called after him appeared in his *Treatise of fluxions* (1742), but the more general Taylor series was published in 1715, and was known to the Scottish mathematician James Gregory (1638–1675). The *Treatise* contains also the method of deciding the maximum/minimum question by investigating the sign of a higher derivative.

Substitution into (7.19) then gives

$$f(x) = f(0) + \frac{x}{1!} f'(0) + \frac{x^2}{2!} f''(0) + \frac{x^3}{3!} f'''(0) + \quad = \sum_{n=0}^{\infty} \frac{x^n}{n!} f^{(n)}(0) \qquad (7.20)$$

This power series is called the **MacLaurin series expansion** of the function $f(x)$.

The MacLaurin series is an expansion of the function $f(x)$ about the point $x = 0$ (Figure 7.2); that is, the value of the function at point x is expressed in terms of the values of the function and its derivatives at $x = 0$. For this to be possible the function and its derivatives must exist at $x = 0$ *and* throughout the interval 0 to x. In addition, the expansion is valid only within the radius of convergence of the series.

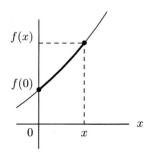

Figure 7.2

Examples of MacLaurin series

1.   The binomial series[11]

The binomial series is the MacLaurin expansion of the function $(1 + x)^a$ for arbitrary values of a. We have

$$f(x) \quad = (1 + x)^a$$
$$f'(x) \quad = a(1 + x)^{a-1}$$
$$f''(x) \quad = a(a - 1)(1 + x)^{a-2}$$
$$f'''(x) \quad = a(a - 1)(a - 2)(1 + x)^{a-3}$$
$$\dots$$
$$f^{(n)}(x) = a(a - 1)(a - 2)\dots(a - n + 1)(1 + x)^{a-n}$$
$$\dots$$

When $x = 0$, the factor $(1 + x)^{a-n}$ is replaced by 1,

$$f(0) = 1, \quad f'(0) = a, \quad f''(0) = a(a - 1), \quad f'''(0) = a(a - 1)(a - 2), \quad \dots$$

and the MacLaurin expansion is

$$(1 + x)^a = 1 + a x + \frac{a(a - 1)}{2!} x^2 + \frac{a(a - 1)(a - 2)}{3!} x^3 + \qquad (7.21)$$

---

[11] The binomial theorem (series) was discovered by Newton in 1665 and described in 1676 in letters written to Henry Oldenburg, secretary of the Royal Society, for transmission to Leibniz. The theorem was published by Wallis in his *Algebra* of 1685.

with general term

$$\frac{a(a-1)(a-2)...(a-n+1)}{n!}x^n$$

The series (7.21) is called the **binomial series**, and is the generalization of the binomial expansion, equation (7.11), to arbitrary powers. In fact, when a is a positive integer n, the $(n+1)$th derivative and all higher derivatives are zero, and (7.21) reduces to the binomial expansion. When a is not a positive integer, the series has radius of convergence $R=1$. Thus, applying the ratio test,

$$R = \lim_{r\to\infty}\left|\frac{c_r}{c_{r+1}}\right| = \lim_{r\to\infty}\left|\frac{r}{a-r}\right| = 1$$

and the binomial series converges for values $-1<x<+1$. It may also converge when $x=\pm1$.

The more general form, the expansion of $(x+y)^a$, is obtained from (7.21) by factorizing out the term x^a if $|x|>|y|$ or y^a if $|y|>|x|$. Thus, if $|x|>|y|$,

$$(x+y)^a = x^a\left(1+\frac{y}{x}\right)^a$$

(7.22)

$$= x^a\left(1+a\left(\frac{y}{x}\right)+\frac{a(a-1)}{2!}\left(\frac{y}{x}\right)^2+\frac{a(a-1)(a-2)}{3!}\left(\frac{y}{x}\right)^3+\right)$$

and the series converges since $|y/x|<1$. Equation (7.22) can also be written as

$$(x+y)^a = x^a + ax^{a-1}y + \frac{a(a-1)}{2!}x^{a-2}y^2 + \frac{a(a-1)(a-2)}{3!}x^{a-3}y^3 + \quad (7.23)$$

EXAMPLES 7.12

(i) $\dfrac{1}{1-x} = (1-x)^{-1} = 1+(-1)(-x)+\dfrac{(-1)(-2)}{2!}(-x)^2 + \dfrac{(-1)(-2)(-3)}{3!}(-x)^3 +$

$\qquad = 1+x+x^2+x^3+\cdots$

(ii) $(1+x)^{1/2} = 1+\left(\dfrac{1}{2}\right)(x)+\dfrac{\left(\frac{1}{2}\right)\left(-\frac{1}{2}\right)}{2!}(x)^2 + \dfrac{\left(\frac{1}{2}\right)\left(-\frac{1}{2}\right)\left(-\frac{3}{2}\right)}{3!}(x)^3 +$

$\qquad = 1+\dfrac{x}{2}-\dfrac{x^2}{8}+\dfrac{x^3}{16}-\dfrac{5x^4}{128}+$

(iii) $(8+x)^{1/2} = \sqrt{8}\left(1+\dfrac{x}{8}\right)^{1/2} = \sqrt{8}\left(1+\left(\dfrac{1}{2}\right)\left(\dfrac{x}{8}\right)+\dfrac{\left(\frac{1}{2}\right)\left(-\frac{1}{2}\right)}{2!}\left(\dfrac{x}{8}\right)^2\right.$

$$+\dfrac{\left(\frac{1}{2}\right)\left(-\frac{1}{2}\right)\left(-\frac{3}{2}\right)}{3!}\left(\dfrac{x}{8}\right)^3+\cdots\Bigg)$$

$$= \sqrt{8}\left(1+\dfrac{x}{16}-\dfrac{x^2}{512}+\dfrac{x^3}{8192}-\dfrac{5x^4}{524288}+\cdots\right)$$

For example, if $x = 1$, using the first five terms of the series,

$$\sqrt{9} = \sqrt{8}\left(1+\dfrac{1}{16}-\dfrac{1}{512}+\dfrac{1}{8192}-\dfrac{5}{524288}+\cdots\right)$$

$$\approx \sqrt{8}\times 1.06065941 \approx 2.999998$$

---

2.   Trigonometric functions

The trigonometric functions $\sin x$, $\cos x$ and $\tan x$ have continuous derivatives at $x = 0$, and can be expanded as MacLaurin series. For example, for the sine function,

$$f(x) = \sin x, \quad f'(x) = \cos x, \quad f''(x) = -\sin x, \quad f'''(x) = -\cos x, \quad \cdots$$
$$f(0) = 0, \qquad f'(0) = 1, \qquad f''(0) = 0, \qquad f'''(0) = -1, \qquad \cdots$$

Then

$$\sin x = x - \dfrac{x^3}{3!} + \dfrac{x^5}{5!} - \dfrac{x^7}{7!} + \cdots$$

Comparison with the exponential series shows that the sine series converges for all values of x.

3.   The logarithmic function

The function $\ln x$ *cannot* be expanded as a power series in x because the function and all its derivatives are discontinuous at $x = 0$. However, the function $\ln(1+x)$ is well behaved at $x = 0$:

$$f(x) \ = \ln(1+x) \qquad\qquad f(0) \ = \ 0$$

$$f'(x) \ = \dfrac{1}{1+x} \qquad\qquad f'(0) \ = \ 1$$

$$f''(x) \ = \dfrac{-1}{(1+x)^2} \qquad\qquad f''(0) \ = -1$$

$$f'''(x) \ = \dfrac{2}{(1+x)^3} \qquad\qquad f'''(0) \ = \ 2$$

$$f''''(x) = \dfrac{-6}{(1+x)^4} \qquad\qquad f''''(0) = -3!$$

$$\cdots \qquad\qquad\qquad\qquad \cdots$$

The nth derivative at $x = 0$ is $f^{(n)}(0) = (-1)^{n-1}(n-1)!$, and

$$\ln(1+x) = x - \frac{x^2}{2} + \frac{x^3}{3} - \frac{x^4}{4} + \cdots$$

The series converges when $|x| < 1$. Thus, $c_r/c_{r+1} = (r+1)/r \rightarrow 1$ as $r \rightarrow \infty$ and by the ratio test (7.18), the radius of convergence is $R = 1$. In addition, the series is the (convergent) alternating harmonic series when $x = 1$:

$$\ln 2 = 1 - \frac{1}{2} + \frac{1}{3} - \frac{1}{4} + \cdots$$

4. A list of some useful series

1. $(1+x)^a = 1 + a\,x + \dfrac{a(a-1)}{2!}x^2 + \dfrac{a(a-1)(a-2)}{3!}x^3 + \cdots$ $|x| < 1$

2. $\sin x = x - \dfrac{x^3}{3!} + \dfrac{x^5}{5!} - \dfrac{x^7}{7!} + \cdots$ all x

3. $\cos x = 1 - \dfrac{x^2}{2!} + \dfrac{x^4}{4!} - \dfrac{x^6}{6!} + \cdots$ all x

4. $\tan x = x + \dfrac{x^3}{3} + \dfrac{2x^5}{15} + \dfrac{17x^7}{315} + \cdots$ $-\dfrac{\pi}{2} < x < \dfrac{\pi}{2}$

5. $\ln(1+x) = x - \dfrac{x^2}{2} + \dfrac{x^3}{3} - \dfrac{x^4}{4} + \cdots$ $-1 < x \le +1$

6. $e^x = 1 + x + \dfrac{x^2}{2!} + \dfrac{x^3}{3!} + \dfrac{x^4}{4!} + \cdots$ all x

7. $\sinh x = x + \dfrac{x^3}{3!} + \dfrac{x^5}{5!} + \dfrac{x^7}{7!} + \cdots$ all x

8. $\cosh x = 1 + \dfrac{x^2}{2!} + \dfrac{x^4}{4!} + \dfrac{x^6}{6!} + \cdots$ all x

➤ Exercises 58–67

The Taylor series[12]

The MacLaurin series, the expansion of a function $f(x)$ about the point $x = 0$, is a special case of the more general expansion of the function about the point $x = a$:

---

[12] Brook Taylor (1685–1731), secretary of the Royal Society. The series called after him appeared in his *Methodus incrementorum* (1715), but had been known to James Gregory.

$$f(x) = f(a) + \frac{(x-a)}{1!} f'(a) + \frac{(x-a)^2}{2!} f''(a) + \frac{(x-a)^3}{3!} f'''(a) + \cdots$$

$$= \sum_{n=0}^{\infty} \frac{(x-a)^n}{n!} f^{(n)}(a)$$

(7.24)

This power series in $(x-a)$ is called a **Taylor series expansion** of the function $f(x)$. The verification and conditions of existence of the Taylor series are as for the MacLaurin series.

EXAMPLE 7.13 Taylor series

(i) Expand x^4 about the point $x = 1$.
We have

$$f(x) = x^4, \ f'(x) = 4x^3, \ f''(x) = 4 \times 3x^2, \ f'''(x) = 4!x, \ f''''(x) = 4!,$$
$$f^{(n)}(x) = 0 \text{ if } n > 4$$

$$f(1) = 1, \ f'(1) = \frac{4!}{3!}, \ f''(1) = \frac{4!}{2!}, \ f'''(1) = \frac{4!}{1!}, \ f''''(1) = 4!$$

Therefore,

$$x^4 = 1 + \binom{4}{1}(x-1) + \binom{4}{2}(x-1)^2 + \binom{4}{3}(x-1)^3 + (x-1)^4$$

(ii) Expand $\cos x$ about $x = \pi/2$.

$$f(x) = \cos x, \ f'(x) = -\sin x, \ f''(x) = -\cos x, \ f'''(x) = \sin x, \ \ldots$$
$$f(\pi/2) = 0, \ f'(\pi/2) = -1, \ f''(\pi/2) = 0, \ f'''(\pi/2) = 1, \ \ldots$$

Therefore

$$\cos x = -(x - \pi/2) + \frac{1}{3!}(x - \pi/2)^3 - \frac{1}{5!}(x - \pi/2)^5 + \frac{1}{7!}(x - \pi/2)^7 + \cdots$$

▶ Exercises 68–71

7.7 Approximate values and limits

The MacLaurin and Taylor series provide a systematic tool for approximating functions in the form of polynomials. Consider, for example, the logarithmic series

$$\ln(1+x) = x - \frac{x^2}{2} + \frac{x^3}{3} - \frac{x^4}{4} + \cdots, \qquad -1 < x \le +1$$

If the series is terminated after a finite number of terms, the result is a polynomial approximation to the function. The series therefore provides a sequence of such approximations:

$$u_1 = x, \qquad u_2 = x - \frac{x^2}{2}, \qquad u_3 = x - \frac{x^2}{2} + \frac{x^3}{3}, \cdots$$

Some values of these are shown in Table 7.2.

Table 7.2   Values of $\ln(1+x)$

$u_1 = x$	$u_2 = x - x^2/2$	$u_3 = x - x^2/2 + x^3/3$	$\ln(1+x) = \lim\limits_{n \to \infty} (u_n)$
0	0	0	0
0.0001	0.0000 9999 5	0.0000 9999 5000	0.0000 9999 5000
0.001	0.0009 995	0.0009 9950 0333	0.0009 9950 0333
0.01	0.0099 5	0.0099 5033 33	0.0099 5033 08
0.1	0.095	0.0953 333	0.0953 310
0.2	0.18	0.1826 66	0.1823 21
1.0	0.5	0.83	0.69

The table shows that $u_1 = x$ is a good approximation to $\ln(1+x)$ when $x \le 0.1$ and that the series converges rapidly for these small values of x, each term providing at least one additional figure of accuracy. Convergence is less good for larger values of x, and eight terms are needed to give 10% accuracy when $x = 1$. The theoretical basis for this use of the series is Taylor's theorem.

Taylor's theorem

Let $f(x)$ be a continuous single-valued function of x with continuous derivatives $f'(x), f''(x), \ldots, f^{(n)}(x)$ in the interval a to x, and let $f^{(n+1)}(x)$ exist within the interval. Then

$$f(x) = f(a) + \frac{(x-a)}{1!} f'(a) + \frac{(x-a)^2}{2!} f''(a) + \cdots$$
$$+ \frac{(x-a)^n}{n!} f^{(n)}(a) + R_n(x) \qquad (7.25)$$

where

$$R_n(x) = \frac{(x-a)^{n+1}}{(n+1)!} f^{(n+1)}(b) \qquad (7.26)$$

and $a < b < x$ is some point in the interval. The term $R_n(x)$ is called the **remainder term** and is the error involved in approximating the function by a polynomial of degree n. The smallest and largest values of $R_n(x)$ are lower and upper bounds to the error. The infinite series is obtained in the limit $n \to \infty$ if $R_n(x) \to 0$ as $n \to \infty$.

EXAMPLE 7.14 The exponential series

By Taylor's theorem (for $a = 0$),

$$e^x = 1 + x + \frac{x^2}{2!} + \cdots + \frac{x^n}{n!} + R_n(x)$$

where

$$R_n = \frac{x^{n+1}}{(n+1)!} e^b$$

for some point $0 < b < x$. When $x > 0$, the smallest and largest values of R_n are given by

$$\frac{x^{n+1}}{(n+1)!} < R_n < \frac{x^{n+1}}{(n+1)!} e^x$$

Then

$$1 + x + \frac{x^2}{2!} + \cdots + \frac{x^n}{n!} + \frac{x^{n+1}}{(n+1)!} < e^x < 1 + x + \frac{x^2}{2!} + \cdots + \frac{x^n}{n!} + \frac{x^{n+1}}{(n+1)!} e^x$$

For example, when $x = 0.2$ and $n = 3$ (and rounding all numbers to six decimal places), the cubic approximation has value

$$e^{0.2} \approx 1 + 0.2 + \frac{(0.2)^2}{2} + \frac{(0.2)^3}{6} = 1.221333$$

and the error bounds are given by

$$0.000067 < R_3 < 0.000067 e^{0.2}$$

so that

$$1.221333 + 0.000067 < e^{0.2} < 1.221333 + 0.000067 e^{0.2}$$

The lower bound is 1.221400. For the upper bound,

$$e^{0.2} < 1.221333 + 0.000067 e^{0.2}$$

$$e^{0.2} - 0.000067 e^{0.2} = 0.999933 e^{0.2} < 1.221333$$

Therefore

$$e^{0.2} < 1.221333/0.999933 = 1.221415$$

and

$$1.221400 < e^{0.2} < 1.221415$$

The exact value is 1.221403.

▸ Exercise 72

Limits

The MacLaurin series shows how a function behaves when the variable is very small. Thus $\ln(1+x) \approx x$ when x is small enough, so that in the immediate vicinity of $x=0$ the function $y = \ln(1+x)$ can be approximated by the straight line $y=x$. Similarly, when x is small enough,

$$\sin x = x - \frac{x^3}{3!} + \frac{x^5}{5!} - \cdots \approx x$$

$$\cos x = 1 - \frac{x^2}{2!} + \frac{x^4}{4!} - \cdots \approx 1 - \frac{x^2}{2}$$

Another way of expressing the same results is in terms of limits. Thus (Figure 7.3)

$$\frac{\sin x}{x} = 1 - \frac{x^2}{3!} + \frac{x^4}{5!} - \cdots$$

so that in the limit $x \to 0$,

$$\lim_{x \to 0} \left(\frac{\sin x}{x} \right) = 1 \qquad (7.27)$$

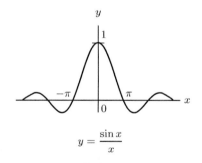

$$y = \frac{\sin x}{x}$$

Figure 7.3

Similarly,

$$\lim_{x \to 0} \left(\frac{1 - \cos x}{x^2} \right) = \frac{1}{2} \qquad (7.28)$$

More generally, the use of power series provides a systematic way of determining the limit of the quotient of two functions,

$$\lim_{x \to a} \frac{f(x)}{g(x)}$$

when $f(x) \to 0$ and $g(x) \to 0$ as $x \to a$. It is not possible to substitute $x=a$ in this case because the result would be the indeterminate $0/0$. However, if both functions are expanded as Taylor series, by equation (7.24), we have

$$\frac{f(x)}{g(x)}=\frac{f(a)+(x-a)f'(a)+\dfrac{(x-a)^2}{2!}f''(a)+\cdots}{g(a)+(x-a)g'(a)+\dfrac{(x-a)^2}{2!}g''(a)+\cdots}$$

Then because $f(a)=g(a)=0$,

$$\frac{f(x)}{g(x)}=\frac{f'(a)+\dfrac{(x-a)}{2!}f''(a)+\cdots}{g'(a)+\dfrac{(x-a)}{2!}g''(a)+\cdots}$$

and, if $g'(a)$ is not zero,

$$\lim_{x\to a}\frac{f(x)}{g(x)}=\frac{f'(a)}{g'(a)}=\lim_{x\to a}\frac{f'(x)}{g'(x)} \qquad (7.29)$$

This method of finding limits is called **l'Hôpital's rule**.[13] If $g'(a)=0$ but $f'(a)\neq0$ the limit is infinite. If $f'(a)$ and $g'(a)$ are both zero the process is repeated to give

$$\lim_{x\to a}\frac{f(x)}{g(x)}=\frac{f''(a)}{g''(a)} \qquad (7.30)$$

and so on.

EXAMPLE 7.15 Find the limits.

(i) $\displaystyle\lim_{x\to0}\left(\frac{\sin x-x}{x^3}\right).$

$$\frac{\sin x-x}{x^3}=\frac{(x-x^3/3!+x^5/5!-\cdots)-x}{x^3}$$

$$=\frac{-x^3/3!+x^5/5!-\cdots}{x^3}=-\frac{1}{3!}+\frac{x^2}{5!}-\cdots\to-\frac{1}{6}\quad\text{as}\quad x\to0$$

(ii) $\displaystyle\lim_{x\to a}\left(\frac{x^2e^x}{(e^x-1)^2}\right).$

[13] Guillaume François Antoine de l'Hôpital (1661–1704). French nobleman and amateur mathematician, he was tutored by Johann Bernoulli in the new calculus. The rule ascribed to him was contained in a letter from Bernoulli in 1694 and appeared in l'Hôpital's influential textbook on the calculus, *Analyse des infiniment petits* (1696).

The numerator is $x^2(1+x+x^2/2+\cdots)$. The denominator is

$$[(1+x+x^2/2+\cdots)-1]^2=(x+x^2/2+\cdots)^2=x^2+x^3+\cdots$$

Therefore,

$$\frac{x^2e^x}{(e^x-1)^2}=\frac{x^2(1+x+\cdots)}{x^2(1+x+\cdots)}\to1\quad\text{as}\quad x\to0$$

This example is important in the statistical mechanics of solids. In Einstein's theory of the heat capacity of simple atomic solids, each atom in the solid is assumed to vibrate with the same frequency v. The molar heat capacity of the solid is then

$$C_v=3R\frac{x^2e^x}{(e^x-1)^2}$$

with $x=hv/kT$, where h is Planck's constant, k is Boltzmann's constant, R is the gas constant and T is the temperature. The taking of the limit $x\to0$ corresponds to letting $T\to\infty$. Then $C_v\to3R$.

▸ Exercises 73–77

7.8 Operations with power series

Let

$$A(x)=\sum_{r=0}^{\infty}a_rx^r\quad\text{and}\quad B(x)=\sum_{r=0}^{\infty}b_rx^r$$

be two power series with radii of convergence R_A and R_B, respectively.

(i) Addition and subtraction.

The power series may be added or subtracted *term by term* to give a power series

$$C(x)=A(x)\pm B(x)=\sum_{r=1}^{\infty}(a_r\pm b_r)x^r \tag{7.31}$$

whose radius of convergence is at least as large as the smaller of R_A and R_B.

(ii) Multiplication.

The product of $A(x)$ and $B(x)$ is the double infinite sum

$$C(x)=A(x)B(x)=\sum_{r=0}^{\infty}\sum_{s=0}^{\infty}a_rb_sx^{r+s}$$

and this can be rearranged as the **Cauchy product**

$$C(x) = a_0 b_0 + (a_0 b_1 + a_1 b_0)x + (a_0 b_2 + a_1 b_1 + a_2 b_0)x^2 + \cdots = \sum_{r=0}^{\infty} c_r x^r \qquad (7.32)$$

where $c_r = \sum_{i=0}^{r} a_i b_{r-i}$

The radius of convergence of the product is equal to the smaller of R_A and R_B.

EXAMPLE 7.16 Multiplication of power series

$$e^{3x} \times e^{-2x} = \left[1 + 3x + \frac{(3x)^2}{2!} + \cdots \right] \left[1 + (-2x) + \frac{(-2x)^2}{2!} + \cdots \right]$$

$$= 1 + \left[1 \times (-2) + 3 \times 1 \right] x + \left[1 \times \frac{(-2)^2}{2!} + 3 \times (-2) + \frac{3^2}{2!} \times 1 \right] x^2 + \cdots$$

$$= 1 + x + \frac{x^2}{2!} + \cdots = e^x$$

▸ Exercise 78

(iii) Differentiation and integration.

A power series may be differentiated or integrated term by term to give a power series whose radius of convergence is the same as that of the original series. Thus, term by term differentiation of

$$A(x) = \sum_{r=0}^{\infty} a_r x^r = a_0 + a_1 x + a_2 x^2 + a_3 x^3 + \cdots$$

gives

$$\frac{dA}{dx} = \sum_{r=1}^{\infty} r a_r x^{r-1} = a_1 + 2a_2 x + 3a_3 x^2 + \cdots \qquad (7.33)$$

Similarly, termwise integration of the series $A(x)$ gives

$$\int A(x)\, dx = \sum_{r=0}^{\infty} a_r \int x^r\, dx = \sum_{r=0}^{\infty} \frac{a_r}{r+1} x^{r+1} + c$$

$$(7.34)$$

$$= c + a_0 x + \frac{a_1}{2} x^2 + \frac{a_2}{3} x^3 + \cdots$$

The radius of convergence of both the derived series and integrated series is $R = R_A$.

EXAMPLE 7.17 Differentiation and integration of power series

$$\frac{d}{dx}\left(\frac{1}{1-x}\right) = \frac{d}{dx}(1+x+x^2+x^3+x^4+\cdots)$$

$$= 1+2x+3x^2+4x^3+\cdots = \frac{1}{(1-x)^2}$$

$$\int\left(\frac{1}{1-x}\right)dx = \int(1+x+x^2+x^3+x^4+\cdots)dx$$

$$= c+x+\frac{x^2}{2}+\frac{x^3}{3}+\frac{x^4}{4}+\frac{x^5}{5}+\cdots = -\ln(1-x)+c$$

➤ Exercises 79, 80

7.9 Exercises

Section 7.2

Find (a) the general term and (b) the recurrence relation for the sequences:

1. 1, 4, 7, 10, ...   **2.** 1, 3, 9, 27, ...   **3.** $1, -\dfrac{1}{5}, \dfrac{1}{25}, -\dfrac{1}{125}, \ldots$

Find the first 6 terms of the sequences:

4. $u_{r+1} = u_r + \dfrac{1}{2}; \quad u_1 = 0$ **5.** $v_n = \left(\dfrac{2}{3}\right)^n; \quad n = 0, 1, 2, \ldots$

6. $u_x = \dfrac{1}{x(x+2)}; \quad x = 1, 2, 3, \ldots$ **7.** $w_{n+1} = \dfrac{w_n}{n}; \quad w_1 = 1$

8. $u_{n+2} = u_{n+1} + 2u_n; \quad u_0 = 1, u_1 = 3$ **9.** $u_{n+2} = 3u_{n+1} - 2u_n; \quad u_0 = 1, u_1 = 1/2$

10. $u_{n+2} = 3u_{n+1} - 2u_n; \quad u_0 = u_1$

Find the limit $r \to \infty$ for:

11. $\dfrac{1}{3^r}$ **12.** 2^r **13.** $\dfrac{1}{r+2}$ **14.** $\dfrac{r}{r+2}$ **15.** $\dfrac{r}{r^2+r+1}$ **16.** $\dfrac{3r^2+3r+1}{5r^2-6r-1}$

17. Find the limit of the sequence $\{u_{n+1}/u_n\}$ for $u_{n+2} = u_{n+1} + 2u_n; \quad u_0 = 1, \quad u_1 = 3$ (see Exercise 8).

Section 7.3

Find the sum of **(i)** the first n terms, **(ii)** the first 10 terms:

18. $1+5+9+13+\cdots$ **19.** $3-2-7-12-\cdots$ **20.** $1+3+9+27+\cdots$

21. $1+\dfrac{1}{3}+\dfrac{1}{9}+\dfrac{1}{27}+\cdots$

Find the sum of the first n terms:

22. $x^3+x^5+x^7+\cdots$ **23.** $x+2x^2+4x^3+\cdots$

Use equation (7.11) to expand in powers of x:

24. $(1+x)^5$ **25.** $(1+x)^7$

Calculate the binomial coefficients $\begin{pmatrix} n \\ r \end{pmatrix}, r = 0, 1, \ldots, n,$ for

26. $n=3$ **27.** $n=4$ **28.** $n=7$

Use equation (7.13) or (7.14) to expand in powers of x:

29. $(1-x)^3$ **30.** $(1+3x)^4$ **31.** $(1-4x)^5$ **32.** $(3-2x)^4$ **33.** $(3+x)^6$

34. **(i)** Calculate the distinct trinomial coefficients $\dfrac{4!}{n_1!\,n_2!\,n_3!}$. **(ii)** Use the coefficients to expand $(a+b+c)^4$.

35. **(i)** Calculate the distinct coefficients $\dfrac{3!}{n_1!\,n_2!\,n_3!\,n_4!}$. **(ii)** Use the coefficients to expand $(a+b+c+d)^3$.

36. Find $\displaystyle\sum_{n=1}^{10} \frac{1}{n(n+1)}$.

37. **(i)** Verify that $\dfrac{1}{r(r+2)} = \dfrac{1}{2}\left(\dfrac{1}{r} - \dfrac{1}{r+2}\right)$, then **(ii)** find the sum of the series $\displaystyle\sum_{r=1}^{n} \frac{1}{r(r+2)}$.

38. **(i)** Express $\dfrac{1}{r(r+1)(r+2)}$ in partial fractions, then **(ii)** show that

$$\sum_{r=1}^{n} \frac{1}{r(r+1)(r+2)} = \frac{1}{4} - \frac{1}{2(n+1)(n+2)}$$

39. **(i)** Verify that $(1+r)^3 - r^3 = 3r^2 + 3r + 1$, then **(ii)** show that

$$\sum_{r=1}^{n} r^2 = \frac{1}{6} n(n+1)(2n+1)$$

40. **(i)** Expand $(1+r)^6 - r^6$, then **(ii)** use the series in Table 7.1 to find the sum of the series

$$\sum_{r=1}^{n} r^5.$$

Section 7.4

(i) Expand in powers of x to terms in x^6. **(ii)** Find the values of x for which the series converge:

41. $\dfrac{1}{1-3x}$ **42.** $\dfrac{1}{1+5x^2}$ **43.** $\dfrac{1}{2+x}$

44. **(i)** Use the geometric series to express the number $1/(10^6 - 1)$ as a decimal fraction.
(ii) Show that the decimal representation of $1/7$ can be written as $142857/(10^6 - 1)$ (see Section 1.4).

45. The vibrational partition function of a harmonic oscillator is given by the series

$$q_v = \sum_{n=0}^{\infty} e^{-n\theta_v/T}$$

where $\theta_v = hv_e/k$ is the vibrational temperature. Confirm that the series is a convergent geometric series, and find its sum.

Section 7.5

Examine the following series for convergence by

Comparison test (use $\ln n < n$):   **46.** $\displaystyle\sum_{n=2}^{\infty} \frac{1}{\ln n}$ **47.** $\displaystyle\sum_{r=1}^{\infty} \frac{\ln r}{r^3}$

D'Alembert ratio test: **48.** $\displaystyle\sum_{s=0}^{\infty} \frac{s^a}{(s+1)!}$ **49.** $\displaystyle\sum_{r=1}^{\infty} \frac{1}{r^a}$

Cauchy integral test: **50.** $\displaystyle\sum_{r=1}^{\infty} \frac{1}{r^a}$ **51.** $\displaystyle\sum_{n=2}^{\infty} \frac{1}{n \ln n}$

Section 7.6

Find the radius of convergence of each of the following series:

52. $\displaystyle\sum_{m=0}^{\infty} \frac{x^m}{4^m}$ **53.** $\displaystyle\sum_{r=0}^{\infty} (-1)^r x^{2r}$ **54.** $\displaystyle\sum_{n=1}^{\infty} n x^n$ **55.** $\displaystyle\sum_{n=1}^{\infty} \frac{x^n}{n^2}$

56. $\displaystyle\sum_{m=1}^{\infty} m^m x^m$ **57.** $\displaystyle\sum_{n=0}^{\infty} \frac{(-1)^n x^{2n}}{3^n}$

Write down the first 5 terms of the MacLaurin series of the following functions:

58. $(1+x)^{1/3}$ **59.** $\dfrac{1}{1+x^2}$ **60.** $(1-x)^{-1/2}$ **61.** $\dfrac{1}{3+x}$

62. $\sin 2x^2$ **63.** $\dfrac{\ln(1-2x)+2x}{x^2}$ **64.** e^{-3x} **65.** $\dfrac{e^{x^2}-1}{x}$

66. A body with rest mass m_0 and speed v has relativistic energy

$$E = mc^2 = \frac{m_0 c^2}{\sqrt{1 - v^2/c^2}}$$

and kinetic energy $T = E - m_0 c^2$. Express T as a power series in v and show that the series reduces to the nonrelativistic kinetic energy in the limit $v/c \to 0$.

67. The equation of state of a gas can be expressed in terms of the series

$$pV = nRT \sum_{i=0}^{\infty} B_i(T) \left(\frac{n}{V}\right)^i$$

where the B_i are called virial coefficients. Find the first three coefficients for

(i) the van der Waals equation, $\left(p + \dfrac{n^2 a}{V^2}\right)(V - nb) = nRT$

(ii) the Dieterici equation, $p(V - nb) = nRTe^{-an/RTV}$

(i) Expand each of the following functions as a Taylor series about the given point, and (ii) find the values of x for which the series converges:

68. $\dfrac{1}{x}, 1$ **69.** $e^x, 2$ **70.** $\sin x, \pi/2$ **71.** $\ln x, 2$

Section 7.7

72. **(i)** Find the MacLaurin expansion of the function $(8+x)^{1/3}$ up to terms in x^4. **(ii)** Use this expansion to find an approximate value of $\sqrt[3]{9}$. **(iii)** Use this value and Taylor's theorem for the remainder to compute upper and lower bounds to the value of $\sqrt[3]{9}$.

Find the limits:

73. $\lim\limits_{x\to 0}\dfrac{e^x-1}{x}$ **74.** $\lim\limits_{x\to 0}\dfrac{\tan x-\sin x}{x^3}$ **75.** $\lim\limits_{x\to 0}\dfrac{e^x+e^{-x}-2}{\cos x-1}$ **76.** $\lim\limits_{x\to 1}\dfrac{\ln x}{x^2-1}$

77. The energy density of black-body radiation at temperature T is given by the Planck formula

$$\rho(\lambda)=\frac{8\pi hc}{\lambda^5}[e^{hc/\lambda kT}-1]^{-1}$$

where λ is the wavelength. Show that the formula reduces to the classical Rayleigh–Jeans law $\rho=8\pi kT/\lambda^4$ **(i)** for long wavelengths $(\lambda\to\infty)$, **(ii)** if Planck's constant is set to zero $(h\to 0)$.

Section 7.8

78. Find the Cauchy product of the power series expansions of $\sin x$ and $\cos x$, and show that it is equal to $\frac{1}{2}\sin 2x$.

79. Differentiate the power series expansion of $\sin x$ and show that the result is $\cos x$.

80. Integrate the power series expansion of $\sin x$ and show that the result is $C-\cos x$, where C is a constant.

8 Complex numbers

8.1 Concepts

We saw in Section 1.7 and in Chapter 2 that the solutions of algebraic equations are not always real numbers; in particular, the solutions of the equation $x^2 = -1$ are $x = \pm\sqrt{-1}$, and the square root of a negative number is not a real number.[1] Such numbers are incorporated into the system of numbers by defining the square root of -1 as a new number which is usually denoted by the symbol i (or j in engineering mathematics) with the property

$$i^2 = -1 \qquad (8.1)$$

A number containing $i = \sqrt{-1}$ is called a **complex number**; examples of complex numbers are $2i$, $-3i$, and $2 + 5i$. The general complex number has the form (the letter z is usually used to denote a complex number)

$$z = x + iy \qquad (8.2)$$

where x and y are real numbers. The number x is called the **real part** of z, and y is called the **imaginary part** of z:

$$x = \text{Re}\,(z), \qquad y = \text{Im}\,(z) \qquad (8.3)$$

If $x = 0$ then $z = iy$ is called **pure imaginary**. If $y = 0$ then $z = x$ is real, so that the set of complex numbers includes the real numbers as subset.

Powers of i

Every integer power of i is one of the numbers i, $-i$, 1, -1; for example

$$i^3 = i^2 \times i = -i, \qquad i^4 = (i^2)^2 = 1, \qquad i^{-1} = \frac{1}{i} = \frac{i}{i^2} = \frac{i}{-1} = -i$$

In general, for integers $n = 0, \pm 1, \pm 2, \ldots$,

$$i^{4n} = +1, \qquad i^{4n+1} = +i, \qquad i^{4n+2} = -1, \qquad i^{4n+3} = -i \qquad (8.4)$$

---

[1] Square roots of negative numbers are mentioned in Cardano's *Ars Magna* of 1545 in connection with the solution of quadratic and cubic equations. Cardano called such a result 'as subtle as it is useless'. Rafael Bombelli (1526–1572), Italian engineer, called $\pm\sqrt{-1}$ 'più di meno' (plus of minus) and 'meno di meno' (minus of minus), and presented the arithmetic of complex numbers in his *Algebra* of 1572. The first serious consideration of complex numbers was by Albert Girard (1595–1632) who in his *L'invention nouvelle en l'algèbre* (New discovery in algebra), 1629, published the first statement of the fundamental theorem of algebra. He called the complex solutions of equations 'impossible' but 'good for three things: for the certainty of the general rule (the theorem), for the fact that there are no other solutions, and for their use'. Descartes used the words 'real' and 'imaginary'. Leibniz factorized $x^4 + a^4 = (x + a\sqrt{i})(x - a\sqrt{i})(x + a\sqrt{-i})(x - a\sqrt{-i})$. Euler proposed the symbol i for $\sqrt{-1}$ in 1777, and this was adopted by Gauss in his *Disquisitiones arithmeticae* of 1801.

Apart from their role in extending the concept of number, complex numbers occur in several branches of mathematics that are important in the physical sciences; for example, the solutions of the differential equations of motion in both classical and quantum mechanics often involve complex numbers. Complex numbers also occur in the formulation of physical theory; the basic equations of quantum mechanics necessarily involve $i = \sqrt{-1}$.

8.2   Algebra of complex numbers

Complex numbers can be added, subtracted, multiplied, and divided in much the same way as ordinary real numbers. It is only necessary to remember to replace i^2 by -1 whenever it occurs. Let two complex numbers be

$$z_1 = x_1 + iy_1, \qquad z_2 = x_2 + iy_2 \tag{8.5}$$

Equality

Two complex numbers are equal if their real parts are equal *and* if their imaginary parts are equal:

$$z_1 = z_2 \quad \text{if} \quad x_1 = x_2 \quad \text{and} \quad y_1 = y_2 \tag{8.6}$$

Addition

$$z_1 + z_2 = (x_1 + x_2) + i(y_1 + y_2) \tag{8.7}$$

The real parts of z_1 and z_2 are added to give the real part of the sum, the imaginary parts are added to give the imaginary part of the sum.

EXAMPLES 8.1   Addition and subtraction

(i) $(3 + 2i) + (4 - 3i) = (3 + 4) + (2 - 3)i = 7 - i$

(ii) $(3 + 2i) - (3 - 2i) = (3 - 3) + (2 + 2)i = 4i$

(iii) $(3 + 2i) + (4 - 2i) = 7$

➤ Exercises 1–3

Multiplication

$$z_1 z_2 = (x_1 + iy_1)(x_2 + iy_2)$$
$$= x_1(x_2 + iy_2) + iy_1(x_2 + iy_2) = (x_1 x_2 + ix_1 y_2) + (iy_1 x_2 + i^2 y_1 y_2) \tag{8.8}$$
$$= (x_1 x_2 - y_1 y_2) + i(x_1 y_2 + y_1 x_2)$$

using $i^2 = -1$.

EXAMPLES 8.2 Multiplication

(i) $(2+3i)(3+4i) = 2(3+4i) + 3i(3+4i) = 6 + 8i + 9i + 12i^2 = -6 + 17i$

(ii) $i(2-3i) = 2i - 3i^2 = 3 + 2i$

> Exercises 4–7

The complex conjugate

If $z = x + iy$ is an arbitrary complex number then the number obtained from it by replacing i by $-i$ is

$$z* = x - iy \qquad (8.9)$$

and is called the **complex conjugate** of z (sometimes $\bar{z}$ is used instead of $z*$). z is then also the complex conjugate of $z*$. The conjugate pair of complex numbers z and $z*$ has the following properties:

(i) $\dfrac{1}{2}(z + z*) = \dfrac{1}{2}\left[(x + iy) + (x - iy)\right] = x = \mathrm{Re}(z) \qquad (8.10)$

(ii) $\dfrac{1}{2}(z - z*) = \dfrac{1}{2}\left[(x + iy) - (x - iy)\right] = iy = i\,\mathrm{Im}(z) \qquad (8.11)$

(iii) $zz* = (x + iy)(x - iy) = x^2 + y^2$ (real and positive) $\qquad (8.12)$

EXAMPLES 8.3 Conjugate pairs of complex numbers

(i) If $z = 2 + 3i$ then $z* = 2 - 3i$ and

$$\frac{1}{2}(z + z*) = 2, \qquad \frac{1}{2}(z - z*) = 3i, \qquad zz* = 2^2 + 3^2 = 13$$

(ii) If $z = 1 - i$ then $z* = 1 + i$ and

$$\frac{1}{2}(z + z*) = 1, \qquad \frac{1}{2}(z - z*) = -i, \qquad zz* = 1 + 1 = 2$$

(iii) Solve the quadratic equation $z^2 - 3z + 4 = 0$ (see Example 2.18).

$$z = \frac{3 \pm \sqrt{9-16}}{2} = \frac{1}{2}\left[3 \pm \sqrt{-7}\right] = \frac{1}{2}\left[3 \pm i\sqrt{7}\right]$$

and the roots of the quadratic are the complex conjugate pair

$$z = \frac{1}{2}\left[3 + i\sqrt{7}\right], \qquad z^* = \frac{1}{2}\left[3 - i\sqrt{7}\right].$$

> Exercises 8–11

Division

$$z_1 \div z_2 = \frac{z_1}{z_2} = \frac{(x_1 + iy_1)}{(x_2 + iy_2)}$$

The division can be performed by the rules of ordinary long division. The simpler method is to make use of property (8.12) of conjugate pairs to transform the denominator into a real number. Thus, multiplying top and bottom by $z_2^* = x_2 - iy_2$, the complex conjugate of the denominator, we have

$$\frac{z_1}{z_2} = \frac{z_1 z_2^*}{z_2 z_2^*} = \frac{(x_1 + iy_1)(x_2 - iy_2)}{(x_2 + iy_2)(x_2 - iy_2)} = \frac{(x_1 + iy_1)(x_2 - iy_2)}{x_2^2 + y_2^2}$$

$$= \left(\frac{x_1 x_2 + y_1 y_2}{x_2^2 + y_2^2}\right) + i\left(\frac{y_1 x_2 - x_1 y_2}{x_2^2 + y_2^2}\right)$$

(8.13)

The division is defined only if $z_2 \neq 0$; that is, $x_2 \neq 0$ and $y_2 \neq 0$.

EXAMPLES 8.4 Division

(i) $\dfrac{2 + 3i}{3 + 4i} = \dfrac{(2 + 3i)(3 - 4i)}{(3 + 4i)(3 - 4i)} = \dfrac{18 + i}{3^2 + 4^2} = \dfrac{18}{25} + \dfrac{i}{25}$

(ii) $\dfrac{1 + i}{1 - i} = \dfrac{(1 + i)(1 + i)}{(1 - i)(1 + i)} = \dfrac{0 + 2i}{1 + 1} = i$

> Exercises 12–15

8.3 Graphical representation

The complex number $z = x + iy$ is represented graphically by a point in a plane, with coordinates (x, y), as in Figure 8.1.[2] The plane is called the **complex plane**. Real

---

[2] John Wallis (1616–1703) first suggested that pure imaginary numbers might be represented on a line perpendicular to the axis of real numbers. Caspar Wessel (1745–1818), Norwegian surveyor, discussed the graphical representation of complex numbers in his *On the analytical representation of direction* of 1797, and Jean Robert Argand (1768–1822), Swiss bookkeeper, in his *Essai* of 1806. Gauss used the same interpretation of complex numbers in his fourth and final proof of the fundamental theorem of algebra in 1848, by which time he believed mathematicians were comfortable enough with complex numbers to accept it. The complex plane is also called the Gaussian plane.

numbers, with $y=0$, are represented by points on the x or **real axis** and pure imaginary numbers, with $x=0$, lie on the y or **imaginary axis**. The representation is called the **Argand diagram**.

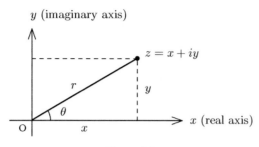

y (imaginary axis)

$z = x + iy$

x (real axis)

Figure 8.1

The distance of point z from the origin, $r = \sqrt{x^2 + y^2}$, is called the **modulus** or **absolute value** of z and is written

$$r = \text{mod } z = |z| \tag{8.14}$$

Complex numbers with the same modulus $r = |z|$ lie on the circle of radius r in the plane. When $|z| = 1$ the point lies on the **unit circle**.

The polar representation

The position of the point $z = x + iy$ in the plane can be specified in terms of the polar coordinates r and θ, as in Figure 8.1. Then

$$x = r \cos \theta, \qquad y = r \sin \theta$$

and the complex number can be written in the polar form

$$z = r(\cos \theta + i \sin \theta) \tag{8.15}$$

The angle θ is called the **argument** or **angle** of z,

$$\theta = \arg z \tag{8.16}$$

The angle can have any real value but, as discussed in Chapter 3, the trigonometric functions in (8.15) are periodic functions of θ so that the number z is unchanged when a multiple of 2π is added to θ. A unique value can be computed from x and y by the prescription given in Section 3.5 for the transformation from cartesian to polar coordinates; thus, given that $\tan \theta = y/x$,

$$\arg z = \tan^{-1}\left(\frac{y}{x}\right) \qquad \text{if} \quad x > 0$$

$$= \tan^{-1}\left(\frac{y}{x}\right) + \pi \qquad \text{if} \quad x < 0 \tag{8.17}$$

in which the inverse tangent has its principal value.

EXAMPLES 8.5 Express $z = x + iy$ in polar form.

(i) $z = 1 + i$

We have $x = 1$ and $y = 1$ so that

$$r = |z| = \sqrt{x^2 + y^2} = \sqrt{2}, \quad \tan^{-1}\left(\frac{y}{x}\right) = \tan^{-1}(1)$$

The principal value of $\tan^{-1}(1)$ is $\pi/4$, and $\arg z = \pi/4$ because the point lies in the first quadrant. Therefore

$$z = \sqrt{2}\left(\cos\frac{\pi}{4} + i\sin\frac{\pi}{4}\right)$$

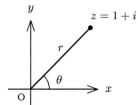

Figure 8.2

(ii) $z = -\frac{1}{2} - \frac{\sqrt{3}}{2}i$

We have $x = -1/2$ and $y = -\sqrt{3}/2$ so that

$$r = \sqrt{\left(\frac{1}{2}\right)^2 + \left(\frac{\sqrt{3}}{2}\right)^2} = 1$$

$$\tan^{-1}\left(\frac{y}{x}\right) = \tan^{-1}(\sqrt{3})$$

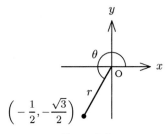

Figure 8.3

The principal value of $\tan^{-1}(\sqrt{3})$ is $\pi/3$ and, because $x < 0$, it follows that $\arg z = \tan^{-1}(\sqrt{3}) + \pi = 4\pi/3$. Therefore

$$z = \cos\frac{4\pi}{3} + i\sin\frac{4\pi}{3}$$

▸ Exercises 16–22

Representation of arithmetic operations

Addition and subtraction

In Figure 8.4, the numbers $z_1 = x_1 + iy_1$ and $z_2 = x_2 + iy_2$ are represented by points P and Q, respectively. The representation of the sum

$$z_1 + z_2 = (x_1 + x_2) + i(y_1 + y_2)$$

is obtained by completing the parallelogram OPSQ. Point S has coordinates $(x_1 + x_2, y_1 + y_2)$ and therefore represents the sum. Similarly for subtraction; the difference $z_1 - z_2$ is the sum of z_1 and $-z_2$.

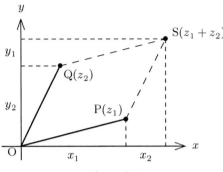

Figure 8.4

Multiplication and division

The operations of multiplication and division are the more easily described when the numbers are expressed in polar form. If

$$z_1 = r_1(\cos \theta_1 + i \sin \theta_1), \qquad z_2 = r_2(\cos \theta_2 + i \sin \theta_2) \qquad (8.18)$$

then

$$
\begin{aligned}
z_1 z_2 &= r_1 r_2 (\cos \theta_1 + i \sin \theta_1)(\cos \theta_2 + i \sin \theta_2) \\
&= r_1 r_2 [(\cos \theta_1 \cos \theta_2 - \sin \theta_1 \sin \theta_2) + i(\cos \theta_1 \sin \theta_2 + \sin \theta_1 \cos \theta_2)] \quad (8.19) \\
&= r_1 r_2 [\cos(\theta_1 + \theta_2) + i \sin(\theta_1 + \theta_2)]
\end{aligned}
$$

It follows that the product of two complex numbers has modulus equal to the product of the moduli of the numbers, and has argument equal to the sum of the arguments:

$$|z_1 z_2| = |z_1| \times |z_2|, \qquad \arg(z_1 z_2) = \arg(z_1) + \arg(z_2) \qquad (8.20)$$

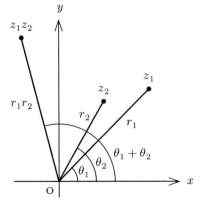

Figure 8.5

If the two numbers are a complex conjugate pair, $z = x + iy$ and $z^* = x - iy$, then

$$|z| = |z^*| = \sqrt{x^2 + y^2}, \qquad \arg z = -\arg z^*$$

and

$$zz^* = |z|^2 = x^2 + y^2, \qquad \arg zz^* = 0$$

In the case of division,

$$\left| \frac{z_1}{z_2} \right| = \frac{|z_1|}{|z_2|}, \qquad \arg \left(\frac{z_1}{z_2} \right) = \arg(z_1) - \arg(z_2) \tag{8.21}$$

and

$$\frac{z_1}{z_2} = \frac{r_1}{r_2} \left[\cos(\theta_1 - \theta_2) + i \sin(\theta_1 - \theta_2) \right] \tag{8.22}$$

For the inverse of a complex number, it follows from (8.22) that

$$\frac{1}{z} = \frac{1}{r} \left[\cos(-\theta) + i \sin(-\theta) \right] \tag{8.23}$$

and, because $\cos(-\theta) = \cos\theta$ and $\sin(-\theta) = -\sin\theta$,

$$\frac{1}{z} = \frac{1}{r}(\cos\theta - i\sin\theta) \tag{8.24}$$

EXAMPLES 8.6 Express each of $z_1 z_2$, z_1/z_2 and z_2/z_1 as a single complex number for

$$z_1 = \sqrt{2}\left(\cos\frac{\pi}{4} + i\sin\frac{\pi}{4} \right), \qquad z_2 = \cos\frac{4\pi}{3} + i\sin\frac{4\pi}{3}$$

(see Examples 8.5)

We have $r_1 = \sqrt{2}$, $r_2 = 1$, $\theta_1 = \pi/4$, and $\theta_2 = 4\pi/3$. Therefore

(i) $r_1 r_2 = \sqrt{2}$, $\theta_1 + \theta_2 = 19\pi/12$ and, by equation (8.19),

$$z_1 z_2 = r_1 r_2 \left[\cos(\theta_1 + \theta_2) + i\sin(\theta_1 + \theta_2) \right] = \sqrt{2} \left[\cos\frac{19\pi}{12} + i\sin\frac{19\pi}{12} \right]$$

(ii) $r_1/r_2 = \sqrt{2}$, $\theta_1 - \theta_2 = -13\pi/12$ and, by equation (8.22),

$$\frac{z_1}{z_2} = \frac{r_1}{r_2}\left[\cos(\theta_1 - \theta_2) + i\sin(\theta_1 - \theta_2)\right]$$

$$= \sqrt{2}\left[\cos\left(-\frac{13\pi}{12}\right) + i\sin\left(-\frac{13\pi}{12}\right)\right]$$

$$= \sqrt{2}\left[\cos\frac{13\pi}{12} - i\sin\frac{13\pi}{12}\right]$$

(iii) $z_2/z_1 = (z_1/z_2)^{-1}$ and, by equation (8.24),

$$\frac{z_2}{z_1} = \frac{1}{\sqrt{2}}\left[\cos\frac{13\pi}{12} + i\sin\frac{13\pi}{12}\right]$$

▸ Exercises 23, 24

de Moivre's formula

It follows from the relations (8.20) that the product of three or more complex numbers has modulus equal to the product of the moduli of the numbers and has argument equal to the sum of the arguments. For example, for the product of three numbers,

$$|z_1z_2z_3| = |z_1z_2| \times |z_3| = |z_1| \times |z_2| \times |z_3|$$

$$\arg(z_1z_2z_3) = \arg(z_1z_2) + \arg(z_3) = \arg(z_1) + \arg(z_2) + \arg(z_3)$$

Therefore

$$z_1z_2z_3 = r_1r_2r_3[\cos(\theta_1 + \theta_2 + \theta_3) + i\sin(\theta_1 + \theta_2 + \theta_3)]$$

and, in general,

$$z_1z_2\cdots z_n = r_1r_2\cdots r_n[\cos(\theta_1 + \theta_2 + \cdots + \theta_n) + i\sin(\theta_1 + \theta_2 + \cdots + \theta_n)] \quad (8.25)$$

In the special case, when all the numbers are equal to $z = r(\cos\theta + i\sin\theta)$,

$$z^n = r^n(\cos n\theta + i\sin n\theta) \quad (8.26)$$

For a number on the unit circle in the complex plane $(r = 1)$, $z = \cos\theta + i\sin\theta$ and

$$(\cos\theta + i\sin\theta)^n = \cos n\theta + i\sin n\theta \quad (8.27)$$

This is **de Moivre's formula** for positive integers n.[3] The formula is also valid for other values of n. For example, by equations (8.27) and (8.23),

$$\frac{1}{(\cos \theta + i \sin \theta)^n} = \frac{1}{\cos n\theta + i \sin n\theta} = \cos(-n\theta) + i\sin(-n\theta)$$

so that

$$(\cos \theta + i \sin \theta)^{-n} = \cos(-n\theta) + i \sin(-n\theta) = \cos n\theta - i \sin n\theta \qquad (8.28)$$

de Moivre's formula can be used to derive many of the formulas of trigonometry (see Section 3.4). For example, equation (8.27) with $n = 2$ is

$$(\cos \theta + i \sin \theta)^2 = \cos 2\theta + i \sin 2\theta$$

Expansion of the left side of this gives

$$(\cos^2 \theta - \sin^2 \theta) + i(2 \sin \theta \cos \theta) = \cos 2\theta + i \sin 2\theta$$

A single equation between two complex numbers is equivalent to two equations between real numbers; two complex numbers are equal only when the real parts are equal and the imaginary parts are equal. Therefore,

$$\cos^2 \theta - \sin^2 \theta = \cos 2\theta, \qquad 2 \sin \theta \cos \theta = \sin 2\theta$$

(see equations (3.23) and (3.24)). Similar formulas, expressing $\sin n\theta$ and $\cos n\theta$ in terms of powers of $\sin \theta$ and $\cos \theta$, are obtained in this way for any positive integer n; the expression on the left side of equation (8.27) is expanded by means of the binomial formula, equation (7.14), and its real and imaginary parts equated to the corresponding terms on the right side of the equation.

The generalization and significance of de Moivre's formula are discussed in Section 8.5.

EXAMPLE 8.7 Express $\cos 5\theta$ and $\sin 5\theta$ in terms of $\sin \theta$ and $\cos \theta$.

By the binomial expansion (7.14), or by using Pascal's triangle,

$$(a + b)^5 = a^5 + 5a^4 b + 10a^3 b^3 + 10a^2 b^3 + 5ab^4 + b^5$$

so that

$$(\cos \theta + i \sin \theta)^5 = (\cos^5 \theta - 10 \cos^3 \theta \sin^2 \theta + 5 \cos \theta \sin^4 \theta)$$
$$+ i(5 \cos^4 \theta \sin \theta - 10 \cos^2 \theta \sin^3 \theta + \sin^5 \theta)$$

---

[3] Abraham de Moivre (1667–1754), fled to England in 1688 from the persecution of the French Huguenots. The first form of the formula occurs in a *Philosophical Transactions* paper of 1707. de Moivre was a friend of Newton. In his later years, Newton would tell visitors who came to him with questions on mathematics to 'go to Mr. de Moivre, he knows these things better than I do'.

Therefore,

$$\cos 5\theta = \cos^5\theta - 10\cos^3\theta\sin^2\theta + 5\cos\theta\sin^4\theta = 16\cos^5\theta - 20\cos^3\theta + 5\cos\theta$$

$$\sin 5\theta = \sin^5\theta - 10\sin^3\theta\cos^2\theta + 5\sin\theta\cos^4\theta = 16\sin^5\theta - 20\sin^3\theta + 5\sin\theta$$

and the final expressions have been obtained by using the formula $\sin^2\theta + \cos^2\theta = 1$.

> Exercises 25–27

8.4   Complex functions

Let $g(x)$ and $h(x)$ be (real) functions of the real variable x. A complex function of x is then

$$f(x) = g(x) + ih(x) \qquad (8.29)$$

where $i = \sqrt{-1}$. Such a function differs in no essential way from a real function of one variable, except that the possible values of the function are complex numbers. The discussion of the properties of complex numbers applies to complex functions; for example, the complex conjugate function $f*(x)$ is defined by

$$f*(x) = g(x) - ih(x) \qquad (8.30)$$

with property

$$f(x)f*(x) = |f(x)|^2 = g(x)^2 + h(x)^2 \qquad (8.31)$$

where $|f(x)|$ is the modulus of the function. The quantity $ff*$ plays an important role in wave theories, when the wave function is complex.

EXAMPLE 8.8

(i) Express the complex function $f(x) = 2x^2 + (7 + 2i)x - (4 + i)$ in the form $f(x) = g(x) + ih(x)$, where $g(x)$ and $h(x)$ are real. (ii) Solve $g(x) = 0$, $h(x) = 0$, then $f(x) = 0$. (iii) Find $|f(x)|^2$

(i) $f(x) = (2x^2 + 7x - 4) + i(2x - 1)$
(ii) $g(x) = 2x^2 + 7x - 4 = (2x - 1)(x + 4) = 0$ when $x = 1/2$ and $x = -4$
$h(x) = 2x - 1 = 0$ when $x = 1/2$
Therefore $f(x) = 0$ when $x = 1/2$
(iii) $|f(x)|^2 = g(x)^2 + h(x)^2 = 4x^4 + 28x^3 + 37x^2 - 60x + 17$

> Exercise 28

A more advanced application of complex numbers is in the extension of the concepts of variable and function to include complex variables and functions of a complex variable.[4] Thus, if x and y are real variables then $z = x + iy$ is a complex variable, and $f(z)$ is a function of the complex variable z. For example, the function

$$f(z) = z^2 + z + 1$$

can be written as

$$
\begin{aligned}
f(z) &= (x + iy)^2 + (x + iy) + 1 = (x^2 - y^2 + x + 1) + i(2xy + y) \\
&= g(x, y) + ih(x, y)
\end{aligned}
$$

where $g(x, y)$ and $h(x, y)$, the real and imaginary parts of $f(z)$, are real functions of x and y. The properties of $f(z)$ as a function of the single complex variable z are more general than the properties of real functions. For many purposes in the physical sciences only one such function is of importance, and is discussed in the following section.

➤ Exercise 29

8.5 Euler's formula

It is known from the theory of functions of a complex variable that the exponential function e^z, where z is a complex number, can be expanded in the familiar infinite series

$$e^z = 1 + z + \frac{z^2}{2!} + \frac{z^3}{3!} + \cdots \qquad (8.32)$$

If z is the imaginary number $z = i\theta$ then

$$e^{i\theta} = 1 + (i\theta) + \frac{(i\theta)^2}{2!} + \frac{(i\theta)^3}{3!} + \cdots$$

$$= \left(1 - \frac{\theta^2}{2!} + \frac{\theta^4}{4!} - \frac{\theta^6}{6!} + \cdots\right) + i\left(\theta - \frac{\theta^3}{3!} + \frac{\theta^5}{5!} - \frac{\theta^7}{7!} + \cdots\right)$$

The real and imaginary parts of this function are the series expansions of the trigonometric functions $\cos\theta$ and $\sin\theta$, respectively (see Section 7.6), so that

$$e^{i\theta} = \cos\theta + i\sin\theta \qquad (8.33)$$

---

[4] The first discussion of functions of a complex variable appeared in a letter by Gauss to Bessel in 1811, together with a description of the geometric interpretation of complex numbers. The theory was developed independently by Cauchy from about 1814. Cauchy was the most prolific mathematician of the 19th century. He swamped the weekly bulletin of the Paris Academy of Sciences, *Comptes Rendus*, forcing it to introduce a rule, still in force, restricting publications to four pages.

This relation between the exponential function and the trigonometric functions is called **Euler's formula**.[5] It forms the basis for the unified theory of the elementary functions, and it is one of the important relations in mathematics.

The function $z = e^{i\theta}$ has modulus $|z| = \sqrt{\cos^2\theta + \sin^2\theta} = 1$ and argument $\arg z = \theta$. For each value of θ, the number lies on the unit circle of the complex plane, and as θ varies from 0 to 2π the function defines the unit circle (Figure 8.6). The complex conjugate of $z = e^{i\theta}$ is

$$z^* = e^{-i\theta} = \cos\theta - i\sin\theta \qquad (8.34)$$

and this is also the inverse, $z^{-1} = z^* = e^{-i\theta}$. Then

$$zz^* = zz^{-1} = e^{i\theta}e^{-i\theta} = e^0 = 1$$

The pair of equations (8.33) and (8.34) can be inverted to give relations for the trigonometric functions in terms of the exponentials:

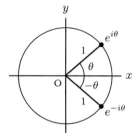

Figure 8.6

$$\cos\theta = \frac{1}{2}\left(e^{i\theta} + e^{-i\theta}\right) \qquad (8.35)$$

$$\sin\theta = \frac{1}{2i}\left(e^{i\theta} - e^{-i\theta}\right) \qquad (8.36)$$

From the polar form of a complex number, equation (8.15), it now follows that every complex number can be written as

$$z = x + iy = re^{i\theta} \qquad (8.37)$$

where $r = |z|$ and $\theta = \arg z$. The complex conjugate and inverse functions of z are

$$z^* = x - iy = re^{-i\theta}, \qquad z^{-1} = \frac{1}{x+iy} = \frac{1}{r}e^{-i\theta} \qquad (8.38)$$

EXAMPLE 8.9 The number $z = 1 + i$.

It was shown in Example 8.5 that

$$z = 1 + i = \sqrt{2}\left(\cos\frac{\pi}{4} + i\sin\frac{\pi}{4}\right)$$

[5] The formula, and others for the logarithmic and trigonometric functions of z, appeared in Euler's *Introductio* of 1748.

The number can therefore be written as

$$z = \sqrt{2}\, e^{i\pi/4}$$

with complex conjugate and inverse

$$z^* = \sqrt{2}\, e^{-i\pi/4} = \sqrt{2}\left(\cos\frac{\pi}{4} - i\sin\frac{\pi}{4}\right)$$

$$z^{-1} = \frac{1}{\sqrt{2}}\, e^{-i\pi/4} = \frac{1}{\sqrt{2}}\left(\cos\frac{\pi}{4} - i\sin\frac{\pi}{4}\right)$$

Also,

$$\cos\frac{\pi}{4} = \frac{1}{2}\left(e^{i\pi/4} + e^{-i\pi/4}\right), \qquad \sin\frac{\pi}{4} = \frac{1}{2i}\left(e^{i\pi/4} - e^{-i\pi/4}\right)$$

➤ Exercises 30–33

EXAMPLE 8.10 Express in cartesian form $x + iy$:

(i) $2e^{i\pi/3} = 2\left[\cos\pi/3 + i\sin\pi/3\right] = 2\left[1/2 + i\sqrt{3}/2\right] = 1 + \sqrt{3}\,i$

(ii) $e^{-i\pi/2} = \cos(-\pi/2) + i\sin(-\pi/2) = -i$

➤ Exercises 34–36

EXAMPLE 8.11 The number $e^{i\pi}$.

By Euler's formula,

$$e^{i\pi} = \cos\pi + i\sin\pi$$

Because $\cos\pi = -1$ and $\sin\pi = 0$ it follows that

$$e^{i\pi} = -1 \qquad\qquad\qquad (8.39)$$

This relation, involving the transcendental numbers e and π, the negative unit -1, and the imaginary unit i, is probably the most remarkable relation in mathematics.

➤ Exercises 37–39

➤ Exercises 40–42

de Moivre's formula

When θ in Euler's formula, (8.33), is replaced by $n\theta$ where n is an arbitrary number, the result is

$$e^{in\theta} = \cos n\theta + i \sin n\theta$$

But because $e^{in\theta} = (e^{i\theta})^n$, it follows that

$$(e^{i\theta})^n = (\cos \theta + i \sin \theta)^n = \cos n\theta + i \sin n\theta \qquad (8.40)$$

This is de Moivre's formula, (8.27), generalized for arbitrary numbers n (that can themselves be complex).

EXAMPLE 8.12 The square root of i:

We have $i = e^{i\pi/2}$. Therefore $\sqrt{i} = \pm e^{i\pi/4} = \pm \left[\cos \pi/4 + i \sin \pi/4 \right] = \pm \frac{1}{\sqrt{2}} \left[1 + i \right]$

Check: $\left(\pm \frac{1}{\sqrt{2}} \left[1 + i \right] \right)^2 = \frac{1}{2}(1 + i^2 + 2i) = i$

▸ Exercise 43

Rotation operators

When a complex number $z = re^{i\alpha}$ is multiplied by $e^{i\theta}$ the product is a number with the same modulus as z but with argument increased by θ:

$$e^{i\theta}z = e^{i\theta}re^{i\alpha} = re^{i(\alpha+\theta)}$$

Graphically, as shown in Figure 8.7, the multiplication corresponds to the (anti-clockwise) rotation of the representative point through angle θ about the origin of the complex plane, from

$$z = x + iy = r \cos \alpha + ir \sin \alpha$$

with cartesian coordinates $x = r \cos \alpha$, $y = r \sin \alpha$ to

$$e^{i\theta}z = z' = x' + iy' = r \cos(\alpha + \theta) + ir \sin(\alpha + \theta)$$

with coordinates

$$x' = r \cos(\alpha + \theta) = r(\cos \alpha \cos \theta - \sin \alpha \sin \theta)$$

$$y' = r \sin(\alpha + \theta) = r(\sin \alpha \cos \theta + \cos \alpha \sin \theta)$$

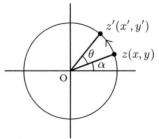

Figure 8.7

The coordinates of the rotated point are therefore related to the coordinates of the unrotated point by

$$x' = x \cos \theta - y \sin \theta$$
$$y' = x \sin \theta + y \cos \theta$$

(8.41)

The function $e^{i\theta}$ can be regarded as a representation of the rotation operator $\mathcal{R}_\theta$ which transforms the coordinates (x, y) of a point z into the coordinates (x', y') of the rotated point z':

$$\mathcal{R}_\theta(x, y) = (x', y')$$

(8.42)

Equations (8.41) play an important role in the mathematical formulation of rotations (see Chapter 18).

▸ Exercise 44

8.6 Periodicity

The trigonometric functions $\cos \theta$ and $\sin \theta$ are periodic functions of θ with period 2π (see Section 3.2). It follows that the exponential function $e^{i\theta}$ is also periodic with period 2π. Thus, if θ is increased by 2π,

$$e^{i(\theta + 2\pi)} = e^{i\theta} \times e^{2\pi i} = e^{i\theta}(\cos 2\pi + i \sin 2\pi)$$

Therefore, because $\cos 2\pi = 1$ and $\sin 2\pi = 0$, it follows that $e^{2\pi i} = 1$ and

$$e^{i(\theta + 2\pi)} = e^{i\theta}$$

More generally, the function is unchanged when a multiple of 2π is added to θ:

$$e^{i(\theta + 2\pi n)} = e^{i\theta}, \qquad n = 0, \pm 1, \pm 2, \dots$$

(8.43)

Graphically, changing the argument θ by $2\pi n$ corresponds to moving the representative point on the unit circle through n full rotations back to its original position (n anticlockwise rotations if n is positive, $|n|$ clockwise rotations if n is negative).

The function $e^{i\theta}$ occurs in the physical sciences whenever periodic motion is described or when a system has periodic structure. We consider here three important representative classes of physical situations exhibiting periodic behaviour.

Periodicity on a circle. The n nth roots of 1

Figure 8.8 shows three equidistant points on the unit circle, at the vertices of an equilateral triangle. The points correspond to complex numbers

$$z_k = e^{(2\pi k/3)i}, \qquad k = 0, 1, 2$$

with unit modulus and arguments $2\pi k/3$. These numbers have the property

$$(z_k)^3 = \left(e^{(2\pi k/3)i}\right)^3 = e^{2\pi ki} = 1$$

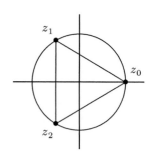

Figure 8.8

so that each is a cube root of the number 1:

$$z_0 = e^0 = 1$$

$$z_1 = e^{2\pi i/3} = \cos\frac{2\pi}{3} + i \sin\frac{2\pi}{3} = -\frac{1}{2} + \frac{\sqrt{3}}{2}i$$

$$z_2 = e^{4\pi i/3} = \cos\frac{4\pi}{3} + i \sin\frac{4\pi}{3} = -\frac{1}{2} - \frac{\sqrt{3}}{2}i$$

We note that z_1 and z_2 are a complex conjugate pair, with $e^{4\pi i/3} = e^{-2\pi i/3}$. Because of the periodicity of the exponential, the three roots can be specified by any three consecutive values of k; conveniently as

$$z_k = e^{2\pi k i/3}, \qquad k = 0, \pm 1$$

such that $z_0 = 1$, $z_{\pm 1} = e^{\pm 2\pi i/3}$.

In general the n nth roots of the number 1 are[6]

$$z_k = e^{2\pi k i/n} \quad \text{for} \quad k = \begin{cases} 0, \pm 1, \pm 2, \ldots, \pm(n-1)/2 & \text{if } n \text{ is odd} \\ 0, \pm 1, \pm 2, \ldots, \pm(n/2-1), n/2 & \text{if } n \text{ is even} \end{cases} \tag{8.44}$$

Thus when n is odd, the only real root is $+1$ (for $k=0$) and the rest occur as complex conjugate pairs. When n is even, two of the roots are real, ± 1 (for $k=0$, $n/2$). The n representative points lie on the vertices of a regular n-sided polygon.

EXAMPLE 8.13 The six sixth roots of 1.

The six roots are

$$z_k = e^{2\pi k i/6}, \qquad k = 0, \pm 1, \pm 2, 3$$

or $\quad z_0 = e^0 = 1$

$$z_{\pm 1} = e^{\pm \pi i/3} = \cos\frac{\pi}{3} \pm i \sin\frac{\pi}{3} = \frac{1}{2} \pm \frac{\sqrt{3}}{2}i$$

$$z_{\pm 2} = e^{\pm 2\pi i/3} = \cos\frac{2\pi}{3} \pm i \sin\frac{2\pi}{3} = -\frac{1}{2} \pm \frac{\sqrt{3}}{2}i$$

$$z_3 = e^{\pi i} = \cos\pi + i \sin\pi = -1$$

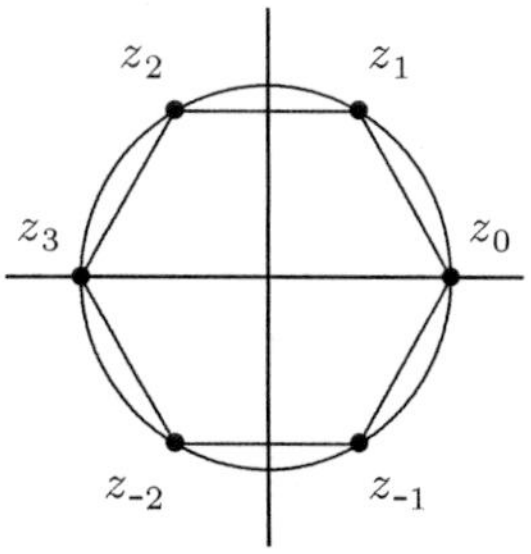

Figure 8.9

> Exercises 45–47

---

[6] The nth roots of a complex number were discussed by de Moivre in a *Philosophical Transactions* paper of 1739.

A function that has the same circular periodicity as the figure with n equidistant points on a circle is

$$f(\theta) = e^{in\theta} \tag{8.45}$$

This function is periodic in θ with period $2\pi/n$. Thus

$$f(\theta + 2\pi/n) = e^{in(\theta + 2\pi/n)} = e^{in\theta} \times e^{2\pi i} = e^{in\theta} = f(\theta)$$

Such functions are important for the description of systems with circular periodicity.

Periodicity on a line

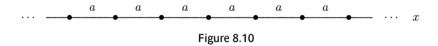

Figure 8.10

Figure 8.10 shows a simple linear array of equidistant points representing, for example, a linear lattice. A function of x that has the same periodicity as the lattice must satisfy the periodicity condition

$$f(x + a) = f(x) \tag{8.46}$$

The simplest such function is

$$f(x) = e^{2\pi x i/a} \tag{8.47}$$

Thus,

$$f(x + a) = e^{2\pi(x+a)i/a} = e^{2\pi x i/a} \times e^{2\pi i} = f(x)e^{2\pi i} = f(x)$$

Functions like (8.47) are important for the description of the properties of periodic systems such as crystals. The functions are readily generalized for three-dimensional periodic systems: the function

$$f(x, y, z) = e^{2\pi x i/a} e^{2\pi y i/b} e^{2\pi z i/c} \tag{8.48}$$

has period a in the x-direction, b in the y-direction, and c in the z-direction.

Rotation in quantum mechanics

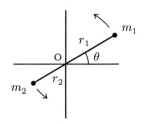

Figure 8.11 shows a system of two masses, m_1 and m_2, joined by a rigid rod (of negligible mass) rotating about the centre of mass at O. As discussed in Section 5.6 (Example 5.13) the system has moment of inertia $I = \mu r^2$ where $\mu = m_1 m_2/(m_1 + m_2)$ is the reduced mass and $r = r_1 + r_2$ is the distance between the masses. Such a

Figure 8.11

system is called a 'rigid rotor' and the equation of motion in quantum mechanics (the Schrödinger equation) for a rigid rotor in a plane is

$$-\frac{\hbar^2}{2I}\frac{d^2\psi}{d\theta^2} = E\psi \tag{8.49}$$

where the wave function $\psi = \psi(\theta)$ is a function of the orientation variable θ, and E is the (positive) kinetic energy of rotation. The rigid rotor is used in chemistry to model the rotational motion of a molecule.

Equation (8.49) can be written

$$\frac{d^2\psi}{d\theta^2} = -a^2\psi \tag{8.50}$$

where $a^2 = 2IE/\hbar^2 > 0$, and it is readily verified that a solution of this equation is

$$\psi(\theta) = Ce^{ia\theta} \tag{8.51}$$

where C is an arbitrary constant (see Section 12.7 for a more complete discussion). Thus,

$$\frac{d\psi}{d\theta} = iaCe^{ia\theta}, \qquad \frac{d^2\psi}{d\theta^2} = (ia)^2 Ce^{ia\theta} = -a^2 Ce^{ia\theta} = -a^2\psi$$

For the solution (8.51) to be physically significant, and represent rotation, it is necessary that the wave function be unchanged when θ is replaced by $\theta + 2\pi$; it must satisfy the periodicity condition

$$\psi(\theta + 2\pi) = \psi(\theta) \tag{8.52}$$

For the function (8.51),

$$\psi(\theta + 2\pi) = Ce^{ia(\theta+2\pi)} = Ce^{ia\theta} \times e^{2\pi ai} = \psi(\theta) \times e^{2\pi ai}$$

and the periodicity condition is satisfied if $2\pi a$ is a multiple of 2π; that is, if $a = n$ for $n = 0, \pm 1, \pm 2, \dots$ The physically significant solutions of the Schrödinger equation are therefore

$$\psi_n(\theta) = Ce^{in\theta}, \qquad n = 0, \pm 1, \pm 2, \dots \tag{8.53}$$

where the 'quantum number' n has been used to label the solutions. The corresponding values of the energy $E = \hbar^2 a^2/2I$ are

$$E_n = \frac{\hbar^2 n^2}{2I} \tag{8.54}$$

We note that the quantization of the energy of the system has arisen as a consequence of applying the periodicity condition (periodic boundary condition) to the solutions. We note also that the set of wave functions includes all the functions (8.45) for the possible periodicities around a circle.

▸ Exercise 48

8.7 Evaluation of integrals

Integration with respect to a complex variable is an important part of the theory of functions of a complex variable, but is used only in advanced applications in the physical sciences. Ordinary integration over complex functions obeys the same rules as integration over real functions. In addition, complex functions can be used to simplify the evaluation of certain types of integral. For example, it is shown in Example 6.13, how the integral

$$\int e^{-ax} \cos x \, dx$$

can be evaluated by the method of integration by parts. An alternative, more elegant, method is to express the trigonometric function in terms of the (complex) exponential function. We consider the general form

$$I = \int e^{ax} \cos bx \, dx$$

Because $\cos bx$ is the real part of e^{ibx} it follows that the integral I is the real part of the integral obtained from I by replacing $\cos bx$ by e^{ibx}:

$$I = \mathrm{Re} \int e^{ax} e^{ibx} \, dx = \mathrm{Re} \int e^{(a+ib)x} \, dx$$

The complex integral is evaluated by means of the ordinary rule for the integration of an exponential function. Thus (ignoring the constant of integration),

$$\int e^{(a+ib)x} \, dx = \frac{e^{(a+ib)x}}{a+ib} = e^{ax} \frac{e^{ibx}}{a+ib}$$

and this can be resolved into its real and imaginary parts:

$$\int e^{(a+ib)x} \, dx = e^{ax} \left(\frac{\cos bx + i \sin bx}{a+ib} \right)$$

$$= \frac{e^{ax}}{(a^2 + b^2)} \left[(a \cos bx + b \sin bx) + i(a \sin bx - b \cos bx) \right]$$

The integral I is the real part of this:

$$\int e^{ax} \cos bx \, dx = \frac{e^{ax}(a \cos bx + b \sin bx)}{a^2 + b^2} \tag{8.55}$$

The imaginary part is a bonus:

$$\int e^{ax} \sin bx \, dx = \frac{e^{ax}(a \sin bx - b \cos bx)}{a^2 + b^2} \tag{8.56}$$

▸ Exercises 49, 50

8.8 Exercises

Section 8.2

Express as a single complex number:

1. $(2+3i)+(4-5i)$ **2.** $(2+3i)+(2-3i)$ **3.** $(2+3i)-(2-3i)$ **4.** $(5+3i)(3-i)$
5. $(1-3i)^2$ **6.** $(1+2i)^5$ **7.** $(1-3i)(1+3i)$
8. If $z=3-2i$, find **(i)** z^* and **(ii)** zz^*. **(iii)** Express the real and imaginary parts of z in terms of z and z^*.
9. Find z such that $zz^* + 4(z-z^*) = 5 + 16i$.

Solve the equations:
10. $z^2 - 2z + 4 = 0$ **11.** $z^3 + 8 = 0$

Express as a single complex number:

12. $\dfrac{1-i}{1+i}$ **13.** $\dfrac{1}{5+3i}$ **14.** $\dfrac{3+2i}{3-2i}$ **15.** $\dfrac{1}{5} - \dfrac{3-4i}{3+4i}$

Section 8.3

(i) Plot as a point in the complex plane, **(ii)** find the modulus and argument, **(iii)** Express in polar form $r(\cos \theta + i \sin \theta)$:

16. $2i$ **17.** -3 **18.** $1-i$ **19.** $\sqrt{3}+i$ **20.** $-6+6i$ **21.** $-2-\sqrt{12}i$
22. $1/i$

Given z_1 and z_2, express **(i)** $z_1 z_2$, **(ii)** z_1/z_2, **(iii)** z_2/z_1 as a single complex number for

23. $z_1 = 2\left(\cos\dfrac{\pi}{2} + i \sin\dfrac{\pi}{2}\right)$, $z_2 = 3\left(\cos\dfrac{\pi}{3} + i \sin\dfrac{\pi}{3}\right)$

24. $z_1 = 5\left(\cos\dfrac{3\pi}{4} + i \sin\dfrac{3\pi}{4}\right)$, $z_2 = \cos\dfrac{2\pi}{3} + i \sin\dfrac{2\pi}{3}$

25. For $z = 3\left(\cos\dfrac{\pi}{8} + i \sin\dfrac{\pi}{8}\right)$ find **(i)** z^4, **(ii)** z^{-4}.

26. Use de Moivre's formula to show that
    (i) $\cos 4\theta = \cos^4 \theta - 6\cos^2 \theta \sin^2 \theta + \sin^4 \theta$
    (ii) $\sin 4\theta = 4 \sin \theta \cos \theta (\cos^2 \theta - \sin^2 \theta)$
27. Use de Moivre's formula to expand $\cos 8x$ as a polynomial in $\cos x$.

Section 8.4

28. (i) Express the complex function $f(x) = 3x^2 + (1 + 2i)x + 2(i - 1)$ in the form
$f(x) = g(x) + ih(x)$, where $g(x)$ and $h(x)$ are real. **(ii)** solve $g(x) = 0$, $h(x) = 0$, then
$f(x) = 0$. **(iii)** find $|f(x)|^2$.

29. (i) Express the complex function $f(z) = z^2 - 2z + 3$ in the form $f(z) = g(x, y) + ih(x, y)$
where $g(x, y)$ and $h(x, y)$ are real functions of the real variables x and y. **(ii)** Find the
(real) solutions of the pair of equations $g(x, y) = 0$ and $h(x, y) = 0$, and hence of $f(z) = 0$,
(iii) Solve $f(z) = 0$ directly in terms of z to confirm the results of **(ii)**.

Section 8.5

Express **(i)** z, **(ii)** $z*$, **(iii)** z^{-1} in exponential form $re^{i\theta}$:

30. $z = 1 - i$ **31.** $z = \sqrt{3} + i$ **32.** $z = 2i$ **33.** $z = -3$

Express in cartesian form $x + iy$:

34. $3e^{i\pi/4}$ **35.** $e^{-i\pi/3}$ **36.** $2e^{\pi i/6}$ **37.** $e^{i\pi/2}$ **38.** $e^{3\pi i/2}$ **39.** $e^{3\pi i}$

40. Use Euler's formulas for $\cos x$ and $\sin x$ to show that

    (i) $\cos ix = \cosh x$,   **(ii)** $\sin ix = i \sinh x$,   **(iii)** $\tan ix = i \tanh x$

41. Express $\cos(a + ib)$ in the form $x + iy$.

42. Show that $\ln z = \ln|z| + i \arg z$

43. Use de Moivre's formula to find the square roots of $-i$. Locate them on the complex
plane.

44. Find the number obtained from $z = 3 + 2i$ by

    (i) anticlockwise rotation through 30°,

    (ii) clockwise rotation through 30° about the origin of the complex plane.

Section 8.6

Find all the roots and plot them in the complex plane:

45. $\sqrt[4]{1}$ **46.** $\sqrt[5]{1}$ **47.** $\sqrt[8]{1}$

48. The wave functions for the quantum mechanical rigid rotor in a plane are
$\psi_n(\theta) = Ce^{in\theta}$, $n = 0, \pm1, \pm2, \ldots$

    (i) Calculate the 'normalization constant' C for which $\displaystyle\int_0^{2\pi} |\psi_n(\theta)|^2\, d\theta = 1$.

    (ii) Show that $\displaystyle\int_0^{2\pi} \psi_m^*(\theta)\,\psi_n(\theta)\, d\theta = 0$ if $m \neq n$.

Section 8.7

Use complex numbers to integrate:

49. $\displaystyle\int_0^\infty e^{-x} \cos 2x\, dx$ **50.** $\displaystyle\int_0^\infty e^{-2x} \sin^3 x\, dx$

9 Functions of several variables

9.1 Concepts

When the equation of state of the ideal gas is written in the form

$$V = f(p, T, n) = \frac{nRT}{p}$$

it is implied that the volume V of the gas is determined by the values of the pressure p, the temperature T, and the amount of substance n; that is, V is a function of the *three* variables p, T, and n. Functions of more than one variable occur widely in the physical sciences; examples are the thermodynamic functions of state, as in the above example, and all those physical properties of a system whose values depend on position. For example, mass density and potential energy were discussed in Chapter 5 as functions of one variable only, the position along a line. More generally, **functions of position** are functions of the three coordinates of a point in ordinary three-dimensional space.

Let the variable z be a function of the two variables x and y. For example, the equation

$$z = x^2 - 2xy - 3y^2$$

gives z as a particular function of x and y. The expression on the right of the equation defines the function

$$f(x, y) = x^2 - 2xy - 3y^2 \tag{9.1}$$

whose value for a given pair of values of x and y is to be assigned to the variable z. The variables x and y are called **independent variables** if no relation exists between them such that the value of one depends on the value of the other.

EXAMPLE 9.1 The values of the function (9.1) when $(x, y) = (2, 1)$, $(x, y) = (1, 0)$, and $(x, y) = (0, 1)$ are

$$f(2, 1) = 2^2 - 2 \times 2 \times 1 - 3 \times 1^2 = -3$$
$$f(1, 0) = 1^2 - 2 \times 1 \times 0 - 3 \times 0^2 = 1$$
$$f(0, 1) = 0^2 - 2 \times 0 \times 1 - 3 \times 1^2 = -3$$

▸ Exercises 1, 2

9.2 Graphical representation

We saw in Section 2.2 that a (real) function of one variable defines a *curve* in a plane; for example, the graph of $y = x^2 - 2x - 3$ is shown in Figure 2.1. A function of *two* independent variables, $z = f(x, y)$, defines a *surface* in a three-dimensional space.[1]

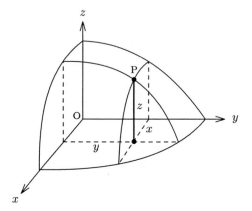

Figure 9.1

In Figure 9.1, Ox, Oy, and Oz are three perpendicular axes, the pair of values (x, y) specifies a point in the (horizontal) xy-plane, and the value of the function is represented by the point P at height z above the plane (coordinate systems in three dimensions are discussed in Chapter 10). As the point (x, y) moves in the xy-plane, the locus of the point P maps out a surface; that is, the point P moves on the surface, and the surface is the representation of the function.

It is possible to draw beautiful three-dimensional representations of functions of two variables by means of modern computer graphics, but such complete physical representations are not possible for functions of three or more variables. In the general case, a function of several variables can be visualized, at least in part, by assigning values to all the variables except one, and plotting the resulting function of the one variable only. Examples of such plots are Figures 4.1 and 4.2 for the volume of the perfect gas, $V = f(p, T, n) = nRT/p$. Figure 4.1 is the graph of V as a function of T with p and n held constant, whilst Figure 4.2 is the graph of V as a function of p at constant T and n. Such simple graphs are often the most useful representation of a function. Figure 9.2 shows graphs of the function (9.1) as a function of x for several values of y.

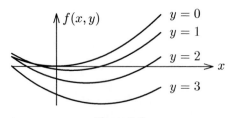

Figure 9.2

---

[1] The representation of surfaces by functions of two variables and the concept of partial derivatives were first considered by Leibniz in the 1690's.

Each of these graphs is a planar 'cut' through the three-dimensional surface. In the general case, a function of n variables defines a 'surface' in an $(n+1)$-dimensional space, and its graph as a function of one of the variables is obtained by taking a planar cut through the representative $(n+1)$-dimensional surface.

9.3 Partial differentiation

We saw in Chapter 4 that the first derivative of a function of one variable is interpreted graphically as the slope of a tangent line to its graph, and dynamically as the rate of change of the function with respect to the variable. For a function of two or more variables there exist as many independent first derivatives as there are independent variables. For example, the function

$$z = f(x, y) = x^2 - 2xy - 3y^2$$

can be differentiated with respect to variable x, with y treated as a constant, to give the **partial derivative** of the function with respect to x

$$\frac{\partial z}{\partial x} = 2x - 2y$$

(read as 'partial dz by dx'),[2] and with respect to y at constant x for the partial derivative with respect to y

$$\frac{\partial z}{\partial y} = -2x - 6y$$

The existence of partial derivatives and the validity of the operation of partial differentiation are subject to the same conditions of continuity and smoothness as for the ordinary (total) derivative. If these conditions are satisfied then the partial derivatives of a function of two variables are defined by the limits (compare equation (4.8))

$$\frac{\partial z}{\partial x} = \lim_{\Delta x \to 0} \left\{ \frac{f(x + \Delta x, y) - f(x, y)}{\Delta x} \right\} \qquad (9.2)$$

$$\frac{\partial z}{\partial y} = \lim_{\Delta y \to 0} \left\{ \frac{f(x, y + \Delta y) - f(x, y)}{\Delta y} \right\} \qquad (9.3)$$

The geometric interpretation of these quantities is shown in Figure 9.3. The plane ABC is parallel to the xz-plane, so that $y = $ constant in the plane and the values of $\partial z/\partial x$ for this value of y are the slopes of the tangent lines to the curve APB. In the same way, the plane DEF is parallel to the yz-plane, and the values of $\partial z/\partial y$ are

---

[2] The notation $\partial z/\partial x$ was first used by Legendre in 1788, but began to be accepted only after Jacobi used it in his theory of determinants in 1841.

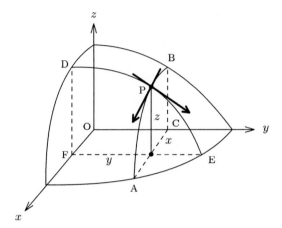

Figure 9.3

the slopes of the tangent lines to the curve DPE. The two tangent lines at point P in Figure 9.3 therefore represent the rates of change of the function $z = f(x, y)$, with $\partial z/\partial x$ for the rate of change along the x-direction and $\partial z/\partial y$ for the rate of change along the y-direction.

The two tangent lines at P in Figure 9.3 define the **tangent plane** at P; that is, the plane that touches the representative surface of the function only at the point P. Every other line through P in the tangent plane is then also a tangent line and its slope is the derivative of the function in some direction. Such a derivative can be expressed in terms of the 'standard' derivatives $\partial z/\partial x$ and $\partial z/\partial y$.

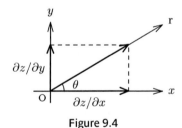

Figure 9.4

In Figure 9.4, the distance r along the direction at angle θ to the x-direction is given by the pair of 'parametric equations'

$$x = r \cos \theta, \qquad y = r \sin \theta \qquad (9.4)$$

and the derivative of the function $z = f(x, y)$ along this r-direction is

$$\frac{\partial z}{\partial r} = \frac{\partial z}{\partial x} \cos \theta + \frac{\partial z}{\partial y} \sin \theta \qquad (9.5)$$

(see Example 9.17).

EXAMPLES 9.2 **Partial differentiation**

(i) $f(x, y, z) = x^2 + 2y^2 + 3z^2 + 4xy + 5xz + 6yz$

$$\frac{\partial f}{\partial x} = 2x + 4y + 5z, \qquad \frac{\partial f}{\partial y} = 4y + 4x + 6z, \qquad \frac{\partial f}{\partial z} = 6z + 5x + 6y$$

(ii) $f(x, y) = (x^2 + 2y^2)^{1/2}$

Let $f = u^{1/2}$ where $u = x^2 + 2y^2$. Then, by the chain rule,

$$\frac{\partial f}{\partial x} = \frac{df}{du} \times \frac{\partial u}{\partial x} = \frac{1}{2}u^{-1/2} \times 2x = x(x^2 + 2y^2)^{-1/2}$$

$$\frac{\partial f}{\partial y} = \frac{df}{du} \times \frac{\partial u}{\partial y} = \frac{1}{2}u^{-1/2} \times 4y = 2y(x^2 + 2y^2)^{-1/2}$$

(iii) $f(x, y) = y \sin(x^2 + y^2)$

By the chain rule, $\dfrac{\partial f}{\partial x} = 2xy \cos(x^2 + y^2)$

To find $\dfrac{\partial f}{\partial y}$, let $f = u \times v$ where $u = y$ and $v = \sin(x^2 + y^2)$. Then, by the product rule,

$$\frac{\partial f}{\partial y} = u \times \frac{\partial v}{\partial y} + v \times \frac{\partial u}{\partial y} = y \times 2y \cos(x^2 + y^2) + \sin(x^2 + y^2) \times 1$$

$$= 2y^2 \cos(x^2 + y^2) + \sin(x^2 + y^2)$$

▸ Exercises 3–7

Higher derivatives

Like the derivative of a function of one variable (Section 4.9), the partial derivative of a function of more than one variable can itself be differentiated if it satisfies the necessary conditions of continuity and smoothness. For example, the cubic function in two variables

$$z = x^3 + 2x^2y + 3xy^2 + 4y^3$$

has partial first derivatives

$$\frac{\partial z}{\partial x} = 3x^2 + 4xy + 3y^2, \qquad \frac{\partial z}{\partial y} = 2x^2 + 6xy + 12y^2$$

and each of these can be differentiated with respect to either variable to give four partial second derivatives. Differentiation of $\partial z/\partial x$ gives

$$\frac{\partial}{\partial x}\left(\frac{\partial z}{\partial x}\right) = \frac{\partial^2 z}{\partial x^2} = 6x + 4y, \qquad \frac{\partial}{\partial y}\left(\frac{\partial z}{\partial x}\right) = \frac{\partial^2 z}{\partial y \partial x} = 4x + 6y$$

and differentiation of $\partial z/\partial y$ gives

$$\frac{\partial}{\partial y}\left(\frac{\partial z}{\partial y}\right) = \frac{\partial^2 z}{\partial y^2} = 6x + 24y, \qquad \frac{\partial}{\partial x}\left(\frac{\partial z}{\partial y}\right) = \frac{\partial^2 z}{\partial x \partial y} = 4x + 6y$$

For a function of two variables, there are a possible 8 partial third derivatives, 16 fourth derivatives, and so on; in general a possible 2^n nth derivatives. In terms of Figure 9.3, the first derivative $\partial z/\partial x$ is the gradient at a point, P say, on the curve APB, and $\partial^2 z/\partial x^2$ is the rate of change of this gradient as the point P moves along the curve. On the other hand, the 'mixed' second derivative $\partial^2 z/\partial y \partial x = \partial(\partial z/\partial x)/\partial y$ is the rate of change of the gradient $\partial z/\partial x$ (in the x-direction) as the point P moves along the curve DPE (in the perpendicular y-direction).

We note that, for the cubic function, the two mixed second derivatives are identical. This is true for functions whose first derivatives are continuous, and is therefore (very nearly) always true in practice:

$$\frac{\partial^2 z}{\partial x \partial y} = \frac{\partial^2 z}{\partial y \partial x} \tag{9.6}$$

Similar results are obtained for higher derivatives.

Alternative notations

The above symbols for partial derivatives become unwieldy for the higher derivatives, and the following more compact notation is often used:

$$f_x = \frac{\partial f}{\partial x}, \quad f_{xx} = \frac{\partial^2 f}{\partial x^2}, \quad f_{xy} = \frac{\partial^2 f}{\partial x \partial y}, \quad f_{xyz} = \frac{\partial^3 f}{\partial x \partial y \partial z}, \quad \dots \tag{9.7}$$

In this notation, equation (9.6) becomes $f_{xy} = f_{yx}$ and, for example, $f_{xxy} = f_{xyx} = f_{yxx}$ (subject to the relevant continuity conditions).

In some applications, particularly in thermodynamics, it is necessary to specify explicitly which variables (or combinations of variables) are to be kept constant. This is achieved by adding the constant variables as subscripts to the ordinary symbol for the partial derivative. For example, for a function of three variables $f(x, y, z)$, the symbol

$$\left(\frac{\partial f}{\partial x}\right)_{y, z} \tag{9.8}$$

means the derivative of f with respect to x at constant y and z.

EXAMPLES 9.3 More partial derivatives

(i) The nonzero partial derivatives of $u = x + y^2 + 2y^3$ are

$$\frac{\partial u}{\partial x} = 1, \quad \frac{\partial u}{\partial y} = 2y + 6y^2, \quad \frac{\partial^2 u}{\partial y^2} = 2 + 12y, \quad \frac{\partial^3 u}{\partial y^3} = 12$$

(ii) The first and second partial derivatives of $u = \sin x \cos y + x/y$ are

$$u_x = \frac{\partial u}{\partial x} = \cos x \cos y + \frac{1}{y}, \qquad u_y = \frac{\partial u}{\partial y} = -\sin x \sin y - \frac{x}{y^2}$$

$$u_{xx} = \frac{\partial^2 u}{\partial x^2} = -\sin x \cos y, \qquad u_{yy} = \frac{\partial^2 u}{\partial y^2} = -\sin x \cos y + \frac{2x}{y^3}$$

$$u_{xy} = \frac{\partial^2 u}{\partial x \partial y} = -\cos x \sin y - \frac{1}{y^2}, \qquad u_{yx} = \frac{\partial^2 u}{\partial y \partial x} = -\cos x \sin y - \frac{1}{y^2}$$

and $u_{xy} = u_{yx}$

(iii) For the ideal gas, $V = nRT/p$ and

$$\left(\frac{\partial V}{\partial p} \right)_{T,n} = -\frac{nRT}{p^2}, \quad \left(\frac{\partial V}{\partial T} \right)_{p,n} = \frac{nR}{p}, \quad \left(\frac{\partial V}{\partial n} \right)_{p,T} = \frac{RT}{p}$$

▸ Exercises 8–20

9.4 Stationary points

We saw in Section 4.10 that a function $f(x)$ of one variable has a stationary value at point $x = a$ if its derivative at that point is zero; that is, if $f'(a) = 0$. Geometrically, the graph of the function has zero slope at the stationary point; its tangent line is 'horizontal'. The corresponding condition for a function of two variables is that the *tangent plane* be horizontal. A function $f(x, y)$ then has stationary point at $(x, y) = (a, b)$ if its partial first derivatives are zero:

$$\frac{\partial f}{\partial x} = \frac{\partial f}{\partial y} = 0 \quad \text{at } (a, b) \tag{9.9}$$

or $f_x(a, b) = f_y(a, b) = 0$. In view of equation (9.5), these are sufficient for all the first derivatives of a continuous function to be zero at a point.

EXAMPLE 9.4 Find the stationary points of the function

$$f(x, y) = x^3 + 6xy^2 - 2y^3 - 12x$$

We have

$$\frac{\partial f}{\partial x} = 3x^2 + 6y^2 - 12, \qquad \frac{\partial f}{\partial y} = 12xy - 6y^2$$

and these are zero when

$$x^2 + 2y^2 = 4, \qquad y(2x - y) = 0$$

The second equation is satisfied when either $y = 0$ or $y = 2x$, and substitution of these in the first equation gives the four stationary points

$$(2, 0), \quad (-2, 0), \quad (2/3, 4/3), \quad (-2/3, -4/3)$$

▸ Exercises 21–23

For a function of one variable, a stationary point at $x = a$ is a local maximum if the second derivative is negative, $f''(a) < 0$, a local minimum if $f''(a) > 0$, and may be a point of inflection if $f''(a) = 0$. The corresponding conditions for a function of two variables are

$$f_{xx} < 0 \quad \text{and} \quad f_{yy} < 0 \quad \text{for a maximum} \tag{9.10a}$$

$$f_{xx} > 0 \quad \text{and} \quad f_{yy} > 0 \quad \text{for a minimum} \tag{9.10b}$$

and

$$f_{xx}f_{yy} - f_{xy}^2 > 0 \quad \text{for either a maximum or a minimum} \tag{9.10c}$$

If the quantity $\Delta = f_{xx}f_{yy} - f_{xy}^2$ is negative then the point is a **saddle point**; a maximum in one direction and a minimum in another. If $\Delta = 0$ then further tests are required to determine the nature of the point. The corresponding conditions are more complicated for functions of more than two variables.

EXAMPLE 9.5 The nature of the stationary points of Example 9.4.

The stationary points of the function $f(x, y) = x^3 + 6xy^2 - 2y^3 - 12x$ are

$$(2, 0), \quad (-2, 0), \quad (2/3, 4/3), \quad (-2/3, -4/3)$$

The determination of the nature of these points is summarized in the following table.

Table 9.1

x	y	f_{xx}	f_{yy}	f_{xy}	$f_{xx}f_{yy} - f_{xy}^2$	nature
2	0	12	24	0	>0	minimum
-2	0	-12	-24	0	>0	maximum
$\dfrac{2}{3}$	$\dfrac{4}{3}$	4	-8	16	<0	saddle point
$-\dfrac{2}{3}$	$-\dfrac{4}{3}$	-4	8	-16	<0	saddle point

➤ Exercises 24–26

Optimization with constraints

The finding of the maxima and minima (**extremum values**) of a function is called **optimization**. In Example 9.4 the variables x and y are independent variables, with no constraints on their values. In many applications in the physical sciences, however, the optimization may be subject to one or more constraints; we have **constrained optimization**. These constraints usually take the form of one or more relations amongst the variables.

EXAMPLE 9.6 Find the extremum value of the function

$$f(x, y) = 3x^2 - 2y^2$$

subject to the constraint $x + y = 2$.

In the absence of the constraint, the function has a saddle point at $x = y = 0$, and no maxima or minima. The constraint is a relation between the variables x and y that reduces the number of independent variables to 1. In the present case, the search for a stationary value is restricted to the line $y = 2 - x$. Thus, substituting $y = 2 - x$ in the function gives

$$f(x, y) = F(x) = 3x^2 - 2(2 - x)^2 = x^2 + 8x - 8$$

Then, for a stationary value,

$$F'(x) = 2x + 8 = 0, \quad F''(x) = 2$$

so that $F(x)$ has minimum value for $x = -4$. The extremum point of $f(x, y)$ is therefore a minimum point at $(x, y) = (-4, 6)$, and the extremum value of the function is $f(-4, 6) = -24$.

➤ Exercises 27(i), 28(i)

The optimization problem in this example has been simplified by using the constraint to eliminate one of the variables. In general, however, such a simplification is either difficult or impossible, particularly for functions of more than two variables or when there are several constraint relations. A general procedure for solving many of the constrained optimization problems in the physical sciences is the **method of Lagrange multipliers.**[3]

We consider a function of three variables, $f(x, y, z)$, and the constraint relation $g(x, y, z) = \text{constant}$. By the method of Lagrangian multipliers there exists a number λ such that

$$\frac{\partial f}{\partial x} - \lambda \frac{\partial g}{\partial x} = 0, \quad \frac{\partial f}{\partial y} - \lambda \frac{\partial g}{\partial y} = 0, \quad \frac{\partial f}{\partial z} - \lambda \frac{\partial g}{\partial z} = 0 \qquad (9.11)$$

at the stationary points. These three equations and the constraint relation are sufficient to determine the values of the stationary points and of the corresponding values of the **Lagrange multiplier** λ.

EXAMPLE 9.7 Use the method of Lagrange multipliers to find the extremum value of the function $f = 3x^2 - 2y^2$ subject to the constraint $x + y = 2$ (Example 9.6).

We have $g = x + y$ and $f - \lambda g = 3x^2 - 2y^2 - \lambda(x + y)$, so that the three equations to be solved are

$$\frac{\partial}{\partial x}(f - \lambda g) = 6x - \lambda = 0, \quad \frac{\partial}{\partial y}(f - \lambda g) = -4y - \lambda = 0, \quad g = x + y = 2$$

The solution is $x = -4$, $y = 6$, $\lambda = -24$, and $f(-4, 6) = -24$ as in Example 9.6.

▸ Exercise 27(ii), 28(ii)

In the general case, the optimization is that of a function of n variables,

$$f(x_1, x_2, x_3, \ldots, x_n) \qquad (9.12a)$$

subject to m $(< n)$ constraint relations

$$g_k(x_1, x_2, x_3, \ldots, x_n) = a_k, \qquad k = 1, 2, 3, \ldots, m \qquad (9.12b)$$

---

[3] Joseph-Louis Lagrange (1736–1813), born in Turin, made important contributions to many branches of mathematics, and has been called the greatest mathematician of the 18th century ('Lagrange is the lofty pyramid of the mathematical sciences', Napoleon Bonaparte). His greatest achievement was the development of the calculus of variations, and in his *Mécanique analytique* of 1788 he extended the mechanics of Newton and Euler. He emphasized that problems in mechanics can generally be solved by reducing them to differential equations. Lagrange invented the name 'derived function' (hence 'derivative') and notation $f'(x)$ for the derivative of $f(x)$.

where $a_1, a_2, ..., a_m$ are constants. The procedure is to construct the **auxiliary function**

$$\phi = f - \lambda_1 g_1 - \lambda_2 g_2 - \lambda_3 g_3 - \cdots - \lambda_m g_m = f - \sum_{k=1}^{m} \lambda_k g_k \qquad (9.13)$$

and to solve the $n+m$ equations

$$\frac{\partial \phi}{\partial x_i} = \frac{\partial f}{\partial x_i} - \sum_{k=1}^{m} \lambda_k \frac{\partial g_k}{\partial x_i} = 0, \qquad i = 1, 2, 3, ..., n$$

$$g_k = a_k, \qquad k = 1, 2, 3, ..., m \qquad (9.14)$$

for the n variables x_i and the m multipliers λ_k.

The following examples demonstrate one application of the method of Lagrangian multipliers in geometry (or packaging) and one important application in chemistry. Another application, the derivation of the Boltzmann distribution, is discussed in Chapter 21.

EXAMPLE 9.8 Find the dimensions of the rectangular box of largest volume for given surface area.

The volume of a box of sides x, y, and z is $V=xyz$ and its surface area is $A = 2(xy + yz + zx)$. The problem is therefore to find the maximum value of V subject to the constraint $A = $ constant. By the method of Lagrangian multipliers we form the auxiliary function $\phi = V - \lambda A$, and solve the set of equations

$$\frac{\partial \phi}{\partial x} = yz - 2\lambda(y+z) = 0, \quad \frac{\partial \phi}{\partial y} = xz - 2\lambda(x+z) = 0,$$

$$\frac{\partial \phi}{\partial z} = xy - 2\lambda(x+y) = 0$$

Multiplication of the first equation by x, the second by y, and the third by z gives

$$xyz - 2\lambda(xy + xz) = xyz - 2\lambda(xy + yz) = xyz - 2\lambda(xz + yz) = 0$$

It follows that $x=y=z$, and the box is a cube of side $\sqrt{A/6}$.

▶ Exercise 29

EXAMPLE 9.9 Secular equations

Variation principles in the physical sciences often lead to the problem of finding the stationary values of a 'quadratic form' in the n variables, $x_1, x_2, ..., x_n$,

$$f(x_1, x_2, ..., x_n) = \sum_{i=1}^{n} \sum_{j=1}^{n} C_{ij} x_i x_j$$

subject to the constraint

$$g(x_1, x_2, ..., x_n) = \sum_{i=1}^{n} x_i^2 = 1$$

where the C_{ij} are constants (with $C_{ij} = C_{ji}$). For example, for $n = 3$,

$$f(x_1, x_2, x_3) = C_{11}x_1^2 + 2C_{12}x_1x_2 + 2C_{13}x_1x_3 + C_{22}x_2^2 + 2C_{23}x_2x_3 + C_{33}x_3^2$$

with constraint $g = x_1^2 + x_2^2 + x_3^2 = 1$.

By the method of Lagrange multipliers the stationary values of the function are obtained by forming the auxiliary function $\phi = f - \lambda g$ and solving equations (9.14). For the case $n = 3$, we have

$$\phi = (C_{11} - \lambda)x_1^2 + 2C_{12}x_1x_2 + 2C_{13}x_1x_3 + (C_{22} - \lambda)x_2^2 + 2C_{23}x_2x_3 + (C_{33} - \lambda)x_3^2$$

Differentiation with respect to x_1, x_2, and x_3 and setting each derivative to zero then gives the set of simultaneous equations

$$(C_{11} - \lambda)x_1 + C_{12}x_2 + C_{13}x_3 = 0$$
$$C_{21}x_1 + (C_{22} - \lambda)x_2 + C_{23}x_3 = 0$$
$$C_{31}x_1 + C_{32}x_2 + (C_{33} - \lambda)x_3 = 0$$

Equations of this kind are often called **secular equations**. They occur, for example, in the 'method of linear combinations' in quantum chemistry, when the quadratic form represents the energy of the system (or an orbital energy in molecular orbital theory) and the numbers x_1, x_2, ..., x_n provide a representation of the state (of an orbital in molecular orbital theory). The significance and solution of such systems of equations are discussed in Chapters 17 and 19.

▸ Exercise 30

9.5 The total differential

Let $z = f(x, y)$ be a function of the variables x and y, and let the values of the variables change continuously from (x, y), at point p in Figure 9.5, to $(x + \Delta x, y + \Delta y)$ at point q. The corresponding change in the function is

$$\Delta z = z_q - z_p = f(x + \Delta x, y + \Delta y) - f(x, y)$$

and is shown in the figure as the displacement P to Q on the representative surface of the function; Δz is the change of 'height' on the surface.

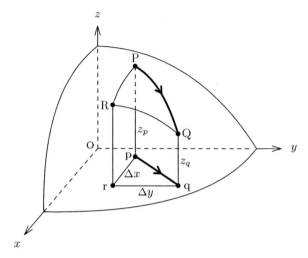

Figure 9.5

For example, for the quadratic function $ax^2 + bxy + cy^2$,

$$z_p = f(x, y) = ax^2 + bxy + cy^2$$

$$z_q = f(x + \Delta x, y + \Delta y) = a(x + \Delta x)^2 + b(x + \Delta x)(y + \Delta y) + c(y + \Delta y)^2$$

and

$$\Delta z = z_q - z_p = (2ax + by)\Delta x + (bx + 2cy)\Delta y + a(\Delta x)^2 + b(\Delta x)(\Delta y) + c(\Delta y)^2$$

Now, because

$$\frac{\partial z}{\partial x} = 2ax + by, \qquad \frac{\partial z}{\partial y} = bx + 2cy$$

$$\frac{1}{2}\frac{\partial^2 z}{\partial x^2} = a, \qquad \frac{\partial^2 z}{\partial x \partial y} = b, \qquad \frac{1}{2}\frac{\partial^2 z}{\partial y^2} = c$$

the change in the function can be written as

$$\Delta z = \left(\frac{\partial z}{\partial x}\right)\Delta x + \left(\frac{\partial z}{\partial y}\right)\Delta y$$

$$+ \frac{1}{2}\left(\frac{\partial^2 z}{\partial x^2}\right)(\Delta x)^2 + \left(\frac{\partial^2 z}{\partial x \partial y}\right)(\Delta x)(\Delta y) + \frac{1}{2}\left(\frac{\partial^2 z}{\partial y^2}\right)(\Delta y)^2$$

(9.15)

This expression is exact for the present case, and for all functions whose third and higher derivatives are zero. For a general function $f(x, y)$, (9.15) represent the first few terms of a Taylor expansion of the function at point $(x + \Delta x, y + \Delta y)$ about the point (x, y) (compare with equation (7.24) for a function of one variable).

If Δx and Δy are small enough, the terms quadratic in Δ are small compared with the linear terms, and an approximate value of Δz is

$$\Delta z \approx \left(\frac{\partial z}{\partial x}\right)_y \Delta x + \left(\frac{\partial z}{\partial y}\right)_x \Delta y \qquad (9.16)$$

This result, valid for *all* continuous functions of two variables, shows that when the changes Δx and Δy are small enough the total change in z is approximately equal to the change in z due to change Δx alone (the first term of the expression on the right of (9.16)) plus the change in z due to change Δy alone (the second term). In addition, the accuracy of the expression improves as Δx and Δy approach zero. As in Section 4.12, we consider infinitesimal changes dx and dy, and define the quantity

$$dz = \left(\frac{\partial z}{\partial x}\right)_y dx + \left(\frac{\partial z}{\partial y}\right)_x dy \qquad (9.17)$$

as the limiting case of (9.16). This quantity is called the **total differential** of z with respect to x and y.[4]

The concept of the total differential is readily generalized for functions of any number of variables; for a function of n variables,

$$z = f(x_1, x_2, x_3, \ldots, x_n)$$

the total differential is

$$dz = \left(\frac{\partial z}{\partial x_1}\right) dx_1 + \left(\frac{\partial z}{\partial x_2}\right) dx_2 + \cdots + \left(\frac{\partial z}{\partial x_n}\right) dx_n = \sum_{i=1}^n \left(\frac{\partial z}{\partial x_i}\right) dx_i \qquad (9.18)$$

where, for example, $\partial z/\partial x_1$ is the partial derivative with respect to variable x_1 with all other variables, $x_2, x_3, \ldots, x_n$, kept constant.

EXAMPLES 9.10 Find the total differential:

(i) $z = \dfrac{x}{y}$

$$\frac{\partial z}{\partial x} = \frac{1}{y}, \quad \frac{\partial z}{\partial y} = -\frac{x}{y^2}, \quad dz = \left(\frac{\partial z}{\partial x}\right)_y dx + \left(\frac{\partial z}{\partial y}\right)_x dy = \frac{1}{y}dx - \frac{x}{y^2}dy$$

---

[4] The total differential and the equality of the mixed second derivatives were discovered in 1719 by Nicolaus (II) Bernoulli (1687–1759). The nephew of Johann (I) and Jakob (John and James, or Jean et Jacques), he is not to be confused with Nicolaus (I), his father, nor with Nicolaus (III), the son of Johann (I) and brother of Daniel and Johann (II). It is Daniel Bernoulli (1700–1782) who wrote the *Hydrodynamica* of 1738 which contains the concept of 'Bernoulli's Theorem' and a development of the kinetic theory of gases.

(ii) $u = (x^2 + y^2 + z^2)^{1/2}$

$$\frac{\partial u}{\partial x} = x(x^2 + y^2 + z^2)^{-1/2} = \frac{x}{u}, \quad \frac{\partial u}{\partial y} = \frac{y}{u}, \quad \frac{\partial u}{\partial z} = \frac{z}{u}$$

$$du = \left(\frac{\partial u}{\partial x}\right)_{y,z} dx + \left(\frac{\partial u}{\partial y}\right)_{z,x} dy + \left(\frac{\partial u}{\partial z}\right)_{x,y} dz = \frac{1}{u}(x\,dx + y\,dy + z\,dz)$$

(iii) $x = r \sin\theta \cos\phi$

$$\frac{\partial x}{\partial r} = \sin\theta \cos\phi, \quad \frac{\partial x}{\partial \theta} = r\cos\theta\cos\phi, \quad \frac{\partial x}{\partial \phi} = -r\sin\theta\sin\phi$$

$$dx = \left(\frac{\partial x}{\partial r}\right)_{\theta,\phi} dr + \left(\frac{\partial x}{\partial \theta}\right)_{r,\phi} d\theta + \left(\frac{\partial x}{\partial \phi}\right)_{r,\theta} d\phi$$

$$= \sin\theta\cos\phi\,dr + r\cos\theta\cos\phi\,d\theta - r\sin\theta\sin\phi\,d\phi$$

> Exercises 31–35

EXAMPLE 9.11 Differential volume

The volume of a one-component thermodynamic system is a function of pressure p, temperature T, and amount of substance n: $V = V(p, T, n)$. The total differential volume is

$$dV = \left(\frac{\partial V}{\partial T}\right)_{p,n} dT + \left(\frac{\partial V}{\partial p}\right)_{T,n} dp + \left(\frac{\partial V}{\partial n}\right)_{p,T} dn$$

$$= \alpha V\,dT - \kappa V\,dp + V_m\,dn$$

where

$$\alpha = \frac{1}{V}\left(\frac{\partial V}{\partial T}\right)_{p,n} \quad \text{the thermal expansivity (coefficient of thermal expansion)}$$

$$\kappa = -\frac{1}{V}\left(\frac{\partial V}{\partial p}\right)_{T,n} \quad \text{the isothermal compressibility}$$

$$V_m = \left(\frac{\partial V}{\partial n}\right)_{p,T} \quad \text{the molar volume}$$

> Exercise 36

One of the principal uses of the total differential in the physical sciences is in the formulation of the laws of thermodynamics (see Examples 9.22 and 9.27). In the following sections we use (9.16) and its limiting form (9.17) to derive a number of differential and integral properties of functions.

9.6 Some differential properties

The total derivative

In the function of two variables, $z=f(x, y)$, let x and y be functions of a third variable t:

$$x=x(t), \qquad y=y(t) \tag{9.19}$$

Then $z=f(x(t), y(t))$ is essentially a function of the single variable t, and there exists an ordinary derivative dz/dt. This derivative is called the **total derivative** of z with respect to t, and can be obtained directly by substituting the functions $x(t)$ and $y(t)$ for the variables in $f(x, y)$ and differentiating the resulting function of t. It can also be obtained indirectly, by dividing the expression (9.16) by Δt and taking the limit $\Delta t \to 0$. Thus, division of

$$\Delta z \approx \left(\frac{\partial z}{\partial x}\right)_y \Delta x + \left(\frac{\partial z}{\partial y}\right)_x \Delta y$$

by Δt gives

$$\frac{\Delta z}{\Delta t} \approx \left(\frac{\partial z}{\partial x}\right)_y \frac{\Delta x}{\Delta t} + \left(\frac{\partial z}{\partial y}\right)_x \frac{\Delta y}{\Delta t} \tag{9.20}$$

and, in the limit $\Delta t \to 0$,

$$\frac{dz}{dt} = \left(\frac{\partial z}{\partial x}\right)_y \frac{dx}{dt} + \left(\frac{\partial z}{\partial y}\right)_x \frac{dy}{dt} \tag{9.21}$$

Alternatively, the same result is obtained by dividing the total differential (9.17) by (infinitesimal) dt.

Equation (9.21) is a generalization of the chain rule (see Section 4.6). For a function of n variables, $u=f(x_1, x_2, x_3, \ldots, x_n)$, in which the variables are all functions of t,

$$\frac{du}{dt} = \left(\frac{\partial u}{\partial x_1}\right)\frac{dx_1}{dt} + \left(\frac{\partial u}{\partial x_2}\right)\frac{dx_2}{dt} + \cdots + \left(\frac{\partial u}{\partial x_n}\right)\frac{dx_n}{dt}$$

$$= \sum_{i=1}^{n} \left(\frac{\partial u}{\partial x_i}\right)\frac{dx_i}{dt} \tag{9.22}$$

EXAMPLE 9.12 Given $z = x^2 + y^3$, where $x = e^t$ and $y = e^{-t}$, find dz/dt.

(i) By substitution:

$$z = e^{2t} + e^{-3t}, \qquad \frac{dz}{dt} = 2e^{2t} - 3e^{-3t} = 2x^2 - 3y^3$$

(ii) By the chain rule (9.21), we have

$$\frac{\partial z}{\partial x} = 2x, \quad \frac{\partial z}{\partial y} = 3y^2 \quad \text{and} \quad \frac{dx}{dt} = e^t = x, \quad \frac{dy}{dt} = -e^{-t} = -y$$

Therefore,

$$\frac{dz}{dt} = \frac{\partial z}{\partial x}\frac{dx}{dt} + \frac{\partial z}{\partial y}\frac{dy}{dt} = (2x) \times (x) + (3y^2) \times (-y) = 2x^2 - 3y^3$$

and this is identical to the result obtained by substitution.

▸ Exercises 37–39

EXAMPLE 9.13 Walking on a circle

The equation of a circle in the xy-plane with centre at the origin and radius a is $x^2 + y^2 = a^2$. A displacement on the circle is most easily described when the equation of the circle is expressed in terms of the polar coordinates $r\, (= a)$ and θ:

$$x = a\cos\theta, \qquad y = a\sin\theta$$

These have the form of the pair of equations (9.19), with t replaced by θ. In general, such equations are called the **parametric equations** of a curve. Let $z = f(x, y)$. Then, by equation (9.21), since $dx/d\theta = -y$ and $dy/d\theta = x$,

$$\frac{dz}{d\theta} = \left(\frac{\partial z}{\partial x}\right)_y \frac{dx}{d\theta} + \left(\frac{\partial z}{\partial y}\right)_x \frac{dy}{d\theta} = -y\left(\frac{\partial z}{\partial x}\right)_y + x\left(\frac{\partial z}{\partial y}\right)_x$$

For example, if $z = xy$ then $\dfrac{\partial z}{\partial x} = y$, $\dfrac{\partial z}{\partial y} = x$, and $\dfrac{dz}{d\theta} = x^2 - y^2$

▸ Exercise 40

A special case of the total derivative (9.21) is obtained if t is replaced by x. Then $y = y(x)$ is an explicit function of x and $z = f(x, y(x))$ can be treated as a function of the single variable x. Then, from (9.21),

$$\frac{dz}{dx} = \left(\frac{\partial z}{\partial x}\right)_y + \left(\frac{\partial z}{\partial y}\right)_x \frac{dy}{dx} \qquad (9.23)$$

We note that if $x = x(t)$ then this total derivative with respect to x is related to the total derivative (9.21) with respect to t by the chain rule:

$$\frac{dz}{dt} = \frac{dz}{dx}\frac{dx}{dt} = \left(\frac{\partial z}{\partial x}\right)_y \frac{dx}{dt} + \left(\frac{\partial z}{\partial y}\right)_x \frac{dy}{dx} \frac{dx}{dt} \qquad (9.24)$$

EXAMPLE 9.14

(i) Given $z = x^2 + y^3$, where $y = 1/x$, find dz/dx. Then, (ii) if $x = e^t$, find dz/dt.

(i) By equation (9.23), since $\dfrac{dy}{dx} = -\dfrac{1}{x^2} = -y^2$,

$$\frac{dz}{dx} = \frac{\partial z}{\partial x} + \frac{\partial z}{\partial y}\frac{dy}{dx} = 2x + (3y^2) \times (-y^2) = 2x - 3y^4$$

(ii) If $x = e^t$ then $\dfrac{dx}{dt} = e^t = x$, and

$$\frac{dz}{dt} = \frac{dz}{dx}\frac{dx}{dt} = 2x^2 - 3xy^4$$

This is identical to the result of Example 9.12, since $x = 1/y$.

▸ Exercise 41

Walking on a contour

Consider changes in x and y that leave the value of the function $z = f(x, y)$ unchanged. In Figure 9.6, the plane ABC is parallel to the xy-plane, so that all points on the curve APB on the representative surface are at constant value of z. The displacement P to Q therefore lies on a contour of the surface. Then, by (9.16),

$$\Delta z = 0 \approx \left(\frac{\partial z}{\partial x}\right)_y \Delta x + \left(\frac{\partial z}{\partial y}\right)_x \Delta y \qquad (9.25)$$

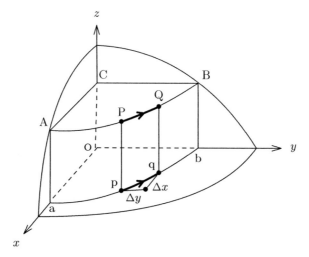

Figure 9.6

and, after division by Δx,

$$0 \approx \left(\frac{\partial z}{\partial x}\right)_y + \left(\frac{\partial z}{\partial y}\right)_x \frac{\Delta y}{\Delta x}$$

In the limit $\Delta x \to 0$, the ratio $\Delta y / \Delta x$ approaches the derivative of y with respect to x *at constant z*, and

$$0 = \left(\frac{\partial z}{\partial x}\right)_y + \left(\frac{\partial z}{\partial y}\right)_x \left(\frac{\partial y}{\partial x}\right)_z \qquad (9.26)$$

The same result is obtained directly from the total differential (9.17) by considering infinitesimal changes dx and dy such that $dz = 0$:

$$dz = 0 = \left(\frac{\partial z}{\partial x}\right)_y dx + \left(\frac{\partial z}{\partial y}\right)_x dy$$

To obtain (9.26) we divide (formally) by dx and, because the process is at constant z, we replace dy/dx by the partial derivative at constant z.

Equation (9.26) can be rearranged as

$$\left(\frac{\partial y}{\partial x}\right)_z = -\left(\frac{\partial z}{\partial x}\right)_y \bigg/ \left(\frac{\partial z}{\partial y}\right)_x \qquad (9.27)$$

and this is the gradient of the graph of the function $y(x)$ for which z is constant; that is, the gradient of the curve ab in the xy-plane in Figure 9.6 for which AB is a contour. In addition, because

$$\left(\frac{\partial z}{\partial x}\right)_y = 1 \bigg/ \left(\frac{\partial x}{\partial z}\right)_y \tag{9.28}$$

the equation can also be written as

$$-1 = \left(\frac{\partial x}{\partial y}\right)_z \left(\frac{\partial y}{\partial z}\right)_x \left(\frac{\partial z}{\partial x}\right)_y \tag{9.29}$$

This form is sometimes called the '–1 rule'.

The presence of the – sign in equations (9.27) and (9.29) sometimes causes unease, but it can be explained geometrically. Thus, for the particular z-surface shown in Figures 9.3 and 9.6 the slopes along the x and y directions are both negative at P (the gradient lines slope 'downward'), so that $(\partial z/\partial x)_y$ and $(\partial z/\partial y)_x$ are negative and their quotient in equation (9.27) is positive. However, Figure 9.6 shows that for motion along the contour from P to Q, Δy is necessarily positive but Δx negative (or vice versa for motion Q to P) so that $(\partial y/\partial x)_z$ is negative. Hence the – sign. Similar considerations apply to the other three possible pairs of signs of the slopes.

We note that the distinction between dependent and independent variables has disappeared from equations (9.26) to (9.29); any one variable can be regarded as a function of the other two. We note also that whereas (9.26) is readily generalized for sets of more than three variables, equations (9.27) and (9.29) are true for only three variables at a time (all others being kept constant).

EXAMPLES 9.15

(i) Given $z = x^2 y^3$, find $\left(\dfrac{\partial y}{\partial x}\right)_z$.

$$\left(\frac{\partial z}{\partial x}\right)_y = 2xy^3, \qquad \left(\frac{\partial z}{\partial y}\right)_x = 3x^2 y^2$$

By equation (9.27),

$$\left(\frac{\partial y}{\partial x}\right)_z = -\left(\frac{\partial z}{\partial x}\right)_y \bigg/ \left(\frac{\partial z}{\partial y}\right)_x = -\frac{2xy^3}{3x^2 y^2} = -\frac{2y}{3x}$$

(ii) For $r = \left(x^2 + y^2\right)^{1/2}$, find $\left(\dfrac{\partial y}{\partial x}\right)_r$.

$$\frac{\partial r}{\partial x} = \frac{x}{r}, \quad \frac{\partial r}{\partial y} = \frac{y}{r}, \quad \text{and} \quad \left(\frac{\partial y}{\partial x}\right)_r = -\frac{x}{y}$$

The representative surface of the function $r=(x^2+y^2)^{1/2}$ is a vertical cone whose contours ($r=$ constant) are circles of radius r parallel to the xy-plane. The quantity $-x/y$ is the gradient at (x, y) on the circle.

(iii) The differential volume of a thermodynamic system is (see Example 9.11)

$$dV = \left(\frac{\partial V}{\partial T}\right)_p dT + \left(\frac{\partial V}{\partial p}\right)_T dp \quad \text{(for fixed amount } n\text{)}.$$

By equation (9.29), $-1 = \left(\frac{\partial V}{\partial p}\right)_T \left(\frac{\partial p}{\partial T}\right)_V \left(\frac{\partial T}{\partial V}\right)_p$

so that

$$\left(\frac{\partial p}{\partial T}\right)_V = -\left(\frac{\partial V}{\partial T}\right)_p \Big/ \left(\frac{\partial V}{\partial p}\right)_T = \frac{\alpha}{\kappa} = \frac{\text{expansivity}}{\text{compressibility}}$$

▸ Exercises 42, 43

Change of constant variable

Let $z=f(x, y)$ be a function of the variables x and y, and let $u=g(x, y)$ be some *other* function of x and y. Consider changes in the variables such that u is constant. Then, division of the total differential dz by infinitesimal change dx at *constant u* gives

$$\left(\frac{\partial z}{\partial x}\right)_u = \left(\frac{\partial z}{\partial x}\right)_y + \left(\frac{\partial z}{\partial y}\right)_x \left(\frac{\partial y}{\partial x}\right)_u \qquad (9.30)$$

This equation shows how the partial derivative with respect to x at constant y is related to the partial derivative with respect to x when some function $u(x, y)$ of x and y is kept constant.

EXAMPLE 9.16 Let $z=f(x, y)$ and $u=x^2+y^2$. Then, by equation (9.27),

$$\left(\frac{\partial y}{\partial x}\right)_u = -\left(\frac{\partial u}{\partial x}\right)_y \Big/ \left(\frac{\partial u}{\partial y}\right)_x = -\frac{x}{y}$$

and, therefore, by equation (9.30),

$$\left(\frac{\partial z}{\partial x}\right)_u = \left(\frac{\partial z}{\partial x}\right)_y - \frac{x}{y}\left(\frac{\partial z}{\partial y}\right)_x$$

This expression is closely related to that in Example 9.13 (walking on a circle) for $dz/d\theta$. Thus, when u is constant $(= a^2$, say), $u = x^2 + y^2 = a^2$ is the equation of a circle of radius a, and displacements at constant u are therefore constrained to the circle, with $x = a \cos \theta$ and $dx/d\theta = -y$. Then,

$$\frac{dz}{d\theta} = \left(\frac{\partial z}{\partial x}\right)_u \frac{dx}{d\theta} = -y \left(\frac{\partial z}{\partial x}\right)_u$$

For example, if $z = xy$ then $\left(\dfrac{\partial z}{\partial x}\right)_u = y - x^2/y$ and $\dfrac{dz}{d\theta} = x^2 - y^2$

> Exercises 44–46

_____

Change of independent variables

Let $z = f(x, y)$ be a function of the independent variables x and y with total differential of z with respect to x and y,

$$dz = \left(\frac{\partial z}{\partial x}\right)_y dx + \left(\frac{\partial z}{\partial y}\right)_x dy \tag{9.17}$$

Let the variables x and y be themselves functions of two other independent variables, u and v:

$$x = x(u, v), \qquad y = y(u, v) \tag{9.31}$$

Then z can be treated as a function of u and v, and its total differential with respect to the new variables is

$$dz = \left(\frac{\partial z}{\partial u}\right)_v du + \left(\frac{\partial z}{\partial v}\right)_u dv \tag{9.32}$$

The relationships between the partial derivatives in (9.32) and those in (9.17) can be obtained in the following way. Divide the total differential (9.17) by du at constant v to give equation (9.33a), and by dv at constant u to give (9.33b):

$$\left(\frac{\partial z}{\partial u}\right)_v = \left(\frac{\partial z}{\partial x}\right)_y \left(\frac{\partial x}{\partial u}\right)_v + \left(\frac{\partial z}{\partial y}\right)_x \left(\frac{\partial y}{\partial u}\right)_v \tag{9.33a}$$

$$\left(\frac{\partial z}{\partial v}\right)_u = \left(\frac{\partial z}{\partial x}\right)_y \left(\frac{\partial x}{\partial v}\right)_u + \left(\frac{\partial z}{\partial y}\right)_x \left(\frac{\partial y}{\partial v}\right)_u \tag{9.33b}$$

The inverse relationships are obtained in the same way from (9.32):

$$\left(\frac{\partial z}{\partial x}\right)_y = \left(\frac{\partial z}{\partial u}\right)_v \left(\frac{\partial u}{\partial x}\right)_y + \left(\frac{\partial z}{\partial v}\right)_u \left(\frac{\partial v}{\partial x}\right)_y \tag{9.34a}$$

$$\left(\frac{\partial z}{\partial y}\right)_x = \left(\frac{\partial z}{\partial u}\right)_v \left(\frac{\partial u}{\partial y}\right)_x + \left(\frac{\partial z}{\partial v}\right)_u \left(\frac{\partial v}{\partial y}\right)_x \tag{9.34b}$$

EXAMPLE 9.17 From cartesian to polar coordinates

Let $z = f(x, y)$ be a function of the cartesian coordinates of a point in the xy-plane. The position of the point is specified equally well in terms of the polar coordinates r and θ, where $x = x(r, \theta) = r \cos \theta$ and $y = y(r, \theta) = r \sin \theta$. Then, replacing u by r and v by θ in equations (9.33),

$$\left(\frac{\partial z}{\partial r}\right)_\theta = \left(\frac{\partial z}{\partial x}\right)_y \left(\frac{\partial x}{\partial r}\right)_\theta + \left(\frac{\partial z}{\partial y}\right)_x \left(\frac{\partial y}{\partial r}\right)_\theta = \left(\frac{\partial z}{\partial x}\right)_y \cos \theta + \left(\frac{\partial z}{\partial y}\right)_x \sin \theta$$

$$\left(\frac{\partial z}{\partial \theta}\right)_r = \left(\frac{\partial z}{\partial x}\right)_y \left(\frac{\partial x}{\partial \theta}\right)_r + \left(\frac{\partial z}{\partial y}\right)_x \left(\frac{\partial y}{\partial \theta}\right)_r = -y \left(\frac{\partial z}{\partial x}\right)_y + x \left(\frac{\partial z}{\partial y}\right)_x$$

$$\tag{9.35}$$

The first of these is identical to equation (9.5) and has the same graphical interpretation. The second is identical to the result obtained in Example 9.13 for motion around a circle. The inverse relationships are, by equations (9.34),

$$\left(\frac{\partial z}{\partial x}\right)_y = \left(\frac{\partial z}{\partial r}\right)_\theta \left(\frac{\partial r}{\partial x}\right)_y + \left(\frac{\partial z}{\partial \theta}\right)_r \left(\frac{\partial \theta}{\partial x}\right)_y = \cos \theta \left(\frac{\partial z}{\partial r}\right)_\theta - \frac{\sin \theta}{r} \left(\frac{\partial z}{\partial \theta}\right)_r$$

$$\tag{9.36}$$

$$\left(\frac{\partial z}{\partial y}\right)_x = \left(\frac{\partial z}{\partial r}\right)_\theta \left(\frac{\partial r}{\partial y}\right)_x + \left(\frac{\partial z}{\partial \theta}\right)_r \left(\frac{\partial \theta}{\partial y}\right)_x = \sin \theta \left(\frac{\partial z}{\partial r}\right)_\theta + \frac{\cos \theta}{r} \left(\frac{\partial z}{\partial \theta}\right)_r$$

▸ Exercises 47, 48

Laplace's equation in two dimensions

The Laplace equation in two dimensions is

$$\frac{\partial^2 f}{\partial x^2} + \frac{\partial^2 f}{\partial y^2} = 0 \tag{9.37}$$

where $f = f(x, y)$ is a function of the cartesian coordinates of a point in a plane. The Laplace equation occurs in many branches of the physical sciences, and is the fundamental equation in 'potential theory', when a physical system is described in terms of a potential function; for example, the gravitational and electrostatic potential functions in a region free of matter satisfy the Laplace equation in three dimensions (see Chapter 10). The equation in two dimensions is important in flow theories; for example in the theory of fluid flow and of heat conduction.[5]

As in Example 9.17, the position of a point in a plane can be specified in terms of the polar coordinates r and θ, where $x = r \cos \theta$ and $y = r \sin \theta$. The function f can therefore be treated as a function of r and θ, $f = f(r, \theta)$, and equation (9.37) can be transformed from an equation in cartesian coordinates to an equation in polar coordinates by the method described in Example 9.17. Example 9.18 shows how this is done, with result

$$\frac{\partial^2 f}{\partial x^2} + \frac{\partial^2 f}{\partial y^2} = \frac{\partial^2 f}{\partial r^2} + \frac{1}{r}\frac{\partial f}{\partial r} + \frac{1}{r^2}\frac{\partial^2 f}{\partial \theta^2} = 0$$

The differential operator

$$\begin{aligned}\nabla^2 &= \frac{\partial^2}{\partial x^2} + \frac{\partial^2}{\partial y^2} \\ &= \frac{\partial^2}{\partial r^2} + \frac{1}{r}\frac{\partial}{\partial r} + \frac{1}{r^2}\frac{\partial^2}{\partial \theta^2}\end{aligned}$$

(9.38)

(read as 'del-squared' or 'nabla-squared') is called the **Laplacian operator** (although the symbol and the name are usually reserved for the three-dimensional form; see Chapter 10). The Laplace equation is then

$$\nabla^2 f = 0 \qquad\qquad (9.39)$$

and a solution f of the equation as known as a **harmonic function**.

EXAMPLE 9.18 The two-dimensional Laplacian in polar coordinates

By the first of equations (9.36) in Example 9.17,

$$\frac{\partial f}{\partial x} = \cos \theta \frac{\partial f}{\partial r} - \frac{\sin \theta}{r}\frac{\partial f}{\partial \theta}$$

---

[5] Pierre Simon de Laplace (1749–1827). His *Traité de mécanique céleste* (Treatise on celestial mechanics) in 5 volumes (1799–1825) marked the culmination of the Newtonian view of gravitation. Legend has it that whilst at the École Militaire, where he taught elementary mathematics to the cadets, he examined, and passed, Napoleon in 1785.

Then

$$\frac{\partial^2 f}{\partial x^2} = \left(\cos\theta\,\frac{\partial}{\partial r} - \frac{\sin\theta}{r}\,\frac{\partial}{\partial\theta}\right)\left(\cos\theta\,\frac{\partial f}{\partial r} - \frac{\sin\theta}{r}\,\frac{\partial f}{\partial\theta}\right)$$

$$= \cos\theta\,\frac{\partial}{\partial r}\left(\cos\theta\,\frac{\partial f}{\partial r} - \frac{\sin\theta}{r}\,\frac{\partial f}{\partial\theta}\right) - \frac{\sin\theta}{r}\,\frac{\partial}{\partial\theta}\left(\cos\theta\,\frac{\partial f}{\partial r} - \frac{\sin\theta}{r}\,\frac{\partial f}{\partial\theta}\right)$$

$$= \cos\theta\left(\cos\theta\,\frac{\partial^2 f}{\partial r^2} + \frac{\sin\theta}{r^2}\,\frac{\partial f}{\partial\theta} - \frac{\sin\theta}{r}\,\frac{\partial^2 f}{\partial r\partial\theta}\right)$$

$$- \frac{\sin\theta}{r}\left(-\sin\theta\,\frac{\partial f}{\partial r} + \cos\theta\,\frac{\partial^2 f}{\partial\theta\partial r} - \frac{\cos\theta}{r}\,\frac{\partial f}{\partial\theta} - \frac{\sin\theta}{r}\,\frac{\partial^2 f}{\partial\theta^2}\right)$$

$$= \cos^2\theta\,\frac{\partial^2 f}{\partial r^2} + \frac{\sin^2\theta}{r}\,\frac{\partial f}{\partial r} + \frac{\sin^2\theta}{r^2}\,\frac{\partial^2 f}{\partial\theta^2} + \frac{2\sin\theta\cos\theta}{r}\left(\frac{1}{r}\,\frac{\partial f}{\partial\theta} - \frac{\partial^2 f}{\partial r\partial\theta}\right)$$

Similarly,

$$\frac{\partial^2 f}{\partial y^2} = \sin^2\theta\,\frac{\partial^2 f}{\partial r^2} + \frac{\cos^2\theta}{r}\,\frac{\partial f}{\partial r} + \frac{\cos^2\theta}{r^2}\,\frac{\partial^2 f}{\partial\theta^2} - \frac{2\sin\theta\cos\theta}{r}\left(\frac{1}{r}\,\frac{\partial f}{\partial\theta} - \frac{\partial^2 f}{\partial r\partial\theta}\right)$$

Therefore

$$\frac{\partial^2 f}{\partial x^2} + \frac{\partial^2 f}{\partial y^2} = \frac{\partial^2 f}{\partial r^2} + \frac{1}{r}\,\frac{\partial f}{\partial r} + \frac{1}{r^2}\,\frac{\partial^2 f}{\partial\theta^2}$$

▸ Exercise 49

EXAMPLE 9.19 Show that the function $f = x^2 - y^2$ satisfies the Laplace equation

(i) In cartesian coordinates,

$$\frac{\partial f}{\partial x} = 2x, \quad \frac{\partial^2 f}{\partial x^2} = 2 \quad \text{and} \quad \frac{\partial f}{\partial y} = -2y, \quad \frac{\partial^2 f}{\partial y^2} = -2$$

Therefore, $\dfrac{\partial^2 f}{\partial x^2} + \dfrac{\partial^2 f}{\partial y^2} = 0$

(ii) In polar coordinates, $f = r^2(\cos^2\theta - \sin^2\theta) = r^2\cos 2\theta$,

$$\frac{\partial f}{\partial r} = 2r\cos 2\theta = \frac{2f}{r}, \qquad \frac{\partial^2 f}{\partial r^2} = 2\cos 2\theta = \frac{2f}{r^2}$$

$$\frac{\partial f}{\partial\theta} = -2r^2\sin 2\theta, \qquad \frac{\partial^2 f}{\partial\theta^2} = -4r^2\cos 2\theta = -4f$$

Therefore, $\dfrac{\partial^2 f}{\partial r^2} + \dfrac{1}{r}\dfrac{\partial f}{\partial r} + \dfrac{1}{r^2}\dfrac{\partial^2 f}{\partial \theta^2} = \dfrac{2f}{r^2} + \dfrac{2f}{r^2} - \dfrac{4f}{r^2} = 0$

▸ Exercises 50–52

EXAMPLE 9.20 Show that the function $f = \ln r$, where $r = \sqrt{x^2 + y^2}$, satisfies the Laplace equation.

Because the function depends on r only, $\partial f/\partial \theta = 0$ and the derivatives with respect to r are total derivatives:

$$\nabla^2 f(r) = \dfrac{d^2 f}{dr^2} + \dfrac{1}{r}\dfrac{df}{dr}$$

Now $df/dr = 1/r$ and $d^2f/dr^2 = -1/r^2$. Therefore $\nabla^2 f = 0$.

This example is important because it can be shown that the only solution of the Laplace equation in two dimensions that depends on $r = \sqrt{x^2 + y^2}$ alone has the general form $f = a \ln r + b$. This function occurs in potential theory in two dimensions.

9.7 Exact differentials

One of the fundamental equations of thermodynamics, combining both the first and second laws, is

$$dU = TdS - pdV \tag{9.40}$$

where U is the internal energy of a thermodynamic system, S is its entropy, and p, V, and T are the pressure, volume, and temperature.* The quantity dU is the total differential of $U = U(S, V)$ as a function of S and V. It can therefore be written as

$$dU = \left(\dfrac{\partial U}{\partial S}\right)_V dS + \left(\dfrac{\partial U}{\partial V}\right)_S dV \tag{9.41}$$

so that, equating (9.40) and (9.41),

$$T = \left(\dfrac{\partial U}{\partial S}\right)_V, \qquad -p = \left(\dfrac{\partial U}{\partial V}\right)_S \tag{9.42}$$

The expression on the right side of (9.40) is called an **exact differential**. In general, a differential

$$F(x, y)\, dx + G(x, y)\, dy \tag{9.43}$$

---

* For a single closed phase with constant composition. More generally, the equation applies to each separate phase, with additional terms if the amounts of substance are not constant.

is exact when there exists a function $z = z(x, y)$ such that

$$F = \left(\frac{\partial z}{\partial x}\right)_y \quad \text{and} \quad G = \left(\frac{\partial z}{\partial y}\right)_x \tag{9.44}$$

The differential can then be equated to the total differential of z:

$$F dx + G dy = dz = \left(\frac{\partial z}{\partial x}\right)_y dx + \left(\frac{\partial z}{\partial y}\right)_x dy \tag{9.45}$$

The general condition that a differential in two variables be exact is that the functions F and G satisfy

$$\left(\frac{\partial F}{\partial y}\right)_x = \left(\frac{\partial G}{\partial x}\right)_y \tag{9.46}$$

By (9.44), each of these partial derivatives is equal to the mixed second derivative of z, $\partial^2 z/\partial x \partial y$. The condition (9.46) is sometimes called the **Euler reciprocity relation**, and is used in thermodynamics to derive a number of relations, called Maxwell relations, amongst thermodynamic properties (see Example 9.22).[6] The significance of exactness is discussed in the following section.

EXAMPLE 9.21 Test of exactness

(i) $F dx + G dy = (x^2 - y^2) dx + 2xy dy$.

We have $\left(\frac{\partial F}{\partial y}\right)_x = -2y$ and $\left(\frac{\partial G}{\partial x}\right)_y = +2y$, and the differential is not exact.

(ii) $F dx + G dy = (2ax + by) dx + (bx + 2cy) dy$.

We have $\left(\frac{\partial F}{\partial y}\right)_x = b = \left(\frac{\partial G}{\partial x}\right)_y$, and the differential is exact.

It is readily verified that it is the total differential of $ax^2 + bxy + cy^2$.

▶ Exercises 53–55

---

[6] The equality of the mixed second partial derivatives was used by Clairaut in 1739 to test a differential for exactness (and also by Euler at about the same time). Alexis Claude Clairaut (1713–1765) was one of a family of 20, only one of whom survived the father. He read a paper on geometry to the Académie des Sciences at the age of 13, and was a member at 18 (a younger brother, known as 'le cadet Clairaut', published a book on the calculus in 1731 at the age of 15, and died of smallpox a year later). His *Recherches sur les courbes à double courbure* (Research on curves of double curvature) in 1731 marked the beginning of the development of a cartesian geometry of three dimensions.

EXAMPLE 9.22 **Maxwell relations**

It follows from equations (9.42) that

$$\left(\frac{\partial T}{\partial V}\right)_S = \frac{\partial^2 U}{\partial V \partial S}, \qquad -\left(\frac{\partial p}{\partial S}\right)_V = \frac{\partial^2 U}{\partial S \partial V}$$

so that

$$\left(\frac{\partial T}{\partial V}\right)_S = -\left(\frac{\partial p}{\partial S}\right)_V$$

This is one of four Maxwell relations of importance in thermodynamics. The other three are derived from the differentials of the enthalpy $H = U + pV$, the Helmholtz energy $A = U - TS$, and the Gibbs energy $G = H - TS$. For example, for the enthalpy (and using (9.40) for dU),

$$dH = dU + pdV + Vdp = TdS + Vdp$$

The quantity dH is the total differential of H as a function of S and p, so that

$$T = \left(\frac{\partial H}{\partial S}\right)_p, \qquad V = \left(\frac{\partial H}{\partial p}\right)_S$$

and the corresponding Maxwell relation is

$$\left(\frac{\partial T}{\partial p}\right)_S = \left(\frac{\partial V}{\partial S}\right)_p$$

Similarly,

$$dA = -SdT - pdV \quad \longrightarrow \quad \left(\frac{\partial S}{\partial V}\right)_T = \left(\frac{\partial p}{\partial T}\right)_V$$

$$dG = -SdT + Vdp \quad \longrightarrow \quad \left(\frac{\partial S}{\partial p}\right)_T = -\left(\frac{\partial V}{\partial T}\right)_p$$

▶ Exercise 56

9.8   Line integrals

Consider the function $F(x)$ and the integral

$$\int_a^b F(x)\, dx \qquad (9.47)$$

The integral was interpreted in Section 5.4 as the 'area under the curve' of the graph of $F(x)$ between $x = a$ and $x = b$. An alternative interpretation is obtained by considering the function $F(x)$ as some property associated with the points x on a line. For example, let $F(x)$ be the mass density of a straight rod of matter of length $b - a$ (see Section 5.6). The differential mass in element dx is then $dm = F(x)\,dx$ and the total mass is $M = \int_a^b F(x)\, dx$. Similarly (see Section 5.7), if $F(x)$ is the force acting on a body at point x on a line, the differential work is $dW = F(x)\,dx$ and the total work from a to b is again the definite integral (9.47).

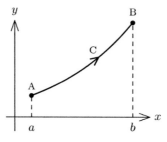

Figure 9.7

In these examples, a function is defined on a straight line, chosen to be the x-axis. More generally, let $y = f(x)$ represent a curve C in the xy-plane in the interval $a \le x \le b$, as shown in Figure 9.7, and let the function $F(x, y)$ be some property associated with the points on the curve. The quantity

$$\int_C F(x, y)\, dx$$

is called a **curvilinear** or **line integral**, and the curve C is called the **path of integration**.[7] If $G(x, y)$ is a second function defined on the curve, the general line integral in the plane is

$$I = \int_C \left[F(x, y)\, dx + G(x, y)\, dy \right] \qquad (9.48)$$

---

[7] The concept and notation of the line integral was used by Maxwell in 1855 in his studies of electric fields. James Clerk Maxwell (1831–1879), born in Edinburgh, ranks with Newton and Einstein in pre-quantal theoretical physics. Building on the work of Michael Faraday, Maxwell presented his field equations in his *Dynamical theory of the electromagnetic field* in 1864. The *Dynamical theory of gases* of 1859 describes the Maxwell distribution, with applications to the theory of viscosity, the conduction of heat, and the diffusion of gases.

The line integral notation appeared in a physics text by Charles Delaunay (1816–1872) in 1856 for the work done along a curve, and quickly became standard in physics.

Line integrals are important in several branches of the physical sciences. For example, when generalized to curves in three dimensions, they provide a method for representing and calculating the work done by a force along an arbitrary path in ordinary space; we return to this topic in Chapter 16.

The line integral (9.48) can be converted into an ordinary integral over either variable when the equation of the curve C is known. Thus, given the curve $y = f(x)$, replacement of dy in (9.48) by $\dfrac{dy}{dx}\,dx$ gives

$$I = \int_a^b \left[F(x,y) + G(x,y)\frac{dy}{dx} \right] dx \qquad (9.49)$$

EXAMPLE 9.23 Find the value of the line integral (9.48) when $F = -y$, $G = xy$, and C is the line in Figure 9.8 from A to B ($x = 1$ to $x = 0$).

The equation of the line is $y = 1 - x$. Then $dy = -dx$, and, by equation (9.49),

$$\int_C \left[-y\,dx + xy\,dy \right] = \int_1^0 \left[-(1-x) - x(1-x) \right] dx$$

$$= \int_0^1 (1 - x^2)\,dx = \frac{2}{3}$$

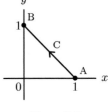

Figure 9.8

▸ Exercises 57, 58

In general, the value of a line integral depends on the path of integration between the end points. This is demonstrated in the following example.

EXAMPLE 9.24 Find the value of the line integral (9.48) when F and G are as in Example 9.23, but C is now the circular arc shown in Figure 9.9.

The equation of the circular arc is $y = +\sqrt{1 - x^2}$. Then

$$dy = \frac{-x}{\sqrt{1-x^2}}\,dx = -\frac{x}{y}\,dx \text{ and}$$

$$I = \int_C \left[-y\,dx + xy\,dy \right]$$

$$= \int_1^0 \left[-\sqrt{1-x^2} - x^2 \right] dx = \frac{\pi}{4} + \frac{1}{3}$$

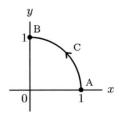

Figure 9.9

(see Example 6.10 for $\int_0^1 \sqrt{1-x^2}\,dx$). The result is different from that obtained in Example 9.23.

> Exercises 59, 60

The dependence of the line integral on the path can be understood, for example, in terms of the work done on a body in moving it from A to B along the path. From the discussion of work and potential energy in Section 5.7, we expect the work done to depend on the path when the forces acting on the body are not conservative forces; for example, when work is done against friction. In particular, net work is done in moving the body around a *closed* path. A dependence on path is also found in thermodynamics.

EXAMPLE 9.25 Work in thermodynamics

When changes in a thermodynamic system are reversible (Section 5.8), the quantity TdS in equation (9.40),

$$dU = TdS - pdV$$

is identified with the heat absorbed by the system, and pdV with the (mechanical) work done by the system. We consider the work done by the ideal gas on (i) expansion from $V_1 = V(p_1, T_1)$, at point A in Figure 9.10, along path $C_1 + C_2$ to $V_2 = V(p_2, T_2)$ at point B, and (ii) the return to A along $C_3 + C_4$.

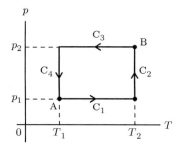

Figure 9.10

(i) Path $C_1 + C_2$

The work done along path C_1 is at constant pressure $p = p_1$ and, by equation (5.71),

$$W_1 = nR(T_2 - T_1).$$

The work done along path C_2 is at constant temperature $T = T_2$ and, by equation (5.72),

$$W_2 = -nRT_2 \ln(p_2/p_1).$$

Therefore

$$\int_{C_1+C_2} pdV = W_1 + W_2 = nR\left[(T_2 - T_1) - T_2 \ln(p_2/p_1)\right]$$

(ii) Path $C_3 + C_4$

The work done along path C_3 is at constant pressure $p = p_2$, and along path C_4 it is at constant temperature $T = T_1$. Therefore $W_3 = nR(T_1 - T_2)$, $W_4 = -nRT_1 \ln(p_1/p_2)$, and

$$\int_{C_3+C_4} pdV = -nR\left[(T_2 - T_1) - T_1 \ln(p_2/p_1)\right]$$

The total work done by the gas around the closed path $C = C_1 + C_2 + C_3 + C_4$ is then

$$\oint_C pdV = -nR(T_2 - T_1)\ln(p_2/p_1)$$

and this is not zero unless either $T_2 = T_1$ or $p_2 = p_1$.

> Exercise 61

When forces are conservative and can be derived from a potential-energy function, as in equation (5.57) of Section 5.7, the work is independent of the path, and no net work is done around a closed path. In general, the value of a line integral is *independent of the path* if the quantity in the square brackets in equation (9.48) is an *exact differential*.[8] By the discussion of Section 9.7, there then exists a function $z(x, y)$ such that

$$F(x, y)dx + G(x, y)dy = dz = \left(\frac{\partial z}{\partial x}\right)_y dx + \left(\frac{\partial z}{\partial y}\right)_x dy \qquad (9.50)$$

and the line integral (9.48) can be written as

$$\int_C dz = \int_C \left[\left(\frac{\partial z}{\partial x}\right)_y dx + \left(\frac{\partial z}{\partial y}\right)_x dy\right] \qquad (9.51)$$

Then, for a path C from A at (x_1, y_1) to B at (x_2, y_2),

$$\int_C dz = \left[z\right]_A^B = z(x_2, y_2) - z(x_1, y_1) \qquad (9.52)$$

and this depends only on the values of z at the end points (we note that the value of the integral changes sign if the direction of integration is changed to B to A). In terms of

[8] This independence of the path for the line integral of an exact differential was observed by Riemann in 1857.

the graphical representation of $z(x, y)$ discussed in Section 9.2, the line integral is the change in 'height' above the xy-plane on the representative surface of z, and this cannot depend on the path between the end points.

EXAMPLE 9.26 Independence of path for an exact differential

Volume is a function of pressure and temperature, $V = V(p, T)$, and the total differential volume is (see Example 9.11 for fixed amount of substance)

$$dV = \left(\frac{\partial V}{\partial T}\right)_p dT + \left(\frac{\partial V}{\partial p}\right)_T dp$$

As in Example 9.25, we consider the paths $C_1 + C_2$ and $C_3 + C_4$ shown in Figure 9.10.

(i) Path $C = C_1 + C_2$:
The change in volume along path C_1 is at constant pressure $p = p_1$ $(dp = 0)$, so that

$$\Delta V_1 = \int_{C_1} \left(\frac{\partial V}{\partial T}\right)_p dT = \int_{T_1}^{T_2} \left(\frac{\partial V}{\partial T}\right)_{p=p_1} dT = \left[V(p_1, T)\right]_{T_1}^{T_2} = V(p_1, T_2) - V(p_1, T_1)$$

Path C_2 is at constant $T = T_2$ $(dT = 0)$, and

$$\Delta V_2 = \int_{C_2} \left(\frac{\partial V}{\partial p}\right)_T dp = \int_{p_1}^{p_2} \left(\frac{\partial V}{\partial p}\right)_{T=T_2} dp = \left[V(p, T_2)\right]_{p_1}^{p_2} = V(p_2, T_2) - V(p_1, T_2)$$

Therefore

$$\Delta V_{A \to B} = \left[V(p_1, T_2) - V(p_1, T_1)\right] + \left[V(p_2, T_2) - V(p_1, T_2)\right]$$
$$= V(p_2, T_2) - V(p_1, T_1)$$

(ii) Path $C = C_3 + C_4$. Path C_3 is at constant p_2 and path C_4 is at constant $T = T_1$. Then, as above,

$$\Delta V_{B \to A} = \left[V(p_2, T_1) - V(p_2, T_2)\right] + \left[V(p_1, T_1) - V(p_2, T_1)\right]$$
$$= V(p_1, T_1) - V(p_2, T_2)$$

Therefore $\Delta V_{B \to A} + \Delta V_{A \to B} = 0$ and the change in volume around the closed path is zero.*

▸ Exercises 62, 63

---

* We note however that the line integral (9.48) around a closed path C is zero in general only if the differential $Fdx + Gdy$ is exact at all points on and *within* the closed path; further consideration of this point is beyond the scope of this book.

EXAMPLE 9.27 Change in entropy in thermodynamics

The change in the entropy when a thermodynamic system goes from a state with pressure and temperature (p_1, T_1) to a state with pressure and temperature (p_2, T_2) can be obtained from calorimetric measurements of the heat capacity and thermal expansivity,

$$C_p = \left(\frac{\partial H}{\partial T}\right)_p, \qquad \alpha = \frac{1}{V}\left(\frac{\partial V}{\partial T}\right)_p$$

in the following way. Treating the entropy as a function of p and T, the total differential of entropy is

$$dS = \left(\frac{\partial S}{\partial T}\right)_p dT + \left(\frac{\partial S}{\partial p}\right)_T dp$$

Then, because entropy is a function of state, the change on going from one state to another is independent of the path:

$$\Delta S_{1\to2} = S(p_2, T_2) - S(p_1, T_1) = \int_C \left[\left(\frac{\partial S}{\partial T}\right)_p dT + \left(\frac{\partial S}{\partial p}\right)_T dp\right]$$

Choosing the path $C = C_1 + C_2$ shown in Figure 9.11,

$$\Delta S_{1\to2} = \Delta S_{1\to3} + \Delta S_{3\to2}$$

where $\Delta S_{1\to3}$ is at constant $p = p_1$:

$$\Delta S_{1\to3} = S(p_1, T_2) - S(p_1, T_1)$$

$$= \int_{C_1} dS = \int_{T_1}^{T_2} \left(\frac{\partial S}{\partial T}\right)_{p=p_1} dT$$

and $\Delta S_{3\to2}$ is at constant $T = T_2$:

$$\Delta S_{3\to2} = S(p_2, T_2) - S(p_1, T_2)$$

$$= \int_{C_2} dS = \int_{p_1}^{p_2} \left(\frac{\partial S}{\partial p}\right)_{T=T_2} dp$$

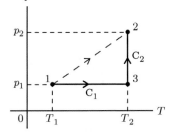

Figure 9.11

Now (Example 9.22), it follows from the differential enthalpy $dH = TdS + Vdp$ that (dividing dH by dT at constant p)

$$\left(\frac{\partial H}{\partial T}\right)_p = T\left(\frac{\partial S}{\partial T}\right)_p = C_p$$

Also, the Maxwell relation derived from the differential Gibbs energy is

$$\left(\frac{\partial S}{\partial p}\right)_T = -\left(\frac{\partial V}{\partial T}\right)_p = -\alpha V.$$

Therefore,

$$\Delta S_{1\to3} = \int_{T_1}^{T_2} \frac{C_p}{T} dT \quad (\text{at } p = p_1), \qquad \Delta S_{3\to2} = -\int_{p_1}^{p_2} \alpha V \, dp \quad (\text{at } T = T_2)$$

Experimentally, the heat capacity is measured at several temperatures between T_1 and T_2 at pressure p_1, and the expansivity is measured at several pressures between p_1 and p_2 at temperature T_2. The integrals are evaluated either by plotting C_p/T against T and αV against p and measuring the area under the curve in each case, or by using a numerical integration method (see Chapter 20).

9.9 Multiple integrals

A function of two independent variables, $f(x, y)$, can be integrated with respect to either, whilst keeping the other constant:

$$\int_a^b f(x, y)\, dx \qquad \text{or} \qquad \int_c^d f(x, y)\, dy$$

EXAMPLE 9.28 Integrate $f(x, y) = 2xy + 3y^2$ with respect to x in the interval $a \le x \le b$, and with respect to y in the interval $c \le y \le d$.

(i) $\displaystyle\int_a^b (2xy + 3y^2)\, dx = \left[x^2 y + 3y^2 x\right]_a^b = y(b^2 - a^2) + 3y^2(b - a)$

(ii) $\displaystyle\int_c^d (2xy + 3y^2)\, dy = \left[xy^2 + y^3\right]_c^d = x(d^2 - c^2) + (d^3 - c^3)$

In addition to these 'partial integrals', the integral with respect to both x and y is also defined:

$$\int_c^d \int_a^b f(x, y)\, dx\, dy = \int_c^d \left\{\int_a^b f(x, y)\, dx\right\} dy = \int_a^b \left\{\int_c^d f(x, y)\, dy\right\} dx \qquad (9.53)$$

This is called a **double integral** and is evaluated by integrating first with respect to one variable and then with respect to the other. The value of the integral does not depend on the order in which the integrations are performed if the function is continuous within the ranges of integration, but some care must be taken when discontinuities are present.

EXAMPLE 9.29 Evaluate the integral (9.53) when $f(x, y) = 2xy + 3y^2$.

(i) Integrating first with respect to x, using the result of Example 9.28(i),

$$\int_c^d \int_a^b (2xy + 3y^2)\, dx\, dy = \int_c^d \left[y(b^2 - a^2) + 3y^2(b - a) \right] dy$$

$$= \frac{1}{2}(d^2 - c^2)(b^2 - a^2) + (d^3 - c^3)(b - a)$$

(ii) Integrating first with respect to y, using the result of Example 9.28(ii),

$$\int_a^b \int_c^d (2xy + 3y^2)\, dy\, dx = \int_a^b \left[x(d^2 - c^2) + (d^3 - c^3) \right] dx$$

$$= \frac{1}{2}(d^2 - c^2)(b^2 - a^2) + (d^3 - c^3)(b - a)$$

▸ Exercises 64, 65

When the order of integration is not given explicitly, some care must be taken to associate each pair of integration limits with the appropriate variable. Thus, if a triple integral is written as

$$I = \int_e^f \int_c^d \int_a^b f(x, y, z)\, dx\, dy\, dz$$

the *convention* is that the integration is to be carried out from the *inside outward*, with (a, b) the limits for x, (c, d) for y, and (e, f) for z:

$$I = \int_e^f \left\{ \int_c^d \left[\int_a^b f(x, y, z)\, dx \right] dy \right\} dz$$

The concepts and methods of integration discussed in Chapters 5 and 6 for integrals over one variable apply to multiple integrals, with some changes in interpretation. The double integral is discussed in some detail in the following sections. The special case of the triple integral and its importance for the description of physical systems in three dimensions are discussed in Chapter 10.

9.10   The double integral

The double integral can be defined as the limit of a (double) sum in the same way as the Riemann integral was defined in Section 5.4. Let $f(x, y)$ by a continuous function of x and y in a rectangular region of the xy-plane, for $a \leq x \leq b$ and $c \leq y \leq d$ (Figure 9.12).

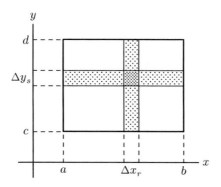

Figure 9.12

Divide the interval $x = a \rightarrow b$ into m subintervals of width $\Delta x_r = x_r - x_{r-1}$ and the interval $y = c \rightarrow d$ into n subintervals of width $\Delta y_s = y_s - y_{s-1}$; that is, divide the rectangle into small rectangles of area $\Delta A_{rs} = \Delta x_r \Delta y_s$. The integral (9.53) is then defined as the limit

$$\int_c^d \int_a^b f(x, y)\, dx\, dy = \lim_{\substack{m \to \infty \\ n \to \infty}} \sum_{s=1}^{n} \sum_{r=1}^{m} f(x_r, y_s) \Delta x_r \Delta y_s \qquad (9.54)$$

(when the limit exists).

The double summation (9.54) can be performed in either order: either as

$$\sum_{r=1}^{m} \left\{ \sum_{s=1}^{n} f(x_r, y_s) \Delta y_s \right\} \Delta x_r = \sum_{r=1}^{m} v_r \Delta x_r \qquad (9.55)$$

where v_r is the sum over the elements in a vertical strip and the second sum is over the vertical strips, or as

$$\sum_{s=1}^{n} \left\{ \sum_{r=1}^{m} f(x_r, y_s) \Delta x_r \right\} \Delta y_s = \sum_{s=1}^{n} h_s \Delta y_s \qquad (9.56)$$

where h_s is the sum over the elements in a horizontal strip and the second sum is over the horizontal strips. These two orders of summation correspond to the two orders of integration in (9.53).

Just as the integral in one variable was interpreted in Section 5.4 as the area under the curve, the double integral can be interpreted as the 'volume under the surface';

that is, the volume between the representative surface of the function $f(x, y)$ and the xy-plane.[9] Alternatively, the integral can be interpreted as a property of the rectangular region in the xy-plane. For example, if $f(x, y)$ is a surface mass density, the differential mass in element of area $dA = dx \, dy$ is $dm = f(x, y) dA$ and the total mass is

$$M = \int_R f(x, y) \, dA = \int_c^d \int_a^b f(x, y) \, dx \, dy \tag{9.57}$$

where $\int_R \cdots dA$ means integration over the (rectangular) region. In the special case of $f(x, y) = 1$, the integral is the area of the rectangle:

$$\int_R dA = \int_c^d \int_a^b dx \, dy = \int_a^b dx \int_c^d dy = (b - a)(d - c)$$

More generally, the region in the xy-plane need not be rectangular. Let $f(x, y)$ be continuous within and on a boundary of a region R of the xy-plane.

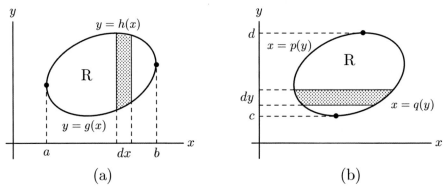

Figure 9.13

The form of the integral over the region depends on how the boundary is defined. In Figure 9.13a the boundary is made up of two sections: $y = g(x)$ 'below' the extreme points $x = a$ and $x = b$, and $y = h(x)$ 'above' these points. Then, within R,

$$g(x) \leq y \leq h(x) \quad \text{for} \quad a \leq x \leq b$$

and the integral over the region (the sum of vertical strips) is

$$I = \int_R f(x, y) \, dA = \int_a^b \left\{ \int_{g(x)}^{h(x)} f(x, y) \, dy \right\} dx \tag{9.58a}$$

In Figure 9.13b the boundary is made up of the two sections: $x = p(y)$ to the 'left' of the extreme points $y = c$ and $y = d$, and $x = q(y)$ to the 'right'. Then, within R,

$$p(y) \leq x \leq q(y) \quad \text{for} \quad c \leq y \leq d$$

[9] The double integral as a volume was discussed by Clairaut in his *Recherches* of 1731. The double integral sign notation ($\iint$) was first used by Lagrange in 1760.

and the integral (the sum of horizontal strips) is

$$I = \int_R f(x, y)\, dA = \int_c^d \left\{ \int_{p(y)}^{q(y)} f(x, y)\, dx \right\} dy \qquad (9.58b)$$

EXAMPLE 9.30 Evaluate the integral of $f(x, y) = 1 + 2xy$ over the region R bounded by the line $y = x$ and the curve $y = x^2$ (or $x = \sqrt{y}$), as in Figure 9.14. Calculate also the area of R.

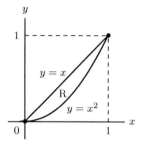

Figure 9.14

(i)   using equation (9.58a) with $g(x) = x^2$ and $h(x) = x$,

$$I = \int_0^1 \left\{ \int_{x^2}^x (1 + 2xy)\, dy \right\} dx$$

$$= \int_0^1 \left\{ \left[y + xy^2 \right]_{x^2}^x \right\} dx = \int_0^1 \left[x + x^3 - x^2 - x^5 \right] dx = \frac{1}{4}$$

(ii)   using equation (9.58b) with $p(y) = y$ and $q(y) = \sqrt{y}$,

$$I = \int_0^1 \left\{ \int_y^{\sqrt{y}} (1 + 2xy)\, dx \right\} dy$$

$$= \int_0^1 \left\{ \left[x + x^2 y \right]_y^{\sqrt{y}} \right\} dy = \int_0^1 \left[\sqrt{y} + y^2 - y - y^3 \right] dy = \frac{1}{4}$$

The area of the region R is

$$A = \int_0^1 \left\{ \int_{x^2}^x dy \right\} dx = \int_0^1 \left(x - x^2 \right) dx = \frac{1}{6}$$

▸ Exercises 66–68

9.11   Change of variables

We saw in Section 6.3 that one of the principal general methods of evaluating integrals is the method of substitution whereby, given a definite integral over the variable x, a new variable of integration can be introduced by setting

$$x = x(u), \qquad dx = \frac{dx}{du}\, du$$

such that

$$\int_a^b f(x)\,dx = \int_{u(a)}^{u(b)} f\big(x(u)\big)\frac{dx}{du}\,du$$

Changes of variable are even more important for double (and higher) integrals because both the function *and* the boundary of the region may be sources of difficulty when expressed in terms of inappropriate variables. Consider, for example, the integral of $f(x,y)=e^{-(x^2+y^2)^{1/2}}$ over a circular region (Figure 9.15).

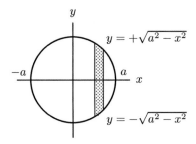

Figure 9.15

By equation (9.58a),

$$I = \int_R f(x,y)\,dA = \int_{-a}^a \left\{\int_{-\sqrt{a^2-x^2}}^{+\sqrt{a^2-x^2}} e^{-(x^2+y^2)^{1/2}}\,dy\right\}dx \qquad (9.59)$$

In this example, both the function and the boundary are expressed more simply in terms of polar coordinates (r, θ). Thus,

$$f(x,y)=f(r\cos\theta, r\sin\theta)=e^{-r}$$

and the equation of the circle of radius a is $r=a$. It is sensible therefore to make the substitution

$$x=r\cos\theta, \qquad y=r\sin\theta$$

to change to polar coordinates. The corresponding element of area, dA, can be deduced from Figure 9.16. The shaded region has area

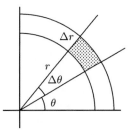

Figure 9.16

$$\Delta A = \Delta r \times r\Delta\theta + \frac{1}{2}(\Delta r)^2\,\Delta\theta$$

so that, for infinitesimal quantities, $dA = r\,dr\,d\theta$. Then

$$\int_R f(x,y)\,dx\,dy = \int_R f(r\cos\theta, r\sin\theta)\,r\,dr\,d\theta \qquad (9.60)$$

The integral (9.59) is then

$$I = \int_0^{2\pi}\int_0^a e^{-r}r\,dr\,d\theta = \int_0^{2\pi} d\theta \times \int_0^a e^{-r}r\,dr = 2\pi\left[1 - e^{-a}(a+1)\right]$$

EXAMPLE 9.31 Evaluate the integral of $f(r,\theta) = e^{-r^2}\sin^2\theta$ over (i) the area of a circle of radius a and (ii) the whole plane.

(i) $$\int_R e^{-r^2}\sin^2\theta\,dA = \int_0^{2\pi}\int_0^a e^{-r^2}\sin^2\theta\,r\,dr\,d\theta$$

Because the integrand is the product of a function of r and a function of θ,

$$e^{-r^2}\sin^2\theta\,r = (e^{-r^2}r)(\sin^2\theta)$$

and because the limits of integration are constants, the double integral can be factorized:

$$I = \int_0^a e^{-r^2}r\,dr \times \int_0^{2\pi}\sin^2\theta\,d\theta$$

Because $\sin^2\theta = \frac{1}{2}(1 - \cos 2\theta)$,

$$\int_0^{2\pi}\sin^2\theta\,d\theta = \frac{1}{2}\int_0^{2\pi}(1 - \cos 2\theta)\,d\theta = \frac{1}{2}\left[\theta - \frac{1}{2}\sin 2\theta\right]_0^{2\pi} = \pi$$

and, using the substitution $u = r^2$, $du = 2r\,dr$,

$$\int_0^a e^{-r^2}r\,dr = \frac{1}{2}\int_0^{a^2}e^{-u}\,du = \frac{1}{2}\left[-e^{-u}\right]_0^{a^2} = \frac{1}{2}\left(1 - e^{-a^2}\right)$$

Therefore, $I = \dfrac{\pi}{2}\left(1 - e^{-a^2}\right)$

(ii) Letting $a \to \infty$,

$$\int_0^{2\pi}\int_0^\infty e^{-r^2}\sin^2\theta\,r\,dr\,d\theta = \frac{\pi}{2}$$

▸ Exercises 69–71

The general case

The transformation from cartesian to polar coordinates is a special case of the general transformation of variables (of coordinate system). It is also the most important in practice, and only a brief discussion of the general case is given here. Let

$$x = x(u, v), \qquad y = y(u, v) \tag{9.61}$$

be continuous and differentiable functions of the variables u and v, such that for each point (x, y) in a region R of the xy-plane there corresponds a unique point (u, v) in the corresponding region R* of the uv-plane. Then

$$\iint_R f(x, y)\, dx\, dy = \iint_{R*} f(x(u, v), y(u, v)) |J|\, du\, dv \tag{9.62}$$

where $|J|$ is the modulus of the **Jacobian** of the transformation,[10]

$$J = \frac{\partial(x, y)}{\partial(u, v)} = \frac{\partial x}{\partial u}\frac{\partial y}{\partial v} - \frac{\partial x}{\partial v}\frac{\partial y}{\partial u} \tag{9.63}$$

For example, if $x = r \cos \theta$ and $y = r \sin \theta$ then

$$J = \frac{\partial x}{\partial r}\frac{\partial y}{\partial \theta} - \frac{\partial x}{\partial \theta}\frac{\partial y}{\partial r} = r \cos^2 \theta + r \sin^2 \theta = r$$

and $dA = dx\,dy \rightarrow dA = r\,dr\,d\theta$, as obtained geometrically from Figure 9.16. The Jacobian is used in advanced formulations of thermodynamics for the transformation of thermodynamic variables. The method can be generalized for three or more variables.

The integral $\displaystyle\int_0^{\infty} e^{-x^2}\, dx$

It was stated in Example 6.16 of Section 6.5 that this integral cannot be evaluated by the methods described in Chapters 5 and 6. It can however by evaluated by means of a 'trick' involving the transformation of a double integral from cartesian to polar coordinates.[11] We have

$$I = \int_0^{\infty} e^{-x^2}\, dx = \frac{1}{2}\int_{-\infty}^{\infty} e^{-x^2}\, dx$$

because the integral is an even function of x (see Section 5.3). Also

$$I = \frac{1}{2}\int_{-\infty}^{\infty} e^{-y^2}\, dy$$

---

[10] Carl Gustave Jacobi (1804–1851), born in Berlin, discussed the quantities called Jacobians in his *De determinantibus functionalibus* (On functional determinants) of 1841. The general transformation of variables in a double integral was given by Euler in 1769.

[11] The integral was first evaluated by de Moivre in 1733. Laplace, following Euler, used the method described here in 1774.

because the value of a definite integral does not depend on the symbol used for the variable of integration. Then

$$I^2 = \frac{1}{4}\int_{-\infty}^{\infty} e^{-x^2}\, dx \int_{-\infty}^{\infty} e^{-y^2}\, dy = \frac{1}{4}\int_{-\infty}^{\infty}\int_{-\infty}^{\infty} e^{-(x^2+y^2)}\, dxdy$$

and the double integral is over the whole xy-plane. A transformation to polar coordinates then gives

$$I^2 = \frac{1}{4}\int_0^{2\pi}\int_0^{\infty} e^{-r^2} r\, drd\theta = \frac{1}{4}\int_0^{2\pi} d\theta \int_0^{\infty} e^{-r^2} r\, dr = \frac{\pi}{4}$$

Therefore,

$$\int_0^{\infty} e^{-x^2}\, dx = \frac{\sqrt{\pi}}{2} \qquad (9.64)$$

9.12 Exercises

Section 9.1

1. Find the value of the function $f(x, y) = 2x^2 + 3xy - y^2 + 2x - 3y + 4$ for
   (i) $(x, y) = (0, 1)$ (ii) $(x, y) = (2, 0)$ (iii) $(x, y) = (3, 2)$
2. Find the value of $f(r, \theta, \phi) = r^2 \sin^2\theta \cos^2\phi + 2\cos^2\theta - r^3 \sin 2\theta \sin\phi$ for
   (i) $(r, \theta, \phi) = (1, \pi/2, 0)$ (ii) $(r, \theta, \phi) = (2, \pi/4, \pi/6)$ (iii) $(r, \theta, \phi) = (0, \pi, \pi/3)$.

Section 9.3

Find $\dfrac{\partial z}{\partial x}$ and $\dfrac{\partial z}{\partial y}$ for

3. $z = 2x^2 - y^2$ 4. $z = x^2 + 2y^2 - 3x + 2y + 3$ 5. $z = e^{2x+3y}$
6. $z = \sin(x^2 - y^2)$ 7. $z = e^{x^2}\cos(xy)$

Find all the nonzero partial derivatives of

8. $z = x^2 - 3x^2 y + 4xy^2$ 9. $u = 3x^2 + y^2 + 2xy^3$

Find all the first and second partial derivatives of

10. $z = 2x^2 y + \cos(x+y)$ 11. $z = \sin(x+y)e^{x-y}$

Show that $f_{xy} = f_{yx}$ for

12. $f = x^3 - 3x^2 y + y^3$ 13. $f = x^2\cos(y-x)$ 14. $f = \dfrac{xy}{x^2 + y^2}$

Show that $f_{xyz} = f_{yzx} = f_{zxy}$ for

15. $f = \cos(x + 2y + 3z)$ 16. $f = xye^{yz}$

17. If $r = (x^2 + y^2 + z^2)^{1/2}$ find $\dfrac{\partial r}{\partial x}, \dfrac{\partial r}{\partial y}, \dfrac{\partial r}{\partial z}$.

18. If $\phi = f(x - ct) + g(x + ct)$, where c is a constant, show that $\dfrac{\partial^2\phi}{\partial x^2} = \dfrac{1}{c^2}\dfrac{\partial^2\phi}{\partial t^2}$.

19. If $xyz + x^2 + y^2 + z^2 = 0$, find $\left(\dfrac{\partial y}{\partial x}\right)_z$.

20. For the van der Waals equation

$$\left(p + \frac{n^2 a}{V^2}\right)(V - nb) - nRT = 0$$

Find **(i)** $\left(\dfrac{\partial V}{\partial T}\right)_{p,n}$, **(ii)** $\left(\dfrac{\partial V}{\partial p}\right)_{T,n}$, **(iii)** $\left(\dfrac{\partial p}{\partial T}\right)_{V,n}$, **(iv)** $\left(\dfrac{\partial p}{\partial V}\right)_{T,n}$.

Section 9.4

Find the stationary points of the following functions:

21. $3 - x^2 - xy - y^2 + 2y$ **22.** $x^3 + y^2 - 3x - 4y + 2$ **23.** $4x^3 - 3x^2y + y^3 - 9y$

Determine the nature of of the stationary points of the functions in Exercises 21–23:

24. $3 - x^2 - xy - y^2 + 2y$ **25.** $x^3 + y^2 - 3x - 4y + 2$ **26.** $4x^3 - 3x^2y + y^3 - 9y$

27. Find the stationary value of the function $f = 2x^2 + 3y^2 + 6z^2$ subject to the constraint $x + y + z = 1$, **(i)** by using the constraint to eliminate z from the function, **(ii)** by the method of Lagrange multipliers.

28. Find the maximum value of the function $f = x^2 y^2 z^2$ subject to the constraint $x^2 + y^2 + z^2 = c^2$, **(i)** by using the constraint to eliminate z from the function, **(ii)** by the method of Lagrange multipliers.

29. **(i)** Find the stationary points of the function $f = (x-1)^2 + (y-2)^2 + (z-2)^2$ subject to the constraint $x^2 + y^2 + z^2 = 1$. **(ii)** Show that these lie at the shortest and longest distances of the point $(1, 2, 2)$ from the surface of the sphere $x^2 + y^2 + z^2 = 1$.

30. **(i)** Show that the problem of finding the stationary values of the function

$$E(x, y, z) = a(x^2 + y^2 + z^2) + 2b(xy + yz)$$

subject to the constraint $x^2 + y^2 + z^2 = 1$ (a and b are constants) is equivalent to solving the secular equations

$$\begin{aligned} (a - \lambda)x + \quad by \qquad\qquad &= 0 \\ bx + (a-\lambda)y + \quad bz &= 0 \\ by + (a-\lambda)z &= 0 \end{aligned}$$

These equations have solutions for three possible values of the Lagrangian multiplier:

$$\lambda_1 = a, \quad \lambda_2 = a + \sqrt{2}b, \quad \lambda_3 = a - \sqrt{2}b$$

(ii) Find the stationary point corresponding to each value of λ (assume x is positive).
(iii) Show that the three stationary values of E are identical to the corresponding values of λ. (This is the Hückel problem for the allyl radical, CH_2CHCH_2; see also Example 17.9).

Section 9.5

Find the total differential df:

31. $f(x, y) = x^2 + y^2$ **32.** $f(x, y) = 3x^2 + \sin(x - y)$ **33.** $f(x, y) = x^3 y^2 + \ln y$

34. $f(x, y, z) = \dfrac{1}{\sqrt{x^2 + y^2 + z^2}}$ **35.** $f(r, \theta, \phi) = r \sin \theta \sin \phi$

36. Write down the total differential of the volume of a two-component system in terms of changes in temperature T, pressure p, and amounts n_A and n_B of the components A and B. Use the full notation with subscripts for constant variables.

Section 9.6

37. Given $z = x^2 + 2xy + 3y^2$, where $x = (1+t)^{1/2}$ and $y = (1-t)^{1/2}$, find $\dfrac{dz}{dt}$ by

   (i) substitution, (ii) the chain rule (9.21).

38. Given $u = e^{x-y}$, where $x = 2\cos t$ and $y = 3t$, use the chain rule to find $\dfrac{du}{dt}$.

39. Given $u = \ln(x+y+z)$, where $x = a\cos t$, $y = b\sin t$, and $z = ct$, find $\dfrac{du}{dt}$.

40. Given $z = \ln(2x+3y)$, $x = a\cos\theta$, $y = a\sin\theta$, use the chain rule to find $\dfrac{dz}{d\theta}$.

41. Given $f = \sin(u+v)$, where $v = \cos u$, (i) find $\dfrac{df}{du}$, (ii) if $u = e^{-t}$, find $\dfrac{df}{dt}$.

42. If $z = x^5 y - \sin y$, find $\left(\dfrac{\partial y}{\partial x}\right)_z$.

43. For the van der Waals gas, use the expressions for $\left(\dfrac{\partial V}{\partial T}\right)_{p,n}$ and $\left(\dfrac{\partial V}{\partial p}\right)_{T,n}$ from Exercise 20

   to find $\left(\dfrac{\partial p}{\partial T}\right)_{V,n}$

44. If $z = x^2 + y^2$ and $u = xy$, find $\left(\dfrac{\partial z}{\partial y}\right)_u$ by (i) substitution, (ii) equation (9.30).

45. Given $z = x\sin y$ and $u = x^2 + 2xy + 3y^2$, find $\left(\dfrac{\partial z}{\partial y}\right)_u$.

46. If $U = U(V, T)$ and $p = p(V, T)$ are functions of V and T and if $H = U + pV$, show that

$$\left(\frac{\partial H}{\partial T}\right)_p - \left(\frac{\partial U}{\partial T}\right)_V = \left[\left(\frac{\partial U}{\partial V}\right)_T + p\right]\left(\frac{\partial V}{\partial T}\right)_p.$$

47. Given $x = au + bv$ and $y = bu - av$, where a and b are constants, (i) if f is a function of x and

   y, express $\left(\dfrac{\partial f}{\partial u}\right)_v$ and $\left(\dfrac{\partial f}{\partial v}\right)_u$ in terms of $\left(\dfrac{\partial f}{\partial x}\right)_y$ and $\left(\dfrac{\partial f}{\partial y}\right)_x$, (ii) if $f = x^2 + y^2$, find $\left(\dfrac{\partial f}{\partial u}\right)_v$

   and $\left(\dfrac{\partial f}{\partial v}\right)_u$ in terms of u and v.

48. Given $u = x^n + y^n$ and $v = x^n - y^n$, where n is a constant,

   (i) show that $\left(\dfrac{\partial x}{\partial u}\right)_v\left(\dfrac{\partial u}{\partial x}\right)_y = \dfrac{1}{2} = \left(\dfrac{\partial y}{\partial v}\right)_u\left(\dfrac{\partial v}{\partial y}\right)_x$. (ii) If f is a function of x and y, express

   $\left(\dfrac{\partial f}{\partial x}\right)_y$ and $\left(\dfrac{\partial f}{\partial y}\right)_x$ in terms of $\left(\dfrac{\partial f}{\partial u}\right)_v$ and $\left(\dfrac{\partial f}{\partial v}\right)_u$. Hence, (iii) if $f = u^2 - v^2$, find $\left(\dfrac{\partial f}{\partial x}\right)_y$

   and $\left(\dfrac{\partial f}{\partial y}\right)_x$ in terms of x and y.

49. If $x = au + bv$, $y = bu - av$, and f is a function of x and y (see Exercise 47(i)), show that

$$(i)\ \frac{\partial^2 f}{\partial u^2} + \frac{\partial^2 f}{\partial v^2} = (a^2 + b^2)\left(\frac{\partial^2 f}{\partial x^2} + \frac{\partial^2 f}{\partial y^2}\right),\ (ii)\ \frac{\partial^2 f}{\partial u \partial v} = ab\left(\frac{\partial^2 f}{\partial x^2} - \frac{\partial^2 f}{\partial y^2}\right) + (b^2 - a^2)\frac{\partial^2 f}{\partial x \partial y}.$$

Show that the following functions of position in a plane satisfy Laplace's equation:

50. $x^5 - 10x^3y^2 + 5xy^4$ **51.** $\left(Ar + \dfrac{B}{r}\right)\sin\theta$ **52.** $r^n \cos n\theta$, $n = 1, 2, 3, \ldots$

Section 9.7

Test for exactness:

53. $(4x + 3y)dx + (3x + 8y)dy$ **54.** $(6x + 5y + 7)dx + (4x + 10y + 8)dy$

55. $y \cos x\, dx + \sin x\, dy$

56. Given the total differential $dG = -S\, dT + V\, dp$, show that $\left(\dfrac{\partial S}{\partial p}\right)_T = -\left(\dfrac{\partial V}{\partial T}\right)_p$.

Section 9.8

57. Evaluate the line integral $\displaystyle\int_C \left[xy\,dx + 2y\,dy\right]$ on the line $y = 2x$ from $x = 0$ to $x = 2$.

58. When the path of integration is given in parametric form $x = x(t)$, $y = y(t)$ from $t = t_A$ to $t = t_B$, the line integral can be evaluated as

$$\int_C \left[F\,dx + G\,dy\right] = \int_{t_A}^{t_B}\left[F\frac{dx}{dt} + G\frac{dy}{dt}\right]dt.$$

Evaluate $\displaystyle\int_C \left[(x^2 + 2y)dx + (y^2 + x)dy\right]$ on the curve with parametric equations $x = t$, $y = t^2$ from $A(0, 0)$ to $B(1, 1)$.

59. Evaluate $\displaystyle\int_C \left[xy\,dx + 2y\,dy\right]$ on the curve $y = x^2$ from $x = 0$ to $x = 2$ (see Exercise 57).

60. Evaluate the line integral $\displaystyle\int_C \left[(x^2 + 2y)dx + (y^2 + x)dy\right]$ on the curve with parametric equations $x = t^2$, $y = t$ from $A(0, 0)$ to $B(1, 1)$ (see Exercise 58).

61. The total differential of entropy as a function of T and p is (Example 9.27)

$$dS = \left(\frac{\partial S}{\partial T}\right)_p dT + \left(\frac{\partial S}{\partial p}\right)_T dp$$

Given that, $\left(\dfrac{\partial S}{\partial T}\right)_p = \dfrac{C_p}{T} = \dfrac{5R}{2T}$ and $\left(\dfrac{\partial S}{\partial p}\right)_T = -\left(\dfrac{\partial V}{\partial T}\right)_p = -\dfrac{R}{p}$ for 1 mole of ideal gas,

show that the (reversible) heat absorbed by the ideal gas round the closed path shown in Figure 9.10 is equal to the work done by the gas; that is, $\displaystyle\oint T\,dS = \oint p\,dV$ (see Example 9.25).

62. **(i)** Show that $F\,dx + G\,dy$ for $F = 9x^2 + 4y^2 + 4xy$ and $G = 8xy + 2x^2 + 3y^2$ is an exact differential. **(ii)** By choosing an appropriate path, evaluate $\displaystyle\int_C [F\,dx + G\,dy]$ from $(x, y) = (0, 0)$ to $(1, 2)$. **(iii)** Show that the result in (ii) is consistent with the differential as the total differential of

$$z(x, y) = 3x^3 + 4xy^2 + 2x^2y + y^3.$$

63. Evaluate $\displaystyle\int_C \left[2xy\,dx + (x^2 - y^2)\,dy \right]$ on the circle with parametric equations $x = \cos\theta$,

$y = \sin\theta$, **(i)** from $A(1, 0)$ to $B(0, 1)$ and **(ii)** around a complete circle $(\theta = 0 \to 2\pi)$. **(iii)** Confirm that the differential $2xy\,dx + (x^2 - y^2)\,dy$ is exact.

Sections 9.9

64. Evaluate the integral $\displaystyle\int_0^3 \int_1^2 (x^2 y + xy^2)\,dx\,dy$ and show that the result is independent of

the order of integration.

65. Evaluate the integral $\displaystyle\int_0^\pi \int_0^R e^{-r} \cos^2\theta \sin\theta\,dr\,d\theta$.

Sections 9.10

Evaluate the integral and sketch the region of integration:

66. $\displaystyle\int_0^2 \int_x^{2x} (x^2 + y^2)\,dy\,dx$ **67.** $\displaystyle\int_0^a \int_0^{\sqrt{a^2-x^2}} xy^2\,dy\,dx$

68. **(i)** Show from a sketch of the region of integration that

$$\int_0^2 \int_{2y-4}^{(2-y)/2} x^3\,dx\,dy = \int_0^1 \int_0^{2-2x} x^3\,dy\,dx + \int_{-4}^0 \int_0^{(x+4)/2} x^3\,dy\,dx,$$

(ii) evaluate the integral.

Section 9.11

Transform to polar coordinates and evaluate:

69. $\displaystyle\int_0^1 \int_0^{\sqrt{1-x^2}} (x^2 + 2xy)\,dy\,dx$ **70.** $\displaystyle\int_{-\infty}^\infty \int_{-\infty}^\infty e^{-2(x^2+y^2)^{1/2}} (x^2 + y^2)^3\,dx\,dy$

71. $\displaystyle\int_0^\infty \int_0^\infty e^{-(x^2+y^2)} x^2\,dx\,dy$

10 Functions in 3 dimensions

10.1 Concepts

A function of three variables, $f(x, y, z)$, in which the variables are the coordinates of a point in ordinary three-dimensional space is called a **function of position**, a **point function**, or a **field**. A great variety of physical quantities are described by such functions. For example, the distribution of mass in a physical body is described by the (volume) mass density, a function of position, the temperature at each point in a body defines a temperature field, and the velocity at each point in a mass of moving fluid is a vector function of position, a velocity field (see Chapter 16). Electric and magnetic fields are vector functions of position, atomic and molecular wave functions are scalar functions of position.

When a function of position is expressed in the form $f(x, y, z)$ it is implied that it is a function of the cartesian coordinates of a point.[1] The value of a function at a given point cannot however depend on the particular system of coordinates used to specify its position. We saw in Chapter 9 that, for points in a plane, a function and the region in which it is defined may sometimes (in fact, often) be expressed more simply in terms of coordinates other than cartesian. The most important and most widely used coordinates, other than the cartesian, are the spherical polar coordinates. They are presented in Section 10.2. Functions of position are discussed in Section 10.3 and volume integrals in 10.4. The Laplacian operator, so important in the physical sciences, is discussed in Section 10.5. In these, the discussion is restricted to the cartesian and spherical polar coordinate systems. The general coordinate system is the subject of Section 10.6. The use of vectors and matrices for the description of systems and processes in three dimensions is discussed in Chapters 16, 18, and 19.

10.2 Spherical polar coordinates

The position of a point in a three-dimensional space is specified uniquely by its three coordinates in a given coordinate system. In the cartesian system shown in Figure 10.1,

---

[1] The analytical geometry of three dimensions has its origins in Clairaut's *Recherches* of 1731, in which he gave $x^2 + y^2 + z^2 = a^2$ as the equation of a sphere of radius a, and in the second volume of Euler's *Introductio* of 1748. The systematic theory was developed by Monge in papers from 1771 and in two influential textbooks written for his students at the École Polytechnique. Gaspar Monge (1746–1818), the son of Jacques Monge, peddler, knife-grinder, and respecter of education, developed a 'descriptive geometry' that formed the basis for modern engineering drawing; the method was classified for a time as a military secret, and published in 1799 in the textbook *Géométrie descriptive*. Monge was the first Director of the École Polytechnique formed in 1794 by the National Convention of the Revolution for the training of engineers and scientists (who show 'a constant love of liberty, equality, and a hatred of tyrants'). The École became a model for colleges throughout Europe and the United States. Teachers at the École included Laplace, Lagrange, and Sylvestre François Lacroix (1765–1843), whose textbook on the calculus was translated into English in 1816 and was influential (with other texts from the École) in bringing European methods to England and the United States. In the textbook *Application de l'analyse à la géométrie*, 1807, Monge discussed the analytical geometry of two and three dimensions. He showed how the coordinates of a point are determined by the perpendiculars from three coordinate planes.

the position of the point P is specified by the ordered triple (x, y, z), the cartesian coordinates. The coordinate x is the distance of P from the yz-plane, y is the distance from the zx-plane, and z is the distance from the xy-plane.

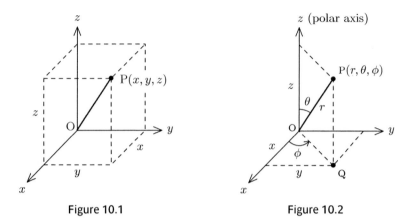

Figure 10.1   Figure 10.2

Other coordinate systems are normally defined in terms of the cartesian system. The system of spherical polar coordinates is shown in Figure 10.2. The distance r of the point from the origin is called the **radial coordinate**; it has possible values from 0 to $+\infty$. The angle θ, the **colatitude**, is the angle between the radial line OP and the z-axis; it has possible values 0 to π. In this context the z-axis is called the **polar axis**. The angle ϕ, the **longitude**, is the angle between the x-axis and the line OQ, the projection of OP in the xy-plane; it has possible values 0 to 2π. Changes in ϕ describe rotation around the polar axis. The spherical polar and cartesian coordinates are related by the equations

$$x = r \sin \theta \cos \phi, \qquad y = r \sin \theta \sin \phi, \qquad z = r \cos \theta \qquad (10.1)$$

As for polar coordinates in a plane (Section 3.5), conversion from spherical polar to cartesian coordinates is straightforward.

EXAMPLE 10.1 Find the cartesian coordinates of the point whose spherical polar coordinates are $(r, \theta, \phi) = (2, \pi/6, \pi/4)$.

$$x = r \sin \theta \cos \phi = 2 \sin(\pi/6) \cos(\pi/4) = 1/\sqrt{2}$$

$$y = r \sin \theta \sin \phi = 2 \sin(\pi/6) \sin(\pi/4) = 1/\sqrt{2}$$

$$z = r \cos \theta = 2 \cos(\pi/6) = \sqrt{3}$$

➤ Exercises 1–3

The conversion from cartesian coordinates to spherical polar coordinates makes use of the following relations:

$$r^2 = x^2 + y^2 + z^2, \quad \theta = \cos^{-1}\left(\frac{z}{r}\right), \quad \phi = \begin{cases} \tan^{-1}\left(\dfrac{y}{x}\right) & \text{if } x > 0 \\[2mm] \tan^{-1}\left(\dfrac{y}{x}\right) + \pi & \text{if } x < 0 \end{cases} \qquad (10.2)$$

in which the inverse functions have their principal values (see Section 3.5).

EXAMPLE 10.2 Find the spherical polar coordinates of the point $(x, y, z) = (-1, 2, -3)$.

$$r^2 = x^2 + y^2 + z^2 = 14, \quad r = \sqrt{14}$$

$$\theta = \cos^{-1}(z/r) = \cos^{-1}(-3/\sqrt{14}) \approx 143.3°$$

$$\phi = \tan^{-1}(y/x) + \pi = \tan^{-1}(-2) + \pi \approx 116.6°$$

▸ Exercises 4–9

10.3 Functions of position

A function of position, or field, is a function of the three coordinates within some region of three-dimensional space. Let the region V (for volume) in Figure 10.3 represent, for example, a body with non-uniform temperature; the temperature is a function of position,

$$T = f(x, y, z)$$

Then, if the cartesian coordinates of the point P are (x_p, y_p, z_p), the temperature at this point is

$$T_p = f(x_p, y_p, z_p)$$

For example, if $f(x, y, z) = z^2 - x^2 - y^2$ then

$$T_p = z_p^2 - x_p^2 - y_p^2$$

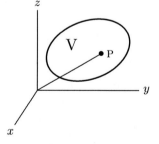

Figure 10.3

The temperature at a point cannot depend on the particular system of coordinates used to specify the position of the point. If the spherical coordinates at P are (r_p, θ_p, ϕ_p) then, by equations (10.1),

$$T_p = r_p^2 \cos^2 \theta_p - r_p^2 \sin^2 \theta_p \cos^2 \phi_p - r_p^2 \sin^2 \theta_p \sin^2 \phi_p = r_p^2(\cos^2 \theta_p - \sin^2 \theta_p)$$

The temperature field is therefore described equally well by the function $g(r, \theta, \phi) = r^2(\cos^2 \theta - \sin^2 \theta)$ such that, at any point in V,

$$T = f(x, y, z) = g(r, \theta, \phi)$$

In this example, the transformation from cartesian to spherical polar coordinates has led to two simplifications of the representation of the field. Firstly, whereas T depends on all three cartesian coordinates, x, y, and z, it only depends on two of the spherical polar coordinates, r and θ, that is, the field is independent of the value of the angle ϕ, and is therefore cylindrically symmetric about the polar (z) axis. The second simplification is that the variables in the function g have been separated; that is, the function $g(r, \theta)$ has been factorized as the product of a function of r and a function of θ. Such simplifications are important for the evaluation of multiple integrals and for the solution of partial differential equations.

EXAMPLE 10.3 Express in spherical polar coordinates: (i) $(x^2 + y^2 + z^2)^{1/2}$,

(ii) $\dfrac{\partial}{\partial x}(x^2 + y^2 + z^2)^{1/2}$, (iii) $\dfrac{\partial}{\partial r}(x^2 + y^2 + z^2)^{1/2}$

(i) $(x^2 + y^2 + z^2)^{1/2} = r$,

(ii) $\dfrac{\partial}{\partial x}(x^2 + y^2 + z^2)^{1/2} = \dfrac{1}{2}(x^2 + y^2 + z^2)^{-1/2} \times 2x = \dfrac{x}{r} = \sin\theta\cos\phi$

(iii) $\dfrac{\partial}{\partial r}(x^2 + y^2 + z^2)^{1/2} = \dfrac{\partial}{\partial r} r = 1$

▸ Exercises 10–13

Density functions

Consider the distribution of mass in a three-dimensional body (see Section 5.6 for a linear distribution of mass). Let P be some point within the body, and let Δm be the mass in a volume Δv surrounding the point P, as shown in Figure 10.4. The ratio $\Delta m/\Delta v$ is then the mass per unit volume, or the average mass density, in Δv. If the mass is not distributed uniformly throughout the body, the value of this ratio depends not only on the position of the volume Δv but also on the size and shape of Δv. As the size of the volume is reduced to zero, $\Delta v \to 0$, the ratio approaches a limit

$$\rho = \lim_{\Delta v \to 0} \left(\frac{\Delta m}{\Delta v} \right) \qquad (10.3)$$

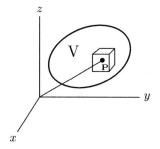

Figure 10.4

and the value of this limit is the density at the point P. We note that although ρ is defined as the limit of a ratio it is not a derivative in the normal sense. We can however consider the differential mass $dm = \rho dv$ in volume element dv. The total mass of the body is then the integral over the volume of the body,

$$M = \int_V dm = \int_V \rho \, dv \tag{10.4}$$

Such volume integrals are triple integrals over the three coordinates of the space and are discussed in Section 10.4.

The concept of density is applicable to any property that is distributed over a region. One important example is the probability density of a distribution in probability theory and statistics, discussed in Chapter 21. Probability densities are also used in quantum mechanics for the interpretation of wave functions and for the computation of the properties of atoms and molecules.

EXAMPLE 10.4 Atomic orbitals and electron probability density

The solutions of the Schrödinger equation for an atom, the atomic orbitals, are nearly always expressed in spherical polar coordinates. Some of the orbitals of the hydrogen atom are listed in Table 10.1 (a_0 is the bohr radius). Each orbital is the product of three functions, one for each coordinate:

$$\psi(r, \theta, \phi) = R(r) \cdot \Theta(\theta) \cdot \Phi(\phi) \tag{10.5}$$

The radial function $R(r)$ determines the size of the orbital, and the radial or 'in-out' motion of the electron in the orbital. The angular functions $\Theta(\theta)$ and $\Phi(\phi)$ determine the shape of the orbital and the angular motion of the electron (its angular momentum). These functions are discussed in greater detail in Chapter 14.

Table 10.1 Atomic orbitals of the hydrogen atom

$$\psi_{1s} = \frac{1}{\sqrt{\pi a_0^3}} e^{-r/a_0}$$

$$\psi_{2s} = \frac{1}{4\sqrt{2\pi a_0^3}} (2 - r/a_0) e^{-r/2a_0}$$

$$\psi_{2p_x} = \frac{1}{4\sqrt{2\pi a_0^5}} r e^{-r/2a_0} \sin\theta \cos\phi$$

$$\psi_{2p_y} = \frac{1}{4\sqrt{2\pi a_0^5}} r e^{-r/2a_0} \sin\theta \sin\phi$$

$$\psi_{2p_z} = \frac{1}{4\sqrt{2\pi a_0^5}} r e^{-r/2a_0} \cos\theta$$

The physical interpretation of an orbital is in terms of an electron probability density; for an electron in orbital ψ the quantity

$$|\psi(r, \theta, \phi)|^2 \, dv \tag{10.6}$$

is interpreted as the probability of finding the electron in the volume element dv at position (r, θ, ϕ). The square modulus, $|\psi|^2 = \psi\psi^*$, is used because wave functions are in general complex functions. The probability of finding the electron in a region V is then the volume integral $\int_V |\psi|^2 \, dv$.

10.4 Volume integrals

A triple, or three-fold, integral has the general form

$$\int_V f(x, y, z) \, dv = \int_{z_1}^{z_2} \int_{y_1}^{y_2} \int_{x_1}^{x_2} f(x, y, z) \, dx \, dy \, dz \tag{10.7}$$

where V is a region in xyz-space. When the variables are the coordinates of a point in ordinary space the integral is often called a volume integral. If the limits of integration in (10.7) are constants then the region V is a rectangular box of sides $x_2 - x_1$, $y_2 - y_1$, $z_2 - z_1$.

EXAMPLE 10.5 Evaluate the integral of the function $f(x, y, z) = 1 + xyz$ over the rectangular box of sides a, b, c shown in Figure 10.5.

The integral (10.7) is

$$\int_V f(x, y, z) \, dv = \int_V dv + \int_V xyz \, dv$$

Then

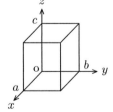

Figure 10.5

(i) $$\int_V dv = \int_0^c \int_0^b \int_0^a dx \, dy \, dz = \int_0^a dx \int_0^b dy \int_0^c dz$$

$$= a \times b \times c = V$$

This is a general result; the integral $\int_V dv$ is the volume of the region V.

(ii) $$\int_V xyz \, dv = \int_0^c \int_0^b \int_0^a xyz \, dx \, dy \, dz = \int_0^a x \, dx \int_0^b y \, dy \int_0^c z \, dz$$

The factorization of the triple integral is allowed because the integrand is the product of three functions, one for each variable, and the limits of integration are constants (see Example 9.31). Then

$$\int_V xyz\, dv = \frac{a^2}{2}\times\frac{b^2}{2}\times\frac{c^2}{2} = \frac{V^2}{8}$$

▸ Exercises 14–17

Spherical polar coordinates

The volume integral in spherical polar coordinates is

$$\int_V f(r,\theta,\phi)\,dv = \int_{\phi_1}^{\phi_2}\int_{\theta_1}^{\theta_2}\int_{r_1}^{r_2} f(r,\theta,\phi)\, r^2 \sin\theta\, dr\, d\theta\, d\phi \qquad (10.8)$$

The form of the volume element

$$dv = r^2 \sin\theta\, dr\, d\theta\, d\phi \qquad (10.9)$$

can be understood in terms of Figure 10.6. The region Δv is a section of spherical shell of thickness Δr between radii r and $r+\Delta r$, angles θ and $\theta+\Delta\theta$, and angles ϕ and $\phi+\Delta\phi$. It can be shown (see Example 10.6) that the volume of the region is

$$\Delta v = \left[r^2\Delta r + r(\Delta r)^2 + \frac{1}{3}(\Delta r)^3 \right]$$

$$\times\left[\cos\theta - \cos(\theta+\Delta\theta) \right]\times\Delta\phi$$

For small Δ-values this volume is approximately equal to the volume of a rectangular box of sides $\Delta r, r\Delta\theta, r\sin\theta\Delta\phi$,

$$\Delta v \approx r^2 \sin\theta\, \Delta r\, \Delta\theta\, \Delta\phi \qquad (10.10)$$

and the volume element (10.9) is obtained for infinitesimal quantities.

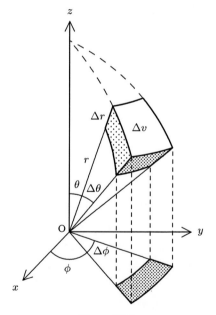

Figure 10.6

EXAMPLE 10.6 Find the volume Δv in Figure 10.6 and show that it reduces to the approximate expression (10.10) for small Δ-values.

$$\Delta v = \int_{\phi}^{\phi+\Delta\phi} \int_{\theta}^{\theta+\Delta\theta} \int_{r}^{r+\Delta r} r^2 \sin\theta \, dr \, d\theta \, d\phi$$

$$= \int_{r}^{r+\Delta r} r^2 \, dr \int_{\theta}^{\theta+\Delta\theta} \sin\theta \, d\theta \int_{\phi}^{\phi+\Delta\phi} d\phi = \left[\frac{r^3}{3}\right]_{r}^{r+\Delta r} \left[-\cos\theta\right]_{\theta}^{\theta+\Delta\theta} \left[\phi\right]_{\phi}^{\phi+\Delta\phi}$$

$$= \left[r^2 \Delta r + r(\Delta r)^2 + \frac{1}{3}(\Delta r)^3\right]\left[\cos\theta - \cos(\theta+\Delta\theta)\right]\Delta\phi$$

When Δr is small enough, the terms in $(\Delta r)^2$ and $(\Delta r)^3$ can be neglected. When $\Delta\theta$ is small,

$$\cos(\theta+\Delta\theta) = \cos\theta\cos\Delta\theta - \sin\theta\sin\Delta\theta \approx \cos\theta - \sin\theta\Delta\theta$$

since $\cos\Delta\theta \to 1$ and $\sin\Delta\theta \to \Delta\theta$ as $\Delta\theta \to 0$. Therefore, for small Δ-values,

$$\Delta v \approx r^2 \sin\theta \Delta r \, \Delta\theta \, \Delta\phi$$

We note that this does not provide a proof of (10.9), since the volume element is used for the construction of the volume integral.

EXAMPLE 10.7 Evaluate the integral of the function $f(r, \theta, \phi) = 1 + r^2 \cos^2\theta \sin^2\phi$ over a sphere of radius a and centre at the origin.

The integral can be evaluated in two parts, as in Example 10.5:

$$\int_V f \, dv = \int_V dv + \int_V r^2 \cos^2\theta \sin^2\phi \, dv$$

The ranges of integration are $r = 0 \to a$, $\theta = 0 \to \pi$, and $\phi = 0 \to 2\pi$. Then

(i) $$\int_V dv = \int_0^{2\pi} \int_0^{\pi} \int_0^{a} r^2 \sin\theta \, dr \, d\theta \, d\phi$$

$$= \int_0^{a} r^2 \, dr \int_0^{\pi} \sin\theta \, d\theta \int_0^{2\pi} d\phi = \frac{a^3}{3} \times 2 \times 2\pi = \frac{4}{3}\pi a^3$$

and this is the volume of the sphere.

(ii) $$\int_V r^2 \cos^2\theta \sin^2\phi \, dv = \int_0^{2\pi} \int_0^{\pi} \int_0^{a} r^4 \cos^2\theta \sin\theta \sin^2\phi \, dr \, d\theta \, d\phi$$

$$= \int_0^{a} r^4 \, dr \int_0^{\pi} \cos^2\theta \sin\theta \, d\theta \int_0^{2\pi} \sin^2\phi \, d\phi$$

$$= \frac{a^5}{5} \times \frac{2}{3} \times \pi$$

> Exercises 18, 19

Integrals over all space

When the region V is the whole three-dimensional space, the volume integral is

$$\int_V f \, dv = \int_{-\infty}^{+\infty} \int_{-\infty}^{+\infty} \int_{-\infty}^{+\infty} f(x,y,z) \, dx \, dy \, dz \qquad \text{in cartesians} \qquad (10.11)$$

$$= \int_0^{2\pi} \int_0^{\pi} \int_0^{\infty} f(r,\theta,\phi) \, r^2 \sin\theta \, dr \, d\theta \, d\phi \qquad \text{in spherical polars} \quad (10.12)$$

These integrals are important, for example, in quantum chemistry for the evaluation of atomic and molecular properties from wave functions obtained as solutions of the Schrödinger equation.

EXAMPLE 10.8 The integral over all space of the electron probability density of the 1s orbital of the hydrogen atom, $\psi_{1s}^2 = (1/\pi a_0^3) e^{-2r/a_0}$ (see Table 10.1), is

$$\int \psi_{1s}^2 \, dv = \frac{1}{\pi a_0^3} \int_0^{2\pi} \int_0^{\pi} \int_0^{\infty} e^{-2r/a_0} r^2 \sin\theta \, dr \, d\theta \, d\phi$$

$$= \frac{1}{\pi a_0^3} \int_0^{\infty} e^{-2r/a_0} r^2 \, dr \int_0^{\pi} \sin\theta \, d\theta \int_0^{2\pi} d\phi$$

The integral over the angles

$$\int_0^{\pi} \sin\theta \, d\theta \int_0^{2\pi} d\phi = 4\pi \qquad (10.13)$$

is the complete solid angle around a point. Then, making use of the standard integral $\int_0^{\infty} e^{-ar} r^n \, dr = n!/a^{n+1}$,

$$\int \psi_{1s}^2 \, dv = \frac{1}{\pi a_0^3} \times \frac{2}{(2/a_0)^3} \times 4\pi = 1$$

This result is consistent with the interpretation of $|\psi|^2 \, dv$ as the probability of finding the electron in element dv. The total probability, the probability of finding the electron somewhere, is the integral over all space, and must by unity. In fact, the coefficients of the orbitals in Table 10.1 have been chosen for this to be true. The orbitals are said to be normalized (to unity).

▶ Exercises 20–22

Average values

We shall see in Chapter 21 that if x is a continuous variable in the interval $a \leq x \leq b$, and if $p(x)\, dx$ is the probability that the variable have value between x and $x + dx$, then the quantity

$$\bar{f} = \int_a^b f(x) p(x)\, dx \tag{10.14}$$

is the average value of the function $f(x)$ in the interval. The generalization to functions of more than one variable involves the corresponding multiple integral. For example, let $f(x, y, z)$ be a function of position in three dimensions, and let $p(x, y, z)\, dx\, dy\, dz$ be the probability that the x-coordinate have value between x and $x + dx$, that the y-coordinate have value between y and $y + dy$, and that the z-coordinate have value between z and $z + dz$. The average value of the function is then the volume integral over all space

$$\bar{f} = \int_{-\infty}^{+\infty} \int_{-\infty}^{+\infty} \int_{-\infty}^{+\infty} f(x,y,z) p(x,y,z)\, dx\, dy\, dz \tag{10.15}$$

The corresponding expression in spherical polar coordinates is

$$\bar{f} = \int_0^{2\pi} \int_0^{\pi} \int_0^{\infty} f(r,\theta,\phi) p(r,\theta,\phi)\, r^2 \sin\theta\, dr\, d\theta\, d\phi \tag{10.16}$$

If the probability density is the modulus square of a wave function in quantum mechanics, $p = |\psi|^2$, the average value

$$\bar{f} = \int f |\psi|^2\, dv \tag{10.17}$$

is usually called the expectation value of f in the state ψ.

EXAMPLE 10.9 Find the average distance of the electron from the nucleus in (i) the $1s$ orbital and (ii) the $2p_z$ orbital of the hydrogen atom (see Table 10.1 in Example 10.4).

In these cases, $f(r, \theta, \phi) = r$ in equation (10.17), and

$$\bar{r} = \int r |\psi|^2\, dv$$

(i) $$\psi_{1s}^2 = \frac{1}{\pi a_0^3} e^{-2r/a_0},$$

$$\bar{r}_{1s} = \frac{1}{\pi a_0^3} \int_0^{2\pi} \int_0^{\pi} \int_0^{\infty} e^{-2r/a_0}\, r^3 \sin\theta\, dr\, d\theta\, d\phi$$

and, after integration over the angles (see equation (10.13)),

$$\bar{r}_{1s} = \frac{4}{a_0^3} \int_0^\infty e^{-2r/a_0} r^3 \, dr = \frac{4}{a_0^3} \times \frac{3!}{(2/a_0)^4} = \frac{3}{2} a_0$$

(ii) $$\psi_{2p_z}^2 = \frac{1}{32\pi a_0^5} r^2 e^{-r/a_0} \cos^2 \theta$$

$$\bar{r}_{2p_z} = \frac{1}{32\pi a_0^5} \int_0^\infty r^5 e^{-r/a_0} \, dr \int_0^\pi \cos^2 \theta \sin \theta \, d\theta \int_0^{2\pi} d\phi$$

$$= \frac{1}{32\pi a_0^5} \times 5! a_0^6 \times \frac{2}{3} \times 2\pi = 5 a_0$$

▸ Exercises 23, 24

10.5   The Laplacian operator

The Laplacian operator in two dimensions was discussed in Section 9.6. The operator in three dimensions is:

in **cartesian coordinates,**

$$\nabla^2 = \frac{\partial^2}{\partial x^2} + \frac{\partial^2}{\partial y^2} + \frac{\partial^2}{\partial z^2} \tag{10.18}$$

in **spherical polar coordinates,**

$$\nabla^2 = \frac{1}{r^2} \frac{\partial}{\partial r} \left(r^2 \frac{\partial}{\partial r} \right) + \frac{1}{r^2 \sin \theta} \frac{\partial}{\partial \theta} \left(\sin \theta \frac{\partial}{\partial \theta} \right) + \frac{1}{r^2 \sin^2 \theta} \frac{\partial^2}{\partial \phi^2} \tag{10.19a}$$

$$= \frac{\partial^2}{\partial r^2} + \frac{2}{r} \frac{\partial}{\partial r} + \frac{1}{r^2} \frac{\partial^2}{\partial \theta^2} + \frac{\cos \theta}{r^2 \sin \theta} \frac{\partial}{\partial \theta} + \frac{1}{r^2 \sin^2 \theta} \frac{\partial^2}{\partial \phi^2} \tag{10.19b}$$

The transformation from cartesian to spherical polar coordinates is achieved in the same way as that described in Example 9.18 for the two-dimensional case. The operator in spherical polar coordinates is usually quoted in the slightly more compact form (10.19a), and the expanded form (10.19b) is obtained by use of the product rule of differentiation. For example,

$$\frac{1}{r^2} \frac{\partial}{\partial r} \left(r^2 \frac{\partial f}{\partial r} \right) = \frac{1}{r^2} \left(r^2 \frac{\partial^2 f}{\partial r^2} + 2r \frac{\partial f}{\partial r} \right) = \frac{\partial^2 f}{\partial r^2} + \frac{2}{r} \frac{\partial f}{\partial r} \tag{10.20}$$

The Laplacian operator occurs in the equations of motion concerned with wave motion and potential theory in both classical mechanics (as in Maxwell's equations of electromagnetic theory) and quantum mechanics (as in Schrödinger's wave mechanics, see Chapter 14).

EXAMPLE 10.10 Evaluate $\nabla^2 f$ for $f(r) = e^{-r}$ in (i) spherical polar coordinates and (ii) cartesian coordinates.

(i) Because $f(r)$ is a function of the radial coordinate only it follows that

$$\frac{\partial f}{\partial \theta} = 0, \qquad \frac{\partial f}{\partial \phi} = 0, \qquad \frac{\partial f}{\partial r} = \frac{df}{dr} = -e^{-r}, \qquad \frac{\partial^2 f}{\partial r^2} = \frac{d^2 f}{dr^2} = e^{-r}$$

Then

$$\nabla^2 f = \frac{d^2 f}{dr^2} + \frac{2}{r}\frac{df}{dr} = e^{-r} - \frac{2}{r}e^{-r} = \left(1 - \frac{2}{r}\right)e^{-r}$$

(ii) In cartesian coordinates,

$$\nabla^2 f = \frac{\partial^2 f}{\partial x^2} + \frac{\partial^2 f}{\partial y^2} + \frac{\partial^2 f}{\partial z^2}$$

$$r = (x^2 + y^2 + z^2)^{1/2}, \qquad \frac{\partial r}{\partial x} = \frac{x}{r}, \qquad \frac{\partial r}{\partial y} = \frac{y}{r}, \qquad \frac{\partial r}{\partial z} = \frac{z}{r}$$

Then, by the chain rule,

$$\frac{\partial f}{\partial x} = \frac{df}{dr}\frac{\partial r}{\partial x} = \frac{x}{r}\frac{df}{dr}, \qquad \frac{\partial^2 f}{\partial x^2} = \frac{\partial}{\partial x}\left(\frac{x}{r}\frac{df}{dr}\right) = \frac{1}{r}\frac{df}{dr} - \frac{x^2}{r^3}\frac{df}{dr} + \frac{x^2}{r^2}\frac{d^2 f}{dr^2}$$

The derivatives with respect to y and z are obtained in the same way. Then

$$\nabla^2 f = \frac{\partial^2 f}{\partial x^2} + \frac{\partial^2 f}{\partial y^2} + \frac{\partial^2 f}{\partial z^2} = \left(\frac{3}{r} - \frac{(x^2 + y^2 + z^2)}{r^3}\right)\frac{df}{dr} + \left(\frac{x^2 + y^2 + z^2}{r^2}\right)\frac{d^2 f}{dr^2}$$

so that, because $x^2 + y^2 + z^2 = r^2$,

$$\nabla^2 f = \frac{2}{r}\frac{df}{dr} + \frac{d^2 f}{dr^2} = \left(-\frac{2}{r} + 1\right)e^{-r}$$

We note that the use of cartesian coordinates in this case naturally involves, via the chain rule, a transformation to spherical polar coordinates.

▸ Exercises 25–27

EXAMPLE 10.11 Evaluate $\nabla^2 f$ for $f(r, \theta, \phi) = re^{-r/2} \sin \theta \cos \phi$.

Using the form (10.19a),

$$\nabla^2 f = \frac{1}{r^2} \frac{\partial}{\partial r}\left(r^2 \frac{\partial f}{\partial r}\right) + \frac{1}{r^2 \sin \theta} \frac{\partial}{\partial \theta}\left(\sin \theta \frac{\partial f}{\partial \theta}\right) + \frac{1}{r^2 \sin^2 \theta} \frac{\partial^2 f}{\partial \phi^2}$$

The function f has the factorized form

$$f(r, \theta, \phi) = (re^{-r/2})(\sin \theta)(\cos \phi) = R(r) \times \Theta(\theta) \times \Phi(\phi)$$

so that

$$\frac{\partial f}{\partial r} = \left(\frac{dR(r)}{dr}\right) \Theta(\theta)\,\Phi(\phi)$$

$$\frac{\partial f}{\partial \theta} = R(r)\left(\frac{d\Theta(\theta)}{d\theta}\right) \Phi(\phi)$$

$$\frac{\partial f}{\partial \phi} = R(r)\,\Theta(\theta)\left(\frac{d\Phi(\phi)}{d\phi}\right)$$

and, therefore,

$$\nabla^2 f = \left[\frac{1}{r^2} \frac{d}{dr}\left(r^2 \frac{dR}{dr}\right)\right]\Theta\Phi + \left[\frac{1}{\sin \theta} \frac{d}{d\theta}\left(\sin \theta \frac{d\Theta}{d\theta}\right)\right]\frac{R\Phi}{r^2} + \left[\frac{d^2\Phi}{d\phi^2}\right]\frac{R\Theta}{r^2 \sin^2 \theta}$$

Now

$$R = re^{-r/2}; \qquad \frac{1}{r^2} \frac{d}{dr}\left(r^2 \frac{dR}{dr}\right) = \left(\frac{r}{4} + \frac{2}{r} - 2\right)e^{-r/2} = \left(\frac{1}{4} + \frac{2}{r^2} - \frac{2}{r}\right)R(r)$$

$$\Theta = \sin \theta; \qquad \frac{1}{\sin \theta} \frac{d}{d\theta}\left(\sin \theta \frac{d\Theta}{d\theta}\right) = \frac{\cos^2 \theta - \sin^2 \theta}{\sin \theta} = \frac{\cos^2 \theta - \sin^2 \theta}{\sin^2 \theta}\Theta(\theta)$$

$$\Phi = \cos \phi; \qquad \frac{d^2\Phi}{d\phi^2} = -\cos \phi = -\Phi(\phi)$$

Therefore,

$$\nabla^2 f = \left(\frac{1}{4} + \frac{2}{r^2} - \frac{2}{r} + \frac{\cos^2 \theta - \sin^2 \theta}{r^2 \sin^2 \theta} - \frac{1}{r^2 \sin^2 \theta}\right)R\Theta\Phi = \left(\frac{1}{4} - \frac{2}{r}\right)f$$

This result can be rearranged to give the equation

$$\left(-\frac{1}{2}\nabla^2 - \frac{1}{r}\right)f = -\frac{1}{8}f$$

and this is essentially the Schrödinger equation of the hydrogen atom, with f the $2p_x$ orbital. The result of Example 10.10 is the same equation for the $1s$ orbital.

➤ Exercises 28–33

EXAMPLE 10.12 Show that the function $f(r) = 1/r$ satisfies the Laplace equation in three dimensions.

The Laplace equation is

$$\nabla^2 f = 0$$

As in Example 10.10, the function does not depend on the angles θ and ϕ. Therefore,

$$\nabla^2 f = \frac{d^2 f}{dr^2} + \frac{2}{r}\frac{df}{dr}$$

Now $df/dr = -1/r^2$, $d^2f/dr^2 = 2/r^3$. Therefore

$$\frac{d^2 f}{dr^2} + \frac{2}{r}\frac{df}{dr} = \frac{2}{r^3} - \frac{2}{r^3} = 0 \qquad (r \neq 0)$$

We note that this demonstration is not valid when $r = 0$, and this singular point requires special treatment in physical applications.

This example shows that the gravitational and Coulomb potential functions satisfy the Laplace equation (see Example 9.20 for the two-dimensional case).

➤ Exercises 34–37

10.6 Other coordinate systems

In addition to the cartesian and spherical polar coordinates, other systems of coordinates have been found useful for the description of physical systems. The most important of these are of the type called **orthogonal curvilinear coordinates**.[2]

In the cartesian system (Figure 10.1), the position of a point $P(x, y, z)$ is defined by the intersection of three mutually perpendicular surfaces (planes); $x = $ constant

[2] Curvilinear coordinates were introduced by Lamé in his *Sur les coordonnées curvilignes et leurs diverses applications*. Gabriel Lamé (1795–1870), engineer and professor at the École Polytechnique, made contributions to the theory of the elasticity of solids and was involved in the construction of the first railways in France.

(parallel to the O*yz*-plane), *y* = constant (parallel to the O*zx*-plane), and *z* = constant (parallel to the O*xy*-plane). The coordinate axes through the point are the lines of intersection of these planes, and are also perpendicular.

In spherical polar coordinates, the position of P(*r*, *θ*, *φ*) is again defined by the intersection of three coordinate surfaces; *r* = constant, *θ* = constant, and *φ* = constant. The surface *r* = constant is that of a sphere of radius *r*. The surface *θ* = constant is that of the right circular cone with apex O and circular base BPB′ in Figure 10.7. The surface *φ* = constant is the (vertical) plane APA′ in the figure.

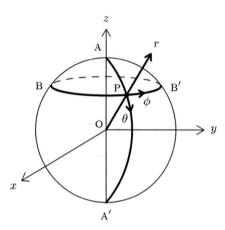

Figure 10.7

The corresponding 'axes' through the point P are the curves of intersection of pairs of these surfaces. As shown in the figure, the *r*-coordinate line is a radial line, the *θ*-coordinate line (curve) is the semicircle APA′, the *φ*-coordinate line is the circle BPB′. These curves (and the surfaces) intersect at right angles; they are perpendicular, or **orthogonal**, at their point of intersection P. This is the characteristic property of the orthogonal curvilinear coordinates.

In the general case, let the cartesian coordinates, *x*, *y*, and *z*, be related to three new quantities, q_1, q_2, and q_3, by the equations

$$x = x(q_1, q_2, q_3), \qquad y = y(q_1, q_2, q_3), \qquad z = z(q_1, q_2, q_3) \qquad (10.21)$$

with inverse relations

$$q_1 = q_1(x, y, z), \qquad q_2 = q_2(x, y, z), \qquad q_3 = q_3(x, y, z) \qquad (10.22)$$

The position of a point is then defined by specifying either *x*, *y*, *z* or q_1, q_2, q_3. Each of the equations (10.22) represents a surface, and the intersection of three such surfaces locates a point. The surfaces q_1 = constant, q_2 = constant, and q_3 = constant are the coordinate surfaces; the curves of intersection in pairs are the coordinate lines. The quantities (q_1, q_2, q_3) are the **curvilinear coordinates** of a point P(*x*, *y*, *z*).

Curvilinear coordinates lines do not necessarily intersect at right angles; they are nonorthogonal in general. Orthogonal curvilinear coordinate systems are those whose coordinate lines (and surfaces) are mutually perpendicular (orthogonal) at their point of intersection. The cartesian and spherical polar systems are the best-known examples. Such coordinate systems have the following properties.

(i) The (infinitesimal) distance between two points on a coordinate line is

$$ds_i = h_i \, dq_i \qquad\qquad (10.23)$$

where

$$h_i^2 = \left(\frac{\partial x}{\partial q_i}\right)^2 + \left(\frac{\partial y}{\partial q_i}\right)^2 + \left(\frac{\partial z}{\partial q_i}\right)^2 \qquad\qquad (10.24)$$

(ii) The volume element is

$$dv = ds_1 \, ds_2 \, ds_3 = h_1 \, h_2 \, h_3 \, dq_1 \, dq_2 \, dq_3 \qquad\qquad (10.25)$$

(iii) The Laplacian operator is

$$\nabla^2 = \frac{1}{h_1 h_2 h_3} \left\{ \frac{\partial}{\partial q_1} \left(\frac{h_2 h_3}{h_1} \frac{\partial}{\partial q_1} \right) + \frac{\partial}{\partial q_2} \left(\frac{h_3 h_1}{h_2} \frac{\partial}{\partial q_2} \right) + \frac{\partial}{\partial q_3} \left(\frac{h_1 h_2}{h_3} \frac{\partial}{\partial q_3} \right) \right\} \qquad (10.26)$$

For the cartesian and spherical polar coordinates,

cartesian: $h_x = h_y = h_z = 1$
spherical polar: $h_r = 1, \quad h_\theta = r, \quad h_\phi = r \sin\theta$

The following are two of the more widely-used alternative coordinate systems.

Cylindrical polar coordinates

These coordinates are the plane polar coordi-nates in the xy-plane plus the z-coordinate, and are useful for the description of systems with cylindrical symmetry.

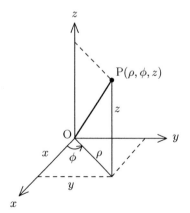

$x = \rho\cos\phi, \quad y = \rho\sin\phi, \quad z = z$

$\rho = 0 \rightarrow \infty, \quad \phi = 0 \rightarrow 2\pi, \quad z = -\infty \rightarrow +\infty$

$h_\rho = 1, \quad h_\phi = \rho, \quad h_z = 1$

$dv = \rho \, d\rho \, d\phi \, dz$

$$\nabla^2 = \frac{1}{\rho}\frac{\partial}{\partial\rho}\left(\rho\frac{\partial}{\partial\rho}\right) + \frac{1}{\rho^2}\frac{\partial^2}{\partial\phi^2} + \frac{\partial^2}{\partial z^2}$$

Figure 10.8

EXAMPLE 10.13 A circular helix lies on the curved surface of a (right) circular cylinder, as shown in Figure 10.9.

The cartesian coordinates, in parametric form, are $x = a \cos t$, $y = a \sin t$, $z = bt$, where a and b are constants, and t is a parameter. In cylindrical polar coordinates, $\rho = a$ is the constant radius of the cylinder, $\phi = t$ describes the rotation around the axis of the cylinder, and $z = bt$ gives the displacement parallel to the axis. The quantity $2\pi b$ is the pitch of the helix, the displacement parallel to the axis made in one revolution about the axis.

Figure 10.9 shows a right-handed helix, with clockwise circulation around the axis.

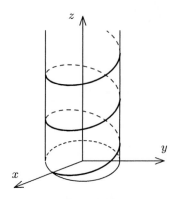

Figure 10.9

> Exercises 38, 39

Confocal elliptic coordinates

These coordinates (also called prolate spheroidal coordinates) are useful for two-centre potential problems.

$$x = a\sqrt{\xi^2 - 1}\,\sqrt{1 - \eta^2}\,\cos\phi, \quad y = a\sqrt{\xi^2 - 1}\,\sqrt{1 - \eta^2}\,\sin\phi, \quad z = a\xi\eta$$

$$\xi = 1 \to \infty, \quad \eta = -1 \to +1, \quad \phi = 0 \to 2\pi$$

$$h_\xi = a\sqrt{\frac{\xi^2 - \eta^2}{\xi^2 - 1}}, \quad h_\eta = a\sqrt{\frac{\xi^2 - \eta^2}{1 - \eta^2}}, \quad h_\phi = a\sqrt{(\xi^2 - 1)(1 - \eta^2)}$$

$$dv = a^3(\xi^2 - \eta^2)\,d\xi d\eta d\phi$$

$$\nabla^2 = \frac{1}{a^2(\xi^2 - \eta^2)}\left\{\frac{\partial}{\partial\xi}\left((\xi^2 - 1)\frac{\partial}{\partial\xi}\right) + \frac{\partial}{\partial\eta}\left((1 - \eta^2)\frac{\partial}{\partial\eta}\right) + \frac{(\xi^2 - \eta^2)}{(\xi^2 - 1)(1 - \eta^2)}\frac{\partial^2}{\partial\phi^2}\right\}$$

The coordinates ξ and η are best visualized, as in Figure 10.10, in terms of the distances r_A and r_B from the foci A and B of an ellipse (the ellipse is defined by $r_A + r_B = $ constant):

$$\xi = \frac{r_A + r_B}{2a}, \quad \eta = \frac{r_A - r_B}{2a}$$

The figure might be a representation of a diatomic molecule with nuclei at A and B, and an electron at P.

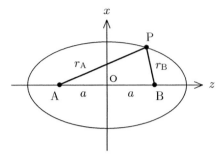

Figure 10.10

EXAMPLE 10.14 The hydrogen molecule

In the method of 'linear combinations of atomic orbitals' (LCAO) for the ground state of the hydrogen molecule, the simplest form of the occupied molecular orbital is $\psi = C(1s_A + 1s_B)$, where (in atomic units) $1s_A = (1/\sqrt{\pi})e^{-r_A}$ and $1s_B = (1/\sqrt{\pi})e^{-r_B}$ are normalized $1s$ orbitals centred on the protons at A and B (see Figure 10.10). C is a constant chosen to normalize the orbital: $\int \psi^2 dv = 1$. Thus

$$1 = \int \left(\frac{1}{\sqrt{\pi}} e^{-r_A} + \frac{1}{\sqrt{\pi}} e^{-r_B} \right)^2 dv = \frac{C^2}{\pi} \int \left(e^{-2r_A} + e^{-2r_B} + 2e^{-(r_A + r_B)} \right) dv$$

$$= C^2 \left[S_{AA} + S_{BB} + 2S_{AB} \right]$$

The integrals S_{AA} and S_{BB} are the normalization integrals for the $1s$ orbital, as in Example 10.8, and equal unity. Then $C = 1/\sqrt{2(1 + S_{AB})}$. To evaluate the overlap integral

$$S_{AB} = \frac{1}{\pi} \int e^{-(r_A + r_B)} dv$$

we use confocal elliptic coordinates (ξ, η, ϕ). Then, if $R = 2a$ is the bond length,

$$\xi = \frac{r_A + r_B}{R}, \quad \eta = \frac{r_A - r_B}{R}$$

and the volume element is

$$dv = \frac{R^3}{8} (\xi^2 - \eta^2) \, d\xi d\eta d\phi$$

The volume integral over all space is then

$$S_{AB} = \frac{1}{\pi} \int e^{-R\xi} dv = \frac{R^3}{8\pi} \int_{\phi=0}^{2\pi} \int_{\eta=-1}^{+1} \int_{\xi=1}^{\infty} e^{-R\xi} (\xi^2 - \eta^2) \, d\xi d\eta d\phi$$

$$= \frac{R^3}{8\pi} \left\{ \int_{\phi=0}^{2\pi} \int_{\eta=-1}^{+1} \int_{\xi=1}^{\infty} e^{-R\xi} \xi^2 \, d\xi d\eta d\phi - \int_{\phi=0}^{2\pi} \int_{\eta=-1}^{+1} \int_{\xi=1}^{\infty} e^{-R\xi} \eta^2 \, d\xi d\eta d\phi \right\}$$

Then, separating variables,

$$S_{AB} = \frac{R^3}{8\pi}\left\{\int_0^{2\pi}d\phi\int_{-1}^{+1}d\eta\int_1^\infty e^{-R\xi}\xi^2 d\xi - \int_0^{2\pi}d\phi\int_{-1}^{+1}\eta^2\,d\eta\int_1^\infty e^{-R\xi}d\xi\right\}$$

$$= \frac{R^3}{8\pi}\left\{\left[2\pi\times2\times\left(\frac{e^{-R}}{R^3}\left(R^2+2R+2\right)\right)\right] - \left[2\pi\times\frac{2}{3}\times\left(\frac{e^{-R}}{R}\right)\right]\right\}$$

$$= \left(1+R+R^2/3\right)e^{-R}$$

▸ Exercises 40, 41

10.7 Exercises

Section 10.2

Find the cartesian coordinates (x, y, z) of the points whose spherical polar coordinates are:
1. $(r, \theta, \phi) = (1, 0, 0)$ 2. $(r, \theta, \phi) = (2, \pi/2, \pi/2)$ 3. $(r, \theta, \phi) = (2, 2\pi/3, 3\pi/4)$
Find the spherical polar coordinates (r, θ, ϕ) of the points:
4. $(x, y, z) = (1, 0, 0)$ 5. $(x, y, z) = (0, 1, 0)$ 6. $(x, y, z) = (1, 2, 2)$
7. $(x, y, z) = (1, -4, -8)$ 8. $(x, y, z) = (-2, -3, 6)$ 9. $(x, y, z) = (-3, 4, -12)$

Section 10.3

Express in spherical polar coordinates:
10. $x^2 - y^2$ 11. $(x^2 + y^2)/z^2$ 12. $2z^2 - x^2 - y^2$
13. Express in spherical polar coordinates:

(i) $(x^2 + y^2 + z^2)^{-1/2}$, (ii) $\dfrac{\partial}{\partial x}(x^2 + y^2 + z^2)^{-1/2}$

Section 10.4

Find the total mass of a mass distribution of density ρ in region V of space:
14. $\rho = x^2 + y^2 + z^2$, V: the cube $0 \le x \le 1, 0 \le y \le 1, 0 \le z \le 1$
15. $\rho = xy^2z^3$, V: the box $0 \le x \le a, 0 \le y \le b, 0 \le z \le c$
16. $\rho = x^2$, V: the region $1 - y \le x \le 1, 0 \le y \le 1, 0 \le z \le 2$
17. $\rho = e^{-(x+y+z)}$, V: the infinite region $x \ge 0, y \ge 0, z \ge 0$
18. $\rho = x^2 + y^2 + z^2$, V: the sphere of radius a, centre at the origin
19. $\rho = \dfrac{\sin^2\theta\cos^2\phi}{r}$, V: the sphere of radius a, centre at the origin
20. $\rho = r^3 e^{-r}$, V: all space

21. The 2s orbital of the hydrogen atom is $\psi_{2s} = \dfrac{1}{4\sqrt{2\pi a_0^3}}(2 - r/a_0)e^{-r/2a_0}$. Show that the

integral of ψ_{2s}^2 over all space is unity.

22. The $3p_z$ orbital of the hydrogen atom is $\psi_{3p_z} = C(6 - r)re^{-r/3}\cos\theta$ (in atomic units) where C is a constant. Find the value of C that normalizes the $3p_z$ orbital.

23. Calculate the average distance from the nucleus of an electron in the $2s$ orbital (see Exercise 21).

24. Calculate the average value of r^2 for the $3p_z$ orbital (see Exercise 22).

Section 10.5

Find $\nabla^2 f$:

25. $f = x^2 y^3 z^4$ **26.** $r^n e^{-ar}$ **27.** $\dfrac{e^{-r}}{r}$ **28.** $e^{-r/3} \sin \theta \sin \phi$ **29.** $xze^{-r/2}$

30. If $f = (2-r)e^{-r/2}$ show that $\nabla^2 f + \dfrac{2f}{r} = \dfrac{f}{4}$.

31. If $f = ze^{-3r/2}$ show that $\nabla^2 f + \dfrac{6f}{r} = \dfrac{9f}{4}$.

32. If $f = xye^{-r}$ show that $\nabla^2 f + \dfrac{6f}{r} = f$.

33. Show that $f(r) = \dfrac{e^{ar}}{r}$, where a is an arbitrary number, satisfies $\nabla^2 f = a^2 f$.

Show that the following functions satisfy the Laplace equation:

34. $2x^3 - 3x(y^2 + z^2)$ **35.** $\dfrac{\cos 2\phi}{r^2 \sin^2 \theta}$ **36.** $r^n \sin^n \theta \cos n\phi, \ n = 1, 2, 3, \ldots$

37. The d'Alembertian operator is

$$\Box^2 = \frac{\partial^2}{\partial x^2} + \frac{\partial^2}{\partial y^2} + \frac{\partial^2}{\partial z^2} - \frac{1}{c^2}\frac{\partial^2}{\partial t^2}$$

where x, y, z are coordinates, t is the time, and c is the speed of light. Show that if the function $f(x, y, z)$ satisfies Laplace's equation then the function $g(x, y, z, t) = f(x, y, z)e^{ikct}$ satisfies the equation $\Box^2 g = k^2 g$.

Section 10.6

38. What are the parametric equations of a left-handed helix in (i) cartesian coordinates, (ii) cylindrical polar coordinates?

39. Integrate the function $f = y^2 z^3$ over the cylindrical region of radius a, between $z = 0$ and $z = 1$, and symmetric about the z-axis.

Use confocal elliptic coordinates, $\xi = \dfrac{r_A + r_B}{R}$, $\eta = \dfrac{r_A - r_B}{R}$, ϕ, to integrate the following functions over all space:

40. $\dfrac{e^{-(r_A + r_B)}}{r_A}$ **41.** $\dfrac{e^{-(r_A + r_B)}}{r_B} \cos^2 \phi$

11 First-order differential equations

11.1 Concepts

A differential equation is an equation that contains derivatives. For example, if y is a function of x then an equation that contains y and x only is an ordinary equation, but an equation that also contains one or more of dy/dx, d^2y/dx^2, and higher derivatives is a differential equation. Examples of differential equations, with representative uses in the physical sciences, are

1. $\dfrac{dx}{dt} = kx$ first-order rate process

2. $\dfrac{dx}{dt} = k(a-x)(b-x)$ second-order rate process

3. $\dfrac{d^2h}{dt^2} = -g$ body falling under the influence of gravity

4. $\dfrac{d^2x}{dt^2} = -\omega^2 x$ classical harmonic oscillator

5. $\dfrac{-\hbar^2}{2m}\dfrac{d^2\psi}{dx^2} + \dfrac{1}{2}kx^2\psi = E\psi$ wave equation for the harmonic oscillator

The five examples describe physical processes.[1] Equations **1** and **2** are used to describe (to model) some simple rate processes in physics and chemistry, and also in biology, engineering, economics, and the social sciences. Equations **3**, **4**, and **5** are equations of motion. In all of these the unknown function is a function of one variable only, $x(t)$ in **1**, **2**, and **4**, $h(t)$ in **3**, $\psi(x)$ in **5**. The derivatives are ordinary derivatives and the equations are **ordinary differential equations**. Equations containing partial derivatives are discussed in Chapter 14.

There exists a system of classification of the possible types of differential equations, but for our purposes we consider only the **order**. The order of a differential equation is the order of the highest derivative in the equation; equations **1** and **2** are first-order equations because they contain a first derivative only, equations **3** to **5** are second-order equations. Almost all the important differential equations in the physical sciences are first order (such as elementary rate processes) or second order (equations of motion such as Newton's second law, the Maxwell equations in electromagnetism, and the Schrödinger equation).

---

[1] Both Newton and Leibniz recognized that physical problems can be formulated in terms of differential equations, and the solution of physical problems provided much of the motivation for the further development of the calculus in the 18th century.

In this chapter we consider only first-order differential equations; second-order equations are discussed in Chapters 12 to 14. We consider first what is meant by *solving* a differential equation.

11.2   Solution of a differential equation

A differential equation containing an unknown function $y(x)$ of the variable x has the general form

$$f\left(x, y, \frac{dy}{dx}, \frac{d^2y}{dx^2}, \ldots\right) = 0 \tag{11.1}$$

in which some combination (a function) of x, y, and the derivatives of y is equal to zero. To **solve**, or **integrate**, the differential equation is meant to find the function (or family of functions) $y(x)$ that satisfies the equation. For example, the first-order equation (see also Section 5.2)

$$\frac{dy}{dx} = 2x \tag{11.2}$$

has the solution

$$y = x^2 + c \tag{11.3}$$

where c is an arbitrary constant, because $d(x^2 + c)/dx = 2x$. The differential equation is solved by performing an (indefinite) integration. Thus, integration of both sides of (11.2) with respect to x gives

$$\int \frac{dy}{dx}\, dx = \int 2x\, dx; \qquad y = x^2 + c \tag{11.4}$$

In the form (11.3), with the arbitrary constant, the solution is called the **general solution**, or **complete integral**, of the differential equation. We saw in Section 5.2 that the general solution (11.3) represents a family of curves, one for each value of c, and that a particular curve of the family is obtained by assigning a particular value to c. The result is a **particular solution** of the differential equation. In the physical context, the general solution represents a whole family of possible physical situations, and the choice of particular solution is determined by the nature and state of the system.

EXAMPLE 11.1 The equation

$$\frac{dx}{dt} = -kx$$

provides a model for a first-order rate process such as radioactive decay or a first-order chemical reaction, where $x(t)$ is the amount of reacting substance at time t. It is shown in Section 11.4 that the general solution of the equation is

$$x(t) = ae^{-kt}$$

where a is the arbitrary constant. In this case, the constant is specified by the amount of substance at any particular time. If $x_0 = x(0)$ is the amount at time $t = 0$ then $a = x_0$ and

$$x(t) = x_0 e^{-kt}$$

is the particular solution. We note that the application of the **initial condition** specifies the significance of the arbitrary constant as well as its value, and is a necessary step in the solution of a physical problem.

In the case of a second-order equation, containing a second derivative, *two* integrations are required to remove the second derivative, so that two arbitrary constants (constants of integration) appear in the general solution. For example, the second-order equation

$$\frac{d^2 y}{dx^2} = 2 \qquad (11.5)$$

is integrated once to give

$$\frac{dy}{dx} = 2x + a \qquad (11.6)$$

and a second time to give

$$y = x^2 + ax + b \qquad (11.7)$$

This is the general solution, and two conditions are required to specify the constants. The general solution of an nth-order differential equation contains n arbitrary constants.*

---

* Some differential equations have no general solution, some have one or more particular solutions only, some have a general solution and some particular solutions not obtained from the general solution. Such cases are not often met in the physical sciences.

EXAMPLE 11.2

(i) Show that $y = e^{-x} + 1$ is a solution of the first-order differential equation

$$\frac{dy}{dx} + y - 1 = 0.$$

We have $\dfrac{dy}{dx} = -e^{-x}$. Therefore

$$\frac{dy}{dx} + y - 1 = \left[-e^{-x} \right] + \left[e^{-x} + 1 \right] - 1 = 0$$

(ii) Find the general solution of the second-order differential equation

$$\frac{d^2 y}{dx^2} = \sin 2x.$$

We need to integrate twice:

$$\frac{dy}{dx} = \int \frac{d^2 y}{dx^2} dx = \int \sin 2x \, dx = -\frac{1}{2} \cos 2x + a$$

$$y = \int \frac{dy}{dx} dx = \int \left[-\frac{1}{2} \cos 2x + a \right] dx = -\frac{1}{4} \sin 2x + ax + b$$

➤ Exercises 1–8

EXAMPLE 11.3 A body falling under the influence of gravity

Let h be the height of the body at time t. The (downward) force acting on the body is $F = -mg$ and, by Newton's second law, $F = md^2 h/dt^2$. The acceleration experienced by the body is therefore

$$\frac{d^2 h}{dt^2} = -g$$

The first integration gives the velocity

$$v(t) = \frac{dh}{dt} = -gt + a$$

and the second integration gives the height at time t

$$h(t) = -\frac{1}{2} gt^2 + at + b$$

A particular solution in this case is given by the initial conditions for the system, the height h_0 and the velocity v_0 at time $t = 0$. Then $a = v_0$, $b = h_0$, and

$$v(t) = -gt + v_0, \quad h(t) = -\frac{1}{2} gt^2 + v_0 t + h_0$$

➤ Exercise 9

All ordinary differential equations can be solved by the numerical methods discussed in Chapter 20. There are however several important standard types whose solutions can be expressed in terms of elementary functions and that frequently occur in mathematical models of physical systems. It is these standard types that are discussed in this and the following chapters.

The general first-order differential equation has the form

$$\frac{dy}{dx} = F(x, y) \tag{11.8}$$

where $F(x, y)$ is a function of x and y, and y is a function of x. Such a differential equation together with an initial condition, $y(x_0) = y_0$ ($y = y_0$ when $x = x_0$), is called an **initial value problem**. We discuss here two important special types of equations that can be solved by elementary methods; separable equations and linear equations.

➤ Exercises 10–12

11.3 Separable equations

Equation (11.8) can be solved by straightforward integration if it has, or if it can be reduced to, the form

$$\frac{dy}{dx} = \frac{f(x)}{g(y)} \tag{11.9}$$

This equation can be written in the differential form[2]

$$g(y) \, dy = f(x) \, dx \tag{11.10}$$

with all the terms involving y on one side of the equal sign and all the terms involving x on the other. Such an equation is called a **separable differential equation** and is solved by (indefinite) integration of each side with respect to the relevant variable:

$$\int g(y) \, dy = \int f(x) \, dx + c \tag{11.11}$$

with the arbitrary constant included on one side. More formally, writing (11.9) as

$$g(y) \frac{dy}{dx} = f(x) \tag{11.12}$$

integration with respect to x gives

$$\int g(y) \frac{dy}{dx} \, dx = \int f(x) \, dx + c \tag{11.13}$$

---

[2] Much of Leibniz's formulation of the calculus involved equations containing differentials (differential equations). The method of separation of variables and the method of reducing a homogeneous equation to separable form discussed in this section were invented by Leibniz in 1691.

and, by the method of substitution for the left side (Chapter 6), this is identical to (11.11). The solution of the differential equation is completed by the evaluation of the integrals in (11.11) by the methods discussed in Chapters 5 and 6 (if possible; otherwise numerical methods must be used) and by the application of the appropriate initial condition. The following examples demonstrate the method. Applications of separable differential equations are discussed in Section 11.4.

EXAMPLE 11.4 Separable equations

Each of the following equations has the form (11.9) with solution (11.11). An initial condition is given in each case.

(i) $\dfrac{dy}{dx} = ax^n, \quad y(0) = 0.$

By equation (11.10), $dy = ax^n\, dx$, and integration gives (when $n \neq -1$)

$$\int dy = a \int x^n\, dx, \qquad y(x) = \frac{ax^{n+1}}{n+1} + c$$

This is the general solution. The required particular solution is obtained by applying the initial condition, that $y = 0$ when $x = 0$. Therefore $c = 0$ and

$$y(x) = \frac{ax^{n+1}}{n+1}$$

(ii) $\dfrac{dy}{dx} = ky, \quad y(x_0) = y_0.$

Separation of the variables and integration gives

$$\frac{dy}{y} = k\, dx, \qquad \int \frac{dy}{y} = k \int dx, \qquad \ln y = kx + c$$

An alternative form of the general solution is obtained by taking the exponential of each side of the equation. On the left side, $e^{\ln y} = y$; on the right side, $e^{kx+c} = e^c\, e^{kx} = ae^{kx}$. Then

$$y = ae^{kx}$$

By the initial condition, $y = y_0$ when $x = x_0$. Therefore, $y(x_0) = y_0 = ae^{kx_0}$ so that $a = y_0 e^{-kx_0}$, and the particular solution is

$$y(x) = y_0 e^{k(x-x_0)}$$

(iii) $\dfrac{dy}{dx} + 3x^2 y^2 = 0, \qquad y(1) = \dfrac{1}{2}.$

Separation of the variables and integration gives

$$-\frac{dy}{y^2} = 3x^2\,dx, \qquad -\int \frac{dy}{y^2} = 3\int x^2\,dx, \qquad \frac{1}{y} = x^3 + c$$

and the general solution is

$$y(x) = \frac{1}{x^3 + c}$$

By the initial condition, $y = 1/2$ when $x = 1$,

$$y(1) = \frac{1}{2} = \frac{1}{1+c}$$

so that $c = 1$ and the required particular solution is

$$y(x) = \frac{1}{x^3 + 1}$$

▸ Exercises 13–22

Reduction to separable form

Some first-order differential equations that are not separable can be made so by means of a suitable change of variables. One such example is when the function $F(x, y)$ on the right of the general form (11.8) has the property

$$F(\lambda x, \lambda y) = F(x, y) \tag{11.14}$$

where λ is an arbitrary number or function; that is, 'scaling' both x and y by the same factor leaves the function unchanged. A function of x and y has this property if it can be expressed as a function of y/x. For example,

$$F(x,y) = \frac{x^2 + y^2}{xy} = \left(\frac{y}{x}\right) + 1 \Big/ \left(\frac{y}{x}\right) = f\left(\frac{y}{x}\right)$$

Then

$$F(\lambda x, \lambda y) = f\left(\frac{\lambda y}{\lambda x}\right) = f\left(\frac{y}{x}\right) = F(x,y) \tag{11.15}$$

The differential equation

$$\frac{dy}{dx} = F(x, y) = f\left(\frac{y}{x}\right) \tag{11.16}$$

is made separable by means of the substitution $u = y/x$. Then

$$y = xu, \qquad \frac{dy}{dx} = u + x\frac{du}{dx} = f(u) \tag{11.17}$$

so that, separating the variables x and u,

$$\frac{du}{f(u) - u} = \frac{dx}{x} \tag{11.18}$$

and the solution of the equation is

$$\int \frac{du}{f(u) - u} = \ln x + c \tag{11.19}$$

An equation of this type is often called a homogeneous equation because a function of type (11.14) is a **homogeneous function**. In general, a function of one or more variables is a homogeneous function of degree n if

$$f(\lambda x, \lambda y, \ldots) = \lambda^n f(x, y, \ldots) \tag{11.20}$$

The function (11.14) is therefore homogeneous of zero degree.

EXAMPLE 11.5 Solve $\dfrac{dy}{dx} = \dfrac{x^2 + y^2}{xy}$.

The substitution $y = xu$ gives

$$\frac{dy}{dx} = u + x\frac{du}{dx} = u + \frac{1}{u}$$

so that

$$x\frac{du}{dx} = \frac{1}{u}.$$

Then, separating the variables and integrating,

$$\int u\, du = \int \frac{dx}{x}, \qquad \frac{u^2}{2} = \ln x + c$$

The general solution of the equation, after substituting $u = y/x$, is therefore

$$\frac{y^2}{2x^2} = \ln x + c$$

The result can be checked by implicit differentiation:

$$\frac{d}{dx}\left(\frac{y^2}{2x^2}\right) = \frac{d}{dx}(\ln x + c), \qquad \frac{y}{x^2}\frac{dy}{dx} - \frac{y^2}{x^3} = \frac{1}{x}$$

and

$$\frac{dy}{dx} = \frac{x^2}{y}\left(\frac{1}{x} + \frac{y^2}{x^3}\right) = \frac{x}{y} + \frac{y}{x} = \frac{x^2 + y^2}{xy}$$

> Exercises 23–25

> Exercises 26–28

11.4   Separable equations in chemical kinetics

A chemical reaction is denoted by a 'stoichiometric' equation of the form

$$aA + bB + \cdots = pP + qQ + \cdots \tag{11.21}$$

in which the symbols A, B, ... (in practice, chemical formulas) represent the chemical species that are reacting to give the products P, Q, ... The presence of the numbers $a, b, \ldots$ and $p, q, \ldots$ in the equation means that a molecules of A are reacting with b molecules of B, ... to give p molecules of P, q molecules of Q, ... For example

$$2H_2 + O_2 = 2H_2O$$

means that two molecules of hydrogen react with one molecule of oxygen to give two molecules of water.

In thermodynamics and in kinetics, the general chemical reaction is written

$$0 = \sum_B v_B\, B \tag{11.22}$$

in which the sum is over all the species present, both reactants and products, and the v_B, called **stoichiometric numbers**, are the numbers $a, b, p, q, \ldots$ in (11.21), but with a change of sign for the reactants. In this form equation (11.21) is

$$0 = -aA - bB + pP + qQ + \cdots$$
$$= v_A\, A + v_B\, B + v_P\, P + v_Q\, Q + \cdots$$

and the H_2/O_2 equation is

$$0 = -2H_2 - O_2 + 2H_2O$$

Rate of reaction

Let n_{A0} be the initial amount of species A, at time $t = 0$, and let n_A be the amount at time t. The 'amount' of reaction that has occurred in time t is given by the **extent of reaction** ξ, defined by

$$n_A = n_{A0} + v_A \xi \qquad (11.23)$$

and a measure of the rate of a reaction is then the rate of change of the extent of reaction, or **rate of conversion**,

$$r = \frac{d\xi}{dt} \qquad (11.24)$$

This rate can also be expressed in terms of the rate of change of the amount of each species taking part in the reaction; differentiating (11.23),

$$\frac{dn_A}{dt} = v_A \frac{d\xi}{dt} \qquad (11.25)$$

(since n_{A0} and v_A are constants), so that

$$r = \frac{d\xi}{dt} = \frac{1}{v_A}\frac{dn_A}{dt} \qquad (11.26)$$

For many purposes in physical chemistry it is more convenient to express the rate in terms of concentrations instead of amounts.* The concentration of species A in volume V is given by $[A] = n_A/V$, and division of equation (11.26) by the volume gives the **rate of reaction** (in terms of the concentration of A)

$$v = \frac{r}{V} = \frac{1}{v_A}\frac{d[A]}{dt} \qquad (11.27)$$

For the general reaction written in the form (11.21) we have

$$v = -\frac{1}{a}\frac{d[A]}{dt} = -\frac{1}{b}\frac{d[B]}{dt} = \frac{1}{p}\frac{d[P]}{dt} = \frac{1}{q}\frac{d[Q]}{dt} = \cdots \qquad (11.28)$$

---

* By convention, the symbol n_A is the amount of species A in the (SI) unit mol, and this is also the unit of ξ. Other quantities used instead of amount are the concentration $[A]$, the partial pressure p_A for a gas, and the dimensionless mole fraction x_A. We follow the normal practice in physical chemistry texts of using concentrations for our discussion of kinetics, but omit the units.

and for the H_2/O_2 reaction,

$$v = -\frac{1}{2}\frac{d[H_2]}{dt} = -\frac{d[O_2]}{dt} = \frac{1}{2}\frac{d[H_2O]}{dt}$$

The mechanism of a chemical reaction is in general complex, proceeding through several elementary kinetic steps, that may be consecutive, competitive, or both, and it is the task of the kineticist to unravel the mechanism and to relate the rate of the overall chemical reaction to the rates of the elementary steps. It is found by experiment that the overall rate may depend on the amounts of all the chemical species present at any time, both reactant and product, as well as on the temperature, the pressure, and the nature of the container. We consider here only the very simplest reactions, those whose rates depend only on the amounts of the reactants; we represent such a reaction by

$$a\mathrm{A} + b\mathrm{B} + \cdots \longrightarrow \text{products} \tag{11.29}$$

It is found from experiment that the rate of the reaction has the form

$$v = k[\mathrm{A}]^\alpha[\mathrm{B}]^\beta\ldots \tag{11.30}$$

where k is called the **rate constant** of the reaction, and the numbers α, β, ..., define the **order** of the reaction. The exponent α is called the order with respect to reactant A, β is the order with respect to B, ..., and the sum $\alpha + \beta + \cdots$ is called the **order of the reaction**. The cases considered are

(1) A $\rightarrow$ products   first order
(2) 2A $\rightarrow$ products   second order
(3) A + B $\rightarrow$ products   second order

(1)   The first-order process: A $\rightarrow$ products

The **rate equation** is

$$v = -\frac{d[\mathrm{A}]}{dt} = k[\mathrm{A}] \tag{11.31}$$

or, if the concentration $[\mathrm{A}]$ is represented by the variable x,

$$\frac{dx}{dt} = -kx \tag{11.32}$$

This is the differential equation discussed in Examples 3.20 and 11.1, and the solution is that given in Example 11.4(ii). Thus, separating the variables and integrating,

$$\frac{dx}{x} = -k\,dt, \qquad \ln x = -kt + c \qquad \text{or} \qquad \ln[\mathrm{A}] = -kt + c \tag{11.33}$$

If the initial concentration of A is $[A]_0$, at time $t = 0$, then $c = \ln[A]_0$ and the solution of the differential rate equation is the **integrated rate equation**

$$\ln[A] = -kt + \ln[A]_0 \qquad (11.34)$$

A plot of $\ln[A]$ against t is a straight line (Figure 11.1) whose slope is $-k$ and whose intercept with the $\ln[A]$ axis is $\ln[A]_0$.

The more conventional expression for the integrated rate equation is obtained from (11.34) by taking the exponential of each side,

$$[A] = [A]_0\, e^{-kt} \qquad (11.35)$$

Figure 11.1

and shows that the first-order process is an example of exponential decay.

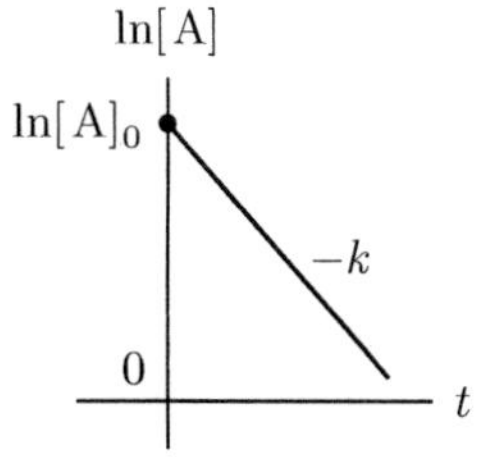

The half-life

The half-life $\tau_{1/2}$ of an exponential decay is the interval in which the amount of reactant is halved:

$$x(t + \tau_{1/2}) = \tfrac{1}{2} x(t)$$

or, from (11.35),

$$[A]_0\, e^{-k(t + \tau_{1/2})} = \frac{1}{2}[A]_0\, e^{-kt}$$

Therefore $e^{-k\tau_{1/2}} = 1/2$ and, taking logs,

$$\tau_{1/2} = \frac{\ln 2}{k}.$$

▸ Exercise 29

(2)   The second-order process: $2A \rightarrow$ products

The rate equation is

$$v = -\frac{1}{2}\frac{d[A]}{dt} = k[A]^2 \qquad (11.36)$$

or,

$$\frac{dx}{dt} = -2kx^2 \qquad (11.37)$$

Separation of the variables and integration gives

$$-\frac{dx}{x^2} = 2k\,dt, \qquad \frac{1}{x} = 2kt + c$$

If x_0 is the initial concentration of reactant then $c = 1/x_0$ and

$$\frac{1}{x} - \frac{1}{x_0} = 2kt \qquad\qquad (11.38)$$

or

$$\frac{1}{[A]} - \frac{1}{[A]_0} = 2kt, \qquad [A] = \frac{[A]_0}{1 + 2kt[A]_0} \qquad (11.39)$$

A plot of $1/[A]$ against t is a straight line (Figure 11.2) whose slope is $2k$ and whose intercept with the vertical axis is $1/[A]_0$.

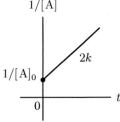

Figure 11.2

(3)   The second-order process: $A + B \rightarrow$ products

The rate equation is

$$v = -\frac{d[A]}{dt} = -\frac{d[B]}{dt} = k[A][B] \qquad\qquad (11.40)$$

Let the initial concentrations of A and B be $[A]_0 = a$ and $[B]_0 = b$, respectively, and let $[A] = a - x$ and $[B] = b - x$ at time t. The rate equation is then

$$-\frac{d(a-x)}{dt} = \frac{dx}{dt} = k(a-x)(b-x) \qquad\qquad (11.41)$$

Two cases need to be distinguished.

(a) The initial concentrations of A and B are equal; $[A]_0 = [B]_0$, $a = b$.

The rate equation (11.41) is

$$\frac{dx}{dt} = k(a-x)^2 \qquad\qquad (11.42)$$

and, separating the variables and integrating,

$$\frac{dx}{(a-x)^2} = k\,dt, \qquad \frac{1}{a-x} = \frac{1}{[A]} = kt + c \qquad (11.43)$$

Then $c = 1/a = 1/[A]_0$ and the integrated rate equation is

$$\frac{1}{[A]} - \frac{1}{[A]_0} = kt \qquad\qquad (11.44)$$

This is the same as in case **(2)** above except for the factor 2, which distinguishes case $A + A$ from case $A + B$.

(b) The initial concentrations of A and B are not equal.

Separating the variables in (11.41) and integrating,

$$\frac{dx}{(a-x)(b-x)} = k\, dt, \qquad \int \frac{dx}{(a-x)(b-x)} = kt + c$$

The integral over the variable x is evaluated by expressing the integrand in partial fractions (see Sections 2.7 and 6.6, and Examples 2.28 and 6.17). Write

$$\frac{1}{(a-x)(b-x)} = \frac{A}{a-x} + \frac{B}{b-x} = \frac{A(b-x) + B(a-x)}{(a-x)(b-x)}$$

It is required that

$$1 = A(b-x) + B(a-x)$$

be true for all values of x. Setting $x = b$ gives $B = 1/(a-b)$ and setting $x = a$ gives $A = 1/(b-a)$. Therefore

$$\frac{1}{(a-x)(b-x)} = \frac{1}{b-a}\left(\frac{1}{a-x} - \frac{1}{b-x} \right)$$

and

$$\int \frac{dx}{(a-x)(b-x)} = \frac{1}{b-a} \int \left(\frac{1}{a-x} - \frac{1}{b-x} \right) dx$$

$$= \frac{1}{b-a}\left(-\ln(a-x) + \ln(b-x) \right) = \frac{1}{b-a} \ln\left(\frac{b-x}{a-x} \right)$$

Therefore

$$\frac{1}{b-a} \ln\left(\frac{b-x}{a-x} \right) = kt + c$$

At time $t = 0$, $x = 0$, and $c = \frac{1}{b-a} \ln\left(\frac{b}{a} \right)$, so that

$$\frac{1}{b-a} \ln \frac{a(b-x)}{b(a-x)} = kt \qquad\qquad (11.45)$$

and the integrated rate equation is

$$\frac{1}{[B]_0 - [A]_0} \ln \frac{[A]_0 [B]}{[B]_0 [A]} = kt \qquad (11.46)$$

A plot of $\ln[B]/[A]$ against t is a straight line.

> Exercises 30–32

11.5 First-order linear equations

A first-order differential equation is called linear if it can be written in the form

$$\frac{dy}{dx} + p(x)y = r(x) \qquad (11.47)$$

The equation is linear in both dy/dx and y. When the right side, $r(x)$, is zero the equation is called **homogeneous**; otherwise it is an **inhomogeneous equation**.

The homogeneous equation

$$\frac{dy}{dx} + p(x)y = 0 \qquad (11.48)$$

The equation is separable:

$$\frac{dy}{y} = -p(x)\,dx, \qquad \ln y = -\int p(x)\,dx + c$$

and, taking the exponential of each side, the general solution is

$$y = a\,e^{-\int p(x)\,dx} \qquad (11.49)$$

where a is the arbitrary constant (if a is chosen as zero then $y = 0$ is called the **trivial solution**).

The inhomogeneous equation

The general linear equation is transformed into a form that can be integrated directly if (11.47) is multiplied throughout by the function

$$F(x) = e^{\int p(x)\,dx} \qquad (11.50)$$

which, given that $\dfrac{d}{dx}\displaystyle\int p(x)\,dx = p(x)$, has the differential property[3]

$$\frac{dF(x)}{dx} = F(x)p(x) \qquad (11.51)$$

---

[3] This method of solving the general first-order differential equation was discovered by Leibniz in 1694.

Then, multiplying (11.47) by $F(x)$,

$$F(x)\frac{dy}{dx} + F(x)p(x)y = F(x)r(x) \qquad (11.52)$$

and, by virtue of (11.51), the left side of this is the derivative of the product $F(x)y$:

$$\frac{d}{dx}\left[F(x)y\right] = F(x)\frac{dy}{dx} + \frac{dF(x)}{dx}y = F(x)\frac{dy}{dx} + F(x)p(x)y$$

Equation (11.52) can therefore be written as

$$\frac{d}{dx}\left[F(x)y\right] = F(x)r(x)$$

and can be integrated to give

$$F(x)y = \int F(x)r(x)\,dx + c \qquad (11.53)$$

This is the general solution of the inhomogeneous equation. The function $F(x)$ is called the **integrating factor** of the differential equation.

EXAMPLES 11.6 Linear equations

(i) $\dfrac{dy}{dx} - y = 3e^{2x}$

In this case, $p(x) = -1$, $\int p(x)\,dx = -x$, and the integrating factor is

$$F(x) = e^{\int p(x)\,dx} = e^{-x}$$

(the constant of integration need not be include in the exponent because it cancels in (11.53)). Then, by formula (11.53),

$$e^{-x}y = \int e^{-x} \times 3e^{2x}\,dx + c = 3\int e^{x}\,dx + c = 3e^{x} + c$$

$$y = e^{x}(3e^{x} + c) = 3e^{2x} + ce^{x}$$

(ii) $\dfrac{dy}{dx} + \dfrac{2}{x}y = 3x^{3}$

The integrating factor is

$$F(x) = e^{\int (2/x)\,dx} = e^{2\ln x} = x^{2}$$

Therefore, by formula (11.53),

$$x^2 y = \int x^2 \times 3x^3 \, dx + c = \frac{x^6}{2} + c$$

$$y = \frac{x^4}{2} + \frac{c}{x^2}$$

▸ Exercises 33–40

11.6   An example of linear equations in chemical kinetics

We saw in Section 11.4 that many (in fact, most) chemical reactions proceed through several elementary steps, and a description of the overall rate process then involves several simultaneous first-order differential equations. Such a system of equations can be solved only in simple cases without resorting to approximations, and numerical methods must otherwise be used. One of the simplest solvable systems is that of two consecutive first-order processes,

$$A \xrightarrow{k_1} B \xrightarrow{k_2} C$$

in which A is transformed into C through the intermediate species B. The process is modelled by the two first-order equations

$$\frac{d[A]}{dt} = -k_1[A] \tag{11.54}$$

$$\frac{d[B]}{dt} = k_1[A] - k_2[B] \tag{11.55}$$

We assume that the initial concentrations, at time $t = 0$, are $[A]_0 = a$, $[B]_0 = 0$, and $[C]_0 = 0$, and that the concentrations at time t are $[A] = a - x$ and $[B] = y$, so that $[C] = x - y$. Then

$$\frac{d(a-x)}{dt} = -k_1(a-x) \tag{11.56}$$

$$\frac{dy}{dt} = k_1(a-x) - k_2 \, y \tag{11.57}$$

The first of these equations is the separable equation (11.32), with solution

$$a - x = ae^{-k_1 t} \tag{11.58}$$

Substitution of this result into equation (11.57) gives

$$\frac{dy}{dt} = ak_1 \, e^{-k_1 t} - k_2 \, y$$

or

$$\frac{dy}{dt} + k_2\, y = ak_1\, e^{-k_1 t} \tag{11.59}$$

This is a linear equation with $p(t) = k_2$, $r(t) = ak_1 e^{-k_1 t}$, and integrating factor

$$F(t) = e^{\int p(t)\, dt} = e^{k_2 t}$$

The solution of equation (11.59) is therefore

$$F(t)y = \int F(t)r(t)\, dt + c$$

$$e^{k_2 t} y = ak_1 \int e^{(k_2 - k_1)t}\, dt + c = \begin{cases} \dfrac{ak_1}{k_2 - k_1} e^{(k_2 - k_1)t} + c & \text{if } k_1 \neq k_2 \\[2mm] ak_1 t + c & \text{if } k_1 = k_2 \end{cases}$$

By the initial conditions, $y = 0$ when $t = 0$. We have

$$c = \begin{cases} -\dfrac{ak_1}{k_2 - k_1} & \text{if } k_1 \neq k_2 \\[2mm] 0 & \text{if } k_1 = k_2 \end{cases}$$

and the required particular solution is

$$y = \begin{cases} \dfrac{ak_1}{k_2 - k_1}\left(e^{-k_1 t} - e^{-k_2 t}\right) & \text{if } k_1 \neq k_2 \\[2mm] ak_1 t e^{-k_1 t} & \text{if } k_1 = k_2 \end{cases} \tag{11.60}$$

The concentration of each species present at time t is therefore

$$[A] = [A]_0 e^{-k_1 t}$$

$$[B] = \begin{cases} \dfrac{[A]_0 k_1}{k_2 - k_1}\left(e^{-k_1 t} - e^{-k_2 t}\right) & \text{if } k_1 \neq k_2 \\[2mm] [A]_0 k_1 t e^{-k_1 t} & \text{if } k_1 = k_2 \end{cases} \tag{11.61}$$

$$[C] = [A]_0 - [A] - [B]$$

Figure 11.3 illustrates the three types of behaviour exhibited by the system. In each case $[A]_0 = 1$ and $k_1 = 1$ (in some appropriate units). Figure (a), with $k_2/k_1 = 10$, shows how the system behaves when $k_2 \gg k_1$; the rate-determining step is $A \rightarrow B$, and the amount of intermediate B remains small throughout the reaction because B is rapidly converted into C. Figure (c), with $k_1/k_2 = 10$, shows the behaviour when $k_1 \gg k_2$; the rate-determining step is $B \rightarrow C$, and the amount of B becomes large because the process $B \rightarrow C$ is slow, and B decreases only when the supply of A has been exhausted.

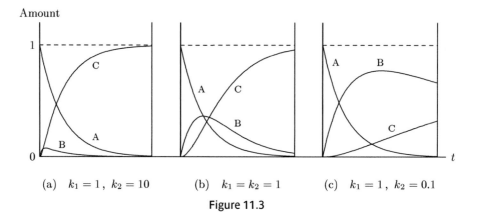

(a) $k_1 = 1$, $k_2 = 10$ (b) $k_1 = k_2 = 1$ (c) $k_1 = 1$, $k_2 = 0.1$

Figure 11.3

➤ Exercises 41, 42

11.7 Electric circuits

The modelling of electric circuits is an important application of linear differential equations. A simple circuit consists of a source of electric energy, such as a generator or a battery, that generates an electromotive force (e.m.f.) E, and one or more elements (Figure 11.4) connected in series; resistors (R), inductors (L), and capacitors (C). The e.m.f. of the source is equal to the difference of electric potential between the two points at which it is attached to the circuit. When the circuit is closed, a current I flows and each element of the circuit causes a drop in voltage (electric potential) in the circuit. The total drop in voltage is equal to the e.m.f. (**Kirchhoff's voltage law**).

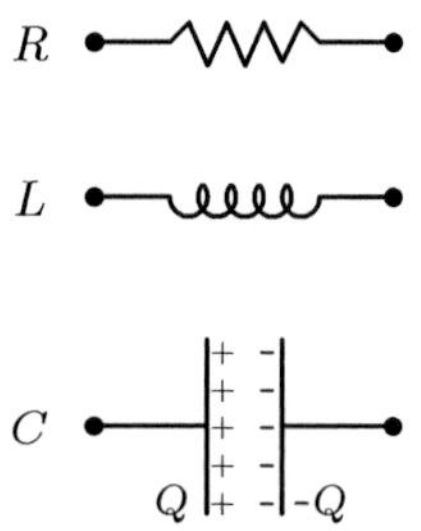

    A resistor is a circuit element that offers resistance to the flow of charge in the circuit. The voltage drop E_R across a resistor is proportional to the current,

$$E_R = RI \quad \textbf{(Ohm's law)} \tag{11.62}$$

and the constant of proportionality R is called the **resistance** of the resistor (see Tables 1.1 and 1.2 for the appropriate SI units). An inductor is a coil which builds up a magnetic field within itself when a current flows. A changing magnetic field develops an induced voltage across the coil that is proportional to the rate of change of current and opposes the current flow:

$$E_L = L\frac{dI}{dt} \tag{11.63}$$

Figure 11.4

where L is the **inductance**. A simple capacitor is made up of two parallel metallic plates separated by a thin layer of insulating material. The plates become charged when a current flows. The rate of charging is $\dfrac{dQ}{dt} = I$, the accumulated charge is $Q = \int I\, dt$, and the potential difference between the plates is

$$E_C = \frac{Q}{C} \tag{11.64}$$

where C is the **capacitance**.

EXAMPLE 11.7 RL circuit

An RL-circuit containing one resistor and one inductor is illustrated in Figure 11.5.
  By Kirchhoff's law,

$$L\frac{dI}{dt} + RI = E \tag{11.65}$$

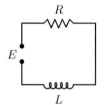

Figure 11.5

and this is a first-order linear differential equation. We have $p = R/L$ (constant) and $r = E/L$ in equation (11.47). The integrating factor is $F(x) = e^{\int (R/L)dt} = e^{Rt/L}$ and the general solution of the equation is

$$I(t) = e^{-Rt/L}\left[\int e^{Rt/L}\frac{E}{L}\,dt + c\right] \tag{11.66}$$

If the e.m.f. is constant, $E = E_0$, then

$$I(t) = e^{-Rt/L}\left[\frac{E_0}{L}\int e^{Rt/L}\,dt + c\right] = \frac{E_0}{R} + ce^{-Rt/L} \tag{11.67}$$

The particular solution for the initial condition $I(0) = 0$ is then

$$I(t) = \frac{E_0}{R}\left(1 - e^{-Rt/L}\right) \tag{11.68}$$

When the circuit is closed the current increases from zero to its steady value E_0/R in a time that depends on the inductive time constant $\tau_L = L/R$.

▸ Exercises 43, 44

11.8 Exercises

Section 11.2

State the order of the differential equation and verify that the given function is a solution:

1. $\dfrac{dy}{dx} - 2y = 2;\quad y = e^{2x} - 1$

2. $\dfrac{d^2 y}{dx^2} + 4y = 0;\quad y = A \cos 2x + B \sin 2x$

3. $\dfrac{d^3 y}{dx^3} = 12;\quad y = 2x^3 + 3x^2 + 4x + 5$

4. $\dfrac{dy}{dx} + \dfrac{3y}{x} = 3x^2;\quad y = \dfrac{x^3}{2} + \dfrac{c}{x^3}$

Find the general solution of the differential equation:

5. $\dfrac{dy}{dx} = x^2$ **6.** $\dfrac{dy}{dx} = e^{-3x}$ **7.** $\dfrac{d^2 y}{dx^2} = \cos 3x$ **8.** $\dfrac{d^3 y}{dx^3} = 24x$

9. A body moves along the x direction under the influence of the force $F(t) = \cos 2\pi t$, where t is the time. **(i)** Write down the equation of motion. **(ii)** Find the solution that satisfies the initial conditions $x(0) = 0$ and $\dot{x}(0) = 1$.

Verify that the given function is a solution of the differential equation, and determine the particular solution for the given initial condition:

10. $x\dfrac{dy}{dx} = 2y;\quad y = cx^2;\quad y = 24$ when $x = 2$

11. $\dfrac{dy}{dx} + 2xy = 0;\quad y = ce^{-x^2};\quad y = 2$ when $x = 2$

12. $\dfrac{dy}{dx} + 2y + 2 = 0;\quad y = ce^{-2x} - 1;\quad y(0) = 4$

Section 11.3

Find the general solution of the differential equation:

13. $\dfrac{dy}{dx} = \dfrac{3x^2}{y}$ **14.** $\dfrac{dy}{dx} = 4xy^2$ **15.** $\dfrac{dy}{dx} = 3x^2 y$

16. $y^2 \dfrac{dy}{dx} = e^x$ **17.** $\dfrac{dy}{dx} = y(y-1)$ **18.** $\dfrac{dy}{dx} = \dfrac{y}{x}$

Solve the initial value problems:

19. $\dfrac{dy}{dx} = \dfrac{y+2}{x-3};\quad y(0) = 1$

20. $\dfrac{dy}{dx} = \dfrac{x^2 - 1}{2y + 1};\quad y(0) = -1$

21. $\dfrac{dy}{dx} = \dfrac{y(y+1)}{x(x-1)};\quad y(2) = 1$

22. $\dfrac{dy}{dx} = e^{x+y};\quad y(0) = 0$

Solve the initial value problems:

23. $\dfrac{dy}{dx} = \dfrac{x+y}{x};\quad y(1) = 2$

24. $2xy\dfrac{dy}{dx} = -(x^2 + y^2);\quad y(1) = 0$

25. $xy^3 \dfrac{dy}{dx} = x^4 + y^4;\quad y(2) = 0$

26. Show that a differential equation of the form

$$\frac{dy}{dx} = f(ax + by + c)$$

is reduced to separable form by means of the substitution $u = ax + by$.

Use the method in Exercise 26 to find the general solution:

27. $\dfrac{dy}{dx} = 2x + y + 3$ **28.** $\dfrac{dy}{dx} = \dfrac{x - y}{x - y + 2}$

Section 11.4

29. Find the interval $\tau_{1/n}$ in which the amount of reactant in a first-order decay process is reduced by factor n.

30. Solve the initial value problem for the nth-order kinetic process A $\rightarrow$ products

$$\frac{dx}{dt} = -kx^n \quad x(0) = a \quad (n > 1)$$

31. The reversible reaction A $\rightleftharpoons$ B, first order in both directions, has rate equation

$$\frac{dx}{dt} = k_1(a - x) - k_{-1}x$$

Find $x(t)$ for initial condition $x(0) = 0$.

32. A third-order process A + 2B $\rightarrow$ products has rate equation

$$\frac{dx}{dt} = k(a - x)(b - 2x)^2$$

where a and b are the initial amounts of A and B, respectively. Show that the solution that satisfies the initial condition $x(0) = 0$ is given by

$$kt = \frac{1}{(2a - b)^2} \ln \frac{a(b - 2x)}{b(a - x)} + \frac{2x}{b(2a - b)(b - 2x)}$$

Section 11.5

Find the general solution:

33. $\dfrac{dy}{dx} + 2y = 4$ **34.** $\dfrac{dy}{dx} - 4xy = x$ **35.** $\dfrac{dy}{dx} + 3y = e^{-3x}$

36. $\dfrac{dy}{dx} + \dfrac{2y}{x} = 2\cos x$ **37.** $\dfrac{dy}{dx} - \dfrac{y}{x^2} = \dfrac{4}{x^2}$ **38.** $\dfrac{dy}{dx} + (2\tan x)y = \sin x$

39. $\dfrac{dy}{dx} + ax^n y = bx^n, \quad (n \neq -1)$ **40.** $\dfrac{dy}{dx} + a\dfrac{y}{x} = x^n$

Section 11.6

41. The system of three consecutive first-order processes A $\xrightarrow{k_1}$ B $\xrightarrow{k_2}$ C $\xrightarrow{k_3}$ D is modelled by the set of equations

$$\frac{d(a - x)}{dt} = -k_1(a - x), \quad \frac{dy}{dt} = k_1(a - x) - k_2 y, \quad \frac{dz}{dt} = k_2 y - k_3 z$$

where $(a - x)$, y, and z are the amounts of A, B, and C, respectively, at time t. Given the initial conditions $x = y = z = 0$ at $t = 0$, find the amount of C as a function of t. Assume $k_1 \neq k_2$, $k_1 \neq k_3$, $k_2 \neq k_3$.

42. The first-order process A $\xrightarrow{k_1}$ B is followed by the parallel first-order processes B $\xrightarrow{k_2}$ C and B $\xrightarrow{k_3}$ D, and the system is modelled by the equations

$$\frac{d(a - x)}{dt} = -k_1(a - x), \quad \frac{dy}{dt} = k_1(a - x) - (k_2 + k_3)y, \quad \frac{dz}{dt} = k_2 y, \quad \frac{du}{dt} = k_3 y$$

where $(a - x)$, y, z and u are the amounts of A, B, C, and D, respectively, at time t. Given the initial conditions $x = y = z = u = 0$ at $t = 0$ find the amount of C as a function of time.

Section 11.7

43. The current in an RL-circuit containing one resistor and one inductor is given by the equation

$$L\frac{dI}{dt} + RI = E$$

Solve the equation for a periodic e.m.f. $E(t) = E_0 \sin \omega t$, with initial condition $I(0) = 0$.

44. Show that the current in an RC-circuit containing one resistor and one capacitor is given by the equation

$$R\frac{dI}{dt} + \frac{I}{C} = \frac{dE}{dt}$$

Solve the equation for **(i)** a constant e.m.f., $E = E_0$, **(ii)** a periodic e.m.f., $E(t) = E_0 \sin \omega t$.

12 Second-order differential equations. Constant coefficients

12.1 Concepts

The second-order differential equations that are important in the physical sciences are linear equations of the general form

$$\frac{d^2 y}{dx^2} + p(x)\frac{dy}{dx} + q(x)y = r(x) \tag{12.1}$$

When the function $r(x)$ is zero the equation

$$\frac{d^2 y}{dx^2} + p(x)\frac{dy}{dx} + q(x)y = 0 \tag{12.2}$$

is called **homogeneous**; otherwise it is **inhomogeneous**.[1] The functions $p(x)$ and $q(x)$ are called the **coefficients** of the equation. When these coefficients are constants the solutions of the equation can be expressed in terms of elementary functions, and it is these linear equations with constant coefficients that are discussed in this chapter. Examples of such equations are the classical equations of motion for the harmonic oscillator and for forced oscillations in mechanical and electrical systems, and the Schrödinger equations for the particle in a box and in a ring. Differential equations with non-constant coefficients are discussed in Chapter 13.

12.2 Homogeneous linear equations

The general homogeneous second-order linear equation with constant coefficients is

$$\frac{d^2 y}{dx^2} + a\frac{dy}{dx} + by = 0 \tag{12.3}$$

where a and b are constants. It is always possible to find a solution of this equation that has the form $e^{\lambda x}$ where λ is a suitable constant.

---

[1] Many of the methods and examples found in textbooks can be traced to Euler's *Institutiones calculi differentialis* (1755) and *Institutiones calculi integralis* (1768–1770, 3 volumes). Euler was responsible for the distinction between homogeneous and inhomogeneous equations, between particular and general solutions, the use of integrating factors, and for the solution of second- and higher-order linear differential equations with constant coefficients.

EXAMPLE 12.1 Show that $y_1 = e^{2x}$ and $y_2 = e^{3x}$ are two solutions of the equation

$$\frac{d^2 y}{dx^2} - 5\frac{dy}{dx} + 6y = 0$$

We have

$$y_1 = e^{2x}, \qquad \frac{dy_1}{dx} = 2e^{2x}, \qquad \frac{d^2 y_1}{dx^2} = 4e^{2x}$$

Therefore

$$\frac{d^2 y_1}{dx^2} - 5\frac{dy_1}{dx} + 6y_1 = 4e^{2x} - 5\times 2e^{2x} + 6e^{2x} = e^{2x}(4-10+6) = 0$$

Similarly,

$$y_2 = e^{3x}, \qquad \frac{dy_2}{dx} = 3e^{3x}, \qquad \frac{d^2 y_2}{dx^2} = 9e^{3x}$$

and

$$\frac{d^2 y_2}{dx^2} - 5\frac{dy_2}{dx} + 6y_2 = 9e^{3x} - 15e^{3x} + 6e^{3x} = 0$$

▸ Exercises 1–3

This example demonstrates the general result that it is always possible to find two particular solutions $y_1(x)$ and $y_2(x)$ of a homogeneous linear equation. If either of these solutions is multiplied by a constant the new function is also a solution. For example, let $y_1(x)$ be a solution of the general homogeneous equation (12.2),

$$\frac{d^2 y_1}{dx^2} + p(x)\frac{dy_1}{dx} + q(x)y_1 = 0$$

and let $y_3(x) = cy_1(x)$, where c is a constant. Then

$$\frac{dy_3}{dx} = \frac{d(cy_1)}{dx} = c\frac{dy_1}{dx}, \qquad \frac{d^2 y_3}{dx^2} = c\frac{d^2 y_1}{dx^2}$$

and

$$\frac{d^2 y_3}{dx^2} + p(x)\frac{dy_3}{dx} + q(x)y_3 = c\left[\frac{d^2 y_1}{dx^2} + p(x)\frac{dy_1}{dx} + q(x)y_1\right] = 0$$

so that $y_3 = cy_1$ is also a solution. The functions y_1 and y_3 are not regarded as distinct solutions however because each is merely a multiple of the other; the functions are said to be **linearly dependent**. In general, two functions, y_1 and y_2, are said to be linearly dependent if there exists a relation

$$a_1 y_1(x) + a_2 y_2(x) = 0 \qquad (12.4)$$

such that a_1 and a_2 are not zero. If (12.4) is true only when $a_1 = a_2 = 0$ then the functions are **linearly independent**, and neither is a multiple of the other. The solutions y_1 and y_2 in Example 12.1 are linearly independent.

We now show that if $y_1(x)$ and $y_2(x)$ are two solutions of a linear homogeneous equation then any **linear combination** of them,

$$y = c_1 y_1 + c_2 y_2 \qquad (12.5)$$

where c_1 and c_2 are arbitrary constants, is also a solution. We have

$$\frac{dy}{dx} = c_1 \frac{dy_1}{dx} + c_2 \frac{dy_2}{dx}, \qquad \frac{d^2 y}{dx^2} = c_1 \frac{d^2 y_1}{dx^2} + c_2 \frac{d^2 y_2}{dx^2}$$

Therefore, substituting into the general homogeneous equation (12.2),

$$\frac{d^2 y}{dx^2} + p(x)\frac{dy}{dx} + q(x)y = \left(c_1 \frac{d^2 y_1}{dx^2} + c_2 \frac{d^2 y_2}{dx^2} \right)$$

$$+ p(x)\left(c_1 \frac{dy_1}{dx} + c_2 \frac{dy_2}{dx} \right) + q(x)\left(c_1 y_1 + c_2 y_2 \right)$$

$$= c_1 \left[\frac{d^2 y_1}{dx^2} + p(x)\frac{dy_1}{dx} + q(x)y_1 \right]$$

$$+ c_2 \left[\frac{d^2 y_2}{dx^2} + p(x)\frac{dy_2}{dx} + q(x)y_2 \right]$$

$$= 0$$

and the result is zero because both sets of terms in square brackets are zero since y_1 and y_2 are solutions. This important property of linear homogeneous equations is called the **principle of superposition** (it is not true for inhomogeneous equations or for nonlinear equations). In particular, when y_1 and y_2 are linearly-independent solutions, the function (12.5), containing two arbitrary constants, is the **general solution** of the homogeneous equation (see Section 11.2).

EXAMPLE 12.2 The general solution of the equation in Example 12.1 is

$$y(x) = c_1 e^{2x} + c_2 e^{3x}$$

Particular solutions are obtained by assigning particular values to c_1 and c_2.

▸ Exercises 4–6

12.3 The general solution

Let one solution of equation (12.3),

$$\frac{d^2 y}{dx^2} + a\frac{dy}{dx} + by = 0 \tag{12.3}$$

be

$$y = e^{\lambda x} \tag{12.6}$$

Then

$$\frac{d}{dx}e^{\lambda x} = \lambda e^{\lambda x}, \qquad \frac{d^2}{dx^2}e^{\lambda x} = \lambda^2 e^{\lambda x}$$

and substitution into (12.3) gives

$$\lambda^2 e^{\lambda x} + a\lambda e^{\lambda x} + b e^{\lambda x} = 0$$

or

$$e^{\lambda x}(\lambda^2 + a\lambda + b) = 0$$

The function (12.6) is therefore a solution of equation (12.3) if

$$\lambda^2 + a\lambda + b = 0 \tag{12.7}$$

This quadratic equation is called the **characteristic equation** or **auxiliary equation** of the differential equation.[2] The possible values of λ are the roots of the quadratic:

$$\lambda_1 = \frac{1}{2}\left(-a + \sqrt{a^2 - 4b}\right), \qquad \lambda_2 = \frac{1}{2}\left(-a - \sqrt{a^2 - 4b}\right) \tag{12.8}$$

[2] In a letter to Johann Bernoulli in 1739 Euler described his discovery of the solution of linear equations with constant coefficients by means of the characteristic equation: 'after treating the problem in many ways, I happened on my solution entirely unexpectedly; before that I had no suspicion that the solution of algebraic equations had so much importance in this matter'.

and the possible solutions of type (12.6) are

$$y_1 = e^{\lambda_1 x}, \qquad y_2 = e^{\lambda_2 x} \tag{12.9}$$

EXAMPLE 12.3 The characteristic equation of the differential equation

$$\frac{d^2 y}{dx^2} - 5\frac{dy}{dx} + 6y = 0$$

is

$$\lambda^2 - 5\lambda + 6 = 0$$

The quadratic can be factorized,

$$\lambda^2 - 5\lambda + 6 = (\lambda - 2)(\lambda - 3)$$

and its roots are $\lambda_1 = 2$ and $\lambda_2 = 3$. Two particular solutions of the differential equation are therefore

$$y_1 = e^{\lambda_1 x} = e^{2x}, \qquad y_2 = e^{\lambda_2 x} = e^{3x}$$

and, because these functions are linearly independent, the general solution is

$$y = c_1 y_1 + c_2 y_2 = c_1 e^{2x} + c_2 e^{3x}$$

➤ Exercises 7, 8

In Example 12.3 the characteristic equation has two distinct real roots so that y_1 and y_2 are linearly independent and the general solution is a linear combination of them. The nature of the roots (12.8) depends on the discriminant $a^2 - 4b$. If a and b are real numbers, the three possible types are

(i) $a^2 - 4b > 0$ two distinct real roots
(ii) $a^2 - 4b = 0$ one real double root
(iii) $a^2 - 4b < 0$ a pair of complex conjugate roots

(i)   Two distinct real roots

The roots are given by (12.8), the particular solutions by (12.9) and, as in Example 12.3, the general solution is

$$y(x) = c_1 y_1(x) + c_2 y_2(x) = c_1 e^{\lambda_1 x} + c_2 e^{\lambda_2 x} \tag{12.10}$$

(ii)   A real double root

When $a^2 - 4b = 0$ the two roots (12.8) are equal, with $\lambda_1 = \lambda_2 = -a/2$. Only one particular solution is therefore obtained from the characteristic equation:

$$y_1(x) = e^{\lambda_1 x} = e^{-ax/2} \tag{12.11}$$

Given one particular solution of a homogeneous equation it is always possible to find a second. In the present case, a second solution, linearly independent of y_1, is

$$y_2(x) = xy_1(x) = xe^{-ax/2} \tag{12.12}$$

This is shown to be a solution by substitution into the differential equation (12.3). Thus

$$\frac{dy_2}{dx} = \left(1 - ax/2\right)e^{-ax/2}, \qquad \frac{d^2 y_2}{dx^2} = \left(-a + a^2 x/4\right)e^{-ax/2}$$

so that

$$\frac{d^2 y_2}{dx^2} + a\frac{dy_2}{dx} + by_2 = \left(-a + \frac{a^2 x}{4}\right)e^{-ax/2} + a\left(1 - \frac{ax}{2}\right)e^{-ax/2} + bxe^{-ax/2}$$

$$= \left(-a + \frac{a^2 x}{4} + a - \frac{a^2 x}{2} + bx\right)e^{-ax/2}$$

$$= \left(-\frac{a^2 x}{4} + bx\right)e^{-ax/2} = -\frac{x}{4}\left(a^2 - 4b\right)e^{-ax/2}$$

$$= 0$$

since $a^2 - 4b = 0$. The general solution in this case is therefore

$$y(x) = c_1 e^{-ax/2} + c_2 xe^{-ax/2} = (c_1 + c_2 x)e^{-ax/2} \tag{12.13}$$

EXAMPLE 12.4 Solve the differential equation $y'' - 4y' + 4y = 0$.

The characteristic equation

$$\lambda^2 - 4\lambda + 4 = (\lambda - 2)^2 = 0$$

has the double root $\lambda = 2$. Two particular solutions are therefore e^{2x} and xe^{2x}, and the general solution is

$$y(x) = (c_1 + c_2 x)e^{2x}$$

‣ Exercises 9, 10

(iii) Complex roots

When the discriminant $a^2 - 4b$ of the characteristic quadratic equation (12.7) is negative, the roots are a pair of complex conjugate numbers.

EXAMPLE 12.5 Solve the differential equation $y'' - 2y' + 2y = 0$.

The characteristic equation is

$$\lambda^2 - 2\lambda + 2 = 0$$

and the roots are $\lambda = \dfrac{1}{2}\left(2 \pm \sqrt{4-8}\right)$, or

$$\lambda_1 = 1 + i, \qquad \lambda_2 = \lambda_1^* = 1 - i$$

Two particular solutions are therefore

$$y_1(x) = e^{\lambda_1 x} = e^{(1+i)x}, \qquad y_2(x) = e^{\lambda_2 x} = e^{(1-i)x}$$

These are linearly independent, and the general solution is

$$y(x) = c_1 e^{(1+i)x} + c_2 e^{(1-i)x}$$
$$= e^x(c_1 e^{ix} + c_2 e^{-ix})$$

This is the exponential form of the solution. A trigonometric form is obtained by expressing $e^{\pm ix}$ in terms of $\sin x$ and $\cos x$ by means of Euler's formula, equations (8.33) and (8.34),

$$e^{\pm ix} = \cos x \pm i \sin x$$

Then

$$y(x) = e^x \left[c_1(\cos x + i \sin x) + c_2(\cos x - i \sin x) \right]$$
$$= e^x \left[(c_1 + c_2)\cos x + i(c_1 - c_2)\sin x \right]$$
$$= e^x(d_1 \cos x + d_2 \sin x)$$

where $d_1 = c_1 + c_2$ and $d_2 = i(c_1 - c_2)$ are (new) arbitrary constants.

▸ Exercises 11, 12

In the general complex case, the roots of the characteristic equation (12.7) are, from (12.8),

$$\lambda = -\frac{a}{2} \pm \sqrt{\left(\frac{a}{2}\right)^2 - b}$$

where $(a/2)^2 - b < 0$. Let $(a/2)^2 - b = -\omega^2$, where ω is real. Then

$$\sqrt{\left(\frac{a}{2}\right)^2 - b} = \sqrt{-\omega^2} = \sqrt{-1}\,\omega = i\omega$$

and the roots are

$$\lambda_1 = -\frac{a}{2} + i\omega, \qquad \lambda_2 = -\frac{a}{2} - i\omega \qquad (12.14)$$

The particular solutions of the differential equation are therefore

$$y_1(x) = e^{(-a/2 + i\omega)x}, \qquad y_2(x) = e^{(-a/2 - i\omega)x} \qquad (12.15)$$

and (when $\omega \neq 0$) the general solution is

$$y(x) = c_1 y_1(x) + c_2 y_2(x)$$
$$= e^{-ax/2}\left(c_1 e^{i\omega x} + c_2 e^{-i\omega x}\right) \qquad (12.16)$$

The trigonometric form of the general solution is (see Example 12.5)

$$y(x) = e^{-ax/2}(d_1 \cos \omega x + d_2 \sin \omega x) \qquad (12.17)$$

where $d_1 = c_1 + c_2$ and $d_2 = i(c_1 - c_2)$.

12.4   Particular solutions

The two arbitrary constants c_1 and c_2 in the general solution,

$$y(x) = c_1 y_1(x) + c_2 y_2(x)$$

of a homogeneous second-order differential equation are normally determined in physical applications by the application of two conditions (two for two constants) in either of the following two forms.

(a)   Initial conditions

The values of the function $y(x)$ *and* of its derivative $y'(x)$ are specified for some value of x:

$$y(x_0) = y_0, \qquad y'(x_0) = y_1 \qquad (12.18)$$

A second-order differential equation with initial conditions is called an **initial value problem**. Initial conditions are usually associated with applications in dynamics, when x is the time variable.

EXAMPLE 12.6 Solve the initial value problem

$$y'' + y' - 6y = 0, \qquad y(0) = 0, \quad y'(0) = 5$$

The characteristic equation is $\lambda^2 + \lambda - 6 = (\lambda - 2)(\lambda + 3) = 0$, with roots $\lambda_1 = 2$ and $\lambda_2 = -3$. The general solution is

$$y(x) = c_1 e^{2x} + c_2 e^{-3x}$$

with first derivative

$$y'(x) = 2c_1 e^{2x} - 3c_2 e^{-3x}$$

Then

$$y(0) = 0 = c_1 + c_2, \qquad y'(0) = 5 = 2c_1 - 3c_2$$

with solution $c_1 = 1, c_2 = -1$. The solution of the initial value problem is therefore

$$y(x) = e^{2x} - e^{-3x}$$

▸ Exercises 13–16

(b) Boundary conditions

In many applications in the physical sciences the variable x is a coordinate and the physical situation is determined by conditions on the value of $y(x)$ at the boundary of the system. The conditions

$$y(x_1) = y_1, \qquad y(x_2) = y_2 \qquad\qquad (12.19)$$

are called **boundary conditions**, and x_1 and x_2 are the end-points of the interval $x_1 \leq x \leq x_2$ within which the differential equation is defined. A differential equation with boundary conditions is called a **boundary value problem**.

EXAMPLE 12.7 Solve the boundary value problem

$$y'' - 2y' + 2y = 0, \qquad y(0) = 1, \quad y(\pi/2) = 2$$

The differential equation is that solved in Example 12.5, with general solution (in the trigonometric form)

$$y(x) = e^x (d_1 \cos x + d_2 \sin x)$$

Application of the boundary conditions gives

$$y(0) = 1 = e^0 (d_1 \cos 0 + d_2 \sin 0) = d_1$$

$$y(\pi/2) = 2 = e^{\pi/2} (d_1 \cos \pi/2 + d_2 \sin \pi/2) = d_2 e^{\pi/2}$$

Therefore, $d_1 = 1$ and $d_2 = 2e^{-\pi/2}$, and the solution of the boundary value problem is

$$y(x) = e^x (\cos x + 2e^{-\pi/2} \sin x)$$

> Exercises 17–19

EXAMPLE 12.8 Solve the boundary value problem

$$y'' + y' - 6y = 0, \qquad y(0) = 1, \quad y(x) \to 0 \text{ as } x \to \infty$$

The equation is that discussed in Example 12.6, with general solution

$$y(x) = c_1 e^{2x} + c_2 e^{-3x}$$

The first boundary condition gives

$$y(0) = 1 = c_1 + c_2$$

The second condition requires that the solution go to zero as x goes to infinity. The function e^{-3x} has this property but the function e^{2x} must be excluded. The condition therefore requires that we set $c_1 = 0$. Then $c_2 = 1$ and the solution of the boundary value problem is

$$y(x) = e^{-3x}$$

> Exercise 20

The equation $y'' + \omega^2 y = 0$

In Examples 12.6 to 12.8, the two conditions on the solution of the differential equation are sufficient to specify a particular solution. In some cases however, when the differential equation itself contains parameters to be determined, one or more additional conditions are required to completely specify the solution. Several examples of such equations are discussed in Chapter 13. One of the most important simple examples in the physical sciences is the equation

$$\frac{d^2 y}{dx^2} + \omega^2 y = 0 \tag{12.20}$$

where ω is a real number. The general solution is given by either (12.16) with $a = 0$,

$$y(x) = c_1 e^{i\omega x} + c_2 e^{-i\omega x} \tag{12.21}$$

or by the equivalent trigonometric form (12.17),

$$y(x) = d_1 \cos \omega x + d_2 \sin \omega x \tag{12.22}$$

If ω is a given constant then two initial or boundary conditions are normally sufficient to specify a particular solution. This is exemplified in Section 12.5 for the classical harmonic oscillator as an initial value problem. However, if ω is a parameter to be determined, then an additional condition is required. This is the case for the quantum-mechanical 'particle in a box', as a boundary value problem, discussed in Section 12.6.

A third possibility arises when the physical situation requires the imposition of a **periodic (cyclic) boundary condition** (see Section 8.6),

$$y(x + \lambda) = y(x) \tag{12.23}$$

where λ is the period. In this case, replacement of x by $x + \lambda$ in the general solution (12.21) gives

$$y(x + \lambda) = c_1 e^{i\omega(x+\lambda)} + c_2 e^{-i\omega(x+\lambda)} = c_1 e^{i\omega x} e^{i\omega\lambda} + c_2 e^{-i\omega x} e^{-i\omega\lambda}$$

The condition (12.23) is then satisfied if both $e^{i\omega\lambda} = 1$ and $e^{-i\omega\lambda} = 1$. As discussed in Section 8.6, this is the case when $\omega\lambda$ is a multiple of 2π,

$$\omega\lambda = 2\pi n, \qquad n = 0, \pm1, \pm2, \pm3, \ldots \tag{12.24}$$

The possible values of ω are then (using n to label the values) $\omega_n = 2\pi n/\lambda$, and the corresponding solutions are

$$y_n(x) = c_1 e^{(2\pi n x/\lambda)i} + c_2 e^{-(2\pi n x/\lambda)i}, \qquad n = 0, \pm1, \pm2, \pm3, \ldots \tag{12.25}$$

The trigonometric form of these solutions is

$$y_n = d_1 \cos \frac{2\pi n x}{\lambda} + d_2 \sin \frac{2\pi n x}{\lambda} \tag{12.26}$$

The constants c_1 and c_2, or d_1 and d_2, are not determined. The case of the periodic boundary condition is exemplified in Section 12.7 for the quantum-mechanical problem of the 'particle in a ring'.

EXAMPLE 12.9 Solve the equation $\dfrac{d^2 y}{d\theta^2} + \omega^2 y = 0$ for the cyclic boundary conditions $y(\theta + \pi) = y(\theta)$.

The general solution is

$$y(\theta) = c_1 e^{i\omega\theta} + c_2 e^{-i\omega\theta}.$$

Then

$$y(\theta+\pi)=c_1 e^{i\omega\theta}e^{i\omega\pi}+c_2 e^{-i\omega\theta}e^{-i\omega\pi}$$

and $y(\theta+\pi)=y(\theta)$ if $\pi\omega=2\pi n$ for integer n. Therefore, $\omega=2n$ for $n=0,\pm1,\pm2,\dots$ and

$$y(\theta)=c_1 e^{2in\theta}+c_2 e^{-2in\theta}$$

▸ Exercise 21

12.5   The harmonic oscillator

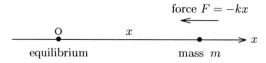

Figure 12.1

The simple (linear) harmonic oscillator (Figure 12.1) consists of a body moving in a straight line under the influence of a force

$$F=-kx \tag{12.27}$$

whose magnitude is proportional to the displacement x of the body from the fixed point O, the point of equilibrium, and whose direction is towards this point. The quantity k is called the force constant and the negative sign ensures that the force acts in the direction opposite to that of the displacement.

By Newton's second law of motion, the acceleration experienced by the body is given by

$$F=m\frac{d^2 x}{dt^2} \tag{12.28}$$

and simple harmonic motion is therefore described by the differential equation

$$m\frac{d^2 x}{dt^2}=-kx \tag{12.29}$$

A variety of physical systems can be modelled in terms of simple harmonic motion; the oscillations of a spring balance (when equation (12.27) is called Hooke's law), the swings of a pendulum, the oscillations of a diving springboard and, important in chemistry, the vibrations of the atoms in a molecule or crystal.

EXAMPLE 12.10 The vibrations of diatomic molecules

The vibrations of a diatomic molecule are often modelled in terms of the Morse potential

$$V(R) = D_e \left[1 - e^{-a(R-R_e)} \right]^2$$

(12.30)

where (Figure 12.2) R is the distance between the nuclei, R_e is the distance at equilibrium (the equilibrium bond length), D_e is the dissociation energy of the molecule and a is a constant (the vibrations of the molecule can be visualized in terms of a ball rolling forwards and backwards in the 'potential well' in Figure 12.2).

A stable molecule in its ground or low-lying excited vibrational states undergoes only small displacements, $R - R_e$, from equilibrium. Then, expanding the potential-energy function $V(R)$ as a power series in $(R - R_e)$,

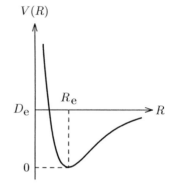

Figure 12.2

$$V = D_e \left[a^2 (R - R_e)^2 - a^3 (R - R_e)^3 + \cdots \right]$$

$$\approx a^2 D_e (R - R_e)^2$$

(12.31)

The force acting between the nuclei of the molecule is (see Section 5.7, equation (5.57)),

$$F = -\frac{dV}{dR}$$

(12.32)

Therefore, for small displacements, differentiation of (12.31) gives

$$F \approx -2a^2 D_e (R - R_e)$$

(12.33)

If $k = 2a^2 D_e$ and $x = (R - R_e)$, the force is $F \approx -kx$, and the vibrations of the molecule are (approximately) simple harmonic.

Equation (12.29) can be written in the standard form (12.3) of a homogeneous linear equation with constant coefficients,

$$\frac{d^2 x}{dt^2} + \frac{k}{m} x = 0$$

(12.34)

or, setting $k/m = \omega^2$, in the form of equation (12.20),

$$\frac{d^2x}{dt^2} + \omega^2 x = 0 \qquad (12.35)$$

with general solution, in trigonometric form,

$$x(t) = d_1 \cos \omega t + d_2 \sin \omega t \qquad (12.36)$$

The state of the system is determined by the initial conditions. For example, let the displacement and velocity at time $t = 0$ be

$$x(0) = A, \qquad x'(0) = 0 \qquad (12.37)$$

Then, for the initial displacement,

$$x(0) = A = d_1 \cos 0 + d_2 \sin 0 = d_1$$

For the velocity

$$x'(t) = -d_1 \omega \sin \omega t + d_2 \omega \cos \omega t$$

$$x'(0) = -d_1 \omega \sin 0 + d_2 \omega \cos 0 = d_2 \omega = 0$$

Therefore $d_1 = A, d_2 = 0$, and the solution of the initial value problem is

$$x(t) = A \cos \omega t \qquad (12.38)$$

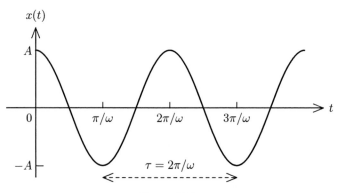

Figure 12.3

The graph of the solution is shown in Figure 12.3. The maximum displacement of the body from equilibrium is the **amplitude** A. The wavelength of the representative curve is the **period** $\tau = 2\pi/\omega$, the time taken for one complete oscillation. The inverse of the period, $v = 1/\tau = \omega/2\pi$, is called the **frequency** of oscillation, the number of oscillations in unit time. The quantity $\omega = 2\pi v$ is called the **angular frequency**. The frequency of oscillation is related to the force constant k by $\omega^2 = k/m$:

$$\omega = \sqrt{\frac{k}{m}}, \qquad v = \frac{1}{2\pi}\sqrt{\frac{k}{m}} \qquad (12.39)$$

Energetics

The harmonic oscillator is a conservative system (see Section 5.7) and the force $F = -kx$ is the derivative of a potential-energy function, $F = -dV/dx$. The potential energy is therefore the integral

$$V(x) = \int \frac{dV}{dx}\, dx = -\int F\, dx = \frac{1}{2}kx^2 + c$$

The potential energy is chosen, by convention, to be zero at the equilibrium position, when $x = 0$. Then

$$V = \frac{1}{2}kx^2 \tag{12.40}$$

The kinetic energy is $T = mv^2/2$, where v is the velocity, and the total energy is $E = T + V$. For the state of the system described by solution (12.38),

$$x = A\cos\omega t, \qquad v = x' = -A\omega\sin\omega t$$

so that

$$V = \frac{1}{2}kx^2 = \frac{1}{2}kA^2\cos^2\omega t$$

and

$$T = \frac{1}{2}mv^2 = \frac{1}{2}m\omega^2 A^2\sin^2\omega t = \frac{1}{2}kA^2\sin^2\omega t$$

since $\omega^2 = k/m$. The total energy is therefore the constant

$$E = T + V = \frac{1}{2}kA^2 \tag{12.41}$$

The relation between the potential and kinetic energies is shown in Figure 12.4. At maximum displacement from equilibrium, $x = A$, the potential energy has its maximum value, $V = E$, and the kinetic energy is zero. As the body approaches the

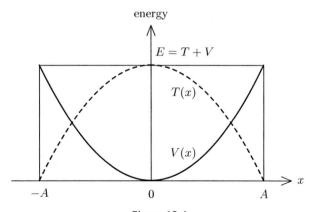

Figure 12.4

equilibrium position, potential energy is converted into kinetic energy and, at $x = 0$, $V = 0$ and the kinetic energy is a maximum, $T = E$.

‣ Exercises 22, 23

12.6   The particle in a one-dimensional box

The 'particle in a one-dimensional box' is the name given to the system consisting of a body allowed to move freely along a line of finite length. In quantum mechanics, this is one of the simplest systems that demonstrate the quantization of energy.

The Schrödinger equation for a particle of mass m moving in the x-direction is

$$-\frac{\hbar^2}{2m}\frac{d^2\psi(x)}{dx^2} + V(x)\psi(x) = E\psi(x) \tag{12.42}$$

where $V(x)$ is the potential energy of the particle at position x, E is the (constant) total energy, and ψ is the wave function. For the present system the potential-energy function is (Figure 12.5)

$$V(x) = \begin{cases} 0 & \text{for} \quad 0 < x < l \\ \infty & \text{for} \quad x \le 0 \ \text{ and } \ x \ge l \end{cases} \tag{12.43}$$

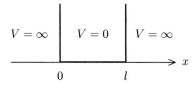

Figure 12.5

The constant value of V inside the box ensures that no force acts on the particle in this region; setting $V = 0$ means that the energy E is the (positive) kinetic energy of the particle. The infinite value of V at the 'walls' and outside the box ensures that the particle cannot leave the box; in quantum mechanics this means that the wave function is zero at the walls and outside the box.

For the particle within the box, we therefore have the boundary value problem

$$-\frac{\hbar^2}{2m}\frac{d^2\psi(x)}{dx^2} = E\psi(x) \tag{12.44}$$

with boundary conditions

$$\psi(0) = \psi(l) = 0 \tag{12.45}$$

Equation (12.44) is identical in form to equation (12.20), or to equation (12.35) for the harmonic oscillator. Thus, setting

$$\omega^2 = \frac{2mE}{\hbar^2} \tag{12.46}$$

we have

$$\frac{d^2\psi}{dx^2} + \omega^2\psi = 0 \tag{12.47}$$

with general solution, in trigonometric form,

$$\psi(x) = d_1 \cos \omega x + d_2 \sin \omega x \tag{12.48}$$

We now apply the boundary conditions (12.45). At $x = 0$,

$$\psi(0) = d_1 \cos 0 + d_2 \sin 0 = d_1 = 0$$

and at $x = l$ (having set $d_1 = 0$ in (12.48)),

$$\psi(l) = d_2 \sin \omega l = 0 \tag{12.49}$$

It follows, because the sine function is zero only when its argument is a multiple of π, that $\omega l = n\pi$ where n is an integer:

$$\omega = \frac{n\pi}{l}, \qquad n = 0, \pm 1, \pm 2, \dots \tag{12.50}$$

The solutions of the boundary value problem are therefore

$$\psi_n(x) = d_2 \sin \frac{n\pi x}{l}, \qquad n = 1, 2, 3, \dots \tag{12.51}$$

where the permitted solutions have been labelled with the **quantum number** n. The value $n = 0$ has been discounted because the trivial solution $\psi_0(x) = 0$ is not a physical solution, and negative values of n have been discounted because $\psi_{-n}(x) = -\psi_n(x)$ is merely ψ_n with a change of sign.

For each unique solution ψ_n there exists an energy

$$E_n = \frac{n^2 h^2}{8ml^2} \tag{12.52}$$

(using $E = \hbar^2 \omega^2 / 2m = h^2 \omega^2 / 8\pi^2 m$). We see that quantization of the energy is a consequence of constraining the motion of the particle to a finite region of space by the application of the boundary conditions; this is a general result in quantum mechanics. We note that these results are not restricted to motion in a straight line; they are valid for any continuous curve of length l if the variable x is measured along the curve.

The wave functions (12.51) are not yet fully specified because the coefficient d_2 is not defined. To determine d_2 we invoke the quantum-mechanical interpretation of the square of the wave function as a probability density (see also Example 10.4):

$$\psi^2(x)dx = \text{probability that the particle be in element } dx \text{ at position } x$$

The total probability that the particle be in the box is unity so that, integrating over the length of the box,

$$1 = \int_0^l \psi^2(x)\, dx = (d_2)^2 \int_0^l \sin^2 \frac{n\pi x}{l}\, dx = (d_2)^2 \frac{l}{2}$$

Therefore $d_2 = \sqrt{2/l}$, and the **normalized** solutions are

$$\psi_n(x) = \sqrt{\frac{2}{l}} \sin \frac{n\pi x}{l} \tag{12.53}$$

The graphs of the first three of these solutions are shown in Figure 12.6. These graphs show that the wave function ψ_n has $(n-1)$ zeros, or **nodes**, between the end points. This is a demonstration of a general property of wave functions; the number of zeros (nodal points, curves or surfaces) increases as the energy of the system increases.

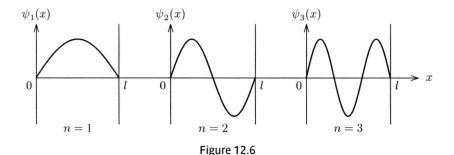

Figure 12.6

▸ Exercises 24, 25

The particle in a box is important in the physical sciences not only as one of the simplest solvable problems that demonstrate the phenomenon of quantization, but also because the system, when generalized to three dimensions, is used in statistical thermodynamics to derive the thermodynamic properties of the ideal gas. We now use the solutions (12.53) to demonstrate an important property of solutions of the Schrödinger equation.

Orthogonality

The Schrödinger equation for the particle in a box can be written as

$$\left\{ -\frac{\hbar^2}{2m} \frac{d^2}{dx^2} \right\} \psi = E\psi$$

or

$$\mathcal{H}\psi = E\psi \tag{12.54}$$

where the differential operator

$$\mathcal{H} = -\frac{\hbar^2}{2m}\frac{d^2}{dx^2} \tag{12.55}$$

is called the **Hamiltonian operator**, or simply the Hamiltonian, of the system. The effect of operating with $\mathcal{H}$ on ψ is to generate a multiple of ψ. The (time-independent) Schrödinger equation can always be written in the form (12.54) as an **eigenvalue equation**, with the system specified by the Hamiltonian operator and appropriate boundary conditions. The permitted wave functions $\psi = \psi_n$ are called the **eigenfunctions** of the Hamiltonian, and the corresponding energies $E = E_n$ are the **eigenvalues** of $\mathcal{H}$.

An important property of eigenfunctions is that of **orthogonality**. In the present case, consider the integral

$$I = \int_0^l \psi_m(x)\,\psi_n(x)\,dx \tag{12.56}$$

where ψ_m and ψ_n are two *different* eigenfunctions (12.53). Then

$$I = \frac{2}{l}\int_0^l \sin\frac{m\pi x}{l}\,\sin\frac{n\pi x}{l}\,dx \tag{12.57}$$

$$= 0 \quad \text{when} \quad m \neq n$$

Proof. The integral is evaluated by making use of the addition properties of trigonometric functions, as discussed in Section 6.2. Thus,

$$\sin\frac{m\pi x}{l}\,\sin\frac{n\pi x}{l} = \frac{1}{2}\left[\cos\left(\frac{(m-n)\pi x}{l}\right) - \cos\left(\frac{(m+n)\pi x}{l}\right)\right]$$

so that

$$I = \frac{1}{2}\int_0^l \cos\left(\frac{(m-n)\pi x}{l}\right)dx - \frac{1}{2}\int_0^l \cos\left(\frac{(m+n)\pi x}{l}\right)dx$$

Now,

$$\int_0^l \cos\left(\frac{(m\pm n)\pi x}{l}\right)dx = \left[\frac{1}{(m\pm n)\pi}\sin\left(\frac{(m\pm n)\,\pi x}{l}\right)\right]_0^l$$

$$= \frac{l}{(m\pm n)\pi}\left[\sin(m\pm n)\pi - \sin 0\right] = 0$$

since $\sin 0 = 0$ and the sine of a multiple, $m\pm n$, of π is also zero.

It follows therefore that

$$\int_0^l \psi_m(x)\psi_n(x)\,dx = 0 \qquad \text{when} \qquad m \neq n \tag{12.58}$$

The functions are said to be **orthogonal**. For *normalized* wave functions, when $\int_0^l \psi_n(x)\psi_n(x)\,dx = 1$, we can write

$$\int_0^l \psi_m(x)\psi_n(x)\,dx = \delta_{mn} = \begin{cases} 1 & \text{if} \quad m = n \\ 0 & \text{if} \quad m \neq n \end{cases} \tag{12.59}$$

The quantity δ_{mn}, equal to 1 if $m = n$ (normalization) and equal to 0 if $m \neq n$ (orthogonality), is called the **Kronecker delta**. Functions that satisfy (12.59) are said to be **orthonormal** (orthogonal and normalized).

▸ Exercise 26

12.7 The particle in a ring

The Schrödinger equation for a particle of mass m moving freely in a circle of radius r is, Figure 12.7,

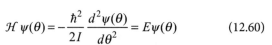

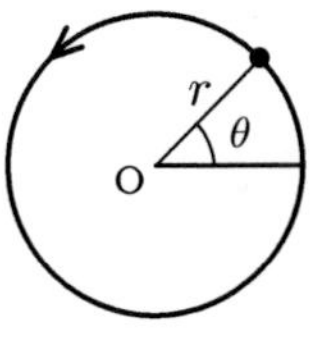

$$\mathcal{H}\,\psi(\theta) = -\frac{\hbar^2}{2I}\frac{d^2\psi(\theta)}{d\theta^2} = E\psi(\theta) \tag{12.60}$$

Figure 12.7

where E is the (positive) kinetic energy of the particle and $I = mr^2$ is its moment of inertia with respect to the centre of the circle. Setting

$$\omega^2 = \frac{2I\,E}{\hbar^2} \tag{12.61}$$

we have

$$\frac{d^2\psi}{d\theta^2} + \omega^2\psi = 0 \tag{12.62}$$

with general solution, in exponential form

$$\psi(\theta) = c_1 e^{i\omega\theta} + c_2 e^{-i\omega\theta} \tag{12.63}$$

For the wave function to be continuous around the circle, it must satisfy the periodic boundary condition

$$\psi(\theta + 2\pi) = \psi(\theta) \tag{12.64}$$

Equations (12.62) and (12.64) are the periodic boundary value problem discussed in Section 12.4, with x replaced by θ and λ by 2π. The allowed values of ω are therefore $\omega_n = n$, and the corresponding solutions are the eigenfunctions

$$\psi_n(\theta) = c_1 e^{in\theta} + c_2 e^{-in\theta}, \qquad n = 0, \pm 1, \pm 2, \ldots \tag{12.65}$$

with eigenvalues

$$E_n = \frac{\hbar^2 n^2}{2I} \tag{12.66}$$

We note that the states of the system with quantum number $n \neq 0$ occur in degenerate pairs, ψ_n and ψ_{-n},

$$\mathcal{H}\psi_n = E_n \psi_n, \qquad \mathcal{H}\psi_{-n} = E_n \psi_{-n}$$

By the principle of superposition (Section 12.2), every linear combination of a pair of degenerate eigenfunctions is itself an eigenfunction with the same eigenvalue,

$$\mathcal{H}(a\psi_n + b\psi_{-n}) = E_n(a\psi_n + b\psi_{-n})$$

where a and b are arbitrary. It is physically possible to distinguish degenerate states of a quantum-mechanical system only by the application of an external force to break the degeneracy. In the absence of such a force, therefore, every choice of coefficients c_1 and c_2 is equally good. It is conventional to choose $c_2 = 0$ in (12.65), to give the set of eigenfunctions

$$\psi_n(\theta) = c_1 e^{in\theta}, \qquad n = 0, \pm 1, \pm 2, \ldots \tag{12.67}$$

and to choose c_1 to normalize these functions. The normalization condition for the *complex* functions is

$$\int_0^{2\pi} \psi_n^*(\theta)\, \psi_n(\theta)\, d\theta = 1 \tag{12.68}$$

where $\psi_n^*(\theta) = c_1^* e^{-in\theta}$ is the complex conjugate function of $\psi_n(\theta)$. Then

$$\int_0^{2\pi} \psi_n^*(\theta)\, \psi_n(\theta)\, d\theta = |c_1|^2 \int_0^{2\pi} e^{-in\theta} e^{in\theta}\, d\theta = |c_1|^2 \int_0^{2\pi} d\theta = 2\pi |c_1|^2$$

and this is unity if $c_1 = 1/\sqrt{2\pi}$. The normalized eigenfunctions are therefore

$$\psi_n(\theta) = \frac{1}{\sqrt{2\pi}} e^{in\theta}, \qquad n = 0, \pm 1, \pm 2, \ldots \tag{12.69}$$

These functions form an **orthonormal** set in the interval $0 \leq \theta \leq 2\pi$; they are orthogonal as well as normalized:

$$\int_0^{2\pi} \psi_n^*(\theta)\, \psi_m(\theta)\, d\theta = 0 \qquad \text{if} \quad n \neq m \tag{12.70}$$

EXAMPLE 12.11 Show that the functions ψ_1 and ψ_2 satisfy the orthogonality condition (12.70).

We have

$$\int_0^{2\pi} \psi_1^*(\theta)\,\psi_2(\theta)\,d\theta = \frac{1}{2\pi} \int_0^{2\pi} e^{-i\theta} e^{2i\theta}\,d\theta = \frac{1}{2\pi}\int_0^{2\pi} e^{i\theta}\,d\theta$$

$$= \frac{1}{2\pi}\left[\frac{1}{i}e^{i\theta}\right]_0^{2\pi} = \frac{1}{2\pi i}\left[e^{2\pi i} - 1\right]$$

and this is zero because $e^{2\pi i} = 1$.

▸ Exercise 27

The functions (12.69) can be interpreted as representing clockwise rotation of the particle when $n > 0$ and anticlockwise rotation when $n < 0$ (see Section 8.6 for the rigid rotor). They are complex functions when $n \neq 0$, and for some purposes it is more convenient to have solutions of the Schrödinger equation that are real functions. The convention is to use the trigonometric functions obtained by means of Euler's formula. We have

$$\psi_{\pm n}(\theta) = \frac{1}{\sqrt{2\pi}}(\cos n\theta \pm i\sin n\theta)$$

and the combinations

$$\frac{1}{\sqrt{2}}(\psi_n + \psi_{-n}) = \frac{1}{2\sqrt{\pi}}\cos n\theta, \qquad \frac{1}{i\sqrt{2}}(\psi_n - \psi_{-n}) = \frac{1}{2\sqrt{\pi}}\sin n\theta \quad (12.71)$$

are the alternative orthonormal set. The symmetries of these functions are illustrated in Figure 12.8.

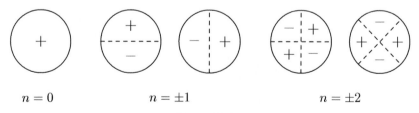

$$n = 0 \qquad\qquad n = \pm 1 \qquad\qquad n = \pm 2$$

Figure 12.8

We note that these symmetries are also the symmetries around the molecular axis of the molecular orbitals of a linear molecule, with $n = 0$ for σ-orbitals, $n = \pm 1$ for π-orbitals, $n = \pm 2$ for δ-orbitals, and so on. They are also the symmetries of the normal modes of vibration of a circular drum.

▸ Exercise 28

The general case

If the end-points of the linear box discussed in Section 12.6 are joined to form a simple closed loop, the Schrödinger equation is unchanged,

$$\frac{d^2\psi}{dx^2} + \omega^2\psi = 0, \qquad \omega^2 = \frac{2mE}{\hbar^2} \tag{12.72}$$

if the variable x is measured along the loop, but the two boundary conditions (12.45) are replaced by the single periodic condition

$$\psi(x+l) = \psi(x) \tag{12.73}$$

Equations (12.72) and (12.73) are the periodic boundary problem discussed in Section 12.4 (with λ replaced by l). The allowed values of ω are $\omega_n = 2\pi n/l$, and the solutions, in trigonometric form, are given by equation (12.26),

$$\psi_n(x) = d_1 \cos\frac{2\pi nx}{l} + d_2 \sin\frac{2\pi nx}{l} \tag{12.74}$$

These results are valid for a simple closed loop with any shape. The results for the circle of radius r are then obtained by replacing l by the circumference $2\pi r$, and the variable x by $r\theta$.

> Exercise 29

12.8 Inhomogeneous linear equations

The general inhomogeneous second-order linear equation with constant coefficients is

$$\frac{d^2y}{dx^2} + a\frac{dy}{dx} + by = r(x) \tag{12.75}$$

where a and b are constants. Particular solutions of this equation can be found by elementary methods for several important types of **inhomogeneity** $r(x)$.

EXAMPLE 12.12 Find a particular solution of the equation $y'' + 3y' + 2y = 2x^2$.

The form of the function on the right of the equation suggests a solution of type

$$y = a_0 + a_1 x + a_2 x^2$$

Then

$$y' = a_1 + 2a_2 x, \qquad y'' = 2a_2$$

and

$$y'' + 3y' + 2y = (2a_2 + 3a_1 + 2a_0) + (6a_2 + 2a_1)x + 2a_2 x^2$$

This is equal to $2x^2$ if

$$(2a_2 + 3a_1 + 2a_0) = 0, \qquad (6a_2 + 2a_1) = 0, \qquad a_2 = 1$$

so that $a_2 = 1$, $a_1 = -3$ and $a_0 = 7/2$. A particular solution is therefore

$$y = \frac{7}{2} - 3x + x^2$$

➤ Exercise 30

The method used in this example is the **method of undetermined coefficients**; a_0, a_1 and a_2 being the coefficients to be determined in this case. We consider first how, given a particular solution, a general solution may be obtained.

Let y_p be a particular solution of the inhomogeneous equation (12.75), so that

$$\frac{d^2 y_p}{dx^2} + a \frac{dy_p}{dx} + by_p = r(x) \qquad (12.76)$$

Let y_h be the general solution (12.5) of the corresponding *homogeneous* equation (12.3), called the **reduced equation** in this context:

$$y_h(x) = c_1 y_1(x) + c_2 y_2(x) \qquad (12.77)$$

$$\frac{d^2 y_h}{dx^2} + a \frac{dy_h}{dx} + by_h = 0 \qquad (12.78)$$

It follows that the sum of the functions $y_h(x)$ and $y_p(x)$,

$$y(x) = y_h(x) + y_p(x) \qquad (12.79)$$

is also a solution of the inhomogeneous equation:

$$\frac{d^2 y}{dx^2} + a \frac{dy}{dx} + by = \left[\frac{d^2 y_h}{dx^2} + a \frac{dy_h}{dx} + by_h \right] + \left[\frac{d^2 y_p}{dx^2} + a \frac{dy_p}{dx} + by_p \right] = r(x)$$

using equations (12.76) and (12.78). The function (12.79) contains two arbitrary constants and is the general solution of the inhomogeneous equation. The function y_h is often called the **complementary function** and y_p the **particular integral**:

general solution = complementary function + particular integral

---

EXAMPLE 12.13 Find the general solution of the equation $y'' + 3y' + 2y = 2x^2$.

By Example 12.12, the particular integral is

$$y_p(x) = \frac{7}{2} - 3x + x^2$$

The reduced (homogeneous) equation is

$$y'' + 3y' + 2y = 0$$

and has characteristic equation $\lambda^2 + 3\lambda + 2 = (\lambda+1)(\lambda+2) = 0$, with roots $\lambda_1 = -1$ and $\lambda_2 = -2$. The general solution of the homogeneous equation, the complementary function, is therefore

$$y_h(x) = c_1 e^{-x} + c_2 e^{-2x}$$

and the general solution of the inhomogeneous equation is

$$y(x) = y_h(x) + y_p(x) = c_1 e^{-x} + c_2 e^{-2x} + \frac{7}{2} - 3x + x^2$$

➤ Exercises 31, 32

The method of undetermined coefficients

This method can be used for many elementary functions $r(x)$ in (12.75), and is summarized in Table 12.1 for some of the more important types.

Table 12.1

	Term in $r(x)$	Choice of y_p
1.	$ce^{\alpha x}$	$ke^{\alpha x}$
2.	$cx^n \quad (n = 0, 1, 2, \dots)$	$a_0 + a_1 x + a_2 x^2 + \cdots + a_n x^n$
3.	$c \cos \omega x \quad$ or $\quad c \sin \omega x$	$k \cos \omega x + l \sin \omega x$
4.	$ce^{\alpha x} \cos \omega x \quad$ or $\quad ce^{\alpha x} \sin \omega x$	$e^{\alpha x}(k \cos \omega x + l \sin \omega x)$

The table gives the initial choice of particular integral y_p corresponding to each function $r(x)$. The coefficients in y_p are determined by substituting y_p into the inhomogeneous equation. Example 12.12 demonstrates the method for case 2 in the table. This is sufficient unless a term in y_p is already a solution of the corresponding homogeneous equation. In that case

(a) if the characteristic equation of the homogeneous equation has two distinct roots, multiply y_p by x before substitution in (12.75), or
(b) if the characteristic equation has a double root, multiply y_p by x^2 before substitution in (12.75). In addition, if $r(x)$ is the sum of two or more terms, the total particular solution is the corresponding sum of the corresponding y_p's.

EXAMPLE 12.14 Find the general solution of the equation $y'' + 3y' + 2y = 3e^{-2x}$.

By Example 12.13, the general solution of the homogeneous equation is

$$y_h(x) = c_1 e^{-x} + c_2 e^{-2x}$$

By Table 12.1, case **1**, the choice of particular integral should be $y_p = ke^{-2x}$, but this is already a solution of the homogeneous equation. By prescription (a) therefore, we use

$$y_p = kxe^{-2x}$$

Then

$$y'_p = ke^{-2x} - 2kxe^{-2x}, \qquad y''_p = -4ke^{-2x} + 4kxe^{-2x}$$

so that

$$y''_p + 3y'_p + 2y_p = -ke^{-2x}$$

and this is equal to $y_p = 3e^{-2x}$ if $k = -3$. Then $y_p = -3xe^{-2x}$ and the general solution is

$$y(x) = y_h(x) + y_p(x) = c_1 e^{-x} + c_2 e^{-2x} - 3xe^{-2x}$$

▸ Exercises 33, 34

EXAMPLE 12.15 Find the general solution of the equation $y'' - 4y' + 4y = e^{2x}$.

By Example 12.4, the characteristic equation of the homogeneous equation has the double root $\lambda = 2$, and the complementary function is

$$y_h(x) = (c_1 + c_2 x)e^{2x}$$

In this case, by prescription (b), the function $y_p = ke^{2x}$ is multiplied by x^2. Then

$$y_p = kx^2 e^{2x}, \qquad y'_p = 2kxe^{2x} + 2kx^2 e^{2x}, \qquad y''_p = 2ke^{2x} + 8kxe^{2x} + 4kx^2 e^{2x}$$

so that

$$y''_p - 4y'_p + 4y_p = 2ke^{2x}$$

and this is equal to e^{2x} when $k = 1/2$. The general solution is therefore

$$y(x) = y_h(x) + y_p(x) = \left(c_1 + c_2 x + \frac{1}{2}x^2 \right) e^{2x}$$

▸ Exercise 35

▸ Exercises 36–38

12.9   Forced oscillations

An important equation in the theory of forced oscillations in mechanical and electrical systems is the inhomogeneous differential equation

$$m\frac{d^2x}{dt^2} + c\frac{dx}{dt} + kx = F_0 \cos \omega t \qquad (12.80)$$

This is the equation of motion for a body moving under the influence of the force

$$F = m\frac{d^2x}{dt^2} = -kx - c\frac{dx}{dt} + F_0 \cos \omega t \qquad (12.81)$$

The term $-kx$ tells us that we have a harmonic oscillator. The term $-c\,dx/dt$ is a 'damping force' proportional to the velocity representing, for example, the drag experienced by the body moving in a fluid. The term $F_0 \cos \omega t$ is an external periodic force that interacts with the motion of the oscillator. For example, a charge q in the presence of an electric field experiences a force qE in the direction of the field. In the case of an alternating field, $E = E_0 \cos \omega t$, the force acting on an electron, with charge $q = -e$, is $-eE_0 \cos \omega t$. If the electron is undergoing simple harmonic motion (in the absence of the field) the total force acting on it is[3]

$$F = -kx - eE_0 \cos \omega t \qquad (12.82)$$

This is an example of (12.81) with no damping, and it is this case that we consider here. When $c = 0$, equation (12.80) has the form

$$\frac{d^2x}{dt^2} + \omega_0^2 x = A \cos \omega t \qquad (12.83)$$

where $\omega_0 = \sqrt{k/m}$ is the angular frequency of the oscillator in the absence of the field; the quantity $\nu_0 = \omega_0/2\pi$ is called the **natural frequency** of the oscillator.

The general solution of the homogeneous equation corresponding to (12.83),

$$\frac{d^2x}{dt^2} + \omega_0^2 x = 0$$

is (see Section 12.4)

$$x_h(t) = d_1 \cos \omega_0 t + d_2 \sin \omega_0 t \qquad (12.84)$$

For the inhomogeneity $r(t) = -A \cos \omega t$ in (12.83), we have the particular integral (case **3** in Table 12.1)

$$x_p(t) = c \cos \omega t + d \sin \omega t$$

---

[3] Euler submitted his solution to the problem of the forced harmonic oscillator to the St Petersburg Academy of Sciences on March 30, 1739.

Then

$$\frac{dx_p}{dt} = -\omega c \sin \omega t + \omega d \cos \omega t, \qquad \frac{d^2 x_p}{dt^2} = -\omega^2 (c \cos \omega t + d \sin \omega t)$$

and

$$\frac{d^2 x_p}{dt^2} + \omega_0^2 x_p = (\omega_0^2 - \omega^2)(c \cos \omega t + d \sin \omega t)$$

This is equal to $A \cos \omega t$ when $d = 0$ and $c = A/(\omega_0^2 - \omega^2)$. The particular integral is therefore

$$x_p(t) = \frac{A \cos \omega t}{\omega_0^2 - \omega^2} \qquad (12.85)$$

and the general solution of equation (12.83) for the oscillator in the external field is

$$x(t) = x_h(t) + x_p(t)$$
$$= d_1 \cos \omega_0 t + d_2 \sin \omega_0 t + \frac{A \cos \omega t}{\omega_0^2 - \omega^2} \qquad (12.86)$$

The solution shows that the behaviour of the system depends strongly on the relative values of ω and ω_0. In particular, when the frequency ω of the external force is close to the natural frequency of the oscillator, the maximum amplitude of oscillation (the maximum value of $x(t)$), is large, and tends to infinity as $\omega \to \omega_0$. This phenomenon of large oscillations is called **resonance**, and is an important factor in the study of vibrating systems in both classical and quantum mechanics.

In the case of resonance, when $\omega = \omega_0$, the function (12.85) is not the required particular integral because it is already a solution of the homogeneous equation. The required function is (by prescription (**a**) in Section 12.8)

$$x_p(t) = t(c \cos \omega_0 t + d \sin \omega_0 t) \qquad (12.87)$$

Substitution in (12.83) then gives $c = 0$ and $d = A/2\pi\omega_0$, and the particular integral in this case is

$$x_p(t) = \frac{A}{2\omega_0} t \sin \omega_0 t \qquad (12.88)$$

The graph of this function, in Figure 12.9, shows how in the case of resonance the amplitude of the oscillation increases indefinitely with time. In mechanical systems resonance may be avoided by the application of a suitable damping force.

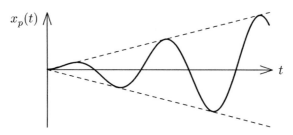

Figure 12.9

‣ Exercise 39

12.10   Exercises

Section 12.2

1. Show that e^{-2x} and $e^{2x/3}$ are particular solutions of the differential equation $3y'' + 4y' - 4y = 0$.
2. Show that e^{3x} and xe^{3x} are particular solutions of the differential equation $y'' - 6y' + 9y = 0$.
3. Show that $\cos 2x$ and $\sin 2x$ are particular solutions of the differential equation $y'' + 4y = 0$.
Write down the general solution of the differential equation in
4. Exercise 1.   **5.** Exercise 2.   **6.** Exercise 3.

Section 12.3

Find the general solutions of the differential equations:
7. $y'' - y' - 6y = 0$ **8.** $2y'' - 8y' + 3y = 0$ **9.** $y'' - 8y' + 16y = 0$
10. $4y'' + 12y + 9y = 0$ **11.** $y'' + 4y' + 5y = 0$ **12.** $y'' + 3y' + 5y = 0$

Section 12.4

Solve the initial value problems:

13. $\dfrac{d^2x}{dt^2} + \dfrac{dx}{dt} - 2x = 0; \quad x(0) = 1, \quad \dfrac{dx}{dt}(0) = 0$

14. $\dfrac{d^2x}{dt^2} + 6\dfrac{dx}{dt} + 9x = 0; \quad x(1) = 0, \quad \dfrac{dx}{dt}(1) = 1$

15. $\dfrac{d^2x}{dt^2} + 9x = 0; \quad x(\pi/3) = 0, \quad \dfrac{dx}{dt}(\pi/3) = -1$

16. $\dfrac{d^2x}{dt^2} - 2\dfrac{dx}{dt} + 2x = 0; \quad x(0) = 1, \quad \dfrac{dx}{dt}(0) = 0$

Solve the boundary value problems:

17. $\dfrac{d^2y}{dx^2} + 4\dfrac{dy}{dx} + 8y = 0; \quad y(\pi/2) = -1, \quad y(3\pi/4) = 1$

18. $\dfrac{d^2y}{dx^2} + 9y = 0; \quad y = 0$ when $x = 0, \quad y = 1$ when $x = \pi/2$

19. $\dfrac{d^2y}{dx^2} + 8\dfrac{dy}{dx} + 16y = 0; \quad y(0) = 0, \quad y(1) = 1$

20. $\dfrac{d^2 y}{dx^2} + \dfrac{dy}{dx} - 2y = 0; \quad y(0) = 2, \; y \to 0 \text{ as } x \to \infty$

21. Solve $\dfrac{d^2 \theta}{dt^2} + a^2 \theta = 0$ subject to the condition $\theta(t + 2\pi\tau) = \theta(t)$.

Section 12.5

22. Given that the general solution of the equation of motion $m\ddot{x} = -kx$ for the harmonic oscillator is $x(t) = a \cos \omega t + b \sin \omega t$, where $\omega = \sqrt{k/m}$, **(i)** show that the solution can be written in the form $x(t) = A \cos(\omega t - \delta)$, where A is the amplitude of the vibration and δ is the phase angle, and express A and δ in terms of a and b; **(ii)** find the amplitude and phase angle for the initial conditions $x(0) = 1, \dot{x}(0) = \omega$.

23. Solve the equation of motion for the harmonic oscillator with initial conditions $x(0) = 0$, $\dot{x}(0) = u_0$.

Section 12.6

24. For the particle in a box, find the nodes and sketch the graph of the wave function ψ_n for **(i)** $n = 4$ and **(ii)** $n = 5$.

25. **(i)** Solve the Schrödinger equation (12.44) for the particle in a box of length l with potential-energy function $V = 0$ for $-l/2 \le x \le +l/2$, $V = \infty$ for $x \le -l/2$ and $x \ge +l/2$. **(ii)** Show that the solutions ψ_n are even functions of x when n is odd and odd functions when n is even. **(iii)** Show that the solutions are the same as those given by (12.53) if x is replaced by $x + l/2$, except for a possible change of sign.

26. For the particle in the box in Section 12.6, show that wave functions $\psi_n(x) = \sqrt{\dfrac{2}{l}} \sin \dfrac{n\pi x}{l}$ for $n = 1$ and $n = 2$ are **(i)** normalized, **(ii)** orthogonal.

Section 12.7

27. For the particle in a ring show that wave functions $\psi_n(\theta) = 1/\sqrt{2\pi} \, e^{in\theta}$ for $n = 3$ and $n = 4$ are **(i)** normalized, **(ii)** orthogonal.

28. The diagrams of Figure 12.8 are maps of the signs and nodes of some real wave functions (12.71) for the particle in a ring. Draw the corresponding diagrams for **(i)** $n = \pm 3$, **(ii)** $n = \pm 4$.

29. Verify that equation (12.72) and its solutions (12.74) are transformed into (12.62) and (12.65) by means of the change of variable $\theta = x/r$.

Section 12.8

30. Find a particular solution of the differential equation $y'' - y' - 6y = 2 + 3x$.

Find the general solutions of the differential equations:

31. $y'' - y' - 6y = 2 + 3x$ **32.** $y'' - 8y' + 16y = 1 - 4x^3$

33. $y'' - y' - 6y = 2e^{-3x}$ **34.** $y'' - y' - 2y = 3e^{-x}$

35. $y'' - 8y' + 16y = e^{4x}$ **36.** $y'' - y' - 6y = 2 \cos 3x$

37. $y'' + 4y = 3 \sin 2x$ **38.** $y'' - y' - 6y = 2 + 3x + 2e^{-3x} + 2 \cos 3x$

Section 12.9

39. An RLC-circuit contains a resistor (resistance R), an inductor (inductance L), and a capacitor (capacitance C) connected in series with a source of e.m.f. E.

(i) Use Kirchhoff's voltage law (Section 11.7) to show that the current $I(t)$ in the circuit is given by the inhomogeneous equation

$$L\frac{d^2I}{dt^2} + R\frac{dI}{dt} + \frac{I}{C} = \frac{dE}{dt}$$

(ii) Find the solution of the homogeneous equation (for $dE/dt = 0$), and confirm that it decays exponentially as $t \to \infty$.

(iii) Show that the particular integral for the periodic e.m.f. $E(t) = E_0 \sin \omega t$ is

$$I_p(t) = I_0 \sin(\omega t - \delta)$$

where $I_0 = \dfrac{E_0}{\sqrt{R^2 + S^2}}$, $\tan\delta = \dfrac{S}{R}$, and $S = \omega L - \dfrac{1}{\omega C}$.

13 Second-order differential equations. Some special functions

13.1 Concepts

We saw in Sections 12.5 to 12.7 how three very different physical problems are modelled in terms of the *same* differential equation: equation (12.35) for the classical harmonic oscillator as an initial value problem, the same equation, (12.47), for the quantum-mechanical particle in a box as a boundary value problem, and (12.62) for the quantum-mechanical particle in a ring as a periodic boundary value problem. This is a common phenomenon in the mathematical modelling of physical systems, and a number of equations are important enough to have been given names. Some of these are listed in Table 13.1.

Table 13.1

Name	Equation
Legendre	$(1-x^2)y'' - 2xy' + l(l+1)y = 0$
Associated Legendre	$(1-x^2)y'' - 2xy' + [l(l+1) - m^2/(1-x^2)]y = 0$
Hermite	$y'' - 2xy' + 2ny = 0$
Laguerre	$xy'' + (1-x)y' + ny = 0$
Associated Laguerre	$xy'' + (m+1-x)y' + (n-m)y = 0$
Bessel	$x^2 y'' + xy' + (x^2 - n^2)y = 0$

The equations in Table 13.1 are all second-order homogeneous linear equations with non-constant coefficients, and they have particular solutions that play an important role in the mathematics of the physical sciences. These solutions are often called **special functions**.

The standard methods used to solve the equations in Table 13.1 and many other linear differential equations in the sciences are the power-series method, described in Section 13.2, and the more general Frobenius method, outlined in Section 13.3. The former is used in Section 13.4 to obtain the physically-significant particular solutions of the Legendre equation. These solutions are called Legendre polynomials and they occur whenever a physical problem in three dimensions is formulated in terms of spherical polar coordinates. Subsequent sections are devoted to brief descriptions of the other special functions that are solutions of the equations listed in Table 13.1.

13.2   The power-series method

Many important second-order linear differential equations

$$y'' + p(x)y' + q(x)y = r(x) \tag{13.1}$$

have at least one particular solution that can be expressed as a power series,

$$y(x) = a_0 + a_1 x + a_2 x^2 + a_3 x^3 + \cdots = \sum_{m=0}^{\infty} a_m x^m \tag{13.2}$$

The series is substituted into the differential equation to determine the numbers a_0, a_1, a_2, ..., and the particular solutions for a physical system are obtained by the application of the appropriate boundary or initial conditions.

The power-series method can be used when $p(x)$, $q(x)$ and $r(x)$ are polynomials or if they can be expanded as power series in x. It can therefore be used to solve the Legendre, associated Legendre and Hermite equations (the first two are transformed into the standard form (13.1) by division by $(1 - x^2)$).

We demonstrate the power-series method in Examples 13.1 and 13.2 by solving two equations for which we already know the solutions; a first-order equation in Example 13.1 and a second-order equation in Example 13.2.

EXAMPLE 13.1 Use the power-series method to solve the equation

$$\frac{dy}{dx} + y = 0$$

By equation (13.2), we express the solution as the power series

$$y = a_0 + a_1 x + a_2 x^2 + a_3 x^3 + \cdots = \sum_{m=0}^{\infty} a_m x^m$$

Then

$$\frac{dy}{dx} = a_1 + 2a_2 x + 3a_3 x^2 + \cdots = \sum_{m=1}^{\infty} m a_m x^{m-1}$$

and, substituting in the differential equation,

$$\frac{dy}{dx} + y = (a_1 + 2a_2 x + 3a_3 x^2 + \cdots) + (a_0 + a_1 x + a_2 x^2 + \cdots)$$

$$= (a_1 + a_0) + (2a_2 + a_1)x + (3a_3 + a_2)x^2 + \cdots$$

For this to be equal to zero for all values of x (within the radius of convergence of the power series) the coefficient of each power of x must be zero:

$$a_1 + a_0 = 0, \quad 2a_2 + a_1 = 0, \quad 3a_3 + a_2 = 0, \ldots$$

so that

$$a_1 = -a_0, \quad a_2 = -\frac{a_1}{2} = +\frac{a_0}{2!}, \quad a_3 = -\frac{a_2}{3} = -\frac{a_0}{3!}, \quad \ldots$$

and, in general, $a_m = (-1)^m a_0/m!$. Therefore,

$$y = \sum_{m=0}^{\infty} a_m x^m = \sum_{m=0}^{\infty} \frac{a_0 (-1)^m x^m}{m!} = a_0 \sum_{m=0}^{\infty} \frac{(-x)^m}{m!}$$

We recognize the infinite series as the power-series expansion of the function e^{-x}:

$$e^{-x} = 1 - x + \frac{x^2}{2!} - \frac{x^3}{3!} + \cdots = \sum_{m=0}^{\infty} \frac{(-x)^m}{m!}$$

Therefore $y = a_0 e^{-x}$ where a_0 is an arbitrary constant (see Example 11.4(ii) with $k = -1$).

▶ Exercises 1, 2

EXAMPLE 13.2 Use the power-series method to solve the equation[1]

$$\frac{d^2 y}{dx^2} + y = 0$$

By equation (13.2), we have

$$y = \sum_{m=0}^{\infty} a_m x^m, \quad y' = \sum_{m=1}^{\infty} m a_m x^{m-1}, \quad y'' = \sum_{m=2}^{\infty} m(m-1) a_m x^{m-2}$$

We can write

$$y'' = \sum_{m=2}^{\infty} m(m-1) a_m x^{m-2} = (2 \times 1)a_2 + (3 \times 2)a_3 x + (4 \times 3)a_4 x^2 + \cdots$$

$$= \sum_{m=0}^{\infty} (m+2)(m+1) a_{m+2} x^m$$

---

[1] Leibniz discovered the differential equation for the sine function in 1693 by a geometric argument. He then solved the equation by a method equivalent to that given in this example to obtain the power-series representation of the function. The ability to represent functions as power series was a vital element in the development of the calculus by both Leibniz and Newton.

Then

$$y'' + y = \sum_{m=0}^{\infty} a_m x^m + \sum_{m=0}^{\infty} (m+2)(m+1)a_{m+2}x^m$$

$$= \sum_{m=0}^{\infty} \left[a_m + (m+2)(m+1)a_{m+2} \right] x^m$$

and this is zero if the coefficient of each power of x is separately zero:

$$a_m + (m+2)(m+1)a_{m+2} = 0$$

For the even values of m,

$$a_2 = -\frac{a_0}{2 \times 1}, \quad a_4 = -\frac{a_2}{4 \times 3} = +\frac{a_0}{4!}, \quad a_6 = -\frac{a_4}{6 \times 5} = -\frac{a_0}{6!}, \cdots$$

and for the odd values of m,

$$a_3 = -\frac{a_1}{3 \times 2} = -\frac{a_1}{3!}, \quad a_5 = -\frac{a_3}{5 \times 4} = +\frac{a_1}{5!}, \quad a_7 = -\frac{a_5}{7 \times 6} = -\frac{a_1}{7!}, \cdots$$

Therefore,

$$y = \left[a_0 + a_2 x^2 + a_4 x^4 + \cdots \right] + \left[a_1 x + a_3 x^3 + a_5 x^5 + \cdots \right]$$

$$= a_0 \left[1 - \frac{x^2}{2!} + \frac{x^4}{4!} - \frac{x^6}{6!} + \cdots \right] + a_1 \left[x - \frac{x^3}{3!} + \frac{x^5}{5!} - \frac{x^7}{7!} + \cdots \right]$$

We recognize the two series in square brackets as the power-series expansions of $\cos x$ and $\sin x$. Therefore

$$y = a_0 \cos x + a_1 \sin x$$

where a_0 and a_1 are arbitrary constants (see equation (12.22) with $\omega = 1$).

▸ Exercises 3–5

13.3 The Frobenius method

The **Frobenius method**[2] is an extended power-series method that is used to solve second-order linear equations that can be written in the form

$$y'' + \frac{b(x)}{x}y' + \frac{c(x)}{x^2}y = 0 \qquad (13.3)$$

[2] Georg Frobenius (1849–1917), German mathematician, is also known for his work in matrix algebra. In a monograph in 1878 he organized the theory of matrices into the form that it has today.

in which $b(x)$ and $c(x)$ are polynomials or can be expanded as power series in x. It includes the simple power-series method discussed in the previous section as a special case and can be used for all the equations in Table 13.1, and for many of the linear differential equations in the sciences.

Every differential equation of the form (13.3) has at least one solution that can be expressed as

$$y(x) = x^r (a_0 + a_1 x + a_2 x^2 + a_3 x^3 + \cdots) = x^r \sum_{m=0}^{\infty} a_m x^m \qquad (13.4)$$

where r (which may be zero) is an **indicial parameter** to be determined, and $a_0 \neq 0$. We consider first the case of $b(x) = b_0$, $c(x) = c_0$, where b_0 and c_0 are constants:

$$x^2 y'' + b_0 x y' + c_0 y = 0 \qquad \text{(Euler–Cauchy equation)} \qquad (13.5)$$

(for convenience, the equation has been multiplied by x^2). We have

$$\begin{aligned}
y &= x^r [a_0 + a_1 x + \cdots] \\
y' &= x^{r-1} [r a_0 + (r+1) a_1 x + \cdots] \\
y'' &= x^{r-2} [r(r-1) a_0 + (r+1) r a_1 x + \cdots]
\end{aligned} \qquad (13.6)$$

and substitution into (13.5) gives

$$x^r [r(r-1) a_0 + (r+1) r a_1 x + \cdots] \\
+ b_0 x^r [r a_0 + (r+1) a_1 x + \cdots] + c_0 x^r [a_0 + a_1 x + \cdots] = 0$$

or, in full,

$$\sum_{m=0}^{\infty} \left[(r+m)^2 + (b_0 - 1)(r+m) + c_0 \right] a_m x^{m+r} = 0 \qquad (13.7)$$

The coefficient of each power of x in this equation must be zero so that, for the coefficient of x^r ($m = 0$),

$$r^2 + (b_0 - 1) r + c_0 = 0 \qquad (13.8)$$

This is called the **indicial equation**, and the roots are the possible values of the parameter r in (13.4). In general, $b(x)$ and $c(x)$ in (13.3) are polynomials or can be expanded as power series in x:

$$b(x) = b_0 + b_1 x + b_2 x^2 + b_3 x^3 + \cdots, \qquad c(x) = c_0 + c_1 x + c_2 x^2 + c_3 x^3 + \cdots$$

but the indicial equation (13.8) remains valid.

EXAMPLE 13.3 Indicial equations

(i) $x^2 y'' - \dfrac{1}{2} x y' + \dfrac{1}{2} y = 0$

We have $b_0 = -1/2$, $c_0 = 1/2$, and the indicial equation is $r^2 - 3r/2 + 1/2 = 0$ with roots $r_1 = 1$ and $r_2 = 1/2$.

(ii) $x^2 y'' + x y' + (x^2 - 1/4) y = 0$
This is the Bessel equation (see Table 13.1) for $n = 1/2$. We have $b(x) = 1 = b_0$ and $c(x) = x^2 - 1/4$ so that $c_0 = -1/4$. Then $r^2 - 1/4 = (r - 1/2)(r + 1/2) = 0$ and the indicial parameters are $r = \pm 1/2$.

(iii) $xy'' + 2y' + 4xy = 0$
We have $x^2 y'' + 2xy' + 4x^2 y = 0$. Therefore, $b(x) = 2 = b_0$ and $c(x) = x^2$ so that $c_0 = 0$. The indicial equation is $r^2 + (b_0 - 1)r + c_0 = r^2 + r = 0$, with roots $r_1 = 0$ and $r_2 = -1$.

▸ Exercises 6–9

Once the values of the indicial parameter have been determined, the solution of the differential equation can proceed as in the power-series method. The general solution is

$$y(x) = c_1 y_1(x) + c_2 y_2(x) \tag{13.9}$$

in which c_1 and c_2 are arbitrary constants, and $y_1(x)$ and $y_2(x)$ are independent particular solutions. At least one of y_1 and y_2 has the form (13.4), and the other depends on the solutions, r_1 and r_2, of the indicial equation. We have the following three cases.

1 Distinct roots not differing by an integer
Both y_1 and y_2 have the form (13.4),

$$y_1(x) = x^{r_1}(a_0 + a_1 x + a_2 x^2 + \cdots) \tag{13.10a}$$

$$y_2(x) = x^{r_2}(A_0 + A_1 x + A_2 x^2 + \cdots) \tag{13.10b}$$

2 Double root
If $r_1 = r_2 = r$, one solution has the form (13.4),

$$y_1(x) = x^r(a_0 + a_1 x + a_2 x^2 + \cdots) \tag{13.11a}$$

and the second solution is

$$y_2(x) = y_1(x) \ln x + x^{r+1}(A_0 + A_1 x + A_2 x^2 + \cdots) \tag{13.11b}$$

3 Roots differ by an integer
One solution has the form (13.4),

$$y_1(x) = x^{r_1}(a_0 + a_1 x + a_2 x^2 + \cdots) \tag{13.12a}$$

and the second solution is

$$y_2(x) = k y_1(x) \ln x + x^{r_2}(A_0 + A_1 x + A_2 x^2 + \cdots) \tag{13.12b}$$

in which $r_1 > r_2$ and the constant k *may be zero.*

▸ Exercises 10–12

Many of the particular solutions of second-order differential equations that are of interest in the physical sciences are generally series of type (13.4), without a logarithmic term.

EXAMPLE 13.4 The Bessel equation $x^2 y'' + xy' + (x^2 - n^2)y = 0$ for $n = \pm 1/2$.

By Example 13.3(ii), the indicial roots are $r = \pm 1/2$ and the solutions are nominally of type 3, equations (13.12). In the present case, however, there is no logarithmic term (see Exercise 13), and we show here that the particular solution with indicial parameter $r = +1/2$ is the Bessel function

$$J_{1/2}(x) = \sqrt{\frac{2}{\pi x}} \sin x$$

We have

$$y = x^{1/2} \sum_{m=0}^{\infty} a_m x^m = \sum_{m=0}^{\infty} a_m x^{m+1/2}$$

Then

$$y' = \sum_{m=0}^{\infty} \left(m + \tfrac{1}{2}\right) a_m x^{m-1/2}, \quad y'' = \sum_{m=0}^{\infty} \left(m^2 - \tfrac{1}{4}\right) a_m x^{m-3/2}$$

and

$$x^2 y'' + xy' + (x^2 - \tfrac{1}{4})y = x^{1/2} \sum_{m=0}^{\infty} a_m \left[m(m+1)x^m + x^{m+2} \right]$$

$$= x^{1/2} \left[2a_1 x + (a_0 + 2\times 3a_2)x^2 + (a_1 + 3\times 4a_3)x^3 + (a_2 + 4\times 5a_4 x^4) + \quad \right]$$

This is zero if the coefficient of each power of x is separately zero. The coefficient of x^1 is $2a_1$ so that $a_1 = 0$ and it follows that $a_m = 0$ for odd values of m. For the even values, as in Example 13.2,

$$a_2 = -\frac{1}{3!}a_0, \quad a_4 = -\frac{1}{5 \times 4}a_2 = +\frac{1}{5!}a_0, \quad a_6 = -\frac{1}{7 \times 6}a_4 = -\frac{1}{7!}a_0, \cdots$$

Therefore,

$$y(x) = a_0 x^{1/2}\left[1 - \frac{x^2}{3!} + \frac{x^4}{5!} - \frac{x^6}{7!} + \cdots\right] = a_0 x^{-1/2}\left[x - \frac{x^3}{3!} + \frac{x^5}{5!} - \frac{x^7}{7!} + \cdots\right]$$

$$= a_0 x^{-1/2}\sin x$$

By convention $a_0 = \sqrt{2/\pi}$ and the function $J_{1/2}(x) = \sqrt{\frac{2}{\pi x}}\sin x$ is the Bessel function

of order $n = 1/2$. The same procedure with $r = -1/2$ gives the Bessel function of order $-1/2$ (Exercise 13).

▸ Exercises 13–15

13.4 The Legendre equation

The Legendre equation is

$$(1 - x^2)y'' - 2xy' + l(l+1)y = 0 \tag{13.13}$$

where l is a real number. This equation arises whenever a physical problem in three dimensions is formulated in terms of spherical polar coordinates, r, θ and ϕ (see Section 10.2), in which case the variable x is replaced by $\cos\theta$. Of interest in the physical sciences are then those solutions that are finite in the interval $-1 \le x \le +1$ (corresponding to $0 \le \theta \le \pi$).

Equation (13.13) can be solved by the series method in the way described in Section 13.2. The series

$$y = \sum_m a_m x^m = a_0 + a_1 x + a_2 x^2 + \cdots \tag{13.14}$$

and its derivatives y' and y'' are substituted into (13.13). Then, setting the coefficient of each power of x equal to zero gives the recurrence relation

$$a_{m+2} = \frac{m(m+1) - l(l+1)}{(m+1)(m+2)}a_m = -\frac{(l-m)(l+m+1)}{(m+1)(m+2)}a_m \tag{13.15}$$

For the even values of m,

$$a_2 = -\frac{l(l+1)}{2}a_0, \quad a_4 = -\frac{(l-2)(l+3)}{3 \times 4}a_2 = +\frac{(l-2)l(l+1)(l+3)}{4!}a_0, \cdots$$

Similarly for the odd values of m,

$$a_3 = -\frac{(l-1)(l+2)}{3!}a_1, \qquad a_5 = +\frac{(l-3)(l-1)(l+2)(l+4)}{5!}a_1, \ldots$$

A power-series solution of the equation is therefore

$$y(x) = a_0 y_1(x) + a_1 y_2(x) \tag{13.16}$$

where a_0 and a_1 are arbitrary constants and

$$y_1(x) = 1 - \frac{l(l+1)}{2!}x^2 + \frac{(l-2)l(l+1)(l+3)}{4!}x^4 - \cdots \tag{13.17}$$

$$y_2(x) = x - \frac{(l-1)(l+2)}{3!}x^3 + \frac{(l-3)(l-1)(l+2)(l+4)}{5!}x^5 - \cdots \tag{13.18}$$

The series y_1 contains only even powers of x, the series y_2 only odd powers.

Convergence

We first consider the case of non-integer values of l. Both series have radius of convergence $R = 1$ for arbitrary values of l. Thus, the ratio of consecutive terms is, by equation (13.15),

$$\frac{a_{m+2}}{a_m} = \frac{m(m+1) - l(l+1)}{(m+1)(m+2)} \to 1 \quad \text{as} \quad m \to \infty$$

so that, by d'Alembert's ratio test (7.18) for power series, both series converge if $|x| < 1$ and diverge if $|x| > 1$ (unless l is an integer, see below). Both series also diverge when $|x| = 1$. The function (13.16), with two arbitrary constants, is therefore the general solution of the Legendre equation in the interval $-1 < x < 1$. It is also possible to find a solution involving inverse powers of x that is valid for $|x| > 1$.

The Legendre polynomials

The function (13.17) reduces to a polynomial of degree l when l is an even integer or zero. For example, when $l = 2$ the series terminates after the second term to give $y_1(x) = 1 - 3x^2$. The choice $a_1 = 0$ in (13.16) therefore gives a particular solution for even values of l that is finite for all values of x. Similarly, the function (13.18) reduces to a polynomial of degree l when l is an odd integer, and the choice $a_0 = 0$ in (13.16) gives a particular solution that is valid for all values of x. The same considerations apply for negative integer values of l, but the resulting set of polynomials is identical to that obtained for positive values. These particular solutions are called **Legendre polynomials** $P_l(x)$, and they are the solutions of the Legendre equation that are of interest in the physical sciences.[3] By convention, the nonzero arbitrary constant in (13.16) is chosen so that $P_l(1) = 1$. Then, for $l = 0, 1, 2, 3, \ldots$,

$$P_l(x) = \frac{1 \cdot 3 \cdot 5 \cdots (2l-1)}{l!}$$

$$\times \left\{ x^l - \frac{l(l-1)}{2(2l-1)} x^{l-2} + \frac{l(l-1)(l-2)(l-3)}{2 \cdot 4(2l-1)(2l-3)} x^{l-4} - \cdots \right\}$$

(13.19)

and the series is to be continued down to the constant term. The first few of these polynomials are

$$P_0(x) = 1 \qquad\qquad P_1(x) = x$$

$$P_2(x) = \frac{1}{2}(3x^2 - 1) \qquad\qquad P_3(x) = \frac{1}{2}(5x^3 - 3x) \qquad (13.20)$$

$$P_4(x) = \frac{1}{8}(35x^4 - 30x^2 + 3) \qquad P_5(x) = \frac{1}{8}(63x^5 - 70x^3 + 15x)$$

EXAMPLE 13.5 Show that the polynomial $P_3(x)$ is a solution of the Legendre equation (13.13) for $l = 3$.

We have

$$y = P_3(x) = \frac{1}{2}(5x^3 - 3x), \qquad y' = \frac{3}{2}(5x^2 - 1), \qquad y'' = 15x$$

Therefore, for $l = 3$,

$$(1 - x^2)y'' - 2xy' + l(l+1)y$$

$$= (1 - x^2) \times 15x - 2x \times \frac{3}{2}(5x^2 - 1) + 3 \times 4 \times \frac{1}{2}(5x^3 - 3x)$$

$$= (-15 - 15 + 30)x^3 + (15 + 3 - 18)x = 0$$

▸ Exercise 16

---

[3] Adrien-Marie Legendre (1752–1833) has the doubtful distinction of the following footnote in E. T. Bell's *Men of mathematics*: 'Considerations of space preclude an account of his life; much of his best work was absorbed or circumvented by younger mathematicians'. The polynomials appeared in Legendre's *Recherches sur l'attraction des sphéroïdes homogènes* of 1785 as the coefficients in the expansion of the potential function $(1 - 2h \cos\theta + h^2)^{-1/2}$ in powers of h. His textbook *Éléments géométrie* of 1794 did much to reform the teaching of geometry (till then based on Euclid), and an American edition, *Davies' Legendre* (1851), was influential in the United States.

The Legendre polynomials satisfy the recurrence relation

$$(l+1)P_{l+1}(x) - (2l+1)xP_l(x) + lP_{l-1}(x) = 0 \qquad (13.21)$$

so that given $P_0(x) = 1$ and $P_1(x) = x$, all higher polynomials can be derived.

EXAMPLE 13.6 Use the recurrence relation (13.21) to find $P_2(x)$ and $P_3(x)$.

For $l = 1$ the relation (13.21) is $2P_2 - 3xP_1 + P_0 = 0$. Therefore,

$$P_2 = \frac{1}{2}(3xP_1 - P_0) = \frac{1}{2}(3x^2 - 1)$$

For $l = 2$ the relation (13.21) is $3P_3 - 5xP_2 + 2P_1 = 0$. Therefore,

$$P_3 = \frac{1}{3}(5xP_2 - 2P_1) = \frac{1}{3}\left(5x \times \frac{1}{2}(3x^2 - 1) - 2x\right) = \frac{1}{2}(5x^3 - 3x)$$

▸ Exercise 17

The associated Legendre functions

The associated Legendre equation is (see Table 3.1)

$$(1 - x^2)y'' - 2xy' + \left[l(l+1) - \frac{m^2}{(1 - x^2)} \right] y = 0 \qquad (13.22)$$

As with the Legendre equation ($m = 0$), the variable x is identified with $\cos\theta$ in physical applications. In this case, the solutions obtained by the power-series method converge in the interval $-1 \le x \le 1$ when both l and m are integers, with $|m| \le l$:

$$l = 0, 1, 2, 3, \ldots \qquad m = 0, \pm 1, \pm 2, \ldots, \pm l \qquad (13.23)$$

The corresponding particular solutions are called the **associated Legendre functions** $P_l^{|m|}(x)$. They are related to the Legendre polynomials by the differential formula

$$P_l^{|m|}(x) = (1 - x^2)^{|m|/2} \frac{d^{|m|}}{dx^{|m|}} P_l(x) \qquad (13.24)$$

with $P_l^0 = P_l$.

EXAMPLE 13.7 Use the formula (13.24) to derive the associated Legendre functions P_3^m for $m = 1, 2$ and 3, and express these as functions of θ when $x = \cos\theta$ and $(1 - x^2)^{1/2} = \sin\theta$.

We have

$$y = P_3(x) = \frac{1}{2}(5x^3 - 3x), \qquad y' = \frac{3}{2}(5x^2 - 1), \qquad y'' = 15x, \qquad y''' = 15$$

Then

$$P_3^1(x) = \frac{3}{2}(1 - x^2)^{1/2}(5x^2 - 1) \qquad\qquad P_3^1(\cos\theta) = \frac{3}{2}\sin\theta(5\cos^2\theta - 1)$$

$$P_3^2(x) = 15(1 - x^2)x \qquad\qquad\qquad P_3^2(\cos\theta) = 15\sin^2\theta\cos\theta$$

$$P_3^3(x) = 15(1 - x^2)^{3/2} \qquad\qquad\qquad P_3^3(\cos\theta) = 15\sin^3\theta$$

➤ Exercise 18

Orthogonality and normalization

We saw in Section 12.6 that the solutions of the Schrödinger equation for the particle in a box are orthogonal functions (equation (12.58)). Orthogonality is a property of the solutions of a class of second-order linear differential equations called Sturm–Liouville equations that includes the Legendre equation. The Legendre polynomials are orthogonal in the interval $-1 \leq x \leq 1$:

$$\int_{-1}^{+1} P_l(x)\, P_{l'}(x)\, dx = 0 \qquad \text{when} \qquad l \neq l' \tag{13.25}$$

EXAMPLE 13.8 Show that P_1 is orthogonal (a) to P_2 and (b) to P_3.

We have

$$P_1(x) = x, \quad P_2(x) = \frac{1}{2}(3x^2 - 1), \quad P_3(x) = \frac{1}{2}(5x^3 - 3x).$$

Then

(a)
$$\int_{-1}^{+1} P_1(x) P_2(x)\, dx = \frac{1}{2}\int_{-1}^{+1}(3x^3 - x)\, dx = \frac{1}{2}\left[\frac{3x^4}{4} - \frac{x^2}{2}\right]_{-1}^{+1}$$

$$= \frac{1}{2}\left[\left(\frac{3}{4} - \frac{1}{2}\right) - \left(\frac{3}{4} - \frac{1}{2}\right)\right] = 0$$

In this case, $P_1(x)$ is an odd function of x whereas $P_2(x)$ is an even function. The integrand is odd and the integral is zero (see Section 5.3).

(b) $\displaystyle\int_{-1}^{+1} P_1(x) P_3(x)\, dx = \frac{1}{2} \int_{-1}^{+1} (5x^4 - 3x^2)\, dx = \frac{1}{2}\Big[x^5 - x^3 \Big]_{-1}^{+1} = \frac{1}{2}\Big[0 - 0\Big] = 0$

In this case, both $P_1(x)$ and $P_3(x)$ are odd functions so that the orthogonality of the functions is a new property, not a consequence of even/odd parity.

▸ Exercise 19

The corresponding property of the associated Legendre functions is

$$\int_{-1}^{+1} P_l^{|m|}(x) P_{l'}^{|m|}(x)\, dx = 0 \qquad \text{when} \qquad l \neq l' \tag{13.26}$$

In addition when $l = l'$,

$$\int_{-1}^{+1} \left(P_l^{|m|}(x) \right)^2 dx = \frac{2}{(2l+1)} \frac{(l+|m|)!}{(l-|m|)!} \tag{13.27}$$

and this result is used to construct the set of *normalized* associated Legendre functions (and normalized Legendre polynomials when $m = 0$),

$$\Theta_{l,m}(x) = \sqrt{\frac{(2l+1)}{2} \frac{(l-|m|)!}{(l+|m|)!}}\; P_l^{|m|}(x) \tag{13.28}$$

with property

$$\int_{-1}^{+1} \Theta_{l,m}(x) \Theta_{l',m}(x)\, dx = \delta_{l,l'} = \begin{cases} 1 & \text{if} \quad l = l' \\ 0 & \text{if} \quad l \neq l' \end{cases} \tag{13.29}$$

When $x = \cos\theta$ these functions form part of the solutions of the Schrödinger equation for the hydrogen atom (Section 14.6).

EXAMPLE 13.9 Show that (i) P_1^1 is orthogonal to P_3^1, (ii) P_2^2 is orthogonal to P_3^2.

(i) $P_1^1(x) = (1-x^2)^{1/2}, \quad P_3^1(x) = \frac{3}{2}(1-x^2)^{1/2}(5x^2 - 1)$

$$I = \int_{-1}^{+1} P_1^1(x) P_3^1(x)\, dx = \frac{3}{2} \int_{-1}^{+1} (1-x^2)(5x^2 - 1)\, dx$$

The integrand is an even function of x in the interval $-1 \le x \le 1$. Therefore (Section 5.3, equation (5.25))

$$I = 3 \int_0^{+1} (1-x^2)(5x^2-1)\,dx = 3 \int_0^{+1} (-5x^4 + 6x^2 - 1)\,dx$$

$$= 3\left[-x^5 + 2x^3 - x \right]_0^1 = 0$$

(ii) $P_2^2(x) = 3(1-x^2)$, $P_3^2(x) = 15x(1-x^2)$

$$I = \int_{-1}^{+1} P_2^2(x) P_3^2(x)\,dx = 45 \int_{-1}^{+1} x(1-x^2)^2\,dx$$

The integrand is an odd function of x in the interval $-1 \le x \le 1$, and $I = 0$ (Section 5.3, equation (5.26))

▸ Exercise 20

13.5   The Hermite equation

The Hermite equation is

$$y'' - 2xy' + 2ny = 0 \qquad (13.30)$$

where n is a real number. The equation arises in the solution of the Schrödinger equation for the harmonic oscillator. It is solved by the power-series method to give a general solution

$$y(x) = a_0 y_1(x) + a_1 y_2(x)$$

in which the particular solution y_1 contains only even powers of x and y_2 only odd powers of x. As in the case of the Legendre equation, y_1 reduces to a polynomial of degree n when (positive) n is an even integer, and similarly for y_2 when n is an odd integer. These polynomials are called **Hermite polynomials** $H_n(x)$, and they are those solutions of the Hermite equation that are of interest in the physical sciences:

$$H_n(x) = (2x)^n - \frac{n(n-1)}{1!}(2x)^{n-2} + \frac{n(n-1)(n-2)(n-3)}{2!}(2x)^{n-4} - \cdots \qquad (13.31)$$

for $n = 0, 1, 2, 3, \ldots$ The first few of these polynomials are

$$H_0(x) = 1 \qquad\qquad H_1(x) = 2x$$

$$H_2(x) = 4x^2 - 2 \qquad H_3(x) = 8x^3 - 12x \qquad\qquad (13.32)$$

The Hermite polynomials satisfy the recurrence relation

$$H_{n+1}(x) - 2xH_n(x) + 2nH_{n-1}(x) = 0 \tag{13.33}$$

so that, given $H_0 = 1$ and $H_1 = 2x$, all higher polynomials can be derived.

EXAMPLE 13.10

(i) Use the series expansion (13.31) to find $H_4(x)$.

$$H_4(x) = (2x)^4 - \frac{4 \times 3}{1!}(2x)^2 + \frac{4 \times 3 \times 2}{2!}(2x)^0 = 16x^4 - 48x^2 + 12$$

(ii) Verify by substitution in (13.30) that $H_4(x)$ is a solution of the Hermite equation.

$$H_4'(x) = 64x^3 - 96x, \qquad H_4''(x) = 192x^2 - 96$$

Therefore, for $n = 4$, the Hermite equation is

$$y'' - 2xy' + 8y = \left[192x^2 - 96\right] - 2x\left[64x^3 - 96x\right] + 8\left[16x^4 - 48x^2 + 12\right] = 0$$

(iii) Use the recurrence relation (13.33) to find $H_5(x)$.

$$H_5 = 2xH_4 - 8H_3 = 2x[16x^4 - 48x^2 + 12] - 8[8x^3 - 12x] = 32x^5 - 160x^3 + 120x$$

▶ Exercise 21

Hermite functions
A differential equation that is related to the Hermite equation is

$$y'' + (1 - x^2 + 2n)y = 0 \tag{13.34}$$

and this is solved by making the substitution

$$y(x) = e^{-x^2/2}u(x) \tag{13.35}$$

with second derivative

$$y'' = e^{-x^2/2}[u'' - 2xu' - (1 - x^2)u]$$

Substitution of y and its second derivative in (13.34) gives

$$e^{-x^2/2}[u'' - 2xu' + 2nu] = 0$$

The expression in square brackets is the left side of the Hermite equation (13.30), so that particular solutions of (13.34) are the **Hermite functions**

$$y_n(x) = e^{-x^2/2} H_n(x), \qquad n = 0, 1, 2, 3, \dots \qquad (13.36)$$

These functions, finite for all values of x, are orthogonal, with property

$$\int_{-\infty}^{+\infty} y_m(x) y_n(x)\, dx = \int_{-\infty}^{+\infty} e^{-x^2} H_m(x) H_n(x)\, dx = 2^n\, n!\sqrt{\pi}\, \delta_{m,n} \qquad (13.37)$$

The Hermite functions are of interest in chemistry because they are the eigenfunctions of the quantum-mechanical problem of the harmonic oscillator. The graphs of the first three functions are shown in Figure 13.1.

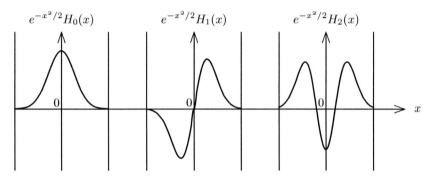

Figure 13.1

▶ Exercise 22

EXAMPLE 13.11 Show that the Hermite function $e^{-\alpha x^2/2} H_n(\sqrt{\alpha}x)$, where $\alpha = \sqrt{km/\hbar^2}$, is a solution of the Schrödinger equation for the harmonic oscillator

$$-\frac{\hbar^2}{2m}\frac{d^2\psi}{dx^2} + \frac{1}{2}kx^2\psi = E\psi$$

Let $z = \sqrt{\alpha}x$. Then, for $\psi = \psi(\sqrt{\alpha}x) = \psi(z)$,

$$\frac{d\psi}{dx} = \frac{d\psi}{dz}\frac{dz}{dx} = \sqrt{\alpha}\frac{d\psi}{dz} = \sqrt{\alpha}\psi', \qquad \frac{d^2\psi}{dx^2} = \alpha\psi''$$

and

$$-\frac{\hbar^2}{2m}\alpha\psi'' + \frac{k}{2\alpha}z^2\psi = E\psi$$

By dividing throughout by $-\hbar^2\alpha/2m$ and noting that $\alpha = m\omega/\hbar$, where $\omega = \sqrt{k/m}$ is the angular frequency of the oscillator, we obtain

$$\psi'' + (1 - z^2 + 2n)\psi = 0 \qquad \text{if} \qquad E = \left(n + \frac{1}{2}\right)\hbar\omega, \qquad n = 0, 1, 2, 3, \ldots$$

This differential equation is identical to equation (13.34). The Hermite functions are therefore the eigenfunctions of the quantum-mechanical harmonic oscillator problem.

13.6 The Laguerre equation

The Laguerre equation[4]

$$xy'' + (1 - x)y' + ny = 0 \tag{13.38}$$

where n is a real number, has a power-series solution that, when n is a positive integer or zero, is a polynomial of degree n called **a Laguerre polynomial** $L_n(x)$:

$$L_n(x) = (-1)^n \left[x^n - \frac{n^2}{1!}x^{n-1} + \frac{n^2(n-1)^2}{2!}x^{n-2} - \cdots + (-1)^n n! \right] \tag{13.39}$$

for $n = 0, 1, 2, 3, \ldots$ The first few of these are

$$L_0(x) = 1 \qquad\qquad L_1(x) = 1 - x$$
$$L_2(x) = 2 - 4x + x^2 \qquad L_3(x) = 6 - 18x + 9x^2 - x^3 \tag{13.40}$$

The Laguerre polynomials satisfy the recurrence relation

$$L_{n+1}(x) - (1 + 2n - x)L_n(x) + n^2 L_{n-1}(x) = 0 \tag{13.41}$$

from which, given L_0 and L_1, all higher polynomials can be found.

▸ Exercises 23, 24

Associated Laguerre functions

The associated Laguerre equation is

$$xy'' + (m + 1 - x)y' + (n - m)y = 0 \tag{13.42}$$

and has polynomial solution when both n and m are positive integers or zero, with $m \le n$. These solutions are the **associated Laguerre polynomials** $L_n^m(x)$. They are related to the Laguerre polynomials by the differential formula

$$L_n^m(x) = \frac{d^m}{dx^m} L_n(x) \tag{13.43}$$

---

[4] Edmond Laguerre (1834–1886).

The associated Laguerre polynomials arise in the solution of the radial part of the Schrödinger equation for the hydrogen atom (Section 14.6), and they occur there in the form of **associated Laguerre functions**

$$f_{n,l}(x) = e^{-x/2} x^l L_{n+l}^{2l+1}(x) \tag{13.44}$$

for $n = 0, 1, 2, 3, \ldots$, $l = 0, 1, 2, \ldots, (n-1)$. These functions satisfy the differential equation

$$f'' + \frac{2}{x} f' + \left(\frac{n}{x} - \frac{l(l+1)}{x^2} - \frac{1}{4} \right) f = 0 \tag{13.45}$$

and they are orthogonal with respect to the **weight function** x^2 in the interval $0 \le x \le \infty$:

$$\int_0^\infty f_{n,l}(x) f_{n',l}(x)\, x^2 dx = \int_0^\infty e^{-x} x^{2l} L_{n+l}^{2l+1}(x) L_{n'+l}^{2l+1}(x)\, x^2 dx \tag{13.46}$$

$$= \frac{2n\left[(n+l) \right]^3}{(n-l-1)!} \delta_{n,n'}$$

13.7 Bessel functions

The Bessel equation is[5]

$$x^2 y'' + xy' + (x^2 - n^2) y = 0 \tag{13.47}$$

where n is a real number. This equation ranks with the Legendre equation in its importance in the physical sciences, although it is met less frequently in chemistry than in physics and engineering. Bessel functions are involved, for example, in the solution of the classical wave equation for the vibrations of circular and spherical membranes, and the same solutions are found for the Schrödinger equation for the particle in a circular box and in a spherical box. The functions are important in the formulation of the theory of scattering processes.

Equation (13.47), when divided by x^2, is of type (13.3) and is solved by the Frobenius method; that is, by expressing the solution in the form (13.4)

$$y(x) = x^r (a_0 + a_1 x + a_2 x^2 + \cdots)$$

---

[5] 'At this junction
It is time to wrestle
With a well-known function
Due to Herr Bessel'.

Friedrich Wilhelm Bessel (1784–1864), German astronomer. Examples of Bessel functions were discussed by Daniel Bernoulli, Euler, and Lagrange, but the first systematic study appeared in 1824 in a paper by Bessel on perturbations of planetary orbits. Bessel is known as the recipient of numerous letters from his close friend Gauss. In 1810 Gauss wrote: 'This winter I am giving two courses of lectures to three students, of whom one is only moderately prepared, the other is less than moderately, and the third lacks both preparation and ability. Such are the burdens …'

The solutions of the indicial equation (13.8) are $r = \pm n$, and the two particular solutions for these values of r are

$$y_1(x) = a_0 x^n \left(1 - \frac{x^2}{2(2n+2)} + \frac{x^4}{2 \cdot 4(2n+2)(2n+4)} - \cdots \right) \qquad (13.48)$$

$$y_2(x) = a_0 x^{-n} \left(1 + \frac{x^2}{2(2n-2)} + \frac{x^4}{2 \cdot 4(2n-2)(2n-4)} + \cdots \right) \qquad (13.49)$$

Important values of the parameter n are the integer and half-integer values.

Bessel functions $J_n(x)$ for integer n

When n is a positive integer or zero, the particular solution (13.48) is

$$J_n(x) = \left(\frac{x}{2} \right)^n \sum_{m=0}^{\infty} \frac{(-1)^m}{m!(n+m)!} \left(\frac{x}{2} \right)^{2m} \qquad (13.50)$$

in which the constant a_0 has been given its conventional value of $a_0 = 1/(2^n n!)$. This is the **Bessel function of the first kind of order n**. The function is finite and converges for all values of x; it converges very fast because of the pair of factorials in the denominator. The functions for $n = 0$ and $n = 1$ are

$$J_0(x) = 1 - \frac{1}{(1!)^2} \left(\frac{x}{2} \right)^2 + \frac{1}{(2!)^2} \left(\frac{x}{2} \right)^4 - \frac{1}{(3!)^2} \left(\frac{x}{2} \right)^6 + \cdots \qquad (13.51)$$

$$J_1(x) = \frac{x}{2} - \frac{1}{1!2!} \left(\frac{x}{2} \right)^3 + \frac{1}{2!3!} \left(\frac{x}{2} \right)^5 - \frac{1}{3!4!} \left(\frac{x}{2} \right)^7 + \cdots \qquad (13.52)$$

and their graphs are shown in Figure 13.2.

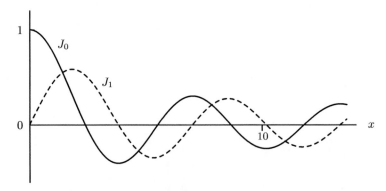

Figure 13.2

Both the expansions and the graphs show that the Bessel functions have properties similar to the trigonometric functions. However, the values of x for which $J_n(x) = 0$, the zeros of the functions, are not equally spaced so that the functions do not have a fixed wavelength. Some of the zeros are (to 3 decimal places)

$$J_0(x) = 0 \quad \text{for} \quad x = 2.405,\, 5.520,\, 8.654,\, 11.792,\, 14.931,\, \ldots \tag{13.53}$$

$$J_1(x) = 0 \quad \text{for} \quad x = 0,\, 3.832,\, 7.016,\, 10.173,\, 13.324,\, \ldots$$

In addition, the amplitude of the waves decreases as x increases. The behaviour for large values of x is that of a damped sine function,

$$J_n(x) \sim \sqrt{\frac{2}{\pi x}} \sin\left(x - \frac{n\pi}{2} + \frac{\pi}{4}\right), \qquad \text{for } x \text{ large} \tag{13.54}$$

For positive values of n the particular solution (13.49) is $J_{-n}(x)$ and it can be shown that this is related to $J_n(x)$ by $J_{-n}(x) = (-1)^n J_n(x)$. The power-series method therefore gives only one independent solution for integral values of n. A second solution can be found; this is the **Bessel function of the second kind**, $Y_n(x)$, of less importance for physical applications.

▶ Exercise 25

Bessel functions $J_{l+1/2}(x)$ of half-integer order

Bessel functions of half-integral order can be expressed in terms of elementary functions. The functions for $n = \pm 1/2$ are (Example 13.4)

$$J_{1/2}(x) = \sqrt{\frac{2}{\pi x}}\, \sin x, \qquad J_{-1/2}(x) = \sqrt{\frac{2}{\pi x}}\, \cos x \tag{13.55}$$

and all others can be obtained by means of the recurrence relation (true for Bessel functions in general)

$$J_{n+1}(x) - \frac{2n}{x} J_n(x) + J_{n-1}(x) = 0 \tag{13.56}$$

EXAMPLE 13.12 Use the recurrence relation (13.56) and the formulas (13.55) to derive the Bessel functions $J_{\pm 3/2}(x)$.

We have, from (13.56) with $n = 1/2$,

$$J_{3/2}(x) - \frac{1}{x} J_{1/2}(x) + J_{-1/2} = 0$$

so that

$$J_{3/2}(x) = \frac{1}{x}J_{1/2}(x) - J_{-1/2}(x) = \sqrt{\frac{2}{\pi x}}\left(\frac{\sin x}{x} - \cos x\right)$$

Similarly, with $n = -1/2$ in (13.56),

$$J_{-3/2}(x) = -\frac{1}{x}J_{-1/2}(x) - J_{1/2}(x) = -\sqrt{\frac{2}{\pi x}}\left(\frac{\cos x}{x} + \sin x\right)$$

▶ Exercise 26

Series expansions for $J_{l+1/2}(x)$ and $J_{-l-1/2}(x)$ are given by equations (13.48) and (13.49), respectively. Thus

$$J_{l+1/2}(x) = \sqrt{\frac{2}{\pi}}x^{l+1/2}\left(1 - \frac{x^2}{2(2l+3)} + \frac{x^4}{2\cdot4(2l+3)(2l+5)} - \cdots\right) \qquad (13.57)$$

$$J_{-l-1/2}(x) = \sqrt{\frac{2}{\pi}}x^{-l-1/2}\left(1 + \frac{x^2}{2(2l-1)} + \frac{x^4}{2\cdot4(2l-1)(2l-3)} + \cdots\right) \qquad (13.58)$$

EXAMPLE 13.13 For $l = 0$,

$$J_{1/2}(x) = \sqrt{\frac{2}{\pi}}x^{1/2}\left(1 - \frac{x^2}{3!} + \frac{x^4}{5!} - \cdots\right) = \sqrt{\frac{2}{\pi x}}\left(x - \frac{x^3}{3!} + \frac{x^5}{5!} - \cdots\right) = \sqrt{\frac{2}{\pi x}}\sin x$$

$$J_{-1/2}(x) = \sqrt{\frac{2}{\pi x}}\left(1 - \frac{x^2}{2!} + \frac{x^4}{4!} - \cdots\right) = \sqrt{\frac{2}{\pi x}}\cos x$$

Equations (13.57) and (13.58) show that whereas $J_{l+1/2}(x) = 0$ at the origin $(x = 0)$ and is therefore finite for all values of x, $J_{-l-1/2}(x) \to \infty$ as $x \to 0$.

The functions of half-integral order are the Bessel functions that occur in the partial wave method in the theory of scattering processes. They occur there in the forms, for $l \geq 0$,

$$j_l(x) = \sqrt{\frac{\pi}{2x}}J_{l+1/2}(x), \qquad \eta_l(x) = (-1)^{l+1}\sqrt{\frac{\pi}{2x}}J_{-l-1/2}(x) \qquad (13.59)$$

The functions $j_l(x)$ are called **spherical Bessel functions of order *l*;** the functions η_l **are the spherical Neumann functions.** They satisfy the differential equation

$$x^2 y'' + 2xy' + [x^2 - l(l+1)]y = 0 \qquad (13.60)$$

These functions often occur in conjunction with the Legendre polynomials when a physical system is formulated in spherical polar coordinates.

▸ Exercise 27

13.8 Exercises

Section 13.2

Use the power-series method to solve the equations:

1. $y' - 3x^2 y = 0$
2. $(1-x)y' - y = 0$. Confirm the solution can be expressed as $y = a/(1-x)$ when $|x| < 1$.
3. $y'' - 9y = 0$. Confirm that the solution can be expressed as $y = ae^{3x} + be^{-3x}$.
4. $(1-x^2)y'' - 2xy' + 2y = 0$ (the Legendre equation (13.13) for $l = 1$).

    Show that the solution can be written as $y = a_1 x + a_0 \left[1 + \dfrac{x}{2} \ln\left(\dfrac{1-x}{1+x}\right) \right]$ when $|x| < 1$.

5. $y'' - xy = 0$ (Airy equation).

Section 13.3

For each of the following, find and solve the indicial equation

6. $x^2 y'' + 3xy' + y = 0$ 7. $x^2 y'' + xy' + (x^2 - n^2)y = 0$ (Bessel equation)
8. $xy'' + (1 - 2x)y' + (x - 1)y = 0$ 9. $x^2 y'' + 6xy' + (6 - x^2)y = 0$
10. **(i)** Find the general solution of the Euler–Cauchy equation $x^2 y'' + b_0 xy' + c_0 y = 0$ for distinct indicial roots, $r_1 \neq r_2$. **(ii)** Show that for a double initial root r, the general solution is $y = (a + b \ln x)x^r$.

Solve the differential equations:

11. $x^2 y'' - \frac{1}{2} xy' + \frac{1}{2} y = 0$ 12. $x^2 y'' - xy' + y = 0$
13. **(i)** Solve the Bessel equation $x^2 y'' + xy' + (x^2 - \frac{1}{4})y = 0$ for indicial root $r = -1/2$ (see Example 13.4 for $r = 1/2$). **(ii)** Confirm that the solution can be written as

    $$y(x) = \dfrac{a_0}{\sqrt{x}} \cos x + \dfrac{a_1}{\sqrt{x}} \sin x = a J_{-1/2}(x) + b J_{1/2}(x)$$

14. **(i)** Use the expansion method to find a particular solution $y_1(x)$ of

    $$xy'' + (1 - 2x)y' + (x - 1)y = 0.$$

    (ii) confirm that $y_2(x) = y_1(x) \ln x$ is a second solution.
15. Find the general solution of $xy'' + 2y' + 4xy = 0$. Assume that there is no logarithmic term in the solution.

Section 13.4

16. Show that the polynomial $P_l(x)$ is a solution of the Legendre equation (13.13) for
    (i) $l = 2$ and **(ii)** $l = 5$.

17. Find the Legendre polynomial $P_6(x)$ **(i)** by means of the recurrence relation (13.21), **(ii)** from the general expression (13.19) for $P_l(x)$.

18. Use the formula (13.24) to find the associated Legendre functions **(i)** P_1^1, **(ii)** $P_4^m(x)$ for $m = 1, 2, 3, 4$. Express the functions in terms of $\cos\theta = x$ and $\sin\theta = (1 - x^2)^{1/2}$.

19. Show that **(i)** P_1 is orthogonal to P_4 and P_5, **(ii)** P_2 is orthogonal to P_0 and P_3.

20. Show that P_2^1 is orthogonal to P_1^1 and P_4^1.

Section 13.5

21. **(i)** Use the series expansion (13.31) to find $H_5(x)$. **(ii)** Verify by substitution in (13.30) that $H_5(x)$ is a solution of the Hermite equation. **(iii)** Use the recurrence relation (13.33) to find $H_6(x)$.

22. Sketch the graph of the Hermite function $e^{-x^2/2}H_3(x)$.

Section 13.6

23. **(i)** Use the power series method to find a solution of the Laguerre equation (13.38). **(ii)** Show that this solution reduces to the polynomial (13.39) when n is a positive integer or zero and when the arbitrary constant is given its conventional value $n!$.

24. Find $L_4(x)$ **(i)** from equation (13.39), **(ii)** from $L_2(x)$ and $L_3(x)$ by means of the recurrence relation (13.41).

Section 13.7

25. **(i)** Find the Bessel function $J_2(x)$ **(i)** from the series expansion (13.50); **(ii)** from $J_0(x)$ and $J_1(x)$ by means of the recurrence relation (13.56).

26. Use the recurrence relation (13.56) to find $J_{5/2}(x)$ and $J_{-5/2}(x)$.

27. Confirm that the spherical Bessel function $j_l(x)$ satisfies equation (13.60).

14 Partial differential equations

14.1 Concepts

An equation that contains partial derivatives is a **partial differential equation**. For example, if f is a function of the independent variables x and y then an equation that contains one or more of $\partial f/\partial x$, $\partial f/\partial y$ and higher partial derivatives, as well as f, x and y, is a partial differential equation. Examples of such equations that are important in the physical sciences are

1. $\dfrac{\partial^2 f}{\partial x^2} = \dfrac{1}{v^2}\dfrac{\partial^2 f}{\partial t^2}$ 1-dimensional wave equation

2. $\dfrac{\partial^2 f}{\partial x^2} = \dfrac{1}{D}\dfrac{\partial f}{\partial t}$ 1-dimensional diffusion equation

3. $\dfrac{\partial^2 f}{\partial x^2} + \dfrac{\partial^2 f}{\partial y^2} + \dfrac{\partial^2 f}{\partial z^2} = \nabla^2 f = 0$ 3-dimensional Laplace equation

4. $\nabla^2 f = g(x, y, z)$ 3-dimensional Poisson equation

5. $-\dfrac{\hbar^2}{2m}\nabla^2 \psi + V(x, y, z)\psi = E\psi$ time-independent Schrödinger equation

6. $-\dfrac{\hbar^2}{2m}\nabla^2 \psi + V(x, y, z, t)\psi = i\hbar\dfrac{\partial\psi}{\partial t}$ time-dependent Schrödinger equation

These are all second-order linear equations. Equation **4** is inhomogeneous, the others are homogeneous equations. In equations **1** and **2**, the unknown function f is a function of the coordinate x and of the time t; $f = f(x, t)$. In **3**, **4** and **5**, the function depends on the three coordinates of a point in ordinary space, and in **6** it is also a function of the time.

As for ordinary differential equations, there are several important standard types of partial differential equation that frequently occur in mathematical models of physical systems, and whose solutions can be expressed in terms of known functions. The equations discussed in this chapter are equation **5** above for the particle in a rectangular box (Section 14.4) and in a circular box (Section 14.5), and for the hydrogen atom (Section 14.6), and equation **1** as applied to the vibrations of an elastic string, such as a guitar string (Section 14.7). These examples demonstrate a number of important principles in the solution of partial differential equations. They show that different boundary and initial conditions can lead to very different types of particular solution, that the symmetry properties of the system being modelled can lead to the phenomenon of 'degeneracy', and that, in some cases, the solutions are expressed in the form of the 'orthogonal expansions' discussed in Chapter 15.

14.2   General solutions

We have seen that the general solution of an ordinary differential equation contains a number of arbitrary constants, usually n for an nth-order equation, and that, in an application, the values of these constants are obtained by the imposition of appropriate initial or boundary conditions. The general solution of a partial differential equation, on the other hand, contains a number of *arbitrary functions*, often n such functions for an nth-order equation. For example, the 1-dimensional wave equation, equation 1 in the list given in Section 14.1,

$$\frac{\partial^2 f}{\partial x^2} = \frac{1}{v^2}\frac{\partial^2 f}{\partial t^2} \tag{14.1}$$

has solution

$$f(x, t) = F(x + vt) \tag{14.2}$$

where F is an arbitrary function of the variable $u = x + vt$. Thus,

$$\frac{\partial f}{\partial x} = \frac{dF}{du}\frac{\partial u}{\partial x} = \frac{dF}{du}, \qquad \frac{\partial^2 f}{\partial x^2} = \frac{d^2 F}{du^2}$$

$$\frac{\partial f}{\partial t} = \frac{dF}{du}\frac{\partial u}{\partial t} = v\frac{dF}{du}, \qquad \frac{\partial^2 f}{\partial t^2} = v^2\frac{d^2 F}{du^2}$$

and equation (14.1) is satisfied. The general solution of the equation is

$$f(x, t) = F(x + vt) + G(x - vt) \tag{14.3}$$

where F and G are both arbitrary functions.[1] The particular functions in any application are determined by appropriate initial and boundary conditions, and one important example is that discussed in Section 14.7.

EXAMPLE 14.1 Verify that the function

$$f(x, t) = 3x^2 - 2xvt + 3v^2t^2$$

is a solution of the wave equation (14.1), and has the general form (14.3).

---

[1] This form of the general solution was first given by d'Alembert in 1747. The origins of the wave equation lie in the discussion *De motu nervi tensi* (On the motion of a tense string) by Brook Taylor in 1713. In 1727 Johann Bernoulli suggested to his son Daniel that he take up Taylor's problem again: 'Of a musical string, of given length and weight, stretched by a given weight, to find its vibrations'. The equation was derived by d'Alembert in *Recherches sur la courbe que forme une corde tendue mise en vibration*, 1747, by considering the string to be composed of infinitesimal masses and applying Newton's force law to each element of mass. Euler published his own solution in *Sur la vibration des cordes* in 1750, and Daniel Bernoulli explored the idea of the superposition of normal modes in his *Réflexions et éclaircissements* in 1755. The debate over the kinds of functions acceptable as solutions of the wave equation was continued for about thirty years by these three, without any one of them being convinced by the others. The problem was resolved through the work of Fourier, Dirichlet, Riemann, and Weierstrass over the next hundred years, and involved a reconsideration of the meaning of function, continuity, and convergence.

The partial derivatives are

$$\frac{\partial f}{\partial x} = 6x - 2vt, \quad \frac{\partial^2 f}{\partial x^2} = 6; \quad \frac{\partial f}{\partial t} = -2xv + 6v^2 t, \quad \frac{\partial^2 f}{\partial t^2} = 6v^2$$

Therefore,

$$6 = \frac{\partial^2 f}{\partial x^2} = \frac{1}{v^2}\frac{\partial^2 f}{\partial t^2}$$

as required. The function can be written as

$$f(x, t) = (x + vt)^2 + 2(x - vt)^2 = F(x + vt) + G(x - vt)$$

➤ Exercises 1, 2

EXAMPLE 14.2 Verify that the function

$$f(x, t) = a \exp\left[-b(x - vt)^2\right]$$

is a solution of the wave equation (14.1).
   We have

$$\frac{\partial^2 f}{\partial x^2} = 2ab\left[-1 + 2b(x - vt)^2\right]\exp\left[-b(x - vt)^2\right]$$

$$\frac{\partial^2 f}{\partial t^2} = 2abv^2\left[-1 + 2b(x - vt)^2\right]\exp\left[-b(x - vt)^2\right]$$

Therefore $\dfrac{\partial^2 f}{\partial x^2} = \dfrac{1}{v^2}\dfrac{\partial^2 f}{\partial t^2}$.

➤ Exercise 3

14.3 Separation of variables

When a partial differential equation, involving two or more independent variables, can be reduced to a set of ordinary differential equations, one for each variable, the equation is called **separable**. The solutions of the partial differential equation are then products of the solutions of the ordinary equations. All the examples discussed in this chapter are of this type.

   We demonstrate the essential principles of the **method of separation of variables** by considering the simplest first-order equation in two variables

$$\frac{\partial f}{\partial x} + \frac{\partial f}{\partial y} = 0 \qquad (14.4)$$

We show that a solution of this equation can be written as

$$f(x, y) = X(x) \times Y(y) \qquad (14.5)$$

in which the solution, a function of the two variables x and y, is expressed as the product of a function of x only and a function of y only. We have

$$\frac{\partial f}{\partial x} = \frac{\partial (XY)}{\partial x} = Y \frac{dX}{dx}$$

because $Y(y)$ does not depend on x. Similarly,

$$\frac{\partial f}{\partial y} = X \frac{dY}{dy}$$

because $X(x)$ is a constant with respect to y. Substitution of $f = XY$ and its derivatives into the differential equation (14.4) then gives

$$Y(y) \frac{dX(x)}{dx} + X(x) \frac{dY(y)}{dy} = 0 \qquad (14.6)$$

and, dividing throughout by $f = X(x) \times Y(y)$,

$$\left[\frac{1}{X(x)} \frac{dX(x)}{dx} \right] + \left[\frac{1}{Y(y)} \frac{dY(y)}{dy} \right] = 0 \qquad (14.7)$$

The first set of terms in square brackets on the left side of (14.7) depends only on the variable x, the second set only on y. If the total is constant (zero here) it follows that each of these sets of terms must be separately constant if x and y are independent variables. Thus a change in the value of x cannot lead to a change in the value of the terms that depend on y only, and vice-versa. Therefore, if the first set of terms equals the constant C then the second set is equal to $-C$ (for the total to be zero):

$$\frac{1}{X} \frac{dX}{dx} = C, \qquad \frac{1}{Y} \frac{dY}{dy} = -C \qquad (14.8)$$

and C is called a **separation constant**. Then

$$\frac{dX}{dx} = CX \qquad (14.9)$$

$$\frac{dY}{dy} = -CY \qquad (14.10)$$

and the partial differential equation (14.4) in two variables has been reduced to two ordinary differential equations, both of which we can solve. The equations, (14.9) in the variable x and (14.10) in the variable y, are separable first-order equations of the kind discussed in Section 11.3, and they have general solutions

$$X(x) = Ae^{Cx}, \qquad Y(y) = Be^{-Cy} \qquad (14.11)$$

A solution of the equation (14.4) in two variables is then the product

$$f(x, y) = X(x) \times Y(y) = Ae^{Cx} \times Be^{-Cy} = De^{C(x-y)} \qquad (14.12)$$

> Exercises 4–7

Particular solutions, including the possible values of the separation constant, can often be obtained by the application of initial and boundary conditions, as exemplified by the important problems discussed in Sections 14.4 to 14.6. In other cases, as discussed in Section 14.7 for the vibrating string, it may be necessary to consider more general solutions that are linear combinations of products.

14.4 The particle in a rectangular box

The Schrödinger equation for a particle of mass m moving in the xy-plane is

$$-\frac{\hbar^2}{2m}\nabla^2\psi(x, y) + V(x, y)\psi(x, y) = E\psi(x, y) \qquad (14.13)$$

where

$$\nabla^2 = \frac{\partial^2}{\partial x^2} + \frac{\partial^2}{\partial y^2} \qquad (14.14)$$

is the two-dimensional Laplacian operator (see Section 9.6). For the present system, the potential energy function is (Figure 14.1)

$$V(x, y) = \begin{cases} 0 & \text{for } 0 < x < a \text{ and } 0 < y < b \\ \infty & \text{elsewhere} \end{cases} \qquad (14.15)$$

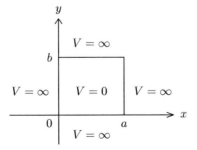

Figure 14.1

As for the one-dimensional case, the constant value of V inside the box ensures that no force acts on the particle in this region; setting $V=0$ means that E is the kinetic energy of the particle. The infinite value of V at the 'walls' and outside the box ensures that the particle cannot leave the box; $\psi=0$ at the walls and outside the box.

For the particle within the box, we have the boundary value problem

$$-\frac{\hbar^2}{2m}\left(\frac{\partial^2\psi}{\partial x^2}+\frac{\partial^2\psi}{\partial y^2}\right)=E\psi \tag{14.16}$$

with conditions

$$\begin{array}{ll}\psi(0,y)=\psi(a,y)=0 & (\psi=0 \text{ when } x=0 \text{ or } x=a)\\[6pt]\psi(x,0)=\psi(x,b)=0 & (\psi=0 \text{ when } y=0 \text{ or } y=b)\end{array} \tag{14.17}$$

The partial differential equation is reduced to two ordinary differential equations by writing the wave function as the product

$$\psi(x,y)=X(x)\times Y(y) \tag{14.18}$$

Then, as in Section 14.3, substitution of this product and its derivatives into (14.13), and division by $\psi=XY$, gives

$$\frac{1}{X}\frac{d^2 X}{dx^2}+\frac{1}{Y}\frac{d^2 Y}{dy^2}=-\frac{2mE}{\hbar^2} \tag{14.19}$$

The term involving x only must be constant and the term involving y only must also be constant. Therefore,

$$\frac{1}{X}\frac{d^2 X}{dx^2}=C_x, \qquad \frac{1}{Y}\frac{d^2 Y}{dy^2}=C_y \tag{14.20}$$

where C_x and C_y are constants, and $C_x+C_y=-2mE/\hbar^2$.

The two-dimensional boundary value problem has been reduced to two one-dimensional boundary value problems:

$$\frac{d^2 X}{dx^2}=C_x X, \qquad X(0)=X(a)=0 \tag{14.21a}$$

$$\frac{d^2 Y}{dy^2}=C_y Y, \qquad Y(0)=Y(b)=0 \tag{14.21b}$$

Both of these are the same problem as that of the particle in a one-dimensional box discussed in Section 12.6. Equation (14.21a) describes the motion of the particle along the x-direction, and has normalized solutions given by (12.53), with appropriate changes of symbols:

$$X_p(x)=\sqrt{\frac{2}{a}}\sin\left(\frac{p\pi x}{a}\right), \qquad p=1,2,3,\ldots \tag{14.22a}$$

and $C_x = -p^2\pi^2/a^2$. Similarly, (14.21b) describes the motion of the particle along the y-direction, and has normalized solutions

$$Y_q(y) = \sqrt{\frac{2}{b}}\sin\left(\frac{q\pi y}{b}\right), \qquad q = 1, 2, 3, \tag{14.22b}$$

and $C_y = -q^2\pi^2/b^2$. The total solutions (eigenfunctions) of the two-dimensional Schrödinger equation (14.13) are the products

$$\psi_{p,q}(x, y) = X_p(x) \times Y_q(y)$$

$$= \sqrt{\frac{2}{a}}\sin\left(\frac{p\pi x}{a}\right) \times \sqrt{\frac{2}{b}}\sin\left(\frac{q\pi y}{b}\right), \qquad p, q = 1, 2, 3, \tag{14.23}$$

and the corresponding total energies (eigenvalues) are

$$E_{p,q} = \frac{h^2}{8m}\left(\frac{p^2}{a^2} + \frac{q^2}{b^2}\right) \tag{14.24}$$

We note that the quantities

$$E_p = \frac{h^2 p^2}{8ma^2} \qquad \text{and} \qquad E_q = \frac{h^2 q^2}{8mb^2} \tag{14.25}$$

are to be identified with the kinetic energies of motion along the x and y directions, respectively.

A square box. Degeneracy

When the sides of the box are not equal and neither is an integer multiple of the other, the eigenvalues (14.24) are all distinct; the states of the system are then said to be **nondegenerate**. For a square box however, with $a = b$,

$$E_{p,q} = \frac{h^2}{8ma^2}(p^2 + q^2) \tag{14.26}$$

and the eigenvalues for $p \neq q$ occur in pairs with $E_{p,q} = E_{q,p}$; for example, $E_{1,2} = E_{2,1} = 5h^2/8ma^2$. States of the system with equal energies are called **degenerate states**.

The occurrence of degeneracy for the square box is a consequence of the symmetry of the system. The eigenfunctions (14.23) are

$$\psi_{p,q}(x, y) = \frac{2}{a}\sin\left(\frac{p\pi x}{a}\right)\sin\left(\frac{q\pi y}{a}\right) \tag{14.27}$$

and an interchange of the x and y coordinate axes gives

$$\psi_{p,q}(y, x) = \frac{2}{a}\sin\left(\frac{p\pi y}{a}\right)\sin\left(\frac{q\pi x}{a}\right) = \psi_{q,p}(x, y) \tag{14.28}$$

that is, the degenerate eigenfunctions are interchanged (when $p \neq q$).* The symmetries of some of the wave functions are illustrated in Figure 14.2.

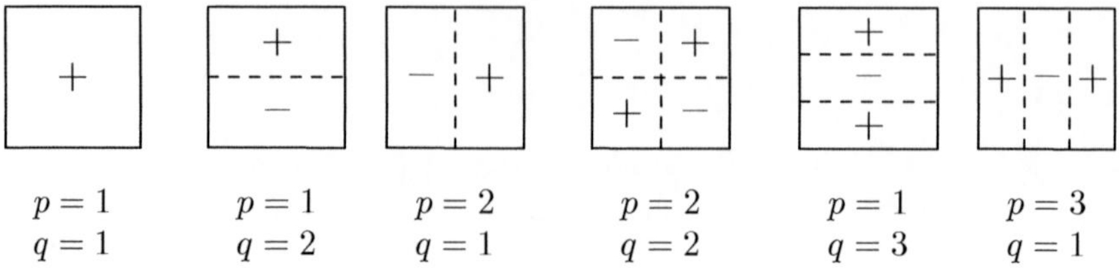

$$p = 1 \qquad p = 1 \qquad p = 2 \qquad p = 2 \qquad p = 1 \qquad p = 3$$
$$q = 1 \qquad q = 2 \qquad q = 1 \qquad q = 2 \qquad q = 3 \qquad q = 1$$

Figure 14.2

The dashed lines are **nodal lines**, where the function is zero. We see that, for example, $\psi_{1,2}$ and $\psi_{2,1}$ are identical except for orientation.

▸ Exercises 8–10

14.5 The particle in a circular box

The motion of a particle in a circular box of radius a is described by the Schrödinger equation

$$-\frac{\hbar^2}{2m}\nabla^2\psi(x, y) + V(x, y)\psi(x, y) = E\psi(x, y) \tag{14.29}$$

as in Section 14.4, but with potential-energy function

$$V(x, y) = \begin{cases} 0 & \text{for } r = \sqrt{x^2 + y^2} < a, \text{ (inside the box)} \\ \infty & \text{elsewhere} \end{cases} \tag{14.30}$$

where r is the distance from the origin at the centre of the box. The equation is not separable in cartesian coordinates because of the functional form of V at the boundary of the box, but it becomes so when the equation is expressed in the plane polar coordinates r and θ. In these coordinates, the two-dimensional Laplacian operator is (see equation (9.38) and Example 9.18)

$$\nabla^2 = \frac{\partial^2}{\partial r^2} + \frac{1}{r}\frac{\partial}{\partial r} + \frac{1}{r^2}\frac{\partial^2}{\partial \theta^2} \tag{14.31}$$

For the particle within the box, with $V = 0$, equation (14.29) is then

$$-\frac{\hbar^2}{2m}\left(\frac{\partial^2\psi}{\partial r^2} + \frac{1}{r}\frac{\partial\psi}{\partial r} + \frac{1}{r^2}\frac{\partial^2\psi}{\partial \theta^2}\right) = E\psi \tag{14.32}$$

---

* An 'accidental' degeneracy may also occur that is not an obvious consequence of symmetry. For example, the states $(7, 1)$ and $(1, 7)$ are degenerate with the state $(5, 5)$.

or, multiplying throughout by $-2mr^2/\hbar^2$, setting

$$\alpha^2 = \frac{2mE}{\hbar^2} \tag{14.33}$$

and rearranging,

$$r^2 \frac{\partial^2 \psi}{\partial r^2} + r \frac{\partial \psi}{\partial r} + \alpha^2 r^2 \psi + \frac{\partial^2 \psi}{\partial \theta^2} = 0 \tag{14.34}$$

This equation can now be reduced to two ordinary equations by writing the wave function as the product

$$\psi(r, \theta) = R(r) \times \Theta(\theta) \tag{14.35}$$

Substitution in (14.34) and division by $\psi = R\Theta$ then gives

$$\left[\frac{r^2}{R} \frac{d^2 R}{dr^2} + \frac{r}{R} \frac{dR}{dr} + \alpha^2 r^2 \right] + \left[\frac{1}{\Theta} \frac{d^2 \Theta}{d\theta^2} \right] = 0 \tag{14.36}$$

Each set of terms in square brackets must be constant so that, with separation constant C,

$$r^2 \frac{d^2 R}{dr^2} + r \frac{dR}{dr} + \alpha^2 r^2 R = CR \tag{14.37}$$

for the radial motion of the particle in the box, and

$$\frac{d^2 \Theta}{d\theta^2} = -C\Theta \tag{14.38}$$

for the angular motion of the particle. We consider the angular equation first.

The angular equation

The function $\Theta(\theta)$ is defined in the interval $0 \le \theta \le 2\pi$ and must satisfy the periodic boundary condition

$$\Theta(2\pi) = \Theta(0) \tag{14.39}$$

for continuity round the circle. We therefore have the same boundary value problem as that discussed in Section 12.7 for the particle in a ring. The normalized solutions are (equation (12.69))

$$\Theta_n(\theta) = \frac{1}{\sqrt{2\pi}} e^{in\theta}, \qquad n = 0, \pm1, \pm2, \ldots \tag{14.40}$$

and substitution in (14.38) gives the separation constant $C = n^2$.

The radial equation

With $C = n^2$, the radial equation (14.37) is

$$r^2 \frac{d^2 R}{dr^2} + r \frac{dR}{dr} + (\alpha^2 r^2 - n^2)R = 0 \qquad (14.41)$$

The equation is transformed into the Bessel equation (Section 13.7) by means of the change of variable $x = \alpha r$. Then

$$\frac{dR}{dr} = \frac{dR}{dx}\frac{dx}{dr} = \alpha \frac{dR}{dx}, \qquad \frac{d^2 R}{dr^2} = \alpha^2 \frac{d^2 R}{dx^2}$$

and (14.41) becomes the Bessel equation (13.47),

$$x^2 \frac{d^2 R}{dx^2} + x \frac{dR}{dx} + (x^2 - n^2)R = 0$$

When n is a positive integer or zero, the solution of this equation is the Bessel function $J_n(x)$ given by (13.50), so that the solutions of the radial equation are

$$R_n(r) = J_n(\alpha r), \qquad n = 0, 1, 2, 3, \dots \qquad (14.42)$$

These solutions are subject to the condition that the wave function vanish at the boundary of the box, when $r = a$. Therefore

$$R_n(a) = J_n(\alpha a) = 0 \qquad (14.43)$$

and the possible values of α are determined by the zeros of the Bessel function, examples of which are given in (13.53) of Section 13.7. If the zeros of $J_n(x)$ are labelled $x_{n,1}, x_{n,2}, x_{n,3}, \dots$, the allowed values of α are

$$\alpha_{n,k} = \frac{x_{n,k}}{a}, \qquad k = 1, 2, 3, \dots \qquad (14.44)$$

and the solutions of the radial equation that satisfy the boundary condition are

$$R_{n,k}(r) = J_n(\alpha_{n,k} r) \qquad (14.45)$$

By equation (14.33), the energy of the system is given by $E = \alpha^2 \hbar^2 / 2m$, so that the energy is quantized, with values

$$E_{n,k} = \frac{\alpha_{n,k}^2 \hbar^2}{2m} = \frac{x_{n,k}^2 \hbar^2}{2ma^2}, \qquad n = 0, \pm 1, \pm 2, \dots \qquad (14.46)$$

and the corresponding total wave functions are

$$\psi_{n,k}(r, \theta) = R_{|n|,k}(r)\Theta_n(\theta) \qquad (14.47)$$

(the radial function depends only on the modulus of n). The symmetries of these wave functions, using the real forms (12.71) of the angular functions, are illustrated in Figure 14.3.

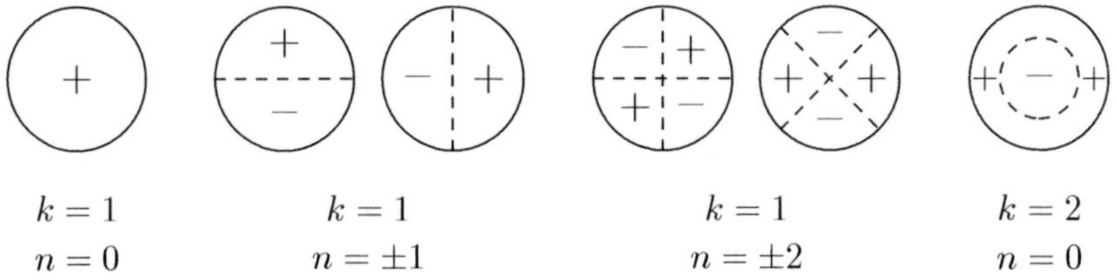

$$k = 1 \qquad\qquad k = 1 \qquad\qquad\qquad k = 1 \qquad\qquad k = 2$$
$$n = 0 \qquad\qquad n = \pm 1 \qquad\qquad\quad n = \pm 2 \qquad\qquad n = 0$$

Figure 14.3

We note that these diagrams also represent the normal modes of motion, the standing waves, of a circular membrane.

▸ Exercise 11

14.6   The hydrogen atom

A hydrogen-like atom is a system of two charges, a nucleus with charge $+Ze$ ($Z = 1$ for the hydrogen atom itself) and an electron with charge $-e$, interacting through a Coulomb force. The potential energy of the system is (see Examples 1.21 and 5.17)

$$V = \frac{-Ze^2}{4\pi\varepsilon_0 r} \tag{14.48}$$

where r is the distance between the charges. Let the nucleus be fixed at the origin of a coordinate system, with the electron at position (x, y, z). The Schrödinger equation for the motion of the electron about the stationary nucleus is then

$$-\frac{\hbar^2}{2m_e}\nabla^2\psi + V\psi = E\psi \tag{14.49}$$

where ∇^2 is the three-dimensional Laplacian and, in terms of cartesian coordinates, the potential energy function is

$$V(x, y, z) = \frac{-Ze^2}{4\pi\varepsilon_0(x^2 + y^2 + z^2)^{1/2}} \tag{14.50}$$

We first simplify the equation by expressing all physical quantities in atomic units (see Section 1.8); the Schrödinger equation 'in atomic units' is

$$-\frac{1}{2}\left(\frac{\partial^2\psi}{\partial x^2} + \frac{\partial^2\psi}{\partial y^2} + \frac{\partial^2\psi}{\partial z^2}\right) - \frac{Z}{(x^2 + y^2 + z^2)^{1/2}}\psi = E\psi \tag{14.51}$$

Separation of variables

The first step in the solution of the partial differential equation in three variables is to separate the variables; that is, to reduce it to three ordinary equations. This is *not* possible in cartesian coordinates because the potential function V cannot then be expressed as a sum of terms, one in each variable only. There exist a number of other coordinate systems, however, in terms of which the separation is possible, and one of these is the system of spherical polar coordinates discussed in Chapter 10. The Laplacian operator in these coordinates is (Section 10.5)

$$\nabla^2 = \frac{1}{r^2}\frac{\partial}{\partial r}\left(r^2\frac{\partial}{\partial r}\right) + \frac{1}{r^2\sin\theta}\frac{\partial}{\partial\theta}\left(\sin\theta\frac{\partial}{\partial\theta}\right) + \frac{1}{r^2\sin^2\theta}\frac{\partial^2}{\partial\phi^2}$$

and the Schrödinger equation is then (after multiplication by -2 and rearrangement)

$$\frac{1}{r^2}\frac{\partial}{\partial r}\left(r^2\frac{\partial\psi}{\partial r}\right) + \frac{1}{r^2\sin\theta}\frac{\partial}{\partial\theta}\left(\sin\theta\frac{\partial\psi}{\partial\theta}\right) + \frac{1}{r^2\sin^2\theta}\frac{\partial^2\psi}{\partial\phi^2} + \frac{2Z}{r}\psi + 2E\psi = 0$$

$$(14.52)$$

We first separate the radial terms from the angular terms by writing the wave function as the product

$$\psi(r,\,\theta,\,\phi) = R(r) \times Y(\theta,\,\phi) \tag{14.53}$$

Substitution into (14.52), division throughout by $\psi = RY$, and multiplication by r^2 then gives

$$\left[\frac{1}{R}\frac{d}{dr}\left(r^2\frac{dR}{dr}\right) + 2Zr + 2Er^2\right] + \frac{1}{Y}\left[\frac{1}{\sin\theta}\frac{\partial}{\partial\theta}\left(\sin\theta\frac{\partial Y}{\partial\theta}\right) + \frac{1}{\sin^2\theta}\frac{\partial^2 Y}{\partial\phi^2}\right] = 0$$

$$(14.54)$$

We call the separation constant $l(l+1)$ for reasons that will become clear. Then

$$\frac{1}{R}\frac{d}{dr}\left(r^2\frac{dR}{dr}\right) + 2Zr + 2Er^2 = l(l+1)$$

or

$$\frac{1}{r^2}\frac{d}{dr}\left(r^2\frac{dR}{dr}\right) + \left(-\frac{l(l+1)}{r^2} + \frac{2Z}{r} + 2E\right)R = 0 \tag{14.55}$$

is the **radial equation** of the hydrogen atom, and

$$\frac{1}{\sin\theta}\frac{\partial}{\partial\theta}\left(\sin\theta\frac{\partial Y}{\partial\theta}\right) + \frac{1}{\sin^2\theta}\frac{\partial^2 Y}{\partial\phi^2} + l(l+1)Y = 0 \tag{14.56}$$

is the **angular equation**.

To separate the angular variables we now write

$$Y(\theta, \phi) = \Theta(\theta) \times \Phi(\phi) \qquad (14.57)$$

Substitution of this product and its derivatives in the angular equation, division by $Y = \Theta\Phi$, and multiplication by $\sin^2 \theta$ gives

$$\left[\frac{\sin \theta}{\Theta} \frac{d}{d\theta} \left(\sin \theta \frac{d\Theta}{d\theta} \right) + l(l+1) \sin^2 \theta \right] + \left[\frac{1}{\Phi} \frac{d^2\Phi}{d\phi^2} \right] = 0 \qquad (14.58)$$

so that, with separation constant $-m^2$

$$\frac{d^2\Phi(\phi)}{d\phi^2} = -m^2 \Phi(\phi) \qquad (14.59)$$

$$\frac{1}{\sin \theta} \frac{d}{d\theta} \left(\sin \theta \frac{d\Theta}{d\theta} \right) + \left(l(l+1) - \frac{m^2}{\sin^2 \theta} \right) \Theta = 0 \qquad (14.60)$$

The separation of the variables is complete. It is now only necessary to solve the three boundary value problems represented by equation (14.55), (14.59), and (14.60), with appropriate boundary conditions.

The Φ equation

$$\frac{d^2\Phi(\phi)}{d\phi^2} = -m^2 \Phi(\phi) \qquad (14.59)$$

The function Φ is defined in the interval $0 \le \phi \le 2\pi$ and must satisfy the condition

$$\Phi(2\pi) = \Phi(0) \qquad (14.61)$$

for continuity round the circle. We therefore have the same boundary value problem as that discussed in Section 12.7 for the particle in a ring and in 14.5 for angular motion of the particle in a circular box. The normalized solutions are

$$\Phi_m(\phi) = \frac{1}{\sqrt{2\pi}} e^{im\phi}, \qquad m = 0, \pm 1, \pm 2, \ldots \qquad (14.62)$$

or, in real form (equations (12.71)),

$$\frac{1}{\sqrt{2}} (\Phi_m + \Phi_{-m}) = \frac{1}{2\sqrt{\pi}} \cos m\phi, \qquad \frac{1}{i\sqrt{2}} (\Phi_m - \Phi_{-m}) = \frac{1}{2\sqrt{\pi}} \sin m\phi \qquad (14.63)$$

for $m = 0, 1, 2, \ldots$. The functions form an orthonormal set, with property

$$\int_0^{2\pi} \Phi_m^*(\phi) \Phi_{m'}(\phi) \, d\phi = \delta_{m,m'} \qquad (14.64)$$

The Θ equation

$$\frac{1}{\sin\theta}\frac{d}{d\theta}\left(\sin\theta\frac{d\Theta}{d\theta}\right)+\left(l(l+1)-\frac{m^2}{\sin^2\theta}\right)\Theta=0 \qquad (14.60)$$

The equation is transformed into the associated Legendre equation (13.22) by means of the substitution $x=\cos\theta$. We have

$$\frac{d\Theta}{d\theta}=\frac{d\Theta}{dx}\frac{dx}{d\theta}=-\sin\theta\frac{d\Theta}{dx}$$

$$\frac{d^2\Theta}{d\theta^2}=-\cos\theta\frac{d\Theta}{dx}+\sin^2\theta\frac{d^2\Theta}{dx^2}=(1-x^2)\frac{d^2\Theta}{dx^2}-x\frac{d\Theta}{dx}$$

Therefore,

$$\frac{1}{\sin\theta}\frac{d}{d\theta}\left(\sin\theta\frac{d\Theta}{d\theta}\right)=\frac{d^2\Theta}{d\theta^2}+\frac{\cos\theta}{\sin\theta}\frac{d\Theta}{d\theta}=(1-x^2)\frac{d^2\Theta}{dx^2}-2x\frac{d\Theta}{dx}$$

and equation (14.60) becomes

$$(1-x^2)\frac{d^2\Theta}{dx^2}-2x\frac{d\Theta}{dx}+\left(l(l+1)-\frac{m^2}{(1-x^2)}\right)\Theta=0 \qquad (14.65)$$

and this is identical to the associated Legendre equation. The finite solutions in the interval $-1\le x\le 1$ $(0\le\theta\le\pi)$ are the (normalized) associated Legendre functions (equation (13.28))

$$\Theta_{l,m}(\cos\theta)=\sqrt{\frac{(2l+1)}{2}\frac{(l-|m|)!}{(l+|m|)!}}\,P_l^{|m|}(\cos\theta), \qquad \begin{cases}l=0,1,2,3,\\ m=0,\pm1,\pm2,\quad,\pm l\end{cases} \qquad (14.66)$$

These functions form an orthonormal set, with property (equation (13.29))

$$\int_0^\pi \Theta_{l,m}(\cos\theta)\Theta_{l',m}(\cos\theta)\sin\theta\,d\theta=\delta_{l,l'} \qquad (14.67)$$

Spherical harmonics

The products of the angular functions $\Theta_{l,m}$ and Φ_m,

$$Y_{l,m}(\theta,\phi)=\Theta_{l,m}(\theta)\Phi_m(\phi)=\sqrt{\frac{2l+1}{4\pi}\frac{(l-|m|)!}{(l+|m|)!}}\,P_l^{|m|}(\cos\theta)\,e^{im\phi} \qquad (14.68)$$

are called **spherical harmonics**. They occur whenever a physical problem in three dimensions is formulated in spherical polar coordinates.* Some of these functions are listed in Table 14.1. The functions are complex when $m \neq 0$ and (see equations (14.63)) it is sometimes more convenient to use the corresponding real functions

$$\frac{1}{\sqrt{2}}(Y_{l,m} + Y_{l,-m}), \qquad \frac{1}{i\sqrt{2}}(Y_{l,m} - Y_{l,-m}), \qquad m > 0 \qquad (14.69)$$

Spherical harmonics multiplied by the factor r^l are called **solid harmonics**, and the real forms of these are also listed in the table, with their conventional names in atomic structure theory.

By virtue of the orthonormality relations (14.64) and (14.67), the spherical harmonics form an orthonormal set over a complete solid angle ($\theta = 0 \to \pi$, $\phi = 0 \to 2\pi$):

$$\int_0^{2\pi} \int_0^{\pi} Y_{l,m}^*(\theta, \phi) Y_{l',m'}(\theta, \phi) \sin\theta \, d\theta \, d\phi = \delta_{l,l'} \, \delta_{m,m'} \qquad (14.70)$$

Table 14.1

Spherical harmonics	Solid harmonics (real)
$Y_{0,0} = \left(\dfrac{1}{4\pi}\right)^{1/2}$	$s = \left(\dfrac{1}{4\pi}\right)^{1/2}$
$Y_{1,0} = \left(\dfrac{3}{4\pi}\right)^{1/2} \cos\theta$	$p_z = \left(\dfrac{3}{4\pi}\right)^{1/2} z$
$Y_{1,\pm1} = \left(\dfrac{3}{8\pi}\right)^{1/2} \sin\theta e^{\pm i\phi}$	$p_x = \left(\dfrac{3}{4\pi}\right)^{1/2} x, \quad p_y = \left(\dfrac{3}{4\pi}\right)^{1/2} y$
$Y_{2,0} = \left(\dfrac{5}{16\pi}\right)^{1/2} (3\cos^2\theta - 1)$	$d_{z^2} = \left(\dfrac{5}{8\pi}\right)^{1/2} (3z^2 - r^2)$
$Y_{2,\pm1} = \left(\dfrac{15}{8\pi}\right)^{1/2} \sin\theta \cos\theta e^{\pm i\phi}$	$d_{xz} = \left(\dfrac{15}{4\pi}\right)^{1/2} xz, \quad d_{yz} = \left(\dfrac{15}{4\pi}\right)^{1/2} yz$
$Y_{2,\pm2} = \left(\dfrac{15}{32\pi}\right)^{1/2} \sin^2\theta e^{\pm 2i\phi}$	$d_{x^2-y^2} = \left(\dfrac{15}{16\pi}\right)^{1/2} (x^2 - y^2), \quad d_{xy} = \left(\dfrac{15}{4\pi}\right)^{1/2} xy$

Angular momentum

The angular equation (14.56) can be written as the eigenvalue equation (with appropriate units)

$$- {}^2\left[\frac{1}{\sin\theta}\frac{\partial}{\partial\theta}\left(\sin\theta\frac{\partial}{\partial\theta}\right) + \frac{1}{\sin^2\theta}\frac{\partial^2}{\partial\phi^2}\right] Y_{l,m} = l(l+1)\, {}^2 Y_{l,m} \qquad (14.71)$$

* The spherical harmonics are often defined with an additional 'phase factor' $(-1)^{(m+|m|)/2}$ that multiplies the function by (-1) when m is odd and positive.

or

$$\mathcal{L}^2 Y_{l,m} = l(l+1)\hbar^2 Y_{l,m} \qquad (14.72)$$

in which $\mathcal{L}^2$ is the quantum-mechanical operator for the square of angular momentum. The spherical harmonics are therefore the eigenfunctions of $\mathcal{L}^2$. They describe the possible states of angular momentum of the system, and the eigenvalues $l(l+1)\hbar^2$ are the allowed values of the square of angular momentum. In addition, it follows from (14.62) that

$$-i\frac{d\Phi_m}{d\phi} = m\Phi_m$$

and, therefore, that

$$-i\hbar\frac{\partial Y_{l,m}}{\partial\phi} = m\hbar Y_{l,m} \qquad (14.73)$$

or

$$\mathcal{L}_z Y_{l,m} = m\hbar Y_{l,m} \qquad (14.74)$$

$\mathcal{L}_z$ is the quantum-mechanical operator representing the component of angular momentum in the z-direction, and the eigenvalues $m\hbar$ of $\mathcal{L}_z$ are the allowed values of this component. The number l is called the angular momentum quantum number and m the component of angular momentum (or magnetic) quantum number.

The radial equation

$$\frac{1}{r^2}\frac{d}{dr}\left(r^2\frac{dR}{dr}\right) + \left(-\frac{l(l+1)}{r^2} + \frac{2Z}{r} + 2E\right)R = 0 \qquad (14.55)$$

The equation has two sets of solutions for the hydrogen atom. In the bound states of the atom the electron is effectively confined to the vicinity of the nucleus by application of the boundary condition $R(r) \to 0$ as $r \to \infty$. The energies of these states are negative (the zero of energy is for the two charges at rest and at infinite separation) and energy must be supplied to the system in order to ionize the atom. States with positive energy are unbound states (continuum states) in which the electron moves freely in the presence of the nucleus but is not bound to it. We consider here only the bound states, with $E < 0$. We set

$$\alpha^2 = -2E, \qquad \lambda = \frac{Z}{\alpha} \qquad (14.75)$$

and introduce the new variable

$$\rho = 2\alpha r \qquad (14.76)$$

Then

$$\frac{dR}{dr} = 2\alpha \frac{dR}{d\rho}, \qquad \frac{d^2 R}{dr^2} = 4\alpha^2 \frac{d^2 R}{d\rho^2}$$

and the radial equation becomes

$$\frac{d^2 R}{d\rho^2} + \frac{2}{\rho}\frac{dR}{d\rho} + \left(-\frac{l(l+1)}{\rho^2} + \frac{\lambda}{\rho} - \frac{1}{4} \right) R = 0 \tag{14.77}$$

This is identical to equation (13.45) for the associated Laguerre functions when $\lambda = n$, a positive integer. The solutions of the radial equation are therefore given by equation (13.44)

$$e^{-\rho/2}\rho^l L_{n+l}^{2l+1}(\rho)$$

These functions are finite and continuous for all positive values of ρ, and therefore of r, and they satisfy the boundary condition for bound states. The functions may be normalized, using equation (13.46), and the resulting normalized radial wave functions are*

$$R_{n,l}(r) = -\left\{ \left(\frac{2Z}{n} \right)^3 \frac{(n-l-1)!}{2n\{(n+l)!\}^3} \right\}^{1/2} e^{-\rho/2}\rho^l L_{n+l}^{2l+1}(\rho) \tag{14.78}$$

where, because $\alpha = Z/n$ when $\lambda = n$,

$$\rho = \frac{2Z}{n}r \tag{14.79}$$

The allowed values of the quantum numbers are

$$n = 1, 2, 3, \ldots \qquad l = 0, 1, 2, \ldots, (n-1) \tag{14.80}$$

The radial functions form an orthonormal set with respect to the weight function r^2 in the interval $0 \le r \le \infty$:

$$\int_0^\infty R_{n,l}(r) R_{n',l}(r) \, r^2 dr = \delta_{n,n'} \tag{14.81}$$

Some of the radial functions are listed in Table 14.2.

---

* By convention the radial functions (14.78) are defined with a − sign as part of the normalization constant.

Table 14.2 Radial functions $R_{n,l}(\rho)$, $\rho = 2Zr/n$

n	l	Name	$R_{n,l}(\rho)$
1	0	1s	$2Z^{3/2}e^{-\rho/2}$
2	0	2s	$\dfrac{1}{2\sqrt{2}}Z^{3/2}(2-\rho)\,e^{-\rho/2}$
2	1	2p	$\dfrac{1}{2\sqrt{6}}Z^{3/2}\rho e^{-\rho/2}$
3	0	3s	$\dfrac{1}{9\sqrt{3}}Z^{3/2}(6-6\rho+\rho^2)\,e^{-\rho/2}$
3	1	3p	$\dfrac{1}{9\sqrt{6}}Z^{3/2}(4-\rho)\rho e^{-\rho/2}$
3	2	3d	$\dfrac{1}{9\sqrt{30}}Z^{3/2}\rho^2 e^{-\rho/2}$

The graphs of these are shown in Figure 14.4. The number of radial nodes (spherical nodal surfaces of the total wave function) is $n - l - 1$, excluding the zero at $r = 0$ when $l > 0$.

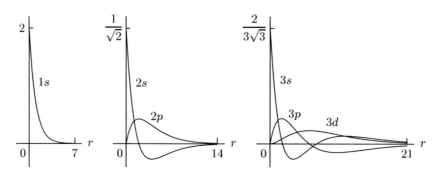

Figure 14.4

The energy

The allowed values of the energy are obtained from (14.75); when $\lambda = n$, $\alpha = Z/n$ and $E = -\alpha^2/2$. Therefore

$$E_n = -\frac{Z^2}{2n^2}, \qquad n = 1, 2, 3, \ldots \tag{14.82}$$

The number n is called the principal quantum number.

The total wave function

The total solutions of the Schrödinger equation for the bound states of the hydrogen-like atom are

$$\psi_{n,l,m}(r, \theta, \phi) = R_{n,l}(r)\Theta_{l,m}(\theta)\Phi_m(\phi) \tag{14.83}$$

and the numbers n, l and m are called the principal quantum number, (n), the angular momentum quantum number, (l), and the component of angular momentum (or magnetic) quantum number (m). From the orthonormality relations (14.64) for Φ_m, (14.67) for $\Theta_{l,m}$ and (14.81) for $R_{n,l}$ it follows that the total wave functions form an orthonormal set in three-dimensional space:

$$\int \psi_{n,l,m}^* \, \psi_{n',l',m'} \, dv = \int_0^{2\pi} \int_0^{\pi} \int_0^{\infty} \psi_{n,l,m}^*(r, \theta, \phi) \, \psi_{n',l',m'}(r, \theta, \phi) \, r^2 \sin\theta \, dr \, d\theta \, d\phi$$

$$= \delta_{n,n'}\delta_{l,l'}\delta_{m,m'} \tag{14.84}$$

Some of the total wave functions of the hydrogen atom, the atomic orbitals, are illustrated in Figure 14.5 by contour diagrams in an appropriate plane containing the nucleus.* Solid contours represent positive values of the orbitals, dashed contours represent negative values, dotted contours are nodes. We note that wave functions $\psi_{n,l,m}$ with energy E_n have a total of $(n-1)$ nodal surfaces. It is these diagrams that have led to the conventional pictorial representation of atomic orbitals.

➤ Exercises 12–15

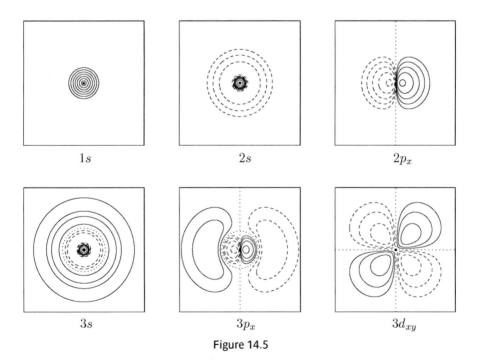

Figure 14.5

* The diagrams in Figure 14.5 are to scale within boxes of side 20 (a_0). The contour values are $\pm 0.002 \times 2^i, i = 0, 1, 2, 3, \ldots$

14.7   The vibrating string

We consider an elastic string, such as a guitar string, of length l and uniform linear density ρ that is stretched and fixed at both ends under tension T. The string is distorted transversely (pulled sideways), released and allowed to vibrate (Figure 14.6). The transverse displacement of the string is then a function of position x and time t,

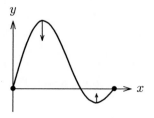

Figure 14.6

$$y = y(x, t) \qquad (14.85)$$

and when the vibrations are small the motion of the string is described by the wave equation

$$\frac{\partial^2 y}{\partial x^2} = \frac{1}{v^2} \frac{\partial^2 y}{\partial t^2} \qquad (14.86)$$

where $v^2 = T/\rho$. The boundary and initial conditions on the solutions of the equation are

$$y(0, t) = y(l, t) = 0 \qquad (14.87)$$

for no displacement at the ends of the string, and

$$y(x, 0) = f(x), \qquad \left(\frac{\partial y}{\partial t}\right)_{t=0} = g(x) \qquad (14.88)$$

for the initial displacement and velocity (functions).

Separation of variables

We write the displacement function (a wave function) as the product

$$y(x, t) = F(x) \times G(t) \qquad (14.89)$$

Substitution in the wave equation and division by $y = FG$ then gives

$$\frac{1}{F} \frac{d^2 F}{dx^2} = \frac{1}{Gv^2} \frac{d^2 G}{dt^2} \qquad (14.90)$$

Both sides of the equation must be constant so that, with separation constant $-\lambda^2$, the problem in two variables reduces to the *boundary value problem*

$$\frac{d^2 F}{dx^2} + \lambda^2 F = 0, \qquad F(0) = F(l) = 0 \qquad (14.91)$$

and the *initial value problem*

$$\frac{d^2G}{dt^2} + \lambda^2 v^2 G = 0 \tag{14.92}$$

with initial conditions (14.88), where $f(x)$ and $g(x)$ are given functions.

The boundary value problem (14.91) is identical to that discussed in Section 12.6 for the particle in a box. The allowed values of the separation constant are given by

$$\lambda_n = \frac{n\pi}{l}, \qquad n = 1, 2, 3, \ldots \tag{14.93}$$

and the corresponding (unnormalized) particular solutions are

$$F_n(x) = \sin \frac{n\pi x}{l} \tag{14.94}$$

For each value of λ_n, equation (14.92) is

$$\frac{d^2G}{dt^2} + \omega_n^2 G = 0 \tag{14.95}$$

where

$$\omega_n = \lambda_n v = \frac{n\pi v}{l} \tag{14.96}$$

and has solution

$$G_n(t) = A_n \cos \omega_n t + B_n \sin \omega_n t \tag{14.97}$$

where A_n and B_n are constants determined by the initial conditions. A set of solutions of the wave equation for the vibrating string is therefore

$$y_n(x, t) = \sin \frac{n\pi x}{l} \left[A_n \cos \omega_n t + B_n \sin \omega_n t \right], \qquad n = 1, 2, 3, \ldots \tag{14.98}$$

These solutions are called the **eigenfunctions** of the system. The quantities $\omega_n = n\pi v/l$ are called the **eigenvalues**. The set of values $\{\omega_1, \omega_2, \omega_3, \ldots\}$ is called the **eigenvalue spectrum**.

Normal modes of motion

Each eigenfunction $y_n(x, t)$ is a periodic function of time with period $2\pi/\omega_n$; the motion is transverse harmonic motion with frequency $v_n = \omega_n/2\pi$. This motion is called the nth **normal mode** of the vibrating string. The first mode, with $n = 1$, is called the fundamental, the second, with $n = 2$, is the first overtone, and so on. The motions in space of the first few of these normal modes are illustrated in Figure 14.7 (see also Figure 12.6 for the particle in a box).

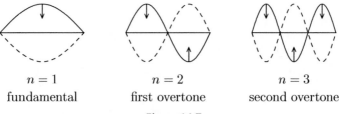

$n = 1$	$n = 2$	$n = 3$
fundamental	first overtone	second overtone

Figure 14.7

The nth mode has $n-1$ **nodes** between the end points; that is, points of zero displacement that do not move; the solutions (14.98) represent **standing waves**.

The complete solution

Each of the normal modes contains two constants, A_n and B_n, that are to be determined by the initial conditions (14.88); that is, by the way the motion is initiated. It is possible to choose the initial conditions so that the actual motion is one of the normal modes; for example, the nth mode is produced if

$$y(x,0) = f(x) = A_n \sin \frac{n\pi x}{l}$$

$$\left(\frac{\partial y}{\partial t}\right)_{t=0} = g(x) = B_n \omega_n \sin \frac{n\pi x}{l}$$

$$(14.99)$$

In general however, a single mode will not satisfy the initial conditions; the motion is not a pure normal mode but is a mixture or superposition of normal modes.

To obtain the solution for the general case, we invoke the principle of superposition discussed in Section 12.2; if $y_1, y_2, y_3, \ldots$ are solutions of a homogeneous linear equation then any linear combination of them is also a solution. Thus, for each normal mode we have

$$\frac{\partial^2 y_n}{\partial x^2} = \frac{1}{v^2} \frac{\partial^2 y_n}{\partial t^2} \tag{14.100}$$

so that, if

$$y = c_1 y_1 + c_2 y_2 + c_3 y_3 + \cdots \tag{14.101}$$

then

$$\frac{\partial^2 y}{\partial x^2} = \frac{1}{v^2} \frac{\partial^2 y}{\partial t^2} \tag{14.102}$$

The general solution of the wave equation that satisfies the boundary conditions is therefore a superposition of normal modes

$$y(x, t) = \sum_{n=1}^{\infty} \sin \frac{n\pi x}{l} \left[A_n \cos \omega_n t + B_n \sin \omega_n t \right] \qquad (14.103)$$

(the constants $c_1, c_2, c_3, \ldots$ in (14.101) have been absorbed in the A's and B's in (14.103)). This solution satisfies the initial conditions when the A's and B's are determined from the equations

$$y(x, 0) = f(x) = \sum_{n=1}^{\infty} A_n \sin \frac{n\pi x}{l}$$

$$\qquad (14.104)$$

$$\left(\frac{\partial y}{\partial t} \right)_{t=0} = g(x) = \sum_{n=1}^{\infty} B_n \omega_n \sin \frac{n\pi x}{l}$$

The two series in (14.104) are examples of the Fourier series discussed in Chapter 15. We return to this problem in Section 15.5 to determine the A's and B's for typical initial functions $f(x)$ and $g(x)$.

> Exercises 16–18

14.8 Exercises

Section 14.2

1. Show that the function $f(x, t) = a \sin (bx) \cos (vbt)$ (i) satisfies the one-dimensional wave equation (14.1), (ii) has the form $f(x, t) = F(x + vt) + G(x - vt)$.

2. The diffusion equation

$$\frac{\partial f}{\partial t} = D \frac{\partial^2 f}{\partial x^2}$$

provides a model of, for example, the transfer of heat from a hot region of a system to a cold region by conduction when $f(x, t)$ is a temperature field, or the transfer of matter from a region of high concentration to one of low concentration when f is the concentration. Find the functions $V(x)$ for which $f(x, t) = V(x)e^{ct}$ is a solution of the equation.

3. (i) It is shown in Example 14.2 that the function $f(x, t) = a \exp[-b(x - vt)^2]$ is a solution of the wave equation (14.1). Sketch graphs of $f(x, t)$ as a function of x at times $t = 0, t = 2/v, t = 4/v$ (use, for example, $a = b = 1$), to demonstrate that the function represents a wave travelling to the right (in the positive x-direction) at constant speed v. (ii) Verify that $g(x, t) = a \exp[-b(x + vt)^2]$ is also a solution of the wave equation, and hence that every superposition $F(x, t) = f(x, t) + g(x, t)$ is a solution. (iii) Sketch appropriate graphs of $f(x, t) + g(x, t)$ to demonstrate how this function develops in time.

Section 14.3

Find solutions of the following equations by the method of separation of variables:

4. $2\dfrac{\partial f}{\partial x} + \dfrac{\partial f}{\partial t} = 0$ **5.** $y\dfrac{\partial f}{\partial x} - x\dfrac{\partial f}{\partial y} = 0$ **6.** $\dfrac{\partial^2 f}{\partial x^2} + \dfrac{\partial^2 f}{\partial y^2} = 0$ **7.** $\dfrac{\partial^2 f}{\partial x \partial y} + f = 0$

Section 14.4

8. Show that the wave functions (14.23) satisfy the orthonormality conditions

$$\int_0^b \int_0^a \psi_{p,q}(x,y)\,\psi_{r,s}(x,y)\,dx\,dy = \begin{cases} 1 & \text{if } p = r \text{ and } q = s \\ 0 & \text{otherwise} \end{cases}$$

9. (i) Find the energies (in units of $h^2/8ma^2$) of the lowest 11 states of the particle in a square box of side a, and sketch an appropriate energy-level diagram. (ii) The six diagrams in Figure 14.2 are maps of the signs and nodes of the wave functions (14.27) for the lowest six states, using the real forms of the angular functions. Sketch the corresponding diagrams for the next five lowest states.

10. (i) Solve the Schrödinger equation for the particle in a three-dimensional rectangular box with potential energy function

$$V(x,y,z) = \begin{cases} 0 & \text{for } 0 < x < a, \quad 0 < y < b, \quad 0 < z < c \\ \infty & \text{elsewhere} \end{cases}$$

(ii) What are the possible degeneracies of the eigenvalues for a cubic box?

Section 14.5

11. Some zeros of the Bessel functions $J_n(x)$ are:

$$
\begin{aligned}
J_0(x) = 0 \quad &\text{for} \quad x = 2.4048,\ 5.5201,\ 8.6537 \\
J_1(x) = 0 \quad & \qquad\quad x = 3.8317,\ 7.0156,\ 10.1736 \\
J_2(x) = 0 \quad & \qquad\quad x = 5.1356,\ 8.4172 \\
J_3(x) = 0 \quad & \qquad\quad x = 6.3802,\ 9.7610 \\
J_4(x) = 0 \quad & \qquad\quad x = 7.5883
\end{aligned}
$$

(i) Find the energies (in units of $\hbar^2/2ma^2$) of the lowest 10 states of the particle in a circular box of radius a, and sketch an appropriate energy-level diagram. (ii) The six diagrams in Figure 14.3 are maps of the signs and nodes of the wave functions (14.47) for the lowest six states, using the real forms of the angular functions. Sketch the corresponding diagrams for the next four states.

Section 14.6

12. (i) Make use of Tables 14.1 and 14.2 to write down the total wave function $\psi_{1,0,0}$ for the hydrogen-like atom. (ii) Substitute this wave function into the Schrödinger equation (14.52), and confirm that it is a solution of the equation with E given by (14.82).

13. Repeat Exercise 12 for the wave function $\psi_{2,1,0}$.

14. Show that the radial functions $R_{1,0}$ and $R_{2,0}$ in Table 14.2 satisfy the orthogonality condition (14.81).

15. The Schrödinger equation for the particle in a spherical box of radius a is $-(\hbar^2/2m)\nabla^2\psi + V\psi = E\psi$, with potential energy function $V(x,y,z) = 0$ for $r = \sqrt{x^2 + y^2 + z^2} < a$ and ∞ elsewhere.

(i) Show that the equation is separable in spherical polar coordinates, with the same angular wave functions, the spherical harmonics (14.68), as for the hydrogen atom. (ii) Show that the radial equation reduces to the Bessel equation (13.60) for spherical Bessel functions $j_l(x)$ where, as in Section 14.5 for the particle in a circular box, $x = \sqrt{2mE/\hbar^2}\, r$. (iii) Use the boundary condition to find an expression for the quantized energy in terms of the zeros of the Bessel functions. (iv) Find the wave function and energy of the ground state.

Section 14.7

16. Find the solution of the wave equation for the vibrating string that satisfies the initial conditions

$$y(x, 0) = 3 \sin \pi x/l, \quad (\partial y/\partial t)_{t=0} = 0.$$

17. A homogeneous thin bar of length l and constant cross-section is perfectly insulated along its length with the ends kept at constant temperature $T = 0$ (on some temperature scale). The temperature profile of the bar is a function $T(x, t)$ of position x $(0 \le x \le l)$ and of time t, and satisfies the heat-conduction (diffusion) equation $\dfrac{\partial T}{\partial t} = D\dfrac{\partial^2 T}{\partial x^2}$ where D is the thermal diffusivity of the material. The boundary conditions are $T(0, t) = T(l, t) = 0$. Find the solution of the equation for initial temperature profile $T(x, 0) = 3 \sin \pi x/l$.

18. (i) Find the general solution of the Laplace equation $\dfrac{\partial^2 u}{\partial x^2} + \dfrac{\partial^2 u}{\partial y^2} = 0$ in the rectangle $0 \le x \le a, 0 \le y \le b$ subject to the boundary conditions

$$u(0, y) = 0, \quad u(a, y) = 0$$
$$u(x, 0) = 0, \quad u(x, b) = f(x)$$

where $f(x)$ is an arbitrary function of x. (ii) Find the particular solution for

$$f(x) = \sin\frac{3\pi x}{a}.$$

15 Orthogonal expansions. Fourier analysis

15.1 Concepts

We saw in Section 7.6 that many functions can be expanded as power series and that, indeed, some functions are defined by such series. In general, a function $f(x)$ of the variable x can be expanded in powers of x as a MacLaurin series if the function and its derivatives exist at $x=0$ and throughout the interval 0 to x. The expansion is then valid within the radius of convergence of the series.

The power series expansion of a function is a special case of a more general type of expansion. Consider the function

$$f(x) = a_0 + a_1 x + a_2 x^2 + \cdots = \sum_{l=0}^{\infty} a_l x^l \qquad (15.1)$$

We can regard each power of x as the function $g_l(x) = x^l$; that is, $g_0(x) = x^0 = 1$, $g_1(x) = x$, $g_2(x) = x^2$ and so on. The expansion is then

$$f(x) = a_0 g_0(x) + a_1 g_1(x) + a_2 g_2(x) + \cdots = \sum_{l=0}^{\infty} a_l g_l(x) \qquad (15.2)$$

This formalism suggests that it may be possible to find other sets of functions $\{g_l\}$ that can be used instead of simple powers, and that it may be possible in this way to expand functions that cannot be expanded as power series.

The sets of functions that are of particular importance for expansions in series consist of functions with the property of orthogonality. The theory of orthogonal expansions is introduced in Section 15.2. Two examples of expansions in Legendre polynomials, important in potential theory and in scattering theory, are discussed in Section 15.3, with an application in electrostatics. Fourier series are developed in Section 15.4, and used in Section 15.5 to solve the wave equation for the vibrating string for a given set of initial conditions. Fourier transforms, essential for the analysis of the results of diffraction experiments and in Fourier transform spectroscopy, are discussed in Section 15.6.

15.2 Orthogonal expansions

We introduce the concept of expansions in sets of orthogonal functions by demonstrating that the power series (15.1) can be written as a linear combination of Legendre polynomials,

$$f(x) = c_0 P_0(x) + c_1 P_1(x) + c_2 P_2(x) + \cdots = \sum_{l=0}^{\infty} c_l P_l(x) \qquad (15.3)$$

and that, when the value of the variable is restricted to the interval $-1 \leq x \leq +1$, the orthogonality of the Legendre polynomials can be used to derive a general formula for the coefficients c_l in (15.3) in terms of the coefficients a_l in (15.1). The Legendre polynomials were discussed in Section 13.4, and the first few are

$$P_0(x) = 1 \qquad\qquad\qquad P_1(x) = x$$

$$P_2(x) = \frac{1}{2}(3x^2 - 1) \qquad\qquad P_3(x) = \frac{1}{2}(5x^3 - 3x) \tag{15.4}$$

$$P_4(x) = \frac{1}{8}(35x^4 - 30x^2 + 3) \qquad P_5(x) = \frac{1}{8}(63x^5 - 70x^3 + 15x)$$

The function $P_l(x)$ is a polynomial of degree l in x, and it is possible to express every power x^l as a linear combination of Legendre polynomials of degree up to l. For example, it follows from (15.4) that

$$x^0 = P_0 \qquad\qquad\qquad x^1 = P_1$$

$$x^2 = \frac{1}{3}(2P_2 + P_0) \qquad\qquad x^3 = \frac{1}{5}(2P_3 + 3P_1) \tag{15.5}$$

$$x^4 = \frac{1}{35}(8P_4 + 20P_2 + 7P_0) \qquad x^5 = \frac{1}{63}(8P_5 + 28P_3 + 27P_1)$$

The power series (15.1) can therefore be written (using only terms from (15.5)),

$$f(x) = a_0 P_0 + a_1 P_1 + \frac{a_2}{3}(2P_2 + P_0) + \frac{a_3}{5}(2P_3 + 3P_1)$$

$$+ \frac{a_4}{35}(8P_4 + 20P_2 + 7P_0) + \frac{a_5}{63}(8P_5 + 28P_3 + 27P_1) + \cdots$$

$$= \left(a_0 + \frac{a_2}{3} + \frac{a_4}{5} + \cdots \right) P_0 + \left(a_1 + \frac{3a_3}{5} + \frac{3a_5}{7} + \cdots \right) P_1 \tag{15.6}$$

$$+ \left(\frac{2a_2}{3} + \frac{4a_4}{7} + \cdots \right) P_2 + \left(\frac{3a_3}{5} + \frac{4a_5}{9} + \cdots \right) P_3$$

$$+ \left(\frac{8a_4}{35} + \cdots \right) P_4 + \left(\frac{8a_5}{63} + \cdots \right) P_5 + \cdots$$

and this has the required form (15.3). The Legendre polynomials occur in the physical sciences with the variable $x = \cos\theta$, so that we are interested only in the interval $-1 \leq x \leq +1$. The polynomials are orthogonal in this interval (see equations (13.25) and (13.27)),

$$\int_{-1}^{+1} P_k(x) P_l(x)\, dx = \frac{2}{2l+1}\delta_{k,l} = \begin{cases} 0 & \text{if} \quad k \neq l \\[2mm] \dfrac{2}{2l+1} & \text{if} \quad k = l \end{cases} \tag{15.7}$$

and this can be used to obtain a general formula for the coefficients c_l in (15.3). We multiply equation (15.3) by $P_k(x)$ and integrate. Then

$$\int_{-1}^{+1} P_k(x)f(x)\,dx = \int_{-1}^{+1} P_k(x)\left[\sum_{l=0}^{\infty} c_l P_l(x)\right] dx = \sum_{l=0}^{\infty} c_l \int_{-1}^{+1} P_k(x)P_l(x)\,dx$$

and because, by (15.7), each integral on the right is zero except for the one with $l = k$, only that term contributes to the sum:

$$\int_{-1}^{+1} P_k(x)f(x)\,dx = c_k \int_{-1}^{+1} P_k(x)P_k(x)\,dx = \frac{2}{2k+1}c_k$$

Therefore

$$c_k = \frac{2k+1}{2}\int_{-1}^{+1} P_k(x)f(x)\,dx \tag{15.8}$$

This important formula gives the values of the coefficients c_l in the expansion of a finite continuous function of x, for $|x| \le 1$, in terms of the Legendre polynomials. Then, for the power series $f(x) = \sum_{l=0}^{\infty} a_l x^l$,

$$c_k = \frac{2k+1}{2}\sum_{l=0}^{\infty} a_l \int_{-1}^{+1} P_k(x)x^l\,dx \tag{15.9}$$

Replacement of $P_k(x)$ in the integrals by its expansion in powers of x, from tabulations like (15.4) or from the general formula (13.19), then gives the coefficient c_k in terms of the coefficients a_l.

EXAMPLE 15.1 Use equation (15.9) for $k = 0$ to find c_0.

We have $P_0 = 1$ so that equation (15.9) is

$$c_0 = \frac{1}{2}\sum_{l=0}^{\infty} a_l \int_{-1}^{+1} x^l\,dx$$

where

$$\int_{-1}^{+1} x^l\,dx = \left[\frac{x^{l+1}}{l+1}\right]_{-1}^{+1} = \begin{cases} \dfrac{2}{l+1} & \text{if } l \text{ is even} \\ 0 & \text{if } l \text{ is odd} \end{cases}$$

Therefore

$$c_0 = a_0 + \frac{a_2}{3} + \frac{a_4}{5} + \frac{a_6}{7} + \cdots = \sum_{l=0}^{\infty} \frac{a_{2l}}{2l+1}$$

and this is in agreement with the coefficient of P_0 shown in (15.6).

▸ Exercises 1, 2

The general case

Let $g_n(x)$, $n = 1, 2, 3, \ldots$, be a set of functions, possibly complex, that are orthogonal in the interval $a \le x \le b$ with respect to the weight function $w(x)$:

$$\int_a^b g_m^*(x) g_n(x) w(x) \, dx = 0 \quad \text{if} \quad m \ne n \tag{15.10}$$

($g_m^* = g_m$ for real functions). Let $f(x)$ be an arbitrary function, defined in the interval $a \le x \le b$, that can be expanded in the set $\{g_n(x)\}$,

$$f(x) = \sum_{n=0}^{\infty} c_n g_n(x) \tag{15.11}$$

Then, multiplication by $g_m^*(x)$ and integration with respect to weight function $w(x)$ in the interval a to b gives

$$\int_a^b g_m^*(x) f(x) w(x) \, dx = \sum_{n=0}^{\infty} c_n \int_a^b g_m^*(x) g_n(x) w(x) \, dx$$

$$= c_m \int_a^b g_m^*(x) g_m(x) w(x) \, dx$$

so that (replacing m by n),

$$c_n = \frac{\displaystyle\int_a^b g_n^*(x) f(x) w(x) \, dx}{\displaystyle\int_a^b g_n^*(x) g_n(x) w(x) \, dx} \tag{15.12}$$

The denominator in this expression is the normalization integral of the function $g_n(x)$; it is the square of the **norm**

$$\|g_n\| = \sqrt{\int_a^b g_n^*(x) g_n(x) w(x) \, dx} \tag{15.13}$$

The norm of a function is sometimes interpreted as the 'magnitude' or 'length' of the function. The coefficients in the expansion (15.11) of $f(x)$ are then given by

$$c_n = \frac{1}{\|g_n\|^2} \int_a^b g_n^*(x) f(x) w(x)\, dx \tag{15.14}$$

It is often more convenient if the functions of the expansion set are normalized, as well as orthogonal. Normalization is achieved by dividing each function by its norm,

$$\bar{g}_n(x) = \frac{1}{\|g_n\|} g_n(x) \tag{15.15}$$

The resulting set of functions is an *orthonormal* set,

$$\int_a^b \bar{g}_n^*(x)\bar{g}_n(x) w(x)\, dx = \delta_{m,n} \tag{15.16}$$

The expansion (15.11) then becomes

$$f(x) = \sum_{n=0}^{\infty} c_n\, \bar{g}_n(x) \tag{15.17}$$

and (15.14) for the expansion coefficients becomes

$$c_n = \int_a^b \bar{g}_n^*(x)\, f(x)\, w(x)\, dx \tag{15.18}$$

The concept of orthogonal expansions is readily generalized for functions of more than one variable.

Completeness of orthogonal sets

A number of orthogonal sets have been discussed in previous chapters; in particular, the several 'special functions' in Chapter 13 and the solutions of the Schrödinger equation (eigenfunctions of the Hamiltonian) in Chapters 12 and 14. These sets share the property of **completeness**:

An orthonormal set $\{g(x)\}$ defined in the interval $a \le x \le b$ with respect to weight function $w(x)$ is called complete if 'every' function $f(x)$ defined in the interval can be represented by a linear combination of the functions of the set,

$$f(x) = \sum_{n=0}^{\infty} c_n g_n(x), \qquad c_n = \int_a^b g_n^*(x)\, f(x) w(x)\, dx \tag{15.19}$$

This expression of completeness is sufficient for most purposes. We consider two qualifications.

(1) By *every* function is meant not only continuous functions but the more general class of **piecewise continuous** functions that have a finite number of finite discontinuities and a finite number of maxima and minima in the interval.[1]

(2) By the *representation* of the function is meant that the function can be approximated arbitrarily closely by a series

$$f(x) \approx f_k(x) = c_0 g_0(x) + c_1 g_1(x) + \cdots + c_k g_k(x) \tag{15.20}$$

and the limit of the series is such that

$$\int_a^b \left[f(x) - f_k(x) \right]^2 w(x) dx \to 0 \text{ as } k \to \infty \tag{15.21}$$

The series is said to converge to $f(x)$ *in the mean*, and the function can in practice be replaced by the series even though (15.21) means that the series need not converge to $f(x)$ for every x. It is this convergence in the mean that allows the representation of some discontinuous functions, but care must sometimes be taken when equating derivatives and integrals of the series to the derivatives and integrals of the function it represents.

15.3   Two expansions in Legendre polynomials

We saw in the previous section that a power series in x can be represented as a linear combination of Legendre polynomials. Two examples in the physical sciences are

$$(1 - 2xt + t^2)^{-1/2} = \sum_{l=0}^{\infty} t^l P_l(x), \qquad -1 \le x \le +1 \tag{15.22}$$

of importance in potential theory (electrostatics and gravitational theory), and

$$e^{itx} = \sum_{l=0}^{\infty} (2l + 1) i^l j_l(t) P_l(x) \tag{15.23}$$

of importance in scattering theory, where the $j_l(t)$ are spherical Bessel functions (equation (13.59)). The series (15.22) converges when $|x| < 1$ and $|t| < 1$, and is derived in Example 15.2. The series (15.23) converges when $|x| < 1$ for all values of t, and its derivation is discussed in Example 15.3.

---

[1] This was first proved by Dirichlet for the Fourier series ('Dirichlet conditions'). Gustav Peter Lejeune-Dirichlet (1805–1859), German mathematician, was influenced by the work of Fourier whilst a student in Paris in the early 1820's.

EXAMPLE 15.2 Derive the expansion (15.22).

We treat the function to be expanded as a function of t, and expand it as a MacLaurin series. We have

$$
\begin{aligned}
f(t) &= (1 - 2xt + t^2)^{-1/2} & f(0) &= 1 \\
f'(t) &= (x - t)(1 - 2xt + t^2)^{-3/2} & f'(0) &= x \\
f''(t) &= 3(x - t)^2(1 - 2xt + t^2)^{-5/2} - (1 - 2xt + t^2)^{-3/2} & f''(0) &= 3x^2 - 1
\end{aligned}
$$

and similarly, $f'''(0) = 15x^3 - 9x$, $f''''(0) = 105x^4 - 90x^2 + 9$, and so on.

Comparison with the Legendre polynomials listed in (15.4) shows that, for the lth derivative, $f^{(l)}(0) = l!P_l(x)$. The MacLaurin series

$$
f(t) = f(0) + tf'(0) + \frac{t^2}{2!}f''(0) + \frac{t^3}{3!}f'''(0) + \frac{t^4}{4!}f''''(0) + \cdots
$$

is then

$$
f(t) = (1 - 2xt + t^2)^{-1/2} = P_0 + tP_1(x) + t^2 P_2(x) + t^3 P_3(x) + t^4 P_4(x) + \cdots
$$

and this is equation (15.22).

EXAMPLE 15.3 Find the first term in the expansion (15.23).

If $e^{itx} = \displaystyle\sum_{l=0}^{\infty} c_l P_l(x)$ then $c_l = \dfrac{2l+1}{2} \displaystyle\int_{-1}^{+1} P_l(x) e^{itx}\, dx$. Therefore

$$
c_0 = \frac{1}{2}\int_{-1}^{+1} e^{itx}\, dx = \frac{1}{2}\left[\frac{e^{itx}}{it}\right]_{-1}^{+1} = \frac{1}{2it}\left[e^{it} - e^{-it}\right]_{-1}^{+1} = \frac{1}{t}\sin t
$$

It follows from equations (13.55) and (13.59) for Bessel functions of half-integral order that

$$
c_0 = \sqrt{\frac{\pi}{2t}} J_{1/2}(t) = j_0(t)
$$

▸ Exercises 3, 4

The expansion of electrostatic potential

The potential of a single charge

The electrostatic potential at point P due to the presence of a point charge q is (see Figure 15.1)

$$
V = \frac{q}{4\pi\varepsilon_0 R_q} \tag{15.24}
$$

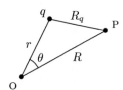

Figure 15.1

where R_q is the distance of point P from the charge (if a charge q' is placed at P, the potential energy of the system of two charges is $Vq' = qq'/4\pi\varepsilon_0 R_q$). By the cosine rule for the triangle OPq, we have $R_q^2 = r^2 + R^2 - 2rR\cos\theta$, so that (15.24) can be written as

$$V = \frac{q}{4\pi\varepsilon_0}(R^2 - 2rR\cos\theta + r^2)^{-1/2} \tag{15.25}$$

We consider the case $R > r$. We write (15.25) as

$$V = \frac{q}{4\pi\varepsilon_0 R}\left[1 - 2\left(\frac{r}{R}\right)\cos\theta + \left(\frac{r}{R}\right)^2\right]^{-1/2}$$

and, by equation (15.22) with $t = r/R$ and $x = \cos\theta$, this can be expanded in Legendre polynomials as

$$V = \frac{q}{4\pi\varepsilon_0 R}\sum_{l=0}^{\infty}\left(\frac{r}{R}\right)^l P_l(\cos\theta), \qquad \frac{r}{R} < 1 \tag{15.26}$$

When $r > R$, the same expansion is valid, but with r and R interchanged.

The potential of a distribution of charges

We consider a system of N charges,

$$q_1 \text{ at } (x_1, y_1, z_1), q_2 \text{ at } (x_2, y_2, z_2), \ldots, q_n \text{ at } (x_n, y_n, z_n)$$

as illustrated in Figure 15.2 (for 2 charges).

Each charge makes its individual contribution to the electrostatic potential at point P, and the total potential at P is the sum of these contributions,

$$V = \sum_{i=1}^{N}\frac{q_i}{4\pi\varepsilon_0 R_i} \tag{15.27}$$

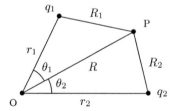

Figure 15.2

If the point P is *exterior* to the distribution of charges, so that $R > r_i$, for all i, the potential can be expanded term by term as in (15.26):

$$V = \sum_{i=1}^{N}\frac{q_i}{4\pi\varepsilon_0 R}\sum_{l=0}^{\infty}\left(\frac{r_i}{R}\right)^l P_l(\cos\theta_i)$$

$$\tag{15.28}$$

$$= \frac{1}{4\pi\varepsilon_0}\sum_{l=0}^{\infty}\frac{1}{R^{l+1}}\left[\sum_{i=1}^{N}q_i r_i^l P_l(\cos\theta_i)\right]$$

This analysis of the potential is important for the description of the electrostatic interaction of charge distributions, for example in the theory of intermolecular forces. The quantities

$$Q_l = \sum_i q_i r_i^l P_l (\cos \theta_i) \qquad (15.29)$$

in (15.28) are called the electric **multipole moments** of order l of the charge distribution (strictly, Q_l is the component of the multipole in the direction OP, see Chapter 16). The expansion (15.28) is called the **multipole expansion** of the potential. The first few moments are

$$Q_0 = \sum_{i=1}^{N} q_i \qquad\qquad \text{total charge}$$

$$Q_1 = \sum_{i=1}^{N} q_i r_i \cos \theta_i \qquad\qquad \text{dipole moment} \qquad (15.30)$$

$$Q_2 = \sum_{i=1}^{N} q_i r_i^2 \frac{1}{2}(3 \cos^2 \theta_i - 1) \qquad \text{quadrupole moment}$$

EXAMPLE 15.4 The field of a dipole

For the system shown in Figure 15.3 of two unlike charges $+q$ and $-q$ separated by distance r:

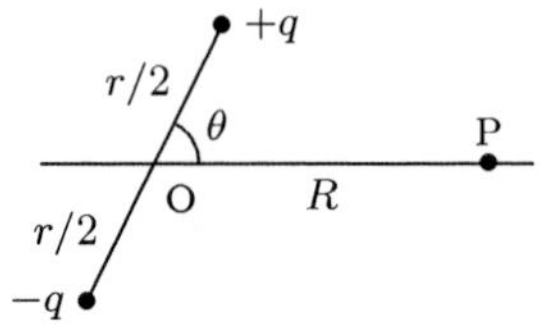

1. $Q_0 = +q - q = 0$

and the total charge is zero.

Figure 15.3

2. $Q_1 = \dfrac{1}{2} qr \cos \theta - \dfrac{1}{2} qr \cos(\pi + \theta) = qr \cos \theta = \mu \cos \theta$

where $\mu = qr$ is the (scalar) dipole moment of the pair of charges, and $\mu \cos \theta$ is the component of the dipole moment in the direction OP.

3. $Q_2 = q \left(\dfrac{r}{2}\right)^2 \cdot \dfrac{1}{2}(3 \cos^2 \theta - 1) - q \left(\dfrac{r}{2}\right)^2 \cdot \dfrac{1}{2}(3 \cos^2(\pi + \theta) - 1) = 0$

and, because $P_l(\cos \theta)$ is an even function of $\cos \theta$, every even-order multipole moment is zero.

The potential at point P is then

$$V = \frac{\mu \cos \theta}{4\pi\varepsilon_0 R^2} + \frac{Q_3}{4\pi\varepsilon_0 R^4} + \frac{Q_5}{4\pi\varepsilon_0 R^6} + \cdots$$

When R is large enough ($R \gg r$), the expansion of the potential can be truncated after the leading term,

$$V \approx \frac{\mu \cos \theta}{4\pi\varepsilon_0 R^2}$$

This is the potential that dominates the long-range interactions of a neutral polar molecule.

▸ Exercises 5, 6

15.4 Fourier series

The Fourier-series representation of a function $f(x)$ is the expansion of the function in terms of the set of trigonometric functions

$$\begin{aligned} \cos nx, && n = 0, 1, 2, \ldots \\ \sin nx, && n = 1, 2, 3, \ldots \end{aligned} \tag{15.31}$$

in the interval $-\pi \le x \le \pi$ (we consider intervals of arbitrary width later in this section). These functions are orthogonal in the interval (and in every interval of width 2π),

$$\int_{-\pi}^{+\pi} \cos mx \cos nx \, dx = 0, \quad m \ne n \tag{15.32}$$

$$\int_{-\pi}^{+\pi} \sin mx \sin nx \, dx = 0, \quad m \ne n \tag{15.33}$$

$$\int_{-\pi}^{+\pi} \cos mx \sin nx \, dx = 0, \quad \text{all } m, n \tag{15.34}$$

and the corresponding normalization integrals are

$$\int_{-\pi}^{+\pi} \cos nx \cos nx \, dx = \begin{cases} 2\pi & \text{if } n = 0 \\ \pi & \text{if } n > 0 \end{cases} \tag{15.35}$$

$$\int_{-\pi}^{+\pi} \sin nx \sin nx \, dx = \pi \quad \text{if } n > 0 \tag{15.36}$$

The (trigonometric) Fourier series is usually written in the form[2]

$$f(x) = \frac{a_0}{2} + a_1 \cos x + a_2 \cos 2x + a_3 \cos 3x + \cdots$$

$$+ b_1 \sin x + b_2 \sin 2x + b_3 \sin 3x + \cdots \qquad (15.37)$$

$$= \frac{a_0}{2} + \sum_{n=1}^{\infty} (a_n \cos nx + b_n \sin nx)$$

and, because of the orthogonality of the expansion functions, the **Fourier coefficients** are given by

$$a_n = \frac{1}{\pi} \int_{-\pi}^{+\pi} f(x) \cos nx \, dx \qquad (15.38)$$

$$b_n = \frac{1}{\pi} \int_{-\pi}^{+\pi} f(x) \sin nx \, dx \qquad (15.39)$$

EXAMPLE 15.5 Confirm the relations (i) (15.32) and (ii) (15.35).

(i) By the trigonometric relations (3.22), when $m \neq n$,

$$\cos mx \cos nx = \frac{1}{2} \Big[\cos(m+n)x + \cos(m-n)x \Big]$$

so that

$$\int_{-\pi}^{+\pi} \cos mx \cos nx \, dx = \frac{1}{2} \int_{-\pi}^{+\pi} \Big[\cos(m+n)x + \cos(m-n)x \Big] dx$$

$$= \frac{1}{2} \left[\frac{\sin(m+n)x}{m+n} + \frac{\sin(m-n)x}{m-n} \right]_{-\pi}^{+\pi} = 0$$

and the zero is obtained because the sine of an integer multiple of π is zero.

(ii) For $m = n = 0$, when $\cos mx = \cos mx = 1$,

$$\int_{-\pi}^{+\pi} dx = \Big[x \Big]_{-\pi}^{+\pi} = 2\pi$$

---

[2] Jean-Baptiste Joseph Fourier (1768–1830) studied the series in connection with his work on the diffusion of heat. The work was first presented to the French Academy in 1807, and published in final form in 1822 in the *Théorie analytique de la chaleur* (Analytic theory of heat). It generated new developments in the mathematical theory of functions, and provided a new and powerful tool for the analysis and solution of physical problems.

For $m = n > 0$,

$$\int_{-\pi}^{+\pi} \cos^2 nx \, dx = \int_{-\pi}^{+\pi} \frac{1}{2}(1 + \cos 2nx) \, dx = \left[\frac{1}{2}\left(x + \frac{\sin 2nx}{2n} \right) \right]_{-\pi}^{+\pi} = \pi$$

> Exercise 7

Periodicity

The trigonometric functions (15.31) are periodic functions of x with period 2π and, therefore, every linear combination of them is also periodic. This means that the function $f(x)$ can be extended to values outside the base interval $-\pi \le x \le \pi$ by means of the periodicity relation

$$f(x + 2\pi) = f(x) \qquad (15.40)$$

For example, the function

$$f(x) = \begin{cases} \sin x, & \text{for} \quad 0 \le x \le \pi \\ 0 & \text{for} \quad -\pi \le x \le 0 \end{cases} \qquad (15.41)$$

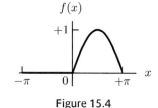

Figure 15.4

shown in Figure 15.4 can be extended by means of (15.40) to give the periodic function shown in Figure 15.5.

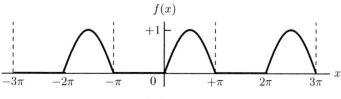

Figure 15.5

Periodic functions occur in the mathematical modelling of physical systems that exhibit periodic phenomena. For example, the function shown in Figure 15.5, with $f(x)$ as current and x as time, might represent the result of passing an alternating electric current through a rectifier that allows current to flow in one direction only (a half-wave rectifier). Despite the discontinuities (of the gradient) exhibited by the function, it is one of an important class of functions that can be represented as Fourier series. An example of such a function is discussed in Example 15.6.

In general, a function that satisfies the periodicity relation (15.40) can be expanded as a Fourier series if it is single-valued and piecewise continuous; that is, continuous except for a finite number of finite discontinuities and only a finite number of maxima and minima in any finite interval (that is, every reasonably well-behaved function).

EXAMPLE 15.6 The Fourier series of the function

$$f(x) = \begin{cases} 1, & \text{for} \quad 0 < x < \pi \\ 0, & \text{for} \quad -\pi < x < 0 \end{cases} \tag{15.42}$$

The graph of the (extended) function, Figure 15.6, shows that the function is discontinuous when x is a multiple of π, and the function can therefore be expanded as the Fourier series (15.37),

$$f(x) = \frac{a_0}{2} + \sum_{n=1}^{\infty} (a_n \cos nx + b_n \sin nx)$$

everywhere other than at these points of discontinuity. It is a property of Fourier series that the value of the series at a point of discontinuity is the mean value of the function at the point. In the present case, this mean value is $1/2$.

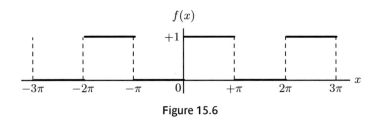

Figure 15.6

The coefficient a_0. By equation (15.38) with $n = 0$,

$$a_0 = \frac{1}{\pi} \int_{-\pi}^{+\pi} f(x) \, dx = \frac{1}{\pi} \int_{-\pi}^{0} f(x) \, dx + \frac{1}{\pi} \int_{0}^{+\pi} f(x) \, dx$$

The function $f(x)$ is equal to zero in the first integral on the right, and is equal to unity in the second. Therefore,

$$a_0 = \frac{1}{\pi} \int_{0}^{+\pi} dx = 1 \tag{15.43}$$

The coefficients a_n, $n > 0$. By equation (15.38),

$$a_n = \frac{1}{\pi} \int_{-\pi}^{+\pi} f(x) \cos nx \, dx = \frac{1}{\pi} \int_{0}^{+\pi} \cos nx \, dx = \frac{1}{\pi} \left[\frac{\sin nx}{n} \right]_{0}^{\pi} = 0 \tag{15.44}$$

All the Fourier coefficients a_n are zero except for $a_0 = 1$.

The coefficients b_n. By equation (15.39),

$$b_n = \frac{1}{\pi}\int_{-\pi}^{+\pi} f(x)\sin nx\, dx = \frac{1}{\pi}\int_0^{+\pi}\sin nx\, dx = \frac{1}{\pi}\left[-\frac{\cos nx}{n}\right]_0^{\pi} = \frac{1}{n\pi}(1-\cos n\pi)$$

We have $\cos n\pi = +1$ when n is an even integer, and $\cos n\pi = -1$ when n is odd. It follows that b_n is nonzero only when n is odd:

$$b_n = \frac{2}{n\pi}, \quad n \text{ odd} \tag{15.45}$$

By the results (15.43) to (15.45), the Fourier series of the function (15.42) is therefore

$$f(x) = \frac{a_0}{2} + b_1\sin x + b_3\sin 3x + b_5\sin 5x + \cdots$$

$$= \frac{1}{2} + \frac{2}{\pi}\left(\sin x + \frac{\sin 3x}{3} + \frac{\sin 5x}{5} + \cdots\right) \tag{15.46}$$

To illustrate the behaviour of the series we consider the n-term approximations (partial sums),

$$S_1 = \frac{1}{2} \qquad\qquad S_2 = \frac{1}{2} + \frac{2}{\pi}\sin x$$

$$S_3 = \frac{1}{2} + \frac{2}{\pi}\left(\sin x + \frac{\sin 3x}{3}\right) \qquad S_4 = \frac{1}{2} + \frac{2}{\pi}\left(\sin x + \frac{\sin 3x}{3} + \frac{\sin 5x}{5}\right)$$

$\cdots$

The graphs of the first four partial sums are shown in Figure 15.7.

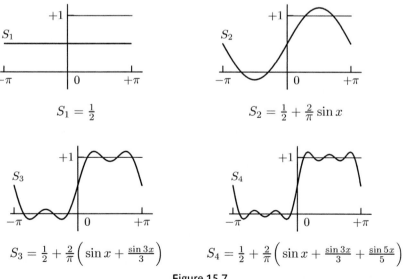

$$S_1 = \tfrac{1}{2} \qquad\qquad S_2 = \tfrac{1}{2} + \tfrac{2}{\pi}\sin x$$

$$S_3 = \tfrac{1}{2} + \tfrac{2}{\pi}\left(\sin x + \tfrac{\sin 3x}{3}\right) \qquad S_4 = \tfrac{1}{2} + \tfrac{2}{\pi}\left(\sin x + \tfrac{\sin 3x}{3} + \tfrac{\sin 5x}{5}\right)$$

Figure 15.7

This example shows how an arbitrary wave is built up by interference of harmonic waves, with constructive (additive) interference in some regions and destructive (subtractive) interference in others. The constant term S_1 is the *mean value* of the function in the interval. The two-term approximation, S_2, includes the first correction to this mean value, and S_3, S_4, ... include successively smaller corrections. The graph of the partial sum S_{50} is shown in Figure 15.8, and demonstrates clearly how the Fourier series, continuous at all points in the interval, deals with the discontinuities of the function.

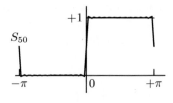

Figure 15.8

▸ Exercises 8–11

Change of period

A periodic function $f(x)$ with period 2π is transformed into a periodic function $g(z)$ with period $2l$ when the variable x in $f(x)$ is replaced by $\pi z/l$. Thus, if

$$f(x) = f\left(\frac{\pi z}{l}\right) = g(z)$$

then

$$f(x + 2\pi) = f\left(\frac{\pi}{l}(z + 2l)\right) = g(z + 2l)$$

For example, the function $\sin x = \sin(\pi z/l)$ is periodic in x with period 2π, and is periodic in z with period $2l$.

This simple change of variable is used to generalize the Fourier method, equations (15.37) to (15.39), for the expansion of functions with periods other than 2π. Thus, a function $f(x)$ that is periodic in x with period $2l$,

$$f(x) = f(x + 2l) \tag{15.47}$$

can be expanded as the Fourier series

$$f(x) = \frac{a_0}{2} + \sum_{n=1}^{\infty}\left(a_n \cos\frac{n\pi x}{l} + b_n \sin\frac{n\pi x}{l}\right) \tag{15.48}$$

with Fourier coefficients

$$a_n = \frac{1}{l}\int_{-l}^{+l} f(x)\cos\frac{n\pi x}{l}\, dx \tag{15.49}$$

$$b_n = \frac{1}{l}\int_{-l}^{+l} f(x)\sin\frac{n\pi x}{l}\, dx \tag{15.50}$$

The following example demonstrates the Fourier series for an arbitrary period, and it also prepares the ground for the completion of the solution of the wave equation for the vibrating string discussed in Section 14.7.

EXAMPLE 15.7 The Fourier series of the function

$$f(x) = \begin{cases} -\dfrac{2A}{l}(l+x) & \text{for} \quad -l \le x \le -\dfrac{1}{2} \\[2mm] \dfrac{2Ax}{l} & \text{for} \quad -\dfrac{1}{2} \le x \le +\dfrac{1}{2} \\[2mm] \dfrac{2A}{l}(l-x) & \text{for} \quad +\dfrac{1}{2} \le x \le +l \end{cases}$$

(15.51)

Figure 15.9 shows that the function is an *odd* function of x, with $f(-x)=-f(x)$. It follows from the discussion of even and odd functions in Section 5.3 that an integral

$$\int_{-l}^{+l} f(x)g(x)\,dx$$

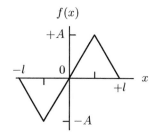

Figure 15.9

is zero unless $g(x)$ is also an odd function of x, or if it contains an odd component. Of the expansion functions in (15.48), the sine functions are odd but the cosine functions are even. All the integrals (15.49) for a_n are therefore zero, and the Fourier series reduces to the Fourier sine series for the expansion of an odd function:

$$f(x) = \sum_{n=1}^{\infty} b_n \sin \frac{n\pi x}{l}$$

(15.52)

with coefficients given by (15.50). It also follows from the discussion of Section 5.3, equation (5.25), that the integral (15.50) over the whole interval $-l \le x \le l$ is equal to twice the integral over the half-interval $0 \le x \le l$:

$$b_n = \frac{2}{l} \int_0^{+l} f(x) \sin \frac{n\pi x}{l}\, dx$$

(15.53)

For the function (15.51), we then have

$$b_n = \frac{2}{l}\left\{ \int_0^{l/2} f(x) \sin \frac{n\pi x}{l}\, dx + \int_{l/2}^{l} f(x) \sin \frac{n\pi x}{l}\, dx \right\}$$

$$= \frac{4A}{l^2}\left\{ \int_0^{l/2} x \sin \frac{n\pi x}{l}\, dx + \int_{l/2}^{l} (l-x) \sin \frac{n\pi x}{l}\, dx \right\}$$

and, after integration by parts,

$$b_n = \begin{cases} 0 & \text{if } n \text{ even} \\ +\dfrac{8A}{n^2\pi^2} & \text{if } n = 1, 5, 9, 13, \ldots \\ -\dfrac{8A}{n^2\pi^2} & \text{if } n = 3, 7, 11, 15, \ldots \end{cases} \qquad (15.54)$$

Therefore

$$f(x) = \frac{8A}{\pi^2}\left[\frac{1}{1^2}\sin\frac{\pi x}{l} - \frac{1}{3^2}\sin\frac{3\pi x}{l} + \frac{1}{5^2}\sin\frac{5\pi x}{l} - \cdots\right] \qquad (15.55)$$

▶ Exercise 12, 13

15.5 The vibrating string

It was shown in Section 14.7 that the solution of the wave equation for the vibrating string of length l is (equation (14.103))

$$y(x, t) = \sum_{n=1}^{\infty} \sin\frac{n\pi x}{l}\left[A_n \cos\omega_n t + B_n \sin\omega_n t\right] \qquad (15.56)$$

in which the coefficients A_n and B_n are determined by the initial conditions (equations (14.104))

$$y(x, 0) = f(x) = \sum_{n=1}^{\infty} A_n \sin\frac{n\pi x}{l}$$

$$\left(\frac{\partial y}{\partial t}\right)_{t=0} = g(x) = \sum_{n=1}^{\infty} B_n\omega_n \sin\frac{n\pi x}{l} \qquad (15.57)$$

where $f(x)$ and $g(x)$ are given displacement and velocity functions that describe the state of the system at time $t = 0$.

We suppose that, at $t = 0$, we pull the string sideways at its centre, and then let go. The initial shape of the string is shown in Figure 15.10, and the displacement function is therefore identical to the function discussed in Example 15.7, except that it is defined only within the half-interval $0 \le x \le l$. The Fourier representation of the function is then given by equation (15.55):

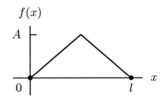

Figure 15.10

$$y(x, 0) = f(x) = \frac{8A}{\pi^2}\left[\frac{1}{1^2}\sin\frac{\pi x}{l} - \frac{1}{3^2}\sin\frac{3\pi x}{l} + \frac{1}{5^2}\sin\frac{5\pi x}{l} - \cdots\right] \qquad (15.58)$$

and the coefficients A_n in (15.56) are identical to the Fourier coefficients b_n given by (15.54). In addition, the initial velocity of the string is zero at all points along its length, so that

$$\left(\frac{\partial y}{\partial t}\right)_{t=0} = g(x) = 0 \qquad (15.59)$$

and all the coefficients B_n in (15.56) are zero. It follows that, for the given initial conditions, the motion of the string is given by the wave function

$$y(x, t) = \frac{8A}{\pi^2}\left[\sin\frac{\pi x}{l}\cos\omega_1 t - \frac{1}{9}\sin\frac{3\pi x}{l}\cos\omega_3 t + \frac{1}{25}\sin\frac{5\pi x}{l}\cos\omega_5 t - \cdots\right]$$

$$(15.60)$$

where, by equation (14.96), $\omega_n = n\pi v/l$ with v constant. The function (15.60) is periodic in t with period $\tau = 2\pi/\omega_1 = 2l/v$, and Figure 15.11 shows how the string behaves over the first quarter of a period; the graphs have been obtained from the 25-term approximation to the wave function (terms up to $n = 49$).

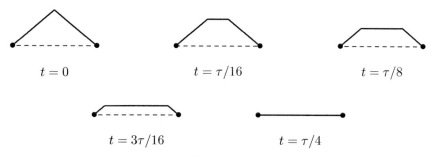

$t = 0$ $t = \tau/16$ $t = \tau/8$

$t = 3\tau/16$ $t = \tau/4$

Figure 15.11

This startling behaviour can be understood by a consideration of the forces acting at each point on the string. Because the tension is uniform throughout, no net force acts at a point on a straight-line section. The essential shape of the string is therefore maintained, with the motion determined by the instantaneous force acting at the two points where the gradient changes (or at the single point at the turning points of the motion).

▸ Exercise 14

15.6 Fourier transforms

In Section 15.4 we were concerned with the use of Fourier series for the representation of functions that are periodic, but several important applications of **Fourier analysis** in the physical sciences involve functions that are not periodic. The Fourier analysis of nonperiodic functions is achieved by letting the width of the base interval become indefinitely large ($l \to \infty$), and by transforming the Fourier series into an infinite integral, called a Fourier integral or a Fourier transform.

The infinite interval

The Fourier series (15.48) for arbitrary interval $2l$,

$$f(x) = \frac{a_0}{2} + \sum_{n=1}^{\infty} \left(a_n \cos \frac{n\pi x}{l} + b_n \sin \frac{n\pi x}{l} \right) \qquad (15.61)$$

can be written as

$$f(x) = \sum_{n=0}^{\infty} c_n \qquad (15.62)$$

where

$$c_0 = \frac{a_0}{2}, \quad c_n = a_n \cos \frac{n\pi x}{l} + b_n \sin \frac{n\pi x}{l}, \quad n > 0$$

A sum like (15.62) can be considered as the sum of the areas of rectangles of width $\Delta n = 1$ and height c_n, as illustrated in Figure 15.12,

$$f(x) = \sum_{n=0}^{\infty} c_n \, \Delta n \qquad (15.63)$$

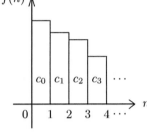

Figure 15.12

We now make the the substitutions

$$y_n = \frac{\pi}{l} n, \quad \Delta y_n = \frac{\pi}{l} \Delta n, \quad u(y_n) = \frac{l}{\pi} a_n, \quad v(y_n) = \frac{l}{\pi} b_n \qquad (15.64)$$

Equation (15.61) is then

$$f(x) = \left\{ \frac{u(y_0)}{2} + \sum_{n=1}^{\infty} \left[u(y_n) \cos x y_n + v(y_n) \sin x y_n \right] \right\} \Delta y_n \qquad (15.65)$$

where, by (15.49) and (15.50) for the Fourier coefficients,

$$u(y_n) = \frac{l a_n}{\pi} = \frac{1}{\pi} \int_{-l}^{+l} f(x) \cos x y_n \, dx \qquad (15.66)$$

$$v(y_n) = \frac{l b_n}{\pi} = \frac{1}{\pi} \int_{-l}^{+l} f(x) \sin x y_n \, dx \qquad (15.67)$$

We are now in a position to let $l \to \infty$ and $\Delta y_n \to 0$. Equation (15.65) has the form

$$f(x) = \sum_{n=0}^{\infty} F(y_n) \Delta y_n \qquad (15.68)$$

so that, by the discussion in Section 5.4 of the Riemann integral as the limit of a sum,

$$f(x) = \lim_{l \to \infty} \sum_{n=0}^{\infty} F(y_n) \, \Delta y_n = \int_0^{\infty} F(y) \, dy$$

Equation (15.65) is then

$$f(x) = \int_0^{\infty} \left[u(y) \cos xy + v(y) \sin xy \right] dy \qquad (15.69)$$

where, after taking the limit in (15.66) and (15.67),

$$u(y) = \frac{1}{\pi} \int_{-\infty}^{+\infty} f(x) \cos xy \, dx \qquad (15.70)$$

$$v(y) = \frac{1}{\pi} \int_{-\infty}^{+\infty} f(x) \sin xy \, dx \qquad (15.71)$$

The representation (15.69) of $f(x)$ is called a **Fourier integral** (the name Fourier transform is normally reserved for a slightly different form, as described below).*

EXAMPLE 15.8 For the square wave

$$f(x) = \begin{cases} A & \text{for} \quad -k < x < +k \\ 0 & \text{otherwise} \end{cases} \qquad (15.72)$$

Find (i) the Fourier series expansion in interval $-l \le x \le +l$ with $k < l$, (ii) the Fourier integral in the limit $l \to \infty$

(i) The Fourier series

Figure 15.13 shows that the function is an *even* function of x, and it can therefore be represented in the interval $-l \le x \le +l$ by the **Fourier cosine series**

$$f(x) = \frac{a_0}{2} + \sum_{n=1}^{\infty} a_n \cos \frac{n\pi x}{l}$$

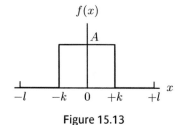

Figure 15.13

where $a_0 = \dfrac{2Ak}{l}$ and, for $n > 0$,

$$a_n = \frac{2}{l} \int_0^l f(x) \cos \frac{n\pi x}{l} \, dx = \frac{2A}{l} \int_0^k \cos \frac{n\pi x}{l} \, dx = \frac{2Ak}{l} \frac{\sin(n\pi k/l)}{(n\pi k/l)}$$

---

* The 'derivation' of (15.69) given here is only an outline, but is sufficient for most purposes; all problems associated with discontinuities of the function and with the existence of the limit have been ignored.

The Fourier-series representation of the function in the finite interval is therefore

$$f(x) = \frac{2Ak}{l}\left[\frac{1}{2} + \sum_{n=1}^{\infty} \frac{\sin(n\pi k/l)}{(n\pi k/l)}\cos\frac{n\pi x}{l}\right]$$

(ii) The Fourier integral

To find the value of this expression in the limit $l \to \infty$, we make the substitution

$$y_n = n\pi/l, \; \Delta y_n = \pi/l$$

Then

$$f(x) = \frac{2Ak}{\pi}\left[\frac{1}{2} + \sum_{n=1}^{\infty} \frac{\sin(ky_n)}{(ky_n)}\cos xy_n\right]\Delta y_n \qquad (15.73)$$

and, in the limit $l \to \infty$,

$$f(x) = \frac{2Ak}{\pi}\int_0^{\infty} \frac{\sin(ky)}{(ky)}\cos xy \, dy \qquad (15.74)$$

(the constant term, corresponding to $n = 0$, does not contribute).

This Fourier cosine integral representation of the function can be written in the form (15.69) as

$$f(x) = \int_0^{\infty} u(y)\cos xy \, dy \qquad (15.75)$$

with $u(y)$ given by (15.70). Thus

$$u(y) = \frac{1}{\pi}\int_{-\infty}^{+\infty} f(x)\cos xy \, dx = \frac{2}{\pi}\int_0^{\infty} f(x)\cos xy \, dx$$

$$= \frac{2A}{\pi}\int_0^{k} \cos xy \, dx = \frac{2Ak}{\pi}\frac{\sin(ky)}{(ky)} \qquad (15.76)$$

as required by (15.74). For the corresponding pair of Fourier transforms see Figure 15.14(a).

The exponential form

The Fourier-integral representation of a function, equations (15.69) to (15.71), can be written in exponential form by making use of the Euler relations

$$\cos xy = \frac{1}{2}(e^{ixy} + e^{-ixy}), \qquad \sin xy = \frac{1}{2i}(e^{ixy} - e^{-ixy}) \qquad (15.77)$$

We define the function

$$w(y) = \frac{1}{2}\Big[u(y) - iv(y)\Big] \tag{15.78}$$

It is shown in Example 15.9 that equation (15.69) can then be written as

$$f(x) = \int_{-\infty}^{+\infty} w(y)e^{ixy}\,dy \tag{15.79}$$

with equations (15.70) and (15.71) replaced by

$$w(y) = \frac{1}{2\pi}\int_{-\infty}^{+\infty} f(x)\,e^{-ixy}\,dx \tag{15.80}$$

EXAMPLE 15.9 Derivation of the exponential form

Substitution of the Euler relations (15.77) into (15.69) gives

$$f(x) = \frac{1}{2}\int_{0}^{\infty}\Big[u(y)(e^{ixy}+e^{-ixy}) - iv(y)(e^{ixy}-e^{-ixy})\Big]dy$$

$$= \frac{1}{2}\int_{0}^{\infty}\Big[u(y)-iv(y)\Big]e^{ixy}\,dy + \frac{1}{2}\int_{0}^{\infty}\Big[u(y)+iv(y)\Big]e^{-ixy}\,dy$$

or, replacing y by $-y$ in the second integral on the right,

$$f(x) = \frac{1}{2}\int_{0}^{\infty}\Big[u(y)-iv(y)\Big]e^{ixy}\,dy + \frac{1}{2}\int_{-\infty}^{0}\Big[u(-y)+iv(-y)\Big]e^{ixy}\,dy$$

From equations (15.70) and (15.71), we have

$$u(-y) = u(y), \qquad v(-y) = -v(y)$$

Therefore,

$$f(x) = \frac{1}{2}\int_{-\infty}^{\infty}\Big[u(y)-iv(y)\Big]e^{ixy}\,dy$$

$$= \int_{-\infty}^{+\infty} w(y)e^{ixy}\,dy$$

where

$$w(y) = \frac{1}{2}\Big[u(y)-iv(y)\Big] = \frac{1}{2\pi}\int_{-\infty}^{+\infty} f(x)e^{-ixy}\,dy$$

Fourier transform pairs

Equations (15.79) and (15.80) are conventionally written in the more symmetrical form by means of the substitution $g(y) = \sqrt{2\pi}\, w(y)$. Then

$$f(x) = \frac{1}{\sqrt{2\pi}} \int_{-\infty}^{+\infty} g(y) e^{ixy}\, dy \tag{15.81}$$

$$g(y) = \frac{1}{\sqrt{2\pi}} \int_{-\infty}^{+\infty} f(x) e^{-ixy}\, dx \tag{15.82}$$

Two functions that are related by equations (15.81) and (15.82) are called a pair of **Fourier transforms**. The function g is called the Fourier transform of the function f, and f is called the (inverse) Fourier transform of g. In general, the Fourier transform of a function $f(x)$ exists if the function is piecewise continuous and if it is absolutely integrable; that is, if

$$\int_{-\infty}^{+\infty} |f(x)|\, dx \qquad \text{exists}$$

Three of the important elementary Fourier transform pairs are illustrated in Figure 15.14.

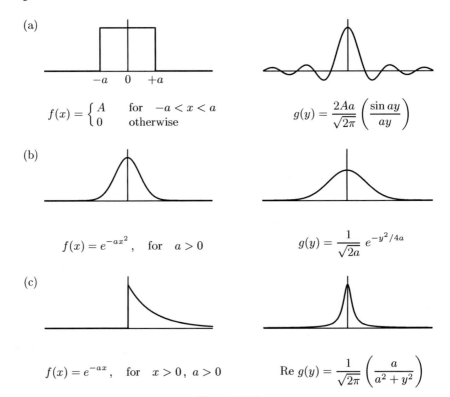

(a)

$$f(x) = \begin{cases} A & \text{for} \quad -a < x < a \\ 0 & \text{otherwise} \end{cases} \qquad g(y) = \frac{2Aa}{\sqrt{2\pi}} \left(\frac{\sin ay}{ay} \right)$$

(b)

$$f(x) = e^{-ax^2}, \quad \text{for} \quad a > 0 \qquad g(y) = \frac{1}{\sqrt{2a}} e^{-y^2/4a}$$

(c)

$$f(x) = e^{-ax}, \quad \text{for} \quad x > 0,\ a > 0 \qquad \text{Re } g(y) = \frac{1}{\sqrt{2\pi}} \left(\frac{a}{a^2 + y^2} \right)$$

Figure 15.14

EXAMPLE 15.10 Find the Fourier transform of the exponential function, Figure 15.14(c),

$$f(x) = e^{-ax} \qquad \text{for } x > 0 \text{ and } a > 0$$

By equation (15.82),

$$g(y) = \frac{1}{\sqrt{2\pi}} \int_{-\infty}^{+\infty} f(x) e^{-ixy} \, dx = \frac{1}{\sqrt{2\pi}} \int_{0}^{\infty} e^{-ax} e^{-ixy} \, dx$$

$$= \frac{1}{\sqrt{2\pi}} \int_{0}^{+\infty} e^{-(a+iy)x} \, dx$$

The presence of a complex exponent does not alter the normal rule for the integration of the exponential. Thus

$$g(y) = \frac{1}{\sqrt{2\pi}} \left[\frac{-e^{-(a+iy)x}}{a+iy} \right]_{0}^{\infty} = \frac{1}{\sqrt{2\pi}} \left(\frac{1}{a+iy} \right)$$

because $e^{-(a+iy)x} = e^{-ax} e^{-iyx} \rightarrow 0$ as $x \rightarrow \infty$. The Fourier transform is complex and to find the real part we write

$$g(y) = \frac{1}{\sqrt{2\pi}} \left(\frac{a-iy}{(a+iy)(a-iy)} \right) = \frac{1}{\sqrt{2\pi}} \left(\frac{a-iy}{a^2+y^2} \right)$$

Then

$$\text{Re } g(y) = \frac{1}{\sqrt{2\pi}} \left(\frac{a}{a^2+y^2} \right) \tag{15.83}$$

This function is called a **Lorentzian**, and gives the shapes of spectral lines in Fourier transform spectroscopy.

▶ Exercises 15–17

In applications in the physical sciences, the variables x and y are usually a pair of **conjugate variables**. The most important of these are the coordinate–momentum pair, such as the x coordinate and its conjugate momentum, the linear momentum p_x, and the time–frequency (or time–energy) pair. A Fourier transformation is then a transformation of the description of a physical system from one in the space (or domain) of one of the variables of a conjugate pair to that of the other. The momentum to coordinate Fourier transformation is an essential tool in the analysis of the results of diffraction experiments; the observed diffraction pattern is the representation of the structure of the system in momentum space and the Fourier transformation gives the structure in ordinary (coordinate) space. Of even greater importance to the chemist is the time to frequency transformation because it forms the basis for Fourier transform magnetic resonance spectroscopy. In such an experiment, the molecules of a sample (or other species with excitable degrees of freedom) are excited by a short pulse of radiation, and the system is observed as it relaxes to its thermodynamically stable state.

Relaxation is essentially a first-order kinetic process and in the simplest case, when the molecules can undergo transitions at only one frequency, ω_0 say, the intensity of the output signal shows exponential decay as a function of time:

$$I(t) = e^{-t/T} \tag{15.84}$$

where T is called the relaxation time. This is an example of case (c) in Figure 15.14, with x as time t and y as frequency ω. The Fourier transform of the decay curve is therefore a Lorentzian curve (see Example 15.10)

$$F(\omega) = \text{Re} \frac{1}{\sqrt{2\pi}} \int_0^\infty I(t) e^{-i\omega t}\, dt = \frac{1}{\sqrt{2\pi}} \left(\frac{T}{1 + \omega^2 T^2} \right) \tag{15.85}$$

The width of this 'spectral line' is inversely proportional to the relaxation time T. Its position (at ω_0) is determined in practice by the fine structure of the output, as illustrated in Example 15.11.

EXAMPLE 15.11 Find the real part of the Fourier transform of the exponentially damped harmonic wave,

$$f(x) = e^{-ax} \cos bx \qquad \text{for } x > 0 \text{ and } a > 0$$

By equation (15.82),

$$g(y) = \frac{1}{\sqrt{2\pi}} \int_{-\infty}^{+\infty} f(x) e^{-ixy}\, dx$$

Figure 15.15

$$= \frac{1}{\sqrt{2\pi}} \int_0^\infty e^{-ax} \cos bx\, e^{-ixy}\, dx$$

or, because $\cos bx = (e^{ibx} + e^{-ibx})/2$,

$$g(y) = \frac{1}{2\sqrt{2\pi}} \int_0^\infty \left[e^{-(a+ib-iy)x} + e^{-(a-ib-iy)x} \right] dx$$

$$= \frac{1}{2\sqrt{2\pi}} \left(\frac{1}{a+ib-iy} + \frac{1}{a-ib-iy} \right)$$

The real part of this is (see Example 15.10)

$$\operatorname{Re} g(y) = \frac{1}{2\sqrt{2\pi}} \left(\frac{a}{a^2 + (b-y)^2} + \frac{a}{a^2 + (b+y)^2} \right) \tag{15.86}$$

If we replace (x, y, a, b) by $(t, \omega, 1/T, \omega_0)$ in (15.86), we obtain the pair of Fourier transforms

$$f(t) = e^{-t/T} \cos \omega_0 t, \quad g(\omega) = \frac{1}{2\sqrt{2\pi}} \left(\frac{T}{1 + (\omega - \omega_0)^2 T^2} + \frac{T}{1 + (\omega + \omega_0)^2 T^2} \right)$$

The second term of $g(\omega)$ is in practice small compared with the first, and the function then reduces to a Lorentzian centred at $\omega = \omega_0$.

15.7 Exercises

Section 15.2

1. Given the power series $f(x) = a_0 + a_1 x + a_2 x^2 + \cdots$, use equation (15.9) to find the coefficient c_1 of $P_1(x)$ in the expansion (15.3) of $f(x)$ in Legendre polynomials.
2. (i) Find the first three terms of the expansion of the function

$$f(x) = \begin{cases} 0 & \text{if} \quad -1 < x < 0 \\ x & \text{if} \quad 0 < x < 1 \end{cases}$$

(Figure 15.16) in Legendre polynomials. (ii) Sketch graphs of the one-term, two-term, and three-term representations of $f(x)$.

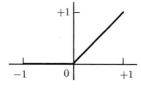

Figure 15.16

Section 15.3

3. Show that the coefficient of $P_1(x)$ in the expansion

$$e^{itx} = \sum_{l=0}^{\infty} c_l P_l(x)$$

is $c_1 = 3i j_1(t)$ where $j_1(t) = \frac{1}{t} \left(\frac{\sin t}{t} - \cos t \right)$.

4. Expand in terms of Legendre polynomials (i) $\cos tx$, (ii) $\sin tx$.
5. Find the first nonzero term in the expansion in powers of $1/R$ of the potential at point P for the system of three charges shown in Figure 15.17.

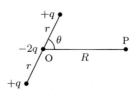

Figure 15.17

6. For the square system of five charges in Figure 15.18, show that the leading term in the expansion in powers of $1/R$ of the electrostatic potential at P (in the plane of the square) is $V = qr^2/4\pi\varepsilon_0 R^3$, independent of orientation.

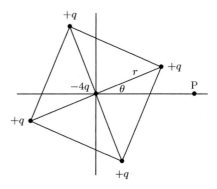

Figure 15.18

Section 15.4

7. Confirm the relations **(i)** (15.33), **(ii)** (15.34) and **(iii)** (15.36).

8. A periodic function with period 2π is defined by

$$f(x) = \begin{cases} 1 & \text{if } -\dfrac{\pi}{2} < x < \dfrac{\pi}{2} \\[2mm] 0 & \text{if } \dfrac{\pi}{2} < |x| < \pi \end{cases}$$

(i) Draw the graph of the function in the interval $-3\pi \le x \le 3\pi$. **(ii)** Find the Fourier series of the function [Hint: $f(x)$ is an even function of x]. **(iii)** Use the series to show that

$$\frac{\pi}{4} = 1 - \frac{1}{3} + \frac{1}{5} - \frac{1}{7} + \cdots$$

[Hint: substitute a suitable value for x in the series].

9. A function with period 2π is defined by

$$f(x) = x, \qquad -\pi < x < \pi$$

(i) Draw the graph of the function in the interval $-3\pi \le x \le 3\pi$. **(ii)** Find the Fourier series of the function. [Hint: $f(x)$ is an odd function of x.] **(iii)** Draw the graphs of the first four partial sums of the series.

10. A function with period 2π is defined by

$$f(x) = x^2, \qquad -\pi < x < \pi$$

(i) Draw the graph of the function in the interval $-3\pi \le x \le 3\pi$. **(ii)** Find the Fourier series of the function. **(iii)** Use the series to show that

$$\frac{\pi^2}{6} = \sum_{n=1}^{\infty} \frac{1}{n^2} = 1 + \frac{1}{4} + \frac{1}{9} + \frac{1}{16} + \cdots$$

$$\frac{\pi^2}{12} = \sum_{n=1}^{\infty} \frac{(-1)^{n+1}}{n^2} = 1 - \frac{1}{4} + \frac{1}{9} - \frac{1}{16} + \cdots$$

11. Show that the Fourier series of the periodic function defined by (see Figure 15.4)

$$f(t) = \begin{cases} \sin t & \text{if } 0 \le t \le \pi \\ 0 & \text{if } -\pi \le t \le 0 \end{cases}$$

is $\dfrac{1}{\pi} + \dfrac{1}{2}\sin t - \dfrac{2}{\pi}\left[\dfrac{\cos 2t}{1 \cdot 3} + \dfrac{\cos 4t}{3 \cdot 5} + \dfrac{\cos 6t}{5 \cdot 7} + \cdots \right]$

12. A function with period $2l$ is defined by

$$f(x) = \begin{cases} x(l-x) & \text{if} \quad 0 < x < l \\ x(l+x) & \text{if} \quad -l < x < 0 \end{cases}$$

(i) Draw the graph of the function in the interval $-l \le x \le l$. (ii) Find the Fourier series of the function.

13. A function with period $2l$ is defined by

$$f(x) = \begin{cases} x(l-x) & \text{if} \quad 0 < x < l \\ -x(l+x) & \text{if} \quad -l < x < 0 \end{cases}$$

(i) Draw the graph of the function in the interval $-l \le x \le l$. (ii) Find the Fourier series of the function.

Section 15.5

14. Find the solution of the diffusion equation

$$\frac{\partial T}{\partial t} = D \frac{\partial^2 T}{\partial x^2}$$

in the interval $0 \le x \le l$ that satisfies the boundary conditions $T(0, t) = T(l, t) = 0$ and initial condition $T(x, 0) = x(l - x)$, and that decreases exponentially with time (See Exercise 17 of Chapter 14).

Section 15.6

Find the Fourier transform:

15. $f(x) = \begin{cases} 1, & \text{if} \quad a < x < b \\ 0, & \text{otherwise} \end{cases}$

16. $e^{-a|x|} \ (a > 0)$

17. Show that the Fourier transform of $e^{-ax^2} \ (a > 0)$ is

$$\frac{1}{\sqrt{2a}} e^{-y^2/4a} \qquad \text{(Figure 15.14b)}$$

$$\left[\text{Hint: change variable to } t = \sqrt{a}\, x + \frac{iy}{2\sqrt{a}} \text{ and use } \int_{-\infty}^{\infty} e^{-\alpha t^2}\, dt = \sqrt{\frac{\pi}{\alpha}} \right]$$

16 Vectors

16.1 Concepts

Physical quantities such as mass, temperature, and distance have values that are specified by single real numbers, in appropriate units; for example, 3 kg, 273 K, and 12 m. Such quantities, having **magnitude** only, are called **scalar** quantities and obey the rules of the algebra of real numbers. Other physical quantities, called **vectors**, require both magnitude and **direction** for their specification. For example, velocity is speed in a given direction, the speed being the magnitude of the velocity vector. Other examples are force, electric field, magnetic field, and displacement. To qualify as vectors these quantities must obey the rules of **vector algebra**.[1] Vector notation and vector algebra are important for the formulation and solution of physical problems in three dimensions; in mechanics, fluid dynamics, electromagnetic theory, and engineering design. Some of these uses of vectors, important in molecular dynamics, spectroscopy, and theoretical chemistry, are discussed in examples throughout this chapter.

A vector is represented graphically by a directed line segment; that is, a segment of line whose length is the magnitude of the vector, in appropriate units, and whose orientation in space, together with an arrowhead, gives the direction.

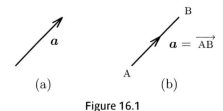

(a)   (b)

Figure 16.1

Figure 16.1 shows two graphical representations of the same vector, and two ways of representing it in print. In (a), the vector a is represented by an arrow; in (b), A and B are the **initial** and **terminal** points of the vector, and the notation $\overrightarrow{AB}$ is particularly useful when the vector is a displacement in space. The length, magnitude, or modulus of a is written as $|a|$ or simply as a (a scalar). A vector of unit length is called a **unit vector**. A vector of zero length is call the **null vector 0**, and no direction is defined in this case.

---

[1] Vector algebra has its origins in the algebra of quaternions discovered by Hamilton in 1843 and in Grassmann's theory of n-dimensional vector spaces of 1844. The modern notation for vectors in three dimensions is due to Gibbs.

William Rowan Hamilton (1805–1865), born in Dublin, is reputed to have been fluent in Latin, Greek, the modern European languages, Hebrew, Persian, Arabic, Sanskrit, and others at the age of ten. He entered Trinity College, Dublin, in 1823 and became Astronomer Royal of Ireland and Professor of Astronomy in 1827 without taking his degree. He is best known for the reformulation and generalization of the mechanics of Newton, Euler, and Lagrange that became important in the formulation of statistical and quantum mechanics.

Figures 16.2 (a) and (b) illustrate several types of vector associated with the motion of a body in space, along curve PQ, under the influence of external forces. In Figure (a), the position of point A on the curve is given by $a = \overrightarrow{OA}$ whose initial point is at the origin of a coordinate system and whose terminal point is at A. The vector a is called a **position vector**, and its value (magnitude and direction) changes as the body moves along the curve. When the body has moved from A to B the position vector has changed from a to b, and the displacement A to B is given by the vector $c = \overrightarrow{AB}$. We will see that vector algebra gives $c = b - a$; that is, the displacement A to B is equal to the change of position vector. We note that a position vector is a type of displacement vector; but one whose initial point is bound to a fixed point, the origin of a coordinate system.

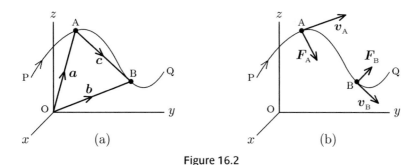

Figure 16.2

At each point along the curve PQ, the body has velocity v along the direction of motion at that point; that is, the direction of v at every point is tangential to the curve. Figure (b) shows the velocities at points A and B, and also forces acting on the body at these points.

16.2 Vector algebra

Equality

Two vectors a and b are equal,

$$a = b \qquad (16.1)$$

if they have the same length and the same direction. The initial points of equal vectors need coincide *only* if they are bound vectors, such as position vectors. Thus two separated bodies moving at the same speed and in the same direction have equal velocities.

Vector addition

The sum, or resultant, $a + b$ of vectors a and b is defined geometrically in Figure 16.3. In (a) the vectors are positioned with the initial point of b coincident with the terminal point of a. The sum $a + b$ is then equal to the vector drawn from the initial point of a to the terminal point of b. Figure (b) also shows that addition is commutative, with

$$a + b = b + a \qquad (16.2)$$

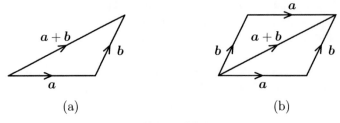

Figure 16.3

In this case a and b define the pairs of parallel sides of a parallelogram, and vector addition is said to obey the **parallelogram law**. If a and b are displacements of a body then their sum $a+b$ is the total displacement. If a and b are two forces acting on a body, then $a+b$ is the total (or resultant) force; the parallelogram law is then called the 'parallelogram of forces'.[2]

Subtraction

If $a+b=0$, where 0 is the null vector, then $b=-a$ is a vector that has the same length as a, $|b|=|a|$, but points in the opposite direction. The subtraction of vectors is then defined by (Figure 16.4)

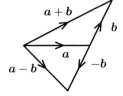

$$a+(-b)=a-b \qquad (16.3)$$

Figure 16.4

Scalar multiplication

The vector $a+a=2a$ has twice the length of a and has the same direction. In general, the product of a scalar (number) c and a vector a is written as ca. It has c times the length of a, has the same direction as a if c is positive, and has opposite direction if c is negative (Figure 16.5).

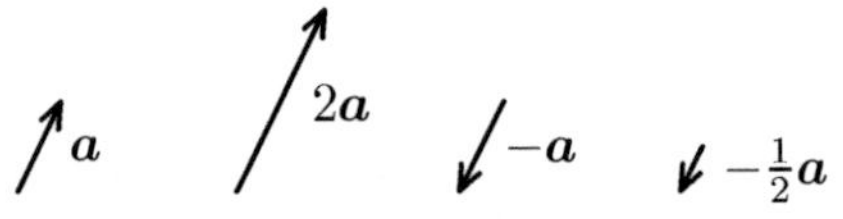

$$\nearrow a \qquad \nearrow 2a \qquad \swarrow -a \qquad \swarrow -\tfrac{1}{2}a$$

Figure 16.5

If $c=0$ then $ca=0$, and the direction is not defined. A vector a divided by its length $|a|$ is the unit vector $\hat{a}$ that has the same direction as a:

$$\hat{a} = \frac{a}{|a|} \qquad (16.4)$$

Unit vectors are often used to define direction.

---

[2] The parallelogram of velocities was described by Heron of Alexandria (1st century AD) in his *Mechanics*, but also appears in a work attributed to Aristotle (384–322 BC).

EXAMPLE 16.1 Show that the diagonals of a parallelogram bisect each other.

In Figure 16.6, C is the midpoint of diagonal OD, and C' is the midpoint of AB. Then

$$\overrightarrow{OC} = \frac{1}{2}\overrightarrow{OD} = \frac{1}{2}(\overrightarrow{OA} + \overrightarrow{AD}) = \frac{1}{2}(a+b)$$

Also

$$\overrightarrow{OC'} = \overrightarrow{OA} + \overrightarrow{AC'} = a + \frac{1}{2}\overrightarrow{AB}$$

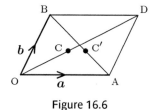

Figure 16.6

But $\overrightarrow{AB} = b - a$. Therefore

$$\overrightarrow{OC'} = a + \frac{1}{2}(b-a) = \frac{1}{2}(a+b) = \overrightarrow{OC}$$

The midpoints C and C' therefore coincide.

EXAMPLE 16.2 Show that the mean of the position vectors of the vertices of a triangle is the position vector of the centroid of the triangle.

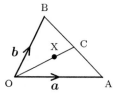

Figure 16.7

In Figure 16.7, the vertices A and B have positions (position vectors) a and b relative to vertex O (with null position vector **0**). The mean of the position vectors of the vertices is therefore

$$\overrightarrow{OX} = \frac{1}{3}(\overrightarrow{OO} + \overrightarrow{OA} + \overrightarrow{OB}) = \frac{1}{3}(0 + a + b) = \frac{1}{3}(a+b)$$

It is shown in Example 16.1 that if C is the mid point of AB then $\overrightarrow{OC} = (a+b)/2$. Therefore

$$\overrightarrow{OX} = \frac{2}{3}\overrightarrow{OC}$$

and the mean lies on the line joining the vertex O to the midpoint of the opposite side. Similarly for the position with respect to the two other vertices, so that X lies at the point of intersection of these lines, and this is the centroid of the triangle.

▸ Exercise 1

16.3   Components of vectors

The component of a vector in a given direction is the length of its projection in that direction. In Figure 16.8 the component of *a* along the direction OP is the length

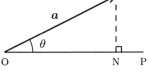

Figure 16.8

$$ON = |a| \cos \theta \qquad (16.5)$$

The concept of component is essential for the practical use of vectors for the solution of physical problems in three dimensions.

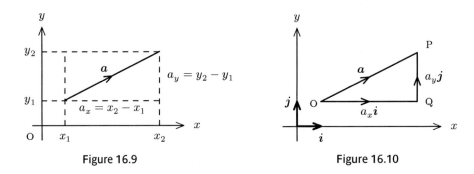

Figure 16.9   **Figure 16.10**

We consider first the simpler case of vectors in a plane. Let the initial and terminal points of *a* in the *xy*-plane be (x_1, y_1) and (x_2, y_2), as shown in Figure 16.9. The (cartesian) component of the vector in the *x*-direction is $a_x = x_2 - x_1$, and the component in the *y*-direction is $a_y = y_2 - y_1$. These two components are sufficient to specify the vector uniquely. Thus the length of the vector is $|a| = \sqrt{a_x^2 + a_y^2}$, and the direction is given by the slope a_y/a_x. We write the vector in terms of its cartesian components as

$$a = (a_x, a_y) \qquad (16.6)$$

Also, if the *x* and *y* directions are described by the unit vectors *i* and *j*, as in Figure 16.10, then $a = \overrightarrow{OP}$ can be expressed as the sum of the two vectors, $\overrightarrow{OQ} = a_x i$ in the *x*-direction (a_x times the unit vector *i* in the *x*-direction) and $\overrightarrow{QP} = a_y j$ in the *y*-direction,

$$a = a_x i + a_y j \qquad (16.7)$$

More generally, a vector in three dimensions (Figure 16.11) can be specified by its components, a_x, a_y and a_z, in the three cartesian directions *i, j*, and *k* (*k* is the unit vector in the *z*-direction).
We write

$$a = (a_x, a_y, a_z) = a_x i + a_y j + a_z k \qquad (16.8)$$

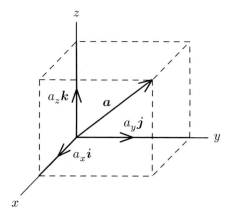

Figure 16.11

The rules of vector algebra can then be formulated in terms of vector components, and it is in this way that vectors are used for the solution of physical problems.[3] We consider the vectors

$$\boldsymbol{a} = (a_x, a_y, a_z) \qquad \boldsymbol{b} = (b_x, b_y, b_z)$$

(i) **Equality.** The vectors are equal when their corresponding components are equal,

$$\boldsymbol{a} = \boldsymbol{b} \quad \text{if} \quad a_x = b_x, \quad a_y = b_y \quad \text{and} \quad a_z = b_z \tag{16.9}$$

(ii) **Addition.** The sum $\boldsymbol{a} + \boldsymbol{b}$ is obtained by adding the corresponding components,

$$\boldsymbol{a} + \boldsymbol{b} = (a_x + b_x, a_y + b_y, a_z + b_z) \tag{16.10}$$

This is illustrated in Figure 16.12 for vectors in a plane.

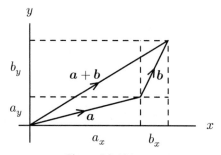

Figure 16.12

---

[3] Hamilton discovered quaternions in his search for an algebra of 3-dimensional complex numbers. They have the form $a + bi + cj + dk$, with rules of combination $i^2 + j^2 + k^2 = -1$, $ij = k = -ji$, $jk = i = -kj$, $ki = j = -ik$. Hamilton called a the scalar part and $bi + cj + dk$ the vector part. He discovered the rules whilst walking with his wife along the Royal Canal on 16 October 1843, a Monday, and he carved them on a stone on Brougham Bridge.

(iii) **Scalar multiplication.** The product ca of the scalar c and the vector a is obtained by multiplying each component of a by c,

$$ca = (ca_x, ca_y, ca_z) \tag{16.11}$$

EXAMPLE 16.3

(i) Find the vector $a = \overrightarrow{PQ}$ whose initial point P is $p = (2, 1, 0)$ and whose terminal point Q is $q = (1, 3, -2)$. (ii) What is the length of a? (iii) Find the unit vector parallel to a.

(i) $a = q - p = (1, 3, -2) - (2, 1, 0) = (-1, 2, -2)$

(ii) $|a| = \sqrt{(-1)^2 + 2^2 + (-2)^2} = 3$

(iii) $\hat{a} = a/|a| = (-1/3, 2/3, -2/3)$

➤ Exercises 2–5

EXAMPLE 16.4 Given $a = (2, 3, 1)$, $b = (1, -2, 0)$, and $c = (5, 2, -1)$, find (a) $d = 2a + 3b - c$ and (b) $|d|$.

(a) If $d = (d_x, d_y, d_z)$ then

$$d_x = 2a_x + 3b_x - c_x = (2 \times 2 + 3 \times 1 - 5) = 2$$
$$d_y = 2a_y + 3b_y - c_y = (2 \times 3 + 3 \times (-2) - 2) = -2$$
$$d_z = 2a_z + 3b_z - c_z = (2 \times 1 + 3 \times 0 - (-1)) = 3$$

and $d = (2, -2, 3)$.

(b) $|d| = \sqrt{d_x^2 + d_y^2 + d_z^2} = \sqrt{2^2 + (-2)^2 + 3^2} = \sqrt{17}$

➤ Exercises 6–10

Equation (16.9) for the equality of vectors and Example 16.4 show that, for vectors in three dimensions, a vector equation is equivalent to three simultaneous scalar equations, one for each component. Conversely, the three scalar equations are expressed by the single vector equation. Vector algebra then provides a powerful method for the formulation and solution of physical problems that involve vector quantities. The resolution into the component equations is often necessary only when the solution of a problem needs to be completed by the insertion of numerical values for the components.

EXAMPLE 16.5 The centre of mass of a system of N masses, m_1 with position vector $\mathbf{r}_1$ ('at position $\mathbf{r}_1$'), m_2 at $\mathbf{r}_2$, ..., m_N at $\mathbf{r}_N$ (Figure 16.13) is

$$\mathbf{R} = \frac{1}{M}(m_1\mathbf{r}_1 + m_2\mathbf{r}_2 + \cdots + m_N\mathbf{r}_N) = \frac{1}{M}\sum_{i=1}^{N} m_i\mathbf{r}_i \qquad (16.12)$$

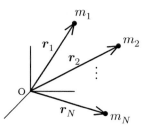

where $M = \sum_{i=1}^{N} m_i$ is the total mass.

The cartesian components of a position vector $\mathbf{r}$ of a point are the cartesian coordinates of the point; $\mathbf{r} = (x, y, z)$. The vector equation (16.12) therefore corresponds to three ordinary (scalar) equations, one for each coordinate of $\mathbf{R} = (X, Y, Z)$:

Figure 16.13

$$X = \frac{1}{M}\sum_{i=1}^{N} m_i x_i \qquad Y = \frac{1}{M}\sum_{i=1}^{N} m_i y_i \qquad Z = \frac{1}{M}\sum_{i=1}^{N} m_i z_i \qquad (16.13)$$

(see equation (5.40) for the one-dimensional case).

▸ Exercise 11

EXAMPLE 16.6 Dipole moments

The system of two charges shown in Figure 16.14, with $-q$ at $\mathbf{r}_1$ and q at $\mathbf{r}_2$, defines an electric dipole with vector dipole moment

$$\boldsymbol{\mu} = -q\mathbf{r}_1 + q\mathbf{r}_2 = q(\mathbf{r}_2 - \mathbf{r}_1) = q\mathbf{r} \qquad (16.14)$$

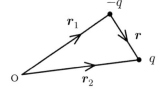

More generally, a system of N charges, q_1 at $\mathbf{r}_1$, q_2 at $\mathbf{r}_2$, ..., q_N at $\mathbf{r}_N$ has dipole moment *with respect to the origin* O as point of reference,

Figure 16.14

$$\boldsymbol{\mu} = (q_1\mathbf{r}_1 + q_2\mathbf{r}_2 + \cdots + q_N\mathbf{r}_N) = \sum_{i=1}^{N} q_i\mathbf{r}_i \qquad (16.15)$$

This quantity depends on the position of the reference point if the total charge $Q = \sum q_i$ is not zero. Thus, the position of charge q_i with respect to some point $\mathbf{R}$ is $\mathbf{r}_i - \mathbf{R}$, and the dipole moment of the system of charges with respect to $\mathbf{R}$ is

$$\boldsymbol{\mu}(\mathbf{R}) = \sum_{i=1}^{N} q_i(\mathbf{r}_i - \mathbf{R}) = \left(\sum_{i=1}^{N} q_i\mathbf{r}_i\right) - \left(\sum_{i=1}^{N} q_i\right)\mathbf{R} = \boldsymbol{\mu}(\mathbf{0}) - Q\mathbf{R} \qquad (16.16)$$

and $\boldsymbol{\mu}(\mathbf{R}) = \boldsymbol{\mu}(\mathbf{0})$ only if $Q = 0$.

A system of electric dipoles with moments $\boldsymbol{\mu}_1, \boldsymbol{\mu}_2, \ldots, \boldsymbol{\mu}_N$, has total dipole moment

$$\boldsymbol{\mu} = \boldsymbol{\mu}_1 + \boldsymbol{\mu}_2 + \cdots + \boldsymbol{\mu}_N = \sum_{i=1}^{N} \boldsymbol{\mu}_i \tag{16.17}$$

The total dipole moment of a molecule is sometimes interpreted as the (vector) sum of 'bond moments'; that is, a dipole moment is associated with each bond. In some cases of high symmetry, these bond moments may cancel to give zero total dipole moment. For example, the methane molecule in its stable state has its four hydrogens at the vertices of a regular tetrahedron, with the carbon at the centre. If one of the vertices is placed at the '111–position', with $r_1 = (a, a, a)$, the positions of the other vertices are $r_2 = (a, -a, -a)$, $r_3 = (-a, a, -a)$, and $r_4 = (-a, -a, a)$. The length of each of these bond vectors is $\sqrt{3}a$, the CH bondlength, and the dipole moment of each bond lies along the direction of the bond, and is therefore a multiple of the bond vector, $\boldsymbol{\mu}_i = k r_i$ $(i = 1, 2, 3, 4)$. The total dipole moment is then

$$\begin{aligned}
\boldsymbol{\mu} &= \boldsymbol{\mu}_1 + \boldsymbol{\mu}_2 + \boldsymbol{\mu}_3 + \boldsymbol{\mu}_4 \\
&= k(r_1 + r_2 + r_3 + r_4) \\
&= k(a + a - a - a, a - a + a - a, a - a - a + a) \\
&= k(0, 0, 0) \\
&= 0
\end{aligned}$$

➤ Exercise 12

Base vectors

The cartesian unit vectors i, j and k lie along the x, y, and z directions and have components,

$$i = (1, 0, 0), \qquad j = (0, 1, 0), \qquad k = (0, 0, 1) \tag{16.18}$$

The rules (16.10) and (16.11) then confirm that every vector in the three-dimensional space can be expressed as a linear combination of these three **base vectors**:

$$\begin{aligned}
a &= a_x i + a_y j + a_z k \\
&= a_x(1, 0, 0) + a_y(0, 1, 0) + a_z(0, 0, 1) \\
&= (a_x, 0, 0) + (0, a_y, 0) + (0, 0, a_z) \\
&= (a_x, a_y, a_z)
\end{aligned} \tag{16.19}$$

EXAMPLE 16.7 Given $a = 2i + 3j + k$ and $b = i - 2j$, find (i) $d = 2a + 3b$, (ii) a vector perpendicular to b, (iii) a vector perpendicular to d.

(i) $d = 2a + 3b = 2(2i + 3j + k) + 3(i - 2j) = (4i + 6j + 2k) + (3i - 6j)$

$\qquad = 7i + 2k$

(ii) Vector $b = i - 2j$ lies in the xy-plane, vector k lies along the z-direction. Therefore λk (any λ) is perpendicular to b.

(iii) Vector d lies in the xz-plane, and λj is perpendicular to d.

➤ Exercises 13, 14

Although the cartesian unit vectors provide the most widely used representation of vectors in three dimensions, other representations are sometimes more useful. Thus, any three noncoplanar vectors can be used as base. If a, b and c are three such vectors, not necessarily unit vectors or perpendicular, then any other vector in the space can be written

$$u = u_a a + u_b b + u_c c \qquad (16.20)$$

The numbers u_a, u_b, and u_c are the components of u along the directions of the base vectors. An example of the use of non-cartesian bases is found in crystallography in the description of the properties of regular lattices. In this case a, b and c define the crystal axes and the unit cell, and every lattice point has position vector (16.20) with components that are integers. This is illustrated in Figure 16.15 for a planar lattice.

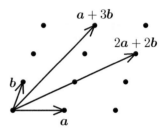

Figure 16.15

16.4 Scalar differentiation of a vector

A vector $a = a(t)$ is a function of the scalar variable t if its magnitude or its direction, or both, depends on the value of t. In Figure 16.16, $\overrightarrow{OA} = a(t)$ and $\overrightarrow{OB} = a(t + \Delta t)$ are the vectors for values t and $t + \Delta t$ of the variable. The change in the vector in interval Δt is

$$\Delta a = a(t + \Delta t) - a(t)$$

and the derivative of the vector with respect to t is defined in the usual way by the limit (if the limit exists)

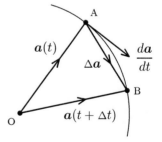

Figure 16.16

$$\frac{da}{dt} = \lim_{\Delta t \to 0} \left(\frac{\Delta a}{\Delta t} \right) = \lim_{\Delta t \to 0} \left\{ \frac{a(t + \Delta t) - a(t)}{\Delta t} \right\} \qquad (16.21)$$

Taking the limit corresponds to letting point B approach point A along the curve defined by $a(t)$. The direction of $\overrightarrow{AB} = \Delta a$ approaches that of the tangent to the curve at A, and this is the direction of the derivative at A.

In terms of the components of the vector, if $a = a_x i + a_y j + a_z k$, where i, j, and k are (*constant*) base vectors, the dependence of a on t is that of the components:

$$a(t) = a_x(t)i + a_y(t)j + a_z(t)k \tag{16.22}$$

and

$$\frac{da}{dt} = \left(\frac{da_x}{dt}\right)i + \left(\frac{da_y}{dt}\right)j + \left(\frac{da_z}{dt}\right)k \tag{16.23}$$

EXAMPLE 16.8 Find da/dt and d^2a/dt^2 for $a(t) = 3t^2 i + 2 \sin t\, j + e^{-t} k$.

$$\frac{da}{dt} = 6t i + 2 \cos t\, j - e^{-t} k, \qquad \frac{d^2a}{dt^2} = 6i - 2 \sin t\, j + e^{-t} k$$

▸ Exercises 15, 16

We note that the base vectors may also depend on t if the coordinate system itself (the frame of reference) is undergoing changes

Parametric representation of a curve

If $r = r(t)$ is the position vector of a point for each value of t in some interval then, given a cartesian coordinate system,

$$r(t) = x(t)i + y(t)j + z(t)k \qquad (16.24)$$

is a **parametric representation** of a curve C in three-dimensional space, and t is the parameter of the representation (Figure 16.17). The sense of increasing values of t is called the positive sense on C, and defines a direction on the curve.

When the parameter t is the time variable, derivatives of position vectors provide a general method for the description of the mechanics of dynamic systems.

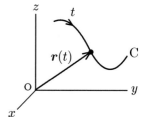

Figure 16.17

EXAMPLE 16.9 Velocity, acceleration, momentum and force

Velocity and acceleration

If $r = r(t)$ is the position vector of a body in space, and the parameter t is the time variable, then dr/dt is the velocity of the body,

$$v = \frac{dr}{dt} = \left(\frac{dx}{dt}\right)i + \left(\frac{dy}{dt}\right)j + \left(\frac{dz}{dt}\right)k = v_x i + v_y j + v_z k \tag{16.25}$$

where v_x, v_y and v_z are the components of velocity in the three cartesian directions. **Velocity** is therefore the **rate of change of position**. The magnitude of the velocity, the speed, is then

$$v = |v| = \sqrt{v_x^2 + v_y^2 + v_z^2} \tag{16.26}$$

and the kinetic energy of the body is

$$T = \frac{1}{2}mv^2 = \frac{1}{2}m\left(v_x^2 + v_y^2 + v_z^2\right) \tag{16.27}$$

The **acceleration** of the body is the **rate of change of velocity**,

$$a = \frac{dv}{dt} = \frac{d^2 r}{dt^2}$$

$$= \left(\frac{dv_x}{dt}\right)i + \left(\frac{dv_y}{dt}\right)j + \left(\frac{dv_z}{dt}\right)k = \left(\frac{d^2x}{dt^2}\right)i + \left(\frac{d^2y}{dt^2}\right)j + \left(\frac{d^2z}{dt^2}\right)k \tag{16.28}$$

The vector form of Newton's second law of motion is then $F = ma$, equivalent to three scalar equations, one for each component,

$$F_x = m\frac{d^2x}{dt^2}, \qquad F_y = m\frac{d^2y}{dt^2}, \qquad F_z = m\frac{d^2z}{dt^2},$$

Linear momentum and force

In mechanics the momentum p of a body of mass m moving with velocity v is defined as $p = mv$. The direction of p lies along the direction of the line of motion, and p is often referred to as the **linear momentum**. By Newton's second law of motion, the force acting on a body is given by the rate of change of momentum that it induces:

$$F = \frac{dp}{dt} \tag{16.29}$$

Equation (16.29) shows that when no external forces act on a system whose linear momentum is p then $dp/dt = 0$, and p is a constant vector. This is Newton's first law of motion; the *principle of conservation of linear momentum*.

EXAMPLE 16.10 A body of mass m moves along the curve $r = 4t\,i + \cos 2t\,j$. Find (i) the velocity of the body, (ii) its acceleration, (iii) the force acting on the body. Describe the motion of the body (iv) in the x-direction, (v) in the y-direction and (vi) overall.

(i) $v = \dfrac{dr}{dt} = 4i - 2\sin 2t\,j$, (ii) $a = \dfrac{dv}{dt} = -4\cos 2t\,j$, (iii) $F = ma = -4m\cos 2t\,j$

(iv) The velocity in the x-direction is $v_x = 4$ so that the body is moving in the positive x-direction with constant speed 4. (v) The acceleration is in the y-direction with $a_y = d^2y/dt^2 = -4y$ and, by equation (12.29) in Section 12.5, the motion is simple harmonic in this direction. (vi) The body moves to the right with constant speed $v_x = 4$ whilst undergoing simple harmonic motion in the perpendicular y-direction, as shown in Figure 16.18.

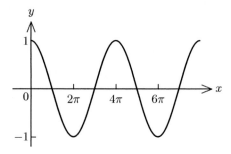

Figure 16.18

> Exercises 17, 18

16.5 The scalar (dot) product

The application of vectors to physical problems has led to the definition of two ways of multiplying a vector by a vector; one way gives a scalar as product, the other gives a vector product (Section 16.6).[4]

The **scalar product** or **dot product** of a and b is defined as

$$a \cdot b = ab \cos \theta \qquad (16.30)$$

(read as 'a dot b') where $a = |a|$ and $b = |b|$ are the lengths of the vectors and θ ($0 \leq \theta \leq \pi$) is the angle between the directions of a and b. In terms of components, if $a = (a_x, a_y, a_z)$ and $b = (b_x, b_y, b_z)$ then

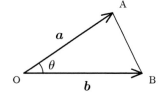

Figure 16.19

$$a \cdot b = a_x b_x + a_y b_y + a_z b_z \qquad (16.31)$$

[4] Clerk Maxwell used Hamilton's quaternions in the *Treatise on Electricity and Magnetism* of 1873, but it was observed, independently, by Heaviside and Gibbs that the full algebra of quaternions was not necessary for the representation of physical quantities.

Oliver Heaviside (1850–1925), English electrical engineer, is noted for his development of Laplace transform techniques. Josiah Willard Gibbs (1839–1903), American engineer, chemist, and 'among the most renowned of theoretical physicists of all time' (Max Planck), was born, lived, and died in New Haven, Connecticut. Gibbs is best known for his work in thermodynamics and statistical mechanics. His introduction of the concept of chemical potential in *On the equilibrium of heterogeneous substances* in 1876–8 marks the foundation of modern chemical thermodynamics. He developed the modern notation and the concepts of scalar and vector products in the early 1880's to simplify the mathematical treatment in Maxwell's *Treatise*. The unpublished work, circulated as lecture notes, was popularized by Edwin B. Wilson's *Vector analysis, founded upon the lectures of J. Willard Gibbs* in 1901.

Proof

To show the equivalence of the definitions (16.30) and (16.31), we apply the cosine rule to the triangle in Figure 16.19:

$$(AB)^2 = (OA)^2 + (OB)^2 - 2(OA)(OB)\cos\theta = a^2 + b^2 - 2ab\cos\theta$$

The vector $\overrightarrow{AB}$ is $\boldsymbol{b} - \boldsymbol{a}$, and its length is given by

$$\begin{aligned}(AB)^2 &= (b_x - a_x)^2 + (b_y - a_y)^2 + (b_z - a_z)^2 \\ &= (a_x^2 + a_y^2 + a_z^2) + (b_x^2 + b_y^2 + b_z^2) - 2(a_x b_x + a_y b_y + a_z b_z) \\ &= a^2 + b^2 - 2(a_x b_x + a_y b_y + a_z b_z)\end{aligned}$$

Therefore $ab\cos\theta = a_x b_x + a_y b_y + a_z b_z$.

EXAMPLE 16.11 Given $\boldsymbol{a} = (3,\ 1,\ -1)$ and $\boldsymbol{b} = (1,\ 2,\ -3)$ find $\boldsymbol{a}\cdot\boldsymbol{b}$, $\boldsymbol{b}\cdot\boldsymbol{a}$ and the angle between the vectors.

By equation (16.31),

$$\boldsymbol{a}\cdot\boldsymbol{b} = 3\times 1 + 1\times 2 + (-1)\times(-3) = 8, \quad \boldsymbol{b}\cdot\boldsymbol{a} = 1\times 3 + 2\times 1 + (-3)\times(-1) = 8$$

This example demonstrates that *scalar multiplication is commutative*,

$$\boldsymbol{a}\cdot\boldsymbol{b} = \boldsymbol{b}\cdot\boldsymbol{a} \tag{16.32}$$

By equation (16.30), $\cos\theta = \dfrac{\boldsymbol{a}\cdot\boldsymbol{b}}{ab}$, and the lengths of the vectors are $a = |\boldsymbol{a}| = \sqrt{11}$ and $b = |\boldsymbol{b}| = \sqrt{14}$. Therefore

$$\cos\theta = \frac{8}{\sqrt{154}}, \quad \theta = \cos^{-1}\left(\frac{8}{\sqrt{154}}\right) \approx 0.8702 \approx 49.86°$$

▸ Exercises 19–21

The sign of the scalar product is the sign of the cosine, so that the scalar product can be positive, zero, or negative:

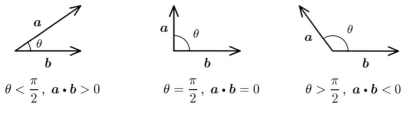

$$\theta < \frac{\pi}{2}, \ \boldsymbol{a}\cdot\boldsymbol{b} > 0 \qquad \theta = \frac{\pi}{2}, \ \boldsymbol{a}\cdot\boldsymbol{b} = 0 \qquad \theta > \frac{\pi}{2}, \ \boldsymbol{a}\cdot\boldsymbol{b} < 0$$

Figure 16.20

For nonzero vectors a and b, the scalar product is zero when the vectors are perpendicular,

$$a \cdot b = 0 \qquad (16.33)$$

The vectors are then said to be **orthogonal**, with a orthogonal to b, and b orthogonal to a.* We note that equation (16.33) shows that it is not possible to cancel vectors in a vector equation in the same way as is possible for scalars. Thus the equation

$$a \cdot b = a \cdot c$$

has the three possible solutions: (i) $a = 0$, (ii) $b = c$, (iii) a is orthogonal to $(b - c)$.

EXAMPLE 16.12 Find the value of λ for which $a = (2, \lambda, 1)$ and $b = (4, -2, -2)$ are orthogonal.

For orthogonality, $a \cdot b = 0 = 2 \times 4 + \lambda \times (-2) + 1 \times (-2) = 6 - 2\lambda$. Therefore $\lambda = 3$.

➤ Exercises 22, 23

When a and b are the same vector, (16.31) gives

$$a \cdot a = a_x^2 + a_y^2 + a_z^2 = |a|^2$$

The length of a vector is therefore given in terms of the scalar product by

$$|a| = \sqrt{a \cdot a} \qquad (16.34)$$

The use of cartesian base vectors

The base vectors i, j, and k are orthogonal and of unit length so that, by equations (16.33) and (16.34),

$$
\begin{array}{llll}
i \cdot j = 0 & j \cdot k = 0 & k \cdot i = 0 & \text{(orthogonality)} \\
i \cdot i = 1 & j \cdot j = 1 & k \cdot k = 1 & \text{(unit length)}
\end{array}
\qquad (16.35)
$$

The expression (16.31) for the scalar product follows from these properties of the base vectors. Thus, expressing a and b in terms of the base vectors,

$$
\begin{aligned}
a \cdot b &= (a_x i + a_y j + a_z k) \cdot (b_x i + b_y j + b_z k) \\
&= a_x b_x\, i \cdot i + a_x b_y\, i \cdot j + a_x b_z\, i \cdot k + a_y b_x\, j \cdot i + a_y b_y\, j \cdot j + \cdots + a_z b_z\, k \cdot k \\
&= a_x b_x + a_y b_y + a_z b_z
\end{aligned}
$$

➤ Exercises 24–27

---

* Orthogonal means perpendicular for vectors in ordinary space, but the definition of orthogonality applies generally to vectors in 'vector spaces' of arbitrary dimensions.

The following examples demonstrate two applications of scalar products in the physical sciences.

EXAMPLE 16.13 Force and work

Let a body be displaced from position $r_1 = (x_1, y_1, z_1)$ to position $r_2 = (x_2, y_2, z_2)$ under the influence of a *constant* force F. The displacement of the body is

$$d = r_2 - r_1 = (x_2 - x_1, y_2 - y_1, z_2 - z_1)$$

and the work done by the force is

$$W = F \cdot d = Fd \cos \theta \qquad (16.36)$$

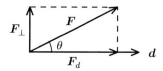

Figure 16.21

The quantity $F \cos \theta$ is the component of the force in the direction of d, and this is the only component of the force that contributes to the work. Thus, as in Figure 16.21, the force can be written as

$$F = F_d + F_\perp$$

where F_d acts along the line of d and $F_\perp$ acts at right angles to d. The latter cannot cause a displacement along d.

In terms of cartesian components, the work (16.36) is

$$W = F_x d_x + F_y d_y + F_z d_z = W_x + W_y + W_z \qquad (16.37)$$

where, for example, F_x is the component of F in the x-direction and $W_x = F_x d_x$ is the work done in moving the body through distance $d_x = x_2 - x_1$ in this direction ('work along the x-direction'). We note that the work done *by* the body *against* the force is $-W = -F \cdot d$.

For example, if the force is $F = (2, 1, 0)$ and the displacement is $d = (2, -3, 1)$, in appropriate units, the work done is

$$W = F \cdot d = 2 \times 2 + 1 \times (-3) + 0 \times 1 = 1$$

In addition, the component of F in the direction of d can be written in terms of the scalar product as

$$F_d = F \cos \theta = \frac{F \cdot d}{d} = F \cdot \hat{d}$$

where $\hat{d} = d/d$ is the unit vector in direction d. In the present case, $d = \sqrt{14}$ so that $F_d = 1/\sqrt{14}$.

➤ Exercise 28

The general case

If the force is *not* constant along the path r_1 to r_2 then the work has the form of a line integral (Section 9.8). Consider a body moving from point A at r_1 to point B at r_2 along the curve C under the influence of a force F whose value varies from point to point on C (Figure 16.22). Let $F(r)$ be the force at point r on the curve. The work done on the body between positions r and $r + \Delta r$ is $\Delta W \approx F(r) \cdot \Delta r$ and, in the limit $|\Delta r| \to 0$ on the curve, the element of work is $dW = F(r) \cdot dr$. The total work done from A to B on C is then

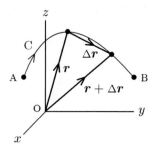

Figure 16.22

$$W_{AB} = \int_C F(r) \cdot dr \tag{16.38}$$

and the integral is a line integral. In terms of components, $F = (F_x, F_y, F_z)$ and $dr = (dx, dy, dz)$, $F \cdot dr = F_x dx + F_y dy + F_z dz$, and

$$W_{AB} = \int_C \left[F_x dx + F_y dy + F_z dz \right] \tag{16.39}$$

This is a generalization of equation (9.48) for a curve in three dimensions, and the discussion of Section 9.8 applies with only minor changes. In particular, if the force is a conservative force then, by a generalization of the discussion of conservative forces in Sections 5.7 and 9.8, the components of the force can be expressed as (partial) derivatives of a potential-energy function V (see equation (5.57)),

$$F_x = -\frac{\partial V}{\partial x} \qquad F_y = -\frac{\partial V}{\partial y} \qquad F_z = -\frac{\partial V}{\partial z} \tag{16.40}$$

Then

$$F_x dx + F_y dy + F_z dz = -\left(\frac{\partial V}{\partial x} dx + \frac{\partial V}{\partial y} dy + \frac{\partial V}{\partial z} dz \right) = -dV \tag{16.41}$$

and dV is the total differential of the potential-energy function $V(r) = V(x, y, z)$. It follows that, for a conservative force, the line integral (16.38) is independent of the path, and is equal to the difference in potential energy between A and B:

$$W_{AB} = -\int_A^B dV = V_A - V_B \tag{16.42}$$

▶ Exercise 29

EXAMPLE 16.14 Charges in an electric field

The force experienced by a charge q in the presence of an electric field E is $F = qE$. If the field is an electrostatic field (constant in time) then this force is conservative so that, by equation (16.40), the components of the field, $E = F/q$, are (minus) the derivatives of a function $\phi = V/q$:

$$E_x = -\frac{\partial \phi}{\partial x}, \qquad E_y = -\frac{\partial \phi}{\partial y}, \qquad E_z = -\frac{\partial \phi}{\partial z} \tag{16.43}$$

The function ϕ is the **electrostatic potential function** (potential energy per unit charge) of the field.

If the field E is a *uniform* field, constant in space, then integration of these equations gives (see (5.58) for a constant force)

$$\phi(r) = -(xE_x + yE_y + zE_z) + C = -r \cdot E + C \tag{16.44}$$

where $\phi(r)$ is the electrostatic potential at position $r = (x, y, z)$ and C is an arbitrary constant. The potential energy of charge q at r in the field is then

$$V = q\phi(r) = -qr \cdot E + qC \tag{16.45}$$

The potential energy of a system of charges, q_1 at r_1, q_2 at $r_2 \ldots, q_N$ at r_N, is the sum of the energies of the individual charges,

$$V = q_1 \phi(r_1) + q_2 \phi(r_2) + \quad + q_N \phi(r_N) = \sum_{i=1}^{N} q_i \phi(r_i)$$

Figure 16.23

For the potential (16.44), we therefore have

$$V = \sum_{i=1}^{N} (-q_i r_i \cdot E + q_i C) = -\left(\sum_{i=1}^{N} q_i r_i\right) \cdot E + \left(\sum_{i=1}^{N} q_i\right) C \tag{16.46}$$

$$= -\mu \cdot E + QC$$

where μ is the dipole moment of the system of charges and Q is the total charge. The term QC is zero for an electrically neutral system or if the potential ϕ is chosen to be zero at the origin (the usual choice). Then

$$V = -\mu \cdot E = -\mu E \cos \theta$$

➤ Exercise 30

16.6   The vector (cross) product

Every pair of non-parallel vectors, a and b, defines a parallelogram, Figure 16.24, whose area is $ab \sin \theta$ (base × perpendicular height).

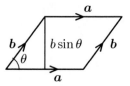

Figure 16.24

In addition, Figures 16.25 show that when the vectors are set head to tail, they define a sense of direction of rotation; anticlockwise, or 'direction up', in (i) for a into b, and clockwise, or 'direction down', in (ii) for b into a.

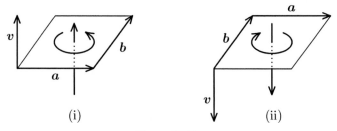

(i)   (ii)

Figure 16.25

These properties of a pair of vectors define a new vector, the **vector product** or **cross product** of a and b,

$$v = a \times b \tag{16.47}$$

(read as 'a cross b'), whose length is

$$|v| = ab \sin \theta \tag{16.48}$$

and whose direction is perpendicular to both a and b (to the parallelogram), and is such that the triple of vectors (a, b, v), in this order, form a right-handed system (as in Figure 16.25(i)). Figure (ii) shows that this direction is reversed when the order of a and b in (16.47) is reversed, so that

$$b \times a = -a \times b \tag{16.49}$$

Unlike the scalar product therefore, the vector product is *not* commutative; it is **anti-commutative**.

It follows from (16.48) that the vector product is zero if $\theta = 0$ or $\theta = \pi$; that is, if a and b have the same or opposite directions. In particular,

$$a \times a = 0 \tag{16.50}$$

In terms of cartesian base vectors

The unit vectors i, j and k form a right-handed system of vectors and have properties

$$i \times j = k \qquad j \times k = i \qquad k \times i = j$$
$$i \times i = 0 \qquad j \times j = 0 \qquad k \times k = 0 \tag{16.51}$$

The vector product $a \times b$ can then be expressed in terms of cartesian components:

$$\begin{aligned}
a \times b &= (a_x i + a_y j + a_z k) \times (b_x i + b_y j + b_z k) \\
&= a_x b_x i \times i + a_x b_y i \times j + a_x b_z i \times k + a_y b_x j \times i + a_y b_y j \times j \\
&\quad + a_y b_z j \times k + a_z b_x k \times i + a_z b_y k \times j + a_z b_z k \times k \\
&= a_x b_y k + a_x b_z (-j) + a_y b_x (-k) + a_y b_z i + a_z b_x j + a_z b_y (-i)
\end{aligned}$$

(remembering that, for example, $j \times i = -i \times j$). Therefore

$$a \times b = (a_y b_z - a_z b_y) i + (a_z b_x - a_x b_z) j + (a_x b_y - a_y b_x) k \tag{16.52}$$

This form of the vector product can be written, and more easily remembered, in the form of a determinant (Chapter 17),

$$a \times b = \begin{vmatrix} i & j & k \\ a_x & a_y & a_z \\ b_x & b_y & b_z \end{vmatrix} \tag{16.53}$$

EXAMPLE 16.15 Given $a = (3, 1, -1)$ and $b = (1, 2, -3)$, find $a \times b$, $b \times a$, and the area $|a \times b|$ of the parallelogram defined by a and b.

By equation (16.52),

$$a \times b = \left[1 \times (-3) - (-1) \times 2 \right] i + \left[(-1) \times 1 - 3 \times (-3) \right] j + \left[3 \times 2 - 1 \times 1 \right] k$$

$$= -i + 8j + 5k = (-1, 8, 5)$$

Also, $b \times a = -a \times b = (1, -8, -5)$

$$|a \times b| = \sqrt{1^2 + 8^2 + 5^2} = \sqrt{90}$$

➤ Exercises 31–40

Vector products are used for the description of surfaces in geometry, and in the vector integral calculus for the evaluation of surface integrals. In mechanics, the vector product is used for the description of properties associated with torque and angular motion.

EXAMPLE 16.16 Moment of force (torque)

In mechanics the magnitude of the moment of force F, or torque T, about a point O is defined as the product $T=|F|d$, where d is the perpendicular distance from O to the line of action of the force (Figure 16.26).

If r is the vector from O to a point A on the line, then $d=|r|\sin\theta$ and

$$T=|r||F|\sin\theta=|r\times F|$$

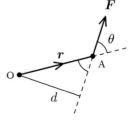

Figure 16.26

The corresponding vector

$$T=r\times F \qquad\qquad (16.54)$$

is perpendicular to r and F, and its direction is that of the axis through O about which the force tends to produce rotation.

▸ Exercise 41

EXAMPLE 16.17 An electric dipole in an electric field

An electric dipole $\mu=qr$ (see Example 16.6) in an electric field E experiences a torque

$$T=\mu\times E \qquad\qquad (16.55)$$

that tends to align the dipole along the direction of the field. Thus, from Figure 16.27, the total torque about O is

$$\begin{aligned} T&=r_1\times F_1+r_2\times F_2\\ &=q(r_1-r_2)\times E=qr\times E=\mu\times E \end{aligned}$$

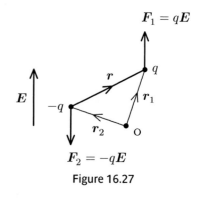

Figure 16.27

▸ Exercises 42, 43

EXAMPLE 16.18 Angular velocity and angular momentum

Angular velocity

A particle moving in a circle of radius r with speed v has angular velocity ω (angular speed ω) about an axis through the centre with magnitude given by

$$v=\omega r \qquad\qquad (16.56)$$

and direction at right angles to the plane of motion (out of the page for the motion shown in Figure 16.28).

More generally, we consider a particle (that may be part of a rotating body) rotating about the axis OA through the fixed point O, as illustrated in Figure 16.29. In this case, equation (16.56) is replaced by

$$v = \omega r \sin \theta \qquad (16.57)$$

and this is just the magnitude of the vector product

$$\boldsymbol{v} = \boldsymbol{\omega} \times \boldsymbol{r} \qquad (16.58)$$

which also correctly defines the direction of each of the three vectors with respect to the others. For a rigid body rotating with angular velocity $\boldsymbol{\omega}$, equation (16.58) gives the velocity at each point $\boldsymbol{r}$ in the body.

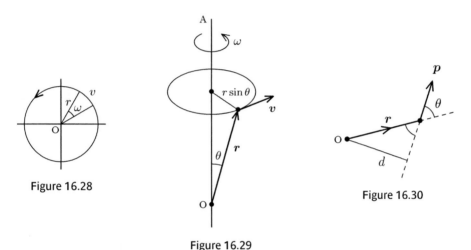

Figure 16.28

Figure 16.29

Figure 16.30

Angular momentum

The angular momentum $\boldsymbol{l}$ of a mass m about a point is defined as the moment of the linear momentum $\boldsymbol{p} = m\boldsymbol{v}$ about the point,

$$\boldsymbol{l} = \boldsymbol{r} \times \boldsymbol{p} \qquad (16.59)$$

Its magnitude about the point O in Figure 16.30 is $l = pd = pr \sin \theta$, and its direction is perpendicular to the plane defined by $\boldsymbol{r}$ and $\boldsymbol{p}$. This direction is not in general the same as that of the angular velocity $\boldsymbol{\omega}$, as can be seen by reference to Figure 16.29 (with $\boldsymbol{v}$ replaced by $\boldsymbol{p}$). It can be shown that (see Exercise 44)

$$\boldsymbol{l} = mr^2 \boldsymbol{\omega} - m(\boldsymbol{r} \cdot \boldsymbol{\omega})\boldsymbol{r} \qquad (16.60)$$

so that $\boldsymbol{l}$ and $\boldsymbol{\omega}$ have the same direction only if $\boldsymbol{r}$ and $\boldsymbol{\omega}$ are perpendicular vectors, when $\boldsymbol{r} \cdot \boldsymbol{\omega} = 0$. This is the case for motion in a circle with the reference point O at the centre of the circle. Then

$$\boldsymbol{l} = mr^2 \boldsymbol{\omega} = I\boldsymbol{\omega} \qquad (16.61)$$

where $I = mr^2$ is the moment of inertia about the axis of rotation. The relation between I and $\boldsymbol{\omega}$ is less simple than that given by (16.61) for arbitrary motion of a particle or for the rotation of a rigid body. The general relation is

$$l = \mathbf{I}\boldsymbol{\omega} \tag{16.62}$$

where $\mathbf{I}$ is a quantity with nine components called the moment of inertia tensor (or matrix), and is discussed in Example 19.15.

➤ Exercise 44

Conservation of angular momentum

As discussed in Example 16.9, Newton's second law of motion can be written as $F = \dfrac{dp}{dt}$. Taking the cross product of r with both sides of this equation gives

$$r \times F = r \times \frac{dp}{dt}$$

The left side is the torque T acting on the system (Example 16.16). The right side is, using the product rule of differentiation (valid for both scalar and vector products),

$$r \times \frac{dp}{dt} = \frac{d}{dt}(r \times p) - \frac{dr}{dt} \times p$$

But $\dfrac{dr}{dt} = v$ is parallel to p, and their vector product is zero. Therefore, since $r \times p = l$,

$$T = \frac{dl}{dt} \tag{16.63}$$

This form of Newton's second law, that the rate of change of angular momentum of a system is equal to the applied torque, shows that the angular momentum of a system is constant in the absence of external torques.

➤ Exercises 45–47

16.7 Scalar and vector fields

A function of the coordinates of a point in space is called a function of position or field. A *scalar* function of position, a **scalar field**,

$$f = f(r) = f(x, y, z) \tag{16.64}$$

has a value at each point $r = (x, y, z)$; that is, a scalar (a number) is associated with each point. Examples of scalar fields have been discussed in Chapter 10. A *vector* function of position, a **vector field**,

$$v = v(r) = v(x, y, z) \tag{16.65}$$

defines a vector associated with each point $\mathbf{r}$. An example of a vector field is the velocity field used to describe fluid flow in hydrodynamics; a velocity is associated with every point. Electric and magnetic fields are vector fields.

The theory of vector fields is an essential tool in hydrodynamics and electromagnetism. A basic feature of vector field theory is the use of vector differential operators to describe how a field changes from point to point in space. The concept of the gradient of a scalar field, with some applications, is discussed in the following section. The divergence and the curl of a vector field are introduced in Section 16.9. These quantities are used in more advanced applications of the vector calculus in the theory of fields, and only a very brief description is given here.

16.8 The gradient of a scalar field

The **gradient** of a scalar function of position $f = f(x, y, z)$ is defined as the vector

$$\text{grad } f = \frac{\partial f}{\partial x}\mathbf{i} + \frac{\partial f}{\partial y}\mathbf{j} + \frac{\partial f}{\partial z}\mathbf{k} \tag{16.66}$$

This quantity can be interpreted as the result of operating on the function f with the vector differential operator

$$\nabla = \frac{\partial}{\partial x}\mathbf{i} + \frac{\partial}{\partial y}\mathbf{j} + \frac{\partial}{\partial z}\mathbf{k} \tag{16.67}$$

(read as 'del' or 'nabla'), so that

$$\text{grad } f = \nabla f = \left(\frac{\partial}{\partial x}\mathbf{i} + \frac{\partial}{\partial y}\mathbf{j} + \frac{\partial}{\partial z}\mathbf{k} \right) f$$
$$= \frac{\partial f}{\partial x}\mathbf{i} + \frac{\partial f}{\partial y}\mathbf{j} + \frac{\partial f}{\partial z}\mathbf{k} \tag{16.68}$$

($\mathbf{i}, \mathbf{j}$, and $\mathbf{k}$ are assumed to be constant vectors).[5]

EXAMPLE 16.19 Find grad V for $V = x + 2yz + 3z^2$.

We have $\dfrac{\partial V}{\partial x} = 1$, $\dfrac{\partial V}{\partial y} = 2z$, and $\dfrac{\partial V}{\partial z} = 2y + 6z$. Therefore $\nabla V = \mathbf{i} + 2z\mathbf{j} + (2y + 6z)\mathbf{k}$.

---

[5] The symbol ∇ was introduced by Hamilton, and (possibly) called nabla after a harp-like musical instrument used in Palestine in Biblical times.

EXAMPLE 16.20 Force and potential energy

By a generalization to three dimensions of the discussion of conservative forces in Section 5.7 (see also Example 16.13), the components of a conservative force are derivatives of a potential-energy function V,

$$F_x = -\frac{\partial V}{\partial x}, \qquad F_y = -\frac{\partial V}{\partial y}, \qquad F_z = -\frac{\partial V}{\partial z}$$

The force is therefore (minus) the gradient of V,

$$F = -\nabla V = -\left(\frac{\partial V}{\partial x} i + \frac{\partial V}{\partial y} j + \frac{\partial V}{\partial z} k \right) \tag{16.69}$$

▸ Exercises 48–50

EXAMPLE 16.21 Coulomb forces

The potential energy of interaction of charges q_1 and q_2 separated by distance r is $V = q_1 q_2 / 4\pi\varepsilon_0 r$ (see Example 5.19). If r is the position of q_2 relative to q_1 (Figure 16.31) then the force acting on q_2 due to the presence of q_1 is

Figure 16.31

$$F = -\nabla V = -\frac{q_1 q_2}{4\pi\varepsilon_0} \left[\frac{\partial}{\partial x}\left(\frac{1}{r}\right) i + \frac{\partial}{\partial y}\left(\frac{1}{r}\right) j + \frac{\partial}{\partial z}\left(\frac{1}{r}\right) k \right]$$

$$= \frac{q_1 q_2}{4\pi\varepsilon_0} \left(\frac{x}{r^3} i + \frac{y}{r^3} j + \frac{z}{r^3} k \right) = \frac{q_1 q_2}{4\pi\varepsilon_0 r^3}(x i + y j + z k)$$

$$= \frac{q_1 q_2 r}{4\pi\varepsilon_0 r^3}$$

The unit vector from q_1 to q_2 is $\hat{r} = \dfrac{r}{r}$. Therefore

$$F = \frac{q_1 q_2 r}{4\pi\varepsilon_0 r^3} = \frac{q_1 q_2}{4\pi\varepsilon_0 r^2} \hat{r} \tag{16.70}$$

The force has strength $\dfrac{q_1 q_2}{4\pi\varepsilon_0 r^2}$, and acts along the line from q_1 to q_2 for like charges, from q_2 to q_1 for unlike charges (see also Example 5.17). In addition, the force per unit charge acting at point r due to the presence of charge q_1 is $E = F/q_2$. Then

$$E = \frac{q_1}{4\pi\varepsilon_0 r^2}\,\hat{\boldsymbol{r}} = -\nabla\phi \qquad (16.71)$$

is the electrostatic field of charge q_1, and $\phi = V/q_2 = q_1/4\pi\varepsilon_0 r$ is the corresponding (scalar) electrostatic potential field.

The meaning of the gradient of a function of position $f(\boldsymbol{r})$ is clarified by considering how the value of the function changes from point to point in space. We consider a differential (infinitesimal) displacement $\boldsymbol{r} \rightarrow \boldsymbol{r} + d\boldsymbol{r}$ or, in cartesian components, $(x, y, z) \rightarrow (x + dx, y + dy, z + dz)$:

$$d\boldsymbol{r} = (dx, dy, dz) = \boldsymbol{i}\,dx + \boldsymbol{j}\,dy + \boldsymbol{k}\,dz \qquad (16.72)$$

The corresponding change in the function, $df = f(\boldsymbol{r} + d\boldsymbol{r}) - f(\boldsymbol{r})$, is the total differential

$$df = \frac{\partial f}{\partial x}\,dx + \frac{\partial f}{\partial y}\,dy + \frac{\partial f}{\partial z}\,dz \qquad (16.73)$$

This can be written as the scalar product of the gradient ∇f and the displacement $d\boldsymbol{r}$ (Equations (16.68) and (16.72)),

$$\left(\frac{\partial f}{\partial x}\boldsymbol{i} + \frac{\partial f}{\partial y}\boldsymbol{j} + \frac{\partial f}{\partial z}\boldsymbol{k}\right)\cdot(\boldsymbol{i}\,dx + \boldsymbol{j}\,dy + \boldsymbol{k}\,dz) = \frac{\partial f}{\partial x}\,dx + \frac{\partial f}{\partial y}\,dy + \frac{\partial f}{\partial z}\,dz$$

so that

$$df = \nabla f \cdot d\boldsymbol{r} \qquad (16.74)$$

The quantity ∇f is therefore the generalization to three dimensions of the ordinary derivative df/dx; the vector operator ∇ is the generalization of the differential (gradient) operator d/dx, and is sometimes written as $d/d\boldsymbol{r}$.

16.9 Divergence and curl of a vector field

The divergence

The **divergence** of a vector field $\boldsymbol{v}(\boldsymbol{r})$ is defined as the scalar quantity

$$\operatorname{div}\boldsymbol{v} = \nabla\cdot\boldsymbol{v} = \frac{\partial v_x}{\partial x} + \frac{\partial v_y}{\partial y} + \frac{\partial v_z}{\partial z} \qquad (16.75)$$

In hydrodynamics, the vector field is the velocity field of a fluid, and the divergence is a flux density; the value of $\operatorname{div}\boldsymbol{v}(\boldsymbol{r})\,dV$ at a point $\boldsymbol{r}$ is a measure of the net flux, or rate of flow, of fluid out of a volume element dV at the point.

We consider here only the special case of a vector field that is the gradient of a scalar field, $v = \nabla f$. Then

$$\text{div } v = \nabla \cdot \nabla f = \left(\frac{\partial}{\partial x} i + \frac{\partial}{\partial y} j + \frac{\partial}{\partial z} k \right) \cdot \left(\frac{\partial f}{\partial x} i + \frac{\partial f}{\partial y} j + \frac{\partial f}{\partial z} k \right)$$

$$= \frac{\partial^2 f}{\partial x^2} + \frac{\partial^2 f}{\partial y^2} + \frac{\partial^2 f}{\partial z^2}$$

(16.76)

It follows that $\nabla \cdot \nabla = \nabla^2$ is the Laplacian operator. When a scalar field $f(r)$ satisfies Laplace's equation $\nabla^2 f = 0$ at a point then the derived vector field ∇f has zero divergence at the point, and there is no net flux out of a volume element at the point. This is the case for an incompressible fluid in a region containing no sources (of fluid) or sinks and for an electrostatic field in a region free of charge.

The curl

The **curl** (or **rotation** rot v) of a vector field $v(r)$ is defined as the vector

$$\text{curl } v = \nabla \times v = \left(\frac{\partial v_z}{\partial y} - \frac{\partial v_y}{\partial z} \right) i + \left(\frac{\partial v_x}{\partial z} - \frac{\partial v_z}{\partial x} \right) j + \left(\frac{\partial v_y}{\partial x} - \frac{\partial v_x}{\partial y} \right) k$$

$$= \begin{vmatrix} i & j & k \\ \dfrac{\partial}{\partial x} & \dfrac{\partial}{\partial y} & \dfrac{\partial}{\partial z} \\ v_x & v_y & v_z \end{vmatrix}$$

(16.77)

(see Equations (16.52) and (16.53)). In hydrodynamics, the curl of the velocity field at a point is a measure of the circulation of fluid around the point.

A vector field that is the gradient of a scalar field has zero curl,

$$\text{curl } v = 0 \quad \text{if} \quad v = \text{grad} f \tag{16.78}$$

Thus, the electrostatic field is the gradient of the scalar electrostatic potential, $E = -\nabla \phi$, so that $\nabla \times E = 0$. On the other hand, a vector field that is itself the curl of a vector field has zero divergence,

$$\text{div } v = 0 \quad \text{if} \quad v = \text{curl } w \tag{16.79}$$

Thus, the magnetic field can always be expressed as the curl of a vector function, $B = \nabla \times A$, where A is called the magnetic vector potential. Then $\nabla \cdot B = 0$.

▸ Exercises 51–55

16.10   Vector spaces

The vectors discussed in this chapter are three-dimensional vectors, or vectors in a three-dimensional **vector space**. The concept of vector can be extended to any number of dimensions by defining vectors in n dimensions as quantities that have n components and that obey the laws of vector algebra described in Sections 16.2 and 16.3. In particular, an n–dimensional vector space can be defined by means of n orthogonal unit vectors,

$$\boldsymbol{e}_1 = (1, 0, 0, \ldots, 0) \quad \boldsymbol{e}_2 = (0, 1, 0, \ldots, 0) \quad \cdots \quad \boldsymbol{e}_n = (0, 0, 0, \ldots, 1) \quad (16.80)$$

Every vector in the space can then be expressed as a linear combination of these,

$$\boldsymbol{a} = (a_1, a_2, a_3, \ldots, a_n) = a_1\boldsymbol{e}_1 + a_2\boldsymbol{e}_2 + a_3\boldsymbol{e}_3 + \cdots + a_n\boldsymbol{e}_n \quad (16.81)$$

These quantities obey the rules of vector algebra.[6]

For the vector space so defined, an **inner (scalar) product** is associated with each pair of vectors,

$$\boldsymbol{a} \cdot \boldsymbol{b} = a_1 b_1 + a_2 b_2 + a_3 b_3 + \cdots + a_n b_n = \boldsymbol{b} \cdot \boldsymbol{a} \quad (16.82)$$

The inner product is often denoted by $(\boldsymbol{a}, \boldsymbol{b})$. Two vectors are orthogonal ('perpendicular') if their inner product is zero, and the length or **norm** of a vector is

$$|a| = \sqrt{\boldsymbol{a} \cdot \boldsymbol{a}} = \sqrt{a_1^2 + a_2^2 + a_3^2 + \cdots + a_n^2} \quad (16.83)$$

A vector space with these properties is called an **inner product space**. The particular space described in this section is the n–**dimensional Euclidean space** R^n.

16.11   Exercises

Section 16.2

1. The position vectors of the points A, B, C and D are $\boldsymbol{a}$, $\boldsymbol{b}$, $\boldsymbol{c}$, and $\boldsymbol{d}$. Express the following quantities in terms of $\boldsymbol{a}$, $\boldsymbol{b}$, $\boldsymbol{c}$, and $\boldsymbol{d}$: (i) $\overrightarrow{AB}$ and $\overrightarrow{BA}$, (ii) the position of the centroid of the points, (iii) the position of the midpoint of $\overrightarrow{BC}$, (iv) the position of an arbitrary point on the line $\overrightarrow{BC}$ (the equation of the line).

Section 16.3

2. Two sides of the triangle ABC are $\overrightarrow{AB} = (2, 1, -1)$ and $\overrightarrow{AC} = (3, 2, 0)$. Find $\overrightarrow{BC}$

Find (i) the vector $\boldsymbol{a} = (a_1, a_2, a_3)$ with the given initial point $P(x_1, y_1, z_1)$ and terminal point $Q(x_2, y_2, z_2)$, (ii) the length of $\boldsymbol{a}$, (iii) the unit vector parallel to $\boldsymbol{a}$.

3. $P(1, -2, 0), Q(4, 2, 0)$ 4. $P(-3, 2, 1), Q(-1, -3, 2)$ 5. $P(0, 0, 0), Q(2, 3, 1)$

---

[6] A detailed treatment of n-dimensional vector spaces was given by Hermann Günther Grassmann (1809–1877), German mathematician, in his *Die lineale Ausdehnungslehre, ein neuer Zweig der Mathematik* (The theory of linear extension, a new branch of mathematics) of 1862. The work included the algebras of Hamilton's quaternions and Gibbs' vectors as special cases, but was largely unknown during his lifetime.

For $a = (1, 2, 3)$, $b = (-2, 3, -4)$, $c = (0, 4, -1)$, find

6. $a + b, b + a$ **7.** $3a, -a, a/3$ **8.** $3a + 2b - 3c$ **9.** $3a - 3c, 3(a - c)$

10. $|a + b|, |a| + |b|$

11. Three masses, $m_1 = 2, m_2 = 3, m_3 = 1$, have position vectors $r_1 = (3, -2, 1), r_2 = (2, -1, 0),$ $r_3 = (0, 1, -2)$, respectively. Find **(i)** the position vector of the centre of mass, **(ii)** the position vectors of the masses with respect to the centre of mass.

12. Three charges, $q_1 = 3, q_2 = -2, q_3 = 1$, have position vectors $r_1 = (2, 2, 1), r_2 = (2, -2, 3),$ $r_3 = (0, -4, -3)$, respectively. Find **(i)** the dipole moment of the system of charges with respect to the origin, **(ii)** the position of the point with respect to which the dipole moment is zero.

13. Forces are said to be in equilibrium if the total force is zero. Find f such that f, $f_1 = 2i - 3j + k$ and $f_2 = 2j - k$ are in equilibrium.

14. For the vectors $a = 2i - j + 3k, b = 2j - 2k, c = -k$, find **(i)** $x = 2a + b + 4c$, **(ii)** a vector perpendicular to c and x, **(iii)** a vector perpendicular to b and c.

Section 16.4

Differentiate with respect to t.

15. $2ti + 3t^2 j$ **16.** $(\cos 2t, 3 \sin t, 2t)$

17. A body of mass m moves along the curve $r(t) = x(t)i + y(t)j$, where $x = at$ and $y = (at - \frac{1}{2}gt^2)$ at time t. **(i)** Find the velocity and acceleration at time t. **(ii)** Find the force acting on the body. Describe the motion of the body **(iii)** in the x-direction, **(iv)** in the y-direction, **(v)** overall.

18. A body of mass m moves along the curve $r(t) = x(t)i + y(t)j + z(t)k$, where $x = 2 \cos 3t$, $y = 2 \sin 3t$ and $z = 3t$ at time t. **(i)** Find the velocity and acceleration at time t. **(ii)** Find the force acting on the body. Describe the motion of the body **(iii)** in the x- and y- directions, **(iv)** in the xy-plane, **(v)** in the z-direction, **(vi)** overall.

Section 16.5

For $a = (1, 3, -2)$, $b = (0, 3, 1)$, $c = (1, -1, -3)$, find:

19. $a \cdot b, b \cdot a$ **20.** $(a - b) \cdot c, a \cdot c - b \cdot c$ **21.** $(a \cdot c)b$

22. Show that $a = (1, 2, 3)$, $b = (0, -3, 2)$ and $c = (-13, 2, 3)$ are orthogonal vectors.

23. Find the value of λ for which $a = (\lambda, 3, 1)$ and $b = (2, 1, -1)$ are orthogonal.

For $a = i, b = j, c = 2i - 3j + k$, find

24. $a \cdot b$ **25.** $b \cdot c$ **26.** $a \cdot c$

27. Find the angles between the direction of $a = i - j + \sqrt{2}k$ and the x-, y-, and z- directions.

28. A body undergoes the displacement d under the influence of a force $f = 3i + 2j$. Calculate the work done **(i)** by the force on the body when $d = 2i - j$, **(ii)** by the body against the force when $d = i - 3k$, **(iii)** by the body against the force when $d = 2k$.

29. A body undergoes a displacement from $r_1 = (0, 0, 0)$ to $r_2 = (2, 3, 1)$ under the influence of the conservative force $F = xi + 2yj + 3zk$. **(i)** Calculate the work $W(r_1 \rightarrow r_2)$ done on the body. **(ii)** Find the potential-energy function $V(r)$ of which the components of the force are $(-)$ the partial derivatives. **(iii)** Confirm that $W(r_1 \rightarrow r_2) = V(r_1) - V(r_2)$.

30. Calculate the energy of interaction between the system of charges $q_1 = 2, q_2 = -3$ and $q_3 = 1$ at positions $r_1 = (3, -2, 1), r_2 = (0, 1, 2)$ and $r_3 = (0, 2, 1)$, respectively, and the applied electric field $E = -2k$.

Section 16.6

For $a = (1, 3, -2)$, $b = (0, 3, 1)$, $c = (0, -1, 2)$, find:

31. $a \times b, b \times a$ **32.** $b \times c, |b \times c|$ **33.** $(a + b) \times c, a \times c + b \times c$

34. $a \times c + c \times a$ **35.** $(a \times c) \cdot b$ **36.** $a \times (b \times c), (a \times b) \times c$

37. Show that $a \times b$ is orthogonal to a and b.

38. The quantity $a \cdot (b \times c)$ is called a **triple scalar product**. Show that

(i) $a \cdot (b \times c) = c \cdot (a \times b) = b \cdot (c \times a)$ (ii) $a \cdot (b \times c) = \begin{vmatrix} a_x & a_y & a_z \\ b_x & b_y & b_z \\ c_x & c_y & c_z \end{vmatrix}$ (determinant)

39. The quantity $a \times (b \times c)$ is called a **triple vector product**.
    (i) By expanding in terms of components, show that $a \times (b \times c) = (a \cdot c)b - (a \cdot b)c$
    (ii) Confirm this formula for the vectors $a = (1, 3, -2)$, $b = (0, 3, 1)$, $c = (0, -1, 2)$.

40. Find the area of the parallelogram whose vertices (in the xy-plane) have coordinates $(1, 2), (4, 3), (8, 6), (5, 5)$.

41. The force F acts on a line through the point A. Find the moment of the force about the point O for
    (i) $F = (1, -3, 0)$, A(2, 1, 0), O(0, 0, 0)   (ii) $F = (0, 1, -1)$, A(1, 1, 0), O(1, 0, 2)
    (iii) $F = (1, 0, -2)$, A(0, 0, 0), O(1, 0, 3)

42. Calculate the torque experienced by the system of charges $q_1 = 2$, $q_2 = -3$ and $q_3 = 1$ at positions $r_1 = (3, -2, 1)$, $r_2 = (0, 1, 2)$, and $r_3 = (0, 2, 1)$, respectively, in the electric field $E = -k$.

43. A charge q moving with velocity v in the presence of an electric field E and a magnetic field **B** experiences a total force $F = qE + qv \times B$ called the **Lorentz force**. Calculate the force acting on the charge $q = 3$ moving with velocity $v = (2, 3, 1)$ in the presence of the electric field $E = 2i$ and magnetic field $B = 3j$.

44. Use the property of the triple vector product (Exercise 39) to derive equation (16.60) from equations (16.58) and (16.59).

45. The position of a particle of mass m moving in a circle of radius R about the z-axis with angular speed ω is given by the vector function $r(t) = x(t)i + y(t)j + zk$ where $x(t) = R \cos \omega t$, $y(t) = R \sin \omega t$, $z = $ constant. (i) What is the angular velocity ω about the z-axis? (ii) Find the velocity v of the particle in terms of ω, x, y, and z. (iii) Find the angular momentum of the particle in terms of ω, x, y, and z. (iv) Confirm equation (16.60) in this case.

46. For the system in Exercise 45, show that $l = I\omega$ when $z = 0$, where I is the moment of inertia about the axis of rotation.

47. The total angular momentum of a system of particles is the sum of the angular momenta of the individual particles. If the system in Exercise 45 is replaced by a system of two particles of mass m with positions $r_1 = x(t)i + y(t)j + zk$ and $r_2 = -x(t)i - y(t)j + zk$, find the total angular momentum $l = l_1 + l_2$, and show that $l = I\omega$ where I is the total moment of inertia about the axis of rotation. This example demonstrates that $l = I\omega$ when the axis of rotation is an axis of symmetry of the system.

Section 16.8

Find the gradient ∇f for
48. $f = 2x^2 + 3y^2 - z^2$ 49. $f = xy + zx + yz$ 50. $f = (x^2 + y^2 + z^2)^{-1/2}$

Section 16.9

Find div v and curl v for
51. $v = xi + yj + zk$ 52. $v = zi + xj + yk$ 53. $v = yzi + zxj + xyk$
54. Show that curl $v = 0$ if $v = \text{grad} f$.   55. Show that div $v = 0$ if $v = \text{curl } w$.

17 Determinants

17.1 Concepts

Many problems in the physical sciences, in engineering, and in statistics give rise to systems of simultaneous linear equations. The methods of elementary algebra are adequate when the number of such equations is small; two or three, as discussed in Section 2.8. In some cases however the number of equations can be large, and alternative methods are then required both for the numerical solution of the large 'linear systems', and for the formulation and theoretical analysis of the problems that give rise to them. Some of the practical methods of solution, the 'numerical methods', are discussed in Chapter 20. The branch of mathematics concerned with the theory of linear systems is **matrix algebra**, the subject of Chapters 18 and 19, but several of the more important and useful results in the theory of linear equations can be derived independently from quantities called determinants. The theory of determinants is discussed in this chapter as a separate topic, partly in preparation for the more general matrix algebra of Chapters 18 and 19, and partly because determinants have certain symmetry properties that have made them an important tool in quantum mechanics. They are used in quantum chemistry to construct electronic wave functions that are consistent with the requirements of the Pauli Exclusion Principle.

The concept of determinants has its origin in the solution of simultaneous linear equations.[1] We consider the pair of equations

$$(1) \quad a_1 x + b_1 y = c_1$$
$$(2) \quad a_2 x + b_2 y = c_2$$

$$(17.1)$$

where a_1, b_1, c_1, a_2, b_2, and c_2 are constants. The equations are linear in the 'unknowns' x and y, and can be solved by the elementary methods of algebra. Thus, to solve for x, we multiply equation (1) by b_2 and equation (2) by b_1 to give

$$(1') \quad a_1 b_2 x + b_1 b_2 y = c_1 b_2$$
$$(2') \quad b_1 a_2 x + b_1 b_2 y = b_1 c_2$$

so that, subtracting (2') from (1'),

$$(a_1 b_2 - b_1 a_2) x = c_1 b_2 - b_1 c_2 \qquad (17.2)$$

Similarly for y,

$$(a_1 b_2 - b_1 a_2) y = a_1 c_2 - c_1 a_2 \qquad (17.3)$$

---

[1] The earliest descriptions of the method of solving sets of linear equations by determinants, known as Cramer's method, were by the Japanese Seki Kowa (1642–1708) in a manuscript of 1683, and by Leibniz in a letter to l'Hôpital in 1693 (published in 1850) in which he also gave the condition for the consistency of the equations. The first published account appeared in MacLaurin's *Treatise of algebra* (posthumously in 1748).

When $a_1b_2 - b_1a_2$ is not zero, the required (unique) values of x and y are

$$x = \frac{c_1 b_2 - b_1 c_2}{a_1 b_2 - b_1 a_2}, \qquad y = \frac{a_1 c_2 - c_1 a_2}{a_1 b_2 - b_1 a_2} \tag{17.4}$$

The quantity in the denominators of (17.4) is a property of the coefficients in equations (17.1), and is written in the form

$$\begin{vmatrix} a_1 & b_1 \\ a_2 & b_2 \end{vmatrix} = a_1 b_2 - b_1 a_2 \tag{17.5}$$

The symbol on the left is called a **determinant**;[2] the expression on the right side defines its value.

The solution (17.4) of the system of equations (17.1) can now be written as

$$x = \frac{D_1}{D}, \qquad y = \frac{D_2}{D} \tag{17.6}$$

where

$$D = \begin{vmatrix} a_1 & b_1 \\ a_2 & b_2 \end{vmatrix}, \quad D_1 = \begin{vmatrix} c_1 & b_1 \\ c_2 & b_2 \end{vmatrix}, \quad D_2 = \begin{vmatrix} a_1 & c_1 \\ a_2 & c_2 \end{vmatrix} \tag{17.7}$$

EXAMPLE 17.1 Use determinants to solve the equations

$$2x - 3y = 5$$
$$x + 5y = 9$$

We have

$$D = \begin{vmatrix} 2 & -3 \\ 1 & 5 \end{vmatrix} = 2 \times 5 - (-3) \times 1 = 13$$

$$D_1 = \begin{vmatrix} 5 & -3 \\ 9 & 5 \end{vmatrix} = 5 \times 5 - (-3) \times 9 = 52, \quad D_2 = \begin{vmatrix} 2 & 5 \\ 1 & 9 \end{vmatrix} = 2 \times 9 - 5 \times 1 = 13$$

Therefore $x = D_1/D = 4$ and $y = D_2/D = 1$.

▸ Exercises 1–4

---

[2] The name determinant was coined by Cauchy in 1812 in the first of a long series of papers on the subject of a class of alternating symmetric functions such as $a_1 b_2 - b_1 a_2$. In the general case, he arranged the n^2 different quantities in a square array, and used the abbreviation $(a_{1,n})$ to represent the array with which a determinant is associated. Cauchy used determinants for problems in geometry and in physics, and for the quantity now called the Jacobian. He introduced the concept of minors and the expansion along any row or column.

The determinant (17.5) is a property of a square array of $4 = 2^2$ **elements**, the coefficients of x and y in the system of equations (17.1). It is a determinant of **order** 2. In the general case, a determinant of order n is a property of a square array of n^2 elements, and is written

$$
\begin{vmatrix}
a_{11} & a_{12} & a_{13} & \cdots & a_{1n} \\
a_{21} & a_{22} & a_{23} & \cdots & a_{2n} \\
a_{31} & a_{32} & a_{33} & \cdots & a_{3n} \\
\vdots & \vdots & \vdots & & \vdots \\
a_{n1} & a_{n2} & a_{n3} & \cdots & a_{nn}
\end{vmatrix}
\tag{17.8}
$$

In this notation, the element a_{ij} lies in the ith row and jth column of the array; for example, a_{23} lies in the second row and third column. The determinant is sometimes denoted by the symbol $|a_{ij}|$. Determinants of order n arise from the consideration of systems of n linear equations in n unknowns.

17.2   Determinants of order 3

The system of three linear equations in three unknowns, x_1, x_2 and x_3,

$$
\begin{aligned}
(1) & \quad a_{11}x_1 + a_{12}x_2 + a_{13}x_3 = b_1 \\
(2) & \quad a_{21}x_1 + a_{22}x_2 + a_{23}x_3 = b_2 \\
(3) & \quad a_{31}x_1 + a_{32}x_2 + a_{33}x_3 = b_3
\end{aligned}
\tag{17.9}
$$

can be solved for x_1 by multiplying

$$
\text{equation} \quad (1) \quad \text{by} \quad a_{22}a_{33} - a_{23}a_{32} = \begin{vmatrix} a_{22} & a_{23} \\ a_{32} & a_{33} \end{vmatrix}
$$

$$
(2) \quad \text{by} \quad -(a_{12}a_{33} - a_{13}a_{32}) = -\begin{vmatrix} a_{12} & a_{13} \\ a_{32} & a_{33} \end{vmatrix}
$$

$$
(3) \quad \text{by} \quad a_{12}a_{23} - a_{13}a_{22} = \begin{vmatrix} a_{12} & a_{13} \\ a_{22} & a_{23} \end{vmatrix}
$$

and adding the three results. The coefficient of x_1 in the resulting expression defines the determinant of order 3:

$$
D = \begin{vmatrix}
a_{11} & a_{12} & a_{13} \\
a_{21} & a_{22} & a_{23} \\
a_{31} & a_{32} & a_{33}
\end{vmatrix}
\tag{17.10}
$$

$$
= a_{11}\begin{vmatrix} a_{22} & a_{23} \\ a_{32} & a_{33} \end{vmatrix} - a_{21}\begin{vmatrix} a_{12} & a_{13} \\ a_{32} & a_{33} \end{vmatrix} + a_{31}\begin{vmatrix} a_{12} & a_{13} \\ a_{22} & a_{23} \end{vmatrix}
$$

and, expanding the second-order determinants,

$$D = a_{11}a_{22}a_{33} - a_{11}a_{23}a_{32} - a_{21}a_{12}a_{33} + a_{21}a_{13}a_{32} + a_{31}a_{12}a_{23} - a_{31}a_{13}a_{22}$$

$$(17.11)$$

The solution of the system of three equations can then be expressed in terms of third-order determinants as

$$x_1 = \frac{D_1}{D}, \qquad x_2 = \frac{D_2}{D}, \qquad x_3 = \frac{D_3}{D} \qquad (17.12)$$

where $D \neq 0$ is the determinant of the coefficients, (17.10), and

$$D_1 = \begin{vmatrix} b_1 & a_{12} & a_{13} \\ b_2 & a_{22} & a_{23} \\ b_3 & a_{32} & a_{33} \end{vmatrix}, \quad D_2 = \begin{vmatrix} a_{11} & b_1 & a_{13} \\ a_{21} & b_2 & a_{23} \\ a_{31} & b_3 & a_{33} \end{vmatrix}, \quad D_3 = \begin{vmatrix} a_{11} & a_{12} & b_1 \\ a_{21} & a_{22} & b_2 \\ a_{31} & a_{32} & b_3 \end{vmatrix}$$

$$(17.13)$$

EXAMPLE 17.2 Use determinants to solve the equations

$$2x - 3y + 4z = 8$$

$$y - 3z = -7$$

$$x + 2y + 2z = 11$$

The determinant of the coefficients is, by equation (17.10),

$$D = \begin{vmatrix} 2 & -3 & 4 \\ 0 & 1 & -3 \\ 1 & 2 & 2 \end{vmatrix} = 2 \begin{vmatrix} 1 & -3 \\ 2 & 2 \end{vmatrix} - 0 \begin{vmatrix} -3 & 4 \\ 2 & 2 \end{vmatrix} + 1 \begin{vmatrix} -3 & 4 \\ 1 & -3 \end{vmatrix}$$

$$= 2 \times 8 - 0 \times (-14) + 1 \times 5 = 21$$

The determinants D_1, D_2, and D_3 are, by equations (17.13),

$$D_1 = \begin{vmatrix} 8 & -3 & 4 \\ -7 & 1 & -3 \\ 11 & 2 & 2 \end{vmatrix} = 21, \quad D_2 = \begin{vmatrix} 2 & 8 & 4 \\ 0 & -7 & -3 \\ 1 & 11 & 2 \end{vmatrix} = 42, \quad D_3 = \begin{vmatrix} 2 & -3 & 8 \\ 0 & 1 & -7 \\ 1 & 2 & 11 \end{vmatrix} = 63$$

Therefore $x = D_1/D = 1$, $y = D_2/D = 2$, $z = D_3/D = 3$.

▸ Exercise 5

Minors and cofactors

The **minor** M_{ij} of element a_{ij} of a determinant D is the determinant obtained by deleting row i and column j of D. For example, (17.14) shows the result of deleting row 2 and column 3 of a third-order determinant.

$$\begin{vmatrix} a_{11} & a_{12} & a_{13} \\ a_{21} & a_{22} & a_{23} \\ a_{31} & a_{32} & a_{33} \end{vmatrix} = \begin{vmatrix} a_{11} & a_{12} \\ a_{31} & a_{32} \end{vmatrix} = M_{23} \tag{17.14}$$

In general, the minors of a determinant of order n are determinants of order $n-1$. They are important because they are used for the expansion of a determinant in terms of its elements. Thus, equation (17.10) can be written as

$$\begin{vmatrix} a_{11} & a_{12} & a_{13} \\ a_{21} & a_{22} & a_{23} \\ a_{31} & a_{32} & a_{33} \end{vmatrix} = a_{11} M_{11} - a_{21} M_{21} + a_{31} M_{31} \tag{17.15}$$

This is called **expansion along the first column**; each element of the first column is multiplied by its minor, and the products are added with appropriate signs. The sign associated with element a_{ij} is

$$(-1)^{i+j} = \begin{cases} +1 & \text{if} \quad i+j \quad \text{is even} \\ -1 & \text{if} \quad i+j \quad \text{is odd} \end{cases} \tag{17.16}$$

The signs for the third-order determinant are

$$\begin{vmatrix} + & - & + \\ - & + & - \\ + & - & + \end{vmatrix} \tag{17.17}$$

A determinant can be expanded along *any* row or column.

EXAMPLE 17.3 Expand a determinant of order 3 along the second row.

The elements of the second row are a_{21}, a_{22} and a_{23}. Therefore, making use of (17.17) for the signs,

$$\text{row 2} \longrightarrow \begin{vmatrix} a_{11} & a_{12} & a_{13} \\ a_{21} & a_{22} & a_{23} \\ a_{31} & a_{32} & a_{33} \end{vmatrix} = -a_{21} M_{21} + a_{22} M_{22} - a_{23} M_{23}$$

The minor M_{21} is

$$M_{21} = \begin{vmatrix} a_{11} & a_{12} & a_{13} \\ a_{21} & a_{22} & a_{23} \\ a_{31} & a_{32} & a_{33} \end{vmatrix} = \begin{vmatrix} a_{12} & a_{13} \\ a_{32} & a_{33} \end{vmatrix} = a_{12}a_{33} - a_{13}a_{32}$$

Similarly,

$$M_{22} = \begin{vmatrix} a_{11} & a_{13} \\ a_{31} & a_{33} \end{vmatrix} = a_{11}a_{33} - a_{13}a_{31}, \quad M_{23} = \begin{vmatrix} a_{11} & a_{12} \\ a_{31} & a_{32} \end{vmatrix} = a_{11}a_{32} - a_{12}a_{31}$$

The complete expansion of the determinant in terms of its elements is therefore

$$-a_{21} M_{21} + a_{22} M_{22} - a_{33} M_{33} = -a_{21}(a_{12}a_{33} - a_{13}a_{32})$$
$$+ a_{22}(a_{11}a_{33} - a_{13}a_{31}) - a_{23}(a_{11}a_{32} - a_{12}a_{31})$$

and this is identical to the result (17.11) obtained by expansion along the first column.

EXAMPLE 17.4 Find the value of the following determinant by expansion along (a) the first row and (b) the third column:

$$D = \begin{vmatrix} 2 & 1 & 3 \\ 4 & -2 & 0 \\ -1 & 1 & 0 \end{vmatrix}$$

(a) The expansion along the first row is

$$D = 2 \times \begin{vmatrix} -2 & 0 \\ 1 & 0 \end{vmatrix} - 1 \times \begin{vmatrix} 4 & 0 \\ -1 & 0 \end{vmatrix} + 3 \times \begin{vmatrix} 4 & -2 \\ -1 & 1 \end{vmatrix}$$

$$= 2 \times 0 - 1 \times 0 + 3 \times 2 = 6$$

(b) The expansion along the third column is

$$D = 3 \begin{vmatrix} 4 & -2 \\ -1 & 1 \end{vmatrix} - 0 + 0 = 3 \times 2 = 6$$

> Exercises 6–9

The **cofactor** C_{ij} of element a_{ij} is the minor M_{ij} multiplied by the appropriate sign,

$$C_{ij} = (-1)^{i+j} M_{ij} \tag{17.18}$$

The expansion (17.15) along the first column is then

$$\begin{vmatrix} a_{11} & a_{12} & a_{13} \\ a_{21} & a_{22} & a_{23} \\ a_{31} & a_{32} & a_{33} \end{vmatrix} = a_{11}C_{11} + a_{21}C_{21} + a_{31}C_{31} = \sum_{i=1}^{3} a_{i1} C_{i1}$$

More generally, the expansion along the jth column is

$$\begin{vmatrix} a_{11} & a_{12} & a_{13} \\ a_{21} & a_{22} & a_{23} \\ a_{31} & a_{32} & a_{33} \end{vmatrix} = a_{1j}C_{1j} + a_{2j}C_{2j} + a_{3j}C_{3j} = \sum_{i=1}^{3} a_{ij} C_{ij} \tag{17.19}$$

and along the ith row is

$$\begin{vmatrix} a_{11} & a_{12} & a_{13} \\ a_{21} & a_{22} & a_{23} \\ a_{31} & a_{32} & a_{33} \end{vmatrix} = a_{i1} C_{i1} + a_{i2} C_{i2} + a_{i3}C_{i3} = \sum_{j=1}^{3} a_{ij} C_{ij} \tag{17.20}$$

An expansion in terms of cofactors is called a **Laplace expansion** of the determinant

---

EXAMPLE 17.5 Find the cofactors of the elements of the third column of

$$\begin{vmatrix} 2 & 1 & 3 \\ 4 & -2 & 0 \\ -1 & 1 & 0 \end{vmatrix}$$

We have

$$a_{13} = 3 \quad M_{13} = \begin{vmatrix} 4 & -2 \\ -1 & 1 \end{vmatrix} = 2 \quad C_{13} = +M_{13} = 2$$

$$a_{23} = 0 \quad M_{23} = \begin{vmatrix} 2 & 1 \\ -1 & 1 \end{vmatrix} = 3 \quad C_{23} = -M_{23} = -3$$

$$a_{33} = 0 \quad M_{33} = \begin{vmatrix} 2 & 1 \\ 4 & -2 \end{vmatrix} = -8 \quad C_{33} = +M_{33} = -8$$

▶ Exercise 10

17.3   The general case

A determinant of order n is a property of a square array of n^2 elements, written

$$D = \begin{vmatrix} a_{11} & a_{12} & a_{13} & \cdots & a_{1n} \\ a_{21} & a_{22} & a_{23} & \cdots & a_{2n} \\ a_{31} & a_{32} & a_{33} & \cdots & a_{3n} \\ \vdots & \vdots & \vdots & & \vdots \\ a_{n1} & a_{n2} & a_{n3} & \cdots & a_{nn} \end{vmatrix} \tag{17.21}$$

The determinant is defined for $n = 1$ by

$$D = |a_{11}| = a_{11} \tag{17.22}$$

It is defined for $n \geq 2$ by the expansion along any row i $(i = 1, 2, \ldots, n)$

$$D = a_{i1} C_{i1} + a_{i2} C_{i2} + \cdots + a_{in} C_{in} = \sum_{j=1}^{n} a_{ij} C_{ij} \tag{17.23}$$

or, equivalently, by the expansion along any column j $(j = 1, 2, \ldots, n)$

$$D = a_{1j} C_{1j} + a_{2j} C_{2j} + \cdots + a_{nj} C_{nj} = \sum_{i=1}^{n} a_{ij} C_{ij} \tag{17.24}$$

The quantity

$$C_{ij} = (-1)^{i+j} M_{ij} \tag{17.25}$$

is the cofactor of element a_{ij}, and M_{ij} is the minor of a_{ij}, the determinant of order $n-1$ obtained from D by deletion of row i and column j.

EXAMPLE 17.6   Find the value of

$$D = \begin{vmatrix} 2 & 1 & 1 & 3 \\ 4 & -2 & 0 & 1 \\ -1 & 1 & 0 & 2 \\ 0 & 3 & 1 & -1 \end{vmatrix}$$

Expansion along the first row gives

$$D = 2\begin{vmatrix} -2 & 0 & 1 \\ 1 & 0 & 2 \\ 3 & 1 & -1 \end{vmatrix} - 1\begin{vmatrix} 4 & 0 & 1 \\ -1 & 0 & 2 \\ 0 & 1 & -1 \end{vmatrix} + 1\begin{vmatrix} 4 & -2 & 1 \\ -1 & 1 & 2 \\ 0 & 3 & -1 \end{vmatrix} - 3\begin{vmatrix} 4 & -2 & 0 \\ -1 & 1 & 0 \\ 0 & 3 & 1 \end{vmatrix}$$

Each of the third-order determinants can be evaluated by the method described in Section 17.2; that is, by expansion in minors of order 2. Then, expanding each along the first row,

$$D = 2\left\{-2\begin{vmatrix} 0 & 2 \\ 1 & -1 \end{vmatrix} - 0\begin{vmatrix} 1 & 2 \\ 3 & -1 \end{vmatrix} + 1\begin{vmatrix} 1 & 0 \\ 3 & 1 \end{vmatrix}\right\}$$

$$-1\left\{4\begin{vmatrix} 0 & 2 \\ 1 & -1 \end{vmatrix} - 0\begin{vmatrix} -1 & 2 \\ 0 & -1 \end{vmatrix} + 1\begin{vmatrix} -1 & 0 \\ 0 & 1 \end{vmatrix}\right\}$$

$$+1\left\{4\begin{vmatrix} 1 & 2 \\ 3 & -1 \end{vmatrix} - (-2)\begin{vmatrix} -1 & 2 \\ 0 & -1 \end{vmatrix} + 1\begin{vmatrix} -1 & 1 \\ 0 & 3 \end{vmatrix}\right\}$$

$$-3\left\{4\begin{vmatrix} 1 & 0 \\ 3 & 1 \end{vmatrix} - (-2)\begin{vmatrix} -1 & 0 \\ 0 & 1 \end{vmatrix} + 0\begin{vmatrix} -1 & 1 \\ 0 & 3 \end{vmatrix}\right\}$$

$$= 2 \times 5 - 1 \times (-9) + 1 \times (-29) - 3 \times 2 = -16$$

> Exercises 11–13

EXAMPLE 17.7 Find the value of the 'triangular' determinant

$$D = \begin{vmatrix} 1 & 2 & 3 & 4 & 5 \\ 0 & 6 & 7 & 8 & 9 \\ 0 & 0 & 10 & 11 & 12 \\ 0 & 0 & 0 & 13 & 14 \\ 0 & 0 & 0 & 0 & 15 \end{vmatrix}$$

Expansion along the first column gives

$$D = 1\begin{vmatrix} 6 & 7 & 8 & 9 \\ 0 & 10 & 11 & 12 \\ 0 & 0 & 13 & 14 \\ 0 & 0 & 0 & 15 \end{vmatrix} = 1 \times 6\begin{vmatrix} 10 & 11 & 12 \\ 0 & 13 & 14 \\ 0 & 0 & 15 \end{vmatrix}$$

$$= 1 \times 6 \times 10\begin{vmatrix} 13 & 14 \\ 0 & 15 \end{vmatrix} = 1 \times 6 \times 10 \times 13 \times 15$$

The value of a triangular determinant is equal to the product of the elements on the diagonal,

$$D = a_{11} \times a_{22} \times a_{33} \times \cdots \times a_{nn}.$$

> Exercise 14

NOTE: on the evaluation of determinants and the solution of linear equations.
The expansion of a determinant in terms of its elements consists of $n!$ products of n elements at a time, and involves $n!(n-1)$ multiplications (see (17.11) for $n=3$). It follows therefore that such expansions do *not* provide a practical (or accurate) method for the evaluation of large determinants. The same is true of the use of Cramer's rule for the solution of linear equations discussed in the following section. Except for very small values of n, numerical methods such as those described in Section 17.6 and Chapter 20 must always be used.

17.4 The solution of linear equations

We have seen in Sections 17.1 and 17.2 that the solutions of two linear equations, (17.6), and of three linear equations, (17.12), can be written as ratios of determinants when the determinant of the coefficients of the unknowns is not zero. In general, the system of n linear equations,

$$\begin{aligned}
a_{11}x_1 + a_{12}x_2 + a_{13}x_3 + \cdots + a_{1n}x_n &= b_1 \\
a_{21}x_1 + a_{22}x_2 + a_{23}x_3 + \cdots + a_{2n}x_n &= b_2 \\
a_{31}x_1 + a_{32}x_2 + a_{33}x_3 + \cdots + a_{3n}x_n &= b_3 \\
\vdots \quad \vdots \quad \vdots \qquad \vdots \quad \vdots \\
a_{n1}x_1 + a_{n2}x_2 + a_{n3}x_3 + \cdots + a_{nn}x_n &= b_n
\end{aligned} \tag{17.26}$$

has a single unique solution if the determinant of the coefficients,

$$D = \begin{vmatrix}
a_{11} & a_{12} & a_{13} & \cdots & a_{1n} \\
a_{21} & a_{22} & a_{23} & \cdots & a_{2n} \\
a_{31} & a_{32} & a_{33} & \cdots & a_{3n} \\
\vdots & \vdots & \vdots & & \vdots \\
a_{n1} & a_{n2} & a_{n3} & \cdots & a_{nn}
\end{vmatrix} \tag{17.27}$$

is not zero. This solution is given by **Cramer's rule**,[3]

$$x_1 = \frac{D_1}{D}, \quad x_2 = \frac{D_2}{D}, \quad \cdots \quad x_n = \frac{D_n}{D} \tag{17.28}$$

where D_k is obtained from D by replacement of the kth column of D by the column with elements $b_1, b_2, \ldots, b_n$. For example,

$$x_2 = \frac{D_2}{D} = \frac{1}{D} \begin{vmatrix}
a_{11} & b_1 & a_{13} & \cdots & a_{1n} \\
a_{21} & b_2 & a_{23} & \cdots & a_{2n} \\
a_{31} & b_3 & a_{33} & \cdots & a_{3n} \\
\vdots & \vdots & \vdots & & \vdots \\
a_{n1} & b_n & a_{n3} & \cdots & a_{nn}
\end{vmatrix} \tag{17.29}$$

[3] Gabriel Cramer (1704–1752), Swiss, published the rule in his *Introduction à l'analyse des lignes courbes algébriques* (Introduction to the analysis of algebraic curves) in 1750.

This result can be derived by a generalization of the method described in Section 17.2 for the system of three equations. Thus, for x_2, multiplication of each equation of (17.26) by the cofactor of the coefficient of x_2 followed by addition gives $Dx_2 = D_2$. Then $x_2 = D_2/D$ if $D \neq 0$. We note that Cramer's rule applies even when all the quantities b_k on the right sides of equations (17.26) are zero. In this case, all the determinants D_k are also zero, and the solution is

$$x_1 = x_2 = x_3 = \cdots = x_n = 0 \tag{17.30}$$

Examples of the use of Cramer's rule are Examples 17.1 and 17.2.

▸ Exercises 15, 16

The case $D = 0$

Cramer's rule provides the unique solution of the system of linear equations (17.26) so long as the determinant of the coefficients (17.27) is not zero. The rule does not apply however when $D = 0$ because division by $D = 0$ in (17.28) is not allowed. There is then in general no unique solution or, in some cases, no solution of the equations at all. For example, the system

$$
\begin{align}
(1) \quad & 2x + 2y + z = 10 \\
(2) \quad & x + 2y - 2z = -3 \\
(3) \quad & 3x + 2y + 4z = 20
\end{align}
\tag{17.31}
$$

has determinant

$$
D = \begin{vmatrix} 2 & 2 & 1 \\ 1 & 2 & -2 \\ 3 & 2 & 4 \end{vmatrix} = 2 \begin{vmatrix} 2 & -2 \\ 2 & 4 \end{vmatrix} - 2 \begin{vmatrix} 1 & -2 \\ 3 & 4 \end{vmatrix} + \begin{vmatrix} 1 & 2 \\ 3 & 2 \end{vmatrix} = 24 - 20 - 4 = 0
$$

and no solution exists because the equations are **inconsistent**. Thus, addition of equations (2) and (3) gives

$$4x + 4y + 2z = 17$$

whereas twice equation (1) is

$$4x + 4y + 2z = 20$$

The system (17.31) is made consistent by, for example, changing the right side of (3) to 23:

$$
\begin{align}
(1) \quad & 2x + 2y + z = 10 \\
(2) \quad & x + 2y - 2z = -3 \\
(3) \quad & 3x + 2y + 4z = 23
\end{align}
\tag{17.32}
$$

The sum of (2) and (3) is now equal to twice (1), but this means that equation (1) contains no information not already contained in the other two equations. The equations are said to be **linearly dependent**, and each equation can be expressed as a linear combination of the others. We have effectively only two equations in three unknowns. For example, solving (1) and (2), or any pair, for x and y in terms of z gives

$$x = 13 - 3z, \qquad y = \frac{1}{2}(5z - 16) \qquad (17.33)$$

and this is a solution of the system (17.32) for every value of z.

▸ Exercise 17

Homogeneous equations

When at least one of the quantities b_k on the right sides of equations (17.26) is not zero, the equations are called **inhomogeneous equations**. When all the b_k are zero, they are called **homogeneous equations**:

$$\begin{aligned}
a_{11}x_1 + a_{12}x_2 + a_{13}x_3 + \cdots + a_{1n}x_n &= 0 \\
a_{21}x_1 + a_{22}x_2 + a_{23}x_3 + \cdots + a_{2n}x_n &= 0 \\
a_{31}x_1 + a_{32}x_2 + a_{33}x_3 + \cdots + a_{3n}x_n &= 0 \\
\vdots \qquad \vdots \qquad \vdots \qquad\quad \vdots \qquad \vdots \\
a_{n1}x_1 + a_{n2}x_2 + a_{n3}x_3 + \cdots + a_{nn}x_n &= 0
\end{aligned} \qquad (17.34)$$

Only the zero solution (17.30) exists if $D \neq 0$, but other solutions also exist when $D = 0$. For example, the system

$$\begin{aligned}
(1) \quad & 2x + 2y + z = 0 \\
(2) \quad & x + 2y - 2z = 0 \\
(3) \quad & 3x + 2y + 4z = 0
\end{aligned} \qquad (17.35)$$

has $D = 0$ and, like (17.32), the equations are linearly dependent. One solution is **the trivial (zero) solution** $x = y = z = 0$. *Nonzero* solutions are obtained by solving any pair of the equations for x and y in terms of z:

$$x = -3z, \qquad y = \frac{5z}{2} \qquad (17.36)$$

for all values of z. A *unique* solution is obtained only if a further, independent, relation amongst x, y, and z is known.

This example demonstrates one of the most important theorems of systems of linear equations:

> A system of homogeneous linear equations has nontrivial
> solution only if the determinant of the coefficients is zero.

The general condition that a determinant be zero is discussed in Section 17.5. Systems of linear equations are discussed in greater detail as matrix equations in Chapter 19.

> Exercise 18

Secular equations

A number of problems in the physical sciences give rise to systems of equations of the form

$$a_{11}x_1 + a_{12}x_2 + a_{13}x_3 + \cdots + a_{1n}x_n = \lambda x_1$$
$$a_{21}x_1 + a_{22}x_2 + a_{23}x_3 + \cdots + a_{2n}x_n = \lambda x_2$$
$$a_{31}x_1 + a_{32}x_2 + a_{33}x_3 + \cdots + a_{3n}x_n = \lambda x_3 \qquad (17.37)$$
$$\vdots \qquad \vdots \qquad \vdots \qquad \vdots \qquad \vdots$$
$$a_{n1}x_1 + a_{n2}x_2 + a_{n3}x_3 + \cdots + a_{nn}x_n = \lambda x_n$$

where λ is a parameter to be determined. For example, in molecular-orbital theory, the Schrödinger equation is replaced by such a set of linear equations in which the quantities $x_1, x_2, \ldots, x_n$ represents an orbital and λ the corresponding orbital energy. The equations can be written as

$$(a_{11} - \lambda)x_1 + \quad a_{12}x_2 \quad + \quad a_{13}x_3 \quad + \cdots + \quad a_{1n}x_n \quad = 0$$
$$a_{21}x_1 \quad + (a_{22} - \lambda)x_2 + \quad a_{23}x_3 \quad + \cdots + \quad a_{2n}x_n \quad = 0$$
$$a_{31}x_1 \quad + \quad a_{32}x_2 \quad + (a_{33} - \lambda)x_3 + \cdots + \quad a_{3n}x_n \quad = 0 \qquad (17.38)$$
$$\vdots \qquad \qquad \vdots \qquad \qquad \vdots \qquad \qquad \vdots \qquad \vdots$$
$$a_{n1}x_1 \quad + \quad a_{n2}x_2 \quad + \quad a_{n3}x_3 \quad + \cdots + (a_{nn} - \lambda)x_n = 0$$

These homogeneous equations, called **secular equations**, have non-trivial solution only if the determinant of the coefficients is zero,

$$\begin{vmatrix} (a_{11} - \lambda) & a_{12} & a_{13} & \cdots & a_{1n} \\ a_{21} & (a_{22} - \lambda) & a_{23} & \cdots & a_{2n} \\ a_{31} & a_{32} & (a_{33} - \lambda) & \cdots & a_{3n} \\ \vdots & \vdots & \vdots & & \vdots \\ a_{n1} & a_{n2} & a_{n3} & \cdots & (a_{nn} - \lambda) \end{vmatrix} = 0 \qquad (17.39)$$

The determinant is called a **secular determinant** in this context. It is zero only for some values of the parameter λ and these are obtained by solving equation (17.39). Because the expansion of the secular determinant is a polynomial of degree n in λ, the required values of λ are the n roots of the polynomial.

EXAMPLE 17.8 Find the values of λ for which the following system of equations has nonzero solution:

$$-2x + y + z = \lambda x$$
$$-11x + 4y + 5z = \lambda y$$
$$-x + y = \lambda z$$

The equations can be written

$$(-2 - \lambda)x + y + z = 0$$
$$-11x + (4 - \lambda)y + 5z = 0$$
$$-x + y + (-\lambda)z = 0$$

and have nonzero solution when

$$D = \begin{vmatrix} -2 - \lambda & 1 & 1 \\ -11 & 4 - \lambda & 5 \\ -1 & 1 & -\lambda \end{vmatrix} = 0$$

Therefore

$$D = (-2 - \lambda) \begin{vmatrix} 4 - \lambda & 5 \\ 1 & -\lambda \end{vmatrix} - \begin{vmatrix} -11 & 5 \\ -1 & -\lambda \end{vmatrix} + \begin{vmatrix} -11 & 4 - \lambda \\ -1 & 1 \end{vmatrix}$$

$$= (-2 - \lambda)[-\lambda(4 - \lambda) - 5] - [11\lambda + 5] + [-11 + (4 - \lambda)]$$

$$= -\lambda^3 + 2\lambda^2 + \lambda - 2 = -(\lambda - 1)(\lambda + 1)(\lambda - 2)$$

and $D = 0$ when $\lambda = 1, -1$, and 2.

For each of the n roots of the secular determinant there exists a solution of the secular equations (17.38).

EXAMPLE 17.9 Solve the Hückel molecular-orbital problem for the allyl radical $CH_2 \, CH \, CH_2$ in terms of the Hückel parameters α and β:

$$(1) \quad (\alpha - E)c_1 + \beta c_2 = 0$$
$$(2) \quad \beta c_1 + (\alpha - E)c_2 + \beta c_3 = 0$$
$$(3) \quad \beta c_2 + (\alpha - E)c_3 = 0$$

The equations have nonzero solution when the determinant of the coefficients of c_1, c_2, and c_3 is zero:

$$\begin{vmatrix} \alpha - E & \beta & 0 \\ \beta & \alpha - E & \beta \\ 0 & \beta & \alpha - E \end{vmatrix} = (\alpha - E)\left[(\alpha - E)^2 - 2\beta^2 \right] = 0$$

The roots are $E_1 = \alpha$, $E_2 = \alpha + \sqrt{2}\beta$ and $E_3 = \alpha - \sqrt{2}\beta$.

The corresponding solutions of the equations are obtained by replacing E in the secular equations by each root in turn.

For $E = E_1 = \alpha$:   (1) $\beta c_2 = 0$ $\longrightarrow$ $c_2 = 0$

(2) $\beta c_1 + \beta c_3 = 0$ $\longrightarrow$ $c_1 = -c_3$

(3) $\beta c_2 = 0$ $\longrightarrow$ $c_2 = 0$

We see that equations (1) and (3) are identical, so that only two of the three equations are independent. We solve for c_1 and c_2 in terms of (arbitrary) c_3. Similarly,

$E = E_2 = \alpha + \sqrt{2}\beta$:   (3) $\beta c_2 - \sqrt{2}\beta c_3 = 0$ $\longrightarrow$ $c_2 = \sqrt{2}c_3$

(1) $-\sqrt{2}\beta c_1 + \beta c_2 = 0$ $\longrightarrow$ $c_1 = c_2/\sqrt{2} = c_3$

$E = E_3 = \alpha - \sqrt{2}\beta$:   (3) $\beta c_2 + \sqrt{2}\beta c_3 = 0$ $\longrightarrow$ $c_2 = -\sqrt{2}c_3$

(1) $\sqrt{2}\beta c_1 + \beta c_2 = 0$ $\longrightarrow$ $c_1 = -c_2/\sqrt{2} = c_3$

Setting $c_3 = 1$ for convenience, the three solutions of the secular problem are therefore

E	c_1	c_2	c_3
α	-1	0	1
$\alpha + \sqrt{2}\beta$	1	$\sqrt{2}$	1
$\alpha - \sqrt{2}\beta$	1	$-\sqrt{2}$	1

> Exercises 19–21

17.5   Properties of determinants

The following are the more important general properties of determinants.

1. Transposition

Because the same value of a determinant is obtained by expansion along any row *or* column,

the value of a determinant is unchanged if its rows and columns are interchanged:

$$\begin{vmatrix} a_1 & b_1 & c_1 \\ a_2 & b_2 & c_2 \\ a_3 & b_3 & c_3 \end{vmatrix} = \begin{vmatrix} a_1 & a_2 & a_3 \\ b_1 & b_2 & b_3 \\ c_1 & c_2 & c_3 \end{vmatrix} \qquad (17.40)$$

2. Multiplication by a constant

If all the elements of any row (or column) are multiplied by the same factor λ, the value of the new determinant is λ times the value of the old determinant:

$$\begin{vmatrix} \lambda a_1 & \lambda b_1 & \lambda c_1 \\ a_2 & b_2 & c_2 \\ a_3 & b_3 & c_3 \end{vmatrix} = \lambda \begin{vmatrix} a_1 & b_1 & c_1 \\ a_2 & b_2 & c_2 \\ a_3 & b_3 & c_3 \end{vmatrix} \qquad (17.41)$$

EXAMPLE 17.10 Examples of Property 2.

$$\begin{vmatrix} 2 & 4 & 6 \\ 1 & 8 & 9 \\ 3 & 12 & 27 \end{vmatrix} = 2 \begin{vmatrix} 1 & 2 & 3 \\ 1 & 8 & 9 \\ 3 & 12 & 27 \end{vmatrix} = 2 \times 2 \begin{vmatrix} 1 & 1 & 3 \\ 1 & 4 & 9 \\ 3 & 6 & 27 \end{vmatrix}$$

$$= 2 \times 2 \times 3 \begin{vmatrix} 1 & 1 & 1 \\ 1 & 4 & 3 \\ 3 & 6 & 9 \end{vmatrix} = 2 \times 2 \times 3 \times 3 \begin{vmatrix} 1 & 1 & 1 \\ 1 & 4 & 3 \\ 1 & 2 & 3 \end{vmatrix} = 144$$

A special case is $\lambda = 0$: if all the elements of a row (or column) are zero, the value of the determinant is zero:

$$\begin{vmatrix} 0 & 0 & 0 \\ a_2 & b_2 & c_2 \\ a_3 & b_3 & c_2 \end{vmatrix} = 0 \qquad (17.42)$$

3. Addition rule

If all the elements of any row (or column) are written as the sum of two terms the determinant can be written as the sum of two determinants:

$$\begin{vmatrix} a_1 + d_1 & b_1 & c_1 \\ a_2 + d_2 & b_2 & c_2 \\ a_3 + d_3 & b_3 & c_3 \end{vmatrix} = \begin{vmatrix} a_1 & b_1 & c_1 \\ a_2 & b_2 & c_2 \\ a_3 & b_3 & c_3 \end{vmatrix} + \begin{vmatrix} d_1 & b_1 & c_1 \\ d_2 & b_2 & c_2 \\ d_3 & b_3 & c_3 \end{vmatrix} \qquad (17.43)$$

This follows from the expansion of the three determinants along the relevant row or column; column 1 in (17.43). Thus,

$$
\begin{vmatrix} a_1+d_1 & b_1 & c_1 \\ a_2+d_2 & b_2 & c_2 \\ a_3+d_3 & b_3 & c_3 \end{vmatrix} = (a_1+d_1)\begin{vmatrix} b_2 & c_2 \\ b_3 & c_3 \end{vmatrix} - (a_2+d_2)\begin{vmatrix} b_1 & c_1 \\ b_3 & c_3 \end{vmatrix} + (a_3+d_3)\begin{vmatrix} b_1 & c_1 \\ b_2 & c_2 \end{vmatrix}
$$

$$
= a_1\begin{vmatrix} b_2 & c_2 \\ b_3 & c_3 \end{vmatrix} - a_2\begin{vmatrix} b_1 & c_1 \\ b_3 & c_3 \end{vmatrix} + a_3\begin{vmatrix} b_1 & c_1 \\ b_2 & c_2 \end{vmatrix}
$$

$$
+ d_1\begin{vmatrix} b_2 & c_2 \\ b_3 & c_3 \end{vmatrix} - d_2\begin{vmatrix} b_1 & c_1 \\ b_3 & c_3 \end{vmatrix} + d_3\begin{vmatrix} b_1 & c_1 \\ b_2 & c_2 \end{vmatrix}
$$

$$
= \begin{vmatrix} a_1 & b_1 & c_1 \\ a_2 & b_2 & c_2 \\ a_3 & b_3 & c_3 \end{vmatrix} + \begin{vmatrix} d_1 & b_1 & c_1 \\ d_2 & b_2 & c_2 \\ d_3 & b_3 & c_3 \end{vmatrix}
$$

4.   Interchange of rows (or columns). Antisymmetry

If two rows (or two columns) of a determinant are interchanged the value of the determinant is multiplied by (–1):

$$
\begin{vmatrix} a_1 & b_1 & c_1 \\ a_2 & b_2 & c_2 \\ a_3 & b_3 & c_3 \end{vmatrix} = - \begin{vmatrix} a_2 & b_2 & c_2 \\ a_1 & b_1 & c_1 \\ a_3 & b_3 & c_3 \end{vmatrix} \quad \text{(interchange of rows 1 and 2)} \qquad (17.44)
$$

$$
\begin{vmatrix} a_1 & b_1 & c_1 \\ a_2 & b_2 & c_2 \\ a_3 & b_3 & c_3 \end{vmatrix} = - \begin{vmatrix} a_1 & c_1 & b_1 \\ a_2 & c_2 & b_2 \\ a_3 & c_3 & b_3 \end{vmatrix} \quad \text{(interchange of columns 2 and 3)} \quad (17.45)
$$

The determinant is said to be **antisymmetric** with respect to the interchange of rows or columns. It is this property of determinants that has made them so useful for the construction of electronic wave functions.

EXAMPLE 17.11 Determinant of order 2.

For interchange of columns,

$$
\begin{vmatrix} a_1 & b_1 \\ a_2 & b_2 \end{vmatrix} = a_1 b_2 - b_1 a_2 = -(b_1 a_2 - a_1 b_2) = - \begin{vmatrix} b_1 & a_1 \\ b_2 & a_2 \end{vmatrix}
$$

For interchange of rows

$$D_1 = \begin{vmatrix} 1 & 2 \\ 3 & 4 \end{vmatrix} = 1 \times 4 - 2 \times 3 = -2$$

$$D_2 = \begin{vmatrix} 3 & 4 \\ 1 & 2 \end{vmatrix} = 3 \times 2 - 4 \times 1 = +2 = -D_1$$

EXAMPLE 17.12 Determinant of order 3.

By Example 17.2,

$$D = \begin{vmatrix} 2 & -3 & 4 \\ 0 & 1 & -3 \\ 1 & 2 & 2 \end{vmatrix} = 21$$

Interchange of rows 1 and 3 gives

$$\begin{vmatrix} 1 & 2 & 2 \\ 0 & 1 & -3 \\ 2 & -3 & 4 \end{vmatrix} = \begin{vmatrix} 1 & -3 \\ -3 & 4 \end{vmatrix} - 2 \begin{vmatrix} 0 & -3 \\ 2 & 4 \end{vmatrix} + 2 \begin{vmatrix} 0 & 1 \\ 2 & -3 \end{vmatrix}$$

$$= -5 - 12 - 4 = -21 = -D$$

5. **Two rows or columns equal**

The value of a determinant is *zero* if two rows (or two columns) are equal:

$$\begin{vmatrix} a_1 & b_1 & c_1 \\ a_1 & b_1 & c_1 \\ a_3 & b_3 & c_3 \end{vmatrix} = 0 \qquad\qquad (17.46)$$

This follows from the antisymmetry property of the determinant. By Property 4, the sign of the determinant is changed when two rows (or columns) are interchanged. If the rows (or columns) are identical then such an interchange must leave the determinant unchanged. Therefore $-D = D$ and this is possible only if $D = 0$.

EXAMPLE 17.13 Example of Property 5: rows 1 and 2 are equal.

$$\begin{vmatrix} 1 & 1 & 1 \\ 1 & 1 & 1 \\ 2 & 3 & 4 \end{vmatrix} = \begin{vmatrix} 1 & 1 \\ 3 & 4 \end{vmatrix} - \begin{vmatrix} 1 & 1 \\ 2 & 4 \end{vmatrix} + \begin{vmatrix} 1 & 1 \\ 2 & 3 \end{vmatrix} = 1 - 2 + 1 = 0$$

6. Proportional rows or columns

The value of a determinant is *zero* if one row (or column) is a multiple of another row (or column):

$$\begin{vmatrix} \lambda a_2 & \lambda b_2 & \lambda c_2 \\ a_2 & b_2 & c_2 \\ a_3 & b_3 & c_3 \end{vmatrix} = \lambda \begin{vmatrix} a_2 & b_2 & c_2 \\ a_2 & b_2 & c_2 \\ a_3 & b_3 & c_3 \end{vmatrix} = 0 \qquad (17.47)$$

This follows from Property 2 for the multiple of a determinant, and Property 5 for two equal rows.

EXAMPLE 17.14 Example of Property 6: column 1 is a multiple of column 2.

$$\begin{vmatrix} 4 & 2 & 3 \\ 2 & 1 & 4 \\ 6 & 3 & 5 \end{vmatrix} = 2 \begin{vmatrix} 2 & 2 & 3 \\ 1 & 1 & 4 \\ 3 & 3 & 5 \end{vmatrix} = 0$$

---

7. Addition of rows or columns

The value of a determinant is *unchanged* when a multiple of any row (or column) is added to any other row (or column):

$$\begin{vmatrix} a_1 + \lambda b_1 & b_1 & c_1 \\ a_2 + \lambda b_2 & b_2 & c_2 \\ a_3 + \lambda b_3 & b_3 & c_3 \end{vmatrix} = \begin{vmatrix} a_1 & b_1 & c_1 \\ a_2 & b_2 & c_2 \\ a_3 & b_3 & c_3 \end{vmatrix} + \begin{vmatrix} \lambda b_1 & b_1 & c_1 \\ \lambda b_2 & b_2 & c_2 \\ \lambda b_3 & b_3 & c_3 \end{vmatrix}$$

$$\qquad (17.48)$$

$$= \begin{vmatrix} a_1 & b_1 & c_1 \\ a_2 & b_2 & c_2 \\ a_3 & b_3 & c_3 \end{vmatrix}$$

This follows from Property 3 for the addition of determinants and Property 6 for one column equal to a multiple of an other.

EXAMPLE 17.15 Example of Property 7. Three times column 2 has been added to column 1.

$$\begin{vmatrix} 4 & 1 & 1 \\ 8 & 3 & 4 \\ 15 & 5 & 6 \end{vmatrix} = \begin{vmatrix} (1+3\times1) & 1 & 1 \\ (2+3\times3) & 3 & 4 \\ (0+3\times5) & 5 & 6 \end{vmatrix} = \begin{vmatrix} 1 & 1 & 1 \\ 2 & 3 & 4 \\ 0 & 5 & 6 \end{vmatrix} + 3 \begin{vmatrix} 1 & 1 & 1 \\ 3 & 3 & 4 \\ 5 & 5 & 6 \end{vmatrix} = \begin{vmatrix} 1 & 1 & 1 \\ 2 & 3 & 4 \\ 0 & 5 & 6 \end{vmatrix}$$

---

8. Linearly-dependent rows or columns

The value of a determinant is *zero* if the rows (or columns) are linearly dependent; that is, if a row (or column) is a linear combination of the others.

$$\begin{vmatrix} \lambda b_1 + \mu c_1 & b_1 & c_1 \\ \lambda b_2 + \mu c_2 & b_2 & c_2 \\ \lambda b_3 + \mu c_3 & b_3 & c_3 \end{vmatrix} = \lambda \begin{vmatrix} b_1 & b_1 & c_1 \\ b_2 & b_2 & c_2 \\ b_3 & b_3 & c_3 \end{vmatrix} + \mu \begin{vmatrix} c_1 & b_1 & c_1 \\ c_2 & b_2 & c_2 \\ c_3 & b_3 & c_3 \end{vmatrix} = 0 \quad (17.49)$$

This follows from Properties 5, 6 and 7. It is the general condition for the value of a determinant to be zero.

➤ Exercises 22–24

9. Derivative of a determinant

If the elements of a determinant D are differentiable functions, the derivative D' of D can be written

$$D' = D_1 + D_2 + D_3 + \cdots + D_n \quad (17.50)$$

where D_i is obtained from D by differentiation of the elements of the ith row.

EXAMPLE 17.16 Differentiation of a third-order determinant.

If the elements are functions of x,

$$\frac{d}{dx} \begin{vmatrix} a_1 & a_2 & a_3 \\ b_1 & b_2 & b_3 \\ c_1 & c_2 & c_3 \end{vmatrix} = \begin{vmatrix} \frac{da_1}{dx} & \frac{da_2}{dx} & \frac{da_3}{dx} \\ b_1 & b_2 & b_3 \\ c_1 & c_2 & c_3 \end{vmatrix} + \begin{vmatrix} a_1 & a_2 & a_3 \\ \frac{db_1}{dx} & \frac{db_2}{dx} & \frac{db_3}{dx} \\ c_1 & c_2 & c_3 \end{vmatrix} + \begin{vmatrix} a_1 & a_2 & a_3 \\ b_1 & b_2 & b_3 \\ \frac{dc_1}{dx} & \frac{dc_2}{dx} & \frac{dc_3}{dx} \end{vmatrix}$$

➤ Exercise 25

17.6 Reduction to triangular form

It is demonstrated in Example 17.7 that a 'triangular' determinant has value equal to the product of its diagonal elements,

$$\begin{vmatrix} a_{11} & a_{12} & a_{13} & a_{14} & \cdots & a_{1n} \\ 0 & a_{22} & a_{23} & a_{24} & \cdots & a_{2n} \\ 0 & 0 & a_{33} & a_{34} & \cdots & a_{3n} \\ 0 & 0 & 0 & a_{44} & \cdots & a_{4n} \\ \vdots & \vdots & \vdots & \vdots & & \vdots \\ 0 & 0 & 0 & 0 & \cdots & a_{nn} \end{vmatrix} = a_{11} \times a_{22} \times a_{33} \times \cdots \times a_{nn} \qquad (17.51)$$

Every determinant can be reduced to triangular form by means of a systematic application of Property 7 in Section 17.5. The method is an example of the elimination methods discussed in Chapter 20, and is illustrated in Example 17.17 for a third-order determinant.

EXAMPLE 17.17 Example of reduction to triangular form.

$$\begin{vmatrix} 1 & 2 & 3 \\ 3 & 2 & 5 \\ 2 & 3 & 6 \end{vmatrix} \qquad \text{subtract } (3 \times \text{row 1}) \text{ from row 2}$$

$$= \begin{vmatrix} 1 & 2 & 3 \\ 0 & -4 & -4 \\ 2 & 3 & 6 \end{vmatrix} \qquad \text{subtract } (2 \times \text{row 1}) \text{ from row 3}$$

$$= \begin{vmatrix} 1 & 2 & 3 \\ 0 & -4 & -4 \\ 0 & -1 & 0 \end{vmatrix} \qquad \text{subtract } (\tfrac{1}{4} \times \text{row 2}) \text{ from row 3}$$

$$= \begin{vmatrix} 1 & 2 & 3 \\ 0 & -4 & -4 \\ 0 & 0 & 1 \end{vmatrix} = 1 \times (-4) \times 1 = -4$$

▸ Exercises 26–27

17.7   Alternating functions

A function $f(x_1, x_2, x_3, \ldots, x_n)$ of n variables is called an **alternating function**, or **totally antisymmetric**, if the interchange of any two of the variables has the effect of multiplying the value of the function by (-1). For the interchange of x_1 and x_2,

$$f(x_2, x_1, x_3, \ldots, x_n) = -f(x_1, x_2, x_3, \ldots, x_n) \qquad (17.52)$$

If two variables are equal the function is zero,

$$f(x_1, x_1, x_3, \ldots, x_n) = 0 \qquad (17.53)$$

A determinant is an alternating function of its rows (or columns). More importantly, a determinant that is an alternating function of n variables, $x_1, x_2, x_3 \ldots, x_n$, has the form

$$\begin{vmatrix} f_1(x_1) & f_1(x_2) & f_1(x_3) & \cdots & f_1(x_n) \\ f_2(x_1) & f_2(x_2) & f_2(x_3) & \cdots & f_2(x_n) \\ f_3(x_1) & f_3(x_2) & f_3(x_3) & \cdots & f_3(x_n) \\ \vdots & \vdots & \vdots & & \vdots \\ f_n(x_1) & f_n(x_2) & f_n(x_3) & \cdots & f_n(x_n) \end{vmatrix} \qquad (17.54)$$

where $f_1, f_2, \ldots, f_n$ are arbitrary functions. The interchange of any pair of variables leads to the interchange of two columns and, therefore, to a change of sign.

For $n = 2$,

$$\begin{vmatrix} f_1(x_1) & f_1(x_2) \\ f_2(x_1) & f_2(x_2) \end{vmatrix} = f_1(x_1)f_2(x_2) - f_1(x_2)f_2(x_1) \qquad (17.55)$$

For $n = 3$,

$$\begin{vmatrix} f_1(x_1) & f_1(x_2) & f_1(x_3) \\ f_2(x_1) & f_2(x_2) & f_2(x_3) \\ f_3(x_1) & f_3(x_2) & f_3(x_3) \end{vmatrix}$$

$$= \begin{cases} f_1(x_1)f_2(x_2)f_3(x_3) - f_1(x_1)f_2(x_3)f_3(x_2) \\ + f_1(x_2)f_2(x_3)f_3(x_1) - f_1(x_2)f_2(x_1)f_3(x_3) \\ + f_1(x_3)f_2(x_1)f_3(x_2) - f_1(x_3)f_2(x_2)f_3(x_1) \end{cases} \qquad (17.56)$$

The expansion of the determinant has $n!$ products of the functions $f_1, f_2, \ldots, f_n$, each with a distinct ordering of the n variables $x_1, x_2, \ldots, x_n$; these orderings are the $n!$ permutations of n objects. Thus, in (17.56), the $3! = 6$ permutations of x_1, x_2, and x_3 are

$$x_1 x_2 x_3, \quad x_1 x_3 x_2, \quad x_2 x_3 x_1, \quad x_2 x_1 x_3, \quad x_3 x_1 x_2, \quad x_3 x_2 x_1 \quad (17.57)$$

Each term contributes to the sum with $+$ sign if the permutation is obtained from $x_1 x_2 x_3$ by an *even* number of transpositions, and with $-$ sign for an *odd* number of transpositions.

▸ Exercise 29

Alternating functions in the form of single determinants or sums of determinants are important in quantum mechanics for the construction of electronic wave functions. The electron is a member of the class of particles called **fermions**, particles with half-integral spin (a particle with zero or integral spin is call a **boson**). The wave function of a system of identical fermions is totally antisymmetric with respect to the interchange of the coordinates (including spin) of the fermions; that is, the interchange of the coordinates of any pair of fermions results in the change of sign of the wave function.* This is just the property of an alternating function. Thus, if the functions $f_1, f_2, \dots, f_n$ in the determinant (17.54) represent the occupied states of the n electrons of a system, and if $x_1, x_2, \dots, x_n$ represent the n sets of coordinates (including spin) of the electrons, then the functions are called **spin-orbitals** and the determinant (17.54) is called a **Slater determinant**. Because the Slater determinant is antisymmetric, the interchange (of the coordinates and spin) of any pair of electrons results in a change of sign of the determinant. If two of the functions (spin-orbitals) are the same then two rows of the determinant are equal and the determinant is zero. This is an expression of the Pauli exclusion principle, that no two electrons can be in the same state (spin-orbital).

17.8 Exercises

Section 17.1

1. Use determinants to solve the pair of equations

$$4x + y = 11$$
$$3x + 2y = 12$$

Evaluate:

2. $\begin{vmatrix} 2 & 0 \\ 0 & 3 \end{vmatrix}$ 3. $\begin{vmatrix} 0 & 1 \\ -2 & 3 \end{vmatrix}$ 4. $\begin{vmatrix} \cos n\theta & -\sin n\theta \\ \sin n\theta & \cos n\theta \end{vmatrix}$

Section 17.2

5. Use determinants to solve the equations

$$x + y + z = 6$$
$$x + 2y + 3z = 14$$
$$x + 4y + 9z = 36$$

Evaluate the determinants by expansion along **(i)** the first row, **(ii)** the second column:

6. $\begin{vmatrix} 2 & 3 & 5 \\ 0 & 1 & 2 \\ 3 & 4 & 1 \end{vmatrix}$ 7. $\begin{vmatrix} 1 & 1 & 1 \\ 1 & -1 & 0 \\ 1 & 1 & -2 \end{vmatrix}$ 8. $\begin{vmatrix} 1 & 3 & -2 \\ 0 & -1 & 2 \\ 0 & 0 & 4 \end{vmatrix}$ 9. $\begin{vmatrix} 0 & 3 & 2 \\ 2 & 0 & 1 \\ 2 & 6 & 0 \end{vmatrix}$

10. **(i)** Find the cofactors of all the elements of

$$\begin{vmatrix} 1 & 2 & 3 \\ 2 & 0 & -1 \\ 1 & -1 & 1 \end{vmatrix}$$

(ii) Confirm that the same value of the determinant is obtained by expansion along every row and every column

---

* The wave function of a system of bosons is totally symmetric, and the interchange of the coordinates of any pair of identical bosons leaves the wave function unchanged.

Section 17.3

Evaluate:

11.
$$\begin{vmatrix} 2 & 0 & 1 & 3 \\ 3 & 1 & 0 & 4 \\ 1 & -1 & 2 & 3 \\ 2 & 2 & 1 & 0 \end{vmatrix}$$

12.
$$\begin{vmatrix} 3 & 4 & 0 & 0 \\ 1 & 2 & 0 & 0 \\ 0 & 0 & 3 & 1 \\ 0 & 0 & 4 & 2 \end{vmatrix}$$

13.
$$\begin{vmatrix} 1 & 1 & 0 & 0 \\ 3 & 2 & 2 & -3 \\ 2 & 1 & -1 & 2 \\ 5 & 3 & 1 & -1 \end{vmatrix}$$

14.
$$\begin{vmatrix} -2 & 6 & 17 & -5 \\ 0 & 3 & 22 & -17 \\ 0 & 0 & 4 & 12 \\ 0 & 0 & 0 & -6 \end{vmatrix}$$

Section 17.4

Use Cramer's rule to solve the systems of equations:

15. $3x - 2y - 2z = 0$
$x + y - z = 0$
$2x + 2y + z = 0$

16. $w + 2x + 3y + z = 5$
$2w + x + y + z = 3$
$w + 2x + y = 4$
$x + y + 2z = 0$

17. **(i)** Show that the following equations have no solution unless $k = 3$, **(ii)** solve for this value of k.

$2x - y + z = 2$
$3x + y - 2z = 1$
$x - 3y + 4z = k$

18. **(i)** Find k for which the following equations have a nontrivial solution, **(ii)** solve for this value of k.

$kx + 5y + 3z = 0$
$5x + y - z = 0$
$kx + 2y + z = 0$

Find **(i)** the values of λ for which the following systems of equations have nontrivial solutions, **(ii)** the solutions for these values of λ.

19. $2x + y = \lambda x$
$x + 2y = \lambda y$

20. $3x + y = \lambda x$
$x + 3y + z = \lambda y$
$y + 3z = \lambda z$

21. $x + 2y - 3z = \lambda x$
$2x + 4y - 6z = \lambda y$
$-x - 2y + 3z = \lambda z$

Section 17.5

Use the properties of determinants to show that:

22.
$$\begin{vmatrix} 3 & 6 & -3 \\ 2 & 1 & 5 \\ 1 & 2 & -1 \end{vmatrix} = 0$$

23.
$$\begin{vmatrix} 2 & 2 & -1 \\ 3 & 2 & 2 \\ -1 & 0 & -3 \end{vmatrix} = 0$$

24.
$$\begin{vmatrix} a & b \\ c & d \end{vmatrix} = \frac{1}{2} \begin{vmatrix} a-b & a+b \\ c-d & c+d \end{vmatrix}$$

25. Differentiate the following determinant with respect to x:

$$\begin{vmatrix} 1 & 2x & 3x^2 \\ 4x^3 & 5x^4 & 6x^5 \\ 7x^6 & 8x^7 & 9x^8 \end{vmatrix}$$

Section 17.6

Evaluate the following determinants by reduction to triangular form:

26. $\begin{vmatrix} 3 & 2 & -2 \\ 6 & 1 & 5 \\ -9 & 3 & 4 \end{vmatrix}$ 27. $\begin{vmatrix} 1 & 0 & 1 & 1 \\ 0 & 1 & 1 & 0 \\ 1 & 0 & 0 & 1 \\ 1 & 1 & 1 & 0 \end{vmatrix}$ 28. $\begin{vmatrix} 2 & 4 & 6 & 3 \\ 8 & 15 & 14 & 13 \\ 7 & 9 & 13 & 9 \\ 1 & 2 & 5 & 5 \end{vmatrix}$

Section 17.7

29. Expand the determinant

$$\begin{vmatrix} \psi_1(x_1) & \psi_1(x_2) & \psi_1(x_3) & \psi_1(x_4) \\ \psi_2(x_1) & \psi_2(x_2) & \psi_2(x_3) & \psi_2(x_4) \\ \psi_3(x_1) & \psi_3(x_2) & \psi_3(x_3) & \psi_3(x_4) \\ \psi_4(x_1) & \psi_4(x_2) & \psi_4(x_3) & \psi_4(x_4) \end{vmatrix}$$

18 Matrices and linear transformations

18.1 Concepts

Matrix algebra is the branch of mathematics concerned with the theory of systems of linear equations. Just as ordinary algebra is the essential tool for the manipulation and solution of single equations (or small numbers of equations), matrix algebra is used for the manipulation of *systems* of equations, and for the construction of numerical methods of solution of the problems that give rise to such systems in the physical sciences, in engineering, and in statistics. The economy and simplicity of the matrix formalism also makes it the ideal tool for the theoretical analysis of the properties and structure of systems of equations, and of the physical problems that lead to them. One of the important applications in chemistry makes use of the concept of linear transformation for the matrix representation of symmetry operations in the description of the symmetry properties of molecules, of molecular wave functions, and of normal modes of vibration. In this chapter, the elements of matrix algebra are presented in Sections 18.2 to 18.4, and the matrix theory of linear transformations in Sections 18.5 and 18.6. Symmetry operations are discussed in Section 18.7, with a brief introduction to symmetry groups.

A matrix consists of $m \times n$ quantities, or elements, arranged in a rectangular array made up of m rows and n columns, and enclosed in parentheses.[1] Examples are

$$\begin{pmatrix} 2 & -1 \\ 0 & 3 \end{pmatrix}, \quad \begin{pmatrix} a & b & c \\ d & e & f \end{pmatrix}, \quad \begin{pmatrix} x_1 \\ x_2 \\ x_3 \\ x_4 \end{pmatrix}, \quad \begin{pmatrix} 1 & 0 & 0 \\ 0 & 1 & 0 \\ 0 & 0 & 1 \end{pmatrix} \tag{18.1}$$

and the general notation for an $m \times n$ matrix (read as 'm by n matrix'), with mn elements, is (see Section 17.1 for determinants)

$$\mathbf{A} = \begin{pmatrix} a_{11} & a_{12} & a_{13} & \cdots & a_{1n} \\ a_{21} & a_{22} & a_{23} & \cdots & a_{2n} \\ a_{31} & a_{32} & a_{33} & \cdots & a_{3n} \\ \vdots & \vdots & \vdots & & \vdots \\ a_{m1} & a_{m2} & a_{m3} & \cdots & a_{mn} \end{pmatrix} \tag{18.2}$$

---

[1] The term matrix was coined by Sylvester in 1850 to denote 'an oblong arrangement of terms consisting, suppose, of m lines and n columns' because from it 'we may form various systems of determinants'. James Joseph Sylvester (1814–1897), studied mathematics at Cambridge but was ineligible for a Cambridge degree on religious grounds (he was a Jew); he received his degree from Trinity College, Dublin. His appointments include the chair of mathematics at the newly founded Johns Hopkins University in Baltimore (1876–1883). He was the founding Editor of the American Journal of Mathematics (1878). He is best known for his collaboration with his friend Cayley on the theory of matrices and forms.

In this notation, the element a_{ij} lies in the ith row and jth column. Although curved brackets are used in this text, matrices are often written with square brackets, and are also denoted by symbols such as (a_{ij}) and $[a_{ij}]$. To qualify as matrices, arrays must obey the laws of matrix algebra (Section 18.3).

Matrix algebra has its origins in the description of coordinate transformations.[2] Let the pair of linear equations

$$
\begin{aligned}
x' &= a_1 x + b_1 y \\
y' &= a_2 x + b_2 y
\end{aligned}
\tag{18.3}
$$

describe a change of coordinates in the xy-plane from (x, y) to (x', y'). This **coordinate transformation** is completely characterized by the four coefficients a_1, b_1, a_2, b_2; that is, by the array

$$
\begin{pmatrix} a_1 & b_1 \\ a_2 & b_2 \end{pmatrix}
\tag{18.4}
$$

In matrix algebra, the equations (18.3) are written as the single matrix equation

$$
\begin{pmatrix} x' \\ y' \end{pmatrix} = \begin{pmatrix} a_1 & b_1 \\ a_2 & b_2 \end{pmatrix} \begin{pmatrix} x \\ y \end{pmatrix}
\tag{18.5}
$$

or, assigning symbols to the arrays,

$$
\mathbf{r}' = \mathbf{A}\mathbf{r}
\tag{18.6}
$$

where

$$
\mathbf{r} = \begin{pmatrix} x \\ y \end{pmatrix}, \qquad \mathbf{r}' = \begin{pmatrix} x' \\ y' \end{pmatrix}, \qquad \mathbf{A} = \begin{pmatrix} a_1 & b_1 \\ a_2 & b_2 \end{pmatrix}
\tag{18.7}
$$

The quantities $\mathbf{r}$ and $\mathbf{r}'$ are **column matrices** (or **column vectors**) whose elements are the coordinates before and after the coordinate transformation, and $\mathbf{A}$ is the **square matrix** of the coefficients that represents and determines the transformation. Coordinate transformations are examples of the linear transformations discussed in Section 18.5.

---

[2] The algebra of coordinate transformations was discussed by Gauss in 1801, and developed into an algebra by Cayley in his *Memoir on the theory of matrices* of 1858, in which he introduced the single-letter notation for a matrix and derived the rules of addition and multiplication.

Arthur Cayley (1821–1895), graduated from Trinity College, Cambridge, in 1842, was Fellow for seven years, then practiced law for 14 years during which time he published several hundreds of papers. His output is rivalled only by Euler and Cauchy, with 967 published papers in his Collected Works. Best known for his work on matrices and forms, he also developed the analytical geometry of n-dimensional space in 1843, gave the first definition of an abstract group in 1854, and made contributions to graph theory, with an application to the study of chemical isomers in 1874. He was in great demand as a referee because of his encyclopaedic knowledge of mathematics.

EXAMPLE 18.1 Rotation as a coordinate transformation

Consider a point P in the xy-plane, with coordinates (x, y) with respect to the coordinate system Oxy. As discussed in Section 8.5, an anticlockwise rotation through angle θ about the Oz axis moves the point to position P′ with coordinates (x', y'), as in Figure 18.1, such that (equations (8.41))

$$x' = x \cos \theta - y \sin \theta$$
$$y' = x \sin \theta + y \cos \theta$$

(18.8)

An alternative way of interpreting these equations (the 'passive' instead of 'active' interpretation) is as the change of coordinates of the *fixed* point P when the coordinate system itself undergoes a *clockwise* rotation through angle θ, from Oxy to $Ox'y'$ as in Figure 18.2.

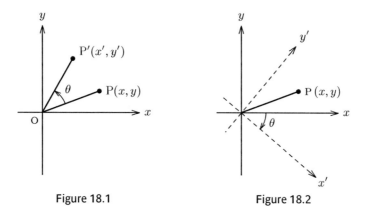

Figure 18.1   Figure 18.2

The corresponding matrix equation in both cases is

$$\begin{pmatrix} x' \\ y' \end{pmatrix} = \begin{pmatrix} \cos \theta & -\sin \theta \\ \sin \theta & \cos \theta \end{pmatrix} \begin{pmatrix} x \\ y \end{pmatrix}$$

(18.9)

and the transformation matrix

$$\begin{pmatrix} \cos \theta & -\sin \theta \\ \sin \theta & \cos \theta \end{pmatrix}$$

(18.10)

completely characterizes the coordinate transformation.

➤ Exercises 1, 2

18.2   Some special matrices

Square matrices

If the number of rows of a matrix is equal to the number of columns, $m = n$, the matrix is called a **square matrix** of **order** n. For example, a square matrix of order 3 is

$$\begin{pmatrix} a_{11} & a_{12} & a_{13} \\ a_{21} & a_{22} & a_{23} \\ a_{31} & a_{32} & a_{33} \end{pmatrix} \qquad \text{(square matrix)} \qquad (18.11)$$

In such a matrix, the diagonal containing the elements $a_{11}, a_{22}, \ldots, a_{nn}$ is called the **principal diagonal** (the word 'principal' is often omitted), and the elements a_{ii} on this diagonal are called the **diagonal elements** of the matrix. Those elements, a_{ij} for $i \neq j$, not on the diagonal are called the **off-diagonal elements**; the off-diagonal elements of the matrix (18.11) are $a_{12}, a_{13}, a_{23}, a_{21}, a_{31}$, and a_{32}.

A matrix whose off-diagonal elements are all zero is called a **diagonal matrix**; for example,

$$\begin{pmatrix} a_{11} & 0 & 0 \\ 0 & a_{22} & 0 \\ 0 & 0 & a_{33} \end{pmatrix} \qquad \text{(diagonal matrix)} \qquad (18.12)$$

The diagonal matrix of order n whose diagonal elements are all unity is called the **unit matrix I** (or $\mathbf{I}_n$) of order n. For order 3,

$$\mathbf{I} = \begin{pmatrix} 1 & 0 & 0 \\ 0 & 1 & 0 \\ 0 & 0 & 1 \end{pmatrix} \qquad \text{(unit matrix)} \qquad (18.13)$$

Two important scalar properties of a square matrix $\mathbf{A}$ are the **determinant** of the matrix, denoted by $|\mathbf{A}|$ or $\det \mathbf{A}$,

$$|\mathbf{A}| = \det \mathbf{A} = \begin{vmatrix} a_{11} & a_{12} & a_{13} & \cdots & a_{1n} \\ a_{21} & a_{22} & a_{23} & \cdots & a_{2n} \\ a_{31} & a_{32} & a_{33} & \cdots & a_{3n} \\ \vdots & \vdots & \vdots & & \vdots \\ a_{n1} & a_{n2} & a_{n3} & \cdots & a_{nn} \end{vmatrix} \qquad (18.14)$$

and the sum of the diagonal elements of the matrix, called the **trace** (or sometimes **spur**) of the matrix,

$$\text{tr } \mathbf{A} = a_{11} + a_{22} + a_{33} + \cdots + a_{nn} = \sum_{i=1}^{n} a_{ii} \qquad (18.15)$$

EXAMPLE 18.2 Find the determinant and trace of the matrix

$$\mathbf{A} = \begin{pmatrix} 2 & 1 & 1 & 3 \\ 4 & -2 & 0 & 1 \\ -1 & 1 & 0 & 2 \\ 0 & 3 & 1 & -1 \end{pmatrix}$$

By Example 17.6, the determinant of **A** is

$$\det \mathbf{A} = \begin{vmatrix} 2 & 1 & 1 & 3 \\ 4 & -2 & 0 & 1 \\ -1 & 1 & 0 & 2 \\ 0 & 3 & 1 & -1 \end{vmatrix} = -16$$

The trace of **A** is

$$\text{tr } \mathbf{A} = 2 + (-2) + 0 + (-1) = -1$$

➤ Exercises 3–6

Vectors

A matrix containing a single column only is called a **column matrix** or **column vector**; a matrix containing one row only is a **row matrix** or **row vector**. The elements of a vector are called **components**. Although matrices are normally denoted by uppercase boldface letters, vectors are denoted by lowercase letters;

$$\mathbf{a} = (a_1 \ a_2 \ a_3 \ \cdots \ a_n), \qquad \mathbf{b} = \begin{pmatrix} b_1 \\ b_2 \\ b_3 \\ \vdots \\ b_n \end{pmatrix} \qquad (18.16)$$

(only the single subscript is required for the components of a vector). Matrix algebra includes vector algebra as a special case.

The transpose matrix

The transpose $\mathbf{A}^T$ (or $\tilde{\mathbf{A}}$) of an $m \times n$ matrix $\mathbf{A}$ is the $n \times m$ matrix obtained from $\mathbf{A}$ by the interchange of rows and columns; the first column of $\mathbf{A}^T$ is the first row of $\mathbf{A}$, the second column of $\mathbf{A}^T$ is the second row of $\mathbf{A}$, and so on. For example,

$$\text{if} \quad \mathbf{A} = \begin{pmatrix} a_1 & b_1 \\ a_2 & b_2 \\ a_3 & b_3 \end{pmatrix} \quad \text{then} \quad \mathbf{A}^T = \begin{pmatrix} a_1 & a_2 & a_3 \\ b_1 & b_2 & b_3 \end{pmatrix} \quad (18.17)$$

EXAMPLES 18.3 Transpose matrices

(i) if $\quad \mathbf{A} = \begin{pmatrix} 1 & 2 & 0 \\ -1 & 4 & 2 \end{pmatrix} \quad$ then $\quad \mathbf{A}^T = \begin{pmatrix} 1 & -1 \\ 2 & 4 \\ 0 & 2 \end{pmatrix}$

(ii) if $\quad \mathbf{A} = \begin{pmatrix} 1 & 2 \\ -3 & 4 \end{pmatrix} \quad$ then $\quad \mathbf{A}^T = \begin{pmatrix} 1 & -3 \\ 2 & 4 \end{pmatrix}$

(iii) if $\quad \mathbf{a} = \begin{pmatrix} 3 \\ 2 \\ -1 \end{pmatrix} \quad$ then $\quad \mathbf{a}^T = (3 \quad 2 \quad -1)$

(iv) if $\quad \mathbf{A} = \begin{pmatrix} 1 & 3 & 0 \\ 3 & 2 & -1 \\ 0 & -1 & 0 \end{pmatrix} \quad$ then $\quad \mathbf{A}^T = \mathbf{A}$

> Exercises 7–12

Case (iv) above is an example of a **symmetric matrix**, whose transpose is equal to the matrix:

$$\mathbf{A}^T = \mathbf{A} \quad (18.18)$$

The determinant

The value of the determinant of a square matrix is unchanged when the matrix is transposed,

$$\det \mathbf{A}^T = \det \mathbf{A} \quad (18.19)$$

18.3 Matrix algebra

Equality of matrices

Two matrices, $\mathbf{A} = (a_{ij})$ and $\mathbf{B} = (b_{ij})$, are equal if they have the same dimensions (the same number of rows and the same number of columns), and if the corresponding elements are equal:

$$(a_{ij}) = (b_{ij}) \quad \text{if} \quad a_{ij} = b_{ij} \quad \text{for all} \quad i, j \qquad (18.20)$$

EXAMPLE 18.4 Equality of matrices

If $\quad \mathbf{A} = \begin{pmatrix} a_{11} & a_{12} & a_{13} \\ a_{21} & a_{22} & a_{23} \end{pmatrix} \quad$ and $\quad \mathbf{B} = \begin{pmatrix} 1 & 2 & 0 \\ -3 & 4 & 2 \end{pmatrix}$

then $\quad \mathbf{A} = \mathbf{B}$ if $a_{11} = 1$, $a_{12} = 2$, $a_{13} = 0$, $a_{21} = -3$, $a_{22} = 4$, $a_{23} = 2$.

Addition of matrices

The sum of two matrices is defined only when the matrices have the same dimensions. If $\mathbf{A} = (a_{ij})$ and $\mathbf{B} = (b_{ij})$ are both $m \times n$ matrices, their sum is an $m \times n$ matrix obtained by adding the corresponding elements of $\mathbf{A}$ and $\mathbf{B}$:

$$\mathbf{A} + \mathbf{B} = (a_{ij} + b_{ij}) \qquad (18.21)$$

EXAMPLES 18.5 Addition of matrices

$$\begin{pmatrix} 1 & 2 & 3 \\ 4 & 5 & 6 \end{pmatrix} + \begin{pmatrix} 1 & -2 & 0 \\ 3 & 0 & -1 \end{pmatrix} = \begin{pmatrix} 2 & 0 & 3 \\ 7 & 5 & 5 \end{pmatrix}$$

$$(1 \quad -2 \quad 3) + (-3 \quad 1 \quad 2) = (-2 \quad -1 \quad 5)$$

$$\begin{pmatrix} a_1 \\ a_2 \\ a_3 \end{pmatrix} + \begin{pmatrix} b_1 \\ b_2 \\ b_3 \end{pmatrix} = \begin{pmatrix} a_1 + b_1 \\ a_2 + b_2 \\ a_3 + b_3 \end{pmatrix}$$

> Exercises 13–18

Multiplication of a matrix by a scalar

The product of an $m \times n$ matrix $\mathbf{A} = (a_{ij})$ and a scalar (number) c is the $m \times n$ matrix whose elements are obtained by multiplying each element of $\mathbf{A}$ by c:

$$c\mathbf{A} = (ca_{ij}) \qquad (18.22)$$

EXAMPLES 18.6 Multiplication by a scalar

If $\quad \mathbf{A} = \begin{pmatrix} 1 & 2 \\ -3 & 0 \\ 5 & -1 \end{pmatrix}$

then $\quad -\mathbf{A} = \begin{pmatrix} -1 & -2 \\ 3 & 0 \\ -5 & 1 \end{pmatrix}, \qquad 3\mathbf{A} = \begin{pmatrix} 3 & 6 \\ -9 & 0 \\ 15 & -3 \end{pmatrix}, \qquad 0\mathbf{A} = \begin{pmatrix} 0 & 0 \\ 0 & 0 \\ 0 & 0 \end{pmatrix}$

▸ Exercise 19

It follows from the rules of addition and multiplication by a scalar that a linear combination of $m \times n$ matrices is an $m \times n$ matrix whose elements are the linear combinations of corresponding elements. If $\mathbf{A} = (a_{ij})$, $\mathbf{B} = (b_{ij})$, and $\mathbf{C} = (c_{ij})$ then

$$\alpha \mathbf{A} + \beta \mathbf{B} + \gamma \mathbf{C} = (\alpha a_{ij} + \beta b_{ij} + \gamma c_{ij}) \tag{18.23}$$

where α, β, and γ are scalars.

EXAMPLES 18.7 Linear combinations of matrices

$$2\begin{pmatrix} 1 & 2 & 3 \\ 4 & 5 & 6 \end{pmatrix} - 3\begin{pmatrix} 1 & -2 & 0 \\ 3 & 0 & -1 \end{pmatrix} = \begin{pmatrix} -1 & 10 & 6 \\ -1 & 10 & 15 \end{pmatrix}$$

$$2\begin{pmatrix} 1 & -1 \\ 2 & 0 \\ -2 & 3 \end{pmatrix} + \begin{pmatrix} -2 & 2 \\ -4 & 0 \\ 4 & -6 \end{pmatrix} = \begin{pmatrix} 0 & 0 \\ 0 & 0 \\ 0 & 0 \end{pmatrix}$$

▸ Exercise 20

The $m \times n$ matrix whose elements are all zero is called the $m \times n$ **null matrix** (or **zero matrix**) **0**. It follows from the above rules that

$$\text{if} \quad \alpha \mathbf{A} + \beta \mathbf{B} = \mathbf{0} \quad \text{then} \quad \mathbf{A} = -\left(\frac{\beta}{\alpha}\right)\mathbf{B} \tag{18.24}$$

Matrix multiplication

The matrix product $\mathbf{C} = \mathbf{AB}$ (in this order, with $\mathbf{A}$ to the left of $\mathbf{B}$) is defined only if

$$\text{the number of columns of } \mathbf{A} = \text{the number of rows of } \mathbf{B} \qquad (18.25)$$

Then, if $\mathbf{A}$ is an $m \times n$ matrix with elements a_{ij} and $\mathbf{B}$ is an $n \times p$ matrix with elements b_{ij}, the product $\mathbf{C} = \mathbf{AB}$ is an $m \times p$ matrix whose elements are

$$c_{ij} = a_{i1}b_{1j} + a_{i2}b_{2j} + a_{i3}b_{3j} + \cdots + a_{in}b_{nj} = \sum_{k=1}^{n} a_{ik}b_{kj} \qquad (18.26)$$

In the simplest case, if $\mathbf{a}$ is a row vector (matrix) with n components a_i and $\mathbf{b}$ is a column vector (matrix) with n components b_i, the product $\mathbf{ab}$ is

$$\mathbf{ab} = (a_1 \ a_2 \ a_3 \ \cdots \ a_n)\begin{pmatrix} b_1 \\ b_2 \\ b_3 \\ \vdots \\ b_n \end{pmatrix} = a_1b_1 + a_2b_2 + a_3b_3 + \cdots + a_nb_n = \sum_{k=1}^{n} a_kb_k \quad (18.27)$$

The product is a number, a 1×1 matrix. In this case the matrix product corresponds to the scalar product $\mathbf{a} \cdot \mathbf{b}$ of two (n-dimensional) vectors (see sections 16.5 and 16.10).

In the general case, the prescription (18.26) for the ijth element is the 'scalar product' of the ith row of $\mathbf{A}$ and the jth column of $\mathbf{B}$. Thus, with the relevant row and column in boldface,

$$\mathbf{C} = \mathbf{AB} = \begin{pmatrix} a_{11} & a_{12} & a_{13} & \cdots & a_{1n} \\ a_{21} & a_{22} & a_{23} & \cdots & a_{2n} \\ \vdots & \vdots & \vdots & & \vdots \\ \mathbf{a_{i1}} & \mathbf{a_{i2}} & \mathbf{a_{i3}} & \cdots & \mathbf{a_{in}} \\ \vdots & \vdots & \vdots & & \vdots \\ a_{m1} & a_{m2} & a_{m3} & \cdots & a_{mn} \end{pmatrix}\begin{pmatrix} b_{11} & b_{12} & \cdots & \mathbf{b_{1j}} & \cdots & b_{1p} \\ b_{21} & b_{22} & \cdots & \mathbf{b_{2j}} & \cdots & b_{2p} \\ b_{31} & b_{32} & \cdots & \mathbf{b_{3j}} & \cdots & b_{2p} \\ \vdots & \vdots & & \vdots & & \vdots \\ b_{n1} & b_{n2} & \cdots & \mathbf{b_{nj}} & \cdots & b_{np} \end{pmatrix}$$

$$= \begin{pmatrix} c_{11} & c_{12} & \cdots & c_{1j} & \cdots & c_{1p} \\ c_{21} & c_{22} & \cdots & c_{2j} & \cdots & c_{2p} \\ \vdots & \vdots & & \vdots & & \vdots \\ c_{i1} & c_{i2} & \cdots & \mathbf{c_{ij}} & \cdots & c_{ip} \\ \vdots & \vdots & & \vdots & & \vdots \\ c_{m1} & c_{m2} & \cdots & c_{mj} & \cdots & c_{mp} \end{pmatrix} \qquad (18.28)$$

EXAMPLES 18.8 Matrix multiplication

(i) Two square matrices:

$$\begin{pmatrix} c_1 & d_1 \\ c_2 & d_2 \end{pmatrix}\begin{pmatrix} a_1 & b_1 \\ a_2 & b_2 \end{pmatrix} = \begin{pmatrix} c_1a_1 + d_1a_2 & c_1b_1 + d_1b_2 \\ c_2a_1 + d_2a_2 & c_2b_1 + d_2b_2 \end{pmatrix}$$

(ii) Two rectangular matrices:

$$\begin{pmatrix} 2 & 0 & -3 \\ 1 & 1 & -2 \end{pmatrix}\begin{pmatrix} 2 & 3 & 4 & 1 \\ 1 & 2 & 2 & 0 \\ 0 & -1 & 2 & 0 \end{pmatrix} = \begin{pmatrix} 4 & 9 & 2 & 2 \\ 3 & 7 & 2 & 1 \end{pmatrix}$$

dimensions: (2×3) (3×4) (2×4)

(iii) Row vector $\times$ column vector:

$$\begin{pmatrix} 1 & 2 & 3 \end{pmatrix}\begin{pmatrix} 2 \\ -2 \\ 1 \end{pmatrix} = (1) = 1$$

(iv) Column vector $\times$ row vector:

$$\begin{pmatrix} 2 \\ -2 \\ 1 \end{pmatrix}\begin{pmatrix} 1 & 2 & 3 \end{pmatrix} = \begin{pmatrix} 2 & 4 & 6 \\ -2 & -4 & -6 \\ 1 & 2 & 3 \end{pmatrix}$$

(v) Matrix $\times$ column vector:

$$\begin{pmatrix} 2 & 0 & -3 \\ 1 & 1 & -2 \end{pmatrix}\begin{pmatrix} 1 \\ 2 \\ 3 \end{pmatrix} = \begin{pmatrix} -7 \\ -3 \end{pmatrix}$$

(vi) Row vector $\times$ matrix:

$$\begin{pmatrix} 1 & 2 & 3 \end{pmatrix}\begin{pmatrix} 2 & 0 & -3 \\ 1 & 1 & -2 \\ -1 & 2 & 0 \end{pmatrix} = \begin{pmatrix} 1 & 8 & -7 \end{pmatrix}$$

▸ Exercises 21–37

Properties of matrix multiplication

The associative law

$$A(BC) = (AB)C = ABC \qquad (18.29)$$

Matrix multiplication is associative, and the brackets can be omitted.

The distributive law

$$A(B + C) = AB + AC \qquad (18.30)$$

The commutative law

Matrix multiplication is *non-commutative* in general (but not in every case),[3]

$$AB \neq BA \qquad (18.31)$$

If A is an $m \times n$ matrix and B is an $n \times p$ matrix then AB is $m \times p$ but BA is not defined unless $p = m$, in which case AB is a square matrix of order m, and BA is a square matrix of order n. Cases (iii) and (iv) in Examples 18.8 demonstrate the non-commutation for the product of a row matrix and a column matrix. In this case AB and BA have different dimensions. Only square matrices *may* commute.

EXAMPLE 18.9 Commuting matrices

An example of a pair of *commuting* matrices is

$$A = \begin{pmatrix} 0 & 1 \\ 1 & 0 \end{pmatrix}, \qquad B = \begin{pmatrix} 2 & 1 \\ 1 & 2 \end{pmatrix}$$

for which

$$AB = \begin{pmatrix} 0 & 1 \\ 1 & 0 \end{pmatrix}\begin{pmatrix} 2 & 1 \\ 1 & 2 \end{pmatrix} = \begin{pmatrix} 1 & 2 \\ 2 & 1 \end{pmatrix} = \begin{pmatrix} 2 & 1 \\ 1 & 2 \end{pmatrix}\begin{pmatrix} 0 & 1 \\ 1 & 0 \end{pmatrix} = BA$$

➤ Exercise 38

---

[3] Hamilton's algebra of quaternions and Cayley's matrix algebra were the first examples of algebras not restricted by the commutative law of multiplication. They led to the development of general algebras that has continued to this day.

But two matrices need not commute even if they are both square.

EXAMPLE 18.10 Non-commuting matrices

If $\qquad \mathbf{A} = \begin{pmatrix} 1 & 0 \\ 0 & 0 \end{pmatrix}, \qquad \mathbf{B} = \begin{pmatrix} 1 & 1 \\ 1 & 0 \end{pmatrix}$

then $\qquad \mathbf{AB} = \begin{pmatrix} 1 & 0 \\ 0 & 0 \end{pmatrix} \begin{pmatrix} 1 & 1 \\ 1 & 0 \end{pmatrix} = \begin{pmatrix} 1 & 1 \\ 0 & 0 \end{pmatrix}$

but $\qquad \mathbf{BA} = \begin{pmatrix} 1 & 1 \\ 1 & 0 \end{pmatrix} \begin{pmatrix} 1 & 0 \\ 0 & 0 \end{pmatrix} = \begin{pmatrix} 1 & 0 \\ 1 & 0 \end{pmatrix}$

so that $\mathbf{AB} \neq \mathbf{BA}$.

‣ Exercise 39

Because of the possibility of non-commutation, it is essential that the correct order of the factors in a product be observed. In the product **AB**, the matrix **B** is **premultiplied**, or **multiplied from the left**, by **A**; the matrix **A** is **postmultiplied**, or **multiplied from the right**, by **B**.

A quantity that plays an important role in quantum mechanics is the **commutator** of the matrices **A** and **B**,

$$[\mathbf{A}, \mathbf{B}] = \mathbf{AB} - \mathbf{BA} \qquad (18.32)$$

‣ Exercises 40, 41

EXAMPLE 18.11 The Pauli spin matrices

In quantum mechanics, electron spin is sometimes represented by the three Pauli spin matrices, one for each cartesian component of the spin angular momentum,

$$\mathbf{S}_x = \frac{1}{2}\hbar \begin{pmatrix} 0 & 1 \\ 1 & 0 \end{pmatrix}, \qquad \mathbf{S}_y = \frac{1}{2}\hbar \begin{pmatrix} 0 & -i \\ i & 0 \end{pmatrix}, \qquad \mathbf{S}_z = \frac{1}{2}\hbar \begin{pmatrix} 1 & 0 \\ 0 & -1 \end{pmatrix} \qquad (18.33)$$

where $i = \sqrt{-1}$, and $\hbar = h/2\pi$ where h is Planck's constant. The commutation properties of these matrices are

$$[\mathbf{S}_x, \mathbf{S}_y] = \mathbf{S}_x \mathbf{S}_y - \mathbf{S}_y \mathbf{S}_x$$

$$= \frac{1}{4}\hbar^2 \left[\begin{pmatrix} 0 & 1 \\ 1 & 0 \end{pmatrix} \begin{pmatrix} 0 & -i \\ i & 0 \end{pmatrix} - \begin{pmatrix} 0 & -i \\ i & 0 \end{pmatrix} \begin{pmatrix} 0 & 1 \\ 1 & 0 \end{pmatrix} \right]$$

$$= \frac{1}{4}\hbar^2 \left[\begin{pmatrix} i & 0 \\ 0 & -i \end{pmatrix} - \begin{pmatrix} -i & 0 \\ 0 & i \end{pmatrix} \right]$$

$$= \frac{1}{2} i\hbar^2 \begin{pmatrix} 1 & 0 \\ 0 & -1 \end{pmatrix} = i\hbar \mathbf{S}_z$$

and similarly for the other pairs. Therefore

$$[\mathbf{S}_x, \mathbf{S}_y] = i\hbar \mathbf{S}_z, \qquad [\mathbf{S}_y, \mathbf{S}_z] = i\hbar \mathbf{S}_x, \qquad [\mathbf{S}_z, \mathbf{S}_x] = i\hbar \mathbf{S}_y \qquad (18.34)$$

In addition,

$$\mathbf{S}_x^2 = \mathbf{S}_x \mathbf{S}_x = \frac{1}{4}\hbar^2 \begin{pmatrix} 0 & 1 \\ 1 & 0 \end{pmatrix} \begin{pmatrix} 0 & 1 \\ 1 & 0 \end{pmatrix} = \frac{1}{4}\hbar^2 \begin{pmatrix} 1 & 0 \\ 0 & 1 \end{pmatrix} = \frac{1}{4}\hbar^2 \mathbf{I}$$

and similarly for $\mathbf{S}_y^2$ and $\mathbf{S}_z^2$. Therefore

$$\mathbf{S}_x^2 + \mathbf{S}_y^2 + \mathbf{S}_z^2 = \frac{3}{4}\hbar^2 \mathbf{I} \qquad (18.35)$$

represents the square of spin angular momentum. The quantity $\frac{3}{4}\hbar^2$ is the square of the magnitude of the spin angular momentum of an electron, whose (total) spin quantum number is $s = \frac{1}{2}$,

$$s(s+1)\hbar^2 = \frac{3}{4}\hbar^2, \qquad \text{for} \qquad s = \frac{1}{2} \qquad (18.36)$$

➤ Exercise 42

Multiplication by a unit matrix

If $\mathbf{A}$ is an $m \times n$ matrix and if $\mathbf{I}_m$ and $\mathbf{I}_n$ are the unit matrices of orders m and n, respectively, then

$$\mathbf{I}_m \mathbf{A} = \mathbf{A} = \mathbf{A} \mathbf{I}_n \qquad (18.37)$$

EXAMPLE 18.12 Multiplication by a unit matrix

$$\begin{pmatrix} 1 & 0 & 0 \\ 0 & 1 & 0 \\ 0 & 0 & 1 \end{pmatrix}\begin{pmatrix} 2 & 3 \\ 1 & 2 \\ 0 & -1 \end{pmatrix} = \begin{pmatrix} 2 & 3 \\ 1 & 2 \\ 0 & -1 \end{pmatrix} = \begin{pmatrix} 2 & 3 \\ 1 & 2 \\ 0 & -1 \end{pmatrix}\begin{pmatrix} 1 & 0 \\ 0 & 1 \end{pmatrix}$$

➤ Exercise 43

The product is a zero matrix

If $\mathbf{AB} = \mathbf{0}$ then it does *not* necessarily follow that $\mathbf{A} = \mathbf{0}$ or $\mathbf{B} = \mathbf{0}$ or that $\mathbf{BA} = \mathbf{0}$ (even if it exists).

EXAMPLE 18.13 The product is a zero matrix

$$\begin{pmatrix} 1 & 0 \\ 1 & 0 \end{pmatrix}\begin{pmatrix} 0 & 0 \\ 1 & 1 \end{pmatrix} = \begin{pmatrix} 0 & 0 \\ 0 & 0 \end{pmatrix} \quad \text{but} \quad \begin{pmatrix} 0 & 0 \\ 1 & 1 \end{pmatrix}\begin{pmatrix} 1 & 0 \\ 1 & 0 \end{pmatrix} = \begin{pmatrix} 0 & 0 \\ 2 & 0 \end{pmatrix}$$

➤ Exercises 44, 45

The determinant and trace of a matrix product

If $\mathbf{A}$ and $\mathbf{B}$ are square matrices of the same order then the determinant of the product $\mathbf{AB}$, and of $\mathbf{BA}$, is equal to the product of the determinants of $\mathbf{A}$ and $\mathbf{B}$,

$$\det \mathbf{AB} = \det \mathbf{A} \times \det \mathbf{B} = \det \mathbf{BA} \qquad (18.38)$$

If $\mathbf{C} = \mathbf{AB}$ then, by the prescription (18.26), a diagonal element of the product is

$$c_{ii} = \sum_{k=1}^{n} a_{ik} b_{ki}$$

The trace of the product matrix is therefore

$$\mathrm{tr}\,\mathbf{C} = \mathrm{tr}\,\mathbf{AB} = \sum_{i=1}^{n} \sum_{k=1}^{n} a_{ik} b_{ki} \qquad (18.39)$$

This is equal to the trace of the product in reverse order. Thus if $\mathbf{D} = \mathbf{BA}$ then a diagonal element of $\mathbf{D}$ is

$$d_{kk} = \sum_{i=1}^{n} b_{ki} a_{ik}$$

and the trace is

$$\text{tr } \mathbf{D} = \text{tr } \mathbf{BA} = \sum_{k=1}^{n} \sum_{i=1}^{n} b_{ki} a_{ik} \tag{18.40}$$

and this is identical to the sum in (18.39).

The transpose of a matrix product

The transpose of the product of two, or more, matrices is equal to the product of the transpose matrices taken *in reverse order*,

$$(\mathbf{AB})^{\mathsf{T}} = \mathbf{B}^{\mathsf{T}} \mathbf{A}^{\mathsf{T}} \tag{18.41}$$

EXAMPLE 18.14 The transpose of a product

Let $\quad \mathbf{A} = \begin{pmatrix} 2 & 0 & -3 \\ 1 & 1 & -2 \end{pmatrix}, \quad \mathbf{B} = \begin{pmatrix} 2 & 3 & 4 & 1 \\ 1 & 2 & 2 & 0 \\ 0 & -1 & 2 & 0 \end{pmatrix}$

Then $\quad \mathbf{AB} = \begin{pmatrix} 2 & 0 & -3 \\ 1 & 1 & -2 \end{pmatrix} \begin{pmatrix} 2 & 3 & 4 & 1 \\ 1 & 2 & 2 & 0 \\ 0 & -1 & 2 & 0 \end{pmatrix} = \begin{pmatrix} 4 & 9 & 2 & 2 \\ 3 & 7 & 2 & 1 \end{pmatrix}$

and $\quad \mathbf{B}^{\mathsf{T}} \mathbf{A}^{\mathsf{T}} = \begin{pmatrix} 2 & 1 & 0 \\ 3 & 2 & -1 \\ 4 & 2 & 2 \\ 1 & 0 & 0 \end{pmatrix} \begin{pmatrix} 2 & 1 \\ 0 & 1 \\ -3 & -2 \end{pmatrix} = \begin{pmatrix} 4 & 3 \\ 9 & 7 \\ 2 & 2 \\ 2 & 1 \end{pmatrix} = (\mathbf{AB})^{\mathsf{T}}$

▸ Exercise 46

18.4 The inverse matrix

If $\mathbf{A}$ and $\mathbf{B}$ are both *square* matrices of order n then $\mathbf{B}$ is the inverse matrix of $\mathbf{A}$ (and *vice-versa*) if

$$\mathbf{BA} = \mathbf{AB} = \mathbf{I}$$

where $\mathbf{I}$ is the unit matrix of order n. The inverse matrix of $\mathbf{A}$ is denoted by $\mathbf{A}^{-1}$:

$$\mathbf{A}^{-1} \mathbf{A} = \mathbf{A} \mathbf{A}^{-1} = \mathbf{I} \tag{18.42}$$

For order 2 the pair of inverse matrices is (if $ad - bc \neq 0$)

$$\mathbf{A} = \begin{pmatrix} a & b \\ c & d \end{pmatrix}, \qquad \mathbf{A}^{-1} = \frac{1}{ad - bc} \begin{pmatrix} d & -b \\ -c & a \end{pmatrix} \tag{18.43}$$

EXAMPLE 18.15 The inverse matrix of order 2

Show that if $\mathbf{A} = \begin{pmatrix} 1 & 2 \\ 3 & 4 \end{pmatrix}$ then $\mathbf{A}^{-1} = -\frac{1}{2} \begin{pmatrix} 4 & -2 \\ -3 & 1 \end{pmatrix}$

We have

$$\mathbf{A}^{-1}\mathbf{A} = -\frac{1}{2} \begin{pmatrix} 4 & -2 \\ -3 & 1 \end{pmatrix} \begin{pmatrix} 1 & 2 \\ 3 & 4 \end{pmatrix} = -\frac{1}{2} \begin{pmatrix} -2 & 0 \\ 0 & -2 \end{pmatrix} = \begin{pmatrix} 1 & 0 \\ 0 & 1 \end{pmatrix}$$

and similarly for $\mathbf{A}\mathbf{A}^{-1}$.

The denominator in the expression for $\mathbf{A}^{-1}$ in (18.43) is the determinant of the matrix $\mathbf{A}$,

$$\det \mathbf{A} = \begin{vmatrix} a & b \\ c & d \end{vmatrix} = ad - bc \tag{18.44}$$

It follows that the inverse of the matrix $\mathbf{A}$ exists only if the determinant of $\mathbf{A}$ is not zero. This is the general result. A square matrix with nonzero determinant is called **nonsingular**. A matrix whose determinant is zero is called **singular**, and the inverse of a singular matrix is not defined. The inverse of a nonsingular matrix can be obtained by means of the following prescription.

1. Replace each element a_{ij} by its cofactor C_{ij} in the determinant of $\mathbf{A}$ (see Section 17.2):

$$\mathbf{A} = \begin{pmatrix} a_{11} & a_{12} & \cdots & a_{1n} \\ a_{21} & a_{22} & \cdots & a_{2n} \\ \vdots & \vdots & & \vdots \\ a_{n1} & a_{n2} & \cdots & a_{nn} \end{pmatrix} \longrightarrow \begin{pmatrix} C_{11} & C_{12} & \cdots & C_{1n} \\ C_{21} & C_{22} & \cdots & C_{2n} \\ \vdots & \vdots & & \vdots \\ C_{n1} & C_{n2} & \cdots & C_{nn} \end{pmatrix}$$

2. Transpose the matrix of cofactors:

$$\longrightarrow \begin{pmatrix} C_{11} & C_{21} & \cdots & C_{n1} \\ C_{12} & C_{22} & \cdots & C_{n2} \\ \vdots & \vdots & & \vdots \\ C_{1n} & C_{2n} & \cdots & C_{nn} \end{pmatrix} = \hat{\mathbf{A}}$$

The matrix $\hat{\mathbf{A}}$ is called the **adjoint** of $\mathbf{A}$.

3. Divide the adjoint matrix by the determinant of $\mathbf{A}$:

$$\mathbf{A}^{-1} = \frac{\hat{\mathbf{A}}}{\det \mathbf{A}} \qquad (18.45)$$

EXAMPLE 18.16 The inverse of order 2

The matrix

$$\mathbf{A} = \begin{pmatrix} a_{11} & a_{12} \\ a_{21} & a_{22} \end{pmatrix}$$

has determinant $\det \mathbf{A} = a_{11}a_{22} - a_{12}a_{21}$, and the cofactors of its elements are $C_{11} = a_{22}$, $C_{12} = -a_{21}$, $C_{21} = -a_{12}$, $C_{22} = a_{11}$. Then

$$\hat{\mathbf{A}} = \begin{pmatrix} C_{11} & C_{21} \\ C_{12} & C_{22} \end{pmatrix} = \begin{pmatrix} a_{22} & -a_{12} \\ -a_{21} & a_{11} \end{pmatrix}$$

so that

$$\mathbf{A}^{-1} = \frac{1}{a_{11}a_{22} - a_{12}a_{21}} \begin{pmatrix} a_{22} & -a_{12} \\ -a_{21} & a_{11} \end{pmatrix}$$

This is the same as formula (18.43).

▸ Exercise 47

EXAMPLE 18.17 Find the inverse of

$$\mathbf{A} = \begin{pmatrix} 2 & 1 & 3 \\ 4 & -2 & 0 \\ -1 & 1 & 0 \end{pmatrix}.$$

The determinant of this matrix is shown in Example 17.4 to have value $\mathbf{A} = 6$, and the cofactors of several of its elements are computed in Example 17.5. The matrix of the cofactors is

$$\begin{pmatrix} 0 & 0 & 2 \\ 3 & 3 & -3 \\ 6 & 12 & -8 \end{pmatrix}$$

and the inverse is the transpose of this divided by the value of the determinant,

$$\mathbf{A}^{-1} = \frac{1}{6} \begin{pmatrix} 0 & 3 & 6 \\ 0 & 3 & 12 \\ 2 & -3 & -8 \end{pmatrix}$$

Check: $\mathbf{A}^{-1}\mathbf{A} = \dfrac{1}{6} \begin{pmatrix} 0 & 3 & 6 \\ 0 & 3 & 12 \\ 2 & -3 & -8 \end{pmatrix} \begin{pmatrix} 2 & 1 & 3 \\ 4 & -2 & 0 \\ -1 & 1 & 0 \end{pmatrix} = \dfrac{1}{6} \begin{pmatrix} 6 & 0 & 0 \\ 0 & 6 & 0 \\ 0 & 0 & 6 \end{pmatrix} = \begin{pmatrix} 1 & 0 & 0 \\ 0 & 1 & 0 \\ 0 & 0 & 1 \end{pmatrix}$

➤ Exercises 48–52

We note that the above method of finding the inverse of a matrix involves the evaluation of the determinant and of n^2 cofactors. As discussed in the note to Section 17.3, this is not a practical method except for very small matrices (see Chapter 20).

The inverse of a matrix product

The inverse of the product of two, or more, matrices is equal to the product of the inverse matrices taken in reverse order,

$$(\mathbf{AB})^{-1} = \mathbf{B}^{-1}\mathbf{A}^{-1} \tag{18.46}$$

➤ Exercise 53

18.5   Linear transformations

The coordinate transformation discussed in Section 18.1, equations (18.3) to (18.7), is a type of linear transformation. More generally, a linear transformation is a set of linear equations*

---

* The *most* general form is a transformation of a vector with n components into a vector with m components. The transformation matrix is then rectangular with dimensions $m \times n$.

$$x_1' = a_{11}x_1 + a_{12}x_2 + a_{13}x_3 + \cdots + a_{1n}x_n$$
$$x_2' = a_{21}x_1 + a_{22}x_2 + a_{23}x_3 + \cdots + a_{2n}x_n$$
$$x_3' = a_{31}x_1 + a_{32}x_2 + a_{33}x_3 + \cdots + a_{3n}x_n \qquad\qquad (18.47)$$
$$\vdots \qquad \vdots \qquad \vdots \qquad\quad \vdots \qquad\quad \vdots$$
$$x_n' = a_{n1}x_1 + a_{n2}x_2 + a_{n3}x_3 + \cdots + a_{nn}x_n$$

that represents the transformation of the vector with components $(x_1, x_2, \ldots, x_n)$ into the new vector $(x_1', x_2', \ldots, x_n')$. In matrix notation,

$$\mathbf{x}' = \mathbf{A}\mathbf{x} \qquad\qquad (18.48)$$

where

$$\mathbf{x}' = \begin{pmatrix} x_1' \\ x_2' \\ x_3' \\ \vdots \\ x_n' \end{pmatrix}, \quad \mathbf{A} = \begin{pmatrix} a_{11} & a_{12} & a_{13} & \cdots & a_{1n} \\ a_{21} & a_{22} & a_{23} & \cdots & a_{2n} \\ a_{31} & a_{32} & a_{33} & \cdots & a_{3n} \\ \vdots & \vdots & \vdots & & \vdots \\ a_{n1} & a_{n2} & a_{n3} & \cdots & a_{nn} \end{pmatrix}, \quad \mathbf{x} = \begin{pmatrix} x_1 \\ x_2 \\ x_3 \\ \vdots \\ x_n \end{pmatrix}$$

The n-dimensional vectors are represented by the column matrices $\mathbf{x}$ and $\mathbf{x}'$, and $\mathbf{A}$ is called the **transformation matrix** that transforms $\mathbf{x}$ into $\mathbf{x}'$.

EXAMPLE 18.18 Linear transformations in two dimensions

Let (x, y) be the cartesian coordinates of a point in the xy-plane, and let the matrices

$$\mathbf{A} = \begin{pmatrix} \cos\theta & -\sin\theta \\ \sin\theta & \cos\theta \end{pmatrix}, \quad \mathbf{B} = \begin{pmatrix} 0 & 1 \\ 1 & 0 \end{pmatrix}, \quad \mathbf{C} = \begin{pmatrix} 1 & 0 \\ 0 & -1 \end{pmatrix}, \quad \mathbf{D} = \begin{pmatrix} 1 & 0 \\ 0 & a \end{pmatrix}$$

represent transformations in the plane. Then, as illustrated in Figure 18.3,

(a) $\mathbf{A}$ is the rotation through angle θ about the origin discussed in Example 18.1.
(b) $\mathbf{B}$ interchanges the x and y coordinates,

$$\begin{pmatrix} 0 & 1 \\ 1 & 0 \end{pmatrix}\begin{pmatrix} x \\ y \end{pmatrix} = \begin{pmatrix} y \\ x \end{pmatrix}$$

and represents reflection in (or rotation through 180° about) the line $x = y$.

(c) **C** changes the sign of the y coordinate,

$$\begin{pmatrix} 1 & 0 \\ 0 & -1 \end{pmatrix} \begin{pmatrix} x \\ y \end{pmatrix} = \begin{pmatrix} x \\ -y \end{pmatrix}$$

and represents reflection in the x-axis.

(d) **D** multiplies the y coordinate by factor a,

$$\begin{pmatrix} 1 & 0 \\ 0 & a \end{pmatrix} \begin{pmatrix} x \\ y \end{pmatrix} = \begin{pmatrix} x \\ ay \end{pmatrix}$$

and represents a stretch in the y direction if $a > 1$ and a contraction if $0 < a < 1$.

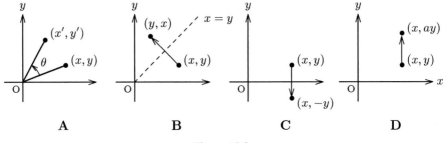

Figure 18.3

▸ Exercises 54, 55

Simultaneous transformations

A coordinate transformation can be applied simultaneously to more than one point if the column matrix of the coefficients of one point is replaced by the rectangular matrix whose columns are the coordinates of several points. Thus, if $x_1, x_2, x_3, \ldots, x_n$ are column vectors for n points in a three-dimensional space, we construct the matrix

$$\mathbf{X} = (\mathbf{x}_1 \ \mathbf{x}_2 \ \mathbf{x}_3 \ \cdots \ \mathbf{x}_n) = \begin{pmatrix} x_1 & x_2 & x_3 & \cdots & x_n \\ y_1 & y_2 & y_3 & \cdots & y_n \\ z_1 & z_2 & z_3 & \cdots & z_n \end{pmatrix} \tag{18.49}$$

If $\mathbf{Ax}_i = \mathbf{x}'_i$ then

$$\begin{aligned} \mathbf{AX} &= \mathbf{A}(\mathbf{x}_1 \ \mathbf{x}_2 \ \mathbf{x}_3 \ \cdots \ \mathbf{x}_n) \\ &= (\mathbf{Ax}_1 \ \mathbf{Ax}_2 \ \mathbf{Ax}_3 \ \cdots \ \mathbf{Ax}_n) \\ &= (\mathbf{x}'_1 \ \mathbf{x}'_2 \ \mathbf{x}'_3 \ \cdots \ \mathbf{x}'_n) \\ &= \mathbf{X}' \end{aligned} \tag{18.50}$$

and the columns of $\mathbf{X}'$ are the coordinates of the transformed points.

EXAMPLE 18.19 Rotate the square whose corners have positions in the xy-plane

$$(x, y) = (2, 1), (3, 1), (3, 2), (2, 2)$$

through $\pi/4$ about the origin as in Figure 18.4.

Because $\cos \pi/4 = \sin \pi/4 = 1/\sqrt{2}$, the transformation matrix is

$$\frac{1}{\sqrt{2}} \begin{pmatrix} 1 & -1 \\ 1 & 1 \end{pmatrix}$$

The coordinates of the corners of the square after rotation are then given by

$$\frac{1}{\sqrt{2}} \begin{pmatrix} 1 & -1 \\ 1 & 1 \end{pmatrix} \begin{pmatrix} 2 & 3 & 3 & 2 \\ 1 & 1 & 2 & 2 \end{pmatrix} = \frac{1}{\sqrt{2}} \begin{pmatrix} 1 & 2 & 1 & 0 \\ 3 & 4 & 5 & 4 \end{pmatrix}$$

$$= \begin{pmatrix} \dfrac{1}{\sqrt{2}} & \dfrac{2}{\sqrt{2}} & \dfrac{1}{\sqrt{2}} & 0 \\ \dfrac{3}{\sqrt{2}} & \dfrac{4}{\sqrt{2}} & \dfrac{5}{\sqrt{2}} & \dfrac{4}{\sqrt{2}} \end{pmatrix}$$

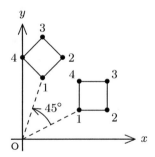

Figure 18.4

> Exercises 56–59

Consecutive transformations

Let the transformation $\mathbf{x'} = \mathbf{Ax}$, equation (18.48), be followed by a second transformation

$$\mathbf{x''} = \mathbf{Bx'} \qquad\qquad (18.51)$$

that transforms $\mathbf{x'}$ into $\mathbf{x''}$. Substitution of (18.48) into (18.51) then gives

$$\mathbf{x''} = \mathbf{BAx} = \mathbf{Cx} \qquad\qquad (18.52)$$

so that **A** followed by **B** is equivalent to the single transformation whose matrix representation is the matrix product **C = BA**.

EXAMPLE 18.20 Consecutive transformations in two dimensions

Let (x, y) be the cartesian coordinates of a point in the xy-plane, and let the matrices

$$\mathbf{A} = \begin{pmatrix} \dfrac{1}{2} & -\dfrac{\sqrt{3}}{2} \\ \dfrac{\sqrt{3}}{2} & \dfrac{1}{2} \end{pmatrix}, \quad \mathbf{B} = \begin{pmatrix} 0 & 1 \\ 1 & 0 \end{pmatrix}, \quad \mathbf{C} = \begin{pmatrix} -1 & 0 \\ 0 & -1 \end{pmatrix}$$

represent transformations in the plane. Matrix **A** represents anticlockwise rotation through $\pi/3$ about the origin, **B** is reflection in the line $x = y$, and **C** is inversion through the origin. The sequence **A** followed by **B** followed by **C** is illustrated in Figure 18.5, and is equivalent to the single transformation

$$\mathbf{D} = \mathbf{CBA} = \begin{pmatrix} -1 & 0 \\ 0 & -1 \end{pmatrix} \begin{pmatrix} 0 & 1 \\ 1 & 0 \end{pmatrix} \begin{pmatrix} \dfrac{1}{2} & -\dfrac{\sqrt{3}}{2} \\ \dfrac{\sqrt{3}}{2} & \dfrac{1}{2} \end{pmatrix} = \begin{pmatrix} -\dfrac{\sqrt{3}}{2} & \dfrac{1}{2} \\ -\dfrac{1}{2} & \dfrac{\sqrt{3}}{2} \end{pmatrix}$$

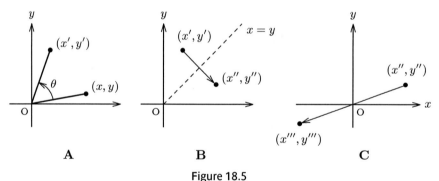

A   B   C

Figure 18.5

The final position of the point is $(x''', y''') = \left(-\dfrac{\sqrt{3}}{2}x - \dfrac{1}{2}y, \ -\dfrac{1}{2}x + \dfrac{\sqrt{3}}{2}y \right)$.

▸ Exercise 60

Inverse transformations

If **A** is a nonsingular square matrix then it has the unique inverse $\mathbf{A}^{-1}$ such that

$$\mathbf{A}^{-1}\mathbf{A} = \mathbf{A}\mathbf{A}^{-1}\mathbf{x} = \mathbf{I}\mathbf{x} = \mathbf{x} \tag{18.53}$$

A nonsingular transformation $\mathbf{A}$ followed by its inverse transformation $\mathbf{A}^{-1}$ (or $\mathbf{A}^{-1}$ followed by $\mathbf{A}$) is therefore equivalent to the identity transformation $\mathbf{I}$; that is, the 'transformation' that leaves every vector unchanged. Thus

$$\text{if} \quad \mathbf{x}' = \mathbf{A}\mathbf{x} \quad \text{and} \quad \mathbf{x}'' = \mathbf{A}^{-1}\mathbf{x}' \quad \text{then} \quad \mathbf{x}'' = \mathbf{A}^{-1}\mathbf{A}\mathbf{x} = \mathbf{x} \qquad (18.54)$$

EXAMPLE 18.21 The matrix

$$\mathbf{A} = \begin{pmatrix} \cos\theta & -\sin\theta \\ \sin\theta & \cos\theta \end{pmatrix}$$

is nonsingular, with determinant $\det \mathbf{A} = \cos^2\theta + \sin^2\theta = 1$, and represents rotation through angle θ. The corresponding inverse transformation is the rotation through angle $-\theta$, with matrix

$$\mathbf{B} = \begin{pmatrix} \cos(-\theta) & -\sin(-\theta) \\ \sin(-\theta) & \cos(-\theta) \end{pmatrix} = \begin{pmatrix} \cos\theta & \sin\theta \\ -\sin\theta & \cos\theta \end{pmatrix}$$

and $\mathbf{B} = \mathbf{A}^{-1}$ is the inverse matrix of $\mathbf{A}$. Thus

$$\mathbf{BA} = \begin{pmatrix} \cos\theta & \sin\theta \\ -\sin\theta & \cos\theta \end{pmatrix} \begin{pmatrix} \cos\theta & -\sin\theta \\ \sin\theta & \cos\theta \end{pmatrix} = \begin{pmatrix} 1 & 0 \\ 0 & 1 \end{pmatrix} = \mathbf{I}$$

In this particular case, the inverse is also equal to the transpose, $\mathbf{A}^{-1} = \mathbf{A}^{\mathsf{T}}$.

18.6   Orthogonal matrices and orthogonal transformations

A nonsingular square matrix is called **orthogonal** when its inverse is equal to its transpose,

$$\mathbf{A}^{-1} = \mathbf{A}^{\mathsf{T}} \qquad \text{(orthogonal matrix)} \qquad (18.55)$$

For example, the matrix discussed in Example 18.21 is orthogonal:

$$\mathbf{A} = \begin{pmatrix} \cos\theta & -\sin\theta \\ \sin\theta & \cos\theta \end{pmatrix}, \qquad \mathbf{A}^{\mathsf{T}} = \begin{pmatrix} \cos\theta & \sin\theta \\ -\sin\theta & \cos\theta \end{pmatrix} = \mathbf{A}^{-1} \qquad (18.56)$$

The characteristic property of an orthogonal matrix is that its columns (and its rows) form a system of orthogonal unit vectors (orthonormal vectors). For order 3, let

$$\mathbf{A} = \begin{pmatrix} a_1 & b_1 & c_1 \\ a_2 & b_2 & c_2 \\ a_3 & b_3 & c_3 \end{pmatrix} = (\mathbf{a} \quad \mathbf{b} \quad \mathbf{c}) \qquad (18.57)$$

where

$$\mathbf{a} = \begin{pmatrix} a_1 \\ a_2 \\ a_3 \end{pmatrix}, \qquad \mathbf{b} = \begin{pmatrix} b_1 \\ b_2 \\ b_3 \end{pmatrix}, \qquad \mathbf{c} = \begin{pmatrix} c_1 \\ c_2 \\ c_3 \end{pmatrix} \tag{18.58}$$

The transpose of **A** is

$$\mathbf{A}^\mathrm{T} = \begin{pmatrix} a_1 & a_2 & a_3 \\ b_1 & b_2 & b_3 \\ c_1 & c_2 & c_3 \end{pmatrix} = \begin{pmatrix} \mathbf{a}^\mathrm{T} \\ \mathbf{b}^\mathrm{T} \\ \mathbf{c}^\mathrm{T} \end{pmatrix} \tag{18.59}$$

where, for example, the row vector $\mathbf{a}^\mathrm{T} = (a_1 \; a_2 \; a_3)$ is the transpose of the column vector **a**. The product $\mathbf{A}^\mathrm{T}\mathbf{A}$ of **A** and its transpose can then by written

$$\mathbf{A}^\mathrm{T}\mathbf{A} = \begin{pmatrix} \mathbf{a}^\mathrm{T} \\ \mathbf{b}^\mathrm{T} \\ \mathbf{c}^\mathrm{T} \end{pmatrix} (\mathbf{a} \; \mathbf{b} \; \mathbf{c}) = \begin{pmatrix} \mathbf{a}^\mathrm{T}\mathbf{a} & \mathbf{a}^\mathrm{T}\mathbf{b} & \mathbf{a}^\mathrm{T}\mathbf{c} \\ \mathbf{b}^\mathrm{T}\mathbf{a} & \mathbf{b}^\mathrm{T}\mathbf{b} & \mathbf{b}^\mathrm{T}\mathbf{c} \\ \mathbf{c}^\mathrm{T}\mathbf{a} & \mathbf{c}^\mathrm{T}\mathbf{b} & \mathbf{c}^\mathrm{T}\mathbf{c} \end{pmatrix} \tag{18.60}$$

where, for example,

$$\mathbf{a}^\mathrm{T}\mathbf{a} = (a_1 \; a_2 \; a_3) \begin{pmatrix} a_1 \\ a_2 \\ a_3 \end{pmatrix} = a_1^2 + a_2^2 + a_3^2 \tag{18.61}$$

$$\mathbf{a}^\mathrm{T}\mathbf{b} = (a_1 \; a_2 \; a_3) \begin{pmatrix} b_1 \\ b_2 \\ b_3 \end{pmatrix} = a_1 b_1 + a_2 b_2 + a_3 b_3 \tag{18.62}$$

We recognize $\mathbf{a}^\mathrm{T}\mathbf{b}$ as the scalar product $\boldsymbol{a} \cdot \boldsymbol{b}$ of the vectors $\boldsymbol{a} = (a_1, \, a_2, \, a_3)$ and $\boldsymbol{b} = (b_1, \, b_2, \, b_3)$ (see Section 16.5), and these vectors are orthogonal if $\mathbf{a}^\mathrm{T}\mathbf{b} = \boldsymbol{a} \cdot \boldsymbol{b} = 0$. Also, the quantity $\mathbf{a}^\mathrm{T}\mathbf{a}$ is the square of the length of the vector $\boldsymbol{a}$, and the vector has unit length if $\mathbf{a}^\mathrm{T}\mathbf{a} = |\boldsymbol{a}|^2 = 1$. It follows that when the columns of **A** form a system of orthonormal vectors, the product $\mathbf{A}^\mathrm{T}\mathbf{A}$ is the unit matrix:

$$\mathbf{A}^\mathrm{T}\mathbf{A} = \begin{pmatrix} \mathbf{a}^\mathrm{T}\mathbf{a} & \mathbf{a}^\mathrm{T}\mathbf{b} & \mathbf{a}^\mathrm{T}\mathbf{c} \\ \mathbf{b}^\mathrm{T}\mathbf{a} & \mathbf{b}^\mathrm{T}\mathbf{b} & \mathbf{b}^\mathrm{T}\mathbf{c} \\ \mathbf{c}^\mathrm{T}\mathbf{a} & \mathbf{c}^\mathrm{T}\mathbf{b} & \mathbf{c}^\mathrm{T}\mathbf{c} \end{pmatrix} = \begin{pmatrix} 1 & 0 & 0 \\ 0 & 1 & 0 \\ 0 & 0 & 1 \end{pmatrix} = \mathbf{I} \tag{18.63}$$

and therefore, $\mathbf{A}^T = \mathbf{A}^{-1}$. It also follows that the determinant of an orthogonal matrix has value ± 1. Thus, because $\det \mathbf{A}^T = \det \mathbf{A}$ and $\det \mathbf{I} = 1$, we have

$$\det(\mathbf{A}^T\mathbf{A}) = \det \mathbf{A}^T \times \det \mathbf{A} = (\det \mathbf{A})^2 = 1 \qquad (18.64)$$

EXAMPLE 18.22 The matrix

$$\mathbf{A} = \begin{pmatrix} \dfrac{2}{3} & \dfrac{1}{3} & -\dfrac{2}{3} \\[2mm] \dfrac{2}{3} & -\dfrac{2}{3} & \dfrac{1}{3} \\[2mm] \dfrac{1}{3} & \dfrac{2}{3} & \dfrac{2}{3} \end{pmatrix}$$

is orthogonal with properties

(i) $\quad \mathbf{A}^T\mathbf{A} = \begin{pmatrix} \dfrac{2}{3} & \dfrac{2}{3} & \dfrac{1}{3} \\[2mm] \dfrac{1}{3} & -\dfrac{2}{3} & \dfrac{2}{3} \\[2mm] -\dfrac{2}{3} & \dfrac{1}{3} & \dfrac{2}{3} \end{pmatrix} \begin{pmatrix} \dfrac{2}{3} & \dfrac{1}{3} & -\dfrac{2}{3} \\[2mm] \dfrac{2}{3} & -\dfrac{2}{3} & \dfrac{1}{3} \\[2mm] \dfrac{1}{3} & \dfrac{2}{3} & \dfrac{2}{3} \end{pmatrix} = \begin{pmatrix} 1 & 0 & 0 \\ 0 & 1 & 0 \\ 0 & 0 & 1 \end{pmatrix}$

and $\mathbf{A}\mathbf{A}^T = (\mathbf{A}^T\mathbf{A})^T = \mathbf{I}$.

(ii) The columns of $\mathbf{A}$ form the vectors

$$a = \left(\frac{2}{3}, \frac{2}{3}, \frac{1}{3}\right), \qquad b = \left(\frac{1}{3}, -\frac{2}{3}, \frac{2}{3}\right), \qquad c = \left(-\frac{2}{3}, \frac{1}{3}, \frac{2}{3}\right)$$

with properties

$$a \cdot a = b \cdot b = c \cdot c = \sqrt{\left(\frac{1}{3}\right)^2 + \left(\frac{2}{3}\right)^2 + \left(\frac{2}{3}\right)^2} = 1$$

$$a \cdot b = \frac{2}{9} - \frac{4}{9} + \frac{2}{9} = 0, \quad b \cdot c = -\frac{2}{9} - \frac{2}{9} + \frac{4}{9} = 0, \quad c \cdot a = -\frac{4}{9} + \frac{2}{9} + \frac{2}{9} = 0$$

Similarly for the rows of $\mathbf{A}$ (the columns of $\mathbf{A}^T$).

(iii) The determinant of $\mathbf{A}$ is

$$\det \mathbf{A} = \frac{1}{3^3} \begin{vmatrix} 2 & 1 & -2 \\ 2 & -2 & 1 \\ 1 & 2 & 2 \end{vmatrix} = -1$$

➤ Exercise 61

Orthogonal transformations

An orthogonal transformation is a linear transformation

$$\mathbf{x}' = \mathbf{A}\mathbf{x} \tag{18.65}$$

whose transformation matrix $\mathbf{A}$ is orthogonal. Orthogonal transformations are important because they preserve the scalar product of vectors; that is, the lengths of vectors and the angles between them are unchanged by an orthogonal transformation.

   All the transformations in Examples 18.18 (except for $\mathbf{D}$), 18.19, and 18.20 are orthogonal. The preservation of lengths and angles is demonstrated in Figure 18.4 of Example 18.19; the size and shape of the figure (square) is not changed by the orthogonal transformation (rotation). All orthogonal transformations in a plane or in a three-dimensional space are either rotations or reflections, or combinations of these, and such transformations are important for the mathematical description of the symmetry properties of molecules.

18.7 Symmetry operations

The symmetry of a physical system is characterized by a set of **symmetry elements**, with each of which is associated one or more transformations called **symmetry operations**. These are transformations that leave the description of the system unchanged. Most important in molecular chemistry are the spatial transformations that result in the interchange of identical nuclei. The possible symmetry elements are then axes of symmetry, planes of symmetry, and a centre of inversion.

EXAMPLE 18.23 Symmetry of the water molecule

The water molecule in its ground state has the nonlinear equilibrium nuclear geometry illustrated in Figure 18.6, with bond angle 105°. The system has three symmetry elements each with one associated operation:

(i) A two-fold axis of symmetry (Oz in the figure). Rotation through 180° about the axis results in the interchange of the hydrogen nuclei.

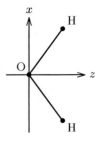

Figure 18.6

(ii) A plane of symmetry (Oyz) that bisects the bond angle. Reflection in the plane again results in the interchange of the hydrogens.
(iii) A plane of symmetry (Oxz) that contains all three nuclei (the molecular plane). Reflection in this plane leaves the nuclei unmoved.

These are distinct symmetry operations and have different effects on the wave functions that describe the states of the molecule.

The mathematical theory of symmetry is **group theory**.[4] We give here only a brief introduction to the concept of symmetry groups and of the matrix representations of groups.

Symmetry groups

We consider the symmetrical plane figure formed by three points at the corners of an equilateral triangle (Figure 18.7), with the figure in the xy-plane of a fixed coordinate system whose origin O lies at the centroid of the triangle.

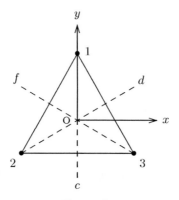

Figure 18.7

The symmetry of the figure can be described in terms of the following set of six symmetry operations:

E the identity operation that leaves every point unmoved
A anticlockwise rotation through 120° about the Oz axis
B anticlockwise rotation through 240° (or clockwise through 120°) about the Oz axis
C rotation through 180° about the Oc axis
D rotation through 180° about the Od axis
F rotation through 180° about the Of axis

---

[4] Group theory has its origins in studies of algebraic equations by Lagrange, Gauss, Abel, and Cauchy. The relation between algebraic equations and the group of permutations was described by Évariste Galois (1811–1832), whose short life included two failures to enter the École Polytechnique (at the second attempt he threw the blackboard eraser at one of the examiners), expulsion from the École Normale, and two arrests, one for threatening the life of the king and one for wearing the uniform of the dissolved National Guard. He was killed in a duel in an 'affair of honour'. His mathematical manuscripts were unread until published by Liouville in 1846. In a paper on quadratic forms in 1882, Heinrich Weber (1842–1913) gave a complete axiomatic description of a finite abstract group, defined Abelian groups, and in 1893 extended the work to infinite groups.

Other symmetry operations (reflections) are possible, but are equivalent to the above six operations for the *plane* figure. The operations may also be interpreted as permutations of the labels (1, 2, 3) of the three points.

The successive application of any two symmetry operations is equivalent to the application of a single operation. The two examples in Figure 18.8 show that the application of operation A followed by C is equivalent to the application of the single operation D, and that C followed by A is equivalent to F. Such combinations of symmetry operations are denoted by the symbolic equations

$$CA = D, \qquad AC = F \tag{18.66}$$

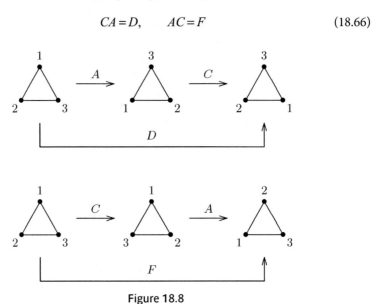

Figure 18.8

The results of the possible combinations of pairs of operations are collected in the **group multiplication table** 18.1.

Table 18.1 A group multiplication table

	E	A	B	C	D	F	(applied first)
E	E	A	B	C	D	F	
A	A	B	E	F	C	D	
B	B	E	A	D	F	C	
C	C	D	F	E	A	B	
D	D	F	C	B	E	A	
F	F	C	D	A	B	E	

The six operations form a closed set called a **group**; we will refer to this group as $G = \{E, A, B, C, D, F\}$. A symmetry group that is made up of rotations, reflections, and inversion is called a **point group** because at least one point is left unmoved by every symmetry operation. There is only one such point in our example, the centroid of the triangle.

‣ Exercise 62

Axioms of group theory

A set of elements $\{E, P, Q, R, ...\}$ forms a group if the following conditions are satisfied.

(i) The combination of any pair of the elements of the group also belongs to the group.

The law of combination depends on the nature of the elements; for example, addition or multiplication if the elements are numbers, matrix multiplication if they are matrices, consecutive application of symmetry or other operations. The combination of two elements, P and Q, is called the product of P and Q and is written as PQ, with some convention about the ordering of the elements. The associative law of combination must hold for all the elements of the group; $P(QR) = (PQ)R = PQR$. The commutative law does not necessarily hold; $PQ \neq QP$ in general; for example, for the group G, $AC \neq CA$. If $PQ = QP$ for all the elements of the group, the group is called an Abelian group.

(ii) One of the elements of the group, denoted by E, has the properties of a unit (or identity) element. For every element P,

$$PE = EP = P$$

(iii) Each element has an inverse which also belongs to the group: if P belongs to the group then its inverse P^{-1} is defined by

$$PP^{-1} = P^{-1}P = E$$

For the group $G = \{E, A, B, C, D, F\}$ the inverse elements are

$$E^{-1} = E, \quad A^{-1} = B, \quad B^{-1} = A, \quad C^{-1} = C, \quad D^{-1} = D, \quad F^{-1} = F$$

Matrix representations of groups

A set of matrices that multiply in accordance with the multiplication table of a group is called a **matrix representation** Γ of the group. Three representations of the group $G = \{E, A, B, C, D, F\}$ are shown in Table 18.2.

Table 18.2 Matrix representations of the group *G*

G	E	A	B	C	D	F
Γ_1	1	1	1	1	1	1
Γ_2	1	1	1	-1	-1	-1
Γ_3	$\begin{pmatrix} 1 & 0 \\ 0 & 1 \end{pmatrix}$	$\begin{pmatrix} -\frac{1}{2} & -\frac{\sqrt{3}}{2} \\ \frac{\sqrt{3}}{2} & -\frac{1}{2} \end{pmatrix}$	$\begin{pmatrix} -\frac{1}{2} & \frac{\sqrt{3}}{2} \\ -\frac{\sqrt{3}}{2} & -\frac{1}{2} \end{pmatrix}$	$\begin{pmatrix} -1 & 0 \\ 0 & 1 \end{pmatrix}$	$\begin{pmatrix} \frac{1}{2} & \frac{\sqrt{3}}{2} \\ \frac{\sqrt{3}}{2} & -\frac{1}{2} \end{pmatrix}$	$\begin{pmatrix} \frac{1}{2} & -\frac{\sqrt{3}}{2} \\ -\frac{\sqrt{3}}{2} & -\frac{1}{2} \end{pmatrix}$

The representations Γ_1 and Γ_2 are called one-dimensional representations (the matrices are the numbers ± 1). The representation Γ_1, in which every symmetry operation is represented by the number +1, clearly satisfies the group multiplication table, Table 18.1, and is called the **trivial** or **totally symmetric** representation of the group. Every group has such a representation. Table 18.3, obtained from the multiplication table 18.1 by replacing each operation by its representative ±1 in Γ_2, shows that Γ_2 is indeed a representation of the group.

Table 18.3 Multiplication table of Γ_2

	+1	+1	+1	−1	−1	−1
+1	+1	+1	+1	−1	−1	−1
+1	+1	+1	+1	−1	−1	−1
+1	+1	+1	+1	−1	−1	−1
−1	−1	−1	−1	+1	+1	+1
−1	−1	−1	−1	+1	+1	+1
−1	−1	−1	−1	+1	+1	+1

In the same way, replacing each operation in Table 18.1 by its representative matrix in the two-dimensional representation Γ_3 confirms that these matrices satisfy the multiplication table.

The matrices of the two-dimensional representation Γ_3 can be derived by considering the result of applying each symmetry operation to the coordinates of a point in the plane. Thus, the operation A is the anticlockwise rotation through $\theta = 120°$ about the origin and its representative matrix is

$$\mathbf{A} = \begin{pmatrix} \cos 120° & -\sin 120° \\ \sin 120° & \cos 120° \end{pmatrix} = \begin{pmatrix} -\dfrac{1}{2} & -\dfrac{\sqrt{3}}{2} \\ \dfrac{\sqrt{3}}{2} & -\dfrac{1}{2} \end{pmatrix} \tag{18.67}$$

The rotation B is the inverse of A, since $AB = BA = E$, and its representative matrix $\mathbf{B}$ is the transpose matrix of $\mathbf{A}$ (all the matrices are orthogonal, with inverse equal to transpose). Similarly, the operation C, rotation through 180° about the Oc axis (the y-axis), transforms a vector $r = (x, y)$ into the vector $r' = (-x, y)$. Its representative matrix is therefore

$$\mathbf{C} = \begin{pmatrix} -1 & 0 \\ 0 & 1 \end{pmatrix} \tag{18.68}$$

since

$$\begin{pmatrix} -1 & 0 \\ 0 & 1 \end{pmatrix}\begin{pmatrix} x \\ y \end{pmatrix} = \begin{pmatrix} -x \\ y \end{pmatrix} \tag{18.69}$$

It is possible to construct any number of representations of all possible dimensions for any group, but it can be shown that only a certain number of these (the 'irreducible representations') are distinct and independent. In the present case, all the possible representations of the group G are either equivalent to or may be reduced to the three representations in Table 18.2.

▸ Exercise 63

18.8 Exercises

Section 18.1

1. Construct transformation matrices that represent the following rotations about the z-axis:
   (i) anticlockwise through 45°,   (ii) anticlockwise through 90°,
   (iii) clockwise through 90°.
2. Construct a transformation matrix that represents the interchange of x and y coordinates of a point.

Section 18.2

For the following matrices,

$$A = \begin{pmatrix} 1 & -2 & 3 \\ 0 & 3 & 4 \end{pmatrix} \quad B = \begin{pmatrix} 0 & 1 & -4 \\ 2 & -3 & 0 \end{pmatrix} \quad C = \begin{pmatrix} -5 & 3 \\ 4 & -1 \\ 2 & -1 \end{pmatrix} \quad D = \begin{pmatrix} 3 & 0 & 0 \\ 0 & 2 & 0 \\ 0 & 0 & -1 \end{pmatrix}$$

$$P = \begin{pmatrix} 1 & -2 \\ 0 & 4 \end{pmatrix} \quad Q = \begin{pmatrix} 3 & 0 \\ 0 & 1 \end{pmatrix} \quad a = \begin{pmatrix} 0 \\ -3 \\ 1 \end{pmatrix} \quad b = (2 \quad 5 \quad -2)$$

find, *if possible*:

3. det A, tr A   4. det D, tr D   5. det P, tr P   6. det a, tr a   7. A^T 8. C^T
9. D^T 10. P^T 11. a^T 12. b^T

Section 18.3

For the above matrices find, *if possible*:

13. A + B   14. A − B   15. B − A   16. C + D   17. $a + b^T$ 18. $a^T + b$
19. 3P   20. 2A + 3B   21. AB   22. BC   23. CB   24. CP
25. PC   26. D^2 27. PQ   28. QP   29. Ba   30. ab
31. ba   32. $a^T b^T$ 33. $b^T a^T$ 34. Ca   35. $a^T C$

36. If $A = \begin{pmatrix} 1 & 1 & 1 \\ 2 & 1 & 2 \\ -2 & 1 & -1 \end{pmatrix}$, show that $A^3 - A^2 - 3A + I = 0$.

37. If $A = \begin{pmatrix} a & b \\ c & d \end{pmatrix}$ and $B = \begin{pmatrix} 2 & 1 \\ 3 & 2 \end{pmatrix}$, find A such that $AB = \begin{pmatrix} 3 & 2 \\ 1 & 4 \end{pmatrix}$.

Given the matrices **A** and **B**, find **AB** and **BA**:

38. $A = \begin{pmatrix} 2 & -1 \\ -1 & 1 \end{pmatrix}$, $B = \begin{pmatrix} -1 & 3 \\ 3 & 2 \end{pmatrix}$ 39. $A = \begin{pmatrix} 1 & 0 \\ 1 & 0 \end{pmatrix}$, $B = \begin{pmatrix} 0 & 1 \\ -1 & 0 \end{pmatrix}$

Find the commutator of the following pairs of matrices:

40. $\begin{pmatrix} 2 & -1 \\ -1 & 1 \end{pmatrix}$, $\begin{pmatrix} -1 & 3 \\ 3 & 2 \end{pmatrix}$ **41.** $\begin{pmatrix} 1 & 0 \\ 1 & 0 \end{pmatrix}$, $\begin{pmatrix} 0 & 1 \\ -1 & 0 \end{pmatrix}$

42. The spin matrices for a nucleus with spin quantum number 1 are

$$\mathbf{I}_x = \frac{\hbar}{\sqrt{2}} \begin{pmatrix} 0 & 1 & 0 \\ 1 & 0 & 1 \\ 0 & 1 & 0 \end{pmatrix}, \quad \mathbf{I}_y = \frac{\hbar}{\sqrt{2}} \begin{pmatrix} 0 & -i & 0 \\ i & 0 & -i \\ 0 & i & 0 \end{pmatrix}, \quad \mathbf{I}_z = \hbar \begin{pmatrix} 1 & 0 & 0 \\ 0 & 0 & 0 \\ 0 & 0 & -1 \end{pmatrix}$$

(i) Find the commutators $\left[\mathbf{I}_x, \mathbf{I}_y \right]$, $\left[\mathbf{I}_y, \mathbf{I}_z \right]$, $\left[\mathbf{I}_z, \mathbf{I}_x \right]$. **(ii)** Find $\mathbf{I}_x^2 + \mathbf{I}_y^2 + \mathbf{I}_z^2$.

43. Given the matrix $\mathbf{A} = \begin{pmatrix} 1 & 2 & 3 & 4 \\ 5 & 6 & 7 & 8 \end{pmatrix}$, find matrices $\mathbf{B}$ and $\mathbf{C}$ for which $\mathbf{BA} = \mathbf{AC} = \mathbf{A}$.

Find the general matrix $\begin{pmatrix} a & b \\ c & d \end{pmatrix}$ for which:

44. $\begin{pmatrix} 1 & 2 \\ 2 & 4 \end{pmatrix}\begin{pmatrix} a & b \\ c & d \end{pmatrix} = \begin{pmatrix} 0 & 0 \\ 0 & 0 \end{pmatrix}$ **45.** $\begin{pmatrix} 1 & 2 \\ 2 & 4 \end{pmatrix}\begin{pmatrix} a & b \\ c & d \end{pmatrix} = \begin{pmatrix} a & b \\ c & d \end{pmatrix}\begin{pmatrix} 1 & 2 \\ 2 & 4 \end{pmatrix} = \begin{pmatrix} 0 & 0 \\ 0 & 0 \end{pmatrix}$

46. Given $\mathbf{A} = \begin{pmatrix} -5 & 3 \\ 4 & -1 \\ 2 & -1 \end{pmatrix}$ and $\mathbf{B} = \begin{pmatrix} 1 & -2 \\ 0 & 4 \end{pmatrix}$, show that $(\mathbf{AB})^\mathsf{T} = \mathbf{B}^\mathsf{T}\mathbf{A}^\mathsf{T}$

Section 18.4

Find the inverse matrix, if possible:

47. $\begin{pmatrix} 2 & -3 \\ 4 & 1 \end{pmatrix}$ **48.** $\begin{pmatrix} 1 & 2 & 3 \\ -2 & 1 & 2 \\ 3 & -1 & -1 \end{pmatrix}$ **49.** $\begin{pmatrix} 2 & -1 & 1 \\ -1 & 1 & -2 \\ -3 & 1 & 0 \end{pmatrix}$

50. $\begin{pmatrix} \frac{1}{\sqrt{2}} & 0 & -\frac{1}{\sqrt{2}} \\ 0 & 1 & 0 \\ \frac{1}{\sqrt{2}} & 0 & \frac{1}{\sqrt{2}} \end{pmatrix}$ **51.** $\begin{pmatrix} 3 & 4 & 0 & 0 \\ 1 & 2 & 0 & 0 \\ 0 & 0 & 3 & 1 \\ 0 & 0 & 4 & 2 \end{pmatrix}$

52. For each matrix $(\mathbf{A})$ of Exercises 47–51, verify (if $\mathbf{A}^{-1}$ exists) that $\mathbf{AA}^{-1} = \mathbf{A}^{-1}\mathbf{A} = \mathbf{I}$.

53. If $\mathbf{A} = \begin{pmatrix} 1 & 2 \\ 3 & 4 \end{pmatrix}$ and $\mathbf{B} = \begin{pmatrix} 5 & 6 \\ 7 & 8 \end{pmatrix}$, show that $(\mathbf{AB})^{-1} = \mathbf{B}^{-1}\mathbf{A}^{-1}$

Section 18.5

54. The linear transformation $\mathbf{r'} = \mathbf{Ar}$, where

$$\mathbf{r'} = \begin{pmatrix} x' \\ y' \\ z' \end{pmatrix}, \quad \mathbf{r} = \begin{pmatrix} x \\ y \\ z \end{pmatrix}, \quad \mathbf{A} = \begin{pmatrix} \cos\theta & -\sin\theta & 0 \\ \sin\theta & \cos\theta & 0 \\ 0 & 0 & 1 \end{pmatrix}$$

represents an anticlockwise rotation through angle θ about the z-axis. **(i)** Write down the corresponding linear equations. **(ii)** Find $\mathbf{A}$ for a clockwise rotation through $\pi/4$ about the z-axis. **(iii)** Show that

$$\mathbf{A}^2 = \begin{pmatrix} \cos 2\theta & -\sin 2\theta & 0 \\ \sin 2\theta & \cos 2\theta & 0 \\ 0 & 0 & 1 \end{pmatrix}$$

and explain its geometric meaning. **(iv)** Explain the geometric meaning of the equation $\mathbf{A}^3 = \mathbf{I}$.

55. For transformations in three dimensions, write down the matrices that represent **(i)** rotation about the x-axis, **(ii)** rotation about the y-axis, **(iii)** reflection in the xy-plane, **(iv)** reflection in the yz-plane, **(v)** reflection in the zx-plane, **(vi)** inversion through the origin.

The coordinates of four points in the xy-plane are given by the columns of the matrix

$$\mathbf{X} = \begin{pmatrix} 2 & 3 & 3 & 2 \\ 1 & 1 & 2 & 2 \end{pmatrix}$$

(see Example 18.19). Find $\mathbf{X}' = \mathbf{AX}$ for each of the following matrices $\mathbf{A}$, and draw appropriate diagrams to illustrate the transformations:

56. $\dfrac{1}{\sqrt{5}}\begin{pmatrix} 1 & -2 \\ 2 & 1 \end{pmatrix}$ **57.** $\begin{pmatrix} 0 & -1 \\ 1 & 0 \end{pmatrix}$ **58.** $\begin{pmatrix} 2 & 0 \\ 0 & 1 \end{pmatrix}$ **59.** $\begin{pmatrix} 3 & 1 \\ -1 & 2 \end{pmatrix}$

60. **(i)** Find the single matrix $\mathbf{A}$ that represents the sequence of consecutive transformations **(a)** anticlockwise rotation through θ about the x-axis, followed by **(b)** reflection in the xy-plane, followed by **(c)** anticlockwise rotation through ϕ about the z-axis. **(ii)** Find $\mathbf{A}$

for $\theta = \pi/3$ and $\phi = -\pi/6$. **(iii)** Find $\mathbf{r}' = \mathbf{Ar}$ for this $\mathbf{A}$ and $\mathbf{r} = \begin{pmatrix} 2 \\ 0 \\ -1 \end{pmatrix}$.

Section 18.6

61. For each of the matrices **(i)**, **(iii)**, and **(vi)** in Exercise 55, **(a)** show that the matrix is orthogonal, **(b)** find its inverse.

Section 18.7

62. The symmetry properties of the plane figure formed by the four points at the corners of a rectangle (not a square), Figure 18.9, can be described in terms of four symmetry operations, the identity operation and three rotations.

(i) Describe these symmetry operations. **(ii)** Construct the group multiplication table.

63. Construct a two-dimensional matrix representation of the group in Exercise 62 by applying each symmetry operation in turn to the coordinates (x, y) of a point in the plane of the figure.

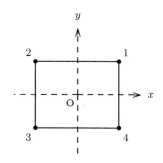

Figure 18.9

19 The matrix eigenvalue problem

19.1 Concepts

The determination of the eigenvalues and eigenvectors of a square matrix is called the eigenvalue problem, and is of importance in many branches of the physical sciences and engineering.[1] For example in quantum chemistry, the application of the variation principle to the Schrödinger equation results in the replacement of the differential eigenvalue equation by an equivalent matrix eigenvalue equation that can be solved by numerical methods. We have already met the eigenvalue problem in Section 9.4, on the use of Lagrangian multipliers for finding the stationary values of a quadratic form (Example 9.9), and in Section 17.4, on the use of determinants for the solution of systems of secular equations. In the present chapter, the matrix eigenvalue problem and the properties of eigenvalues and eigenvectors are discussed in Sections 19.2 and 19.3. The closely related problems of matrix diagonalization and of the reduction of quadratic forms to canonical form are discussed in Sections 19.4 and 19.5. Section 19.6 contains a summary of the complex matrices that are important in advanced group theory and in quantum mechanics.

We first summarize the matrix formulation of the solution of systems of inhomogeneous linear equations, already discussed in Section 17.4 in terms of determinants. A system of n linear equations in n unknowns, $x_1, x_2, x_3, \ldots, x_n$ (equation (17.26)

$$
\begin{aligned}
a_{11}x_1 + a_{12}x_2 + a_{13}x_3 + \cdots + a_{1n}x_n &= b_1 \\
a_{21}x_1 + a_{22}x_2 + a_{23}x_3 + \cdots + a_{2n}x_n &= b_2 \\
a_{31}x_1 + a_{32}x_2 + a_{33}x_3 + \cdots + a_{3n}x_n &= b_3 \\
\vdots \qquad \vdots \qquad \vdots \qquad\qquad \vdots \quad \vdots \\
a_{n1}x_1 + a_{n2}x_2 + a_{n3}x_3 + \cdots + a_{nn}x_n &= b_n
\end{aligned}
\tag{19.1}
$$

can be written as the single matrix equation

$$
\begin{pmatrix}
a_{11} & a_{12} & a_{13} & \cdots & a_{1n} \\
a_{21} & a_{22} & a_{23} & \cdots & a_{2n} \\
a_{31} & a_{32} & a_{33} & \cdots & a_{3n} \\
\vdots & \vdots & \vdots & & \vdots \\
a_{n1} & a_{n2} & a_{n3} & \cdots & a_{nn}
\end{pmatrix}
\begin{pmatrix}
x_1 \\ x_2 \\ x_3 \\ \vdots \\ x_n
\end{pmatrix}
=
\begin{pmatrix}
b_1 \\ b_2 \\ b_3 \\ \vdots \\ b_n
\end{pmatrix}
\tag{19.2}
$$

---

[1] The earliest eigenvalue problems were considered by d'Alembert between 1743 and 1758 in connection with the solution of systems of linear equations with constant coefficients arising from the motions of a loaded string. Cauchy discussed eigenvalues in connection with the forms that describe quadratic surfaces. In 1829 he showed that a quadratic form can be reduced to diagonal form (Section 19.5) by the constrained optimization method involving Lagrangian multipliers described in Example 9.9. The multipliers are the eigenvalues of the associated matrix.

or

$$\mathbf{Ax} = \mathbf{b} \qquad (19.3)$$

We saw in Section 17.4 that the equations have a unique solution when $\det \mathbf{A} \neq 0$, and can then be solved formally in terms of determinants by Cramer's rule. We also saw how solutions may be obtained under certain circumstances when $\mathbf{A}$ is singular, with $\det \mathbf{A} = 0$.

When $\mathbf{A}$ is nonsingular $(\det \mathbf{A} \neq 0)$, $\mathbf{A}^{-1}$ exists, and premultiplication of the left side of (19.3) by $\mathbf{A}^{-1}$ gives

$$\mathbf{A}^{-1}(\mathbf{Ax}) = (\mathbf{A}^{-1}\mathbf{A})\mathbf{x} = \mathbf{Ix} = \mathbf{x}$$

Therefore, premultiplication of both sides of (19.3) by $\mathbf{A}^{-1}$ gives the unique solution

$$\mathbf{x} = \mathbf{A}^{-1}\mathbf{b} \qquad (19.4)$$

This is equivalent to the use of Cramer's rule.

EXAMPLE 19.1 Solve the equations

$$2x - 3y + 4z = 8$$
$$y - 3z = -7$$
$$x + 2y + 2z = 11$$

The matrix of the coefficients and its inverse are

$$\mathbf{A} = \begin{pmatrix} 2 & -3 & 4 \\ 0 & 1 & -3 \\ 1 & 2 & 2 \end{pmatrix}, \qquad \mathbf{A}^{-1} = \frac{1}{21} \begin{pmatrix} 8 & 14 & 5 \\ -3 & 0 & 6 \\ -1 & -7 & 2 \end{pmatrix}$$

Therefore, by equation (19.4),

$$\begin{pmatrix} x \\ y \\ z \end{pmatrix} = \frac{1}{21} \begin{pmatrix} 8 & 14 & 5 \\ -3 & 0 & 6 \\ -1 & -7 & 2 \end{pmatrix} \begin{pmatrix} 8 \\ -7 \\ 11 \end{pmatrix} = \frac{1}{21} \begin{pmatrix} 21 \\ 42 \\ 63 \end{pmatrix} = \begin{pmatrix} 1 \\ 2 \\ 3 \end{pmatrix}$$

This is the the result obtained in Example 17.2 by Cramer's rule.

▶ Exercises 1–3

The matrix equation (19.3) represents the case of inhomogeneous equations, with $\mathbf{b} \neq \mathbf{0}$. In the homogeneous case, we have

$$\mathbf{Ax} = \mathbf{0} \qquad (19.5)$$

and the solution is $\mathbf{x} = \mathbf{0}$ if $\mathbf{A}$ is nonsingular. When $\mathbf{A}$ is *singular*, $\mathbf{x} = \mathbf{0}$ is the trivial solution but, as discussed in Section 17.4, other solutions may exist. In the following section we consider these other solutions for the most important case of the homogeneous system, the matrix eigenvalue problem.

19.2 The eigenvalue problem

A matrix equation of type

$$\mathbf{Ax} = \lambda\mathbf{x} \qquad (19.6)$$

where $\mathbf{A}$ is a square matrix, $\mathbf{x}$ is a column vector, and λ is a number, is a linear transformation in which the matrix $\mathbf{A}$ transforms the vector $\mathbf{x}$ into a multiple of $\mathbf{x}$. The equation can be written as

$$(\mathbf{A} - \lambda\mathbf{I})\mathbf{x} = 0 \qquad (19.7)$$

(since $\mathbf{Ix} = \mathbf{x}$), and therefore represents the system of homogeneous linear equations

$$
\begin{aligned}
(a_{11} - \lambda)x_1 + \quad a_{12}x_2 \quad + \quad a_{13}x_3 \quad + \cdots + \quad a_{1n}x_n \quad &= 0 \\
a_{21}x_1 \quad + (a_{22} - \lambda)x_2 + \quad a_{23}x_3 \quad + \cdots + \quad a_{2n}x_n \quad &= 0 \\
a_{31}x_1 \quad + \quad a_{32}x_2 \quad + (a_{33} - \lambda)x_3 + \cdots + \quad a_{3n}x_n \quad &= 0 \qquad (19.8) \\
\vdots \qquad\qquad \vdots \qquad\qquad \vdots \qquad\qquad \vdots \qquad \vdots \\
a_{n1}x_1 \quad + \quad a_{n2}x_2 \quad + \quad a_{n3}x_3 \quad + \cdots + (a_{nn} - \lambda)x_n &= 0
\end{aligned}
$$

(these are the secular equations discussed in Section 17.4). The equations have the trivial solution $\mathbf{x} = \mathbf{0}$ for all values of λ. The equations also have a nonzero solution if the value of λ can be chosen to make the matrix $(\mathbf{A} - \lambda\mathbf{I})$ singular; that is, if λ is chosen to make the determinant of $(\mathbf{A} - \lambda\mathbf{I})$ zero:

$$\det(\mathbf{A} - \lambda\mathbf{I}) = 0 \qquad (19.9)$$

or

$$
\begin{vmatrix}
(a_{11} - \lambda) & a_{12} & a_{13} & \cdots & a_{1n} \\
a_{21} & (a_{22} - \lambda) & a_{23} & \cdots & a_{2n} \\
a_{31} & a_{32} & (a_{33} - \lambda) & \cdots & a_{3n} \\
\vdots & \vdots & \vdots & & \vdots \\
a_{n1} & a_{n2} & a_{n3} & \cdots & (a_{nn} - \lambda)
\end{vmatrix} = 0 \qquad (19.10)
$$

The left side of this equation is called the **characteristic** (or **secular**) **determinant** of the matrix **A**. It is a polynomial of degree n in λ, and its roots are the values of λ for which (19.9) is true; that is, the values of λ for which the matrix equation (19.6) has nonzero solutions. Equation (19.9) is called the **characteristic equation** of the matrix **A**.

The values of λ for which the **matrix eigenvalue equation** (19.6) has nonzero solutions are called the **eigenvalues** or **characteristic** values of the matrix **A** (they are also called latent roots or proper values). A matrix of order n has n eigenvalues, $\lambda_1, \lambda_2, \ldots, \lambda_n$, and these form the **eigenvalue spectrum** of **A**.

EXAMPLE 19.2 Solve the characteristic equation of the matrix

$$\mathbf{A} = \begin{pmatrix} -2 & 1 & 1 \\ -11 & 4 & 5 \\ -1 & 1 & 0 \end{pmatrix}$$

The characteristic equation of **A** is

$$\det(\mathbf{A} - \lambda\mathbf{I}) = D = \begin{vmatrix} -2-\lambda & 1 & 1 \\ -11 & 4-\lambda & 5 \\ -1 & 1 & -\lambda \end{vmatrix} = 0$$

This equation was solved in Example 17.8:

$$D = -(\lambda+1)(\lambda-1)(\lambda-2) = 0 \quad \text{when} \quad \lambda = -1, +1, \text{ and } +2.$$

Corresponding to each eigenvalue, $\lambda = \lambda_k$, there is a solution $\mathbf{x} = \mathbf{x}_k$ of the eigenvalue equation such that

$$\mathbf{A}\mathbf{x}_k = \lambda_k\mathbf{x}_k \qquad k = 1, 2, 3, \ldots, n \qquad (19.11)$$

These solutions are called the **eigenvectors** or **characteristic vectors** of **A** (they are also called latent vectors, proper vectors, or poles). The vectors are obtained by solving the system of homogeneous equations (19.8) for each value of λ in turn (as in Example 17.9). The problem of finding the eigenvalues and eigenvectors is called **the matrix eigenvalue problem** (or algebraic eigenvalue problem).

EXAMPLE 19.3 Find the eigenvectors of the matrix **A** in Example 19.2.

The secular equations are (see also Example 17.8)

$$
\begin{aligned}
&(1)\ (-2-\lambda)x + y + z = 0 \\
&(2)\ -11x + (4-\lambda)y + 5z = 0 \\
&(3)\ -x + y + (-\lambda)z = 0
\end{aligned}
$$

with eigenvalues $\lambda_1 = -1$, $\lambda_2 = 1$, and $\lambda_3 = 2$. The corresponding solutions of the secular equations are obtained by replacing λ in the equations by each root in turn. We use equations (1) and (2) to solve for x and y in terms of (arbitrary) z:

$$\lambda = \lambda_1 = -1: \quad \begin{matrix} (1) \\ (2) \end{matrix} \quad \left. \begin{matrix} -x + y + z = 0 \\ -11x + 5y + 5z = 0 \end{matrix} \right\} \rightarrow x = 0, \, y = -z$$

$$(3) \qquad -x + y + z = 0$$

We see that equation (3) is identical to (1) and gives no further information. Similarly,

$$\lambda = \lambda_2 = 1: \quad \begin{matrix} (1) \\ (2) \end{matrix} \quad \left. \begin{matrix} -3x + y + z = 0 \\ -11x + 3y + 5z = 0 \end{matrix} \right\} \rightarrow x = z, \, y = 2z$$

$$\lambda = \lambda_3 = 2: \quad \begin{matrix} (1) \\ (2) \end{matrix} \quad \left. \begin{matrix} -4x + y + z = 0 \\ -11x + 2y + 5z = 0 \end{matrix} \right\} \rightarrow x = z, \, y = 3z$$

The three eigenvectors are therefore

$$\mathbf{x}_1 = z \begin{pmatrix} 0 \\ -1 \\ 1 \end{pmatrix}, \quad \mathbf{x}_2 = z \begin{pmatrix} 1 \\ 2 \\ 1 \end{pmatrix}, \quad \mathbf{x}_3 = z \begin{pmatrix} 1 \\ 3 \\ 1 \end{pmatrix}$$

▸ Exercises 4–9

When the eigenvalues, λ_k $(k = 1, 2, \ldots, n)$, are distinct (with no two having the same value, as in Example 19.3) then there exist n distinct (linearly independent) eigenvectors. When two or more eigenvalues are equal (**multiple** or **degenerate eigenvalues**) there may exist *fewer* than n distinct eigenvectors. But *symmetric* (and Hermitian) matrices always have the full complement of n distinct eigenvectors.

EXAMPLES 19.4 Multiple eigenvalues

(i) The characteristic equation of the non-symmetric matrix

$$\begin{pmatrix} 2 & 1 \\ -1 & 4 \end{pmatrix} \text{ is } \begin{vmatrix} 2-\lambda & 1 \\ -1 & 4-\lambda \end{vmatrix} = (\lambda - 3)^2 = 0$$

and the single eigenvalue $\lambda = 3$ is doubly-degenerate. Substitution of this value of λ in either secular equation gives just one eigenvector. Thus,

$$(2 - \lambda)x + y = 0 \xrightarrow{\lambda = 3} -x + y = 0 \rightarrow x = y$$

and the single eigenvector is $\mathbf{x} = y \begin{pmatrix} 1 \\ 1 \end{pmatrix}$, y arbitrary.

(ii) The characteristic equation of the symmetric matrix

$$\begin{pmatrix} 2 & 1 & 1 \\ 1 & 2 & 1 \\ 1 & 1 & 2 \end{pmatrix} \text{ is } \begin{vmatrix} 2-\lambda & 1 & 1 \\ 1 & 2-\lambda & 1 \\ 1 & 1 & 2-\lambda \end{vmatrix} = (4-\lambda)(1-\lambda)^2 = 0$$

with nondegenerate eigenvalue $\lambda_1 = 4$ and the pair of degenerate values $\lambda_2 = \lambda_3 = 1$. The secular equations are

$$\begin{aligned} (2-\lambda)x + \quad y \quad + \quad z \quad &= 0 \\ x \quad + (2-\lambda)y + \quad z \quad &= 0 \\ x \quad + \quad y \quad + (2-\lambda)z &= 0 \end{aligned}$$

For $\lambda = 4$: The solution is $x = y = z$ and the eigenvector is $\mathbf{x}_1 = c_1 \begin{pmatrix} 1 \\ 1 \\ 1 \end{pmatrix}$, c_1 arbitrary.

For $\lambda = 1$: Each secular equation gives $x + y + z = 0$, and every independent pair of vectors that satisfy this condition is a solution for the degenerate eigenvalue; for example,

$$\mathbf{x}_2 = c_2 \begin{pmatrix} 1 \\ 1 \\ -2 \end{pmatrix} \quad \text{and} \quad \mathbf{x}_3 = c_3 \begin{pmatrix} 1 \\ 2 \\ -3 \end{pmatrix}$$

➤ Exercises 10, 11

Two types of matrices whose eigenvalues are of greatest interest in the physical sciences are real symmetric matrices (symmetric matrices whose elements are all real numbers) as in Example 19.4(ii), and the complex Hermitian matrices described in Section 19.6. These matrices have real eigenvalues.

19.3 Properties of the eigenvectors

Property 1. If $\mathbf{x}$ is an eigenvector corresponding to eigenvalue λ, then $k\mathbf{x}$ is also an eigenvector corresponding to the same eigenvalue, for any nonzero value of the number k:

$$\text{if} \quad \mathbf{A}\mathbf{x} = \lambda\mathbf{x} \quad \text{then} \quad \mathbf{A}(k\mathbf{x}) = k(\mathbf{A}\mathbf{x}) = k(\lambda\mathbf{x}) = \lambda(k\mathbf{x}) \tag{19.12}$$

Eigenvectors that differ only in a constant factor are not treated as distinct. It is convenient and conventional to choose the factor k to make the eigenvector a unit vector; that is, to normalize the vector.

EXAMPLE 19.5 Normalization of eigenvectors

The eigenvectors $\mathbf{x}$ of Example 19.3 are normalized (have unit length) if

$$\mathbf{x}^T\mathbf{x} = x^2 + y^2 + z^2 = 1$$

For example,

$$\mathbf{x}_2{}^T\mathbf{x}_2 = z^2(1^2 + 2^2 + 1^2) = 1 \quad \text{if} \quad z = 1/\sqrt{6}$$

and the set of three normalized eigenvectors is

$$x_1 = \frac{1}{\sqrt{2}}\begin{pmatrix} 0 \\ -1 \\ 1 \end{pmatrix}, \quad x_2 = \frac{1}{\sqrt{6}}\begin{pmatrix} 1 \\ 2 \\ 1 \end{pmatrix}, \quad x_3 = \frac{1}{\sqrt{11}}\begin{pmatrix} 1 \\ 3 \\ 1 \end{pmatrix}$$

▶ Exercises 12–15

Property 2. If $\mathbf{A}$ is a (real) symmetric matrix, the eigenvectors corresponding to distinct eigenvalues are orthogonal.

Let $\mathbf{x}_k$ and $\mathbf{x}_l$ be eigenvectors of $\mathbf{A}$ corresponding to eigenvalues λ_k and λ_l, respectively. Then

$$\mathbf{A}\mathbf{x}_k = \lambda_k\mathbf{x}_k \tag{19.13}$$

and premultiplication of both sides by $\mathbf{x}_l{}^T$ gives

$$\mathbf{x}_l{}^T\mathbf{A}\mathbf{x}_k = \lambda_k\mathbf{x}_l{}^T\mathbf{x}_k \tag{19.14}$$

Also

$$\mathbf{A}\mathbf{x}_l = \lambda_l\mathbf{x}_l \tag{19.15}$$

and premultiplication of both sides by $\mathbf{x}_k{}^T$ gives

$$\mathbf{x}_k{}^T\mathbf{A}\mathbf{x}_l = \lambda_l\mathbf{x}_k{}^T\mathbf{x}_l \tag{19.16}$$

Now the transpose of a product of matrices is the product of the transpose matrices in reverse order (equation (18.41)). Therefore, taking the transpose of both sides of (19.16), and remembering that $\mathbf{A}^T = \mathbf{A}$ for a symmetric matrix,

$$\mathbf{x}_l{}^T\mathbf{A}\mathbf{x}_k = \lambda_l\mathbf{x}_l{}^T\mathbf{x}_k \tag{19.17}$$

Subtraction of (19.17) from (19.14) then gives

$$0 = (\lambda_k - \lambda_l)\mathbf{x}_l{}^T\mathbf{x}_k \tag{19.18}$$

so that, when $\lambda_k \neq \lambda_p$,

$$\mathbf{x}_l{}^T\mathbf{x}_k = 0 \tag{19.19}$$

and the vectors are orthogonal.

EXAMPLE 19.6   For the eigenvectors of the symmetric matrix in Example 19.4(ii),

$$\mathbf{x}_1{}^T\mathbf{x}_2 = c_1 c_2 (1 \quad 1 \quad 1) \begin{pmatrix} 1 \\ 1 \\ -2 \end{pmatrix} = c_1 c_2 (1+1-2) = 0$$

$$\mathbf{x}_1{}^T\mathbf{x}_3 = c_1 c_3 (1 \quad 1 \quad 1) \begin{pmatrix} 1 \\ 2 \\ -3 \end{pmatrix} = c_1 c_2 (1+2-3) = 0$$

▸ Exercises 16–18

Property 3. For a symmetric matrix, the eigenvectors corresponding to the *same* eigenvalue are either orthogonal or can be made so.

EXAMPLE 19.7   The eigenvectors $\mathbf{x}_2$ and $\mathbf{x}_3$ of Example 19.4(ii), belonging to the degenerate eigenvalue $\lambda = 1$, are *not* orthogonal. Thus

$$\mathbf{x}_2{}^T\mathbf{x}_3 = c_2 c_3 (1 \quad 1 \quad -2) \begin{pmatrix} 1 \\ 2 \\ -3 \end{pmatrix} = c_2 c_3 (1+2+6) \neq 0$$

If the independent eigenvectors $\mathbf{x}_k$ and $\mathbf{x}_l$ correspond to the same eigenvalue $\lambda_k = \lambda_l = \lambda$, then

$$\mathbf{A}\mathbf{x}_k = \lambda \mathbf{x}_k \quad \text{and} \quad \mathbf{A}\mathbf{x}_l = \lambda \mathbf{x}_l \tag{19.20}$$

so that every linear combination of $\mathbf{x}_k$ and $\mathbf{x}_l$ is also an eigenvector of $\mathbf{A}$ with the same eigenvalue. Thus, if

$$\mathbf{x} = a\mathbf{x}_k + b\mathbf{x}_l \tag{19.21}$$

then

$$\mathbf{A}\mathbf{x} = \mathbf{A}(a\mathbf{x}_k + b\mathbf{x}_l) = a(\mathbf{A}\mathbf{x}_k) + b(\mathbf{A}\mathbf{x}_l)$$
$$= a(\lambda\mathbf{x}_k) + b(\lambda\mathbf{x}_l) = \lambda(a\mathbf{x}_k + b\mathbf{x}_l) = \lambda\mathbf{x} \tag{19.22}$$

It is always possible to find two linear combinations of the vectors that are orthogonal. Let $\mathbf{x}_1$ and $\mathbf{x}_2$ be nonorthogonal vectors, with $\mathbf{x}_1{}^T\mathbf{x}_2 \neq 0$. Let $\mathbf{x}_2'$ be the linear combination

$$\mathbf{x}_2' = \mathbf{x}_2 - c\mathbf{x}_1$$

in which the parameter c is chosen such that $\mathbf{x}_2'$ be orthogonal to $\mathbf{x}_1$; that is, $\mathbf{x}_1{}^T\mathbf{x}_2' = 0$. Then

$$\mathbf{x}_1{}^T\mathbf{x}_2' = \mathbf{x}_1{}^T\mathbf{x}_2 - c\mathbf{x}_1{}^T\mathbf{x}_1 = 0 \quad \text{if} \quad c = \frac{\mathbf{x}_1{}^T\mathbf{x}_2}{\mathbf{x}_1{}^T\mathbf{x}_1}$$

and new vector $\mathbf{x}_2' = \mathbf{x}_2 - \left(\dfrac{\mathbf{x}_1{}^T\mathbf{x}_2}{\mathbf{x}_1{}^T\mathbf{x}_1}\right)\mathbf{x}_1$ is orthogonal to $\mathbf{x}_1$. This is an example of the widely-used **Schmidt orthogonalization** method.

EXAMPLE 19.8 Orthogonalization of vectors

The eigenvectors $\mathbf{x}_2$ and $\mathbf{x}_3$ of Example 19.4(ii), belonging to the degenerate eigenvalue $\lambda = 1$, are not orthogonal. We have, ignoring the arbitrary multipliers c_2 and c_3,

$$\mathbf{x}_2{}^T\mathbf{x}_2 = (1 \quad 1 \quad -2)\begin{pmatrix} 1 \\ 1 \\ -2 \end{pmatrix} = 6, \quad \mathbf{x}_2{}^T\mathbf{x}_3 = (1 \quad 1 \quad -2)\begin{pmatrix} 1 \\ 2 \\ -3 \end{pmatrix} = 9$$

$$\text{Then} \quad \mathbf{x}_3' = \mathbf{x}_3 - \left(\frac{\mathbf{x}_2{}^T\mathbf{x}_3}{\mathbf{x}_2{}^T\mathbf{x}_2}\right)\mathbf{x}_1 = \begin{pmatrix} 1 \\ 2 \\ -3 \end{pmatrix} - \frac{9}{6}\begin{pmatrix} 1 \\ 1 \\ -2 \end{pmatrix} = \begin{pmatrix} -1/2 \\ 1/2 \\ 0 \end{pmatrix}$$

is orthogonal to $\mathbf{x}_2$. Including an arbitrary multiplier, the new vector is $\mathbf{x}_3' = c_3'\begin{pmatrix} 1 \\ -1 \\ 0 \end{pmatrix}$

and, with $\mathbf{x}_2$ (and $\mathbf{x}_1$), can now be normalized. The orthonormal eigenvectors of the real symmetric matrix

$$\begin{pmatrix} 2 & 1 & 1 \\ 1 & 2 & 1 \\ 1 & 1 & 2 \end{pmatrix} \quad \text{are} \quad \mathbf{x}_1 = \frac{1}{\sqrt{3}}\begin{pmatrix} 1 \\ 1 \\ 1 \end{pmatrix}, \quad \mathbf{x}_2 = \frac{1}{\sqrt{6}}\begin{pmatrix} 1 \\ 1 \\ -2 \end{pmatrix}, \quad \mathbf{x}_3' = \frac{1}{\sqrt{2}}\begin{pmatrix} 1 \\ -1 \\ 0 \end{pmatrix}$$

▸ Exercise 19

The following important theorem for the eigenvectors of symmetric matrices follows from Properties 1–3:

> The n eigenvectors of a real symmetric matrix of order n form (or can be chosen to form) a system of n orthogonal unit (orthonormal) vectors:

$$\mathbf{x}_k^T \mathbf{x}_l = \delta_{kl} = \begin{cases} 1 & \text{if } k = l \\ 0 & \text{if } k \neq l \end{cases} \tag{19.23}$$

EXAMPLE 19.9 Hückel theory of cyclobutadiene

In the molecular-orbital theory of π-electron systems, the states of the π electrons are described by a matrix eigenvalue equation

$$\mathbf{HC} = E\mathbf{C}$$

in which the matrix $\mathbf{H}$ represents the 'effective Hamiltonian' for a π electron in the system, the eigenvalues E of $\mathbf{H}$ are the orbital energies of the π electrons, and the eigenvectors $\mathbf{C}$ represent the corresponding molecular orbitals (the components of $\mathbf{C}$ are the coefficients in a 'linear combination of atomic orbitals' (LCAO) description of a molecular orbital). In the Hückel theory of cyclobutadiene (C_4H_4), $\mathbf{H}$ is the real symmetric matrix

$$\mathbf{H} = \begin{pmatrix} \alpha & \beta & 0 & \beta \\ \beta & \alpha & \beta & 0 \\ 0 & \beta & \alpha & \beta \\ \beta & 0 & \beta & \alpha \end{pmatrix}$$

(the Hückel parameters α and β are real negative scalars with the dimensions of energy) with characteristic equation

$$\det(\mathbf{H} - E\mathbf{I}) = \begin{vmatrix} \alpha - E & \beta & 0 & \beta \\ \beta & \alpha - E & \beta & 0 \\ 0 & \beta & \alpha - E & \beta \\ \beta & 0 & \beta & \alpha - E \end{vmatrix}$$

$$= (\alpha - E)^2(\alpha - E + 2\beta)(\alpha - E - 2\beta) = 0$$

The eigenvalues (orbital energies) are therefore

$$E_1 = \alpha + 2\beta, \quad E_2 = E_3 = \alpha, \quad E_4 = \alpha - 2\beta$$

and the eigenvalue spectrum is shown in Figure 19.1.

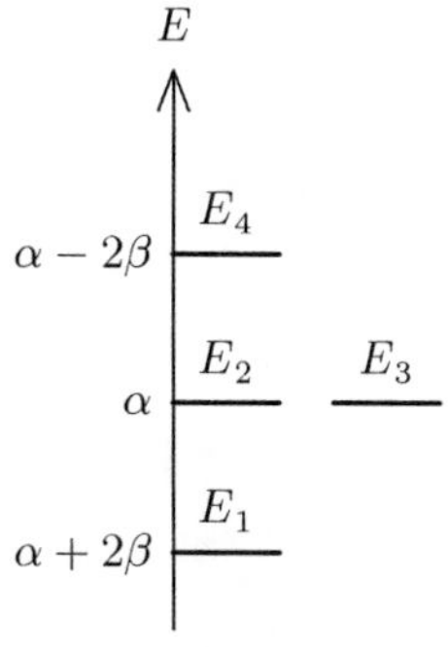

Figure 19.1

The secular equations of the problem are

$$
\begin{aligned}
&(1) \quad (\alpha-E)c_1 + \beta c_2 + \beta c_4 = 0 \\
&(2) \quad \beta c_1 + (\alpha-E)c_2 + \beta c_3 = 0 \\
&(3) \quad \beta c_2 + (\alpha-E)c_3 + \beta c_4 = 0 \\
&(4) \quad \beta c_1 + \beta c_3 + (\alpha-E)c_4 = 0
\end{aligned}
$$

and the eigenvectors are obtained by solving this system of homogeneous equations for each eigenvalue E in turn. We consider first the nondegenerate eigenvalues $E_1 = \alpha + 2\beta$ and $E_4 = \alpha - 2\beta$.

For $E = E_1 = \alpha + 2\beta$,

$$
\begin{aligned}
&(1) \quad -2\beta c_1 + \beta c_2 + \beta c_4 = 0 \\
&(2) \quad \beta c_1 - 2\beta c_2 + \beta c_3 = 0 \\
&(3) \quad \beta c_2 - 2\beta c_3 + \beta c_4 = 0 \\
&(4) \quad \beta c_1 + \beta c_3 - 2\beta c_4 = 0
\end{aligned}
$$

Only three of the four equations are independent; for example, $(1)+(2)+(3)=-(4)$. Solving for c_2, c_3, and c_4 in terms of c_1 gives $c_2 = c_1$, $c_3 = c_1$, $c_4 = c_1$. Similarly, the eigenvector corresponding to eigenvalue $E_4 = \alpha - 2\beta$ has components $c_2 = -c_1$, $c_3 = c_1$, $c_4 = -c_1$. The eigenvectors corresponding to eigenvalues E_1 and E_4 are therefore

$$
\mathbf{C}_1 = c_1 \begin{pmatrix} 1 \\ 1 \\ 1 \\ 1 \end{pmatrix}, \quad
\mathbf{C}_4 = c_4 \begin{pmatrix} 1 \\ -1 \\ 1 \\ -1 \end{pmatrix}
$$

where c_1 and c_4 are arbitrary.

For $E_2 = E_3 = \alpha$, the secular equations are

$$
\begin{aligned}
&(1)=(3) \quad \beta c_2 + \beta c_4 = 0 \\
&(2)=(4) \quad \beta c_1 + \beta c_3 = 0
\end{aligned}
$$

Only two of the four equations are therefore independent, and have solution $c_3 = -c_1$, $c_4 = -c_2$. A pair of eigenvectors corresponding to the doubly-degenerate eigenvalue $E = \alpha$ is therefore

$$
\mathbf{C}_2 = c_2 \begin{pmatrix} 1 \\ 1 \\ -1 \\ -1 \end{pmatrix}, \quad
\mathbf{C}_3 = c_3 \begin{pmatrix} 1 \\ -1 \\ -1 \\ 1 \end{pmatrix}
$$

where c_2 and c_3 are arbitrary. The normalized vectors

$$\mathbf{C}_1 = \frac{1}{2}\begin{pmatrix} 1 \\ 1 \\ 1 \\ 1 \end{pmatrix}, \qquad \mathbf{C}_2 = \frac{1}{2}\begin{pmatrix} 1 \\ 1 \\ -1 \\ -1 \end{pmatrix}, \qquad \mathbf{C}_3 = \frac{1}{2}\begin{pmatrix} 1 \\ -1 \\ -1 \\ 1 \end{pmatrix}, \qquad \mathbf{C}_4 = \frac{1}{2}\begin{pmatrix} 1 \\ -1 \\ 1 \\ -1 \end{pmatrix}$$

are also orthogonal; they form an orthonormal set of four-dimensional vectors.

➤ Exercises 20, 21

➤ Exercise 22

19.4 Matrix diagonalization

Let the square matrix $\mathbf{A}$ have eigenvectors $\mathbf{x}_1, \mathbf{x}_2, \mathbf{x}_3, \ldots, \mathbf{x}_n$ corresponding to eigenvalues $\lambda_1, \lambda_2, \lambda_3, \ldots, \lambda_n$:

$$\mathbf{A}\mathbf{x}_k = \lambda_k \mathbf{x}_k, \qquad k = 1, 2, 3, \ldots, n \tag{19.24}$$

and let $\mathbf{X}$ be the matrix whose columns are the eigenvectors of $\mathbf{A}$,

$$\mathbf{X} = (\mathbf{x}_1 \quad \mathbf{x}_2 \quad \mathbf{x}_3 \quad \cdots \quad \mathbf{x}_n) = \begin{pmatrix} x_{11} & x_{12} & x_{13} & \cdots & x_{1n} \\ x_{21} & x_{22} & x_{23} & \cdots & x_{2n} \\ x_{31} & x_{32} & x_{33} & \cdots & x_{3n} \\ \vdots & \vdots & \vdots & & \vdots \\ x_{n1} & x_{n2} & x_{n3} & \cdots & x_{nn} \end{pmatrix} \tag{19.25}$$

Then

$$\begin{aligned} \mathbf{A}\mathbf{X} &= (\mathbf{A}\mathbf{x}_1 \quad \mathbf{A}\mathbf{x}_2 \quad \mathbf{A}\mathbf{x}_3 \quad \cdots \quad \mathbf{A}\mathbf{x}_n) \\ &= (\lambda_1 \mathbf{x}_1 \quad \lambda_2 \mathbf{x}_2 \quad \lambda_3 \mathbf{A}_3 \quad \cdots \quad \lambda_n \mathbf{x}_n) \\ &= \mathbf{X}\mathbf{D} \end{aligned} \tag{19.26}$$

where $\mathbf{D}$ is the diagonal matrix whose diagonal elements are the eigenvalues of $\mathbf{A}$,

$$\mathbf{D} = \begin{pmatrix} \lambda_1 & 0 & 0 & \cdots & 0 \\ 0 & \lambda_2 & 0 & \cdots & 0 \\ 0 & 0 & \lambda_3 & \cdots & 0 \\ \vdots & \vdots & \vdots & & \vdots \\ 0 & 0 & 0 & \cdots & \lambda_n \end{pmatrix} \tag{19.27}$$

EXAMPLE 19.10 The eigenvectors of the matrix (Examples 19.2 and 19.3)

$$A = \begin{pmatrix} -2 & 1 & 1 \\ -11 & 4 & 5 \\ -1 & 1 & 0 \end{pmatrix}$$

are (ignoring the arbitrary multipliers)

$$\mathbf{x}_1 = \begin{pmatrix} 0 \\ -1 \\ 1 \end{pmatrix}, \qquad \mathbf{x}_2 = \begin{pmatrix} 1 \\ 2 \\ 1 \end{pmatrix}, \qquad \mathbf{x}_3 = \begin{pmatrix} 1 \\ 3 \\ 1 \end{pmatrix}$$

corresponding to eigenvalues $\lambda_1 = -1, \lambda_2 = 1, \lambda_3 = 2$. Then

$$\mathbf{X} = (\mathbf{x}_1 \ \mathbf{x}_2 \ \mathbf{x}_3) = \begin{pmatrix} 0 & 1 & 1 \\ -1 & 2 & 3 \\ 1 & 1 & 1 \end{pmatrix}, \quad \mathbf{D} = \begin{pmatrix} \lambda_1 & 0 & 0 \\ 0 & \lambda_2 & 0 \\ 0 & 0 & \lambda_3 \end{pmatrix} = \begin{pmatrix} -1 & 0 & 0 \\ 0 & 1 & 0 \\ 0 & 0 & 2 \end{pmatrix}$$

and

$$\mathbf{AX} = \begin{pmatrix} -2 & 1 & 1 \\ -11 & 4 & 5 \\ -1 & 1 & 0 \end{pmatrix} \begin{pmatrix} 0 & 1 & 1 \\ -1 & 2 & 3 \\ 1 & 1 & 1 \end{pmatrix} = \begin{pmatrix} 0 & 1 & 2 \\ 1 & 2 & 6 \\ -1 & 1 & 2 \end{pmatrix}$$

$$\mathbf{XD} = \begin{pmatrix} 0 & 1 & 1 \\ -1 & 2 & 3 \\ 1 & 1 & 1 \end{pmatrix} \begin{pmatrix} -1 & 0 & 0 \\ 0 & 1 & 0 \\ 0 & 0 & 2 \end{pmatrix} = \begin{pmatrix} 0 & 1 & 2 \\ 1 & 2 & 6 \\ -1 & 1 & 2 \end{pmatrix}$$

so that $\mathbf{AX} = \mathbf{XD}$

‣ Exercises 23, 24

If the matrix $\mathbf{X}$ of the eigenvectors of $\mathbf{A}$ is nonsingular then premultiplication of both sides of equation (19.26) by the inverse matrix $\mathbf{X}^{-1}$ gives

$$\mathbf{D} = \mathbf{X}^{-1}\mathbf{AX} \tag{19.28}$$

and $\mathbf{A}$ has been reduced to the **diagonal form D**.

EXAMPLE 19.11 Diagonalization

The inverse of the matrix $\mathbf{X}$ of the eigenvectors of $\mathbf{A}$ in Example 19.10 is

$$
\mathbf{X}^{-1} = \begin{pmatrix} -1 & 0 & 1 \\ 4 & -1 & -1 \\ -3 & 1 & 1 \end{pmatrix}
$$

so that $\quad \mathbf{X}^{-1}\mathbf{A}\mathbf{X} = \begin{pmatrix} -1 & 0 & 1 \\ 4 & -1 & -1 \\ -3 & 1 & 1 \end{pmatrix}\begin{pmatrix} -2 & 1 & 1 \\ -11 & 4 & 5 \\ -1 & 1 & 0 \end{pmatrix}\begin{pmatrix} 0 & 1 & 1 \\ -1 & 2 & 3 \\ 1 & 1 & 1 \end{pmatrix} = \begin{pmatrix} -1 & 0 & 0 \\ 0 & 1 & 0 \\ 0 & 0 & 2 \end{pmatrix} = \mathbf{D}$

▸ Exercises 25–28

Similarity transformations

Two square matrices $\mathbf{A}$ and $\mathbf{B}$ are called **similar matrices** or **similarity transforms** if they are related by the **similarity transformation**[2]

$$
\mathbf{B} = \mathbf{C}^{-1}\mathbf{A}\mathbf{C} \tag{19.29}
$$

where $\mathbf{C}$ is a nonsingular matrix. Equation (19.28) is therefore a similarity transformation that reduces $\mathbf{A}$ to diagonal form. When $\mathbf{A}$ is symmetric then, by theorem (19.23), the eigenvectors of $\mathbf{A}$ are orthonormal and $\mathbf{X}$ in (19.28) is an orthogonal matrix, with $\mathbf{X}^{-1} = \mathbf{X}^{\mathrm{T}}$. A symmetric matrix is therefore reduced to diagonal form by the **orthogonal transformation**

$$
\mathbf{D} = \mathbf{X}^{\mathrm{T}}\mathbf{A}\mathbf{X} \tag{19.30}
$$

Two important **invariance properties** of similarity transforms follow from equations (18.38) to (18.40) for the determinant and trace of a matrix product: if $\mathbf{B} = \mathbf{C}^{-1}\mathbf{A}\mathbf{C}$ then

$$
\det \mathbf{B} = \det \mathbf{A} \quad \text{invariance of the determinant} \tag{19.31}
$$

$$
\operatorname{tr} \mathbf{B} = \operatorname{tr} \mathbf{A} \quad \text{invariance of the trace} \tag{19.32}
$$

For example,

$$
\det \mathbf{B} = \det (\mathbf{C}^{-1}\mathbf{A}\mathbf{C}) = \det (\mathbf{A}\mathbf{C}\mathbf{C}^{-1}) = \det (\mathbf{A}\mathbf{I}) = \det \mathbf{A}
$$

---

[2] In his 1878 monograph on the theory of matrices, Frobenius defined similar matrices, discussed the properties of orthogonal matrices and transformations, and showed the relationship between the algebras of matrices and quaternions by determining four 2×2 matrices whose algebra is that of the quaternion quantities $1, i, j, k$.

It follows from these invariance properties that, for a square matrix $\mathbf{A}$,

1. The trace of $\mathbf{A}$ is equal to the sum of the eigenvalues of $\mathbf{A}$,

$$\sum_{k=1}^{n} \lambda_k = \text{tr } \mathbf{A} \tag{19.33}$$

2. The determinant of $\mathbf{A}$ is equal to the product of the eigenvalues of $\mathbf{A}$,

$$\prod_{k=1}^{n} \lambda_k = \det \mathbf{A} \tag{19.34}$$

(the determinant of a diagonal matrix is the product of the diagonal elements).

EXAMPLE 19.12 For the matrices $\mathbf{A}$ and $\mathbf{D}$ of Example 19.11,

Trace: $\text{tr } \mathbf{A} = -2 + 4 + 0 = 2, \qquad \text{tr } \mathbf{D} = -1 + 1 + 2 = 2$

Determinant: $\det \mathbf{A} = -2 \begin{vmatrix} 4 & 5 \\ 1 & 0 \end{vmatrix} - \begin{vmatrix} -11 & 5 \\ -1 & 0 \end{vmatrix} + \begin{vmatrix} -11 & 4 \\ -1 & 1 \end{vmatrix} = 10 - 5 - 7 = -2$

$\det \mathbf{D} = (-1) \times 1 \times 2 = -2$

19.5 Quadratic forms

A quadratic form is a polynomial of the second degree in a set of variables. Examples are

$$x^2 + y^2, \qquad 3x^2 - 4xy + 2y^2, \qquad x_1^2 + 2x_1 x_2 + 3x_2 x_3 - x_2^2 + 3x_3^2$$

The general (real) quadratic form in two variables is

$$\begin{aligned} Q(x, y) &= ax^2 + bxy + byx + cy^2 \\ &= ax^2 + 2bxy + cy^2 \end{aligned} \tag{19.35}$$

in which the coefficients a, b, and c are real numbers. This can be written in matrix form as

$$Q(\mathbf{x}) = (x \ \ y) \begin{pmatrix} a & b \\ b & c \end{pmatrix} \begin{pmatrix} x \\ y \end{pmatrix} = \mathbf{x}^{\mathsf{T}} \mathbf{A} \mathbf{x} \tag{19.36}$$

where $\mathbf{A}$ is the real symmetric matrix of the coefficients, and $\mathbf{x}$ is the vector whose elements are the variables x and y.

EXAMPLE 19.13 Verify that

$$Q = 3x^2 + 2xy + y^2 = (x \; y) \begin{pmatrix} 3 & 1 \\ 1 & 1 \end{pmatrix} \begin{pmatrix} x \\ y \end{pmatrix}$$

We have $\begin{pmatrix} 3 & 1 \\ 1 & 1 \end{pmatrix} \begin{pmatrix} x \\ y \end{pmatrix} = \begin{pmatrix} 3x + y \\ x + y \end{pmatrix}$

and $\quad Q = (x \; y) \begin{pmatrix} 3x + y \\ x + y \end{pmatrix} = x(3x + y) + y(x + y) = 3x^2 + 2xy + y^2$

➤ Exercises 29, 30

The general quadratic form in n variables is

$$Q(x_1, x_2, x_3, \ldots, x_n) = \sum_{i=1}^{n} \sum_{j=1}^{n} a_{ij} x_i x_j$$

$$\begin{aligned} = \; & a_{11}x_1^2 && + a_{12}x_1x_2 && + a_{13}x_1x_3 && + \cdots + a_{1n}x_1x_n \\ & + a_{21}x_2x_1 && + a_{22}x_2^2 && + a_{23}x_2x_3 && + \cdots + a_{2n}x_2x_n \quad (19.37) \\ & + \cdots \\ & + a_{n1}x_nx_1 && + a_{n2}x_nx_2 && + a_{n3}x_nx_3 && + \cdots + a_{nn}x_n^2 \end{aligned}$$

in which the coefficients a_{ij} are real and $a_{ji} = a_{ij}$. In matrix form

$$Q(\mathbf{x}) = \mathbf{x}^T \mathbf{A} \mathbf{x} \qquad (19.38)$$

where

$$\mathbf{x} = \begin{pmatrix} x_1 \\ x_2 \\ x_3 \\ \vdots \\ x_n \end{pmatrix}, \qquad \mathbf{A} = \begin{pmatrix} a_{11} & a_{12} & a_{13} & \cdots & a_{1n} \\ a_{21} & a_{22} & a_{23} & \cdots & a_{2n} \\ a_{31} & a_{32} & a_{33} & \cdots & a_{3n} \\ \vdots & \vdots & \vdots & & \vdots \\ a_{n1} & a_{n2} & a_{n3} & \cdots & a_{nn} \end{pmatrix} \qquad (19.39)$$

and $\mathbf{A}$ is real and symmetric.

➤ Exercise 31

The canonical form

We have seen (equation (19.30)) that a symmetric matrix $\mathbf{A}$ is reduced to diagonal form by the similarity transformation $\mathbf{X}^T\mathbf{AX}$ where $\mathbf{X}$ is the orthogonal matrix whose columns are the orthonormal eigenvectors of $\mathbf{A}$. Because $\mathbf{XX}^T = \mathbf{I}$ (for orthogonal matrix $\mathbf{X}$), we can write (19.38) as

$$
\begin{aligned}
Q &= \mathbf{x}^T(\mathbf{XX}^T)\mathbf{A}(\mathbf{XX}^T)\mathbf{x} \\
&= (\mathbf{x}^T\mathbf{X})\,(\mathbf{X}^T\mathbf{AX})\,(\mathbf{X}^T\mathbf{x}) \\
&= \mathbf{y}^T\mathbf{Dy}
\end{aligned}
\tag{19.40}
$$

where $\mathbf{D} = \mathbf{X}^T\mathbf{AX}$ is the diagonal matrix of the eigenvalues of $\mathbf{A}$, and

$$
\mathbf{y} = \mathbf{X}^T\mathbf{x}
\tag{19.41}
$$

is the vector obtained from $\mathbf{x}$ by the orthogonal transformation $\mathbf{X}^T$.

The quadratic form Q in the n variables $x_1, x_2, x_3, \ldots, x_n$ has been transformed into an equivalent form in the n variables $y_1, y_2, y_3, \ldots, y_n$ that contains only pure square terms:

$$
Q(\mathbf{y}) = \sum_{k=1}^{n} \lambda_k y_k^2 = \lambda_1 y_1^2 + \lambda_2 y_2^2 + \lambda_3 y_3^2 + \cdots + \lambda_n y_n^2
\tag{19.42}
$$

This is the **canonical form** of Q, and the variables y_k are the **canonical variables.**[3]

EXAMPLE 19.14 Transform the following quadratic form into canonical form:

$$
Q = 5x_1^2 + 8x_1x_2 + 5x_2^2
$$

We have

$$
Q = (x_1 \ \ x_2) \begin{pmatrix} 5 & 4 \\ 4 & 5 \end{pmatrix} \begin{pmatrix} x_1 \\ x_2 \end{pmatrix} = \mathbf{x}^T\mathbf{Ax}
$$

and the orthonormal eigenvectors of the symmetric matrix $\mathbf{A}$ are

$$
\mathbf{x}_1 = \frac{1}{\sqrt{2}} \begin{pmatrix} 1 \\ -1 \end{pmatrix}, \quad \mathbf{x}_2 = \frac{1}{\sqrt{2}} \begin{pmatrix} 1 \\ 1 \end{pmatrix}
$$

---

[3] Cayley and Sylvester developed the theory of forms between 1854 and 1878. Sylvester claimed that he discovered and developed the reduction of a quadratic form to canonical form at one sitting 'with a decanter of port wine to sustain nature's flagging energies'. A general description of canonical forms was given by Camille Jordan (1838–1922) in his *Traité des substitutions et des équations algébriques* (Treatise on substitutions and algebraic equations) of 1871, in which he presented many of the modern concepts of group theory within the context of groups of permutations (substitutions).

corresponding to eigenvalues $\lambda_1 = 1$ and $\lambda_2 = 9$. The matrix of the eigenvectors is

$$\mathbf{X} = (\mathbf{x}_1 \quad \mathbf{x}_2) = \frac{1}{\sqrt{2}} \begin{pmatrix} 1 & 1 \\ -1 & 1 \end{pmatrix} \text{ with transpose } \mathbf{X}^T = \frac{1}{\sqrt{2}} \begin{pmatrix} 1 & -1 \\ 1 & 1 \end{pmatrix}$$

Then

$$Q = \mathbf{y}^T \mathbf{D} \mathbf{y} = \lambda_1 y_1^2 + \lambda_2 y_2^2 = y_1^2 + 9 y_2^2$$

where

$$\mathbf{y} = \mathbf{X}^T \mathbf{x} = \frac{1}{\sqrt{2}} \begin{pmatrix} 1 & -1 \\ 1 & 1 \end{pmatrix} \begin{pmatrix} x_1 \\ x_2 \end{pmatrix}, \quad \begin{pmatrix} y_1 \\ y_2 \end{pmatrix} = \frac{1}{\sqrt{2}} \begin{pmatrix} x_1 - x_2 \\ x_1 + x_2 \end{pmatrix}$$

➤ Exercises 32–34

The transformation of a quadratic form into canonical form is also called a **transformation to principal axes**, and has numerous applications in geometry and in the physical sciences. For example, the quadratic form of Example 19.14,

$$Q = 5x_1^2 + 8x_1 x_2 + 5x_2^2$$

represents a family of ellipses in the $x_1 x_2$-plane for positive values of Q. The ellipse for $Q = 9$ is shown in Figure 19.2(a).

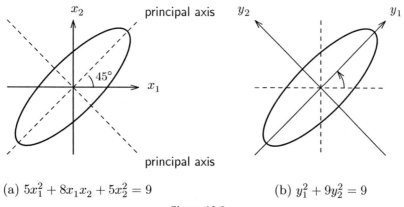

(a) $5x_1^2 + 8x_1 x_2 + 5x_2^2 = 9$ (b) $y_1^2 + 9y_2^2 = 9$

Figure 19.2

The principal axes of the ellipse lie at 45° to the x_1 and x_2 axes, and the transformation to the canonical form

$$Q = y_1^2 + 9 y_2^2$$

then corresponds to a rotation of the coordinate system, from (x_1, x_2) to (y_1, y_2), to bring the coordinate axes into coincidence with the principal axes of the ellipse, as shown in Figure 19.2(b) for $Q = 9$.

EXAMPLE 19.15 The inertia tensor

A body rotating about an axis through its centre of mass with angular velocity ω has kinetic energy of rotation given by the quadratic form

$$T = \frac{1}{2}\omega^T I \omega = \frac{1}{2}(\omega_x \ \omega_y \ \omega_z) \begin{pmatrix} I_{xx} & I_{xy} & I_{xz} \\ I_{yx} & I_{yy} & I_{yz} \\ I_{zx} & I_{zy} & I_{zz} \end{pmatrix} \begin{pmatrix} \omega_x \\ \omega_y \\ \omega_z \end{pmatrix} \qquad (19.43)$$

in which the symmetric matrix I (not to be confused with the unit matrix) is called the **moment of inertia tensor** or, simply, the **inertia tensor**. For example, a mass m at position (x, y, z) has the moment of inertia tensor with components (see Exercise 35)

$$I_{xx} = m(y^2 + z^2), \qquad I_{yy} = m(z^2 + x^2), \qquad I_{zz} = m(x^2 + y^2)$$

$$I_{xy} = -mxy, \qquad I_{yz} = -myz, \qquad I_{zx} = -mzx \qquad (19.44)$$

A principal-axis transformation is that coordinate transformation (x, y, z) to (x', y', z') that brings the coordinate axes into coincidence with the principal axes of inertia of the body. The inertia tensor is diagonal in the new coordinate system, and the kinetic energy of rotation is

$$T = \frac{1}{2}(\omega_{x'} \ \omega_{y'} \ \omega_{z'}) \begin{pmatrix} I_{x'x'} & 0 & 0 \\ 0 & I_{y'y'} & 0 \\ 0 & 0 & I_{z'z'} \end{pmatrix} \begin{pmatrix} \omega_{x'} \\ \omega_{y'} \\ \omega_{z'} \end{pmatrix} \qquad (19.45)$$

$$= \frac{1}{2} I_{x'x'} \omega_{x'}^2 + \frac{1}{2} I_{y'y'} \omega_{y'}^2 + \frac{1}{2} I_{z'z'} \omega_{z'}^2$$

and is the sum of contributions from rotations about the x', y', and z' axes.

Angular momentum

The angular momentum of a rotating body is related to the angular velocity by (see Example 16.18)

$$l = I\omega \qquad (19.46)$$

or, in terms of components,

$$l_x = I_{xx}\omega_x + I_{xy}\omega_y + I_{xz}\omega_z$$
$$l_y = I_{yx}\omega_x + I_{yy}\omega_y + I_{yz}\omega_z \qquad (19.47)$$
$$l_z = I_{zx}\omega_x + I_{zy}\omega_y + I_{zz}\omega_z$$

When the coordinate axes coincide with the principal axes of inertia, $\mathbf{I}$ is diagonal and equations (19.47) reduce to

$$l_x = I_{xx}\omega_x, \qquad l_y = I_{yy}\omega_y, \qquad l_z = I_{zz}\omega_z \qquad (19.48)$$

The kinetic energy of rotation then has the familiar form

$$T = \frac{l_x^2}{2I_{xx}} + \frac{l_y^2}{2I_{yy}} + \frac{l_z^2}{2I_{zz}} \qquad (19.49)$$

that is used in theoretical discussions of the microwave spectroscopy of polyatomic molecules.

▸ Exercise 35

19.6 Complex matrices

Much of the earlier discussion applies equally well to complex matrices as to real matrices and, in this section, we summarize the more important properties of complex matrices, and the ways in which these differ from those of real matrices.

A matrix $\mathbf{A}$ whose elements are complex numbers is called a **complex matrix**, and can be written in the form

$$\mathbf{A} = \mathbf{B} + i\mathbf{C} \qquad (19.50)$$

where $i = \sqrt{-1}$, and $\mathbf{B}$ and $\mathbf{C}$ are real matrices.

The complex conjugate matrix A*

The complex conjugate matrix $\mathbf{A}^*$ is obtained from $\mathbf{A}$ by replacing each element of $\mathbf{A}$ by its complex conjugate:

$$\text{if} \quad \mathbf{A} = (a_{ij}) \quad \text{then} \quad \mathbf{A}^* = (a_{ij}^*) \qquad (19.51)$$

and

$$\mathbf{A}^* = \mathbf{B} - i\mathbf{C} \qquad (19.52)$$

For a *real* matrix, $\mathbf{A}^* = \mathbf{A}$ and $\mathbf{C} = \mathbf{0}$.

The Hermitian conjugate matrix $\mathbf{A}^\dagger$

The Hermitian conjugate of $\mathbf{A}$ is the transpose of the complex conjugate,

$$\mathbf{A}^\dagger = (\mathbf{A}^*)^\mathsf{T} = (\mathbf{A}^\mathsf{T})^* \tag{19.53}$$

Thus, for order 3,

$$\text{if} \quad \mathbf{A} = \begin{pmatrix} a_{11} & a_{12} & a_{13} \\ a_{21} & a_{22} & a_{23} \\ a_{31} & a_{32} & a_{33} \end{pmatrix} \quad \text{then} \quad \mathbf{A}^\dagger = \begin{pmatrix} a_{11}^* & a_{21}^* & a_{31}^* \\ a_{12}^* & a_{22}^* & a_{32}^* \\ a_{13}^* & a_{23}^* & a_{33}^* \end{pmatrix}$$

The Hermitian conjugate is also called the **conjugate transpose matrix**, the **associate matrix**, and (in quantum mechanics) the **adjoint** matrix (not to be confused with the matrix of the same name discussed in Section 18.4).

EXAMPLE 19.16 Find the Hermitian conjugate of the matrix

$$\mathbf{A} = \begin{pmatrix} 3+i & 2+3i & -1 \\ 1-2i & 2 & 2i \end{pmatrix}$$

The complex conjugate $\mathbf{A}^*$ and Hermitian conjugate $\mathbf{A}^\dagger$ are

$$\mathbf{A}^* = \begin{pmatrix} 3-i & 2-3i & -1 \\ 1+2i & 2 & -2i \end{pmatrix}, \quad \mathbf{A}^\dagger = \begin{pmatrix} 3-i & 1+2i \\ 2-3i & 2 \\ -1 & -2i \end{pmatrix}$$

▸ Exercises 36–38

The Hermitian conjugate plays the same role for complex matrices as does the transpose for real matrices. For example, the inner (scalar) product of two vectors $\boldsymbol{a} = (a_1, a_2, a_3)$ and $\boldsymbol{b} = (b_1, b_2, b_3)$ in a complex vector space is defined as (compare equation (18.62))

$$\boldsymbol{a} \cdot \boldsymbol{b} = \mathbf{a}^\dagger \mathbf{b} = \begin{pmatrix} a_1^* & a_2^* & a_3^* \end{pmatrix} \begin{pmatrix} b_1 \\ b_2 \\ b_3 \end{pmatrix} = a_1^* b_1 + a_2^* b_2 + a_3^* b_3 \tag{19.54}$$

The inner product is complex in general, but the quantity

$$\boldsymbol{a} \cdot \boldsymbol{a} = \mathbf{a}^\dagger \mathbf{a} = \begin{pmatrix} a_1^* & a_2^* & a_3^* \end{pmatrix} \begin{pmatrix} a_1 \\ a_2 \\ a_3 \end{pmatrix} = a_1^* a_1 + a_2^* a_2 + a_3^* a_3 \qquad (19.55)$$

$$= |a_1|^2 + |a_2|^2 + |a_3|^2$$

is a positive real number, and defines the length, or norm, of the vector, $|\boldsymbol{a}| = \sqrt{\mathbf{a}^\dagger \mathbf{a}}$.

EXAMPLE 19.17 For the vectors $\boldsymbol{a} = (1 + i, 3, 2i)$ and $\boldsymbol{b} = (2 - i, 4i, -1)$,

$$\mathbf{a}^\dagger \mathbf{a} = \begin{pmatrix} 1-i & 3 & -2i \end{pmatrix} \begin{pmatrix} 1+i \\ 3 \\ 2i \end{pmatrix} = (1-i)(1+i) + 3^2 + (-2i)(2i) = 2 + 9 + 4 = 15$$

$$\mathbf{b}^\dagger \mathbf{b} = \begin{pmatrix} 2+i & -4i & -1 \end{pmatrix} \begin{pmatrix} 2-i \\ 4i \\ -1 \end{pmatrix} = (2+i)(2-i) + (-4i)(4i) + 1 = 5 + 16 + 1 = 22$$

$$\mathbf{a}^\dagger \mathbf{b} = \begin{pmatrix} 1-i & 3 & -2i \end{pmatrix} \begin{pmatrix} 2-i \\ 4i \\ -1 \end{pmatrix} = (1-i)(2-i) + 3(4i) + (-2i)(-1) = 1 + 11i$$

‣ Exercise 39

A vector divided by its norm is a unit vector, and two vectors are orthogonal when their inner product is zero. The complex analogue of an orthonormal system of vectors is then

$$\mathbf{a}_k^\dagger \mathbf{a}_l = \begin{cases} 1 & \text{if } k = l \\ 0 & \text{if } k \neq l \end{cases} \qquad (19.56)$$

Such a system of vectors is called **unitary**.

Hermitian matrices

A complex square matrix that is equal to its Hermitian conjugate is called a Hermitian matrix,

$$\mathbf{A}^\dagger = \mathbf{A} \quad \text{(Hermitian matrix)} \qquad (19.57)$$

A real Hermitian matrix is a symmetric matrix.

EXAMPLE 19.18 Show that the following matrix is Hermitian:

$$\mathbf{A} = \begin{pmatrix} 3 & 2+3i \\ 2-3i & 1 \end{pmatrix}$$

We have

$$\mathbf{A}^* = \begin{pmatrix} 3 & 2-3i \\ 2+3i & 1 \end{pmatrix}, \qquad \mathbf{A}^\dagger = (\mathbf{A}^*)^\mathrm{T} = \begin{pmatrix} 3 & 2+3i \\ 2-3i & 1 \end{pmatrix} = \mathbf{A}$$

▸ Exercise 40

Unitary matrices

A complex square matrix $\mathbf{U}$ is called **unitary** when its Hermitian conjugate is equal to its inverse,

$$\mathbf{U}^\dagger = \mathbf{U}^{-1} \tag{19.58}$$

The characteristic property of a unitary matrix is that its columns (and its rows) form a unitary system of orthonormal vectors as defined by (19.56) (compare the discussion of orthogonal matrices in Section 18.6). For order 3, let

$$\mathbf{U} = (\mathbf{a}\ \mathbf{b}\ \mathbf{c}) = \begin{pmatrix} a_1 & b_1 & c_1 \\ a_2 & b_2 & c_2 \\ a_3 & b_3 & c_3 \end{pmatrix} \tag{19.59}$$

where $\mathbf{a}^\dagger \mathbf{a} = \mathbf{b}^\dagger \mathbf{b} = \mathbf{c}^\dagger \mathbf{c} = 1$ and $\mathbf{a}^\dagger \mathbf{b} = \mathbf{b}^\dagger \mathbf{c} = \mathbf{c}^\dagger \mathbf{a} = 0$. The Hermitian conjugate of $\mathbf{U}$ is

$$\mathbf{U}^\dagger = \begin{pmatrix} \mathbf{a}^\dagger \\ \mathbf{b}^\dagger \\ \mathbf{c}^\dagger \end{pmatrix} = \begin{pmatrix} a_1^* & a_2^* & a_3^* \\ b_1^* & b_2^* & b_3^* \\ c_1^* & c_2^* & c_3^* \end{pmatrix} \tag{19.60}$$

and

$$\mathbf{U}^\dagger \mathbf{U} = \begin{pmatrix} \mathbf{a}^\dagger \\ \mathbf{b}^\dagger \\ \mathbf{c}^\dagger \end{pmatrix} (\mathbf{a}\ \mathbf{b}\ \mathbf{c}) = \begin{pmatrix} \mathbf{a}^\dagger \mathbf{a} & \mathbf{a}^\dagger \mathbf{b} & \mathbf{a}^\dagger \mathbf{c} \\ \mathbf{b}^\dagger \mathbf{a} & \mathbf{b}^\dagger \mathbf{b} & \mathbf{b}^\dagger \mathbf{c} \\ \mathbf{c}^\dagger \mathbf{a} & \mathbf{c}^\dagger \mathbf{b} & \mathbf{c}^\dagger \mathbf{c} \end{pmatrix} = \begin{pmatrix} 1 & 0 & 0 \\ 0 & 1 & 0 \\ 0 & 0 & 1 \end{pmatrix} = \mathbf{I} \tag{19.61}$$

▸ Exercises 41, 42

A real unitary matrix is an orthogonal matrix. Hermitian and unitary matrices play the same roles for complex matrices as do symmetric and orthogonal matrices for real matrices, and they have the same important properties:

(i) The eigenvalues of a Hermitian matrix are real.
(ii) The eigenvectors of a Hermitian matrix of order n form (or can be chosen to form) a unitary system of n orthonormal vectors.
(iii) A Hermitian matrix $\mathbf{A}$ is reduced to diagonal form by means of the similarity (unitary) transformation

$$\mathbf{D} = \mathbf{U}^{\dagger}\mathbf{A}\mathbf{U} \qquad (19.62)$$

where $\mathbf{U}$ is the unitary matrix whose columns are the eigenvectors of $\mathbf{A}$, and $\mathbf{D}$ is the real diagonal matrix whose diagonal elements are the eigenvalues of $\mathbf{A}$.

19.7 Exercises

Section 19.1

Find the inverse of the matrix of the coefficients, and use it to solve the equations:

1. $2x - 3y = 8$
$4x + y = 2$

2. $x + y + z = 6$
$x + 2y + 3z = 14$
$x + 4y + 9z = 36$

3. $w + x + z = 2$
$ x + y + z = 6$
$w + y + z = 3$
$w + x + y = 4$

Section 19.2

Find the eigenvalues and eigenvectors of the following matrices:

4. $\begin{pmatrix} 2 & 2 \\ 1 & 3 \end{pmatrix}$
5. $\begin{pmatrix} 2 & 0 \\ 0 & -3 \end{pmatrix}$
6. $\begin{pmatrix} 3 & 1 \\ -1 & 3 \end{pmatrix}$
7. $\begin{pmatrix} 3 & 1 \\ 1 & 3 \end{pmatrix}$

8. $\begin{pmatrix} 1 & 2 & 0 \\ 2 & 1 & 0 \\ 0 & 2 & 1 \end{pmatrix}$
9. $\begin{pmatrix} 0 & 3 & 0 \\ 3 & 0 & 3 \\ 0 & 3 & 0 \end{pmatrix}$
10. $\begin{pmatrix} 3 & 4 \\ -4 & -5 \end{pmatrix}$
11. $\begin{pmatrix} 4 & 0 & 2 \\ -6 & 1 & -4 \\ -6 & 0 & -3 \end{pmatrix}$

Section 19.3

Normalize the eigenvectors obtained in
12. Exercise 5   **13.** Exercise 7   **14.** Exercise 8   **15.** Exercise 9
Show that the sets of eigenvectors of the symmetric matrices are orthogonal in
16. Exercise 5   **17.** Exercise 7   **18.** Exercise 9
19. Given the three vectors

$$\mathbf{x}_1 = \begin{pmatrix} 1 \\ 1 \\ 1 \end{pmatrix}, \quad \mathbf{x}_2 = \begin{pmatrix} 3 \\ 1 \\ 2 \end{pmatrix}, \quad \mathbf{x}_3 = \begin{pmatrix} 3 \\ 2 \\ 1 \end{pmatrix},$$

use Schmidt orthogonalization to **(i)** find new vectors $\mathbf{x}_2'$ and $\mathbf{x}_3'$ that are orthogonal to $\mathbf{x}_1$, **(ii)** find the new vector $\mathbf{x}_3''$ that is orthogonal to both $\mathbf{x}_1$ and $\mathbf{x}_2'$.

20. The Hückel Hamiltonian matrix of butadiene is

$$\begin{pmatrix} \alpha & \beta & 0 & 0 \\ \beta & \alpha & \beta & 0 \\ 0 & \beta & \alpha & \beta \\ 0 & 0 & \beta & \alpha \end{pmatrix}$$

Find **(i)** the eigenvalues (in terms of α and β), **(ii)** the orthonormal eigenvectors. You may find the following relations useful: $\phi = (\sqrt{5}+1)/2$ (the 'golden section', see Section 7.2), $\phi^2 = (\sqrt{5}+3)/2$, $\phi - 1 = (\sqrt{5}-1)/2 = 1/\phi$, $\phi^2 - 1 = \phi$, $\phi^2 - \phi = 1$.

21. The Hückel Hamiltonian matrix of cyclopropene is

$$\begin{pmatrix} \alpha & \beta & \beta \\ \beta & \alpha & \beta \\ \beta & \beta & \alpha \end{pmatrix}$$

(i) Show that the eigenvalues are $E_1 = \alpha + 2\beta$, $E_2 = E_3 = \alpha - \beta$. **(ii)** Find the normalized eigenvector belonging to eigenvalue E_1. **(iii)** Show that an eigenvector belonging to the doubly-degenerate eigenvalue $\alpha - \beta$ has components x_1, x_2, x_3 that satisfy $x_1 + x_2 + x_3 = 0$. **(iv)** Find two orthonormal eigenvectors corresponding to eigenvalue $\alpha - \beta$ (you may find the results of Examples 19.4(ii) and 19.8 useful).

22. (i) Show that a square matrix $\mathbf{A}$ and its transpose $\mathbf{A}^T$ have the same set of eigenvalues.

(ii) Show that the following two equations are equivalent:

$$\mathbf{A}^T \mathbf{y} = \lambda \mathbf{y}, \qquad \mathbf{y}^T \mathbf{A} = \lambda \mathbf{y}^T$$

The eigenvectors of $\mathbf{A}^T$ are in general different from those of $\mathbf{A}$ (unless $\mathbf{A}$ is symmetric). The vector $\mathbf{y}$ is sometimes called a **left-eigenvector** of $\mathbf{A}$, and an 'ordinary' eigenvector $\mathbf{x}$ of $\mathbf{A}$ is then called a **right-eigenvector**.

(iii) Find the eigenvalues and corresponding normalized right- and left-eigenvectors of

$$\mathbf{A} = \begin{pmatrix} 3 & 2 \\ 0 & 2 \end{pmatrix}.$$

Section 19.4

23. For the matrix

$$\mathbf{A} = \begin{pmatrix} 1 & 2 & 0 \\ 2 & 1 & 0 \\ 0 & 2 & 1 \end{pmatrix}$$

of Exercise 8, construct **(i)** the matrix $\mathbf{X}$ of the eigenvectors and **(ii)** the diagonal matrix $\mathbf{D}$ of the eigenvalues. **(iii)** Show that $\mathbf{AX} = \mathbf{DX}$.

24. Repeat Exercise 23 for the matrix $\begin{pmatrix} 0 & 3 & 0 \\ 3 & 0 & 3 \\ 0 & 3 & 0 \end{pmatrix}$ of Exercise 9.

25. For the matrix $\mathbf{A} = \begin{pmatrix} 3 & 1 \\ 1 & 3 \end{pmatrix}$ of Exercise 7,

(i) construct the matrix $\mathbf{X}$ of the eigenfunctions of $\mathbf{A}$, and find its inverse, $\mathbf{X}^{-1}$,

(ii) calculate $\mathbf{D} = \mathbf{X}^{-1}\mathbf{AX}$ and confirm that $\mathbf{D}$ is the diagonal matrix of the eigenvalues of $\mathbf{A}$.

Repeat Exercise 25 for:

26. $\mathbf{A} = \begin{pmatrix} 2 & 2 \\ 1 & 3 \end{pmatrix}$ of Exercise 4   **27.** $\mathbf{A} = \begin{pmatrix} 1 & 2 & 0 \\ 2 & 1 & 0 \\ 0 & 2 & 1 \end{pmatrix}$ of Exercise 8

28. $\mathbf{A} = \begin{pmatrix} 0 & 3 & 0 \\ 3 & 0 & 3 \\ 0 & 3 & 0 \end{pmatrix}$ of Exercise 9

Section 19.5

Express in matrix form:

29. $5x^2 - 2xy - 3y^2$ **30.** $4xy$ **31.** $3x^2 - 4xy + 2xz - 6yz + y^2 - 2z^2$

Transform the following quadratic forms into canonical form:

32. $7x_1^2 + 6\sqrt{3}x_1x_2 + 13x_2^2$ **33.** $ax^2 + 2bxy + ay^2$

34. $3x_1^2 + 2x_1x_2 + 2x_1x_4 + 3x_2^2 + 2x_2x_3 + 3x_3^2 + 2x_3x_4 + 3x_4^2$

35. Derive equations (19.44) for the components of the inertia tensor.
[Hint: Expand equation (16.60), $\mathbf{l} = mr^2\boldsymbol{\omega} - m(\mathbf{r} \cdot \boldsymbol{\omega})\mathbf{r}$, for the angular momentum in terms of components.]

Section 19.6

Find the complex conjugate and Hermitian conjugate of the following matrices

36. $\begin{pmatrix} 1+i & 2-i \\ 3+i & -i \end{pmatrix}$ **37.** $\begin{pmatrix} 2 & i \\ -i & 1 \end{pmatrix}$ **38.** $\begin{pmatrix} 0 & -i & 0 \\ i & 0 & -i \\ 0 & i & 0 \end{pmatrix}$

39. If $\mathbf{a} = \begin{pmatrix} i \\ 1 \\ -i \end{pmatrix}$ and $\mathbf{b} = \begin{pmatrix} 2i \\ 0 \\ 3 \end{pmatrix}$, find **(i)** $\mathbf{a}^\dagger\mathbf{a}$ **(ii)** $\mathbf{b}^\dagger\mathbf{b}$ **(iii)** $\mathbf{a}^\dagger\mathbf{b}$ **(iv)** $\mathbf{b}^\dagger\mathbf{a}$

40. Which of the matrices in Exercise 36–38 are Hermitian?

41. (i) Show that $\mathbf{A} = \begin{pmatrix} 1/\sqrt{2} & i/\sqrt{2} \\ -i/\sqrt{2} & -1/\sqrt{2} \end{pmatrix}$ is unitary.

(ii) Confirm that both the columns and the rows of **A** form unitary systems of vectors.

42. Repeat Exercise 41 for $\mathbf{A} = \begin{pmatrix} i/\sqrt{3} & i/\sqrt{2} & i/\sqrt{6} \\ i/\sqrt{3} & -i/\sqrt{2} & i/\sqrt{6} \\ i/\sqrt{3} & 0 & -2i/\sqrt{6} \end{pmatrix}$.

20 Numerical methods

20.1 Concepts

A numerical method is a method for obtaining the solution of a mathematical problem in the form of numbers. In many cases the problem can be solved 'exactly' in terms of known functions by means of the analytical methods discussed in the earlier chapters of this book. Numerical methods are then required only if a solution is required in numerical form. Many other problems, on the other hand, cannot be solved analytically, and numerical methods are required for the process of solution itself. Because numerical methods are commonly implemented on computers, they are sometimes described as 'methods for solving problems on a computer'. The development of computers over the past fifty years has been accompanied by the development of sophisticated numerical procedures for the solution of a wide range of problems in the physical sciences, in engineering and in statistics. These procedures have been collected in 'mathematical libraries' or 'numerical packages', available on most standard computers. In this chapter we discuss the general principles underlying some of the more important numerical methods, and treat in detail only the very simplest.

Nearly all numerical operations are necessarily accompanied by errors, and the analysis of these errors is an integral part of any numerical method.

20.2 Errors

Errors in numerical computations arise largely in three ways.

1. **Mistakes.** These include errors in the carrying out of a numerical operation or an error ('bug') in a computer program. Such errors are in principle removable by checking. In the experimental sciences, removable errors also arise from faulty or incorrectly calibrated apparatus. *Mistakes* are not counted as *errors* in numerical analysis.
2. **Mathematical truncation errors.** These are due to the use of approximate representations of functions; for example, as discussed in Section 7.7, when a Taylor series expansion of a function is truncated after a given number of terms. Taylor's theorem then provides a method of determining the error bounds of the result.

   Many numerical methods have their origin in the Taylor expansion and are therefore necessarily accompanied by truncation errors; numerical methods are **approximate methods**.
3. **Rounding errors.** As discussed in Section 1.4, numbers are in practice nearly always approximated by rounding, either to a given number of decimal places in the fixed-point representation, or to a given number of significant figures in the floating-point representation. It is the consequences of this type of error that we consider in this section.

Because of the conventions for rounding (Section 1.4) a number like $a = 1.234$, obtained by rounding to three decimal places, represents all the numbers between

1.2335 and 1.2345. The number a is said to have a maximum absolute error or (**absolute**) **error bound** $\varepsilon_a = 0.0005$, and the bound can be indicated by writing

$$a \pm \varepsilon_a = 1.2340 \pm 0.0005$$

➤ Exercises 1, 2

In fixed-point arithmetic all numbers are rounded to the same number of decimal places, and therefore have the same error bound, ε say. The sum or difference of two numbers then has error bound 2ε. More generally, the sum or difference of numbers a and b, with error bounds ε_a and ε_b, has error bound $\varepsilon_a + \varepsilon_b$.

EXAMPLE 20.1 The numbers $a = 1.234$ and $b = 3.468$ have been rounded to 4 significant figures. Determine the error bound for $a - b$.

The numbers a and b both have error bound $\varepsilon_a = 0.0005$. The largest and smallest possible values of $a - b$ are therefore

$(a + \varepsilon) - (b - \varepsilon) = (a - b) + 2\varepsilon$ or $1.2345 - 3.4675 = -2.2330 = -2.234 + 0.001$

$(a - \varepsilon) - (b + \varepsilon) = (a - b) - 2\varepsilon$ or $1.2335 - 3.4685 = -2.2350 = -2.234 - 0.001$

and the error bound of the difference is $2\varepsilon = 0.001$. Therefore $a - b = -2.234 \pm 0.001$

➤ Exercise 3

For multiplication and division the error propagation is best described in terms of the relative error. A number whose value is given by a and whose true value is a_t has **relative error**

$$r_a = \left| \frac{a - a_t}{a_t} \right| = \frac{\varepsilon_a}{|a_t|} \tag{20.1}$$

When the error is small enough, a_t in the denominator can be replaced by a and (20.1) by $r_a = \varepsilon_a / |a|$.

➤ Exercise 4

The relative error of the product or quotient of a and b, with relative errors r_a and r_b, is $r_a + r_b$. Thus, for multiplication, the largest and smallest possible values of $p = a \times b$ are

$$(a + \varepsilon_a)(b + \varepsilon_b) = ab + a\varepsilon_b + b\varepsilon_a + \varepsilon_a \varepsilon_b$$

$$(a - \varepsilon_a)(b - \varepsilon_b) = ab - a\varepsilon_b - b\varepsilon_a + \varepsilon_a \varepsilon_b$$

The quantity $\varepsilon_a \varepsilon_b$ can be ignored for small enough errors, and the error bound of the product is $\varepsilon_p = a\varepsilon_b + b\varepsilon_a$. Division by $p = ab$ then gives

$$r_p = \frac{\varepsilon_p}{p} = \frac{\varepsilon_a}{a} + \frac{\varepsilon_b}{b} = r_a + r_b \qquad (20.2)$$

EXAMPLE 20.2 If $a = 12.35$ and $b = 2.345$ have been rounded to 4 figures, determine the error bounds for $q = a/b$.

We have $12.35/2.345 = 5.26652$ (to six figures). The absolute and relative error bounds for a are $\varepsilon_a = 0.005$ and $r_a = \varepsilon_a/a \approx 0.0004$; and for b they are $\varepsilon_b = 0.0005$ and $r_b = \varepsilon_b/b \approx 0.0002$. We expect the quotient to have $r_q = r_a + r_b \approx 0.0006$. Thus the largest and smallest values of the quotient are (with results quoted to 6 figures)

$$\frac{12.355}{2.3445} = 5.26978 = 5.26652 + 0.00326,$$

$$\frac{12.345}{2.3455} = 5.26327 = 5.26652 - 0.00325$$

Then $r_q \approx 0.0033/5.26652 \approx 0.0006$, as expected from equation (20.2). This is the relative error for the unrounded product.

The answer may be written as 5.2665 ± 0.0033.

‣ Exercise 5

The bounds computed in the ways described above are examples of **worst-case bounds**. In practice, particularly when floating point arithmetic is used, there is nearly always some cancellation of errors, and the results of sequences of arithmetic operations are expected to have smaller actual errors. Rounding errors are examples of the 'random errors' discussed in Chapter 21 and are amenable to statistical analysis.

Differencing errors

By far the most severe errors arising from the use of rounded numbers occur when two almost equal numbers are subtracted. Thus, $a = 1.234$ and $b = 1.233$, rounded to 4 figures, have $a - b = 0.001$ with error bound 0.001 $(a - b = 0.001 \pm 0.001)$. Differencing errors can however often be minimized by the use of an appropriate numerical procedure (**algorithm**).

EXAMPLE 20.3 Find the solutions of the quadratic equation $x^2 - 36x + 2 = 0$ using 4-figure arithmetic.

The solutions of the general quadratic equation $ax^2 + bx + c = 0$ are

$$x_1 = -\frac{b}{2a} + \sqrt{\left(\frac{b}{2a}\right)^2 - \left(\frac{c}{a}\right)}, \qquad x_2 = -\frac{b}{2a} - \sqrt{\left(\frac{b}{2a}\right)^2 - \left(\frac{c}{a}\right)} \qquad (20.3a)$$

so that, in the present case,

$$x_1 = 18 + \sqrt{322} = 18.00 + 17.94 = 35.94$$

$$x_2 = 18 - \sqrt{322} = 18.00 - 17.94 = 0.06$$

and x_2 has only one significant figure.

However, because the roots of a quadratic have the property $x_1 x_2 = c/a$, an alternative formula for the second root is $x_2 = c/x_1 a$,

$$x_2 = \frac{2.000}{35.94} = 0.05565$$

with error of 1 in the least significant figure.

This example shows that the correct procedure for the solution of a quadratic equation, *irrespective of the number of significant figures used*, is

$$\text{if} \quad b < 0 \quad \text{then} \quad x_1 = -\frac{b}{2a} + \sqrt{\left(\frac{b}{2a}\right)^2 - \left(\frac{c}{a}\right)}, \quad x_2 = \frac{c}{x_1 a}$$

$$\text{(20.3b)}$$

$$\text{if} \quad b > 0 \quad \text{then} \quad x_1 = -\frac{b}{2a} - \sqrt{\left(\frac{b}{2a}\right)^2 - \left(\frac{c}{a}\right)}, \quad x_2 = \frac{c}{x_1 a}$$

▸ Exercise 6

EXAMPLE 20.4 Table 20.1 shows some results obtained for the function $f(x) = (1 - \cos x)/x^2$ with a 10-digit pocket calculator.

Table 20.1 Values of $f(x) = (1 - \cos x)/x^2$

x	Computed $f(x)$	True $f(x)$
0.1	0.4995 8346 3	0.4995 8347 22
0.01	0.4999 95	0.4999 9583 33
0.001	0.4999	0.4999 9995 83
0.0001	0.49	0.4999 9999 96
0.00001	0	0.5000 0000 00

An alternative, and correct, procedure to eliminate the differencing errors is to use the truncated MacLaurin expansion of the function,

$$\frac{1 - \cos x}{x^2} \approx \frac{1}{2!} - \frac{x^2}{4!} + \frac{x^4}{6!} \quad \text{for} \quad |x| \le 0.1$$

which, by Taylor's theorem, has error bound 2.5×10^{-11} for $|x| \le 0.1$.

▸ Exercises 7, 8

20.3   Solution of ordinary equations

An ordinary equation (an equation not involving derivatives or integrals) in one variable can be written in the form

$$f(x) = 0 \qquad (20.4)$$

where $f(x)$ is a function of x. The solutions of the equation are those values of x for which (20.4) is true; they are the **zeros** of the function $f(x)$. For example, a quadratic equation is a special case of the algebraic (polynomial) equation

$$f(x) = a_0 + a_1 x + a_2 x^2 + \cdots + a_n x^n = 0 \qquad (20.5)$$

and the solutions are the roots of the polynomial. The general formula for the roots of the quadratic function was discussed in Chapter 2 and Example 20.3 but, although exact formulas exist for the roots of cubics and quartics, the general algebraic equation must be solved numerically. Similarly, the solutions of a transcendental equation (one that involves transcendental functions) such as

$$e^{-x} = \tan x \quad \text{or} \quad f(x) = e^{-x} - \tan x = 0 \qquad (20.6)$$

cannot normally be written in terms of a finite number of known functions, and the equation must be solved numerically.

All numerical methods of solving equations proceed by **iteration** whereby, given an initial approximate solution, x_0 say, an algorithm exists that uses x_0 to give a new, hopefully more accurate solution x_1. The same algorithm is then applied to x_1 to give x_2, and so on:

$$x_n \xrightarrow[\text{algorithm}]{\text{numerical}} x_{n+1}, \quad n = 0, 1, 2, 3 \ldots \qquad (20.7)$$

The iterative process is terminated when a given accuracy has been achieved; for example, when the change $|x_{n+1} - x_n|$ is less than a predetermined value. The success of a numerical procedure often depends on a good choice of initial approximation, and this can usually be obtained from a graph of the function or from a tabulation of values of the function.

We consider here two simple methods that are widely used, and that also form the basis for more sophisticated methods. We note that methods for equations in one variable can be generalized for several variables and, sometimes, for systems of simultaneous equations.

The bisection method

The starting point for this ancient method are two values of x for which the function is known to lie on opposite sides of zero. As in Figure 20.1, let $f(x_1) < 0$ and $f(x_2) > 0$, and let the function be continuous in the interval x_1 to x_2. We evaluate the function at the midpoint of the interval,

$$x_3 = \frac{1}{2}(x_1 + x_2) \qquad (20.8)$$

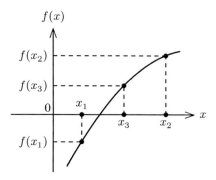

Figure 20.1

Then, if $f(x_3) > 0$, so that the zero lies between x_1 and x_3 as in the figure, repeat for the interval x_1 to x_3, and so on.

EXAMPLE 20.5 Bisection method for $\sqrt{5}$

To find $\sqrt{5}$, write $f(x) = x^2 - 5 = 0$. The root lies between $x_< = 2$ and $x_> = 3$, with $f(x_<) = -1$ and $f(x_>) = 4$. Then $x_3 = (x_< + x_>)/2 = 2.5$. Table 20.2 shows the computation to 4 significant figures on a standard pocket calculator.

Table 20.2 To find the square root of 5 correct to 4 significant figures

n	0	1	2	3	4	5
$x_<$	2	2	2	2.125	2.1875	2.21875
$x_>$	3	2.5	2.25	2.25	2.25	2.25
$\frac{1}{2}(x_< + x_>)$	2.5	2.25	2.125	2.1875	2.21875	2.23438

n	6	7	8	9	10	11
$x_<$	2.23438	2.23438	2.23438	2.23438	2.23535	2.23584
$x_>$	2.25	2.24219	2.23828	2.23633	2.23633	2.23633
$\frac{1}{2}(x_< + x_>)$	2.24219	2.23828	2.23633	2.23535	2.23584	

The values of $x_<$ and $x_>$ after 11 iterations both round to 2.236, and this is the required value $(2.236^2 = 4.999696)$.

▸ Exercises 9–11

The bisection method is an example of a **bracketing method**, with the zero known to lie in an interval of decreasing size. In the present case, the size of the interval is *halved* at every iteration. The convergence is slow, but the method cannot fail. It is the method of last resort when all else fails.

The Newton–Raphson method

The Newton–Raphson method, also called simply Newton's method, is the most famous method of finding the zeros of a function.[1] Unlike the bisection method, it requires the evaluation of the derivative $f'(x)$ as well as the function $f(x)$ itself. The method is illustrated in Figure 20.2.

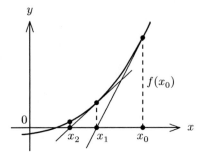

Figure 20.2

Given an approximate solution x_0 of $f(x)=0$, the tangent to the curve at x_0 is extended until it crosses the x-axis, at point x_1. The gradient of the tangent is

$$f'(x_0) = \frac{f(x_0)}{x_0 - x_1} \tag{20.9}$$

so that

$$x_1 = x_0 - \frac{f(x_0)}{f'(x_0)} \tag{20.10}$$

Then x_1 is the new estimate of the zero. The process is repeated with x_0 replaced by x_1 to give x_2, and so on:

$$x_{n+1} = x_n - \frac{f(x_n)}{f'(x_n)} \qquad n = 0, 1, 2, 3, \ldots \tag{20.11}$$

EXAMPLE 20.6 Newton–Raphson for $\sqrt{5}$

To find $\sqrt{5}$, write $f(x) = x^2 - 5 = 0$. Then $f'(x) = 2x$, so that the Newton–Raphson formula is

$$x_{n+1} = x_n - \frac{x_n^2 - 5}{2x_n}$$

Table 20.3 shows the computation using all the figures on a standard 10-digit calculator, starting with $x_0 = 3$.

---

[1] Numerical methods for the solution of equations have their origins in the determination of square and cube roots by the Babylonians and by the Chinese. Chapter 4 of the *Jiuzhang suanshu* (Nine chapters on the mathematical art) of the early Han dynasty, around 200 BC, describes a method of finding square roots that is similar to Newton's method. Jia Xian generalized the method for the solution of polynomial equations in the 11th century (he also described the construction and uses of the Pascal triangle), and the first detailed account was given in the *Shuchu jiuzhang* (Mathematical treatise in nine sections) of 1247 by Qin Jiushao (c. 1202–1261). Newton's method, described in the *Methodus fluxionum* of 1671 but not published until 1736, is essentially the Chinese method for polynomials. Joseph Raphson (1648–1715), who 'was one of the few people whom Newton allowed to see his mathematical papers', published the method in 1690.

Table 20.3 The square root of 5

n	x_n	$f(x_n)/f'(x_n)$	x_{n+1}
0	3	0.66666 6666	2.33333 3333
1	2.33333 3333	0.09523 8095	2.23809 5238
2	2.23809 5238	0.00202 6343	2.23606 8896
3	2.23606 8896	0.00000 0918	2.23606 7978
4	2.23606 7978	0	

The accuracy of the calculator is exhausted after 4 iterations, and the result is in error by only 1 in the 10th significant figure.

▸ Exercises 12–15

Newton–Raphson is derived formally from the Taylor series expansion of a continuous function about a point. Let x be the true value of a zero of $f(x)$, and x_n an approximate value such that $x = x_n + \varepsilon_n$. Then

$$f(x) = f(x_n + \varepsilon_n) = f(x_n) + \varepsilon_n f'(x_n) + \frac{\varepsilon_n^2}{2} f''(x_n) + \cdots \qquad (20.12)$$

The terms in ε_n^2 and higher can be neglected when ε_n is small enough. Then, since $f(x) = 0$,

$$\varepsilon_n = -\frac{f(x_n)}{f'(x_n)} \qquad (20.13)$$

and a new estimate is $x_{n+1} = x_n + \varepsilon_n$, as in (20.11).

It can also be shown from the Taylor series that, when the errors ε_n of x_n and ε_{n+1} of x_{n+1} are small enough,

$$\varepsilon_{n+1} = -\varepsilon_n^2 \frac{f''(x_n)}{2f'(x_n)} \qquad (20.14)$$

This explains the rapid convergence of Newton–Raphson. The error decreases quadratically, and the number of significant figures approximately *doubles* at each iteration. The method is called a **second-order iteration process**, in contrast to the bisection method which, with $\varepsilon_{n+1} \approx \frac{1}{2}\varepsilon_n$, is a first-order process.

The Newton–Raphson method is not a bracketing method and, as shown by equations (20.13) and (20.14), can fail, for example, when $|f'(x_n)|$ is small (or zero). Most problems are avoided either by a suitable choice of initial point x_0 or by combining Newton–Raphson with bisection. Thus, a bisection step is taken whenever Newton–Raphson goes out of bounds or fails to converge.

20.4   Interpolation

Interpolation is the process of finding a function whose graph goes through a number of given points. In Figure 20.3 the points represent the $(n+1)$ number pairs

$$(x_0, y_0), \quad (x_1, y_1), \quad (x_2, y_2), \quad \cdots \quad (x_n, y_n) \tag{20.15}$$

and the dashed curve represents a continuous function $y = f(x)$ such that $y_i = f(x_i)$ for the number pairs (20.15).

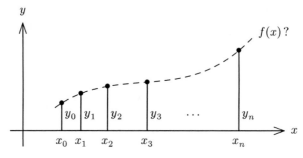

Figure 20.3

The numbers themselves may, for example, be the results of measurements of concentration and time in a kinetics experiment, or of pressure and temperature in a study of the thermodynamic properties of a fluid. Alternatively, they may be the tabulated values of a function that cannot be expressed in simple functional form. The object of interpolation is to find a 'simple' function that can be used to find intermediate points in the interval (x_0, y_0) to (x_n, y_n) (outside this interval the process is extrapolation). Important uses of interpolation are in numerical integration (Section 20.5), whereby a function that cannot be integrated by the methods described in Chapters 5 and 6 is approximated by a simpler function that can be so integrated, and in the numerical solution of differential equations (Section 20.9).

Polynomial interpolation

Given $(n+1)$ points, it is possible to find a unique polynomial p_n of degree n,

$$p_n(x) = a_0 + a_1 x + a_2 x^2 + \cdots + a_n x^n \tag{20.16}$$

that passes through all the points. For example, through any two points there is a unique straight line $(n = 1)$, and through any three points there is a unique quadratic $(n = 2)$.[2]

---

[2] The theoretical basis for polynomial interpolation, and one of the important theorems of modern analysis, is Weierstrass' theorem: 'If $f(x)$ is an arbitrary continuous function defined in the interval $a \le x \le b$, it is always possible to approximate $f(x)$ over the whole interval as closely as we please by a power polynomial of sufficiently high degree'. The essential requirement is continuity; the function may have infinitely many maxima and minima, and need not have a derivative at any point in the interval. Karl Weierstrass (1815–1897), professor at Berlin, made important and influential contributions to analysis.

Linear interpolation

The straight line $y = p_1(x) = a_0 + a_1 x$ through the two points (x_0, y_0) and (x_1, y_1), as in Figure 20.4, satisfies the pair of linear equations

$$y_0 = a_0 + a_1 x_0, \quad y_1 = a_0 + a_1 x_1$$

with solution

$$a_0 = \frac{y_0 x_1 - y_1 x_0}{x_1 - x_0}, \quad a_1 = \frac{y_1 - y_0}{x_1 - x_0}$$

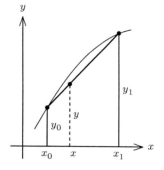

Figure 20.4

Then

$$y = \left(\frac{y_0 x_1 - y_1 x_0}{x_1 - x_0} \right) + \left(\frac{y_1 - y_0}{x_1 - x_0} \right) x$$

and this can be rearranged as

$$y = y_0 + (x - x_0) \left(\frac{y_1 - y_0}{x_1 - x_0} \right) \tag{20.17}$$

The linear interpolation formula is used to find an approximate value of a function at a point x in the interval x_0 to x_1. The method is illustrated in the following example for the (known) function e^x.

EXAMPLE 20.7 Linear interpolation

Two points on the graph of $y = e^x$ are (to 4 decimal places in y) $(x_0, y_0) = (0.80, 2.2255)$ and $(x_1, y_1) = (0.84, 2.3164)$. The linear interpolation formula (20.17) is then

$$e^x \approx y = 2.2255 + (x - 0.80) \left(\frac{2.3164 - 2.2255}{0.84 - 0.80} \right)$$

Then, at $x = 0.832$,

$$e^{0.832} \approx 2.2255 + 0.0727 = 2.2982$$

The true value is 2.2979 (to 4 decimal places).

▸ Exercise 16

When linear interpolation is used between each pair of neighbouring points we have **piecewise linear interpolation**, as illustrated in Figure 20.5 for the points listed in Table 20.4 (compare with Figures 20.6 to 20.8).

Table 20.4

x	y
0.0	1.0000
0.4	1.8902
0.8	1.0517
1.2	0.9577
1.6	0.5207
2.0	1.0000
2.4	2.1184
2.8	1.3809
3.2	1.5690

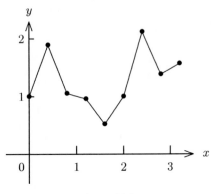

Figure 20.5

Quadratic interpolation

The quadratic function that passes through three successive points, or **nodes**, (x_0, y_0), (x_1, y_1) and (x_2, y_2) can be written as

$$p_2(x) = L_0(x)y_0 + L_1(x)y_1 + L_2(x)y_2 \qquad (20.18)$$

where

$$L_0(x) = \frac{(x - x_1)(x - x_2)}{(x_0 - x_1)(x_0 - x_2)}, \quad L_1(x) = \frac{(x - x_0)(x - x_2)}{(x_1 - x_0)(x_1 - x_2)}$$

$$L_2(x) = \frac{(x - x_0)(x - x_1)}{(x_2 - x_0)(x_2 - x_1)}$$

This is **Lagrange's formula** for quadratic interpolation, and is used to find an approximate value of a function in the interval x_0 to x_2.[3]

EXAMPLE 20.8 Quadratic interpolation

Three points on the graph of e^x are (to 6 decimal places in e^x) $(x_0, y_0) = (0.80, 2.225541)$, $(x_1, y_1) = (0.84, 2.316367)$ and $(x_2, y_2) = (0.88, 2.410900)$. Quadratic interpolation at $x = 0.832$ then gives (compare Example 20.7)

$$e^{0.832} = 0.12y_0 + 0.96y_1 - 0.08y_2 = 2.297905$$

compared with the true value 2.297910 (to 6 decimal places).

▸ Exercise 17

---

[3] Lagrange's general interpolation formula (1795) was anticipated by Euler in the *Institutiones calculi differentialis* of 1755 and given explicitly by Edward Waring (1734–1793), professor at Cambridge, in a Philosophical Transactions paper of 1779.

When quadratic interpolation is used for each triple of neighbouring points we have piecewise quadratic interpolation, as illustrated in Figure 20.6 for the points listed in Table 20.4 (compare with Figures 20.5, 20.7, and 20.8).

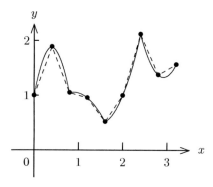

Figure 20.6

Newton's method of divided differences

The interpolating polynomial $p_n(x)$ through $(n+1)$ given points can be expressed in a simple way in terms of **divided differences**. The first and second divided differences are

$$f[x_0, x_1] = \frac{y_1 - y_0}{x_1 - x_0}, \qquad f[x_0, x_1, x_2] = \frac{f[x_1, x_2] - f[x_0, x_1]}{x_2 - x_0} \qquad (20.19)$$

and in general

$$f[x_0, x_1, \ldots, x_k] = \frac{f[x_1, x_2, \ldots, x_k] - f[x_0, x_1, \ldots, x_{k-1}]}{x_k - x_0} \qquad (20.20)$$

Equation (20.17) for $p_1(x)$ is then

$$\begin{aligned} p_1(x) &= y_0 + (x - x_0)f[x_0, x_1] \\ &= p_0(x) + (x - x_0)f[x_0, x_1] \end{aligned} \qquad (20.21)$$

and (20.18) for $p_2(x)$ can be rewritten as

$$\begin{aligned} p_2(x) &= y_0 + (x - x_0)f[x_0, x_1] + (x - x_0)(x - x_1)f[x_0, x_1, x_2] \\ &= p_1(x) + (x - x_0)(x - x_1)f[x_0, x_1, x_2] \end{aligned} \qquad (20.22)$$

In general,

$$p_{k+1}(x) = p_k(x) + (x - x_0)(x - x_1) \cdots (x - x_k)f[x_0, x_1, \ldots, x_{k+1}] \qquad (20.23)$$

The relations amongst the divided differences are demonstrated in Table 20.5 for 5 points. Column D_1 contains the first divided differences, column D_2 the second divided differences, and so on. Each divided difference is the difference of its 'parents' in the column on its left, divided by the difference of the extreme values of x.

Table 20.5 Divided difference interpolation table

x	y	D_1	D_2	D_3	D_4
x_0	y_0				
		$f[x_0, x_1]$			
x_1	y_1		$f[x_0, x_1, x_2]$		
		$f[x_1, x_2]$		$f[x_0, x_1, x_2, x_3]$	
x_2	y_2		$f[x_1, x_2, x_3]$		$f[x_0, x_1, x_2, x_3, x_4]$
		$f[x_2, x_3]$		$f[x_1, x_2, x_3, x_4]$	
x_3	y_3		$f[x_2, x_3, x_4]$		
		$f[x_3, x_4]$			
x_4	y_4				

EXAMPLE 20.9 Newton's method of divided differences for e^x

Five points on the graph of e^x are given in the first two columns of Table 20.6. The divided differences have been computed on a standard 10-digit calculator, and are quoted to 8 decimal places.

Table 20.6 Divided difference table for e^x

x	y	D_1	D_2	D_3	D_4
0.80	2.22554093				
		2.27065125			
0.84	2.31636698		**1.15833750**		
		2.36331825		**0.39393233**	
0.88	2.41089971		1.20560938		**0.10058544**
		2.45976700		0.41002600	
0.92	2.50929039		1.25481250		
		2.56015200			
0.96	2.61169647				

Using the numbers in **boldface** in the table,

$$p_1(x) = 2.22554093 + 2.27065125(x - 0.8) \qquad\qquad \text{for} \quad 0.80 < x < 0.84$$
$$p_2(x) = p_1(x) + 1.15833750(x - 0.8)(x - 0.84) \qquad\qquad 0.80 < x < 0.88$$
$$p_3(x) = p_2(x) + 0.39393233(x - 0.8)(x - 0.84)(x - 0.88) \qquad 0.80 < x < 0.92$$
$$p_4(x) = p_3(x) + 0.10058544(x - 0.8)(x - 0.84)(x - 0.88)(x - 0.92) \quad 0.80 < x < 0.96$$

For $x = 0.832$ the sequence of approximations for $e^{0.832}$ is

$$p_1(x) = 2.29820177 \qquad p_2(x) = 2.29790524$$
$$p_3(x) = 2.29791008 \qquad p_4(x) = 2.29790997$$

and $p_4(x)$ is correct to all the figures quoted (see Examples 20.7 and 20.8).

▸ Exercise 18

Although the x_i are evenly spaced in this example, Newton's formula (20.23) is valid for unevenly spaced points. Modern implementations on computers are essentially adaptations of Newton's method that find the best path through an interpolation table such as Table 20.5 in order to give greatest accuracy.

Spline interpolation

Examples 20.7 to 20.9 show that increasing the degree of the interpolation polynomial leads to increased accuracy for e^x, but this is not generally true when the nodes exhibit maxima and minima as in Figure 20.5. Figure 20.7 shows the result of fitting the polynomial of (maximum) degree 8 to the 9 points given in Table 20.4. This demonstrates that increasing the degree n of the interpolation polynomial can result in spurious oscillations between the nodes, and these oscillations can grow in magnitude indefinitely as n increases. Interpolation should therefore in general be restricted to polynomials of low degree (not more than 4 or 5), but Figure 20.6 shows that, in unfavourable cases, such interpolations can suffer from sharp discontinuities of gradient at the nodes.

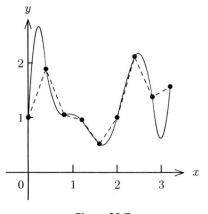
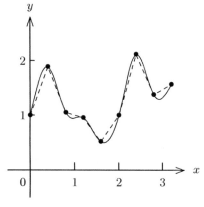

Figure 20.7   Figure 20.8

Both types of disadvantage are avoided in the method of **splines**.[4] This is piecewise polynomial interpolation, but with the polynomial in the interval between each pair of nodes chosen to make the overall interpolation function smooth at the nodes; that

---

[4] Splines are flexible strips of wood long used by designers, engineers and shipwrights to draw smooth curves through given points. The numerical method was described by I. J. Schoenberg in 1946.

is, with continuous first and second (and possibly higher) derivatives at the nodes. In the simplest and most widely used procedure, **cubic spline interpolation**, the function between each pair of nodes is approximated as a cubic. For the jth interval, between (x_j, y_j) and (x_{j+1}, y_{j+1}), the cubic $f_j(x)$ can be written as

$$f_j(x) = a_j + b_j(x - x_j) + c_j(x - x_j)^2 + d_j(x - x_j)^3 \tag{20.24}$$

Then, at the nodes,

$$f_j(x_j) = y_j, \qquad f_j(x_{j+1}) = y_{j+1} \tag{20.25}$$

For continuity of the first derivatives at the nodes, the derivative $f_j'(x)$ is related to the derivatives of the cubics in the adjacent intervals by

$$f_j'(x_j) = f_{j-1}'(x_j), \qquad f_j'(x_{j+1}) = f_{j+1}'(x_{j+1}) \tag{20.26}$$

Similarly for the second derivatives,

$$f_j''(x_j) = f_{j-1}''(x_j), \qquad f_j''(x_{j+1}) = f_{j+1}''(x_{j+1}) \tag{20.27}$$

We consider here only the special case of evenly spaced nodes, with $x_{j+1} - x_j = h$ (all $j = 1$ to n for $(n+1)$ nodes). Then, if the derivative at node j is denoted by k_j, the three sets of conditions (20.25) to (20.27) give the following expressions for the coefficients in the cubic (20.24):

$$a_j = y_j, \qquad b_j = k_j$$

$$c_j = \frac{3}{h^2}(y_{j+1} - y_j) - \frac{1}{h}(k_{j+1} + 2k_j) \tag{20.28}$$

$$d_j = \frac{2}{h^3}(y_j - y_{j+1}) + \frac{1}{h^2}(k_{j+1} + k_j)$$

and the recurrence relation for the gradients,

$$k_{j-1} + 4k_j + k_{j+1} = \frac{3}{h}(y_{j+1} - y_{j-1}) \tag{20.29}$$

These relations are sufficient to determine the interpolation cubics for all the internal intervals ($j = 1$ to n). For the end intervals it is necessary also to specify the slopes k_0 and k_n. If these are chosen by means of the end-point conditions

$$f_0''(x_0) = f_n''(x_n) = 0 \tag{20.30}$$

the cubic splines are called **natural splines**, and these give approximately linear behaviour near the ends. Other choices can be made to alter this behaviour.

Computer programs for cubic spline interpolation exist in all the popular computer programming languages, and are widely used for the display of experimental data and of the results of computational studies in the physical sciences and engineering. The result of natural spline interpolation for the points given in Table 20.4 is illustrated in

Figure 20.8. This demonstrates that curve fitting with splines produces a continuous curve that closely follows the straight lines of piecewise linear interpolation. This does *not* guarantee however that the computed curve is a true representation of the function of which the nodes are a sample. Thus in the present case, the points in Table 20.4 have been computed from the function $f(x) = 1 - (x-1)^2(x-3)\sin\pi x$. This has many local maxima and minima. The 'unrealistic' 8th degree polynomial fit shown in Figure 20.7 is in fact, quite fortuitously, a fairly good representation of the function, whereas the spline fit is only as good as the poor sample of points allows.*

20.5 Numerical integration

Numerical integration (quadrature) is the numerical evaluation of the integral

$$I = \int_a^b f(x)\, dx \tag{20.31}$$

when the integrand $f(x)$ is either given by a table of numbers or when the integral cannot be evaluated by the methods described in Chapters 5 and 6 in terms of a finite number of standard functions. Geometrically, the problem is to estimate the area under the curve. The standard method is to replace the function $f(x)$ by a different function that can be integrated, and one way is to use the interpolation formulas discussed in the previous section. This gives the class of numerical integration methods called **Newton–Cotes quadratures**, of which the trapezoidal rule and Simpson's rule are the simplest examples.[5]

The trapezoidal rule. Linear interpolation

The simplest estimate of an integral (20.31), as an area, is obtained by dividing the interval $a \le x \le b$ into n equal subintervals each of width $h = (b-a)/n$, and approximating the integrand by piecewise linear interpolation; that is, by joining the corresponding adjacent points on the graph by straight lines as shown in Figure 20.9.

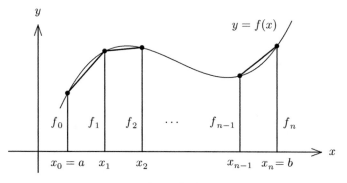

Figure 20.9

* *Garbage in, Garbage out* (Wilf Hey, d. 2007, computer programmer and writer).

[5] The Newton–Cotes formulas first appeared in Newton's letter to Leibniz of 24 October 1676 and in the *Principia*, although particular examples were known long before. They were included in the collection of work by Cotes, the *Harmonia mensurarum*, published posthumously in 1722. Roger Cotes (1682–1716), Cambridge mathematician, spent much of 1709–1713 preparing the second edition of the *Principia*. His work includes one of the earliest discussions of the calculus of logarithmic and trigonometric functions, with tables of integrals.

The area of the trapezium between x_j and x_{j+1} is $(f_j+f_{j+1})h/2$, and the total area is

$$\int_a^b f(x)\,dx \approx T(h) = h\left[\frac{1}{2}f_0 + f_1 + f_2 + \cdots + f_{n-1} + \frac{1}{2}f_n\right] \qquad (20.32)$$

where $f_j = f(a+jh)$.

For an integrand that is continuous and differentiable in the interval $a \le x \le b$, it can be shown that the error in this formula is

$$\int_a^b f(x)\,dx - T(h) = \varepsilon = -\frac{b-a}{12}h^2 f''(c) \qquad (20.33)$$

where $f''(c)$ is the second derivative of the function at some point $x=c$ in the interval. Alternatively, when h is small enough, the error can be approximated as (see the Euler–MacLaurin formula (20.39) below)

$$\varepsilon \approx -\frac{1}{12}h^2\left[f_n' - f_0'\right] \qquad (20.34)$$

where f_0' and f_n' are the derivatives of $f(x)$ at the end points. It follows that halving h (or doubling the number of subintervals n) reduces the error by a factor of four.

EXAMPLE 20.10 Find estimates of the integral $I = \int_0^{1.2} xe^{-x^2}\,dx$ by means of the trapezoidal rule (20.32) and the error formula (20.34) for $n = 3, 6$ and 12.

The required values of the integrand $f(x) = xe^{-x^2}$ are listed in the following table.

Table 20.7 Values of $f(x) = xe^{-x^2}$

x	0.0	0.1	0.2	0.3	0.4	0.5	0.6
$f(x)$	0.000000	0.099005	0.192158	0.274179	0.340858	0.389400	0.418606
x	0.7	0.8	0.9	1.0	1.1	1.2	
$f(x)$	0.428838	0.421834	0.400372	0.367879	0.328017	0.284313	

Then, for example, for $n = 3$ we have

$$h = 0.4, \quad f_0 = 0, \quad f_1 = 0.340858, \quad f_2 = 0.421834, \quad f_3 = 0.284313$$

and

$$T(h) = 0.4[0 + 0.340858 + 0.421834 + \tfrac{1}{2} \times 0.284313] = 0.361939$$

The derivative of the integrand is

$$f'(x) = e^{-x^2}(1 - 2x^2),$$

so that $f_0' = f'(0) = 1$ and $f_n' = f'(1.2) = -0.445424$. Then $\varepsilon = 0.120452h^2$. The results are collected in Table 20.8.

Table 20.8 Values of $I = \int_0^{1.2} xe^{-x^2}\, dx$

n	3	6	12
$T(h)$	0.361939	0.376698	0.380330
ε	0.019272	0.004818	0.001205
$T(h) + \varepsilon$	0.381211	0.381516	0.381535

The exact value of the integral is $\frac{1}{2}(1 - e^{-1.44}) = 0.381536$.

▸ Exercise 19

Simpson's rule. Quadratic interpolation

Simpson's rule has been one of the most popular simple numerical quadrature methods for over two centuries.[6] It is obtained by dividing the interval $a \le x \le b$ into an even number of subintervals, $2n$, each with width $h = (b - a)/2n$, and approximating the integrand by piecewise quadratic interpolation. Then

$$\int_a^b f(x)\, dx \approx \frac{h}{3}\left[f_0 + 4f_1 + 2f_2 + 4f_3 + 2f_4 + \cdots + 4f_{2n-1} + f_{2n}\right] \quad (20.35)$$

or

$$\int_a^b f(x)\, dx \approx \frac{h}{3}\left[\sum(\text{end points}) + 4\sum(\text{odd points}) + 2\sum(\text{even points})\right]$$

The error in Simpson's rule behaves like h^4 when h is small, so that halving h (or doubling the number of subintervals n) reduces the error by a factor of sixteen. Simpson's rule therefore usually converges much more quickly that the (uncorrected) trapezoidal rule.

---

[6] Thomas Simpson (1710–1761), professor of mathematics at the Royal Military Academy at Woolwich, was a self-taught mathematician. He was brought up as a weaver and in 1735 he joined the Mathematical Society at Spitalfields, a weaving community, of which it was the duty of every member 'if he be asked any mathematical or philosophical question by another member, to instruct him in the plainest and easiest manner he is able.' The rule named for him appeared in his *Mathematical dissertations on a variety of physical and analytical subjects* of 1743, but had appeared previously in work by Cavalieri (1639) and Gregory (1668).

EXAMPLE 20.11 Find estimates of the integral $I = \int_0^{1.2} xe^{-x^2}\, dx$ by means of Simpson's rule (20.35) for $2n = 2, 4, 6$, and 12.

We use the values of the integrand in Table 20.7 and construct Table 20.9.

Table 20.9 Simpson's rule for $I = \int_0^{1.2} xe^{-x^2}\, dx$

j	x_j	$f(x_j)$	$2n = 2$	$2n = 4$	$2n = 6$	$2n = 12$
0	0.0	0.000000	×1	×1	×1	×1
1	0.1	0.099005				×4
2	0.2	0.192158			×4	×2
3	0.3	0.274179		×4		×4
4	0.4	0.340858			×2	×2
5	0.5	0.389400				×4
6	0.6	0.418606	×4	×2	×4	×2
7	0.7	0.428838				×4
8	0.8	0.421834			×2	×2
9	0.9	0.400372		×4		×4
10	1.0	0.367879			×4	×2
11	1.1	0.328017				×4
12	1.2	0.284313	×1	×1	×1	×1
		Totals	1.958737	3.819729	5.724269	11.446227
		$\times \dfrac{h}{3} = I$	0.391747	0.381973	0.381618	0.381541

The exact value is 0.381536, and Simpson's rule is accurate enough for many applications even without correction.

▸ Exercises 20–23

The Euler–MacLaurin formula

When the integrand $f(x)$ can be expanded as a Taylor series in the interval $a \le x \le b$, the approximate expression (20.32) for the trapezoidal rule can be replaced by the Euler–MacLaurin formula

$$
\int_a^b f(x)\, dx = h \left[\frac{1}{2} f_0 + f_1 + f_2 + \cdots + f_{n-1} + \frac{1}{2} f_n \right]
$$

$$
- B_2 \frac{h^2}{2!} (f_n' - f_0') - B_4 \frac{h^4}{4!} (f_n''' - f_0''') - B_6 \frac{h^6}{6!} (f_n^{\mathrm{v}} - f_0^{\mathrm{v}}) - \cdots
$$

(20.36)

where $f_0', f_n', f_0''', f_n''', \ldots$ are the odd-order derivatives of $f(x)$ evaluated at the endpoints, and the numbers $B_2 = \frac{1}{6}, B_4 = -\frac{1}{30}, B_6 = \frac{1}{42}, \ldots$ are called **Bernoulli numbers**. When $f(x)$ is a polynomial of degree n the expansion in h^2 terminates after a finite number of terms because all derivatives after the nth are zero. The formula (20.36) is then exact. In other cases, the expansion is not usually a convergent series for any value of h because the Bernoulli numbers B_n, after the first few, increase rapidly in magnitude as n increases. It is instead a special kind of expansion called an **asymptotic expansion**, and it has the property that when truncated at any point the error is less than the magnitude of the last included term.

Apart from its use for the analysis of error in numerical integration, the Euler–MacLaurin formula has some more direct applications in the physical sciences. The formula can be used for evaluating sums of the type $\sum f_m$ over a set of integral values of m. Thus, putting $h = 1$ and inserting the values of the Bernoulli numbers, (20.36) can be rearranged as

$$\sum_{m=0}^{n} f_m = \int_a^b f(x)\,dx + \frac{1}{2}\left[f_0 + f_n\right]$$

$$+ \frac{1}{12}(f_n' - f_0') - \frac{1}{720}(f_n''' - f_0''') + \frac{1}{30240}(f_n^{\mathrm{v}} - f_0^{\mathrm{v}}) -$$

$$(20.37)$$

EXAMPLE 20.12 Evaluate $\displaystyle\sum_{m=0}^{\infty} 1/(10+m)^2$

For $f(x) = 1/(a+x)^2$,

$$\int_0^\infty 1/(a+x)^2\,dx = 1/a, \quad f'(0) = -2/a^3, \quad f'''(0) = -24/a^5, \quad f^{\mathrm{v}}(0) = -720/a^7$$

and the function and all its derivatives tend to 0 as $x \to \infty$. Then

$$\sum_{m=0}^{\infty} \frac{1}{(a+m)^2} = \int_0^\infty f(x)\,dx + \frac{1}{2}f(0) - \frac{1}{12}f'(0) + \frac{1}{720}f'''(0) - \frac{1}{30240}f^{\mathrm{v}}(0) +$$

$$= \frac{1}{a} + \frac{1}{2a^2} + \frac{1}{6a^3} - \frac{1}{30a^5} + \frac{1}{42a^7} -$$

$$\sum_{m=0}^{\infty} \frac{1}{(10+m)^2} = 0.1 + 0.005 + 0.00016666 - 0.00000033 + 0.000000002 -$$

$$= 0.105166$$

to 6 significant figures. We note that a very large number of terms of the sum over m would be required to obtain just the first term of the expansion.

▸ Exercises 24, 25

Formula (20.37) is used in the following examples to derive two results that are important in statistical thermodynamics.

EXAMPLE 20.13 The rotational partition function of the linear rigid rotor

The energy levels of the linear rigid rotor are

$$E_J = \frac{J(J+1)h^2}{8\pi^2 I}, \qquad J = 0, 1, 2, 3, \ldots$$

with degeneracies $g_J = 2J + 1$, where J is the rotational quantum number and I is the moment of inertia. The rotational partition function is then

$$q = \sum_{J=0}^{\infty} g_J e^{-E_J/kT} = \sum_{J=0}^{\infty} (2J+1)\, e^{-J(J+1)\theta/T}$$

where $\theta = h^2/8\pi^2 Ik$ is called the rotational temperature.

Put $f(J) = (2J+1)e^{-J(J+1)\theta/T}$ in the Euler–MacLaurin formula (20.37), with $a = 0$, $b \to \infty$. Then

$$q = \sum_{J=0}^{\infty} f(J) = \int_0^{\infty} f(J)\, dJ + \frac{1}{2}[f(0) + f(\infty)]$$

$$+ \frac{1}{12}[f'(\infty) - f'(0)] - \frac{1}{720}[f'''(\infty) - f'''(0)] + \cdots$$

The integral is evaluated by means of the substitution $u = J(J+1)$. Then

$$\int_0^{\infty} (2J+1)\, e^{-J(J+1)\theta/T}\, dJ = \int_0^{\infty} e^{-u\theta/T}\, du = \frac{T}{\theta}$$

Also, $f(J)$ and all its derivative go to zero as $J \to \infty$, and the values at $J = 0$ can be obtained from the expansion of $f(J)$ as a power series in J,

$$f(J) = 1 + \left(2 - \frac{\theta}{T}\right)J + \left(-\frac{3\theta}{T} + \frac{\theta^2}{2T^2}\right)J^2 + \left(-\frac{2\theta}{T} + \frac{2\theta^2}{T^2} - \frac{\theta^3}{6T^3}\right)J^3 + \cdots$$

Then $f(0) = 1$, $f'(0) = 2 - \theta/T$, $f''(0) = -12\theta/T + 12\theta^2/T^2 - \theta^3/T^3$, and all higher derivatives at $J = 0$ are of order $(\theta/T)^2$ or higher. Then

$$q = \frac{T}{\theta} + \frac{1}{3} + \frac{1}{15}\left(\frac{\theta}{T}\right) + \text{terms in } \left(\frac{\theta}{T}\right)^2 \text{ and higher} \tag{20.38}$$

The largest value of θ for a molecule is $\theta = 87.5\,\text{K}$ for H_2 but most molecules have rotational temperatures close to 0 K. In that case, $T/\theta \gg 1$ except at very low temperatures, so that the first term on the right side of (20.38) is much the largest, and is the rotational partition function normally used in chemistry (after correction for symmetry).

EXAMPLE 20.14 Stirling's approximation for $\ln n!$

The quantity $\ln n!$ for integer n occurs in probability theory and enters statistical thermodynamics with n equal to the number of particles in a mole; that is, with $n \approx 10^{23}$. Stirling's approximation for the logarithm of the factorial of a very large number, $\ln n! \approx n \ln n - n$, can be derived from the Euler–MacLaurin formula. We put $f(x) = \ln x$. Then

$$\ln n! = \ln 1 + \ln 2 + \cdots + \ln n$$

$$= \int_1^n \ln x \, dx + \frac{1}{2}[f(n) + f(1)] + \frac{1}{12}[f'(n) - f'(1)] + \cdots$$

$$\approx \left[x \ln x - x \right]_1^n + \frac{1}{2}(\ln n + \ln 1) + \frac{1}{12}\left(\frac{1}{n} - 1 \right)$$

$$\approx n \ln n - n \quad \text{when} \quad n \gg 1$$

(see also Section 21.6). A more complete treatment gives

$$\ln n! \approx n \ln n - n + \frac{1}{2} \ln 2\pi n + \frac{1}{12n} - \frac{1}{360n^3} + \frac{1}{1260n^5}$$

▸ Exercise 26

Gaussian quadratures

All numerical integration formulas have the form

$$\int_a^b f(x) \, dx \approx \sum_{i=1}^n w_i f(x_i) \tag{20.39}$$

and involve the evaluation of the integrand at n points, $x_1, x_2, \ldots, x_n$, with coefficients, or Gaussian weights, w_i determined by the form of the interpolation formula. In the Newton–Cotes formulas discussed above the points are equally spaced. In Gaussian quadrature formulas the points are chosen to give the greatest possible accuracy for a given interpolation formula, and are not equally spaced.[7]

---

[7] Gauss presented his new method in *Methodus nova integralium valores per approximationem inveniendi* in 1816.

There are a number of Gaussian quadrature formulas appropriate to several kinds of integrand. The simplest is the **Gauss–Legendre** (or, simply, Gaussian) formula

$$\int_{-1}^{+1} f(x)\, dx \approx \sum_{i=1}^{n} w_i f(x_i) \tag{20.40}$$

in which the points x_i are the zeros of the Legendre polynomial $P_n(x)$. The weights w_i are then chosen to make the formula exact if $f(x)$ is a polynomial of degree $2n - 1$. Tabulations of the weights and points to many significant figures (at least 15) exist for many values of n (up to at least 100), and the formula can give very high accuracy for integrals that can be written in the form given by (20.40). Any integral with a finite range of integration a to b can be transformed into one with range -1 to $+1$ by a change of variable to u given by $x = \frac{1}{2}\Big[(b-a)u + (a+b)\Big]$.

EXAMPLE 20.15 Gauss–Legendre quadrature for $I = \int_0^{1.2} xe^{-x^2}\, dx$.

Changing the variable of integration to u given by $x = 0.6(1+u)$ gives

$$I = 0.36 \int_{-1}^{+1} (1+u)e^{-0.36(1+u)^2}\, du \approx \sum_{i=1}^{n} w_i \left[0.36(1+u_i)e^{-0.36(1+u_i)^2} \right]$$

The Gauss–Legendre parameters for $n = 4$ are (to 8 figures)

$$u_1 = -u_4 = -0.86113631, \qquad w_1 = w_4 = 0.34785485$$
$$u_2 = -u_3 = -0.33998104, \qquad w_2 = w_3 = 0.65214515$$

Then $I \approx 0.381532$. The exact value is 0.381536, and Gauss–Legendre integration with 4 points gives greater accuracy than Simpson's rule with 12 intervals.

Other Gaussian quadrature formulas of the more general form

$$\int_a^b f(x)W(x)\, dx \approx \sum_{i=1}^{n} w_i f(x_i) \tag{20.41}$$

exist for a variety of integrand weight functions $W(x)$ and also for infinite integrals. The more important are

Gauss–Chebyshev: $\displaystyle \int_{-1}^{+1} f(x)\frac{1}{\sqrt{1-x^2}}\, dx \approx \sum_{i=1}^{n} w_i f(x_i)$

Gauss–Laguerre: $\displaystyle \int_0^{\infty} f(x)e^{-x}\, dx \approx \sum_{i=1}^{n} w_i f(x_i)$

Gauss–Hermite: $\displaystyle \int_{-\infty}^{+\infty} f(x)e^{-x^2}\, dx \approx \sum_{i=1}^{n} w_i f(x_i)$

Gaussian quadrature formulas are used in applications in which very high accuracy is required; for example, in quantum chemistry.

20.6 Methods in linear algebra

The problems of linear algebra are those associated with systems of linear equations, as described in Chapters 17, 18, and 19:

(a) Solution of sets of simultaneous linear equations, or of the equivalent matrix equation $\mathbf{Ax} = \mathbf{b}$ for the unknown vector $\mathbf{x}$, where $\mathbf{A}$ is a given matrix of coefficients and $\mathbf{b}$ is a known vector.
(b) Calculation of the determinant of a square matrix $\mathbf{A}$.
(c) Calculation of the inverse matrix $\mathbf{A}^{-1}$ of a square matrix $\mathbf{A}$.
(d) Solution of the eigenvalue problem $\mathbf{Ax} = \lambda\mathbf{x}$ for a square matrix $\mathbf{A}$.

There exists a vast and growing body of numerical methods for these problems driven by, and driving, developments of computer technology. Many of these methods are based on, or are related to, the **elimination methods** discussed in the following sections for problems of types (a), (b), and (c).

20.7 Gauss elimination for the solution of linear equations

The Gauss elimination method is the formalization of the elementary method of solving systems of simultaneous linear equations. We consider the three equations in three unknowns (see also Example 2.34).[8]

$$
\begin{array}{lll}
(1) & x + 2y + 3z = 26 & \\
(2) & 2x + 3y + z = 34 & (20.42) \\
(3) & 3x + 2y + z = 39 &
\end{array}
$$

The method proceeds in a sequence of steps.

Step 1. Elimination of x
Equation (1) is called the **pivot equation** and the term in x is called the **pivot** of the step. The elimination of x from all subsequent equations is achieved by subtraction of appropriate multiples of (1). Thus, subtraction of $2 \times (1)$ from (2) and of $3 \times (1)$ from (3) gives the new set of equations

$$
\begin{array}{lll}
(1) & x + 2y + 3z = 26 & \\
(2') & {-y - 5z} = -18 & (20.43) \\
(3') & {-4y - 8z} = -39 &
\end{array}
$$

---

[8] These particular equations arise in a problem on measures of grain in Chapter 8 of the *Jiuzhang suanshu*, and they are solved there by a method essentially identical to Gauss elimination. Gauss described his method in the *Theoria motus corporum celestium* (Theory of motion of the heavenly bodies) of 1809.

Step 2.   Elimination of y

Equation (2') is the new pivot equation, and y is eliminated from subsequent equations by addition or subtraction of appropriate multiples of (2'). In the present case, subtraction of $4 \times (2')$ from (3') gives

$$\begin{array}{llr}
(1) & x + 2y + 3z = & 26 \\
(2') & -y - 5z = & -18 \\
(3'') & 12z = & 33
\end{array}$$
(20.44)

The equations are in **triangular** or **echelon form**, and the solution is completed by **back substitution**.

Step 3.   Back substitution

The equations (20.44) are solved in reverse order. Equation (3'') gives $z = 11/4$, substitution for this in (2') gives $y = 17/4$, and substitution for y and z in (1) gives $x = 37/4$.

➤ Exercises 27–30

The following example demonstrates that Gauss elimination applies to any system of linear equations, even when there is no unique solution or when there is no solution.

───

EXAMPLE 20.16 Use Gauss elimination to solve the equations (see Section 17.4)

$$\begin{array}{ll}
(1) & 2x + 2y + z = 10 \\
(2) & x + 2y - 2z = -3 \\
(3) & 3x + 2y + 4z = \lambda
\end{array}$$

where λ is a number.

Step 1. Subtract $\frac{1}{2} \times (1)$ from (2) and $\frac{3}{2} \times (1)$ from (3):

$$\begin{array}{ll}
(1) & 2x + 2y + z = 10 \\
(2') & y - \dfrac{5}{2}z = -8 \\
(3') & -y + \dfrac{5}{2}z = \lambda - 15
\end{array}$$

Step 2. Add (2') to (3'):

$$\begin{array}{ll}
(1) & 2x + 2y + z = 10 \\
(2') & y - \dfrac{5}{2}z = -8 \\
(3'') & 0 = \lambda - 23
\end{array}$$

When $\lambda = 23$, equation (3″) is redundant, and back substitution gives $y = 5z/2 - 8$ and $x = 13 - 3z$. Because z remains arbitrary, we have infinitely many solutions, one for each possible value of z. On the other hand, no solution exists when $\lambda \neq 23$.

> Exercise 31

Pivoting

The elimination method involves repeated subtraction of a multiple of one equation from another and can lead to serious differencing errors, especially for large systems of equations. To illustrate the problem, we consider the pair of equations

$$(1) \quad 0.0003x_1 + 2.513x_2 = 7.545$$
$$(2) \quad 0.7003x_1 - 2.613x_2 = 6.167 \tag{20.45}$$

of which the solution is $x_1 = 20$, $x_2 = 3$. To solve by the elimination method, we choose (1) as the pivot equation and eliminate x_1 from (2) by subtracting $(0.7003/0.0003) \times$ (1) from (2). If, for example, the calculation is performed using 4-figure arithmetic then $0.7003/0.0003 = 2334$, and the second equation becomes

$$(2') \quad -5868x_2 = -17600$$

Then $x_2 = 2.999$, and equation (1) gives $x_1 = 28.38$. A small error in x_2 has led to a large error in x_1 because the coefficient of x_1 in equation (1) is small compared with that in (2). The resulting differencing errors are avoided if equation (2) is chosen as the pivot equation. Then, multiplication of (2) by $0.0003/0.7003 = 0.0002484$ and subtraction from (1) gives

$$(1') \quad -2.514x_2 = -7.542$$

so that $x_2 = 3.000$. Substitution in (2) then gives $x_1 = 20.0$.

The choice of equation (2) as the pivot equation is an example of **partial pivoting**; that is, choosing the pivot equation in the first step as that equation in which the coefficient of x_1 has largest magnitude. Similarly for x_2 in the second step, and so on.

Partial pivoting is equivalent to a reordering of the equations, but is not always sufficient. For example, equations (20.45) may have been obtained with (1) multiplied by some large number, 3000 say:

$$(1) \quad 0.9000x_1 + 7539x_2 = 22640$$
$$(2) \quad 0.7003x_1 - 2.613x_2 = 6.167 \tag{20.46}$$

The coefficient of x_1 in (1) is now the greater, but choosing (1) as the pivot equation again results in a poor (but different) result. The correct procedure, called **scaled partial pivoting**, is to choose as pivot equation in the first step that equation in which the ratio of the coefficient of x_1 to the largest other coefficient has largest magnitude. This ratio is $0.90000/7539 \approx 0.0001$ in equation (1) and $0.7003/2.613 \approx 0.3$ in equation

(2). The latter is therefore chosen as pivot equation. Similarly for x_2 in step 2, and so on. A further refinement, called **total pivoting**, is to look for the largest relative coefficient of any variable in each step.

Elimination method for the value of a determinant

A determinant can be reduced to triangular form by the first stage of the Gauss elimination method. The value of the determinant is then the product of the diagonal elements of the triangular form, as discussed in Section 17.6.

20.8 Gauss–Jordan elimination for the inverse of a matrix

The inverse matrix $\mathbf{A}^{-1}$ of a nonsingular matrix $\mathbf{A}$ of order n satisfies the matrix equation

$$\mathbf{A}\mathbf{A}^{-1} = \mathbf{I} \tag{20.47}$$

We write

$$\mathbf{A}^{-1} = (\mathbf{x}_1 \quad \mathbf{x}_2 \quad \mathbf{x}_3 \quad \cdots \quad \mathbf{x}_n), \qquad \mathbf{I} = (\mathbf{e}_1 \quad \mathbf{e}_2 \quad \mathbf{e}_3 \quad \cdots \quad \mathbf{e}_n)$$

where $\mathbf{x}_k$ is the column vector formed from the elements of the kth column of $\mathbf{A}^{-1}$, and $\mathbf{e}_k$ is formed from the kth column of the unit matrix $\mathbf{I}$. The matrix equation (20.47) is then equivalent to the n matrix equations

$$\mathbf{A}\mathbf{x}_k = \mathbf{e}_k \qquad (k = 1, 2, \ldots, n) \tag{20.48}$$

each of which represents a system of n simultaneous linear equations, and each of which can therefore be solved by Gauss elimination. The systematic way of doing this is called **Gauss–Jordan elimination**, whereby an **augmented matrix** $(\mathbf{A}|\mathbf{I})$ is transformed into $(\mathbf{I}|\mathbf{A}^{-1})$ by the two stage process demonstrated in the following example.[9]

EXAMPLE 20.17 Gauss–Jordan elimination for the inverse of

$$\mathbf{A} = \begin{pmatrix} 1 & 2 & 3 \\ -2 & 1 & 2 \\ 3 & -1 & -1 \end{pmatrix}$$

The augmented matrix of the problem is

$$(\mathbf{A}|\mathbf{I}) = \left(\begin{array}{ccc|ccc} 1 & 2 & 3 & 1 & 0 & 0 \\ -2 & 1 & 2 & 0 & 1 & 0 \\ 3 & -1 & -1 & 0 & 0 & 1 \end{array} \right)$$

---

[9] Wilhelm Jordan (1842–1899), German geodesist.

The first stage is to reduce **A** to upper triangular form by Gauss elimination:

$$\begin{bmatrix} \text{row } 2 + 2 \times \text{row } 1 \\ \text{row } 3 - 3 \times \text{row } 1 \end{bmatrix} \longrightarrow \left(\begin{array}{ccc|ccc} 1 & 2 & 3 & 1 & 0 & 0 \\ 0 & 5 & 8 & 2 & 1 & 0 \\ 0 & -7 & -10 & -3 & 0 & 1 \end{array}\right)$$

$$\left[\text{row } 3 + \tfrac{7}{5} \times \text{row } 2\right] \longrightarrow \left(\begin{array}{ccc|ccc} 1 & 2 & 3 & 1 & 0 & 0 \\ 0 & 5 & 8 & 2 & 1 & 0 \\ 0 & 0 & \tfrac{6}{5} & -\tfrac{1}{5} & \tfrac{7}{5} & 1 \end{array}\right)$$

The second stage is to reduce the left half of the augmented matrix to diagonal form by further elimination steps:

$$\begin{bmatrix} \text{row } 1 - \tfrac{5}{2} \times \text{row } 3 \\ \text{row } 2 - \tfrac{20}{3} \times \text{row } 3 \end{bmatrix} \longrightarrow \left(\begin{array}{ccc|ccc} 1 & 2 & 0 & \tfrac{3}{2} & -\tfrac{7}{2} & -\tfrac{5}{2} \\ 0 & 5 & 0 & \tfrac{10}{3} & -\tfrac{25}{3} & -\tfrac{20}{3} \\ 0 & 0 & \tfrac{6}{5} & -\tfrac{1}{5} & \tfrac{7}{5} & 1 \end{array}\right)$$

$$\left[\text{row } 1 - \tfrac{2}{5} \times \text{row } 2\right] \longrightarrow \left(\begin{array}{ccc|ccc} 1 & 0 & 0 & \tfrac{1}{6} & -\tfrac{1}{6} & \tfrac{1}{6} \\ 0 & 5 & 0 & \tfrac{10}{3} & -\tfrac{25}{3} & -\tfrac{20}{3} \\ 0 & 0 & \tfrac{6}{5} & -\tfrac{1}{5} & \tfrac{7}{5} & 1 \end{array}\right)$$

The left half is reduced to the unit matrix by dividing row 2 by 5 and row 3 by $\tfrac{6}{5}$. Then

$$\left(\begin{array}{ccc|ccc} 1 & 0 & 0 & \tfrac{1}{6} & -\tfrac{1}{6} & \tfrac{1}{6} \\ 0 & 1 & 0 & \tfrac{2}{3} & -\tfrac{5}{3} & -\tfrac{4}{3} \\ 0 & 0 & 1 & -\tfrac{1}{6} & \tfrac{7}{6} & \tfrac{5}{6} \end{array}\right) = \left(\mathbf{I} \,|\, \mathbf{A}^{-1}\right)$$

and the right half of the augmented matrix is the inverse matrix $\mathbf{A}^{-1}$:

$$\mathbf{A}^{-1} = \frac{1}{6}\begin{pmatrix} 1 & -1 & 1 \\ 4 & -10 & -8 \\ -1 & 7 & 5 \end{pmatrix} \text{ and } \mathbf{A}^{-1}\mathbf{A} = \mathbf{A}\mathbf{A}^{-1} = \mathbf{I}$$

▸ Exercises 32, 33

20.9 First-order differential equations

We saw in Chapters 11 to 14 that some important differential equations can be solved by analytical methods to give solutions that are expressed in terms of known functions. Many differential equations cannot however be solved in this way, either because no appropriate method exists or because there is no analytical solution. It is then necessary to use numerical methods of solution.

An analytical solution is always better than a numerical one. For example, the analytical solution can be evaluated for any numerical values of the independent variables, whereas a numerical solution takes the form of a table of values for a fixed discrete set of values of the variables. An analytical solution will in general contain one or more arbitrary parameters whose values may be determined by the imposition of initial or boundary conditions. In a numerical method, the values of these parameters must be fixed at the outset, before a solution can be obtained. In addition, the functional behaviour of the solution may give additional insights into the nature of the physical problem of which the differential equation is the model.

The first-order differential equations discussed in this section have the general form

$$y'(x) = \frac{dy}{dx} = f(x, y) \qquad (20.49)$$

It is shown in Section 20.10 that an ordinary higher-order equation can always be expressed as a system of simultaneous first-order equations, and the numerical methods for the single equations are also applicable to such systems.

Graphical representation

Equation (20.49) is an expression in the two variables x and y so that a value of $y'(x)$ is defined at each point of the xy-plane. A graphical representation of the differential equation is then obtained by representing the value of $y'(x)$ at each point on a grid in the plane by a line segment with gradient $f(x, y)$, as illustrated in Figure 20.10 for the equation

$$y'(x) = 2(x - 2y)/(x + y).$$

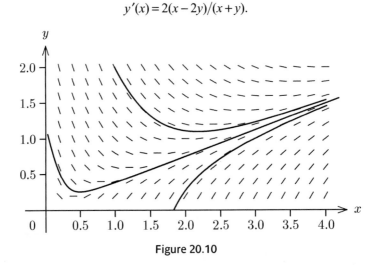

Figure 20.10

Such a diagram is called a **direction field**. A solution of the differential equation then has a graph that is tangential to its direction lines at every point on the curve. The graphs of three solutions are shown in the figure; the upper graph for condition $y(1) = 2$, the middle for $y(0.25) = 0.5$, the lower for $y(3) = 1$.

A numerical method for solving a differential equation is equivalent to starting at a particular point in the direction field and following the field lines. A number of methods are available for the solution of most types of differential equation met in the physical sciences, and most of these have their origin in the Taylor expansion (7.24).

Given the value $y(x_0) = y_0$ of the function $y(x)$ at $x = x_0$, the value at the neighbouring point $x_1 = x_0 + h$ is

$$y(x_1) = y(x_0) + h y'(x_0) + \frac{1}{2} h^2 y''(x_0) + \cdots \qquad (20.50)$$

Then, because $y'(x_0) = f(x_0, y_0)$ by (20.49),

$$y(x_1) = y(x_0) + h f(x_0, y_0) + \frac{1}{2} h^2 y''(x_0) + \cdots$$

or

$$y_1 = y_0 + h F_0 \qquad (20.51)$$

where $F_0 = (y_1 - y_0)/(x_1 - x_0)$ is the slope of the line joining the points 0 and 1 on the graph of $y(x)$ (the average slope of the curve). The different numerical methods use different way of estimating this slope, leading to the recursion relation

$$y_{n+1} = y_n + h F_n \qquad (20.52)$$

where $x_n = x_0 + nh$ and $y_n \approx y(x_n)$. We consider first the simplest possible method, Euler's method.

Euler's method

Euler's method is to truncate the Taylor expansion after the second term,

$$y(x_1) \approx y(x_0) + h y'(x_0)$$

so that F_0 has been approximated by the slope at the initial point (x_0, y_0). An approximate value of $y(x_1)$ is then

$$y_1 = y_0 + h f(x_0, y_0) \qquad (20.53)$$

and this provides us with a simple step-by-step procedure whereby an approximate solution of the equation is obtained by repeated application of the recursion

$$y_{n+1} = y_n + h f(x_n, y_n), \qquad n = 0, 1, 2, \dots \qquad (20.54)$$

where $x_n = x_0 + nh$ and $y_n \approx y(x_n)$. At each step the slope of the line from point n to point $(n+1)$ is that of the direction field line at point n, and the procedure is illustrated in Figure 20.11.

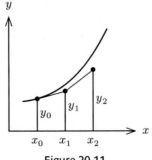

Figure 20.11

Euler's method is called a **first-order** method because only the first power of h is retained in the Taylor expansion. The truncation error at each step is therefore of order h^2, so that halving the step size h reduces the error by a factor of 4. However, the cumulative (global) error for a series of steps is of order h.

EXAMPLE 20.18 Given the initial value problem

$$y'(x) = y + 1, \qquad y(0) = 1$$

use Euler's method to obtain an approximate value of $y(1)$ using step sizes $h = 0.2$ and $h = 0.1$.

The results of applying the recursion

$$y_{n+1} = y_n + h(y_n + 1)$$

are summarized in Table 20.10, and compared with the exact values obtained from the solution $y(x) = 2e^x - 1$.

Table 20.10 Example of Euler's method

h	n	x_n	y_n	exact	error	h	n	x_n	y_n	exact	error
0.2	0	0.0	1.0000	1.0000	0.0000	0.1	0	0.0	1.0000	1.0000	0.0000
							1	0.1	1.2000	1.2103	0.0103
	1	0.2	1.4000	1.4428	0.0428		2	0.2	1.4200	1.4428	0.0228
							3	0.3	1.6620	1.6997	0.0377
	2	0.4	1.8800	1.9836	0.1036		4	0.4	1.9282	1.9836	0.0554
							5	0.5	2.2210	2.2974	0.0764
	3	0.6	2.4560	2.6442	0.1882		6	0.6	2.5431	2.6442	0.1011
							7	0.7	2.8974	3.0275	0.1301
	4	0.8	3.1472	3.4511	0.3039		8	0.8	3.2872	3.4511	0.1639
							9	0.9	3.7159	3.9192	0.2033
	5	1.0	3.9766	4.4366	0.4599		10	1.0	4.1875	4.4366	0.2491

The computed values of $y(1)$ are 3.9766 for step size $h = 0.2$ and 4.1875 for $h = 0.1$, compared with the exact value 4.4366. The table shows that the error is approximately halved when the step size is halved, as expected for a first-order method.

▶ Exercises 34–37

Accuracy increases as the step size is decreased, but the larger number of steps required leads to increasingly larger rounding errors. The practical methods of solving differential equations are procedures for simulating the Taylor expansion to some higher order in h, and the more sophisticated of these involve the use of variable step sizes to control the accumulation of rounding errors and to optimize convergence. The derivation and justification of these methods are beyond the scope of this book, and we give here only a brief description of one popular family of methods.

Runge–Kutta methods

The Runge–Kutta methods are a family of methods of increasing order, with Euler's first-order method as the first member.[10] The second-order Runge–Kutta method is obtained by replacing F_n in (20.52) by an estimate of the gradient halfway between points n and $n+1$. Thus, by Euler's method, an approximate value of y at the midpoint is

$$y_n + \frac{h}{2} f(x_n, y_n) = y_n + \frac{h}{2} k_1 \tag{20.55a}$$

and an approximate slope at the midpoint is

$$k_2 = f\left(x_n + \frac{h}{2}, y_n + \frac{h}{2} k_1\right) \tag{20.55b}$$

The estimated value of $y(x_{n+1})$ is then

$$y_{n+1} = y_n + h k_2 \tag{20.55c}$$

The procedure represented by equations (20.55) is of second order in the step size h.

One of the most widely used of all numerical methods for solving differential equations is the **fourth-order Runge–Kutta method**, defined by the set of equations

$$k_1 = f(x_n, y_n) \qquad\qquad k_2 = f\left(x_n + \frac{h}{2}, y_n + \frac{h}{2} k_1\right)$$

$$k_3 = f\left(x_n + \frac{h}{2}, y_n + \frac{h}{2} k_2\right) \qquad k_4 = f(x_n + h, y_n + h k_3) \tag{20.56}$$

$$y_{n+1} = y_n + \frac{h}{6}[k_1 + 2k_2 + 2k_3 + k_4]$$

---

[10] Carl David Tolmé Runge (1856–1927) and Wilhelm Kutta (1867–1944), German mathematicians, published systematic investigations of methods of 'successive substitutions' in 1895 (Runge) and 1901 (Kutta). This work has been followed by many elaborations and extensions to give some of the most widely used modern methods.

EXAMPLE 20.19 Apply the Runge–Kutta methods of orders 2 and 4 to the initial value problem in Example 20.18, with step size $h = 0.2$.

The result of applying the recursion relations (20.55) for order 2 and (20.56) for order 4 are displayed in Table 20.11.

Table 20.11 Example of Runge–Kutta methods

order	n	x_n	y_n	exact	error	order	n	x_n	y_n	error
2	0	0.0	1.000000	1.000000	0.000000	4	0	0.0	1.000000	0.000000
	1	0.2	1.440000	1.442806	0.002805		1	0.2	1.442800	0.000005
	2	0.4	1.976800	1.983649	0.006849		2	0.4	1.983636	0.000013
	3	0.6	2.631696	2.644238	0.012542		3	0.6	2.644213	0.000025
	4	0.8	3.430669	3.451082	0.020413		4	0.8	3.451042	0.000040
	5	1.0	4.405416	4.436563	0.031147		5	1.0	4.436502	0.000061

The table shows that the second-order method with step size $h = 0.2$ is already considerably more accurate that Euler's first-order method with $h = 0.1$, and that the fourth-order method is very much more accurate still.

▸ Exercises 38–42

20.10 Systems of differential equations

A problem involving second- or higher-order differential equations can always be reduced to one involving a system of simultaneous first-order equations. For example, the general second-order linear equation

$$\frac{d^2 y}{dx^2} + p(x)\frac{dy}{dx} + q(x)y = r(x) \tag{20.57}$$

can be written as the pair of first-order equations

$$\frac{dy}{dx} = z$$

$$\frac{dz}{dx} = r(x) - p(x)z - q(x)y \tag{20.58}$$

for the two functions of x, $y(x)$ and $z(x)$. If (20.57) is the differential equation of an initial value problem, with initial conditions $y(x_0) = y_0$ and $y'(x_0) = z(x_0) = y_1$, then equations (20.58) are a pair of (coupled) first-order initial value problems. The general problem is that of N first-order initial value problems for N functions, $y_1(x)$, $y_2(x)$, ..., $y_N(x)$:

$$y_i'(x) = \frac{dy_i}{dx} = f_i(x, y_1, y_2, ..., y_N), \qquad i = 1, 2, ..., N \qquad (20.59)$$

with given initial values $y_1(x_0), y_2(x_0), \ldots$

Such systems of first-order initial value problems occur in the chemical kinetics of multistep chemical reactions, and are therefore of considerable interest to the chemist (boundary value problems are generally much more difficult). They are solved numerically by applying one of the methods described in Section 20.9 to each equation in turn at each step. For example, for the pair of equations

$$y'(x) = f_1(x, y, z), \qquad z'(x) = f_2(x, y, z) \qquad (20.60)$$

with initial conditions $y(x_0) = y_0$ and $z(x_0) = z_0$, Euler's method leads to the sequence of recursion steps

(1) $y_1 = y_0 + hf_1(x_0, y_0, z_0), \qquad z_1 = z_0 + hf_2(x_0, y_0, z_0)$

(2) $y_2 = y_1 + hf_1(x_1, y_1, z_1), \qquad z_2 = z_1 + hf_2(x_1, y_1, z_1)$

(3) $y_3 = y_2 + hf_1(x_2, y_2, z_2), \qquad$ and so on

The procedure is essentially the same as that for the single equation.

Stiff equations

A stiff differential equation is one whose solution contains terms that differ greatly in their dependence on the independent variable. For example, the second-order initial value problem

$$\frac{d^2 y}{dx^2} = 100\,y, \qquad y(0) = 1, \ \ y'(0) = -10 \qquad (20.61)$$

has the general solution $y = ae^{-10x} + be^{+10x}$ and particular solution $y = e^{-10x}$. A numerical solution of the equation by one of the methods discussed so far gives a solution that behaves correctly for values of x close to $x = 0$ but 'explodes' like e^{+10x} as x increases because of rounding and truncation errors that inevitably lead to a small admixture of the unwanted term; that is, the numerical solution has the form

$$y \approx e^{-10x} + \varepsilon e^{+10x} \qquad (20.62)$$

where ε is not zero.

Problems involving stiff equations occur in kinetics when several elementary processes have very different rate constants. Thus, problem (20.61) is equivalent to the pair of first-order problems

$$\frac{dy}{dx} = z, \qquad \frac{dz}{dx} = 100\,y \qquad (20.63)$$

which can be interpreted as first-order rate processes with rate constants 1 and 100. Stiff problems can sometimes be solved by means of a change of variable, but special numerical methods exist to handle such problems.

20.11   Exercises

Section 20.2

1. Express the following numbers rounded to (a) 3 decimal places, (b) 4 significant figures:
   (i) 1.21271   (ii) 72.0304   (iii) 0.129914   (iv) 0.0024988

2. Find the absolute error bound for each answer of Exercise 1.

3. Compute the values of the following arithmetic expressions. Assuming that all the numbers in the expressions are correctly rounded, find the absolute error bounds of your answers:
   (i) $2.137 + 3.152$,   (ii) $2.137 + 3.152 - 4.672$,   (iii) $12.36 + 14.13 + 16.38$,
   (iv) $12.36 + 14.13 - 16.38$

4. Find the relative error bound for each answer of Exercise 1.

5. Compute the values of the following arithmetic expressions. Assuming that all the numbers in the expressions are correctly rounded, find the absolute error bounds of your answers:
   (i) 22.7×2.59,   (ii) $22.7/2.59$,   (iii) $(17.43 - 12.34)/14.38$

6. Solve $x^2 - 60x + 1 = 0$ by (i) equations (20.3a) and (ii) equations (20.3b), using (a) 4-figure arithmetic, (b) 6-figure arithmetic.

7. (i) Compute $f(x) = x(\sqrt{x+1} - \sqrt{x})$ on a 10-digit calculator (or similar) for $x = 1, 10^2, 10^4, 10^6, 10^8$.

   (ii) Show that the function can be written as $f(x) = \dfrac{x}{\sqrt{x+1} + \sqrt{x}}$. Use this to recompute the function.

8. (i) Compute $\dfrac{e^x - e^{-x}}{2x} - 1$ on a 10-digit calculator (or similar) for $x = 1, 10^{-2}, 10^{-4}, 10^{-6}$.

   (ii) Use the Taylor series to find an expression for the function that is accurate for small values of x. (iii) Use this to recompute the function for $x = 10^{-2}, 10^{-4}, 10^{-6}$.

Section 20.3

Find a solution to 4 significant figures of the following equations by the bisection method, using the given starting values of x:

9. $x^2 - \ln x = 2$; 1.5, 1.6   10. $e^{-x} = \tan x$; 0.5, 0.6   11. $x^3 - 3x^2 + 6x = 5$; 1.0, 1.5

Find a solution to 8 significant figures of the following equations by the Newton–Raphson method starting in every case with $x_0 = 1$ (see Exercises 9 to 11):

12. $x^2 - \ln x = 2$ 13. $e^{-x} = \tan x$ 14. $x^3 - 3x^2 + 6x = 5$

15. Given one root, x_1 say, of a polynomial of degree n, a second root can be obtained by first dividing the polynomial by the factor $(x - x_1)$ to give a polynomial of degree $n - 1$. Show that the computed root of the cubic in Exercise 14 is the only real one.

Section 20.4

Six points on the graph of a function $y = f(x)$ are given by the (x, y) pairs

(0.0, 0.00000)	(0.2, 0.19867)	(0.4, 0.38942)
(0.6, 0.56464)	(0.8, 0.71736)	(1.0, 0.84147)

(the function is $\sin x$).

16. Use linear interpolation to compute $f(0.04), f(0.26), f(0.5), f(0.81)$.

17. Use quadratic interpolation to compute $f(0.04), f(0.26), f(0.81)$.

18. Construct the finite difference interpolation table, and use it to compute $f(0.26)$.

Section 20.5

Given $I = \displaystyle\int_{1.0}^{2.6} \frac{dx}{x}$ in Exercises 19–21:

19. Find estimates of the integral I by means of the trapezoidal rule (20.32) and the error formula (20.34), starting with one strip $(n = 1)$ and doubling the number of strips $(n = 2, 4, 8, …)$ until 4-figure accuracy is obtained.

20. Use Simpson's rule (20.35) for $2n = 2, 4, 8, …$ to find the value of the integral I to 4 decimal places

21. (i) The error in Simpson's rule is Ah^4 when h is small enough, and A is a constant. If $S(2n)$ is the Simpson formula for $2n$ intervals, show that

$$\int_a^b f(x)\,dx \approx \frac{1}{15}[16S(2n) - S(n)] \quad \text{(Richardson's extrapolation)}.$$

(ii) Use this and the results of Exercise 20 for $n = 8$ to estimate a more accurate value of the integral I. Check the accuracy by comparing with the exact value (ln 2.6).

Use Simpson's rule for $2n = 2, 4, 8, …$ to find the values of the following integrals to 5 decimal places:

22. $\displaystyle\int_0^1 \frac{\sin x}{x}\,dx$ $\left[\mathrm{Si}(z) = \displaystyle\int_0^z \frac{\sin x}{x}\,dx \text{ is the Sine integral}\right]$

23. $\displaystyle\int_0^2 e^{-x^2}\,dx$ $\left[\mathrm{erf}(z) = \dfrac{2}{\sqrt{\pi}} \displaystyle\int_0^z e^{-x^2}\,dx \text{ is the error function}\right]$

24. (i) Use the Euler–MacLaurin formula to calculate the sum $S(11) = \sum_{m=0}^{\infty} 1/(11+m)^2$ to six decimal places. (ii) Use the value of $S(10)$ in Example 20.12 to verify that $S(11) = S(10) - 0.01$.

25. Use the results of Exercise 24 to calculate $S(1)$ to six significant figures.

26. Calculate ln 1!, ln 2! and ln 3! (i) using the 'large-number' formula $n \ln n - n$, (ii) from the more accurate formula given in Example 20.14. (iii) Exactly.

Section 20.7

Solve the following systems of equations by the Gauss elimination method:

27. $\begin{aligned} x_1 - 4x_2 &= -2 \\ 3x_1 + x_2 &= 7 \end{aligned}$

28. $\begin{aligned} 2x_1 + x_2 - x_3 &= 6 \\ 4x_1 - x_3 &= 6 \\ -8x_1 + 2x_2 + 2x_3 &= -8 \end{aligned}$

29. $\begin{aligned} x + y + z &= 2 \\ -x + 2y - 3z &= 32 \\ 3x \quad\; - 4z &= 17 \end{aligned}$

30. $\begin{aligned} w + 2x + 3y + z &= 5 \\ 2w + x + y + z &= 3 \\ w + 2x + y &= 4 \\ x + y + 2z &= 0 \end{aligned}$

31. (i) Use Gauss elimination to find the value of λ for which the following equations have a solution.

(ii) Solve the equations for this value of λ:

$$x + y + z = 2$$
$$-x + 2y - 3z = 32$$
$$3x + 5z = \lambda$$

Section 20.8

32. Use Gauss–Jordan elimination to find the inverse of the matrix

$$\begin{pmatrix} -1 & 1 & 2 \\ 3 & -1 & 1 \\ -1 & 3 & 4 \end{pmatrix}$$

33. Given the system of equations,

$$-x - y + 2z = b_1$$
$$3x - y + z = b_2$$
$$-x + 3y + 4z = b_3$$

use the result of Exercise 32 to express x, y, and z in terms of the arbitrary numbers b_1, b_2 and b_3.

Section 20.9

NOTE: The following exercises can be performed using a pocket calculator, but the arithmetic is tedious. You are advised to use a spreadsheet or write your own computer programs to perform the tasks.

34. Apply Euler's method to the initial value problem

$$y'(x) = -y(x), \quad y(0) = 1$$

with step sizes **(i)** $h = 0.2$, **(ii)** $h = 0.1$, **(iii)** $h = 0.05$ to calculate approximate values of $y(x)$ for $x = 0.2, 0.4, 0.6, 0.8, 1.0$. Compare these with the values obtained from the exact solution $y = e^{-x}$.

For initial value problems in Exercises 35 to 37, **(i)** apply Euler's method with step size $h = 0.1$ to compute an approximate value of $y(1)$, **(ii)** confirm the given exact solution and compute the error:

35. $y' = 2 - 2y, \quad y(0) = 0; \quad y = 1 - e^{-2x}$

36. $y' = \dfrac{y^2}{x+1}, \quad y(0) = 1; \quad y = \dfrac{1}{1 - \ln(x+1)}$

37. $y' = \dfrac{y + x^2 - 2}{x+1}, \quad y(0) = 2; \quad y = x^2 + 2x + 2 - 2(x+1)\ln(x+1)$

38. Apply the second-order Runge–Kutta method to the initial value problem in Exercise 34 with step sizes **(i)** $h = 0.2$, **(ii)** $h = 0.1$.

39. Apply the fourth-order Runge–Kutta method to the initial value problem in Exercise 34 with step sizes **(i)** $h = 0.2$, **(ii)** $h = 0.1$.

Apply the fourth-order Runge–Kutta method to the initial value problems in Exercises 35, 36, and 37 with step size $h = 0.1$ to compute **(i)** $y(1)$ and **(ii)** the error:

40. As Exercise 35   **41.** As Exercise 36   **42.** As Exercise 37

21 Probability and statistics

21.1 Concepts

The practice of statistics, the collection and statistical analysis of data, has become a ubiquitous activity of everyday life, and the subject of a vast literature. Three broad areas of statistical activity may be distinguished:

  (i) the practical problems of the design of experiments and of the choice and collection of suitable samples of data ('good experimental practice');

 (ii) the description and presentation of the data and its analysis in terms of appropriate theoretical models;

(iii) the use of the analysis to draw conclusions about the nature of the system under investigation, the quality of the experiment, and the method of collection of data.

Our main concern in this chapter is an introduction to probability theory, the mathematical theory of statistics.[1] Probability theory provides the theoretical models and analytical tools for the organization, interpretation, and analysis of statistical data. Probability theory has additional importance in chemistry in, for example, the description of the collective behaviour of very large numbers of particles in statistical mechanics, the quantum mechanical descriptions of changes of state and of rate processes, the physical interpretation of wave functions, and the enumeration of the ways of assembling basic chemical units to form large molecules, as in the construction of polypeptides from amino acid residues. Conventional ('practical') statistics is represented by brief discussions of descriptive statistics in Section 21.2 and the method of least squares in Section 21.10. There exist several specialist texts on the use of statistical methods in the physical sciences, and the reader should consult one of these for a more comprehensive discussion, and for a proper appreciation of the power and wide range of applications of statistics in the sciences, and in many other fields of study.

21.2 Descriptive statistics

Table 21.1 shows a set of 50 numbers that represent the results of an experiment that involves the counting of events.

---

[1] Probability theory has its origins in ancient times in divination and in games of chance, and has continued to be used for these purposes to the present day. Oresme used a probability argument to conclude that astrology must be false. The modern theory arose from the correspondence of Pascal and Fermat in 1654 following the questions on the results of throwing dice put to Pascal by the gambler de Méré. Pascal's ideas were included in his *Treatise on the arithmetic triangle*. The throwing of dice had been discussed earlier by Cardano in his *Liber de ludo aleae* (Book on games of chance) in 1526, but the earliest systematic account was by Huygens in the *De ratiociniis in aleae ludo* (Calculations in games of chance) of 1657. The first substantial treatment of mathematical probability theory was Jakob Bernoulli's *Ars conjectandi* (The art of conjecture) published in 1713 in which he also proposed the application of probabilities in the social sciences. de Moivre discussed the law of errors and the normal distribution in the second edition of the *Doctrine of chances* (1718, 1738, 1756). Euler and d'Alembert wrote on problems of life expectancy, insurance, lotteries, and others, and the extensive application of statistical methods, particularly in the social sciences, followed the publication of Laplace's definitive *Théorie analytique des probabilités* in 1812 ('at the bottom, the theory of probabilities is only common sense in numbers').

Table 21.1 50 experimental values

4	6	7	3	6	3	5	7	6	6
3	4	4	5	8	6	5	2	6	5
7	7	4	4	3	0	8	5	4	4
6	2	8	2	7	6	5	5	5	4
9	5	2	2	9	3	6	5	7	4

These values have been obtained from a computer simulation of the experiment of tossing a coin. Each entry is the number of heads obtained with 10 tosses, but it might equally well represent the number of faults per day in an industrial process or the number of α-particles emitted each second from a radioactive source.

The first step in the analysis of the data is to determine its structure, both pictorially by means of appropriate graphical displays and numerically by calculating a small number of 'statistics', such as mean value and standard deviation, that summarize the essential properties of the data. In our example, the possible results of the experiment are the 11 integers 0 to 10, and the number of occurrences of each of these is called its **frequency**. The collection of these frequencies is the **frequency distribution** of the experiment, and can be displayed in tabular form as in Table 21.2 or as a bar chart as in Figure 21.1.

Table 21.2

Result	Frequency
0	1
1	0
2	5
3	5
4	9
5	10
6	9
7	6
8	3
9	2
10	0

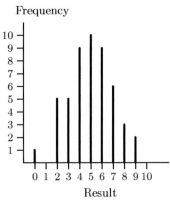

Figure 21.1

▸ Exercise 1

The above distribution is a **discrete** frequency distribution because the possible results of the experiment have discrete values. Many experiments however involve measurement rather than counting. The results can, in principle, then be any values of a continuous range, giving rise to a **continuous** frequency distribution. Table 21.3 shows a set of 50 results of such an experiment, and these may represent, for example, the heights (in appropriate units) of a sample of the population, the weights of a sample of manufactured products, or the results of measurements of a physical quantity such as a rate constant or equilibrium constant.

Table 21.3 50 experimental values

40.6	44.9	47.1	39.5	45.3	38.9	42.9	47.0	45.0	44.2
39.3	40.7	48.4	43.1	48.9	44.9	43.2	37.1	45.3	42.7
47.5	46.5	40.9	40.5	38.9	33.3	49.1	43.7	41.3	41.3
45.3	36.9	49.3	37.3	47.2	44.3	42.9	43.4	43.1	41.1
51.1	43.3	37.9	36.9	53.2	39.3	45.7	42.7	47.1	46.8

For presentation purposes, information of this type is simplified by dividing the range that contains all the sample values into a suitable number of intervals, called **class intervals** or **groups**, and allocating the sample values to their appropriate classes. The number of values in each class is called the **class frequency**. In our example, the smallest and largest values are 33.3 and 53.2, respectively, so that all the values lie in the range 32.0 to 54.0, say, and this is conveniently divided into 11 class intervals, each of width 2.0. A value that falls on a class boundary is assigned to the upper class. The resulting frequency distribution is illustrated by a **frequency histogram** in Figure 21.2 and a **cumulative frequency graph** in Figure 21.3. In the latter, each point on the graph is the frequency of all the values up to and including the value at that point.

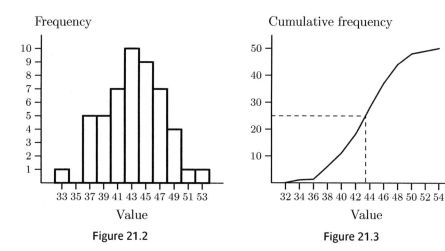

Figure 21.2   Figure 21.3

Both the bar chart in Figure 21.1 and the histogram in Figure 21.2 show distributions of frequencies that are approximately symmetrical about the central value, with frequency decreasing away from the centre. This is the result expected for many different types of experiment. Thus, the heights of members of the (human or other) population are expected to be distributed evenly about some average value, with very small and very large heights less likely than heights near the average. Similarly, 'equally good' measurements of a physical quantity have **random errors**, with positive and negative errors equally likely and with large errors less probable than small errors. The 'S' shape of the cumulative frequency graph is typical of such symmetrical distributions.

> Exercise 2

The pictorial and graphical representations of data provide qualitative information about the centre of a distribution, its width, and its shape. These properties are readily quantified in terms of a small number of computed **distribution statistics** such as the mean, the standard deviation, and the skewness.

   We consider an experiment in which the possible results (**outcomes**) are represented by the discrete variable x whose values form a set of k values $\{x_1, x_2, ..., x_k\}$ (the **population** or **sample space**). If the possible results are any values in a continuous range then the set represents k class intervals. In a particular experiment, let the results of N measurements of x (a sample of the population) consist of

$$n_1 \text{ values of } x_1, \quad n_2 \text{ of } x_2, \quad ..., \quad n_k \text{ of } x_k$$

The total number of measurements (the **sample size**) is then

$$\sum_{i=1}^{k} n_i = N \tag{21.1}$$

In general, different samples of the same (parent) population have different frequency distributions; that is, they have different sets of frequencies $\{n_1, n_2, ..., n_k\}$. Each sample distribution is an approximation to the 'true' distribution of the parent. Small samples differ more than large samples, and a fundamental principle underlying statistics is that the differences between sample distributions are expected to become small when the sample sizes become large enough, and that sample distributions tend to the distribution of the parent population as $N \to \infty$. This is sometimes called the **law of large numbers**.

   The problems associated with finite sample sizes will be touched upon in Section 21.11. We assume in the meantime that N is large enough for these problems to be unimportant.

Mean, mode, and median

The most generally useful measure of the average value of x is the **arithmetic mean**,

$$\bar{x} = \frac{1}{N} \sum_{i=1}^{N} x_i \tag{21.2}$$

in which the sum is over all the N data values, or

$$\bar{x} = \frac{1}{N}(n_1 x_1 + n_2 x_2 + \cdots + n_k x_k) = \frac{1}{N} \sum_{i=1}^{k} n_i x_i \tag{21.3}$$

in which the sum is over the k distinct values (or classes). Other measures of average, seldom used in the sciences, are:

(i) **Mode:** the value of the variable that has the greatest frequency (the most popular value). If this value is unique the distribution is called **unimodal**, but some distributions may have two or more maximum values (**bimodal** or **multimodal** distributions).

(ii) **Median:** the value of the variable that divides the distribution into two equal halves. The values are ordered and the median is the central value if N is odd, and the mean of the two central values when N is even. This quantity is used when order or **rank** is more important than numerical value.

The three measures of average are equal for a symmetrical unimodal distribution.

EXAMPLES 21.1 Mean, mode, and median

(i) The mean value of the data given in Table 21.2 is

$$\frac{1}{50}(1\times 0+0\times 1+5\times 2+5\times 3+9\times 4+10\times 5$$

$$+9\times 6+6\times 7+3\times 8+2\times 9+0\times 10)=4.98$$

This is close to the 'expected' value 5 for the experiment. The same mean is obtained by adding all the values in Table 21.1 and dividing the sum by 50. The mode is 5, with frequency 10, and the median is also 5.

(ii) The mean of the raw data in Table 21.3 is 43.34. The mean obtained from the histogram in Figure 21.2, using the values at the centres of the classes, is

$$\frac{1}{50}(1\times 33+0\times 35+5\times 37+5\times 39+7\times 41+10\times 43$$

$$+9\times 45+7\times 47+4\times 49+51+53)=43.28$$

That the two values of mean are almost the same shows that our allocation of data to classes has not distorted this measure of the distribution. The modal class is 42–44. The median of the raw data is 43.1, and the median class is 42–44, as shown by the dashed lines in Figure 21.3.

▸ Exercises 3, 4

Variance and standard deviation

The mean $\bar{x}$ of a set of data gives the position of the centre of the distribution but no information about the spread or **dispersion** of the data about the mean; two different distributions can have the same mean but very different spreads.

Qualitative measures of spread are the **range**, the difference between the extreme data values, and the **interquartile range**, containing the middle 50% of values, with **upper quartile** and **lower quartile** ranges on either side, each containing 25%. These measures have little use in physical applications. Quantitative measures of spread are obtained from consideration of the deviations $x_i - \bar{x}$ of data values from the mean. The mean deviation (summing over all data elements) is necessarily zero because of the definition of the mean:

$$\frac{1}{N}\sum_{i=1}^{N}(x_i - \bar{x}) = \frac{1}{N}\sum_{i=1}^{N}x_i - \bar{x}\frac{1}{N}\sum_{i=1}^{N}1 = \bar{x} - \bar{x} = 0$$

that is, the positive and negative deviations from the mean cancel. The spread of a distribution is instead nearly always measured in terms of the mean of the *squares* of the deviations,

$$V(x) = \frac{1}{N}\sum_{i=1}^{N}(x_i - \bar{x})^2 \tag{21.4}$$

or, in terms of the k distinct values (or classes),

$$V(x) = \frac{1}{N}\sum_{i=1}^{k}n_i(x_i - \bar{x})^2 \tag{21.5}$$

The quantity $V(x)$ is called the **variance** of the distribution. The square root of the variance is called the **standard deviation**; $V = s^2$ and

$$s = \sqrt{V(x)} = \left\{\frac{1}{N}\sum_{i=1}^{N}(x_i - \bar{x})^2\right\}^{\frac{1}{2}} \tag{21.6}$$

(but see Section 21.11 for corrections to V and s when N is not large). The standard deviation has the same units and dimensions as the data elements.

We note that if the n_i in (21.3) and (21.5) are interpreted as masses at positions x_i on a straight line, then $\bar{x}$ is the position of the centre of mass, $NV(x)$ is the moment of inertia with respect to the centre of mass, and s is the radius of gyration.

EXAMPLES 21.2 Standard deviation

The mean value of the data in Table 21.2 is 4.98 and the variance is

$$V = \frac{1}{50}\sum_{i=1}^{11}n_i(x_i - 4.98)^2 = 3.80$$

The standard deviation is therefore $s = \sqrt{V} = 1.95$, and 70% of the data lies within s of the mean.

▶ Exercise 5

The variance (21.5) can be written in simpler form as 'the mean of the squares minus the square of the mean':

$$V(x) = \frac{1}{N} \sum_{i=1}^{N} (x_i - \bar{x})^2$$

$$= \frac{1}{N} \sum_{i=1}^{N} (x_i^2 - 2\bar{x}x_i + \bar{x}^2) = \frac{1}{N} \sum_{i=1}^{N} x_i^2 - 2\bar{x}\frac{1}{N} \sum_{i=1}^{N} x_i - \bar{x}^2 \frac{1}{N} \sum_{i=1}^{N} 1$$

$$= \frac{1}{N} \sum_{i=1}^{N} x_i^2 - 2\bar{x}\,\bar{x} + \bar{x}^2$$

Therefore, writing $\overline{x^2} = \dfrac{1}{N} \sum_{i=1}^{N} x_i^2$ for the mean of the squares,

$$V(x) = \overline{x^2} - \bar{x}^2 \tag{21.7}$$

Mean absolute deviation

An alternative measure of the spread of a distribution is the mean of the absolute deviations $|x_i - \bar{x}|$,

$$\frac{1}{N} \sum_{i=1}^{N} |x_i - \bar{x}| \tag{21.8}$$

Although this is seldom used, it can be useful for broad distributions with significant numbers of points remote from the centre.

Skewness and kurtosis

The asymmetry of a set of data is measured by the mean cube deviation, with **skewness** (or skew) defined by

$$\gamma = \frac{1}{N} \sum_{i=1}^{N} \left(\frac{x_i - \bar{x}}{s} \right)^3 = \frac{1}{s^3} \left(\overline{x^3} - 3\bar{x}\,\overline{x^2} + 2\bar{x}^3 \right) \tag{21.9}$$

This quantity is zero for a symmetrical distribution, positive if a tail extends further to the right than to the left, and negative if a tail extends further to the left than to the right. It is an important statistic of a distribution, but seldom used in the physical sciences. Even less frequently used is the **kurtosis**, defined in terms of the mean fourth-power deviation.

> Exercise 6

21.3   Frequency and probability

The simplest nontrivial statistical experiment is one that has only two possible outcomes: true or false, on or off, success or failure. Table 21.4 shows some results for the tossing of a coin (an unbiased coin); N is the number of tosses, $n(H)$ is the number, or frequency, of heads, and $f(H) = n(H)/N$ is the fraction, or **relative frequency**, of heads.

Table 21.4

N	$n(\mathrm{H})$	$f(\mathrm{H})$
256	118	0.461
4040	2068	0.512
12000	6062	0.505
24000	11942	0.498

The results are not surprising because we *expect* heads (or tails) to come up about half the time; the two outcomes are said to be 'equally probable'. This experiment is an example of a **random experiment** in which the outcomes depend entirely on 'chance' (the problem of what is meant by the words 'random', 'chance', and 'probability' has generated its own vast literature). Experience shows that random experiments exhibit statistical regularity; that is, the relative frequency of a particular outcome in a long sequence of trials remains about the same when several such sequences are performed. In our example, $f(\mathrm{H}) \approx \frac{1}{2}$ and, for tails, $f(\mathrm{T}) = 1 - f(\mathrm{H}) \approx \frac{1}{2}$. The theoretical value of the relative frequency of an outcome, the value we expect, is the **probability** of the outcome; $P(\mathrm{H}) = P(\mathrm{T}) = \frac{1}{2}$.

Probability distributions

A set of possible outcomes (or **events**) $\{x_1, x_2, \ldots, x_k\}$ with probabilities $P(x_1)$, $P(x_2), \ldots, P(x_k)$ is called a **probability distribution**. Probability distributions are the theoretical models (of populations) with which frequency distributions (of samples) are compared for the analysis of experimental data. The total probability, the probability that there be an outcome, is unity (for certainty), so that

$$\sum_{i=1}^{k} P(x_i) = 1 \tag{21.10}$$

The two most important properties of a probability distribution are its mean and variance (or standard deviation).

The mean or **expectation value** of x is

$$\mu = \sum_{i=1}^{k} x_i P(x_i) \tag{21.11}$$

This is the sum in (21.3) with the relative frequencies n_i/N replaced by probabilities. The symbol μ is that used most frequently, in statistics, for the mean of the population, with $\bar{x}$ reserved for a sample of the population. Other symbols, that emphasize that μ is the expected value, are $E(x)$ and $\langle x \rangle$. If $f(x)$ is a function of x, the mean or expectation value of f is defined as

$$E(f) = \langle f \rangle = \sum_{i=1}^{k} f(x_i) P(x_i) \tag{21.12}$$

The variance V and standard deviation σ of the probability distribution (of the population) are given by*

$$V(x) = \sum_{i=1}^{k} (x_i - \mu)^2 \, P(x_i) = \sigma^2 = \langle x^2 \rangle - \langle x \rangle^2 \qquad (21.13)$$

(see equations (21.5) to (21.7) for the corresponding quantities for a frequency distribution).

21.4   Combinations of probabilities

Exclusive events

The two possible outcomes of the toss of a coin are exclusive events; if one happens then the other does not. Thus, the probability that the outcome is H *and* T in any one toss is zero (impossibility), but the probability that the outcome is H *or* T is unity (certainty):

$$P(\text{H or T}) = P(\text{H}) + P(\text{T}) = 1 \qquad (21.14)$$

In general, if A and B are exclusive events with probabilities $P(A)$ and $P(B)$, the probability that the outcome is event A or event B is

$$P(\text{A or B}) = P(\text{A}) + P(\text{B}) \qquad \text{(exclusive events)} \qquad (21.15)$$

EXAMPLE 21.3 Exclusive events

The possible results of throws of a die are the numbers {1, 2, 3, 4, 5, 6}, each with probability $\frac{1}{6}$ (for a fair die).

(i) The probability that the outcome of throwing the die is greater than 4 (is 5 or 6) is

$$P(>4) = P(5 \text{ or } 6) = P(5) + P(6) = 2 \times \frac{1}{6} = \frac{1}{3}$$

(ii) The probability that the outcome is even is

$$P(\text{even}) = P(2 \text{ or } 4 \text{ or } 6) = P(2) + P(4) + P(6) = 3 \times \frac{1}{6} = \frac{1}{2}$$

> Exercises 7, 8

---

* The symbol σ is always used for the standard deviation of the population, but is sometimes also used instead of s for a sample; see Section 21.11.

Independent events

The possible outcomes of tossing a coin twice are {HH, HT, TH, TT}, each with probability $\frac{1}{4}$. The outcome of the second toss (event B) is independent of the outcome of the first (event A), and the probability of event A *and* B is then the product of the probabilities of A and B separately:

$$P(\text{A and B}) = P(\text{A}) \times P(\text{B}) \quad \text{(independent events)} \qquad (21.16)$$

EXAMPLE 21.4 Find the probability of outcome 10 from two throws of a die.

Of the 36 possible outcomes, those equal to 10 are (5, 5), (4, 6), and (6, 4). The second throw is independent of the first (or, for a single throw of two dice, the outcome of each is independent of that of the other). Each outcome has probability $\frac{1}{6} \times \frac{1}{6} = \frac{1}{36}$ and because they are exclusive,

$$P(10) = P(5 \text{ and } 5) + P(4 \text{ and } 6) + P(6 \text{ and } 4) = 3 \times \frac{1}{36} = \frac{1}{12}$$

> Exercises 9, 10

EXAMPLE 21.5 Independent systems

In statistical thermodynamics the probability $P(E)$ that a system is in a state with energy E is a function of E only. Two noninteracting (independent) systems with energies E_1 and E_2 have combined energy $E_1 + E_2$ and combined probability

$$P(E_1 + E_2) = P(E_1) \times P(E_2) \qquad (21.17)$$

One function that satisfies this equation is $P(E) = e^{\beta E}$, where β is a constant. We have

$$P(E_1) \times P(E_2) = e^{\beta E_1} e^{\beta E_2} = e^{\beta(E_1 + E_2)} = P(E_1 + E_2)$$

It can be shown that this function (multiplied by a constant) is the only one that satisfies (21.17). In statistical thermodynamics the parameter β is inversely proportional to the temperature, $\beta = -1/kT$, where k is Boltzmann's constant, and the exponential probability function describes the Boltzmann distribution.

> Exercise 11

21.5   The binomial distribution

The binomial distribution is the theoretical distribution that describes the results of a given number of independent performances of an experiment that has only two possible outcomes (see Section 21.3). We consider 4 tosses of a coin (or 4 observations of the spin of an electron). Each toss (each 'trial') has two possible outcomes with equal probabilities: $P(H) = P(T) = \frac{1}{2}$. The 4 possible outcomes of 2 tosses are HH, HT, TH, and TT, each with probability $\left(\frac{1}{2}\right)^2 = \frac{1}{4}$. These outcomes are of three types: 2 heads with probability $\frac{1}{4}$, 1 head and 1 tail with probability $\frac{1}{4} + \frac{1}{4} = \frac{1}{2}$, and 2 tails with probability $\frac{1}{4}$. In the same way, the 16 equally probable outcomes of 4 tosses are of 5 types with the following probabilities:

4 heads

$$P(4H) = P(HHHH) = \left(\frac{1}{2}\right)^4 = \frac{1}{16}$$

3 heads and 1 tail

$$P(3H + 1T) = P(HHHT) + P(HHTH) + P(HTHH) + P(THHH) = 4 \times \left(\frac{1}{2}\right)^4$$

2 heads and 2 tails

$$P(2H + 2T) = P(HHTT) + P(HTHT) + P(HTTH)$$
$$+ P(TTHH) + P(THTH) + P(THHT) = 6 \times \left(\frac{1}{2}\right)^4$$

1 head and 3 tails

$$P(1H + 3T) = P(TTTH) + P(TTHT) + P(THTT) + P(HTTT) = 4 \times \left(\frac{1}{2}\right)^4$$

4 tails

$$P(4T) = P(TTTT) = \left(\frac{1}{2}\right)^4$$

We see that the coefficients of the factor $\left(\frac{1}{2}\right)^4$ of the probabilities are the numbers $\{1, 4, 6, 4, 1\}$ that form the fifth row of the Pascal triangle. They are the binomial coefficients $\begin{pmatrix} 4 \\ m \end{pmatrix}$, for m heads and $4 - m$ tails. In the case of n trials, the total number of possible outcomes is 2^n (2 per trial) and the probability of m heads and $n - m$ tails is

$$P(mH + (n - m)T) = P_m = \begin{pmatrix} n \\ m \end{pmatrix} \times \left(\frac{1}{2}\right)^n \qquad (21.18)$$

This collection of probabilities is called a **binomial distribution** (or Bernoulli distribution).

More generally, the binomial distribution applies to any experiment whose outcomes can be treated as two exclusive events. These are often called 'success', with probability p, and 'failure', with probability $q = 1 - p$. The probability of obtaining m successes from n trials is then

$$P_m = \binom{n}{m} p^m q^{n-m} = \frac{n!}{m!(n-m)!} p^m (1-p)^{n-m} \tag{21.19}$$

The mean, variance and standard deviation of the distribution are

$$\mu = \langle m \rangle = np, \qquad V(n) = np(1-p), \qquad \sigma = \sqrt{np(1-p)} \tag{21.20}$$

EXAMPLE 21.6 The probability that the outcome of 10 tosses of a coin consists of m heads (and $10 - m$ tails) is

$$P_m = \binom{10}{m} \times \left(\frac{1}{2}\right)^{10}$$

with $\mu = 5$, $V = 2.5$, and $\sigma = 1.58$. The corresponding binomial distribution is compared in Table 21.5 with the relative frequencies f_m obtained from Table 21.2.

Table 21.5

m	0	1	2	3	4	5	6	7	8	9	10
P_m	0.001	0.01	0.04	0.12	0.21	0.25	0.21	0.12	0.04	0.01	0.001
f_m	0.02	0.00	0.10	0.10	0.18	0.20	0.18	0.12	0.09	0.04	0.00

The mean, variance, and standard deviation of the frequency distribution are $\bar{n} = 4.98$, $V(n) = 3.8$, and $s = 1.95$. The differences between theory and experiment are due to the small size of the sample (50 sets of 10 tosses).

EXAMPLE 21.7 Find the probability of throwing at least 2 'sixes' in 4 throws of a fair die.

The probability of success in one throw is $p = \frac{1}{6}$, so that $q = \frac{5}{6}$. The total probability of at least two successes (two, three or four) is then

$$P = P_2 + P_3 + P_4$$

$$= \binom{4}{2}\left(\frac{1}{6}\right)^2\left(\frac{5}{6}\right)^2 + \binom{4}{3}\left(\frac{1}{6}\right)^3\left(\frac{5}{6}\right) + \binom{4}{4}\left(\frac{1}{6}\right)^4$$

$$= \left(\frac{1}{6}\right)^4 (150 + 20 + 1) = \frac{171}{1294} = 0.132$$

▸ Exercises 12, 13

Multinomial distributions

Multinomial distributions are generalizations of the binomial distribution for experiments that have more than two possible outcomes. Let an experiment have k possible outcomes, $E_1, E_2, ..., E_k$, with respective probabilities $p_1, p_2, ..., p_k$. Then, the joint probability that there are n_i occurrences of outcome E_i ($i = 1, 2, ..., k$), in n independent trials is

$$P(n_1, n_2, ..., n_k) = \begin{pmatrix} n \\ n_1 \ n_2 \ ... \ n_k \end{pmatrix} p_1^{n_1} p_2^{n_2} \cdots p_k^{n_k} \qquad (21.21)$$

where

$$\begin{pmatrix} n \\ n_1 \ n_2 \ \cdots \ n_k \end{pmatrix} = \frac{n!}{n_1! \, n_2! \cdots n_k!}$$

is a multinomial coefficient (Section 7.3).

▸ Exercises 14, 15

21.6 Permutations and combinations

The possible outcomes of tossing a coin or throwing a die are exclusive events with equal probabilities; each of k possible outcomes has probability $1/k$. The counting of equally-probable outcomes is important in several applications of probability theory in the sciences and often involves the counting of the permutations and combinations of sets of objects or events. The following theorems are some of the important results in combinatorial theory.[2]

Permutations

A set of n different (**distinguishable**) objects (real objects, numbers, events, and so on) may be arranged in a row in a number of ways, and each such arrangement is called a **permutation** of the n objects. For example, the three letters A, B, and C can be arranged in the 6 ways

  ABC,   ACB,   BAC,   BCA,   CAB,   CBA

There are $3! = 6$ permutations of 3 different objects. In general:

  1. The number of permutations of n different objects is $n!$

The first object can be put in n different positions, leaving $n-1$ positions for the second object (think of n boxes in a row and n balls of different colours). The first two

---

[2] An early discussion of permutations and combinations is found in the Jewish mystical *Sefer Yetsirah* (Book of creation), possibly of the second century AD, in which the (unknown) author calculates the ways of arranging the 22 letters of the Hebrew alphabet, and the number of combinations taken 2 at a time. The French mathematician, astronomer, and biblical commentator Levi ben Gerson (1288–1344) gave a derivation of the rules of combinations in his *Maasei Hoshev* (Art of the calculator) in 1321. He also invented the Jacob Staff, used for centuries by sailors to measure the angular separation of heavenly bodies.

objects can therefore be arranged in $n(n-1)$ ways. The third object can occupy $n-2$ positions, and the number of ways of arranging 3 different objects is $n(n-1)(n-2)$. In general:

2. The number of permutations of n different objects taken r at a time is

$$^nP_r = n(n-1)(n-2)\cdots(n-r+1) = \frac{n!}{(n-r)!} \qquad (21.22)$$

The total number of permutations of n objects (taken n at a time) is therefore $^nP_n = n!$

EXAMPLE 21.8 The permutations of the 4 objects A, B, C, D taken 2 at a time are

AB and BA, AC and CA, AD and DA, BC and CB, BD and DB, CD and CD

and $^4P_2 = 4!/(4-2)! = 12$.

➤ Exercises 16, 17

Combinations

In a permutation, the order of the selected objects is important. A **combination**, on the other hand, is a selection of objects without regard to order.

3. The number of combinations of n different objects taken r at a time is

$$^nC_r = \frac{n!}{r!(n-r)!} = \binom{n}{r} \qquad (21.23)$$

EXAMPLE 21.9 The number of combinations of 5 objects taken 3 at a time is

$$^5C_3 = \frac{5!}{3!\,2!} = \frac{5\cdot4\cdot3\cdot2\cdot1}{(3\cdot2\cdot1)(2\cdot1)} = 10$$

Thus, for A, B, C, D, E:

ABC, ABD, ABE, ACD, ACE, ADE, BCD, BCE, BDE, CDE

There are a possible 3! permutations for each combination, so that $3! \times {}^nC_3 = {}^nP_3$. In general, as can be seen from (21.22) and (21.23), $^nP_r = r! \times {}^nC_r$.

➤ Exercise 18

The number of combinations of n objects taken r at a time is the same as the number of ways of dividing the n objects into 2 groups of r and $n - r$ (2 boxes, one containing r objects, the other $n - r$). More generally:

4. The number of ways of dividing n different objects into k groups, with n_1 objects in group 1, n_2 in group 2, ..., and n_k in group k, is the multinomial coefficient

$$\frac{n!}{n_1! \, n_2! \, n_3! \cdots n_k!} \tag{21.24}$$

This result is important in statistical thermodynamics when considering the number of ways of distributing molecules amongst the available energy states. It gives rise to the classical or **Boltzmann statistics** that describes the behaviour of many thermodynamic systems.

Distinguishable and indistinguishable objects

The above theorems have been presented for sets of different or distinguishable objects, and they must be modified or reinterpreted if some or all the objects are **indistinguishable**. Two objects are called indistinguishable if their interchange cannot be observed, and therefore does not give a new or distinct permutation or combination. For example, theorem 4 can be reinterpreted for the number of distinct permutations of n objects made up of k groups, the objects in any one group being indistinguishable but different from the objects in all other groups.

EXAMPLE 21.10 The $\begin{pmatrix} 4 \\ 2 \end{pmatrix} = 6$ distinct permutations of the 4 objects A, A, B, B are

AABB, ABAB, ABBA, BAAB, BABA, BBAA

▶ Exercises 19–21

For n indistinguishable objects:

5. The number of permutations of n indistinguishable objects is 1.

6. The number of ways of distributing k indistinguishable objects (for example, electrons) amongst $n \, (\geq k)$ boxes (quantum states) with *not more than one object per box* is $^nC_k = \begin{pmatrix} n \\ k \end{pmatrix}$.

The n boxes are either singly-occupied or are unoccupied, so that theorem 3 applies to the boxes. This result is important in the statistical mechanics of systems of particles that obey the Pauli exclusion principle. It gives rise to the **quantum statistics** called **Fermi–Dirac statistics** that is important for the description of some thermodynamic systems at very low temperatures and, for example, for the description of the properties of the conduction electrons in solids.

> Exercise 22

Large numbers

EXAMPLE 21.11 The number of permutations of 52 different objects (a pack of cards) is $52! \approx 8.1 \times 10^{67}$.

The number of objects considered in some applications of combinatorial theory in the sciences can be very large. For example, we are concerned in statistical thermodynamics with the possible arrangements of the particles of a mole of substance amongst the available energy states, so that $n \approx 10^{23}$, and $n!$ is about $10^{10^{25}}$. The computation and manipulation of such numbers is simplified by means of Stirling's approximation[3]

$$n! \sim \sqrt{2\pi n}\left(\frac{n}{e}\right)^n \qquad (21.25)$$

where the symbol $\sim$ is read as 'asymptotically equal to' and means that the ratio of the two sides of (21.25) approaches unity as $n \to \infty$. In many cases it is $\ln n!$ that is of interest rather than $n!$ itself. Then

$$\ln n! \sim n \ln n - n + \frac{1}{2} \ln 2\pi n \qquad (21.26)$$

Use of this asymptotic formula gives errors of about $1/(12n)$ in $\ln n!$ when n is large (see Example 20.14). When n is large enough, we can put

$$\ln n! \approx n \ln n - n \qquad (21.27)$$

Some values are given in Table 21.6.

Table 21.6 Stirling's approximation

n	$\ln n!$	eq.(21.26)	$n \ln n - n$	$\ln n! - (n \ln n - n)$
10	15.104	15.096	13.026	2.078
52	156.361	156.359	153.465	2.896
10^{10}		2.2×10^{11}	2.2×10^{11}	12.7
10^{23}		5.2×10^{24}	5.2×10^{24}	27.4

---

[3] James Stirling (1692–1770), Scottish mathematician, was, like Cotes and de Moivre, a friend of Newton. The formula named for him appeared in his *Methodus differentialis* in 1730 and, also in 1730, in de Moivre's *Miscellanea analytica*.

Equation (21.26) is already very accurate when $n = 52$, and can be assumed to be exact for the larger values of n in the table. The difference between (21.26) and the 'more approximate' (21.27) is $\frac{1}{2} \ln 2\pi n$, and this increases only very slowly with n. The absolute error of 27.4 when $n = 10^{23}$ corresponds to the fractional error 5×10^{-24} in $\ln n!$

An important use of Stirling's formula is given in Example 21.12 below. We consider first how a distribution, such as the binomial distribution, behaves when the numbers involved are large. Figure 21.4 shows the binomial distribution (21.18) for $n = 10$ and $n = 100$. The graphs are of $P_m/P_{n/2}$ against m/n, with maximum value 1 when $m/n = 1/2$.

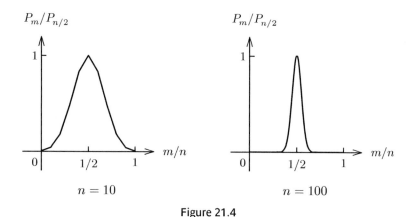

Figure 21.4

They show that when n is large the properties of the distribution are dominated by the probabilities at and close to the maximum. A measure of this relative narrowing of the distribution curve is given by the fractional standard deviation $\sigma/n = 1/(2\sqrt{n})$.

EXAMPLE 21.12 The Boltzmann distribution

In statistical thermodynamics, the Boltzmann distribution for a system of n particles with total energy E is often derived by considering the number of ways the particles can be distributed amongst the particle states. Let there be k available states with energies $\varepsilon_1, \varepsilon_2, \ldots, \varepsilon_k$, and let n_1 particles have energy ε_1, n_2 have energy ε_2, and so on. The number of ways of arranging the n particles in this way (assuming distinguishable particles) is then given by theorem 4, equation (21.24):

$$W = \frac{n!}{n_1! \, n_2! \cdots n_k!} \tag{21.28}$$

When n is large, as it is in a thermodynamic system, the properties of the system are dominated by the distribution for which W has its maximum value. This most probable distribution is obtained by optimizing W with respect to the occupation numbers (treated as continuous variables) subject to the constraints that the total number of particles n is constant and that the total energy E is constant:

$$n = \sum_{i=1}^{k} n_i = \text{constant}, \qquad E = \sum_{i=1}^{k} n_i \varepsilon_i = \text{constant} \qquad (21.29)$$

We first take the logarithm of W and use Stirling's approximation (21.27) to simplify the expression:

$$\ln W = \ln n! - \sum_{i=1}^{k} \ln n_i! = (n \ln n - n) - \sum_{i=1}^{k} (n_i \ln n_i - n_i) \qquad (21.30)$$

The maximum value of $\ln W$ subject to the constraints is then obtained by the method of Lagrange multipliers; that is, we find the maximum value of the auxiliary function

$$\phi = \ln W + \alpha \sum_{i=1}^{k} n_i + \beta \sum_{i=1}^{k} n_i \varepsilon_i \qquad (21.31)$$

where α and β are the multipliers (see Section 9.4). For a stationary value (maximum) with respect to n_i,

$$\frac{\partial \phi}{\partial n_i} = \frac{\partial}{\partial n_i} \ln W + \alpha + \beta \varepsilon_i = 0$$

Therefore, because $\dfrac{\partial}{\partial n_i} \ln W = -\ln n_i$, we have $\ln n_i = \alpha + \beta \varepsilon_i$ so that the most probable distribution of particles is given by the occupation numbers

$$n_i = e^{\alpha + \beta \varepsilon_i} \qquad (21.32)$$

With appropriate interpretations of the quantities α and β, this set of numbers is called the Boltzmann distribution. Thus, $\beta = -1/kT$, so that $n_i = e^{\alpha} e^{-\varepsilon_i/kT}$ and

$$\sum_{i=1}^{k} n_i = n = e^{\alpha} \sum_{i=1}^{k} e^{-\varepsilon_i/kT}$$

The fractional number of particles in state i is then

$$\frac{n_i}{n} = \frac{1}{q} e^{-\varepsilon_i/kT}$$

where $q = \sum_{i=1}^{k} e^{-\varepsilon_i/kT}$ is called the partition function.

21.7 Continuous distributions

The distributions considered so far have been **discrete**, involving the discrete variable x (or a variable that is treated as discrete for practical purposes). Each value of such a variable has an observed relative frequency and a corresponding theoretical probability. Such distributions describe processes that involve the counting of discrete events, and one example is the binomial distribution described in Section 21.5.

Processes that involve the measurement of a continuous quantity are described by **continuous** distributions, for which the variable x can have any value in a continuous range, $a \leq x \leq b$ say. The number of possible outcomes is then infinite, and the probability of a *particular* outcome is not defined (it is effectively zero). We define, instead, the probability that the value of x lies in a specified interval $x_1 \leq x \leq x_2$:

$$P(x_1 \leq x \leq x_2) = \int_{x_1}^{x_2} \rho(x)\, dx \qquad (21.33)$$

where $\rho(x)$ is called the **probability density distribution** (or probability density function, or just probability density). When $x_2 - x_1 = \Delta x$ is small enough, the probability in the interval x to $x + \Delta x$ is

$$P(x \rightarrow x + \Delta x) \approx \rho(x)\Delta x \qquad (21.34)$$

with equal sign for an infinitesimal interval dx. The total probability (the normalization of $\rho(x)$) is

$$P(a \leq x \leq b) = \int_a^b \rho(x)\, dx = 1 \qquad (21.35)$$

Figure 21.5 illustrates the graphical interpretation. The total area under the curve between a and b is equal to 1 and represents the total probability (equation (21.35)). The shaded region has area equal to the probability given by (21.33).

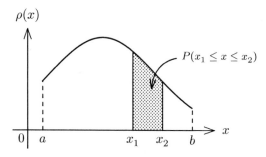

Figure 21.5

The properties of a continuous distribution are obtained by replacing the sums for the discrete distribution by integrals over the range of possible values, and the expectation values are then (see equations (21.11) to (21.23) for the discrete case)

$$\langle f \rangle = \int_a^b f(x)\rho(x)\, dx \qquad (21.36)$$

with mean (expectation value)

$$\mu = \langle x \rangle = \int_a^b x \rho(x)\, dx \tag{21.37}$$

and standard deviation given by

$$\sigma^2 = \int_a^b (x - \mu)^2 \rho(x)\, dx \tag{21.38}$$

An example of a continuous distribution, and probably the most important in statistics, is the Gaussian distribution (normal distribution) discussed in Section 21.8.

EXAMPLE 21.13 The uniform distribution

The simplest continuous distribution has probability

$$\rho(x) = \begin{cases} \dfrac{1}{b-a} & \text{if } a < x < b \\ 0 & \text{otherwise} \end{cases}$$

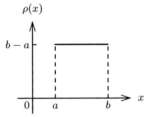

Figure 21.6

The probability density is constant throughout the range, and the probability that the variable x lies in interval $x_1 < x < x_2$ is proportional to the width of the interval:

$$P(x_1 < x < x_2) = \int_{x_1}^{x_2} \rho(x)\, dx = \frac{x_2 - x_1}{b - a}$$

▸ Exercise 23

EXAMPLE 21.14 Radial distribution functions

A continuous distribution in three variables has probability density function such that

$$P(x_1 < x < x_2;\, y_1 < y < y_2;\, z_1 < z < z_2) = \int_{z_1}^{z_2} \int_{y_1}^{y_2} \int_{x_1}^{x_2} \rho(x, y, z)\, dx\, dy\, dz$$

is the probability that the variables lie in the given intervals; that is, (x, y, z) is a point in the rectangular box whose sides are $x_2 - x_1$, $y_2 - y_1$, $z_2 - z_1$. The density function is a function of position in the space of the variables, and $\rho(r)dv$ is the probability that the point is in volume dv at position r. Some continuous three-dimensional distributions of importance in chemistry are the electron probability distributions obtained from solutions of the Schrödinger equation.

If the variables are transformed to spherical polar coordinates in the space then

$$P(r_1 < r < r_2; \theta_1 < \theta < \theta_2; \phi_1 < \phi < \phi_2) = \int_{\phi_1}^{\phi_2} \int_{\theta_1}^{\theta_2} \int_{r_1}^{r_2} \rho(r, \theta, \phi) \, r^2 \sin\theta \, dr \, d\theta \, d\phi$$

is the probability that the point lies in the section of spherical shell between radii r_1 and r_2, angles θ_1 and θ_2, and angles ϕ_1 and ϕ_2 (see the discussion of Figure 10.6). The probability that the point lies *anywhere* in the shell between r_1 and r_2 is then

$$P(r_1 < r < r_2) = P(r_1 < r < r_2; 0 < \theta < \pi; 0 < \phi < 2\pi)$$

$$= \int_{r_1}^{r_2} \left[\int_0^{2\pi} \int_0^{\pi} \rho(r, \theta, \phi) r^2 \sin\theta \, d\theta \, d\phi \right] dr$$

The quantity in square brackets,

$$p(r) = \int_0^{2\pi} \int_0^{\pi} \rho(r, \theta, \phi) r^2 \sin\theta \, d\theta \, d\phi$$

is called a **radial density function**, and $p(r)dr$ is the probability that the variable $r = \sqrt{x^2 + y^2 + z^2}$ has value in the (infinitesimal) interval r to $r+dr$ that is, that (x, y, z) lies in a spherical shell of radius r and thickness dr.

For example, the modulus square of the wave function of an electron, $\rho(r, \theta, \phi) = |\psi(r, \theta, \phi)|^2$ is an electron probability density function. For the electron in the hydrogen atom, the wave function has the form (see equation 14.83)

$$\psi_{n,l,m}(r, \theta, \phi) = R_{n,l}(r) \Theta_{l,m}(\theta) \Phi_m(\phi),$$

and the corresponding *radial* density function is

$$p(r) = r^2 R_{n,l}^2(r)$$

▸ Exercises 24, 25

21.8 The Gaussian distribution

The **Gaussian distribution**, also called the **normal distribution**, is the continuous distribution with mean μ and standard deviation σ whose probability density function is

$$p(x) = \frac{1}{\sigma\sqrt{2\pi}} e^{-(x-\mu)^2/2\sigma^2} \tag{21.39}$$

for all values of x. The graphs for three values of σ are shown in Figure 21.7.

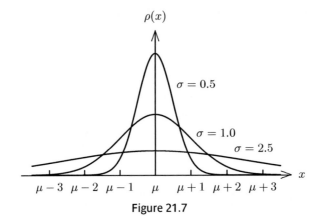

Figure 21.7

The function is symmetric about $x = \mu$. It is broad when σ is large and narrow when σ is small, and it falls to about 0.6 of its maximum value when $x = \pm\sigma$, the points of inflection of the function.

Errors

The Gaussian distribution is important in the sciences because it describes the distribution of errors in a sequence of random experiments.[4] Measurements, however accurately made, are always accompanied by errors. They are of two kinds: determinate and random. Determinate errors may be simply mistakes or they may be systematic errors due to faulty or incorrectly calibrated apparatus. It is usually possible to discover the causes of such errors, and either to eliminate them or make corrections for them. Random errors, on the other hand, are indeterminate. They may be due to the 'shaky hand', the imperfect resolution of an instrument, the unpredictable fluctuations of temperature or voltage.

Errors are in general due to several or many causes. The **central limit theorem** of probability theory tells us that a variable produced by the cumulative effect of several independent variables is approximately Gaussian or 'normal'. Many observed quantities are of this kind; for example, anatomical measurements such as height or length of nose are due to the combined effects of many genetic and environmental factors, and their distributions are Gaussian. In addition, many theoretical distributions behave like Gaussian distributions when the numbers are large. The binomial distribution tends to a Gaussian with $\mu = np$ and $\sigma = \sqrt{np(1-p)}$ as $n \to \infty$. Random numbers have a uniform distribution when considered singly. If taken n at a time, however, they have a distribution that becomes Gaussian as n becomes large.

The distribution function

The probability distribution function $F(x)$ of the Gaussian distribution is defined by

$$F(x) = \int_{-\infty}^{x} \rho(x)\, dx = \frac{1}{\sigma\sqrt{2\pi}} \int_{-\infty}^{x} e^{(x-\mu)^2/2\sigma^2}\, dx \qquad (21.40)$$

---

[4] The first description of the normal law of errors was by de Moivre in 1738. It was derived by Gauss in his *Theory of motion of the heavenly bodies* of 1809 from a principle of maximum probability: 'if any quantity has been determined by several observations, made under the same circumstances and with equal care, the arithmetical mean of the observed values affords the most probable value'. Gauss considered the function (21.39) to be the 'correct' error function because he was able to derive from it the principle of least squares. Laplace gave an alternative derivation of the normal law in 1810, and included it in his 1812 *Analytical theory of probability*.

so that $F(a)$ is the probability that the variable has value $x < a$, and

$$F(x \to \infty) = \int_{-\infty}^{\infty} p(x)\, dx = 1 \tag{21.41}$$

for unit total probability. The probability that the variable has a value in the interval $a < x < b$ is then

$$P(a < x < b) = F(b) - F(a) = \frac{1}{\sigma\sqrt{2\pi}} \int_{a}^{b} e^{(x-\mu)^2/2\sigma^2}\, dx \tag{21.42}$$

and this is the area under the curve between $x = a$ and $x = b$.

The integral in (21.40) cannot be evaluated by the methods of the calculus described in Chapters 5 and 6, but extensive tabulations are found in most statistics texts. These tabulations are of the auxiliary function

$$\Phi(z) = \frac{1}{\sqrt{2\pi}} \int_{-\infty}^{z} e^{-z^2/2}\, dz \tag{21.43}$$

obtained from (21.40) by means of the substitution $z = \dfrac{x - \mu}{\sigma}$. Then $F(x) = \Phi\left(\dfrac{x - \mu}{\sigma}\right)$ and

$$P(a < x < b) = \Phi\left(\frac{b - \mu}{\sigma}\right) - \Phi\left(\frac{a - \mu}{\sigma}\right) \tag{21.44}$$

It is found from these tabulations that the probabilities that x lies within σ, 2σ, and 3σ of the mean are

$$P(\mu - \sigma < x < \mu + \sigma) \quad = 0.6827$$
$$P(\mu - 2\sigma < x < \mu + 2\sigma) = 0.9545 \tag{21.45}$$
$$P(\mu - 3\sigma < x < \mu + 3\sigma) = 0.9973$$

Therefore, about 68% of observed values are expected to lie within one standard deviation of the mean (Figure 21.8), 95% within two standard deviations, and almost all within three standard deviations.

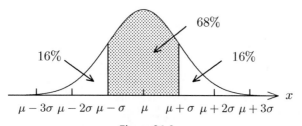

Figure 21.8

21.9   More than one variable

In many experiments in the physical sciences, the outcomes consist of the values of more than one quantity at a time. For example, a kinetics experiment involves the measurement of pairs (c, t) of concentration and time, in thermodynamics we may have the pairs (p, T), in dynamics we may be following the flight of a body by measuring (x, y, z, t) for position and time. Alternatively, the four values may be height, IQ, mark in a maths examination, and personal best time (PB) in the 100 m breaststroke for each member of a class of chemistry students.

Covariance and correlation

We consider an experiment in which each outcome consists of values of a pair of variables (x, y): $(x_1, y_1), (x_2, y_2), \ldots, (x_N, y_N)$ for a sample of N pairs. We can compute the means, $\bar{x}$ and $\bar{y}$, variances, $V(x)$ and $V(y)$, and standard deviations, σ_x and σ_y, of the separate variables. We can also consider the two variables together and determine whether they are dependent on one another. A measure of the dependence of two variables is the **covariance**, defined by

$$\text{cov}(x, y) = \frac{1}{N} \sum_i (x_i - \bar{x})(y_i - \bar{y}) = \overline{xy} - \bar{x}\,\bar{y} \tag{21.46}$$

where $\overline{xy} = \dfrac{1}{N} \sum x_i y_i$. If the variables are *independent* then $\overline{xy} = \bar{x}\bar{y}$ (each value of x can occur with each value of y) and the covariance is zero. If values of x greater than the mean tend to occur with above average values of y, the covariance is positive because the terms $(x_i - \bar{x})(y_i - \bar{y})$ are positive. If large values of x occur with small values of y, the covariance is negative.

An alternative measure of dependence is the **correlation coefficient**

$$\rho_{x,y} = \frac{\text{cov}(x, y)}{\sigma_x \sigma_y} = \frac{\overline{xy} - \bar{x}\,\bar{y}}{\sigma_x \sigma_y} \tag{21.47}$$

This quantity is dimensionless, with values $-1 \le \rho_{x,y} \le +1$. The variables x and y are **uncorrelated** when $\rho_{x,y} = 0$, show **positive correlation** if $\rho_{x,y}$ is positive, and **negative correlation** when it is negative. If $\rho_{x,y} = \pm 1$, x and y are completely (100%) correlated, and a functional relation may exist between the variables. The scatter plots in Figure 21.9 show some examples of correlation.

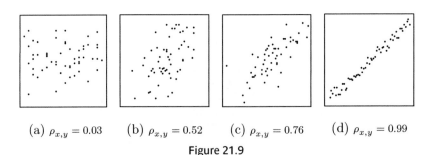

(a) $\rho_{x,y} = 0.03$ (b) $\rho_{x,y} = 0.52$ (c) $\rho_{x,y} = 0.76$ (d) $\rho_{x,y} = 0.99$

Figure 21.9

The first three are for the performance of a class of first-year chemistry students. Figure (a) shows the total chemistry marks (vertically) against the initial letter of the surnames of the students (horizontally). No correlation is expected, and none is found, with $\rho_{x,y} = 0.03$. There is some correlation between the total chemistry marks and the maths marks, with $\rho_{x,y} = 0.52$, and greater correlation ($\rho_{x,y} = 0.76$) between the inorganic and organic chemistry marks. Figure (d) is typical of the results obtained when a linear functional relation is expected to exist between the variables (a high correlation coefficient may also be obtained even when no such direct functional relation exists).

21.10 Least squares

Many experiments in the sciences are performed to measure pairs of physical quantities that are known, or are suspected, to be functionally related; that is, there exists some function $y = f(x)$. One aim of the experiment is then to confirm, or determine, the relation. Such experiments are often designed in such a way that one of the quantities, x say, can be measured precisely (with negligible error), with the errors confined to y; for example, x might be the time in a kinetics experiment and y a concentration.

We consider a sample of N data points, (x_i, y_i), $i = 1, 2, \ldots, N$, in which the values of x are assumed to be precise and with each value of y is associated a measure of precision, σ_i for y_i. If the errors in y are random then σ_i is the standard deviation (or an estimate of the standard deviation) of the normal distribution to which y_i belongs. Then, $y_i \pm \sigma_i$ means that about 68% of measurements of y, when $x = x_i$, lie within σ_i of the mean $\bar{y}_i$. Figure 21.10 shows a typical plot of such a set of data points.

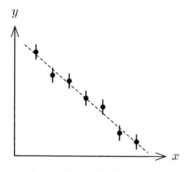

Figure 21.10

Each point is plotted with appropriate error bar, and the dashed line indicates that the points can be fitted to a straight line

$$y = y(x; m, c) = mx + c \tag{21.48}$$

where m and c are parameters that determine the slope and intercept of the line. More generally, the problem is to fit the data points to a (simple) model function

$$y = y(x; a_1, a_2, \ldots, a_k) = y(x; \boldsymbol{a}) \tag{21.49}$$

with k adjustable parameters $\boldsymbol{a} = (a_1, a_2, \ldots, a_k)$ to be determined by some criteria of 'best fit'.

Simple least squares fitting

The simplest and one of the most popular methods of fitting a set of data points to a model function (21.49) is to minimize the quantity

$$D = \sum_{i=1}^{N}\left[y_i - y(x_i; a) \right]^2 = \sum_{i=1}^{N} \varepsilon_i^2 \tag{21.50}$$

with respect to the parameters $a_1, a_2, \ldots, a_k$:

$$\frac{\partial D}{\partial a_1} = 0, \quad \frac{\partial D}{\partial a_2} = 0, \quad \ldots, \quad \frac{\partial D}{\partial a_k} = 0 \tag{21.51}$$

As shown in Figure 21.11, each term of the sum is the square of the difference

$$\varepsilon_i = y_i - y(x_i; a)$$

between the actual and predicted values of y for each value of $x = x_i$.

This method of least squares can be regarded either as an empirical curve-fitting procedure that is justified on the grounds that it often gives sensible results, or it can be derived from the theory of random errors for the special case that the σ_i values are all equal; that is, when all the y_i have the same precision. It is also the method to be used when no information about precision is available.[5]

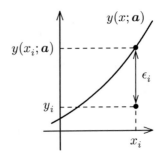

Figure 21.11

The straight-line fit

The quantity to be minimized for a fit to the straight line $y = mx + c$ is

$$D = \sum_{i=1}^{N}\left[y_i - (mx_i + c) \right]^2 \tag{21.52}$$

Then

$$\frac{\partial D}{\partial m} = -2\sum_{i=1}^{N} x_i(y_i - mx_i - c) = 0 \quad \text{or} \quad \sum_{i=1}^{N} x_i y_i - m\sum_{i=1}^{N} x_i^2 - c\sum_{i=1}^{N} x_i = 0$$

$$\frac{\partial D}{\partial c} = -2\sum_{i=1}^{N} (y_i - mx_i - c) = 0 \quad \sum_{i=1}^{N} y_i - m\sum_{i=1}^{N} x_i - c\sum_{i=1}^{N} 1 = 0$$

---

[5] The method of least squares (*la méthode des moindres quarrés*) was developed by Legendre in 1805 in connection with the determination of the orbits of comets, and quickly became a standard method for solving problems in astronomy and geodesy. The method also appeared in 1809 in Gauss' *Theory of motion of the heavenly bodies*. Gauss did not quote Legendre, and claimed that he had been using the method since 1795. Legendre argued that precedence in scientific discoveries could only be established by publication, and was still accusing Gauss in 1827 of appropriating the discoveries of others.

Division by $N = \sum_{i=1}^{N} 1$ then gives the pair of **normal equations**

$$\overline{xy} - m\overline{x^2} - c\overline{x} = 0$$
$$\overline{y} - m\overline{x} - c = 0 \tag{21.53}$$

and these have solution

$$m = \frac{\overline{xy} - \overline{x}\,\overline{y}}{\overline{x^2} - \overline{x}^2}, \qquad c = \overline{y} - m\overline{x} \tag{21.54}$$

The resulting line passes through the centroid $(\overline{x}, \overline{y})$ of the data points.

If the y_i values all have the same precision σ then the estimated parameters m and c have standard deviations given by

$$\sigma_m^2 = \frac{\sigma^2}{N\left(\overline{x^2} - \overline{x}^2\right)}, \qquad \sigma_c^2 = \frac{\sigma^2 \overline{x^2}}{N\left(\overline{x^2} - \overline{x}^2\right)} \tag{21.55}$$

and the linear least-squares fit with estimated errors is $y = mx + c$ with

$$m = \frac{\overline{xy} - \overline{x}\,\overline{y}}{\overline{x^2} - \overline{x}^2} \pm \sigma_m, \qquad c = \overline{y} - m\overline{x} \pm \sigma_c \tag{21.56}$$

EXAMPLE 21.15 Find the linear least-squares fit for the following data points:

Table 21.7

x	0	3	6	9	12	15	18
y	3.3	2.5	2.3	1.7	1.4	0.5	0.2

We have $N = 7$, $\overline{x} = 9$, $\overline{y} = 1.7$, $\overline{x^2} = 117$, $\overline{xy} = 64.5/7 = 9.21429$, $\overline{x^2} - \overline{x}^2 = 36$ and $\overline{xy} - \overline{x}\,\overline{y} = -42.6/7 = -6.08571$. Then

$$\sigma_m = \sigma/\sqrt{252} = 0.0630\sigma, \qquad \sigma_c = \sqrt{\frac{117}{252}}\sigma = 0.6814\sigma$$

so that

$$m = \frac{-42.6}{252} \pm \frac{\sigma}{\sqrt{252}} = -0.1691 \pm 0.0630\sigma$$

$$c = 1.7 + \frac{42.6 \times 9}{252} \pm \sigma\sqrt{\frac{117}{252}} = 3.2214 \pm 0.6814\sigma$$

This is the example shown in Figure 21.10. The error bars correspond to $\sigma = 0.25$ so that the linear least-squares fit in this case is

$$y = mx + c, \quad \text{where} \quad m = -0.1691 \pm 0.0158, \quad c = 3.2214 \pm 0.1704$$

> Exercises 26, 27

Chi-square fitting

When the (known) errors in the values of y are not all the same, with σ_i for y_i, the simple least-squares method is replaced by the more general weighted least-squares method:

$$\chi^2 = \sum_{i=1}^{N} \left[\frac{y_i - f(x_i; \boldsymbol{a})}{\sigma_i} \right]^2 = \text{minimum} \tag{21.57}$$

Each contribution to the sum is weighted by the factor $1/\sigma_i^2$, and this has the effect of giving greater weight to the more precise values of y (for the smaller values of σ_i). This situation arises, for example, when the precision with which the measurements can be made varies over the range of y values, or when the data points come from measurements on several different instruments. This fitting method reduces to simple least squares when all the σ_i are equal.

The minimization of χ^2 for a straight-line fit gives the same equations as before, (21.54) for m and c and (21.55) for σ_m and σ_c, except that the averages are replaced by weighted averages, for example

$$\bar{y} = \frac{1}{N} \sum_{i=1}^{N} y_i \quad \text{is replaced by} \quad \bar{y} = \left(\sum_{i=1}^{N} \frac{y_i}{\sigma_i^2} \right) \Big/ \left(\sum_{i=1}^{N} \frac{1}{\sigma_i^2} \right) \tag{21.58}$$

and, in equations (21.55),

$$\frac{\sigma^2}{N} \quad \text{is replaced by} \quad 1 \Big/ \left(\sum_{i=1}^{N} \frac{1}{\sigma_i^2} \right) \tag{21.59}$$

Fits to other types of function are obtained in the same way, but lead to more complicated formulas. The method can also be generalized for more than two variables.

The justification of the use of equation (21.57) comes from a consideration of the normal distribution (21.39) for random errors. If the errors in y_i are random and if $f(x_i)$ is the (unknown) *true* value of y when $x = x_i$ then

$$p(\varepsilon_i) = \frac{1}{\sigma_i \sqrt{2\pi}} e^{-\varepsilon_i^2 / 2\sigma_i^2} \tag{21.60}$$

is the probability density for the error $\varepsilon_i = y_i - f(x_i)$. If the errors are independent, the total probability density for the N values $\varepsilon_1, \varepsilon_2, \ldots, \varepsilon_k$ is the product

$$p(\varepsilon_1)p(\varepsilon_2)\cdots p(\varepsilon_N) = \left(\sqrt{2\pi}\prod_{i=1}^{N}\sigma_i\right)^{-1}\exp\left[-\frac{1}{2}\sum_{i=1}^{N}\left(\frac{\varepsilon_i}{\sigma_i}\right)^2\right]$$

$$= \left(\sqrt{2\pi}\prod_{i=1}^{N}\sigma_i\right)^{-1}\exp\left[-\frac{1}{2}\chi^2\right]$$

(21.61)

When the true function $f(x)$ is replaced by an approximate function, the expression (21.61) is a measure of the likelihood of the fit, and the likelihood is maximized when χ^2 is as small as possible.

Regression

The fitting of data points to functions is often called **regression** by statisticians; the curves about which points are clustered are called 'regression curves', and their equations are 'regression equations'. For a straight line, the regression is called linear. The name has its origin in Galton's studies on heredity in which he found that, on average, fathers whose heights deviate from the mean height of all fathers have sons whose heights deviate from the mean height of all sons by lesser amounts.[6] He called this 'regression to mediocrity', and the word is widely used, particularly in the social sciences. The type of data of interest in the physical sciences, and the purpose for its statistical analysis, is of a quite different nature. The word regression is entirely inappropriate as a description of the use of statistics in chemistry.

21.11   Sample statistics

In any practical experiment the number of measurements, the sample size, is necessarily finite so that only *estimates* of the parent population or distribution are obtained from the statistics of the sample. Thus, the sample mean defined in equation (21.2) gives an estimate of the population mean,

$$\mu \approx \bar{x} = \frac{1}{N}\sum_{i=1}^{N}x_i$$

(21.62)

and the variance and standard deviation defined in equations (21.5) and (21.6) are estimates of the variance and standard deviation of the population,

$$\sigma^2 \approx s^2 = \frac{1}{N}\sum_{i=1}^{N}(x_i - \bar{x})^2$$

(21.63)

---

[6] Francis Galton (1822–1911), born in Birmingham. He discovered the importance of anti-cyclones in weather systems and was influential in the establishment of the Meteorological Office and the National Physical Laboratory. Best known for his work on heredity, he coined the word 'eugenics' and advocated the application of scientific breeding to human populations.

Considerations of 'best estimates' in sampling theory tell us that although (21.62) is the best estimate of μ obtainable from the sample, (21.63) underestimates the best estimate of variance by the factor $(N-1)/N$, and should be replaced by the corrected estimate

$$\sigma^2 \approx s^2 = \frac{1}{N-1} \sum_{i=1}^{N} (x_i - \bar{x})^2 \tag{21.64}$$

The two estimates of σ given by (21.63) and (21.64) are often distinguished by denoting them by s_N and s_{N-1}, respectively (or by σ_N and σ_{N-1}, as on several popular makes of pocket calculator). The reader is advised always to check which particular version of the standard deviation is being used in any work or application of statistics. The correction has little effect on the estimate of σ when N is large, but can be significant for small samples, with $N < 10$ say, and it is always safer to use (21.64).

EXAMPLE 21.16 A very small sample

For the sample of three values, $x_1 = a - b$, $x_2 = a$ and $x_3 = a + b$, the mean is $\bar{x} = a$ and the two versions of s^2 give

$$s_N^2 = \frac{1}{3}\left[b^2 + 0 + b^2 \right] = \frac{2}{3} b^2, \qquad s_{N-1}^2 = \frac{1}{2}\left[b^2 + 0 + b^2 \right] = b^2$$

21.12 Exercises

Section 21.2

1. The following data consists of the numbers of heads obtained from 10 tosses of a coin:

    5, 5, 4, 4, 7, 4, 3, 7, 6, 4, 2, 5, 6, 4, 5, 3, 5, 4, 2, 6, 7, 2, 4, 5, 6,
    5, 6, 4, 3, 4, 4, 5, 5, 6, 7, 5, 3, 6, 5, 5, 6, 7, 9, 4, 7, 9, 8, 8, 5, 10

    Construct (i) a frequency table, (ii) a frequency bar chart.
2. The following data consists of percentage marks achieved by 60 students in an examination:

    66, 68, 70, 48, 56, 54, 48, 47, 45, 53, 73, 60, 68, 75, 61, 62, 61, 61, 52, 59,
    58, 56, 58, 69, 48, 62, 72, 71, 49, 69, 59, 48, 64, 59, 53, 62, 66, 55, 41, 66,
    60, 38, 54, 69, 60, 53, 60, 64, 57, 54, 73, 46, 73, 58, 50, 66, 37, 60, 47, 70

    Construct (i) a class frequency table for classes of width 5, (ii) the corresponding frequency histogram.
3. Calculate the mean, mode, and median of the data in Exercise 1.
4. Calculate the mean, mode, and median of the data in Exercise 2 (i) using the raw (ungrouped) data, (ii) using the data grouped in classes of width 5.
5. Calculate the variance and standard deviation of the data in Exercise 1.

6. For the data in Exercise 1, (i) calculate $\bar{x}$, $\overline{x^2}$, and $\overline{x^3}$, (ii) use equations (21.7) and (21.9) to compute the standard deviation and the skewness.

Section 21.4

7. A set of 10 balls consists of 6 red balls, 3 blue, and 1 yellow. If a ball is drawn at random, find the probability that it is (i) red, (ii) yellow, (iii) red or yellow, (iv) not blue, (v) not yellow.

8. A set of 50 numbered discs consists of 8 ones, 12 twos, 14 threes, 7 fours, and 9 fives. If one disc is drawn at random, what is the probability that its number is (i) 2, (ii) 4, (iii) 2 or 4, (iv) ≤ 4, (v) odd.

9. Find the probabilities $P(2)$ to $P(12)$ of all the possible outcomes of two throws of a die.

10. Find the probability of the following total scores from three throws of a die: (i) 4, (ii) 8, (iii) 4 or 8, (iv) more than 15.

11. A particle can be in three states with energies ε_0, ε_1, and ε_2 ($\varepsilon_0 < \varepsilon_1 < \varepsilon_2$), and probability distribution $P_i = e^{-\varepsilon_i/kT}/q$ at temperature T. The quantity q is called the particle partition function.
   (i) Express q in terms of T and the energies (use $\sum_i P_i = 1$). (ii) Find the probability distribution in the limit (a) $T \to 0$, (b) $T \to \infty$. (iii) Find the (combined) probability distribution for a system of three independent particles.

Section 21.5

12. Find the probability that at least 3 heads are obtained from 5 tosses of (i) an unbiased coin, (ii) a coin with probability 0.6 of coming up tails.

13. A system that can exist in a number of states with energies $E_0 < E_1 < E_2 < \cdots$ has probability 0.1 of being in an excited state (with $E > E_0$). Find the probability that in 10 independent observations, the system is found in the ground state (with $E = E_0$) (i) every time, (ii) only 5 times, (iii) at least 8 times.

14. Use the probability distribution of the outcomes of throwing a pair of dice, from Exercise 9, to calculate the probability that two throws of a pair of dice have outcome (i) $8 + 12$ (one eight and one twelve), (ii) $9 + 11$, (iii) $10 + 10$. Hence (iv) find the probability of outcome 20 from two throws of two dice.

15. Use the probability distribution of the outcomes of throwing a pair of dice, from Exercise 9, to calculate the probability that three throws of two dice have total outcome 30.

Section 21.6

16. List the permutations of 4 different objects.
17. List the permutations of 5 different objects taken 2 at a time.
18. List the combinations of 5 different objects taken 2 at a time.
19. List the distinct permutations of the 5 objects, A, A, A, B, and B.
20. What is the number of distinct permutations of 8 objects made up of 4 of type A, 3 of B, and 1 of C?
21. Given an inexhaustible supply of objects A, B, and C, what is the number of distinct permutations of these taken 8 at a time?
22. (i) Given 3 distinguishable particles each of which can be in any of 4 states with (different) energies E_1, E_2, E_3, and E_4, (a) what is the total number of ways of distributing the particles amongst the states?, (b) how many states of the system have total energy $E_1 + E_2 + E_4$? (ii) Repeat (i) for 3 electrons instead of distinguishable particles.

Section 21.7

23. The variable x can have any value in the continuous range $0 \leq x \leq 1$ with probability density function $p(x) = 6x(1 - x)$. **(i)** Derive an expression for the probability, $P(x \leq a)$, that the value of x is not greater than a. **(ii)** Confirm that $P(0 \leq x \leq 1) = 1$. **(iii)** Find the mean $\langle x \rangle$ and standard deviation σ. **(iv)** Find the probability that $\langle x \rangle - \sigma \leq x \leq \langle x \rangle + \sigma$.

24. The variable r can have any value in the range $r = 0 \rightarrow \infty$ with probability density function $p(r) = 4r^2 e^{-2r}$. Derive an expression for the probability $P(0 \leq r \leq R)$ that the variable has value not greater than R.

25. Confirm that $p(r)$ in Exercise 24 is the radial density function (in atomic units) of the $1s$ orbital of the hydrogen atom. Find **(i)** the mean value $\langle r \rangle$, **(ii)** the standard deviation σ, **(iii)** the most probable value of r.

Section 21.10

26. **(i)** Find the linear straight-line fit for the following data points. **(ii)** If the errors in y are all equal to $\sigma = 1$, find estimates of the errors in the slope and intercept of the line.

x	1	2	3	4	5	6	7	8	9	10	11	12
y	4.4	4.9	6.4	7.3	8.8	10.3	11.7	13.2	14.8	15.3	16.5	17.2

27. The results of measurements of the rate constant of the second-order decomposition of an organic compound over a range of temperatures are:

T/K	282.3	291.4	304.1	313.6	320.2	331.3	343.8	354.9	363.8	371.7
$k/10^{-3}\,\mathrm{dm^3 mol^{-1} s^{-1}}$	.0249	.0691	0.319	0.921	1.95	5.98	19.4	57.8	114.	212.

The temperature dependence of the rate constant is given by the Arrhenius equation $k = Ae^{-E_a/RT}$, or $\ln k = -E_a/RT + \ln A$, in which the activation energy E_a and pre-exponential factor A may be assumed to be constant over the experimental range of temperature. A plot of $\ln k$ against $1/T$ should therefore be a straight line. **(i)** Construct a table of values of $1/T$ and $\ln k$, and determine the linear least-squares fit to the data assuming only k is in error. **(ii)** Calculate the best values of E_a and A. **(iii)** Assuming that the errors in $\ln k$ are all equal to $\sigma = 0.1$, use equations (21.55) to find estimates of the errors in E_a and A.

Appendix   Standard integrals

Indefinite Integrals

For simplicity, the constant of integration has been omitted from the tabulation.

$$\int x^n \, dx = \frac{x^{n+1}}{n+1} \quad (n \neq -1)$$

$$\int (ax+b)^n \, dx = \frac{(ax+b)^{n+1}}{a(n+1)} \quad (n \neq -1)$$

$$\int \frac{dx}{x} \, dx = \ln x$$

$$\int \frac{dx}{ax+b} = \frac{1}{a} \ln(ax+b)$$

$$\int \frac{\ln x}{x} \, dx = \frac{1}{2}(\ln x)^2$$

$$\int x^n \ln x \, dx = \frac{x^{n+1}}{n+1}\left(\ln x - \frac{1}{n+1}\right)$$
$$(n \neq -1)$$

$$\int e^x \, dx = e^x$$

$$\int e^{ax+b} \, dx = \frac{1}{a} e^{ax+b}$$

$$\int \sin x \, dx = -\cos x$$

$$\int \sin(ax+b) \, dx = -\frac{1}{a}\cos(ax+b)$$

$$\int \cos x \, dx = \sin x$$

$$\int \cos(ax+b) \, dx = \frac{1}{a}\sin(ax+b)$$

$$\int \tan x \, dx = -\ln \cos x$$

$$\int \cot x \, dx = \ln \sin x$$

$$\int \csc x \, dx = \begin{cases} \ln \tan \dfrac{x}{2} \\[2mm] \dfrac{1}{2} \ln \dfrac{1-\cos x}{1+\cos x} \end{cases}$$

$$\int \sec x \, dx = \begin{cases} \ln \tan \left(\dfrac{\pi}{4} + \dfrac{x}{2}\right) \\[2mm] \dfrac{1}{2} \ln \dfrac{1+\sin x}{1-\sin x} \end{cases}$$

$$\int e^{ax} \sin bx \, dx$$
$$= e^{ax}\left(\frac{a \sin bx - b \cos bx}{a^2+b^2}\right)$$

$$\int e^{ax} \cos bx \, dx$$
$$= e^{ax}\left(\frac{a \cos bx + b \sin bx}{a^2+b^2}\right)$$

$$\int \sin^2 x \, dx = \frac{1}{2}(x - \sin x \cos x)$$

$$\int \cos^2 x \, dx = \frac{1}{2}(x + \sin x \cos x)$$

$$\int \tan^2 x \, dx = \tan x - x$$

$$\int \cot^2 x \, dx = -\cot x - x$$

$$\int \sec^2 x \, dx = \tan x$$

$$\int \cosec^2 dx = -\cot x$$

$$\int \sin ax \sin bx \, dx$$

$$= \frac{1}{2}\left(\frac{\sin(a-b)x}{a-b} - \frac{\sin(a+b)x}{a+b} \right)$$

$$(a \neq \pm b)$$

$$\int \cos ax \cos bx \, dx$$

$$= \frac{1}{2}\left(\frac{\sin(a-b)x}{a-b} + \frac{\sin(a+b)x}{a+b} \right)$$

$$(a \neq \pm b)$$

$$\int \sin ax \cos bx \, dx$$

$$= -\frac{1}{2}\left(\frac{\cos(a-b)x}{a-b} + \frac{\cos(a+b)x}{a+b} \right)$$

$$(a \neq \pm b)$$

$$\int \frac{dx}{1+\cos x} = \tan \frac{x}{2}$$

$$\int \frac{dx}{1-\cos x} = -\cot \frac{x}{2}$$

$$\int \sinh x \, dx = \cosh x$$

$$\int \cosh x \, dx = \sinh x$$

$$\int \tanh x \, dx = \ln \cosh x$$

$$\int \coth x \, dx = \ln \sinh x$$

$$\int \frac{dx}{a^2 + x^2} = \frac{1}{a} \tan^{-1}\left(\frac{x}{a} \right)$$

$$\int \frac{dx}{a^2 - x^2} = \frac{1}{2a} \ln \left| \frac{a+x}{a-x} \right|$$

$$\int \frac{dx}{\sqrt{a^2 - x^2}} = \sin^{-1}\left(\frac{x}{a} \right)$$

$$\int \sqrt{a^2 - x^2} \, dx$$

$$= \frac{a^2}{2} \sin^{-1}\left(\frac{x}{a} \right) + \frac{x}{2}\sqrt{a^2 - x^2}$$

$$\int \frac{dx}{\sqrt{x^2 - a^2}} = \cosh^{-1}\left(\frac{x}{a} \right)$$

$$\int \sqrt{x^2 - a^2} \, dx$$

$$= -\frac{a^2}{2} \cosh^{-1}\left(\frac{x}{a} \right) + \frac{x}{2}\sqrt{x^2 - a^2}$$

$$\int \frac{dx}{\sqrt{a^2 + x^2}} = \sinh^{-1}\left(\frac{x}{a} \right)$$

$$\int \sqrt{a^2 + x^2} \, dx$$

$$= \frac{a^2}{2} \sinh^{-1}\left(\frac{x}{a} \right) + \frac{x}{2}\sqrt{a^2 + x^2}$$

$$\int \sin^{-1} x \, dx = x \sin^{-1} x + \sqrt{1-x^2}$$

$$\int \cos^{-1} x \, dx = x \cos^{-1} x - \sqrt{1-x^2}$$

$$\int \tan^{-1} x \, dx = x \tan^{-1} x - \frac{1}{2}\ln(1+x^2)$$

$$\int \cot^{-1} x \, dx = x \cot^{-1} x + \frac{1}{2} \ln(1 + x^2)$$

$$\int \frac{dx}{ax^2 + bx + c}$$

$$\int \frac{dx}{\sqrt{a + bx + x^2}}$$

$$= \begin{cases} \dfrac{1}{2aq} \ln \left| \dfrac{p+q}{p-q} \right| & \text{if } b^2 > 4ac \\[3mm] \dfrac{1}{aq} \tan^{-1} \left(\dfrac{p}{q} \right) & \text{if } b^2 < 4ac \end{cases}$$

$$= \ln \left(x + \frac{b}{2} + \sqrt{a + bx + x^2} \right)$$

$$\int \frac{dx}{\sqrt{a + bx - x^2}} = \sin^{-1} \left(\frac{2x - b}{\sqrt{b^2 + 4a}} \right)$$

$$\text{where } p = x + \frac{b}{2a}, \quad q^2 = \left| \frac{b^2 - 4ac}{4a^2} \right|$$

Definite integrals

$$\int_0^\infty x^n e^{-ax} \, dx = \frac{n!}{a^{n+1}}$$

$$\int_0^a \sqrt{a^2 - x^2} \, dx = \frac{1}{4} \pi a^2$$

$$\int_0^\infty e^{-ax^2} \, dx = \frac{1}{2} \sqrt{\frac{\pi}{a}}$$

$$\int_0^\infty \frac{\sin ax}{x} \, dx = \frac{\pi}{2}, \quad (a > 0)$$

$$\int_0^\infty x^{2n} e^{-ax^2} \, dx = \frac{1 \cdot 3 \cdots (2n-1)}{2^{n+1} a^n} \sqrt{\frac{\pi}{a}}$$

$$\int_0^\infty \frac{\cos ax - \cos bx}{x} \, dx = \ln \frac{b}{a}$$

$$\int_0^\infty x^{2n+1} e^{-ax^2} \, dx = \frac{n!}{2a^{n+1}}$$

$$\int_0^\infty e^{-ax} \sin bx \, dx = \frac{b}{a^2 + b^2}$$

$$\int_0^a \frac{dx}{\sqrt{a^2 - x^2}} = \frac{\pi}{2}$$

$$\int_0^\infty e^{-ax} \cos bx \, dx = \frac{a}{a^2 + b^2}$$

$$\int_0^\pi \sin^2 nx \, dx = \int_0^\pi \cos^2 nx \, dx = \frac{\pi}{2} \quad (n \neq 0, n \text{ integer})$$

$$\int_0^\pi \sin mx \sin nx \, dx = 0 \quad (m \neq n, m \text{ and } n \text{ integers})$$

$$\int_0^\pi \cos mx \cos nx \, dx = 0 \quad (m \neq n, m \text{ and } n \text{ integers})$$

$$\int_0^\pi \sin mx \cos nx \, dx = 0 \quad (\text{all } m \text{ and } n \text{ integers})$$

Reduction formulas

$$\int x^n e^{ax}\,dx = \frac{x^n e^{ax}}{a} - \frac{n}{a}\int x^{n-1} e^{ax}\,dx$$

$$\int \cos^n x\,dx = \frac{\sin x \cos^{n-1} x}{n} + \frac{n-1}{n}\int \cos^{n-2} x\,dx$$

$$\int \sin^n x\,dx = -\frac{\cos x \sin^{n-1} x}{n} + \frac{n-1}{n}\int \sin^{n-2} x\,dx$$

$$\int \sin^m x \cos^n x\,dx = \begin{cases} \dfrac{\sin^{m+1} x \cos^{n-1} x}{m+n} + \dfrac{n-1}{m+n}\displaystyle\int \sin^m x \cos^{n-2} x\,dx \\[2ex] -\dfrac{\sin^{m-1} x \cos^{n+1} x}{m+n} + \dfrac{m-1}{m+n}\displaystyle\int \sin^{m-2} x \cos^n x\,dx \end{cases}$$

Solutions to exercises

Chapter 1

Section 1.2

1. -1	**2.** 7	**3.** 1
4. 12	**5.** -12	**6.** -2
7. 2	**8.** 3/8	**9.** 1/28
10. $-11/18$	**11.** 1/6	**12.** $-1/54$
13. 53/48	**14.** 3/8	**15.** 3/2
16. 5/9	**17.** 1/2	**18.** 15/16
19. 2/5	**20.** 1/6	**21.** 3

Section 1.3

22. 2×3	**23.** $2^4 \times 5$	**24.** 2^8
25. $2 \times 3^4 \times 5$	**26.** 1/6	**27.** 3/7
28. 3/14	**29.** 3/20	**30.** 2
31. 5040	**32.** 3628800	**33.** 3
34. 120	**35.** 10	**36.** 120

Section 1.4

37. 0.01	**38.** 0.002	**39.** 2.000305
40. 0.375	**41.** 0.04	**42.** 0.15625
43. $0.\overline{1}1\ldots$	**44.** $0.\overline{09}09\ldots$	
45. $0.\overline{047619}047619\ldots$		
46. $0.\overline{0588235294117647}05882\ldots$		

47. 0.076923077, 0.07692308, 0.0769231,
0.076923, 0.07692, 0.0769, 0.077, 0.08

48. 1.4142136, 1.414214, 1.41421, 1.4142,
1.414, 1.41, 1.4, 1

49. 3.1415927, 3.141593, 3.14159, 3.1416,
3.142, 3.14, 3.1, 3

Section 1.6

50. a^5	**51.** 1	**52.** a^{-1}	**53.** a
54. a^9	**55.** a^{12}	**56.** a^{-6}	**57.** a^8
58. $a^{5/6}$	**59.** a^3	**60.** $a^2 b^4$	
61. $(a^3 + b^3)^{1/3}$		**62.** 3	**63.** 4
64. 8	**65.** 1/81	**66.** 1	**67.** 1
68. 8	**69.** 14	**70.** 35	**71.** 12
72. 16	**73.** 2	**74.** 19	**75.** 103
76. 37	**77.** 13		

Section 1.7

78. $7 - 2i, 47 - i$ **79.** $-4 - 10i, -29 + 26i$

Section 1.8

80. m^3; volume   **81.** $kg\,m^{-3}$; density

82. $mol\,m^{-3}$; concentration

83. $kg\,m\,s^{-1}$; momentum

84. $kg\,m\,s^{-2} = N$; force

85. $kg\,m^2\,s^{-2} = N\,m = J$; work, energy

86. $kg\,m^{-1}\,s^{-2} = N\,m^{-2} = Pa$; pressure

87. $A\,s = C$; electric charge

88. $kg\,m^2\,A^{-1}\,s^{-3} = J\,C^{-1} = V$; electric potential

89. $kg\,m^2\,s^{-2}\,mol^{-1} = J\,mol^{-1}$; molar energy

90. $kg\,m^2\,s^{-2}\,mol^{-1}\,K^{-1} = J\,mol^{-1}\,K^{-1}$;
heat capacity, molar entropy

91. (i) $26.8224\ m\,s^{-1}$,
(ii) $96.56064\ km\,h^{-1}$

92. (i) in h^{-1}, (ii) $7.0555 \times 10^{-6}\ m\,s^{-1}$,
(iii) $2\ yd\,h^{-1}$, $5.08 \times 10^{-4}\ m\,s^{-1}$
$1.8288 \times 10^{-3}\ km\,h^{-1}$

93. (i) $lbf = 4.448222\ N$,
(ii) $psi = lbf\,in^{-2} = 6894.75729\ Pa$,
(iii) $4.067454\ kJ$

94. (i) $2.3331 \times 10^3\ Pa$,
(ii) $2.3331 \times 10^{-2}\ bar$,
(iii) $2.3026 \times 10^{-2}\ atm$

96. $10^3\ m^{-3}$	**97.** $10^4\ m\,s^{-2}$
98. $kg\,m^{-3}$	**99.** $10^{-6}\ kg\,m\,s^{-2}$
100. $10^{17}\ kg\,m^{-1}\,s^{-2}$	**101.** $10^3\ m\,s^{-1}$
102. $10^2\ kg\,m^2\,s^{-2}$	**103.** $mol\,m^{-3}$

104. (i) 17.0265, (ii) $2.8273 \times 10^{-26}\ kg$,
(iii) $0.01703\ kg\,mol^{-1}$

105. (i) (a) 127.45 pm, (b) 1.2745 Å,
(c) $2.4084\ a_0$, (ii) $1.6266 \times 10^{-27}\ kg$,
(iii) $2.6421 \times 10^{-47}\ kg\,m^2$

106. (i) 86.52 THz, (ii) 3.465 μm,
(iii) 0.3578 eV, $34.52\ kg\,mol^{-1}$

107. (i) $462\ m\,s^{-1}$, (ii) $475\ m\,s^{-1}$

Chapter 2

Section 2.1

1. (i) 2	(ii) -4	(iii) 11	(iv) 0
2. (i) -1	(ii) 4	(iii) -2	
(iv) $-19/9$			

3. (i) 67   **(ii)** -3 **(iii)** -31
(iv) $-197/27$
4. (i) $2a^2 + 4a + 3$ **(ii)** $2y^4 + 4y^2 + 3$
5. (i) $a^2 + 3a - 4$ **(ii)** $a^4 - a^2 - 6$
(iii) $x^2 - x - 6$
(iv) $x^4 - 6x^3 - 2x^2 + 33x + 24$
6. $6x + 1$

Section 2.3

9. $2y^2(3x^2 - xy - 2)$ **10.** $(x+5)(x+1)$
11. $(x+3)(x-2)$ **12.** $(x-5)(x-3)$
13. $(x+2)(x-2)$ **14.** $(2x+3)(2x-3)$
15. $(2x-3)(x+2)$
16. $(x-1)(x+1)(x-3)(x+3)$

17. $\dfrac{1}{3x+2}$ **18.** $\dfrac{x+2}{x+4}$ **19.** $x+2$

20. $x+1$ **21.** $\dfrac{x-3}{x+2}$ **22.** $\dfrac{2x-1}{x-2}$

Section 2.4

23. $y+2$ **24.** $(2y-1)/3$ **25.** $2-3y$

26. $\dfrac{y}{1+y}$ **27.** $\dfrac{2y+3}{3y-2}$ **28.** $\dfrac{1+y}{1-2y}$

29. $\pm\sqrt{\dfrac{1+y}{1-y}}$ **30.** $\pm\sqrt{y^2-1}$

31. $\pm\sqrt{1\pm\sqrt{y}}$ **32.** $B = V_{\mathrm{m}}\left(\dfrac{pV_{\mathrm{m}}}{RT} - 1\right)$

33. $c = \left(\dfrac{\Lambda_{\mathrm{m}}^0 - \Lambda_{\mathrm{m}}}{\mathcal{K}}\right)^2$ **34.** $p = \dfrac{\theta}{\mathcal{K}(1-\theta)}$

Section 2.5

36. $1 + 2x + 3x^2$ **37.** $1 + 2x + 3x^2$

38. $\dfrac{2}{x} + \dfrac{6}{x^2} + \dfrac{12}{x^3}$ **39.** $1 + x + 2x^4 + 6x^9$

40. $y = 3x + 1$ **41.** $y = -2x + 4$

42. Plot Λ_{m} against $\sqrt{c}$ for a straight line.
Then $-\mathcal{K}$ is the slope and Λ_{m}^0 is the intercept.

43. Plot $\dfrac{1}{\rho}\left(\dfrac{\varepsilon_r - 1}{\varepsilon_r + 2}\right)$ against $\dfrac{1}{T}$ for a straight
line. Then μ^2 from the slope and α from the intercept.

44. $1, 2$ **45.** $1/2, -2$ **46.** $\dfrac{1}{2} \pm \sqrt{\dfrac{7}{12}}$

47. 3 (double)   **48.** $-1/2$ (double)

49. $\dfrac{1}{2}\left(-1 \pm i\sqrt{7}\right)$ **50.** $\dfrac{1}{2} \pm \dfrac{i}{\sqrt{12}}$

51. $\dfrac{1}{4}\left[-1 \pm \sqrt{1 + 8\dfrac{y+1}{y-1}}\right]$

52. $\alpha = \dfrac{K_a}{2c}\left\{\sqrt{1 + \dfrac{4c}{K_a}} - 1\right\}$

53. $1, -2, -3$ **54.** 1 (double), 4
55. 1 (triple)   **56.** $-1, 1, 2, 3$

Section 2.6

57. $2 - \dfrac{7}{x+3}$

58. $3x^2 - 8x + 15 - \dfrac{26}{x+2}$

59. $(x+3)(x-2)$

60. $2x^2 + x + 10 + \dfrac{17x + 26}{x^2 - 2x - 2}$

Section 2.7

61. $\dfrac{1}{3}\left[\dfrac{1}{x-1} - \dfrac{1}{x+2}\right]$

62. $\dfrac{1}{3}\left[\dfrac{2}{x} + \dfrac{1}{x+3}\right]$

63. $\dfrac{4}{x+2} - \dfrac{3}{x+1}$

64. $\dfrac{1}{6}\left[-\dfrac{21}{x} + \dfrac{8}{x-1} + \dfrac{25}{x+2}\right]$

65. $\dfrac{1}{9}\left[-\dfrac{1}{x-2} + \dfrac{10}{x-1} + \dfrac{6}{(x-1)^2}\right]$

Section 2.8

66. lines cross at $(x, y) = (2, 1)$
67. lines cross at $(x, y) = (7/13, 4/13)$
68. inconsistent; parallel lines
69. linear dependence; only one line
70. three lines cross at $(x, y, z) = (2, -2, -1)$
71. line crosses ellipse at $(x, y) = (0, -1)$ and
$(x, y) = (1, 1)$
72. line touches (is tangential to) ellipse at
$(x, y) = (1, 0)$
73. complex $(x, y) = \left(3/2 \pm i\sqrt{15}/6, \pm i\sqrt{15}/3\right)$,
line and ellipse do not touch

Chapter 3

Section 3.2

1. $c = 13$,

$$\sin A = \frac{12}{13}, \quad \cos A = \frac{5}{13}, \quad \tan A = \frac{12}{5},$$

$$\operatorname{cosec} A = \frac{13}{12}, \sec A = \frac{13}{5}, \cot A = \frac{5}{12}$$

$$\sin B = \frac{5}{13}, \quad \cos B = \frac{12}{13}, \quad \tan B = \frac{5}{12},$$

$$\operatorname{cosec} B = \frac{13}{5}, \sec B = \frac{13}{12}, \cot B = \frac{12}{5}$$

2. (i) 1   (ii) 1

3. (i) $\pi/36$ (ii) $29\pi/60$ (iii) $2\pi/3$

(iv) $\dfrac{13\pi}{9}$ (v) 3π (vi) 4π

4. (i) $18°$ (ii) $45°$ (iii) $30°$

(iv) $60°$ (v) $67.5°$ (vi) $157.5°$

5. (i) $3/2$ rad $\approx 85.94°$ (ii) $2\pi/5 \approx 1.257$

(iii) $2\pi \approx 6.283$ (iv) $8\pi \approx 25.13$

6. (i) $\sin 3\pi/4 = +1/\sqrt{2}$, $\cos 3\pi/4 = -1/\sqrt{2}$,

  $\tan 3\pi/4 = -1$

(ii) $\sin 5\pi/4 = -1/\sqrt{2}$, $\cos 5\pi/4 = -1/\sqrt{2}$,

  $\tan 5\pi/4 = +1$

(iii) $\sin 7\pi/4 = -1/\sqrt{2}$, $\cos 7\pi/4 = +1/\sqrt{2}$,

  $\tan 7\pi/4 = -1$

10. (i) π (ii) $2\pi/3$

Section 3.3

12. (i) $\pi/6$ (ii) $\pi/2$ (iii) $\pi/3$

(iv) π

13. $\sin^{-1}(1/4) \approx 14.48°$, $\sin^{-1}(1/2) = 30°$,

  $\sin^{-1}(3/4) \approx 48.59°$, $\sin^{-1}(1) = 90°$

Section 3.4

14. $C = 5\pi/12 = 75°$, $b = \sqrt{3/2}$,

  $c = \sqrt{2} \sin 75° \approx 1.3660$

15. $A = \cos^{-1}(0.75) \approx 41.41°$,

  $B = \cos^{-1}(0.5625) \approx 55.77°$,

  $C = \pi - A - B \approx 82.82°$

16. $c \approx 2.8336$, $A \approx 48.47°$, $B \approx 86.53°$

17. $c = \sqrt{5}$, $A = \cos^{-1}(2/\sqrt{5}) \approx 26.57°$,

  $B = \pi - \pi/4 - A \approx 108.43°$

18. (i) $\sin 7\theta = \sin 5\theta \cos 2\theta + \cos 5\theta \sin 2\theta$

(ii) $\sin 3\theta = \sin 5\theta \cos 2\theta - \cos 5\theta \sin 2\theta$

(iii) $\cos 7\theta = \cos 5\theta \cos 2\theta - \sin 5\theta \sin 2\theta$

(iv) $\cos 3\theta = \cos 5\theta \cos 2\theta + \sin 5\theta \sin 2\theta$

19. (i) $\sin 3\theta = 3 \sin \theta - 4 \sin^3 \theta$

(ii) $\cos 3\theta = 4 \cos^3 \theta - 3 \cos \theta$

20. (i) $\cos^2 2x - \sin^2 2x$ (ii) $1 - 2 \sin^2 2x$

(iii) $2\cos^2 2x - 1$

(iv) $1 - 8 \sin^2 x + 8 \sin^4 x$

(v) $1 - 8 \cos^2 x + 8 \cos^4 x$

21. 0.9396

22. (i) $\frac{1}{2}(\sin 8x + \sin 2x)$

(ii) $\frac{1}{2}(\sin 8x - \sin 2x)$

23. (i) $\frac{1}{2}(\cos 2x - \cos 8x)$

(ii) $\frac{1}{2}(\cos 2x + \cos 8x)$

24. (i) $\sin(\pi \pm \theta) = \mp \sin \theta$

(ii) $\cos(\pi \pm \theta) = -\cos \theta$

25. (i) $t = n/2, n = 0, 1, 2, \ldots$

(ii) $t = (2n+1)/4, n = 0, 1, 2, \ldots$

Section 3.5

27. (i) $\left(3/2, 3\sqrt{3}/2\right)$ (ii) $\left(3/2, -3\sqrt{3}/2\right)$

28. (i) $\left(-3/2, 3\sqrt{3}/2\right)$ (ii) $\left(-3/2, -3\sqrt{3}/2\right)$

29. (i) $\sqrt{13}$, $\tan^{-1}(2/3) \approx 33.7°$

(ii) $\sqrt{13}$, $\tan^{-1}(-2/3) + 2\pi \approx 326.3°$

30. (i) $\sqrt{13}$, $\tan^{-1}(-2/3) + \pi \approx 146.3°$

(ii) $\sqrt{13}$, $\tan^{-1}(2/3) + \pi \approx 213.7°$

Section 3.6

32. (i) e^5 (ii) 1   (iii) $1/e$

(iv) e (v) e^9

33. (i) $1 - \dfrac{x}{3} + \dfrac{x^2}{18} - \dfrac{x^3}{162} + \dfrac{x^4}{1944} - \dfrac{x^5}{29160}$

(ii) 0.71653

34. (i) $1 - x^3 + \dfrac{x^6}{2} - \dfrac{x^9}{6} + \dfrac{x^{12}}{24} - \dfrac{x^{15}}{120}$

(ii) 0.999000499833, 4 terms

(iii) 0.999999000001, 3 terms

(iv) 0.999999999000, 2 terms

(v) 0.999999999999, 2 terms

(vi) 1.00000000000, 1 term

36. (i) $n_i/n_j = e^{-(\varepsilon_i - \varepsilon_j)/kT}$ (ii) 1

Section 3.7

37. (i) 2   (ii) 4   (iii) -5 (iv) x^2

(v) $-(ax^2 + bx + c)$ (vi) $-kt$

38. (i) $\ln 6$ **(ii)** $\ln 2/3$ **(iii)** $\ln 32$
(iv) $\ln 2$

39. (i) $\ln x^2$ **(ii)** $\ln(2x-3)$ **(iii)** 0
(iv) x **(v)** x^2

40. $h = -\left(\dfrac{RT}{Mg}\right)\ln\dfrac{p}{p_0}$

41. $p = \left(\dfrac{p^{\ominus}}{\gamma}\right)\exp\left[(\mu - \mu^{\ominus})/RT\right]$

42. (i) $0.2310\ s$ **(ii)** $0.6931 \times 10^5\ s$

Chapter 4

Section 4.2

1. $\Delta y = 3x^2\Delta x + 3x(\Delta x)^2 + (\Delta x)^3$

2. (i) $\dfrac{\Delta y}{\Delta x} = 3x^2 + 3x\Delta x + (\Delta x)^2$

(ii) $\lim_{\Delta x \to 0}\dfrac{\Delta y}{\Delta x} = 3x^2$

3. (i) $\Delta\theta = \dfrac{K\Delta p}{\left[1+Kp\right]\left[1+K(p+\Delta p)\right]}$

(ii) $\lim_{\Delta p \to 0}\dfrac{\Delta\theta}{\Delta p} = \dfrac{K}{(1+Kp)^2}$

Section 4.3

4. $x = -1$, essential **5.** $x = 0$, removable
6. $x = 3$, essential; $x = 0$, removable

Section 4.4

7. 0 **8.** $\pm\infty$ **9.** $1/3$ **10.** $1/2$
11. 1 **12.** 0 **13.** ∞ **14.** -4
15. 2 **16.** $-\ln 2$ **17.** $-\ln 3$

Section 4.5

18. $\dfrac{\Delta y}{\Delta x} = 4x + 3 + 2\Delta x,\quad \dfrac{dy}{dx} = 4x + 3$

19. $\dfrac{\Delta y}{\Delta x} = 4x^3 + 6x^2\Delta x + 4x(\Delta x)^2 + (\Delta x)^3,$
$\dfrac{dy}{dx} = 4x^3$

20. $\dfrac{\Delta y}{\Delta x} = -\dfrac{4x + 2\Delta x}{x^2(x+\Delta x)^2},\quad \dfrac{dy}{dx} = -\dfrac{4}{x^3}$

21. $\dfrac{\Delta y}{\Delta x} = \dfrac{3x^2 + 3x\Delta x + (\Delta x)^2}{(x+\Delta x)^{3/2} + x^{3/2}},\quad \dfrac{dy}{dx} = \dfrac{3}{2}x^{1/2}$

22. $\dfrac{\Delta y}{\Delta x} = -e^{-x} + \dfrac{\Delta x}{2}e^{-x} - \cdots,\quad \dfrac{dy}{dx} = -e^{-x}$

Section 4.6

23. $3x^2$ **24.** $(5/4)x^{1/4}$

25. $(1/3)x^{-2/3}$ **26.** $-3/x^4$

27. $-2 + 6x - 12x^2 + 5\cos x + 6\sin x + 7e^x - 8/x$

28. $nRT(2nB - V)/V^3$

29. $-(1 - 4x^2)\sin x - 8x\cos x$

30. $(5 + 3x)e^x$ **31.** $(\cos x - \sin x)e^x$

32. $1 + \ln x$

33. $\dfrac{6 + 18x - 3x^2 - 4x^3 - 3x^4}{(3 + x^3)^2}$

34. $-\dfrac{8x\sin x + (1 - 4x^2)\cos x}{\sin^2 x}$

35. $-\operatorname{cosec}^2 x$ **36.** $(1 - \ln x)/x^2$

37. $5(1 + x)^4$ **38.** $\dfrac{x}{\sqrt{2 + x^2}}$

39. $\dfrac{2x}{(3 - x^2)^2}$ **40.** $\dfrac{-3(4x - 3)}{2(2x^2 - 3x - 1)^{3/2}}$

41. $4\cos 4x$ **42.** $-2e^{-2x}$

43. $(4x - 3)e^{2x^2 - 3x + 1}$ **44.** $\dfrac{4x - 3}{2x^2 - 3x + 1}$

45. $-(4x - 3)\sin(2x^2 - 3x + 1)$

46. $\cos x e^{\sin x}$ **47.** $-\tan x$

48. $6x\sin(3x^2 + 2)e^{-\cos(3x^2 + 2)}$

49. $\dfrac{1}{2 + x} + \dfrac{1}{3 - x}$ **50.** $\dfrac{\sin 2x + 2\cos 2x}{\sin 2x + \sin^2 x}$

51. $6x(2 + x)^{1/2} + \dfrac{3x^2}{2(2 + x)^{1/2}}$

52. $\cos x \cos 2x - 2\sin\sin 2x$

53. $4\cos^2 2x \sec^2 4x - 2\tan 4x \sin 4x$

54. $2x(1 + 2x^2)e^{2x^2 + 3}$ **55.** $\dfrac{3x(4 + x^2)}{(2 + x^2)^{3/2}}$

56. $\dfrac{1}{4y - 3}$ **57.** $\dfrac{-V^3}{nRT(V + 2nB)}$

58. $-\dfrac{(V - nb)}{p}$

59. $\left[\dfrac{2n^2a}{V^3} - \dfrac{nRT}{(V - nb)^2}\right]^{-1}$ **60.** $\dfrac{2}{\sqrt{1 - 4x^2}}$

61. $\dfrac{2x}{1 + x^4}$ **62.** $\dfrac{1}{\sqrt{x}(1 + x)}$

63. $\dfrac{2}{\sqrt{1 + 4x^2}}$ **64.** $\dfrac{2x}{1 - x^4}$

Section 4.7

65. $-\dfrac{x}{y}$ **66.** $-\dfrac{3 + 2x}{3y^2}$ **67.** $\dfrac{x - y}{x(1 + \ln xy)}$

Section 4.8

68. $\dfrac{2xy^2 - 3}{2y - 2\big/y^2 - 2x^2 y}$

69. $\dfrac{-7}{3(3-x)(4+x)}\left(\dfrac{3-x}{4+x}\right)^{1/3}$

70. $\left(\dfrac{2x}{1+x^2} + \dfrac{1}{2(x-1)} - \dfrac{2}{2x+1}\right.$

$\left. - \dfrac{6x+2}{3(3x^2+2x-1)}\right)$

$\times \dfrac{(1+x^2)(x-1)^{1/2}}{(2x+1)(3x^2+2x-1)^{1/3}}$

71. $\left(\dfrac{1}{2}\cot x - 6x\tan(x^2+1) + \dfrac{4}{3\sin 4x}\right)$

$\times \sin^{1/2} x \cos^3(x^2+1) \tan^{1/3} 2x$

73. 6.1×10^{-4} s

Section 4.9

74. $y' = 15x^4 + 16x^3 - 9x^2 + 2x - 2$

$y'' = 60x^3 + 48x^2 - 18x + 2$

$y''' = 180x^2 + 96x - 18$

$y^{iv} = 360x + 96, \ y^v = 360$

75. $\dfrac{1}{x}, \ -\dfrac{1}{x^2}, \ \dfrac{2}{x^3}, \ -\dfrac{6}{x^4}$ **76.** $3^n e^{3x}$

77. $y^{(n)} = \begin{cases} (-1)^{n/2}\, 2^n \cos 2x & (n \text{ even}) \\ (-1)^{(n+1)/2}\, 2^n \sin 2x & (n \text{ odd}) \end{cases}$

Section 4.10

78. $(x, y) = (3/2, -1/4)$, min

79. $(3, 0)$, min; $(5/3, 32/27)$,

max; $(7/3, 16/27)$, inflection

80. $(0, 3)$, min; $(-1, 5)$,

max; $(-1/2, 4)$, inflection

81. $(1, e^{-1})$, max; $(2, 2e^{-2})$, inflection

83. $(2, -13)$, min; $(0, 3)$,

max; $(3/2, -57/8)$, inflection

84. (i) $A = D_e R_e^{12}, B = 2D_e R_e^6$

(ii) $U(R) = D_e\left[\left(\dfrac{R_e}{R}\right)^{12} - 2\left(\dfrac{R_e}{R}\right)^6\right]$

85. $\sqrt{\dfrac{2kT}{m}}$

86. (i) $t = \dfrac{1}{k_2 - k_1}\ln\left(\dfrac{k_2}{k_1}\right)$

Section 4.11

87. (i) $v = \dot{s} = 4t - 3; a = \ddot{s} = 4$

(iii) $v = 0$ when $s = -9/8$

88. (i) $\theta = (t^3 - 2t^2 - 4t)/2$

(ii) $\omega = (3t^2 - 4t - 4)/2, \dot{\omega} = 3t - 2$

(iv) $\omega = 0$ when $t = 2, \theta = -4$

89. $dy = 2dx$ **90.** $dy = (6x + 2)dx$

91. $dy = \cos x\, dx$ **92.** $dV = 4\pi r^2 dr$

Chapter 5

Section 5.2

1. $2x + C$ **2.** $\dfrac{x^4}{4} + C$ **3.** $\dfrac{3}{5}x^{5/3} + C$

4. $-\dfrac{1}{2x^2} + C$ **5.** $\dfrac{3}{2}x^{2/3} + C$

6. $-\dfrac{1}{4}\cos 4x + C$ **7.** $\dfrac{1}{3}e^{3x} + C$

8. $-\dfrac{1}{2}e^{-2x} + C$ **9.** $\ln A(x - 1)$

10. $\ln A/(3 - x)$ **11.** $\dfrac{x^3}{3} - 9$

12. $\dfrac{1}{4}\sin 4x$ **13.** $x^5 + x^2 + 3x - 38$

14. $3x + 2\ln x - \dfrac{1}{x} + 1$

15. $-4x + 2\sin 2x - \dfrac{1}{4}e^{2x} + \dfrac{1}{4}$

Section 5.3

16. $28/3$ **17.** 2 **18.** $\dfrac{3}{8}$

19. $\ln(5/4)$ **20.** $\dfrac{1}{3}(e^{-3} - e^{-15})$

21. 1 **22.** 0 **23.** $-1/2$

26. $14/3$ **27.** $-2/3\pi$ **28.** 1

31. (ii) $0, 1, 0$ **32.** $40/3$ **33.** 1

34. 0 **35.** -1 **36.** $1/3$

37. 2 **38.** $\ln 2$ **39.** $1 - \ln 2$

40. odd   **41.** even   **42.** odd

43. odd   **44.** even

45. neither; even: $3x^2 + 1$; odd: $2x$

46. neither; even: $\dfrac{1}{2}(e^{-x} + e^x)$;

odd: $\dfrac{1}{2}(e^{-x} - e^x)$

47. neither:

even: $\dfrac{1}{2}\left[(3x^2+1)(e^{-x}+e^{x})+x(e^{-x}-e^{x})\right]$;

odd: $\dfrac{1}{2}\left[(3x^2+1)(e^{-x}-e^{x})+x(e^{-x}+e^{x})\right]$

Section 5.4

48. πab **49.** $61/54$

Section 5.6

50. (i) 1   (ii) 66   (iii) 60

51. (i) $l^3/3$ (ii) $l^2/3$ (iii) $3l/4$

  (iv) $l^5/5 - x_0 l^4/2 + x_0^2 l^3/3$ (v) $l^5/80$

Section 5.7

52. (i) 1   (ii) 7   (iii) 37

53. $\dfrac{1}{2}k(x_B^2 - x_A^2)$

54. (i) $-\dfrac{1}{2}kx^2$ (ii) (a) $-\dfrac{1}{2}k$

  (b) $\dfrac{1}{2}k(x^2-1)$

55. (i) $\dfrac{1}{2}kx^2$ (ii) (a) $\dfrac{1}{2}k$ (b) $\dfrac{1}{2}k(1-x^2)$

Section 5.8

56. (i) $p(V_2 - V_1)$

  (ii) $nRT\ln\left(\dfrac{V_2-nb}{V_1-nb}\right) - n^2 a\left(\dfrac{1}{V_1} - \dfrac{1}{V_2}\right)$

Chapter 6

Section 6.2

1. $\dfrac{1}{12}(6x - \sin 6x) + C$

2. $-\dfrac{1}{12}\cos 6x + C$

3. $-\dfrac{1}{10}(\cos 5x + 5\cos x) + C$

4. $\dfrac{1}{8}(2\cos 2x - \cos 4x) + C$

5. $\dfrac{1}{8}(2\sin 2x - \sin 4x) + C$

6. $\dfrac{1}{42}(3\sin 7x + 7\sin 3x) + C$

7. $\dfrac{\pi}{4}$ **8.** 0   **9.** $-\dfrac{2}{3}$

Section 6.3

11. $\dfrac{1}{18}(3x+1)^6 + C$ **12.** $\dfrac{1}{3}(2x-1)^{3/2} + C$

13. $\dfrac{1}{8}(3x^2 + 2x + 5)^4 + C$

14. $\dfrac{1}{4}(2x^3 + 3x - 1)^{4/3} + C$

15. $-e^{-(x^3+2x)} + C$ **16.** $\dfrac{1}{4}e^{4x-2x^2} + C$

17. $-\dfrac{1}{3}(4 - x^2)^{3/2} + C$ **18.** $\dfrac{1}{2}e^{2\sin x} + C$

19. $\dfrac{2}{3}(1 + e^x)^{3/2} + C$

20. $\dfrac{1}{6}\sin(3x^2 - 1) + C$

21. $\ln(x^2 + x + 2) + C$

22. $\dfrac{1}{2}\ln(2x^3 - x^2 + 3) + C$

23. $-\ln(1 - \sin x) + C$ **24.** $-\ln(\cos x) + C$

25. $-\sqrt{4 - x^2} + C$ **26.** $-\ln[\ln(\cos x)] + C$

27. $\dfrac{1}{4}\sin^4 x + C$ **28.** $\cos[1 - \ln(\cos x)] + C$

29. $\dfrac{1}{2}\tan^{-1}\left(\dfrac{x}{2}\right) + C$

30. $\dfrac{1}{2}\left[\sin^{-1} x - x\sqrt{1 - x^2}\right] + C$

31. $2\left[\sqrt{x} - \tan^{-1}\sqrt{x}\right] + C$ **33.** $\dfrac{1}{6}\ln 10$

34. -4 **35.** $\dfrac{2}{3}$ **36.** $\dfrac{\pi}{4}$ **37.** $\dfrac{1}{2}$

38. $\dfrac{1}{2}$ **39.** $\dfrac{T}{\Theta_r}$

Section 6.4

40. $-x\cos x + \sin x + C$

41. $3(x^2 - 2)\sin x - x(x^2 - 6)\cos x + C$

42. $\dfrac{1}{2}(x+1)^2 \sin 2x + \dfrac{1}{2}(x+1)\cos 2x$

  $-\dfrac{1}{4}\sin 2x + C$

43. $\dfrac{1}{4}(2x^2 - 2x + 1)e^{2x} + C$ **44.** 1

45. $\dfrac{1}{4}$ **46.** $\dfrac{x^2}{4}(2\ln x - 1) + C$

47. $-\left(\dfrac{\ln x + 1}{x}\right) + C$ **48.** $-\dfrac{1}{9}$

49. $-\dfrac{1}{5}(\sin 2x + 2\cos 2x)e^{-x} + C$

50. $\dfrac{e^{ax}}{(a^2 + b^2)}(a\cos bx + b\sin bx) + C$

51. $\dfrac{1}{13}(2 - 3e^{-\pi})$

Section 6.5

52. $\displaystyle\int \sin^n x\, dx = -\dfrac{1}{n}\sin^{n-1}x\cos x$

$+\dfrac{n-1}{n}\displaystyle\int \sin^{n-2}x\, dx$

54. $\dfrac{\sin^6 x\cos^3 x}{9} + \dfrac{\sin^6 x\cos x}{21}$

$-\dfrac{\sin^4 x\cos x}{105} - \dfrac{4\sin^2 x\cos x}{315}$

$-\dfrac{8\cos x}{315} + C$

56. $\dfrac{1}{60}$ **57.** $\dfrac{1}{4}$ **58.** $\dfrac{1}{8}\sqrt{\dfrac{\pi}{2}}$ **59.** $\dfrac{1}{8}$

60. $\left(\dfrac{8kT}{\pi m}\right)^{1/2}$ **61.** $\left(\dfrac{3kT}{m}\right)^{1/2}$ **62.** $\dfrac{1}{T^2}$

Section 6.6

63. $\dfrac{1}{7}\ln\left(\dfrac{2x-1}{x+3}\right) + C$

64. $\ln\dfrac{(x+4)^2}{(x+3)} + C$

65. $\dfrac{1}{2}\Big[7\ln(x+1) - 26\ln(x+2)$

$+ 21\ln(x+3)\Big] + C$

66. $\dfrac{1}{2}\ln(x^2 + 4x + 5) + C$

67. $\dfrac{1}{2}\ln\left(\dfrac{x^2+3}{x^2+4}\right) + C$

68. $\tan^{-1}(x+2) + C$

69. $\dfrac{1}{2}\left(\tan^{-1}(x+2) + \dfrac{x+2}{x^2+4x+5}\right) + C$

70. $\dfrac{1}{2}\ln(x^2 + 4x + 5) - 2\tan^{-1}(x+2) + C$

71. $-\dfrac{1}{2}\left[\dfrac{5x+14}{x^2+4x+5} + 5\tan^{-1}(x+2)\right] + C$

73. $\ln\left(\dfrac{1+\tan\theta/2}{1-\tan\theta/2}\right) + C$

74. $\dfrac{1}{2}\tan^{-1}(2\tan\theta/2) + C$

75. $\ln(1 + \tan\theta/2) + C$

Chapter 7

Section 7.2

1. $u_r = 1 + 3r;\ u_r = u_{r-1} + 3,\ u_0 = 1$

2. $u_r = 3^r;\ u_r = 3u_{r-1},\ u_0 = 1$

3. $u_r = \left(-\dfrac{1}{5}\right)^r;\ u_r = -\dfrac{u_{r-1}}{5},\ u_0 = 1$

4. $0,\ \dfrac{1}{2},\ 1,\ \dfrac{3}{2},\ 2,\ \dfrac{5}{2}$

5. $1,\ \dfrac{2}{3},\ \dfrac{4}{9},\ \dfrac{8}{27},\ \dfrac{16}{81},\ \dfrac{32}{243}$

6. $\dfrac{1}{3},\ \dfrac{1}{8},\ \dfrac{1}{15},\ \dfrac{1}{24},\ \dfrac{1}{35},\ \dfrac{1}{48}$

7. $1,\ \dfrac{1}{2},\ \dfrac{1}{6},\ \dfrac{1}{24},\ \dfrac{1}{120},\ \dfrac{1}{720}$

8. $1, 3, 5, 11, 21, 43$

9. $1,\ \dfrac{1}{2},\ -\dfrac{1}{2},\ -\dfrac{5}{2},\ -\dfrac{13}{2},\ -\dfrac{29}{2}$

10. $u_n = u_0$, all n **11.** 0 **12.** ∞

13. 0 **14.** 1 **15.** 0

16. $3/5$ **17.** 2

Section 7.3

18. (i) $n(2n-1)$ (ii) 190

19. (i) $\dfrac{n}{2}(11 - 5n)$ (ii) -195

20. (i) $\dfrac{1}{2}(3^n - 1)$ (ii) 29524

21. (i) $\dfrac{3}{2}\left[1 - \left(\dfrac{1}{3}\right)^n\right]$

(ii) $\dfrac{1}{2}\left[3 - \dfrac{1}{3^9}\right] \approx 1.499975$

22. $\dfrac{x^3(1 - x^{2n})}{1 - x^2}$ **23.** $x\left[\dfrac{1 - (2x)^n}{1 - 2x}\right]$

24. $1 + 5x + 10x^2 + 10x^3 + 5x^4 + x^5$

25. $1 + 7x + 21x^2 + 35x^3 + 35x^4 + 21x^5$

$+ 7x^6 + x^7$

26. $1, 3, 3, 1$ **27.** $1, 4, 6, 4, 1$

28. $1, 7, 21, 35, 21, 7, 1$

29. $1 - 3x + 3x^2 - x^3$

30. $1 + 12x + 54x^2 + 108x^3 + 81x^4$

31. $1 - 20x + 160x^2 - 640x^3 + 1280x^4 - 1024x^5$

32. $81 - 216x + 216x^2 - 96x^3 + 16x^4$

33. $729 + 1458x + 1215x^2 + 540x^3 + 135x^4$
$+ 18x^5 + x^6$

34. (i) $\dfrac{4!}{4!0!0!} = 1, \quad \dfrac{4!}{3!1!0!} = 4, \quad \dfrac{4!}{2!2!0!} = 6,$

$\dfrac{4!}{2!1!1!} = 12$

(ii) $(a^4 + b^4 + c^4) + 4(a^3b + a^3c + b^3a + b^3c$
$+ c^3a + c^3b) + 6(a^2b^2 + a^2c^2 + b^2c^2)$
$+ 12(a^2bc + b^2ca + c^2ab)$

35. (i) $\dfrac{3!}{3!0!0!0!} = 1, \quad \dfrac{3!}{2!1!0!0!} = 3,$

$\dfrac{3!}{1!1!1!0!} = 6$

(ii) $(a^3 + b^3 + c^3 + d^3) + 3(a^2b + a^2c + a^2d$
$+ b^2a + b^2c + b^2d + c^2a + c^2b + c^2d$
$+ d^2a + d^2b + d^2c) + 6(abc + abd$
$+ acd + bcd)$

36. $10/11$ **37.** $\dfrac{3}{4} - \dfrac{2n+3}{2(n+1)(n+2)}$

40. $\dfrac{n^2}{12}\left[2n^4 + 6n^3 + 5n^2 - 1\right]$

Section 7.4

41. (i) $1 + 3x + 9x^2 + 27x^3 + 81x^4 + 243x^5$
$+ 729x^6 + \cdots$

(ii) $|x| < 3$

42. (i) $1 - 5x^2 + 25x^4 - 125x^6 + \cdots$

(ii) $|x| < 1/\sqrt{5}$

43. (i) $\dfrac{1}{2} - \dfrac{x}{4} + \dfrac{x^2}{8} - \dfrac{x^3}{16} + \dfrac{x^4}{32} - \dfrac{x^5}{64} + \dfrac{x^6}{128} - \cdots$

(ii) $|x| < 2$

44. (i) $0.000001000001000001\cdots$

45. $q_v = \left[1 - e^{-\theta_v/T}\right]^{-1}$

Section 7.5

46. diverges   **47.** converges

48. converges for all a

49. no conclusion (see Exercise 50)

50. diverges if $a \le 1$, converges if $a > 1$

51. diverges

Section 7.6

52. 4   **53.** 1   **54.** 1

55. 1   **56.** 0   **57.** $\sqrt{3}$

58. $1 + \dfrac{x}{3} - \dfrac{x^2}{9} + \dfrac{5x^3}{81} - \dfrac{10x^4}{243}$

59. $1 - x^2 + x^4 - x^6 + x^8$

60. $1 + \dfrac{x}{2} + \dfrac{3x^2}{8} + \dfrac{5x^3}{16} + \dfrac{35x^4}{128}$

61. $\dfrac{1}{3} - \dfrac{x}{9} + \dfrac{x^2}{27} - \dfrac{x^3}{81} + \dfrac{x^4}{243}$

62. $2x^2 - \dfrac{4x^6}{3} + \dfrac{4x^{10}}{15} - \dfrac{8x^{14}}{315} + \dfrac{4x^{18}}{2835}$

63. $-2 - \dfrac{8x}{3} - 4x^2 - \dfrac{32x^3}{5} - \dfrac{32x^4}{3}$

64. $1 - 3x + \dfrac{9x^2}{2} - \dfrac{9x^3}{2} + \dfrac{27x^4}{8}$

65. $x + \dfrac{x^3}{2} + \dfrac{x^5}{6} + \dfrac{x^7}{24} + \dfrac{x^9}{120}$

66. $T = \dfrac{1}{2}m_0v^2\left[1 + \dfrac{3}{4}\left(\dfrac{v}{c}\right)^2 + \dfrac{5}{8}\left(\dfrac{v}{c}\right)^4 + \cdots\right]$

$\to \dfrac{1}{2}m_0v^2$ as $v/c \to 0$

67. (i) $B_0 = 1, B_1 = b - \dfrac{a}{RT}, \quad B_2 = b^2$

(ii) $B_0 = 1, B_1 = b - \dfrac{a}{RT},$

$B_2 = b^2 - \dfrac{ab}{RT} + \dfrac{a^2}{2R^2T^2}$

68. (i) $\displaystyle\sum_{n=0}^{\infty}(1-x)^n$ (ii) $0 < x < 2$

69. (i) $e^2\displaystyle\sum_{n=0}^{\infty}\dfrac{(x-2)^n}{n!}$ (ii) all x

70. (i) $\displaystyle\sum_{n=0}^{\infty}\dfrac{(-1)^n(x-\pi/2)^{2n}}{(2n)!}$ (ii) all x

71. (i) $\ln 2 - \displaystyle\sum_{n=1}^{\infty}\dfrac{1}{n}\left(\dfrac{2-x}{2}\right)^n$ (ii) $0 < x \le 4$

Section 7.7

72. (i) $2 + \dfrac{x}{12} - \dfrac{x^2}{288} + \dfrac{5x^3}{20736} - \dfrac{5x^4}{248832}$

(ii) 2.08008214

(iii) $2.0800832 < \sqrt[3]{9} < 2.0800840$

73. 1   **74.** $1/2$ **75.** -2 **76.** $1/2$

Section 7.8

78. $x - \dfrac{4x^3}{3!} + \dfrac{16x^5}{5!} - \dfrac{64x^7}{7!}$

79. $1 - \dfrac{x^2}{2!} + \dfrac{x^4}{4!} - \dfrac{x^6}{6!} + \cdots$

80. $A + \dfrac{x^2}{2!} - \dfrac{x^4}{4!} + \dfrac{x^6}{6!} - \cdots$

Chapter 8

Section 8.2

1. $6 - 2i$ **2.** 4 **3.** $6i$ **4.** $18 + 4i$
5. $-8 - 6i$ **6.** $41 - 38i$ **7.** 10
8. (i) $3 + 2i$ (ii) 13

   (iii) $3 = \dfrac{1}{2}(z + z^*),\ -2 = \dfrac{1}{2i}(z - z^*)$

9. $z = \pm 1 + 2i$ **10.** $1 \pm i\sqrt{3}$

11. $-2, 1 \pm i\sqrt{3}$ **12.** $-i$

13. $\dfrac{5}{34} - \dfrac{3}{34}i$ **14.** $\dfrac{5}{13} + \dfrac{12}{13}i$

15. $\dfrac{12}{25} + \dfrac{24}{25}i$

Section 8.3

16. (i) 2 (ii) $\pi/2$
   (iii) $2(\cos \pi/2 + i \sin \pi/2)$
17. (i) 3 (ii) π
   (iii) $3(\cos \pi + i \sin \pi)$
18. (i) $\sqrt{2}$ (ii) $-\pi/4$
   (iii) $\sqrt{2}(\cos \pi/4 - i \sin \pi/4)$
19. (i) 2 (ii) $\pi/6$
   (iii) $2(\cos \pi/6 + i \sin \pi/6)$
20. (i) $6\sqrt{2}$ (ii) $3\pi/4$
   (iii) $6\sqrt{2}(\cos 3\pi/4 + i \sin 3\pi/4)$
21. (i) 4 (ii) $4\pi/3$
   (iii) $4(\cos 4\pi/3 + i \sin 4\pi/3)$
22. (i) 1 (ii) $3\pi/2$
   (iii) $\cos 3\pi/2 + i \sin 3\pi/2$

23. (i) $6\left(\cos \dfrac{5\pi}{6} + i \sin \dfrac{5\pi}{6} \right)$

   (ii) $\dfrac{2}{3}\left(\cos \dfrac{\pi}{6} + i \sin \dfrac{\pi}{6} \right)$

   (iii) $\dfrac{3}{2}\left(\cos \dfrac{\pi}{6} - i \sin \dfrac{\pi}{6} \right)$

24. (i) $5\left(\cos \dfrac{17\pi}{12} + i \sin \dfrac{17\pi}{12} \right)$

   (ii) $5\left(\cos \dfrac{\pi}{12} + i \sin \dfrac{\pi}{12} \right)$

   (iii) $\dfrac{1}{5}\left(\cos \dfrac{\pi}{12} - i \sin \dfrac{\pi}{12} \right)$

25. (i) $81i$ (ii) $-i/81$
27. $128 \cos^8 x - 256 \cos^6 x + 160 \cos^4 x$
   $- 32 \cos^2 x + 1$

Section 8.4

28. (i) $f(x) = (3x^2 + x - 2) + i(2x + 2)$
   (ii) $g(x) = 0$ when $x = 2/3$ and $x = -1$,
      $h(x) = 0$ when $x = -1, f(x) = 0$ when
      $x = -1$
   (iii) $(x + 1)^2(9x^2 - 12x + 8)$
29. (i) $f(z) = [x^2 - y^2 - 2x + 3] + i[2y(x - 1)]$
   (ii) $h(x, y) = 0$ when $y = 0$ or $x = 1$,
      $g(x, y) = 0$ when (real) $x = 1,\ y = \pm\sqrt{2}$
      $f(z) = 0$ when $z = 1 \pm i\sqrt{2}$

Section 8.5

30. (i) $\sqrt{2}e^{-i\pi/4}$ (ii) $\sqrt{2}e^{i\pi/4}$

   (iii) $\left(1/\sqrt{2} \right)e^{i\pi/4}$

31. (i) $2e^{i\pi/6}$ (ii) $2e^{-i\pi/6}$
   (iii) $(1/2)e^{-i\pi/6}$
32. (i) $2e^{i\pi/2}$ (ii) $2e^{-i\pi/2}$
   (iii) $(1/2)e^{-i\pi/2}$
33. (i) $3e^{i\pi}$ (ii) $3e^{-i\pi}$ (iii) $(1/3)e^{-i\pi}$

34. $\dfrac{3}{\sqrt{2}} + \dfrac{3}{\sqrt{2}}i$ **35.** $\dfrac{1}{2} - \dfrac{\sqrt{3}}{2}i$

36. $\sqrt{3} + i$ **37.** i **38.** $-i$ **39.** -1
41. $\cos a \cosh b - i \sin a \sinh b$
43. $\pm(1 - i)/\sqrt{2}$

44. (i) $\left(\dfrac{3\sqrt{3}}{2} - 1 \right) + i\left(\dfrac{3}{2} + \sqrt{3} \right)$

   (ii) $\left(\dfrac{3\sqrt{3}}{2} + 1 \right) + i\left(-\dfrac{3}{2} + \sqrt{3} \right)$

Section 8.6

45. $\pm 1, \pm i$
46. $1, \cos 2\pi/5 \pm i \sin 2\pi/5,$
   $\cos 4\pi/5 \pm i \sin 4\pi/5$

47. $1, \dfrac{1}{\sqrt{2}}(1 \pm i), \pm i, \dfrac{1}{\sqrt{2}}(-1 \pm i), -1$

48. (i) $1/\sqrt{2\pi}$

Section 8.7

49. $1/5$ **50.** $6/65$

Chapter 9

Section 9.1

1. (i) 0 (ii) 16 (iii) 36

2. (i) 1 (ii) $-3/2$ (iii) 2

Section 9.3

3. $4x, -2y$ **4.** $2x - 3, 4y + 2$ **5.** $2z, 3z$

6. $2x \cos(x^2 - y^2), -2y \cos(x^2 - y^2)$

7. $(2x \cos xy - y \sin xy)e^{x^2}, \; -xe^{x^2} \sin xy$

8. $\dfrac{\partial z}{\partial x} = 2x - 6xy + 4y^2, \; \dfrac{\partial z}{\partial y} = -3x^2 + 8xy$

$\dfrac{\partial^2 z}{\partial x^2} = 2 - 6y, \; \dfrac{\partial^2 z}{\partial y^2} = 8x$

$\dfrac{\partial^2 z}{\partial x \partial y} = -6x + 8y = \dfrac{\partial^2 z}{\partial y \partial x}$

$\dfrac{\partial^3 z}{\partial x^2 \partial y} = \dfrac{\partial^3 z}{\partial x \partial y \partial x} = \dfrac{\partial^3 z}{\partial y \partial x^2} = -6$

$\dfrac{\partial^3 z}{\partial x \partial y^2} = \dfrac{\partial^3 z}{\partial y \partial x \partial y} = \dfrac{\partial^3 z}{\partial y^2 \partial x} = 8$

9. $u_x = 6x + 2y^3, u_y = 2y + 6xy^2,$

$u_{xx} = 6, u_{yx} = u_{xy} = 6y^2, u_{yy} = 2 + 12xy,$

$u_{yyx} = u_{yxy} = u_{xyy} = 12y, u_{yyy} = 12x,$

$u_{xyyy} = u_{yxyy} = u_{yyxy} = u_{yyyx} = 12$

10. $\dfrac{\partial z}{\partial x} = 4xy - \sin(x + y),$

$\dfrac{\partial z}{\partial y} = 2x^2 - \sin(x + y),$

$\dfrac{\partial^2 z}{\partial x^2} = 4y - \cos(x + y),$

$\dfrac{\partial^2 z}{\partial y^2} = -\cos(x + y),$

$\dfrac{\partial^2 z}{\partial x \partial y} = 4x - \cos(x + y) = \dfrac{\partial^2 z}{\partial y \partial x}$

11. $\dfrac{\partial z}{\partial x} = \left[\sin(x + y) + \cos(x + y) \right] e^{x-y},$

$\dfrac{\partial z}{\partial y} = \left[\cos(x + y) - \sin(x + y) \right] e^{x-y},$

$\dfrac{\partial^2 z}{\partial x^2} = 2 \cos(x + y) e^{x-y},$

$\dfrac{\partial^2 z}{\partial y^2} = -2 \cos(x + y) e^{x-y},$

$\dfrac{\partial^2 z}{\partial x \partial y} = \dfrac{\partial^2 z}{\partial y \partial x} = -2 \sin(x + y) e^{x-y}$

17. $\dfrac{\partial r}{\partial x} = \dfrac{x}{r}, \; \dfrac{\partial r}{\partial y} = \dfrac{y}{r}, \; \dfrac{\partial r}{\partial z} = \dfrac{z}{r}$

19. $-\dfrac{2x + yz}{2y + xz}$

20. (i) $nR \Big/ \left(p - \dfrac{n^2 a}{V^2} + \dfrac{2n^3 ab}{V^3} \right)$

(ii) $-(V - nb) \Big/ \left(p - \dfrac{n^2 a}{V^2} + \dfrac{2n^3 ab}{V^3} \right)$

(iii) $nR/(V - nb)$

(iv) $2n^2 a/V^3 - (p + 2n^2 a/V^2)/(V - nb)$

Section 9.4

21. $(-2/3, 4/3)$ **22.** $(1, 2), (-1, 2)$

23. $(0, \pm\sqrt{3}), (1, 2), (-1, -2)$

24. $(-2/3, 4/3)$: maximum

25. $(1, 2)$: minimum; $(-1, 2)$: saddle point

26. $(0, \pm\sqrt{3})$: saddle points; $(1, 2)$: minimum; $(-1, -2)$: maximum

27. $f = 1$ at $(x, y, z) = (1/2, 1/3, 1/6)$

28. $f = c^6/27$ at $x^2 = y^2 = z^2 = c^2/3$

29. (i) $f = 4$ at $(x, y, z) = (1/3, 2/3, 2/3)$;

$f = 16$ at $(x, y, z) = (-1/3, -2/3, -2/3)$

30. (ii) $\lambda = a : \left(1/\sqrt{2}, 0, -1/\sqrt{2} \right),$

$\lambda = a \pm \sqrt{2}b : \left(1/2, \pm 1/\sqrt{2}, 1/2 \right)$

Section 9.5

31. $2x \, dx + 2y \, dy$

32. $[6x + \cos(x - y)]dx - \cos(x - y)dy$

33. $3x^2 y^2 \, dx + (2x^3 y + 1/y)dy$

34. $-\dfrac{1}{(x^2 + y^2 + z^2)^{3/2}}(x \, dx + y \, dy + z \, dz)$

35. $\sin\theta \sin\phi \, dr + r \cos\theta \sin\phi \, d\theta + r \sin\theta \cos\phi \, d\phi$

36. $dV = \left(\dfrac{\partial V}{\partial T} \right)_{p, n_A, n_B} dT + \left(\dfrac{\partial V}{\partial p} \right)_{T, n_A, n_B} dp$

$+ \left(\dfrac{\partial V}{\partial n_A} \right)_{T, p, n_B} dn_A + \left(\dfrac{\partial V}{\partial n_B} \right)_{T, p, n_A} dn_B$

Section 9.6

37. $-2 - 2t/(1-t^2)^{1/2}$

38. $-(2\sin t + 3)e^{2\cos t - 3t}$

39. $(-a\sin t + b\cos t + c)/(a\cos t + b\sin t + ct)$

40. $(3\cos\theta - 2\sin\theta)/(2\cos\theta + 3\sin\theta)$

41. (i) $(1-\sin u)\cos(u+v),$

    (ii) $-u(1-\sin u)\cos(u+v)$

42. $-5x^4 y/(x^5 - \cos y)$

43. $\left(\dfrac{\partial p}{\partial T}\right)_{V,n} = \dfrac{nR}{V-nb}$

44. $2(y^2 - x^2)/y$

45. $x\cos y - \left(\dfrac{x+3y}{x+y}\right)\sin y$

47. (i) $\dfrac{\partial f}{\partial u} = a\dfrac{\partial f}{\partial x} + b\dfrac{\partial f}{\partial y},\ \dfrac{\partial f}{\partial v} = b\dfrac{\partial f}{\partial x} - a\dfrac{\partial f}{\partial y}$

    (ii) $\dfrac{\partial f}{\partial u} = 2(a^2 + b^2)u,\ \dfrac{\partial f}{\partial v} = 2(a^2 + b^2)v$

48. (ii) $\left(\dfrac{\partial f}{\partial x}\right)_y = nx^{n-1}\left[\left(\dfrac{\partial f}{\partial u}\right)_v + \left(\dfrac{\partial f}{\partial v}\right)_u\right]$

    $\left(\dfrac{\partial f}{\partial y}\right)_x = ny^{n-1}\left[\left(\dfrac{\partial f}{\partial u}\right)_v - \left(\dfrac{\partial f}{\partial v}\right)_u\right]$

    (iii) $\left(\dfrac{\partial f}{\partial x}\right)_y = 4nx^{n-1}y^n,\ \left(\dfrac{\partial f}{\partial y}\right)_x = 4ny^{n-1}x^n$

Section 9.7

53. exact   **54.** not exact   **55.** exact

Section 9.8

57. $64/3$ **58.** 2   **59.** 20

60. $7/3$ **62.** 31

63. (i) $-1/3,$ (ii) 0

Section 9.9

64. 24   **65.** $2(1-e^{-R})/3$

Section 9.10

66. $40/3$ **67.** $a^5/15$

68. (ii) $-51/2$

Section 9.11

69. $\dfrac{1}{4}\left(\dfrac{\pi}{4}+1\right)$ **70.** $315\pi/8$

71. $\pi/8$

Chapter 10

Section 10.2

1. $(0,0,1)$ **2.** $(0,2,0)$

3. $(-\sqrt{3/2},\sqrt{3/2},-1)$ **4.** $(1,\pi/2,0)$

5. $(1,\pi/2,\pi/2)$ **6.** $(3,\cos^{-1}(2/3),\tan^{-1}(2))$

7. $(9,\cos^{-1}(-8/9),\tan^{-1}(-4)+\pi)$

8. $(7,\cos^{-1}(6/7),\tan^{-1}(3/2)+\pi)$

9. $(13,\cos^{-1}(-12/13),\tan^{-1}(-4/3)+\pi)$

Section 10.3

10. $r^2\sin^2\theta\cos 2\phi$ **11.** $\tan^2\theta$

12. $r^2(3\cos^2\theta - 1)$ **13.** (i) $1/r$ (ii) $-x/r^3$

Section 10.4

14. 1   **15.** $a^2 b^3 c^4/24$ **16.** $1/2$

17. 1   **18.** $4\pi a^5/5$ **19.** $2\pi a^2/3$

20. 480π **22.** $\dfrac{1}{81}\sqrt{\dfrac{2}{\pi}}$ **23.** $6\,a_0$

24. 180

Section 10.5

25. $2y^3 z^4 + 6x^2 yz^4 + 12x^2 y^3 z^2$

26. $\left(\dfrac{n(n+1)}{r^2} - \dfrac{2a(n+1)}{r} + a^2\right)r^n e^{-ar}$

27. e^{-r}/r

28. $\left(\dfrac{1}{9} - \dfrac{2}{3r} - \dfrac{2}{r^2}\right)e^{-r/3}\sin\theta\sin\phi$

29. $xze^{-r/2}\left(\dfrac{1}{4} - \dfrac{3}{r}\right)$

Section 10.6

38. (i) $x = a\cos t,\ y = -a\sin t,\ z = bt$

    (ii) $\rho = a,\ \phi = -t,\ z = bt$

39. $\pi a^4/16$ **40.** $\pi(1+R)e^{-R}$

41. $\dfrac{\pi}{2}(1+R)e^{-R}$

Chapter 11

Section 11.2

5. $\dfrac{x^3}{3} + c$ **6.** $-\dfrac{e^{-3x}}{3} + c$

7. $-\dfrac{1}{9}\cos 3x + ax + b$ **8.** $x^4 + ax^2 + bx + c$

9. (i) $m\dfrac{d^2x}{dt^2} = \cos 2\pi t$

(ii) $x(t) = \dfrac{1 - \cos 2\pi t}{4\pi^2 m} + t$

10. $6x^2$ **11.** $2e^{4-x^2}$ **12.** $5e^{-2x} - 1$

Section 11.3

13. $y^2 = 2x^3 + c$

14. $y = \dfrac{-1}{2x^2 + c}$

15. $y = ae^{x^3}$

16. $y^3 = 3e^x + c$

17. $y = \dfrac{1}{1 + ce^x}$

18. $y = cx$

19. $y = 1 - x$

20. $y^2 + y = \dfrac{x^3}{3} - x$

21. $y = x - 1$

22. $y = -\ln(2 - e^x)$

23. $y = x(\ln x + 2)$

24. $y^2 = \dfrac{1}{3}\left(\dfrac{1}{x} - x^2\right)$

25. $y^4 = 4x^4 \ln\left(\dfrac{x}{2}\right)$

27. $y = ae^x - 2x - 5$

28. $(x - y)^2 - 4y = c$

Section 11.4

29. $\tau_{1/n} = \dfrac{\ln n}{k}$

30. $\dfrac{1}{x^{n-1}} - \dfrac{1}{a^{n-1}} = (n-1)kt$

31. $x(t) = \dfrac{k_1 a}{k_1 + k_{-1}}\left[1 - e^{-(k_1 + k_{-1})t}\right]$

Section 11.5

33. $y = 2 + ce^{-2x}$

34. $y = ce^{2x^2} - \dfrac{1}{4}$

35. $y = e^{-3x}(x + c)$

36. $y = \dfrac{2}{x^2}\left[x^2 \sin x + 2x \cos x - 2 \sin x + c\right]$

37. $y = ce^{-1/x} - 4$

38. $y = \cos x + c \cos^2 x$

39. $y = \dfrac{b}{a} + ce^{-ax^{n+1}/(n+1)}$

40. $y = \dfrac{x^{n+1}}{n + a + 1} + cx^{-a}$

Section 11.6

41. $[C] = \dfrac{ak_1 k_2}{k_2 - k_1}\left[\dfrac{e^{-k_1 t} - e^{-k_3 t}}{k_3 - k_1} - \dfrac{e^{-k_2 t} - e^{-k_3 t}}{k_3 - k_2}\right]$

42. $[C] = \dfrac{ak_1 k_2}{k_2 + k_3 - k_1}\left[\dfrac{1}{k_1}(1 - e^{-k_1 t})\right.$

$\left. - \dfrac{1}{k_2 + k_3}(1 - e^{-(k_2 + k_3)t})\right]$

Section 11.7

43. $I(t) = \dfrac{E_0}{R^2 + \omega^2 L^2}\left[\omega L e^{-Rt/L}\right.$

$\left. + R \sin \omega t - \omega L \cos \omega t\right]$

44. (i) $I(t) = Ae^{-t/RC}$

(ii) $I(t) = ce^{-t/RC} + \dfrac{E_0 \omega C}{1 + (\omega RC)^2}$

$\times \left[\cos \omega t + \omega RC \sin \omega t\right]$

Chapter 12

Section 12.2

4. $y = ae^{-2x} + be^{2x/3}$ **5.** $y = (a + bx)e^{3x}$

6. $y = a \cos 2x + b \sin 2x$

Section 12.3

7. $y = ae^{3x} + be^{-2x}$

8. $y = (ae^{2\sqrt{5}x} + be^{-2\sqrt{5}x})e^{2x}$

9. $y = (a + bx)e^{4x}$ **10.** $y = (a + bx)e^{-3x/2}$

11. $y = e^{-2x}(A \cos x + B \sin x)$

12. $y = e^{-3x/2}(ae^{i\sqrt{11}x/2} + be^{-i\sqrt{11}x/2})$

Section 12.4

13. $x = \dfrac{2}{3}e^t + \dfrac{1}{3}e^{-2t}$ **14.** $x = e^{-3(t-1)}(t - 1)$

15. $x = \dfrac{1}{3}\sin 3t$ **16.** $x = (\cos t - \sin t)e^t$

17. $y = e^{\pi - 2x}(\cos 2x - e^{\pi/2}\sin 2x)$

18. $y = -\sin 3x$ **19.** $y = xe^{4(1-x)}$

20. $y = 2e^{-2x}$

21. $\theta_n(t) = Ae^{int/\tau} + Be^{-int/\tau}, n = 0, \pm 1, \pm 2, \ldots$

Section 12.5

22. (i) $A = \sqrt{a^2 + b^2}, \quad \delta = \tan^{-1}(b/a)$

(ii) $A = \sqrt{2}, \quad \delta = \pi/4$

23. $x(t) = (U_0/\omega)\sin \omega t$

Section 12.6

24. (i) $x/l = 0, 1/4, 1/2, 3/4, 1$

(ii) $x/l = 0, 1/5, 2/5, 3/5, 4/5, 1$

25. (i) $\psi_n(x) = \sqrt{\dfrac{2}{l}}\cos\dfrac{n\pi x}{l}$ if n odd,

$\psi_n(x) = \sqrt{\dfrac{2}{l}}\sin\dfrac{n\pi x}{l}$ if n even

Section 12.8

30. $y_p = -\dfrac{1}{4} - \dfrac{x}{2}$

31. $y = ae^{3x} + be^{-2x} - \dfrac{1}{4} - \dfrac{x}{2}$

32. $y = (a + bx)e^{4x} - \dfrac{1}{32}(1 + 9x + 12x^2 + 8x^3)$

33. $y = ae^{3x} + be^{-2x} + \dfrac{1}{3}e^{-3x}$

34. $y = ae^{2x} + (b - x)e^{-x}$

35. $y = (a + bx + x^2/2)e^{4x}$

36. $y = ae^{3x} + be^{-2x} - \dfrac{1}{39}(5\cos 3x + \sin 3x)$

37. $y = (a - 3x/4)\cos 2x + b\sin 2x$

38. $y = ae^{3x} + be^{-2x} - \dfrac{1}{4} - \dfrac{x}{2} + \dfrac{1}{3}e^{-3x}$

$\qquad -\dfrac{1}{39}(5\cos 3x + \sin 3x)$

39. (ii) $I_h(t) = ae^{-(\alpha + \beta)t} + be^{-(\alpha - \beta)t}$
$\qquad$ where $\alpha = R/2L,\ \beta^2 = \alpha^2 - (1/LC)$

Chapter 13

Section 13.2

1. $y = a_0 \sum\limits_{m=0}^{\infty} x^{3m}/m! = a_0 e^{x^3}$

2. $y = a_0 \sum\limits_{m=0}^{\infty} x^m$

3. $y = a_0 \left(1 + \dfrac{(3x)^2}{2!} + \dfrac{(3x)^4}{4!} + \cdots \right)$

$\qquad + \dfrac{a_1}{3}\left(\dfrac{(3x)}{1!} + \dfrac{(3x)^3}{3!} + \dfrac{(3x)^5}{5!} + \cdots \right)$

$\qquad = ae^{3x} + be^{-3x}$ if $a_0 = a + b, a_1 = 3(a - b)$

4. $y = a_1 x + a_0 \left(1 - x^2 - \dfrac{x^4}{3} - \dfrac{x^6}{5} - \dfrac{x^8}{7} - \cdots \right)$

5. $y = a_0 \left(1 + \dfrac{1}{3!}x^3 + \dfrac{1 \cdot 4}{6!}x^6 + \dfrac{1 \cdot 4 \cdot 7}{9!}x^9 + \cdots \right)$

$\qquad + a_1 \left(x + \dfrac{2}{4!}x^4 + \dfrac{2 \cdot 5}{7!}x^7 + \dfrac{2 \cdot 5 \cdot 8}{10!}x^{10} + \cdots \right)$

Section 13.3

6. $r_1 = r_2 = -1$ $\qquad$ **7.** $r = \pm n$

8. $r_1 = r_2 = 0$ $\qquad$ **9.** $r_1 = -2, r_2 = -3$

10. (i) $y = ax^{r_1} + bx^{r_2}$ $\qquad$ (ii) $y = (a + b\ln x)x^r$

11. $y = ax^{1/2} + bx$ $\qquad$ **12.** $y = (a + b\ln x)x$

13. $y = ax^{-1/2}\cos x$ $\qquad$ **14.** $y_1 = ae^x$

15. $y = a\,\dfrac{\cos 2x}{x} + b\,\dfrac{\sin 2x}{x}$

Section 13.4

17. $P_6(x) = \dfrac{1}{16}(231x^6 - 315x^4 + 105x^2 - 5)$

18. (i) $P_1^1 = \sin\theta$

$\qquad$ (ii) $P_4^1 = \dfrac{5}{2}\sin\theta\,(7\cos^3\theta - 3\cos\theta)$

$\qquad\quad P_4^2 = \dfrac{15}{2}\sin^2\theta\,(7\cos^2\theta - 1)$

$\qquad\quad P_4^3 = 105\sin^3\theta\cos\theta,\ P_4^4 = 105\sin^4\theta$

Section 13.5

21. (i) $H_5 = 32x^5 - 160x^3 + 120x$

$\qquad$ (iii) $H_6 = 64x^6 - 480x^4 + 720x^2 - 120$

Section 13.6

23. (i) $y = a_0 \left[1 - \dfrac{n}{(1!)^2}x + \dfrac{n(n-1)}{(2!)^2}x^2 \right.$

$\qquad\qquad\qquad \left. - \dfrac{n(n-1)(n-2)}{(3!)^2}x^3 + \cdots \right]$

24. $L_4 = x^4 - 16x^3 + 72x^2 - 96x + 24$

Section 13.7

25. $J_2 = \left(\dfrac{x}{2} \right)^2 \left[\dfrac{1}{2!} - \dfrac{1}{1!3!}\left(\dfrac{x}{2} \right)^2 + \dfrac{1}{2!4!}\left(\dfrac{x}{2} \right)^4 \right.$

$\qquad\qquad\qquad\qquad \left. - \dfrac{1}{3!5!}\left(\dfrac{x}{2} \right)^6 + \cdots \right]$

26. (i) $J_{5/2} = \sqrt{\dfrac{2}{\pi x}}\left[\left(\dfrac{3}{x^2} - 1 \right)\sin x - \dfrac{3}{x}\cos x \right]$

$\qquad$ (ii) $J_{-5/2} = \sqrt{\dfrac{2}{\pi x}}\left[\left(\dfrac{3}{x^2} - 1 \right)\cos x + \dfrac{3}{x}\sin x \right]$

Chapter 14

Section 14.2

2. $c/D = 0$: $\qquad V = A + Bx$

$\qquad c/D = \lambda^2 > 0$: $\quad V = Ae^{\lambda x} + Be^{-\lambda x}$

$\qquad c/D = -\lambda^2 < 0$: $\quad V = A\cos\lambda x + B\sin\lambda x$

Section 14.3

4. $f(x, t) = Ae^{B(x-2t)}$

5. $f(x, y) = Ae^{B(x^2+y^2)}$

6. For separation constant λ^2;
$\lambda^2 = 0$: $f(x, y) = (a + bx)(c + dy)$
$\lambda^2 > 0$: $f(x, y) = [ae^{\lambda x} + be^{-\lambda x}][c\cos\lambda y$
    $+ d\sin\lambda y]$

7. $f(x, y) = Ae^{(cx-y/c)}$

Section 14.4

9. (i) $E_{1,1} = 2, E_{1,2} = E_{2,1} = 5,$
$E_{2,2} = 8, E_{1,3} = E_{3,1} = 10,$
$E_{2,3} = E_{3,2} = 13,$
$E_{1,4} = E_{4,1} = 17, E_{3,3} = 18$

10. (i) $\psi_{p,q,r}(x, y, z) = \sqrt{\dfrac{2}{a}}\sin\left(\dfrac{p\pi x}{a}\right)$

$\times\sqrt{\dfrac{2}{b}}\sin\left(\dfrac{q\pi y}{b}\right)\times\sqrt{\dfrac{2}{c}}\sin\left(\dfrac{r\pi z}{c}\right)$

$E_{p,q,r} = \dfrac{h^2}{8m}\left(\dfrac{p^2}{a^2} + \dfrac{q^2}{b^2} + \dfrac{r^2}{c^2}\right)$

(ii) Degeneracy if $a = b = c$:
1 if p, q, r all equal (e.g. $E_{2,2,2}$)
3 if two equal (e.g. $E_{1,1,2}, E_{1,2,1}, E_{2,1,1}$)
6 if all different (e.g. $E_{1,2,3}, E_{1,3,2}, E_{2,3,1},$
$E_{2,1,3}, E_{3,1,2}, E_{3,2,1}$)

Section 14.5

11. (i) $E_{0,1} = 5.7831, E_{\pm1,1} = 14.6819,$
$E_{\pm2,1} = 26.3744, E_{0,2} = 30.4715,$
$E_{\pm3,1} = 40.7070, E_{\pm1,2} = 49.2186$

Section 14.6

12. (i) $\psi_{1,0,0} = (Z^3/\pi)^{1/2}e^{-Zr}$
(ii) $E = -Z^2/2$

13. (i) $\psi_{2,1,0} = (Z^5/32\pi)^{1/2}re^{-Zr/2}\cos\theta$
(ii) $E = -Z^2/8$

15. (iii) $E_{n,l} = \dfrac{x_{n,l}^2 h^2}{2ma^2}, n = 1, 2, 3, \ldots,$

$l = 0, 1, 2, \ldots,$ where the allowed
values of $x_{n,l}$ are the zeros of the
spherical Bessel function $j_l(x)$.

(iv) $\psi_{1,0,0} = \sqrt{\dfrac{1}{2\pi a}}\dfrac{\sin(\pi r/a)}{r}$

$E_{1,0} = \dfrac{h^2}{8ma^2}$

Section 14.7

16. $y(x,t) = 3\sin\dfrac{\pi x}{l}\cos\dfrac{\pi vt}{l}$

17. $T = 3\sin\dfrac{\pi x}{l}\exp[-D\pi^2 t/l^2]$

18. (i) $u(x,y) = \displaystyle\sum_{n=1}^{\infty} A_n\sinh(n\pi y/a)\sin(n\pi x/a)$

(ii) $u(x,y) = \dfrac{\sinh(3\pi y/a)}{\sinh(3\pi b/a)}\sin(3\pi x/a)$

Chapter 15

Section 15.2

1. $c_1 = 3\left(\dfrac{a_1}{3} + \dfrac{a_3}{5} + \dfrac{a_5}{7} + \cdots\right)$

2. (i) $\dfrac{1}{4}P_0 + \dfrac{1}{2}P_1 + \dfrac{5}{16}P_2$

Section 15.3

4. (i) $\cos tx = j_0(t)P_0(x) - 5j_2(t)P_2(x)$
    $+ 9j_4(t)P_4(x) - \cdots$
(ii) $\sin tx = 3j_1(t)P_1(x) - 7j_3(t)P_3(x)$
    $+ 11j_5(t)P_5(x) - \cdots$

5. $V = \dfrac{Q_2}{4\pi\varepsilon_0 R^3}, Q_2 = qr^2(3\cos^2\theta - 1)$

Section 15.4

8. (ii) $\dfrac{1}{2} + \dfrac{2}{\pi}\left(\dfrac{\cos x}{1} - \dfrac{\cos 3x}{3} + \dfrac{\cos 5x}{5}\right.$

$\left. - \dfrac{\cos 7x}{7} + \cdots\right)$

9. (ii) $2\left(\dfrac{\sin x}{1} - \dfrac{\sin 2x}{2} + \dfrac{\sin 3x}{3} - \dfrac{\sin 4x}{4} + \cdots\right)$

10. (ii) $\dfrac{\pi^2}{3} - 4\left(\dfrac{\cos x}{1^2} - \dfrac{\cos 2x}{2^2} + \dfrac{\cos 3x}{3^2} - \cdots\right)$

12. (ii) $\dfrac{8l^2}{\pi^3}\left[\sin\left(\dfrac{\pi x}{l}\right) + \dfrac{1}{3^3}\sin\left(\dfrac{3\pi x}{l}\right)\right.$

$\left. + \dfrac{1}{5^3}\sin\left(\dfrac{5\pi x}{l}\right) + \cdots\right]$

13. (ii) $\dfrac{l^2}{6} - \dfrac{4l^2}{\pi^2}\left[\dfrac{1}{2^2}\cos\left(\dfrac{2\pi x}{l}\right)\right.$

$\left. + \dfrac{1}{4^2}\cos\left(\dfrac{4\pi x}{l}\right) + \dfrac{1}{6^2}\cos\left(\dfrac{6\pi x}{l}\right) + \cdots\right]$

Section 15.5

14. $T(x,t) = \dfrac{8l^2}{\pi^3} \displaystyle\sum_{n\ \mathrm{odd}} \dfrac{1}{n^3} \sin\left(\dfrac{n\pi x}{l}\right)$

$\times \exp\left[-n^2\pi^2 Dt/l^2\right]$

Section 15.6

15. $g(y) = \dfrac{e^{-iay} - e^{-iby}}{i\sqrt{2\pi}\, y}$

16. $g(y) = \sqrt{\dfrac{2}{\pi}}\left(\dfrac{a}{a^2 + y^2}\right)$

Chapter 16

Section 16.2

1. (i) $b-a, a-b$ (ii) $\frac{1}{4}(a+b+c+d)$
(iii) $\frac{1}{2}(c+d)$ (iv) $b+\lambda(c-d)$, all λ

Section 16.3

2. $(1, 1, 1)$

3. (i) $(3, 4, 0)$ (ii) 5
(iii) $(3/5, 4/5, 0)$

4. (i) $(2, -5, 1)$ (ii) $\sqrt{30}$
(iii) $(2/\sqrt{30}, -5/\sqrt{30}, 1/\sqrt{30})$

5. (i) $(2, 3, 1)$ (ii) $\sqrt{14}$
(iii) $(2/\sqrt{14}, 3/\sqrt{14}, 1/\sqrt{14})$

6. Both $(-1, 5, -1)$

7. $(3, 6, 9), (-1, -2, -3), (1/3, 2/3, 1)$

8. $(-1, 0, 4)$ **9.** Both $(3, -6, 12)$

10. $\sqrt{27}$, $\sqrt{14} + \sqrt{29}$

11. (i) $(2, -1, 0)$,
(ii) $(1, -1, 1), (0, 0, 0), (-2, 2, -2)$

12. (i) $(2, 6, -6)$ (ii) $(1, 3, -3)$

13. $-2i+j$

14. (i) $4i$ (ii) λj (iii) λi

Section 16.4

15. $2i + 6tj$ **16.** $(-2\sin 2t, 3\cos t, 2)$

17. (i) $v = ai + (a - gt)j,\ a = -gj$
(ii) $F = -mgj$
(iii) to the right with constant speed $v_x = a$
(iv) vertical up and down under the influence of gravity
(v) parabolic up and down to the right

18. (i) $v = -6\sin 3t\, i + 6\cos 3t\, j + 3k$
$a = -18\cos 3t\, i - 18\sin 3t\, j$
$= -9(xi + yj)$
(ii) $-9m(xi + yj)$
(iii) simple harmonic in both directions
(iv) circular in xy-plane at constant angular speed
(v) in z-direction at constant speed
(vi) right-handed circular helix around z-axis

Section 16.5

19. Both 7   **20.** Both 10

21. $(0, 12, 4)$ **23.** -1 **24.** 0

25. -3 **26.** 2   **27.** $\pi/3, 2\pi/3, \pi/4$

28. (i) 4   (ii) -3 (iii) 0

29. (i) $25/2$ (ii) $V = -\dfrac{x^2}{2} - y^2 - \dfrac{3z^2}{2} + C$

30. -6

Section 16.6

31. $9i - j + 3k, -9i + j - 3k$ **32.** $7i, 7$

33. Both $11i - 2j - k$ **34.** 0   **35.** -7

36. $-14j - 21k, i - 18j - 9k$ **40.** 5

41. (i) $-7k$ (ii) i (iii) $-5j$

42. $5i + 6j$ **43.** $-3i + 18k$

45. (i) $\omega = \omega k$ (ii) $v = -\omega yi + \omega xj$
(iii) $l = m\omega[(x^2 + y^2)k - xzi - yzj]$

47. $l = 2m(x^2 + y^2)\omega k$

Section 16.8

48. $4xi + 6yj - 2zk$

49. $(y+z)i + (x+z)j + (x+y)k$

50. $-(x^2 + y^2 + z^2)^{-3/2}(xi + yj + zk)$

51. $3, 0$ **52.** $0, i + j + k$ **53.** $0, 0$

Chapter 17

Section 17.1

1. $x = 2, y = 3$ **2.** 6   **3.** 2   **4.** 1

Section 17.2

5. $x = 1, y = 2, z = 3$ **6.** -11

7. 6   **8.** −4   **9.** 30

10. (i) $C_{11}=-1, C_{12}=-3, C_{13}=-2$
$C_{21}=-5, C_{22}=-2, C_{23}=+3$
$C_{31}=-2, C_{32}=+7, C_{33}=-4$

(ii) −13

Section 17.3

11. −6   **12.** 4   **13.** 0   **14.** 144

Section 17.4

15. $x=y=z=0$ **16.** $w=x=y=1, z=-1$
17. (ii) $x=(z+3)/5, y=(7z-4)/5$, all z
18. (ii) $x=z/3, y=-2z/3$, all z
19. (i) $\lambda=1, 3$
    (ii) $\lambda=1: y=-x, \lambda=3: y=x$, all x
20. (i) $\lambda=3, 3\pm\sqrt{2}$
    (ii) $\lambda=3: x=-z, y=0$, all z
    $\lambda=3\pm\sqrt{2}: x=z, y=\pm\sqrt{2}z$, all z
21. (i) $\lambda=0$ (double), 8
    (ii) $\lambda=0: x=3z-2y$, all y and z
    $\lambda=8: x=-z, y=-2z$, all z

25. $\begin{vmatrix} 0 & 2 & 6x \\ 4x^3 & 5x^4 & 6x^5 \\ 7x^6 & 8x^7 & 9x^8 \end{vmatrix} + \begin{vmatrix} 1 & 2x & 3x^2 \\ 12x^2 & 20x^3 & 30x^4 \\ 7x^6 & 8x^7 & 9x^8 \end{vmatrix}$

$+ \begin{vmatrix} 1 & 2x & 3x^2 \\ 4x^3 & 5x^4 & 6x^5 \\ 42x^5 & 56x^6 & 72x^7 \end{vmatrix}$

Section 17.6

26. $\begin{vmatrix} 3 & 2 & -2 \\ 0 & -3 & 9 \\ 0 & 0 & 25 \end{vmatrix} = -225$

27. $\begin{vmatrix} 1 & 0 & 1 & 1 \\ 0 & 1 & 1 & 0 \\ 0 & 0 & -1 & 0 \\ 0 & 0 & 0 & -1 \end{vmatrix} = 1$

28. $\begin{vmatrix} 2 & 4 & 6 & 3 \\ 0 & -1 & -10 & 1 \\ 0 & 0 & 42 & -13/2 \\ 0 & 0 & 0 & 80/21 \end{vmatrix} = -320$

29. $\psi_1(x_1)\psi_2(x_2)\psi_3(x_3)\psi_4(x_4)$
$-\psi_1(x_1)\psi_2(x_2)\psi_3(x_4)\psi_4(x_3)$
$+\psi_1(x_1)\psi_2(x_3)\psi_3(x_4)\psi_4(x_2)$

$-\psi_1(x_1)\psi_2(x_3)\psi_3(x_2)\psi_4(x_4)$
$+\psi_1(x_1)\psi_2(x_4)\psi_3(x_2)\psi_4(x_3)$
$-\psi_1(x_1)\psi_2(x_4)\psi_3(x_3)\psi_4(x_2)$
$+\psi_1(x_2)\psi_2(x_1)\psi_3(x_4)\psi_4(x_3)$
$-\psi_1(x_2)\psi_2(x_1)\psi_3(x_3)\psi_4(x_4)$
$+\psi_1(x_2)\psi_2(x_3)\psi_3(x_1)\psi_4(x_4)$
$-\psi_1(x_2)\psi_2(x_3)\psi_3(x_4)\psi_4(x_1)$
$+\psi_1(x_2)\psi_2(x_4)\psi_3(x_3)\psi_4(x_1)$
$-\psi_1(x_2)\psi_2(x_4)\psi_3(x_1)\psi_4(x_3)$
$+\psi_1(x_3)\psi_2(x_1)\psi_3(x_2)\psi_4(x_4)$
$-\psi_1(x_3)\psi_2(x_1)\psi_3(x_4)\psi_4(x_2)$
$+\psi_1(x_3)\psi_2(x_2)\psi_3(x_4)\psi_4(x_1)$
$-\psi_1(x_3)\psi_2(x_2)\psi_3(x_1)\psi_4(x_4)$
$+\psi_1(x_3)\psi_2(x_4)\psi_3(x_1)\psi_4(x_2)$
$-\psi_1(x_3)\psi_2(x_4)\psi_3(x_2)\psi_4(x_1)$
$+\psi_1(x_4)\psi_2(x_1)\psi_3(x_3)\psi_4(x_2)$
$-\psi_1(x_4)\psi_2(x_1)\psi_3(x_2)\psi_4(x_3)$
$+\psi_1(x_4)\psi_2(x_2)\psi_3(x_1)\psi_4(x_3)$
$-\psi_1(x_4)\psi_2(x_2)\psi_3(x_3)\psi_4(x_1)$
$+\psi_1(x_4)\psi_2(x_3)\psi_3(x_2)\psi_4(x_1)$
$-\psi_1(x_4)\psi_2(x_3)\psi_3(x_1)\psi_4(x_2)$

Chapter 18

Section 18.1

1. (i) $\begin{pmatrix} \frac{1}{\sqrt{2}} & -\frac{1}{\sqrt{2}} \\ \frac{1}{\sqrt{2}} & \frac{1}{\sqrt{2}} \end{pmatrix}$ **(ii)** $\begin{pmatrix} 0 & -1 \\ 1 & 0 \end{pmatrix}$

(iii) $\begin{pmatrix} 0 & 1 \\ -1 & 0 \end{pmatrix}$ **2.** $\begin{pmatrix} 0 & 1 \\ 1 & 0 \end{pmatrix}$

Section 18.2

3. not possible   **4.** −6, 4   **5.** 4, 5

6. not possible   **7.** $\begin{pmatrix} 1 & 0 \\ -2 & 3 \\ 3 & 4 \end{pmatrix}$

8. $\begin{pmatrix} -5 & 4 & 2 \\ 3 & -1 & -1 \end{pmatrix}$ **9.** $\begin{pmatrix} 3 & 0 & 0 \\ 0 & 2 & 0 \\ 0 & 0 & -1 \end{pmatrix}$

10. $\begin{pmatrix} 1 & 0 \\ -2 & 4 \end{pmatrix}$ **11.** $(0 \ -3 \ 1)$

12. $\begin{pmatrix} 2 \\ 5 \\ -2 \end{pmatrix}$

Section 18.3

13. $\begin{pmatrix} 1 & -1 & -1 \\ 2 & 0 & 4 \end{pmatrix}$

14. $\begin{pmatrix} 1 & -3 & 7 \\ -2 & 6 & 4 \end{pmatrix}$

15. $\begin{pmatrix} -1 & 3 & -7 \\ 2 & -6 & -4 \end{pmatrix}$

16. not possible

17. $\begin{pmatrix} 2 \\ 2 \\ -1 \end{pmatrix}$

18. $(2 \quad 2 \quad -1)$

19. $\begin{pmatrix} 3 & -6 \\ 0 & 12 \end{pmatrix}$

20. $\begin{pmatrix} 2 & -1 & -6 \\ 6 & -3 & 8 \end{pmatrix}$

21. not possible

22. $\begin{pmatrix} -4 & 3 \\ -22 & 9 \end{pmatrix}$

23. $\begin{pmatrix} 6 & -14 & 20 \\ -2 & 7 & -16 \\ -2 & 5 & -8 \end{pmatrix}$

24. $\begin{pmatrix} -5 & 22 \\ 4 & -12 \\ 2 & -8 \end{pmatrix}$

25. not possible

26. $\begin{pmatrix} 9 & 0 & 0 \\ 0 & 4 & 0 \\ 0 & 0 & 1 \end{pmatrix}$

27. $\begin{pmatrix} 3 & -2 \\ 0 & 4 \end{pmatrix}$

28. $\begin{pmatrix} 3 & -6 \\ 0 & 4 \end{pmatrix}$

29. $\begin{pmatrix} -7 \\ 9 \end{pmatrix}$

30. $\begin{pmatrix} 0 & 0 & 0 \\ -6 & -15 & 6 \\ 2 & 5 & -2 \end{pmatrix}$

31. -17

32. -17

33. $\begin{pmatrix} 0 & -6 & 2 \\ 0 & -15 & 5 \\ 0 & 6 & -2 \end{pmatrix}$

34. not possible

35. $(-10 \quad 2)$

37. $\begin{pmatrix} 0 & 1 \\ -10 & 7 \end{pmatrix}$

38. $\begin{pmatrix} -5 & 4 \\ 4 & -1 \end{pmatrix}, \begin{pmatrix} -5 & 4 \\ 4 & -1 \end{pmatrix}$

39. $\begin{pmatrix} 0 & 1 \\ 0 & 1 \end{pmatrix}, \begin{pmatrix} 1 & 0 \\ -1 & 0 \end{pmatrix}$

40. $\begin{pmatrix} 0 & 0 \\ 0 & 0 \end{pmatrix}$

41. $\begin{pmatrix} -1 & 1 \\ 1 & 1 \end{pmatrix}$

42. (i) $[I_x, I_y] = i\hbar I_z,\ [I_y, I_z] = i\hbar I_x,\ [I_z, I_x] = i\hbar I_y$
(ii) $2\hbar^2 \mathbf{1}$

43. $\mathbf{B} = \begin{pmatrix} 1 & 0 \\ 0 & 1 \end{pmatrix}, \mathbf{C} = \begin{pmatrix} 1 & 0 & 0 & 0 \\ 0 & 1 & 0 & 0 \\ 0 & 0 & 1 & 0 \\ 0 & 0 & 0 & 1 \end{pmatrix}$

44. $\begin{pmatrix} -2c & -2d \\ c & d \end{pmatrix}$

45. $d\begin{pmatrix} 4 & -2 \\ -2 & 1 \end{pmatrix}$

Section 18.4

47. $\dfrac{1}{14}\begin{pmatrix} 1 & 3 \\ -4 & 2 \end{pmatrix}$

48. $\dfrac{1}{6}\begin{pmatrix} 1 & -1 & 1 \\ 4 & -10 & -8 \\ -1 & 7 & 5 \end{pmatrix}$

49. singular

50. $\begin{pmatrix} \frac{1}{\sqrt{2}} & 0 & \frac{1}{\sqrt{2}} \\ 0 & 1 & 0 \\ -\frac{1}{\sqrt{2}} & 0 & \frac{1}{\sqrt{2}} \end{pmatrix}$

51. $\dfrac{1}{2}\begin{pmatrix} 2 & -4 & 0 & 0 \\ -1 & 3 & 0 & 0 \\ 0 & 0 & 2 & -1 \\ 0 & 0 & -4 & 3 \end{pmatrix}$

53. both $\dfrac{1}{4}\begin{pmatrix} 50 & -22 \\ -43 & 19 \end{pmatrix}$

Section 18.5

54. (i) $\begin{cases} x' = x\cos\theta - y\sin\theta \\ y' = x\sin\theta + y\cos\theta \\ z' = z \end{cases}$

(ii) $\begin{pmatrix} \frac{1}{\sqrt{2}} & \frac{1}{\sqrt{2}} & 0 \\ -\frac{1}{\sqrt{2}} & \frac{1}{\sqrt{2}} & 0 \\ 0 & 0 & 1 \end{pmatrix}$

55. (i) $\begin{pmatrix} 1 & 0 & 0 \\ 0 & \cos\theta & -\sin\theta \\ 0 & \sin\theta & \cos\theta \end{pmatrix}$

(ii) $\begin{pmatrix} \cos\theta & 0 & \sin\theta \\ 0 & 1 & 0 \\ -\sin\theta & 0 & \cos\theta \end{pmatrix}$

(iii) $\begin{pmatrix} 1 & 0 & 0 \\ 0 & 1 & 0 \\ 0 & 0 & -1 \end{pmatrix}$

(iv) $\begin{pmatrix} -1 & 0 & 0 \\ 0 & 1 & 0 \\ 0 & 0 & 1 \end{pmatrix}$

(v) $\begin{pmatrix} 1 & 0 & 0 \\ 0 & -1 & 0 \\ 0 & 0 & 1 \end{pmatrix}$

(vi) $\begin{pmatrix} -1 & 0 & 0 \\ 0 & -1 & 0 \\ 0 & 0 & -1 \end{pmatrix}$

56. $\dfrac{1}{\sqrt{5}}\begin{pmatrix} 0 & 1 & -1 & -2 \\ 5 & 7 & 8 & 6 \end{pmatrix}$

57. $\begin{pmatrix} -1 & -1 & -2 & -2 \\ 2 & 3 & 3 & 2 \end{pmatrix}$

58. $\begin{pmatrix} 4 & 6 & 6 & 4 \\ 1 & 1 & 2 & 2 \end{pmatrix}$

59. $\begin{pmatrix} 7 & 10 & 11 & 8 \\ 0 & -1 & 1 & 2 \end{pmatrix}$

60. (i) $\begin{pmatrix} \cos\phi & -\cos\theta\sin\phi & \sin\theta\sin\phi \\ \sin\phi & \cos\theta\cos\phi & -\sin\theta\cos\phi \\ 0 & -\sin\theta & -\cos\theta \end{pmatrix}$

(ii) $\begin{pmatrix} \frac{\sqrt{3}}{2} & \frac{1}{4} & -\frac{\sqrt{3}}{4} \\ -\frac{1}{2} & \frac{\sqrt{3}}{4} & -\frac{3}{4} \\ 0 & -\frac{\sqrt{3}}{2} & -\frac{1}{2} \end{pmatrix}$ **(iii)** $\frac{1}{4}\begin{pmatrix} 5\sqrt{3} \\ -1 \\ 2 \end{pmatrix}$

Section 18.6

61. (i) $\mathbf{A}^{-1} = \mathbf{A}^{\mathsf{T}} = \begin{pmatrix} 1 & 0 & 0 \\ 0 & \cos\theta & \sin\theta \\ 0 & -\sin\theta & \cos\theta \end{pmatrix}$

(iii) and (vi) $\mathbf{A}^{-1} = \mathbf{A}^{\mathsf{T}} = \mathbf{A}$

Section 18.7

62. (i) $E =$ identity

$A = 180°$ about Oz-axis

$B = 180°$ about Ox-axis

$C = 180°$ about Oy-axis

(ii)

	E	A	B	C (applied first)
E	E	A	B	C
A	A	E	C	B
B	B	C	E	A
C	C	B	A	E

63. $E\begin{pmatrix} 1 & 0 \\ 0 & 1 \end{pmatrix}$ $A\begin{pmatrix} -1 & 0 \\ 0 & -1 \end{pmatrix}$

$B\begin{pmatrix} 1 & 0 \\ 0 & -1 \end{pmatrix}$ $C\begin{pmatrix} -1 & 0 \\ 0 & 1 \end{pmatrix}$

Chapter 19

Section 19.1

1. $\frac{1}{14}\begin{pmatrix} 1 & 3 \\ -4 & 2 \end{pmatrix}, \begin{pmatrix} 1 \\ -2 \end{pmatrix}$

2. $\frac{1}{2}\begin{pmatrix} 6 & -5 & 1 \\ -6 & 8 & -2 \\ 2 & -3 & 1 \end{pmatrix}, \begin{pmatrix} 1 \\ 2 \\ 3 \end{pmatrix}$

3. $\frac{1}{3}\begin{pmatrix} 1 & -2 & 1 & 1 \\ 1 & 1 & -2 & 1 \\ -2 & 1 & 1 & 1 \\ 1 & 1 & 1 & -2 \end{pmatrix}, \begin{pmatrix} -1 \\ 2 \\ 3 \\ 1 \end{pmatrix}$

Section 19.2

4. $1, c\begin{pmatrix} -2 \\ 1 \end{pmatrix}$; $4, c\begin{pmatrix} 1 \\ 1 \end{pmatrix}$

5. $2, c\begin{pmatrix} 1 \\ 0 \end{pmatrix}$; $-3, c\begin{pmatrix} 0 \\ 1 \end{pmatrix}$

6. $3-i, c\begin{pmatrix} 1 \\ -i \end{pmatrix}$; $3+i, c\begin{pmatrix} 1 \\ i \end{pmatrix}$

7. $2, c\begin{pmatrix} 1 \\ -1 \end{pmatrix}$; $4, c\begin{pmatrix} 1 \\ 1 \end{pmatrix}$

8. $-1, c\begin{pmatrix} 1 \\ -1 \\ 1 \end{pmatrix}$; $1, c\begin{pmatrix} 0 \\ 0 \\ 1 \end{pmatrix}$; $3, c\begin{pmatrix} 1 \\ 1 \\ 1 \end{pmatrix}$

9. $0, c\begin{pmatrix} 1 \\ 0 \\ -1 \end{pmatrix}$; $3\sqrt{2}, c\begin{pmatrix} 1 \\ \sqrt{2} \\ 1 \end{pmatrix}$; $-3\sqrt{2}, c\begin{pmatrix} 1 \\ -\sqrt{2} \\ 1 \end{pmatrix}$

10. -1 (double), $c\begin{pmatrix} 1 \\ -1 \end{pmatrix}$ (one vector only)

11. $0, c\begin{pmatrix} 1 \\ -2 \\ -2 \end{pmatrix}$; 1 (double), $c\begin{pmatrix} 1 \\ a \\ -3/2 \end{pmatrix}$ (all a)

Section 19.3

12. $\begin{pmatrix} 1 \\ 0 \end{pmatrix}, \begin{pmatrix} 0 \\ 1 \end{pmatrix}$ **13.** $\frac{1}{\sqrt{2}}\begin{pmatrix} 1 \\ -1 \end{pmatrix}, \frac{1}{\sqrt{2}}\begin{pmatrix} 1 \\ 1 \end{pmatrix}$

14. $\frac{1}{\sqrt{3}}\begin{pmatrix} 1 \\ -1 \\ 1 \end{pmatrix}, \begin{pmatrix} 0 \\ 0 \\ 1 \end{pmatrix}, \frac{1}{\sqrt{3}}\begin{pmatrix} 1 \\ 1 \\ 1 \end{pmatrix}$

15. $\frac{1}{\sqrt{2}}\begin{pmatrix} 1 \\ 0 \\ -1 \end{pmatrix}, \frac{1}{2}\begin{pmatrix} 1 \\ \sqrt{2} \\ 1 \end{pmatrix}, \frac{1}{2}\begin{pmatrix} 1 \\ -\sqrt{2} \\ 1 \end{pmatrix}$

19. (i) $\mathbf{x}_2' = \begin{pmatrix} 1 \\ -1 \\ 0 \end{pmatrix}, \mathbf{x}_3' = \begin{pmatrix} 1 \\ 0 \\ -1 \end{pmatrix}$

(ii) $\mathbf{x}_3'' = \begin{pmatrix} 1/2 \\ 1/2 \\ -1 \end{pmatrix}$

20. For $\phi = (\sqrt{5}+1)/2$, $c = 1/\sqrt{2(\phi+2)}$

(i) $E_1 = \alpha + \phi\beta$, $E_2 = \alpha + (\phi-1)\beta$,
$E_3 = \alpha - (\phi-1)\beta$, $E_1 = \alpha - \phi\beta$

(ii) $\mathbf{x}_1 = c\begin{pmatrix} 1 \\ \phi \\ \phi \\ 1 \end{pmatrix}$, $\mathbf{x}_2 = c\begin{pmatrix} \phi \\ 1 \\ -1 \\ -\phi \end{pmatrix}$,

$\mathbf{x}_3 = c\begin{pmatrix} \phi \\ -1 \\ -1 \\ \phi \end{pmatrix}$, $\mathbf{x}_4 = c\begin{pmatrix} 1 \\ -\phi \\ \phi \\ -1 \end{pmatrix}$

21. (ii) $\dfrac{1}{\sqrt{3}}\begin{pmatrix} 1 \\ 1 \\ 1 \end{pmatrix}$

(iii) e.g. $\dfrac{1}{\sqrt{2}}\begin{pmatrix} 1 \\ -1 \\ 0 \end{pmatrix}$, $\dfrac{1}{\sqrt{6}}\begin{pmatrix} 1 \\ 1 \\ -2 \end{pmatrix}$

22. (iii) right: $\lambda_1 = 2, \mathbf{x}_1 = \dfrac{1}{\sqrt{5}}\begin{pmatrix} -2 \\ 1 \end{pmatrix}$;

$\lambda_2 = 3, \mathbf{x}_2 = \begin{pmatrix} 1 \\ 0 \end{pmatrix}$

left: $\lambda_1 = 2, \mathbf{y}_1 = \begin{pmatrix} 0 \\ 1 \end{pmatrix}$;

$\lambda_2 = 3, \mathbf{y}_2 = \dfrac{1}{\sqrt{5}}\begin{pmatrix} 1 \\ 2 \end{pmatrix}$

Section 19.4

23. (i) $\begin{pmatrix} 1 & 0 & 1 \\ -1 & 0 & 1 \\ 1 & 1 & 1 \end{pmatrix}$,   (ii) $\begin{pmatrix} -1 & 0 & 0 \\ 0 & 1 & 0 \\ 0 & 0 & 3 \end{pmatrix}$

24. (i) $\begin{pmatrix} 1 & 1 & 1 \\ 0 & \sqrt{2} & -\sqrt{2} \\ -1 & 1 & 1 \end{pmatrix}$,

(ii) $\begin{pmatrix} 0 & 0 & 0 \\ 0 & 3\sqrt{2} & 0 \\ 0 & 0 & -3\sqrt{2} \end{pmatrix}$

25. (i) $\mathbf{X} = \dfrac{1}{\sqrt{2}}\begin{pmatrix} 1 & 1 \\ -1 & 1 \end{pmatrix}$, $\mathbf{X}^{-1} = \dfrac{1}{\sqrt{2}}\begin{pmatrix} 1 & -1 \\ 1 & 1 \end{pmatrix}$

(ii) $\mathbf{D} = \begin{pmatrix} 2 & 0 \\ 0 & 4 \end{pmatrix}$

26. (i) $\mathbf{X} = \begin{pmatrix} \frac{2}{\sqrt{5}} & \frac{1}{\sqrt{2}} \\ \frac{-1}{\sqrt{5}} & \frac{1}{\sqrt{2}} \end{pmatrix}$, $\mathbf{X}^{-1} = \dfrac{1}{3}\begin{pmatrix} \sqrt{5} & -\sqrt{5} \\ \sqrt{2} & 2\sqrt{2} \end{pmatrix}$

(ii) $\mathbf{D} = \begin{pmatrix} 1 & 0 \\ 0 & 4 \end{pmatrix}$

27. (i) $\mathbf{X} = \begin{pmatrix} 1 & 0 & 1 \\ -1 & 0 & 1 \\ 1 & 1 & 1 \end{pmatrix}$,

$\mathbf{X}^{-1} = \dfrac{1}{2}\begin{pmatrix} 1 & -1 & 0 \\ -2 & 0 & 2 \\ 1 & 1 & 0 \end{pmatrix}$

(ii) $\mathbf{D} = \begin{pmatrix} -1 & 0 & 0 \\ 0 & 1 & 0 \\ 0 & 0 & 3 \end{pmatrix}$

28. (i) $\mathbf{X} = \begin{pmatrix} \frac{1}{\sqrt{2}} & \frac{1}{2} & \frac{1}{2} \\ 0 & \frac{1}{\sqrt{2}} & \frac{-1}{\sqrt{2}} \\ \frac{-1}{\sqrt{2}} & \frac{1}{2} & \frac{1}{2} \end{pmatrix}$,

$\mathbf{X}^{-1} = \begin{pmatrix} \frac{1}{\sqrt{2}} & 0 & \frac{-1}{\sqrt{2}} \\ \frac{1}{2} & \frac{1}{\sqrt{2}} & \frac{1}{2} \\ \frac{1}{2} & \frac{-1}{\sqrt{2}} & \frac{1}{2} \end{pmatrix}$

(ii) $\mathbf{D} = \begin{pmatrix} 0 & 0 & 0 \\ 0 & 3\sqrt{2} & 0 \\ 0 & 0 & -3\sqrt{2} \end{pmatrix}$

Section 19.5

29. $\begin{pmatrix} x & y \end{pmatrix}\begin{pmatrix} 5 & -1 \\ -1 & -3 \end{pmatrix}\begin{pmatrix} x \\ y \end{pmatrix}$

30. $\begin{pmatrix} x & y \end{pmatrix}\begin{pmatrix} 0 & 2 \\ 2 & 0 \end{pmatrix}\begin{pmatrix} x \\ y \end{pmatrix}$

31. $\begin{pmatrix} x & y & z \end{pmatrix}\begin{pmatrix} 3 & -2 & 1 \\ -2 & 1 & -3 \\ 1 & -3 & -2 \end{pmatrix}\begin{pmatrix} x \\ y \\ z \end{pmatrix}$

32. $Q = 4y_1^2 + 16y_2^2$, $\begin{pmatrix} y_1 \\ y_2 \end{pmatrix} = \dfrac{1}{2}\begin{pmatrix} \sqrt{3}x_1 - x_2 \\ x_1 + \sqrt{3}x_2 \end{pmatrix}$

33. $Q = (a+b)y_1^2 + (a-b)y_2^2$,

$\begin{pmatrix} y_1 \\ y_2 \end{pmatrix} = \dfrac{1}{\sqrt{2}}\begin{pmatrix} x+y \\ x-y \end{pmatrix}$

34. $Q = 5y_1^2 + 3y_2^2 + 3y_3^2 + y_4^2$,

$$\begin{pmatrix} y_1 \\ y_2 \\ y_3 \\ y_4 \end{pmatrix} = \frac{1}{2} \begin{pmatrix} x_1 + x_2 + x_3 + x_4 \\ x_1 + x_2 - x_3 - x_4 \\ x_1 - x_2 - x_3 + x_4 \\ x_1 - x_2 + x_3 - x_4 \end{pmatrix}$$

Section 19.5

36. $\begin{pmatrix} 1-i & 2+i \\ 3-i & i \end{pmatrix}$, $\begin{pmatrix} 1-i & 3-i \\ 2+i & i \end{pmatrix}$

37. $\begin{pmatrix} 2 & -i \\ i & 1 \end{pmatrix}$, $\begin{pmatrix} 2 & i \\ -i & 1 \end{pmatrix}$

38. $\begin{pmatrix} 0 & i & 0 \\ -i & 0 & i \\ 0 & -i & 0 \end{pmatrix}$, $\begin{pmatrix} 0 & -i & 0 \\ i & 0 & -i \\ 0 & i & 0 \end{pmatrix}$

39. (i) 3   **(ii)** 13   **(iii)** $2 + 3i$
     (iv) $2 - 3i$

40. $\begin{pmatrix} 2 & i \\ -i & 1 \end{pmatrix}$ in Ex. 37, $\begin{pmatrix} 0 & -i & 0 \\ i & 0 & -i \\ 0 & i & 0 \end{pmatrix}$ in Ex. 38

Chapter 20

Section 20.2

1. (i) (a) 1.213   (b) 1.213
   (ii) (a) 72.030   (b) 72.03
   (iii) (a) 0.130   (b) 0.1299
   (iv) (a) 0.002   (b) 0.002499
2. (i) (a) 0.0005   (b) 0.0005
   (ii) (a) 0.0005   (b) 0.005
   (iii) (a) 0.0005   (b) 0.00005
   (iv) (a) 0.0005   (b) 0.0000005
3. (i) 6.289 ± 0.001 **(ii)** 1.617 ± 0.0015
   (iii) 42.87 ± 0.015 **(iv)** 10.11 ± 0.015
4. (i) (a) 0.0004   (b) 0.0004
   (ii) (a) 0.000007   (b) 0.00007
   (iii) (a) 0.004   (b) 0.00004
   (iv) (a) 0.3   (b) 0.0002
5. (i) 58.793 ± 0.243 **(ii)** 8.764 ± 0.036
   (iii) 0.3540 ± 0.0008
6. (i) (a) $x_1 = 60.0, x_2 = 0.015$
       (b) $x_1 = 59.9835, x_2 = 0.01665$
   (ii) (a) $x_1 = 60.0, x_2 = 0.01667$
       (b) $x_1 = 59.9835, x_2 = 0.0166713$

7. (i) 0.414213562; 4.9875621; 49.9987;
       500; 4900
   (ii) 0.414213562; 4.98756211; 49.9987501;
       499.999875; 4999.99999
8. (i) 0.175201194; 1.66585×10^{-5};
       5×10^{-8}; 0
   (ii) $\dfrac{x^2}{6}\left(1 + \dfrac{x^2}{20} + \dfrac{x^4}{840}\right)$
   (iii) $1.666675000 \times 10^{-5}$;
       $1.666666668 \times 10^{-9}$;
       $1.666666667 \times 10^{-13}$

Section 20.3

9. 1.564   **10.** 0.5314   **11.** 1.322
12. 1.5644623   **13.** 0.53139086
14. 1.3221854

Section 20.4

16. 0.03973; 0.25590; 0.47703; 0.72357
17. 0.04037 (from x_0, x_1, x_2);
     0.25673 (from x_0, x_1, x_2);
     0.72425 (from x_3, x_4, x_5)
18. 0.257081 (from x_0 to x_5)

Section 20.5

19. $n = 8$: $0.95834 - 0.00284 = 0.95550$;
     $n = 16$: $0.95622 - 0.00071 = 0.95551$
20. $n = 4$: 0.955559; $n = 8$: 0.9555146
21. (ii) 0.9555117
22. 0.94608
23. 0.88208
24. (i) 0.095166
25. 1.64493
26.

n	(i) $n \ln n - n$	(ii) 'accurate'	(iii) $\ln n!$
1	-1	0.0003	0
2	-0.6137	0.693151	0.693147
3	0.2958	1.7917597	1.7917595

Section 20.7

27. $x_1 = 2, x_2 = 1$
28. $x_1 = 1, x_2 = 2, x_3 = -2$
29. $x = -1, y = 8, z = -5$
30. $w = x = y = 1, z = -1$

31. (i) $\lambda = -28$

(ii) $x = -(28 + 5z)/3, y = (34 + 2z)/3$

Section 20.8

32. $\begin{pmatrix} -0.7 & 0.2 & 0.3 \\ -1.3 & -0.2 & 0.7 \\ 0.8 & 0.2 & -0.2 \end{pmatrix}$

33. $x = -0.7b_1 + 0.2b_2 + 0.3b_3$
$y = -1.3b_1 - 0.2b_2 + 0.7b_3$
$z = 0.8b_1 + 0.2b_2 - 0.2b_3$

Section 20.9

34. (i) 0.800000; 0.640000; 0.512000;
0.409600; 0.327680

(ii) 0.810000; 0.656100; 0.531441;
0.430467; 0.348678

(iii) 0.814506; 0.663420; 0.540360;
0.440127; 0.358486

35. (i) 0.892626   (ii) 0.027961

36. (i) 2.845387   (ii) −0.413504

37. (i) 2.191160   (ii) −0.036251

38. (i) 0.370740   (ii) 0.368541

39. (i) 0.367885   (ii) 0.367880

40. (i) 0.864660   (ii) −0.000004

41. (i) 3.258821   (ii) −0.00009

42. (i) 2.2274120   (ii) 0.0000007

Chapter 21

Section 21.2

1. (i)

result	0	1	2	3	4	5	6	7	8	9	10
frequency	0	0	3	4	11	13	8	6	2	2	1

2. (i)

class	36 −40	41 −45	46 −50	51 −55	56 −60	61 −65	66 −70	71 −75
frequency	2	2	9	8	14	8	11	6

3. 5.22, 5, 5

4. (i) 58.68, 60, 59.5

(ii) 58.67, (56−60), (56−60)

5. 3.172, 1.781

6. (i) 5.22, 30.42, 194.6

(ii) 1.781, 0.474

Section 21.4

7. (i) 0.6   (ii) 0.1   (iii) 0.7
(iv) 0.7   (v) 0.9

8. (i) 0.24   (ii) 0.14   (iii) 0.38
(iv) 0.82   (v) 0.62

9. 1/36, 2/36, 3/36, 4/36, 5/36, 6/36, 5/36,
4/36, 3/36, 2/36, 1/36

10. (i) 3/216   (ii) 21/216
(iii) 24/216   (iv) 10/216

11. (i) $q = e^{-\varepsilon_0/kt} + e^{-\varepsilon_1/kt} + e^{-\varepsilon_2/kt}$

(ii) (a) $P_0 = 1, P_1 = P_2 = 0$
(b) $P_0 = P_1 = P_2 = 1/3$

(ii) $P_{i,j,k} = e^{-(\varepsilon_i + \varepsilon_j + \varepsilon_k)/kT} / q^3$

Section 21.5

12. (i) 0.5   (ii) 0.31754

13. (i) 0.3487   (ii) 0.001488
(iii) 0.9298

14. (i) $10/36^2$ (ii) $16/36^2$
(iii) $9/36^2$ (iv) $35/36^2$

15. $456/36^3$

Section 21.6

16. ABCD, ABDC, ACDB, ACBD,
ADBC, ADCB, BACD, BADC,
BCDA, BCAD, BDAC, BDCA,
CABD, CADB, CBDA, CBAD,
CDAB, CDBA, DABC, DACB,
DBCA, DBAC, DCAB, DCBA

17. AB, AC, AD, AE, BA, BC, BD,
BE, CA, CB, CD, CE, DA, DB,
DC, DE, EA, EB, EC, ED

18. AB, AC, AD, AE, BC, BD, BE,
CD, CE, DE

19. AAABB, AABAB, AABBA,
ABAAB, ABABA, ABBAA,
BAAAB, BAABA, BABAA,
BBAAA

20. $\dfrac{8!}{4!3!1!} = 280$ **21.** 3^8

22. (i) (a) $4^3 = 64$ (b) $3! = 6$

(ii) (a) $\begin{pmatrix} 4 \\ 3 \end{pmatrix} = 4$ (b) 1

Section 21.7

23. (i) $3a^2 - 2a^3$ (iii) $\dfrac{1}{2}$, $\dfrac{1}{2\sqrt{5}}$

   (iv) $\dfrac{7}{5\sqrt{5}}$

24. $1 - e^{-2R}(1 + 2R + 2R^2)$

25. (i) $\dfrac{3}{2}$ (ii) $\dfrac{\sqrt{3}}{2}$ (iii) 1

Section 21.10

26. (i) $y = 1.257x + 2.727$
   (ii) $\sigma_m = 0.084$, $\sigma_c = 0.615$

27. (i) $\ln(k/\mathrm{dm}^3\,\mathrm{mol}^{-1}\,\mathrm{s}^{-1})$
      $= -1.076 \times 10^4 (\mathrm{K}/T) + 27.38$
   (ii) $E_a = 89.5\,\mathrm{kJ\,mol}^{-1}$,
      $A = 7.8 \times 10^{11}\,\mathrm{dm}^3\,\mathrm{mol}^{-1}\,\mathrm{s}^{-1}$
   (iii) $\sigma_m = 114.9$; $E_a = 89.5 \pm 1.0\,\mathrm{kJ\,mol}^{-1}$
      $\sigma_c = 0.355$;
      $5 \times 10^{11} < A/\mathrm{dm}^3\,\mathrm{mol}^{-1}\,\mathrm{s}^{-1} < 11 \times 10^{11}$

Index

A

Abel 50, 525
Abelian group 525, 527
abscissa 34
absolute error bound 559
absolute value of complex number 229
acceleration 21, 118, 454–5
addition
  of complex numbers 19, 226
  of matrices 505
  of real numbers 3
  of vectors 445, 449
adjoint matrix 515
adjoint (Hermitian conjugate) matrix 552
Airy equation 389
Alembert, *see* d'Alembert
algebra
  of complex numbers 226–8
  fundamental theorem of 47
  matrix 474, 505–16
  of real numbers 14–19
  vector 444, 445–7
algebraic
  division 51
  equation 6
  fraction 36, 50
  functions 31–61 (chapter), 50, 179
  number 6
Al-Kashi 9, 200
Al-Khwarizmi 9, 43
all space, integral over 302
allyl radical 487
alternating function 475, 494
alternating series 208
amplitude 71, 350
angle, unit of 65
Ångström 24
angular
  frequency 72, 350
  motion 119
  velocity 119, 464, 550
angular momentum 405, 465, 550
  conservation of 466
  operator 406
anticommutation 462
antisymmetric function 140, 494
antisymmetry of determinant 490
Archimedean number (π) 7, 11

Archimedes 11, 127
area under the curve 132
Argand 228
Argand diagram 229
argument of complex number 229
Aristotle 446
arithmetic 14
  expression 17
  fundamental theory of 7
  mean 598
  operation 14, 17
  operator 14
  progression 192
  rules of 14
  series 196
Arrhenius 1, 626
associated Laguerre functions 385, 407
associated Legendre functions 378, 404
associate matrix 552
associative law
  of addition 14
  of matrix multiplication 509
  of multiplication 14
asymptote 51
asymptotic expansion 577
atomic orbital 82, 298
atomic units 27
augmented matrix 584
auxiliary equation 340
auxiliary function 257
average density 150
average value 303
  of function 134
average velocity 126
Avogadro's constant 23
axis of symmetry 524

B

Babbage 119
Babylonians 11, 43
back substitution 582
Barrow 96, 127
base 9, 16, 80
base units 20
base vectors 452, 458, 463
Bell 377
Bernoullis 260
Bernoulli, Daniel 260, 392

Bernoulli, Jakob 202, 260, 595
Bernoulli, Johann 14, 202, 218, 260, 340, 392
Bernoulli distribution 605
Bernoulli numbers 577
Bessel 236, 385
Bessel equation 368, 374, 385, 400, 415
Bessel functions 385–9, 414, 415
    of first kind 386, 400
    of half-integer order 387–9
    spherical 389, 421
    zeros of 387, 400, 415
binomial
    coefficient 199, 605
    distribution 604–6
    expansion 198
    series 210
bisection method 562
bohr 24
Boltzmann constant 125, 189, 604
Boltzmann distribution 91, 604, 611
Boltzmann statistics 609
Bombelli 225
boson 496
boundary conditions 345, 352, 396, 400, 410
    periodic 244, 347, 356, 399, 403
boundary value problem 345, 352, 357, 396
bracketing method 563
Bragg 1, 90
Briggs 83
butadiene 556

C
calculus, fundamental theorem of 144
canonical form 548
    variable 548
capacitance 333
Cardano 50, 225, 595
cartesian coordinates 33, 269, 295
cartesian unit vectors 448, 452
Cauchy 206, 207, 236, 475, 500, 525, 532
Cauchy product 220
Cauchy's integral test 207
Cavalieri 127, 575
Cayley 499, 500, 509, 548
central limit theorem 616
centre
    of gravity 149
    of inversion 524
    of mass 148, 151, 451
chain rule 103, 105–8, 262
change 94
    infinitesimal 120, 122
    rate of 93

change of
    constant variable 267
    independent variables 268
    of period 430
    variables in integration 166, 285–289
change of coordinates
    cartesian to polar 269, 286
    cartesian to spherical polar 304
characteristic
    determinant 535
    value 535
    vector 535
characteristic equation
    of differential equation 340
    of matrix 535
chi-square fitting 622
circuits, electric 332
Clairaut 273, 284, 294
    le cadet 273
class frequency 597
Clausius–Clapeyron equation 124, 160
coefficients
    binomial 199, 605
    expansion 209, 420
    Fourier 426, 430
    multinomial 200, 607
    virial 223
cofactor 478, 480
colatitude 295
column vector (matrix) 500, 503
combination 608
combinations of probabilities 603
common denominator 5
common factor 7, 37
commutation
    of scalar product 457
    of vector addition 445
commutative law
    of addition 14
    for matrix multiplication 509
    of multiplication 14
commutator 510
commuting matrices 509
comparison test of convergence 205
complementary function 360
completeness 420
complex conjugate 227
    matrix 551
complex function 235
complex matrices 551–5
complex number 19, 225
    absolute value of 229
    argument (angle) of 229

graphical representation of
   228–35
   imaginary part of 225
   modulus of 229
   polar representation of 229
   real part of 225
complex numbers 225–46 (chapter)
complex plane 228
components of vector 448–53, 503
   of force 459
compound-angle identities 74
confocal elliptic coordinates 310
conjugate transpose matrix 552
consecutive transformations 519
conservation
   of angular momentum 466
   of energy 156
   of linear momentum 455
conservative force 155, 460, 468
conservative system 156
constant 3,13
   coefficients 337
   of integration 128
   variable, change of 267
constrained optimization 255
constructive interference 77, 430
continuity 97
continuous distribution 596, 613–7
   variable 13
   function 97
continuum of numbers 7
convergence in the mean 421
convergence of series 204–8
   Cauchy's integral test of 207
   comparison test of 205
   d'Alembert's ratio test of 206
   radius of 208
convergent integral 139
conversion factor 21, 86
coordinate axes 34
coordinates
   cartesian 33, 295
   confocal elliptic 310
   curvilinear 307–12
   cylindrical polar 309
   origin of 34
   orthogonal 308
   polar 77
   spherical polar 294–6, 300
coordinate transformation 500, 501
Copernicus 63
correlation coefficient 618
cosecant 64

cosh 88
cosine 63
cosine rule 73
cotangent 64
Cotes 573, 610
Coulomb force 401, 468
Coulomb's law 153
covariance 618
Cramer 483
Cramer's rule 474, 483, 533
cross product, see vector product
cubic
   equation 6
   factorization of 48
   spline 572
cumulative frequency graph 597
curl of vector field 470
curvilinear coordinates 307–12
curvilinear integral, see line integral
cusp 100
cyclic boundary condition 347
cyclobutadiene 541
cyclopropene 556
cylindrical polar coordinates 309

D
d'Alembert 206, 207, 392, 532, 595
d'Alembertian operator 313
d'Alembert's ratio test 206
Dalton 23
Debye equation 60
decimal
   fraction 9
   point 9
   places 10
   system 9
definite integral 127, 132–42
   in method of substitution 172
degree of polynomial 40
Delaunay 275
del Ferro 50
degeneracy 397
degenerate eigenvalues 536
degenerate states 397
de Moivre 234, 241, 288, 595, 610,
   616
de Moivre's formula 233, 239
density 150
   average 150
   function 151, 297
   at a point 298
   probability 298, 353, 613
dependent variable 3

derivative 96
  higher 113, 251
  second 113
  total 262
  see also partial derivative
Descartes 33, 225
descriptive statistics 595–601
destructive interference 430
determinant 475
  addition rule for 489
  antisymmetry of 490
  characteristic 535
  cofactors of 480
  derivative of 493
  elements of 476
  expansion of 478, 480
  of matrix 502
  of matrix product 512
  minors of 478
  multiplication by a constant of 489
  order of 476
  properties of 488–93
  reduction to triangular form of 482, 493
  secular 486, 535
  Slater 496
  transpose of 488, 504
determinants 474–498 (chapter), 532
diagonal
  element 502
  form 544
  matrix 502
diagonalization of matrix 543–51
Dieterici equation 223
differences, method of 201
differencing errors 560
differentiable 100
differential 119–122, 145
  area 145
  calculus 94
  coefficient 96, 113
  exact 272–4, 278
  mass 298
  operator 97, 129
  total 258–262
  volume 261
differential equations
  first-order 314–36 (chapter)
  homogeneous 337–59, 368
  inhomogeneous 359–65
  linear first-order 328–33
  numerical methods for 585–92
    Euler's method 585–9
    Runge–Kutta methods 589–90

second-order
  with constant coefficients 337–367 (chapter)
  special functions 368–90 (chapter)
  separable 318–28
  see also partial differential equations
differentiation 93–125 (chapter)
  by chain rule 103, 105–10
  of determinant 493
  of exponential function 102
  from first principles 100–2
  of hyperbolic functions 102
  of implicit function 110
  of inverse hyperbolic functions 109–10
  by inverse rule 103, 108
  of inverse trigonometric functions 108–9
  of integrals, parametric 184–6
  logarithmic 111
  of logarithmic function 102
  of multiple of a function 103
  parametric, of integrals 184–6
  of power series 220
  by product rule 103
  by quotient rule 103
  by rule 102–110
  scalar, of vector 453
  successive 113
  of sum of functions 103
  of trigonometric functions 102
  see also partial differentiation
diffusion equation 391, 413, 415, 443
dimensional analysis 22
dimensions 19
dipole, field of 424
dipole moment 424, 451, 461
directed line segment 444
direction field 586
Dirichlet 392, 421
Dirichlet conditions 421
discontinuity 97
  essential 98
  finite 98, 136
  infinite 98, 137
  removable 98
discontinuous derivative 136
discontinuous function, integration of 136
discrete distribution 595
discrete variable 13
discriminant 44
dispersion 599
displacement vector 445
dissipative force 156

distinguishable objects 607
distribution
  binomial 604–6
  Boltzmann 91, 604, 611
  continuous 596, 613–7
  expectation value of 602
  frequency 596
  Gaussian (normal) 82, 615–7
  kurtosis of 601
  Maxwell–Boltzmann 125, 189
  mean of 598, 602, 614
  mean absolute deviation of 601
  median of 598
  mode of 598
  multinomial 607
  normal, *see* Gaussian
  probability 602
  probability density 613
  range of 600
  skewness of 601
  standard deviation of 600, 603, 614
  statistics 598
  uniform 614
  variance of 603
distributive law 14
  of matrix multiplication 509
divergence of vector field 469
divergent integral 139
divided differences, method of 569
division
  of complex numbers 228
  long 10, 51
dot notation 119
dot product, *see* scalar product
double integrals 282, 283–9
double-valued function 38

E

echelon form 582
eigenfunction 355, 397, 411
eigenvalue 355, 411, 535
  degenerate 536
  equation 355, 535
  matrix, problem 532–57 (chapter)
  spectrum 411
eigenvectors 535
  left- 556
  normalization of 537–8
  orthogonal 538
  properties of 537–43
  right- 556
electric circuits 332
electron volt 25

electrostatic energy 157
electrostatic potential 157, 461, 468
  multipole expansion of 422–5
element
  symmetry 524
  volume 298, 300, 309
elimination
  Gauss 581–4
  Gauss–Jordan 584–5
elimination methods 494, 581–5
energy 22, 154, 351
  conservation of 156
  kinetic 47, 154, 351
  molar 25
  potential 47, 154, 155, 351, 460, 468
  quantization of 353
  relativistic 223
enthalpy 274
entropy 272, 280
equality
  of complex numbers 226
  of matrices 505
  of vectors 445, 449
equation(s)
  algebraic 6
  auxiliary 340
  characteristic 340, 535
  cubic 6, 48, 114
  eigenvalue 355, 535
  homogeneous 321, 485, 534
  inconsistent 56, 484
  indicial 372
  inhomogeneous 485, 532
  linear 41
  linearly dependent 57
  matrix eigenvalue 535
  normal 621
  numerical solution of 562–5
  parametric 263
  pivot 583
  polynomial 42
  quadratic 6, 43, 560
  reduced 360
  secular 257, 290, 486, 534
  separable 318
  simultaneous 55, 474
  stiff 591
  see also differential equations
error 558, 616
  bound 559
  differencing 560
  function 593
  random 560, 597

relative 559
rounding 558
truncation 558
essential discontinuity 98
ethene 116
Euclid 7, 36, 64, 195, 197
Euclidean space 471
Eudoxus 127
Euler 7, 80, 237, 256, 288, 294, 337, 340,
    363, 392, 444, 500, 568, 595
Euler formula 236–40, 358
    method 587
    number (e) 7, 12, 80
    reciprocity relation 273
Euler–Cauchy equation 372
Euler–MacLaurin formula 576–9
even function 140
events 602
    exclusive 603
    independent 604
exact differential 272–4, 278
exactness, test of 273
exclusion (Pauli) principle 496, 609
expansion
    asymptotic 575
    binomial 198
    coefficients 199, 209, 420
    of electrostatic potential 422–5
    of a gas 159
    isobaric 159
    isothermal 159
    Laplace 480
    in Legendre polynomials 416–9, 421–5
    multinomial 200
    multipole 424
    in orthogonal functions 416–21
expansions, orthogonal 416–43 (chapter)
expectation value 303, 602, 613
explicit function 39
exponent 16, 80
exponential function 17, 80–2
    decay 81
    differentiation of 101, 102
    growth 81
extent of reaction 323
extremum 255

F
factorial 8
factorization 7, 34
Faraday 275
Fermat 33, 96, 117, 127, 200, 595
Fermi–Dirac statistics 609

fermion 496
Ferrari 50
Fibonacci 9, 192
Fibonacci sequence (series) 193, 195
field 294, 466
    direction 586
    scalar 466
    vector 466
    see also function of position
figures, significant 12
finite series 191, 196–203
first-order differential equations 314–36
    (chapter)
    numerical methods for 585–92
first-order linear differential equations
    328–33
first-order reaction 324
fixed point 12, 558
floating point 12, 558
force 22, 152, 455, 459, 468
    components of 459
    conservative 155, 460, 468
    Coulomb 401, 468
    dissipative 156
    Lorentz 473
    moment of 149, 464
    work done by 459
force constant 46
forced oscillations 363
form
    canonical 548–51
    echelon 582
    quadratic 546–51
    triangular 493, 582
Fourier 392, 421, 426
    analysis 425–41
    coefficients 426, 430
    cosine series 435
    integral 433
    series 425–33
    sine series 431
    transform 433–41
        exponential form 436
        pair of 438
fraction(s)
    addition of 4
    algebraic 36, 50
    decimal 9
    division of 4, 6
    multiplication of 4, 5
    partial 52
frame of reference 34
frequency 21, 25, 71, 596

angular 72, 350
  class 597
  graph, cumulative 597
  histogram 597
  natural 363
  relative 601
  of vibration 72, 350, 363
frequency distribution 596
  continuous 596
  discrete 596
Frobenius 371, 545
  method of 371–5
function(s) 2, 31
  algebraic 31–61 (chapter), 50
  alternating 475, 494
  antisymmetric 140, 494
  associated Laguerre 385
  associated Legendre 378
  auxiliary 257
  average value of 134
  Bessel 385–9, 400, 421
  complementary 360
  complex 235
  continuous 97
  discontinuous 136
  double-valued 38
  even 140
  explicit 39
  exponential 17, 80–2
  graphical representation of 32, 248
  harmonic 270
  Hermite 382
  homogeneous 321
  hyperbolic 87–9
  implicit 39, 110
  inverse 37
  inverse hyperbolic 88
  inverse trigonometric 72–3
  linear 35, 41
  logarithmic 83–6, 212
  odd 140
  orthogonal 355
  orthogonal set of 419
  orthonormal 356
  periodic 70
  point 294
  of position 247, 294, 296–9, 466
  quadratic 35, 43
  rational 50
  rational trigonometric 183
  single-valued 37, 427
  symmetric 140
  transcendental 50, 62–92 (chapter)

  trigonometric 63–79, 212
  vector, of position 466
functions of several variables 247–93
    (chapter)
  differentiation of, *see* partial
    differentiation
  graphical representation of 248
  stationary points of 253–8
functions in three dimensions 294–313
    (chapter)
fundamental theorem
  of algebra 47
  of arithmetic 7
  of the calculus 144

G

Galileo 22, 127
Galois 525
Galton 623
Gauss 47, 49, 225, 228, 236, 385, 500, 525,
    579, 581, 616, 620
Gauss elimination 581–4
Gaussian
  (complex) plane 228
  distribution 82, 615–7
  quadratures 579–81
    Gauss–Chebyshev 580
    Gauss–Hermite 580
    Gauss–Laguerre 580
    Gauss–Legendre 580
Gauss–Jordan elimination 584–5
general solution
  of differential equation 315, 339, 340–4,
    360, 373, 525
  of partial differential equation 392–3
geometric
  progression 192
  series 197, 203
Gibbs 456, 471
Gibbs energy 274
Girard 225
golden section (ratio) 195, 556
gradient 93
gradient operator (grad) 467
gradient (vector) of scalar field 467
graph 33
graphical representation
  of complex numbers 228–35
  of first-order differential equation 586
  of functions 32, 248
Grassmann 444, 471
Gregory, James 127, 209, 213, 575
Gregory of St Vincent 127

group 526
   Abelian 525, 527
   matrix representation of 527
   multiplication table 526
   point 526
   symmetry 525
group theory 525–9
   axioms of 527

H

half-life 325
Hamilton 444, 449, 467, 471, 509
Hamiltonian operator 47, 355
harmonic function 270
harmonic oscillator 46, 71, 348–52
   in quantum mechanics 47, 383–4
harmonic sequence 193
harmonic series 204
harmonic wave 70, 71, 76
hartree 27
Heaviside 456
Helmholtz energy 274
Hermite 11, 12, 50
   equation 368, 381
   functions 382
   polynomials 381
Hermitian conjugate matrix 552
Hermitian matrix 553
Heron 446
Herschel 119
Hey 573
Hipparchus of Nicaea 63
histogram 597
homogeneous
   function 321
   differential equations 321, 328, 337–59
   linear equations 485, 534
Hooke's law 348
Hôpital, *see* l'Hôpital
Hückel (theory) 116, 290, 487, 541, 556
Huygens 96, 595
hydrogen atom 14, 30, 82, 298, 302, 303,
   401–9, 414
hyperbolic functions 87–9
   differentiation of 102
   expansion in series of 213
   inverse 88
hyperbolic substitution 170
hypotenuse 63

I

identity transformation 521
imaginary axis 229
implicit function 39

differentiation of 110
improper integral 137
improper rational function 51
inconsistent equations 56, 484
increment 94
indefinite integral 127–31
independent events 604
independent variables 3, 247
   change of 268
index rule (law) 16
indicial equation 372
   parameter 372
indistinguishable objects 609
inductance 333
inertia tensor 466, 550–1
infinite integral 138
infinitely small 122
infinitesimal change 120, 122
infinite series 191, 196, 203–21
infinity 15
inflection, point of 115
inhomogeneity 359
inhomogeneous differential equations 328,
   337, 359–65
inhomogeneous linear equations 485, 532
initial conditions 316, 344, 410, 432
initial value problem 318, 344
inner product 471
integer 3
integer variable 4
integral
   calculus 127, 142–7
   convergent 139
   definite 127, 132–42, 172
   divergent 139
   double 282, 283–9
   Fourier 433
   improper 137
   indefinite 127–31
   infinite 138
   as a length 146
   line (curvilinear) 275–81
   multiple 281
   operator 129
   over all space 302
   parametric differentiation of 184
   partial 281
   particular 360
   sign 128
   standard 128, 163, 627–30 (Appendix)
   triple (three-fold) 299
   volume 299–304
integral test of convergence 207
integrand 128

integrated rate equation 325
integrating factor 329
integration 126–162, 163–190 (chapters)
   by change of variable 166
   use of complex numbers 244
   constant of 128, 130
   of discontinuous function 136
   of even function 141
   by hyperbolic substitution 170
   limits of 133
   numerical 573–81
   of odd function 141
   by partial fractions, method of 179–84
   by parts 173–6
   path of 275
   of power series 220
   range of 133
   of rational functions 179–84
   by reduction formula 176–9
   by substitution, method of 165–72
   by use of trigonometric relations 163–5
   by trigonometric substitution 170
   variable of 128
internal energy 272
interpolation 566–73
   linear 567, 573
   by method of divided differences 569
   piecewise linear 567
   piecewise quadratic 569
   polynomial 566
   quadratic 568
   spline 571
inverse
   matrix 513, 584
   of matrix product 516
   rule 103, 108
   transformation 520
inverse function 37, 108
   hyperbolic 88, 109–10
   trigonometric 72, 108–9
inverse-square law 153
irrational number 6
irreversible process 159
isobaric expansion 159
isothermal expansion 159
iteration 562
   second-order 565
IUPAC 20

J
Jacobi 249, 288
Jacobian 288, 475
Jia Xian 564
Jones 7

Jordan, Camille 548
Jordan, Wilhelm 584

K
Kepler 127
kinetic energy 47, 154, 351
kinetics
   linear equations in 330–3
   separable equations in 322–7
Kirchhoff's law 332
Kohlrausch's law 59
Kronecker delta 356
kurtosis 601
Kutta 589

L
Lacroix 294
Lagrange 256, 284, 294, 444, 525, 568
   interpolation formula 568
   multipliers 256, 290, 532
Laguerre 384
   equation 368, 384
   equation, associated 368, 384, 407
   function, associated 385, 407
   polynomial 384
     associated 384
Lambert 11
Lamé 307
Langmuir isotherm 59, 122
Laplace 270, 288, 294, 595, 616
Laplace equation 269, 307, 391, 415
Laplace expansion of determinant 480
Laplacian operator 270, 304, 309, 401
law of large numbers 598
least squares 619–23
   linear (straight-line fit) 620–2
left-eigenvector 556
Legendre 249, 377, 620
   equation 368, 375–81
   equation, associated 368, 404
   functions, associated 378, 404
   polynomials 376
     expansion in 416–9, 421–5
Leibniz 14, 31, 96, 102, 120, 127, 129, 144, 202, 210, 225, 248, 314, 318, 328, 370, 474
Lennard-Jones potential 125
Leonardo of Pisa (Fibonacci) 9
Leonardo da Vinci 195
Levi ben Gerson 607
l'Hôpital 218, 474
l'Hôpital's rule 218
limit 98, 217
limit of sequence 193

Lindemann 11
linear
  combination of solutions 339
  equation 6
  function 35, 41
  momentum 455
  motion 118
  velocity 118
linear differential equations 328–33
  homogeneous 328, 337–59, 368
  inhomogeneous 328, 359–65
  integrating factor for 329
linear equations, system of 474, 483–8, 534
  solution by Gauss elimination 581
linear interpolation 567
  piecewise 567, 573
linearly dependent
  equations 57
  solutions 339
linearly independent solutions 339
linear least squares 620–2
linear transformations 516–24, 534
line (curvilinear) integrals 275–81
  dependence on path of 277
  independence of path of 278
Liouville 525
logarithm 83
logarithmic differentiation 111
logarithmic function 83–6
  combination properties of 84
  conversion factors for 86
  differentiation of 102, 111–2
  expansion in series 212–3
logarithmic plot 112
long division 10, 51
longitude 295
Lorentz force 473
Lorentzian 188, 439, 440

M
MacLaurin 207, 209, 474
Maclaurin series 208–14, 416
magnitude 19
  of vector 444
mass 20, 148
  centre of 148, 151, 451
  molar 24
  moment of 148
  reduced 24, 150
  relative 24
mathematical notation 7
mathematical truncation error 558
matrices 499–531, 532–57 (chapters)

  addition of 505
  complex 551–5
  commutator of 510
  equality of 505
  similar 545
matrix 499
  adjoint 515
  adjoint (Hermitian conjugate) 552
  associate 552
  augmented 584
  column 500, 503
  complex conjugate 551
  conjugate transpose 552
  determinant of 502
  diagonal 502
  diagonalization 543–51
  Hermitian 563
  Hermitian conjugate 552
  Hückel 541
  inverse 513–6, 584
  multiplication of by scalar 505
  nonsingular 514, 521
  null 506, 512
  orthogonal 521–4
  rectangular 499, 516
  representation of group 527
  row 503
  similarity transformation of 545
  singular 514
  spin 510
  square 500, 502
  symmetric 504
  trace of 503
  transformation 517
  transpose 504
  unit 502
  unitary 554
  zero 506, 512
matrix algebra 474, 505–16
matrix eigenvalue problem 532–57 (chapter)
matrix multiplication 507
  associative law of 509
  commutative law for 509
  distributive law of 509
matrix product
  determinant of 512
  inverse of 516
  trace of 512
  transpose of 513
maximum 114, 209
Maxwell 275, 456
Maxwell–Boltzmann distribution 125, 189
Maxwell equations 304

Maxwell relations 273, 274
mean absolute deviation 601
mean of distribution 598, 602, 614, 623
median of distribution 598
minimum 43, 114, 209
minor of determinant 478
minus-one rule 266
mistakes 558
mode of distribution 598
modulus 15
    of complex number 229
Moivre, see de Moivre
molar energy 25
molar mass 24
mole 23
moment
    of force 149, 464
    of inertia 24, 149, 151, 466
    of inertia tensor 466, 550–1
    of mass 148
momentum
    angular 405, 465, 550
    linear 455
Monge 294
Morse potential 349
Müller (Regiomontanus) 63
multinomial 200
    expansion 200
    coefficient 200, 607
    distribution 607
multiple eigenvalues 536
multiple integral 281
multiplication
    of complex numbers 19, 226
    of real numbers 4
    matrix 507
    of matrix by scalar 505
    of vector by scalar 446, 450
    of vector by vector 456
multipole expansion 424
multipole moment 424

N

nabla 467
Napier 9, 83
Napoleon 256, 270
natural frequency 363
natural logarithm 83
natural spline 572
negative area 135
Nernst 1
Newton 78, 96, 102, 119, 127, 144, 202, 210,
    234, 256, 314, 370, 444, 564, 573

Newton–Cotes quadrature 573
Newton–Raphson method 564
Newton's method of divided differences 569
Newton's first law 147, 466
Newton's second (force) law 71, 147, 317,
    348, 392, 455, 466
nodal lines 397
node 354, 412, 568
    stationary 77
nondegeneracy 397
nonsingular matrix 514, 521
norm 419, 471
normal (Gaussian) distribution 82, 615–8
normal equations 621
normalization 302, 354, 379
    of eigenvectors 537
normal modes 411
    superposition of 412
notation
    mathematical 7
    for partial derivatives 249, 252
null matrix 506, 512
null vector 444
number
    algebraic 6
    Archimedean (∂) 7, 11
    Bernoulli 577
    complex 19, 225
    continuum of 7
    Euler (e) 7, 12
    imaginary 19, 225
    irrational 6
    natural 3
    prime 7
    rational 4
    real 3, 7
    trancendental 6
number system, positional 9
numerals, Hindu-Arabic 9
numerical integration 573–81
numerical methods 558–94 (chapter)

O

odd function 140
Ohm's law 332
Oldenburg 210
operation, symmetry 524
operator
    angular momentum 406
    arithmetic 14
    d'Alembertian 313
    differential 97, 129
    gradient 467

Hamiltonian 47, 355
    integral 129
    Laplacian 270, 304, 309, 401
    rotation 239
optimization 255
    with constraints 255
orbital, atomic 82, 298
order
    of differential equation 314
    of magnitude 26
    of reaction 324
ordinary differential equation 314
ordinary logarithm 83
ordinate 24
Oresme 126, 204, 595
orthogonal
    curvilinear coordinates 307–12
    eigenvectors 538
    functions 355
        expansion in 416–21
    matrix 521–4
    set, complete 420
    transformation 521–4, 545
    vectors 458
orthogonal expansions 416–43 (chapter)
orthogonality 354
orthogonalization, Schmidt 540
orthonormal
    functions 356
    set 357
oscillation
    forced 363
    period of 350
oscillator, harmonic 46, 47, 348–52
outcome 598

P

Pacioli 195
parallelogram of forces 446
parallelogram law 446
parametric differentiation of integrals 184
parametric equations 263
parametric representation of curve 454
parity 140
partial derivative 249
    higher 251
    notation for 249, 252
partial differential equations 391–415
        (chapter)
partial differentiation 249–53
partial fraction 52
    method of 179–84
partial integral 281

partial pivoting 583
partial sum (series) 196
particle in
    a circular box 398–401
    a one-dimensional box 352–6
    a rectangular box 395–8
    a ring 356–9
    a spherical box 414
    a three-dimensional box 414
particular integral 360
particular solution of differential equation
        315, 344–8, 373
partition function 612
    rotational 188, 578
    vibrational 222
Pascal 127, 200, 595
Pascal triangle 200, 564, 605
path of integration 275
Pauli principle 496, 609
Pauli spin matrices 510
Peacock 119
period 70
    change of 430
    of oscillation 350
periodic boundary condition 244, 347, 356,
        399
periodic function 70
periodicity 427
    on a circle 240
    of complex numbers 240–4
    condition 243
    on a line 242
permutation 607
piecewise continuous function 421, 427
piecewise interpolation 567
Piero della Francesco 195
pH 85
pivot equation 583
pivoting
    scaled partial 583
    partial 583
    total 584
Planck formula 224
Planck's constant 25
plane of symmetry 524
point function, *see* function of position
point group 526
point of singularity 51
Poisson equation 391
polar axis 295
polar coordinates 77–9
    change from cartesian 78, 269
polar coordinates, spherical 294–6, 300

polynomial 40–50
  degree of 40
  equation 42
  Hermite 381
  interpolation 566
  Laguerre 384
  Legendre 376–8
  roots of 42
population 598
  mean 602
  standard deviation 603
  variance 603
position, function of 247, 294, 296–9, 466
position vector 445
positional number system 9
potential energy 47, 154, 155, 351, 460, 468
  electrostatic 157, 461, 468
  of a distribution of charges 422–5
potential theory 270, 304, 421
power series 208, 416
  operations with 219–21
  method for differential equations 369–71
precedence, rules of 17
pressure 22
pressure–volume work 157–9
prime number 7
principal axes 549, 550
principal diagonal 502
principal value 73
principle of least time 117
principle of superposition 339, 412
probabilities, combinations of 603
probability density 298, 353, 613
probability distribution 602
probability and statistics 595–626 (chapter)
product rule 103, 104
progression
  arithmetic 192
  geometric 192
proper rational function 51
Ptolemy 63
Pythagoras 64

Q
Qin Jiushao 564
quadrant 67
quadratic
  equation 6, 43, 560
  form 546–51
  function 35, 43
  interpolation 568, 575
  roots of 43

quadrature 127, 573, 579
quadrupole moment 424
quantum number 243, 353, 406, 511
quartic 35, 49, 116
quaternion 449
quotient rule 103, 105

R
radial coordinate 295
radial distribution function 614
radian 65
radius of convergence 208
random errors 560, 598, 616
random experiment 602, 616
Raphson 564
rate
  of change 93, 455
  of conversion 323
  of reaction 323
rate constant 82, 324
rate equation 324
rational
  function 50, 179
  integrands 179, 183
  number 4
  trigonometric function 183
ratio test of convergence 206
Rayleigh–Jeans law 224
reaction
  extent of 323
  first-order 324
  order of 324
  rate of 323
  rate of conversion of 323
  second-order 325
real
  axis 229
  number 3, 7
  variable 13
reciprocity relation 273
Recorde 1
recurrence relations (integration) 176
reduced mass 24, 150
reduced equation 360
reduction formulas (integration) 173–5
reduction to separable form 320
Regiomontanus 63
regression 623
relative error 559
relative frequency 601
relativistic energy 223
remainder term (Taylor) 215
removable discontinuity 98

representation, matrix 527
  totally symmetric (trivial) 528
resistance 332
resonance 364
reversible process 159
Rheticus 63
Rhind papyrus 11, 192
Richardson's extrapolation 593
Riemann 144, 278, 392
Riemann integral 144
right-eigenvector 556
rigid body 466
rigid rotor 243, 578
Roberval 69, 127
Robert of Chester 9
Robinson 122
roots
  of characteristic (secular) determinant
    486–8, 535
  of complex number 240
  of polynomial 42
rotational partition function 188, 578
rotation operator 239
rounding 13
rounding error 558
row vector (matrix) 503
Runge 589
Runge–Kutta methods 589–90

S

saddle point 254
sample 598, 623
  mean 623
  size 598
  space 598
  standard deviation 623
  statistics 623
  variance 623, 624
scalar 444
  differentiation of vector 453–6
  field 466
  multiplication of vector 446, 450
  (dot) product 456–61
  triple product 473
scaled partial pivoting 583
Schmidt orthogonalization 540
Schoenberg 571
Schrödinger equation 27, 298, 352, 391, 395,
    398, 401, 414
secant 64
second-order differential equations
  with constant coefficients 337–67
    (chapter)
  special functions 368–90 (chapter)

second-order iteration process 565
second-order partial differential equations
    391–415 (chapter)
  general solution of 392–3
  separation of variables 393–5
second-order reaction 325
secular determinant 486, 535
secular equations 257, 290, 486, 534
Seki Kowa 474
separable differential equations 318–28
  in kinetics 322–7
separable equations 318, 393
separable form, reduction to 320
separation constant 394
separation of variables 393, 402, 410
sequence(s) 191–5
  arithmetic (progression) 192
  geometric (progression) 192
  harmonic 193
  Fibonacci (series) 193, 195
  limit of 193
series 191–224 (chapter)
  alternating 208
  arithmetic 196
  binomial 198, 210
  finite 191, 196–203
  Fourier 416, 425–32
  geometric 197, 203
  harmonic 204
  infinite 191, 196, 203–21
  MacLaurin 208–14, 416
  power 208, 219, 369, 416
  sum of 196
  Taylor 213
series expansion
  of hyperbolic functions 213
  of logarithmic function 212–3
  of trigonometric functions 212–3
significant figures 12
similarity transformation 545–51
similar matrices 545
Simpson 575
Simpson's rule 575–6
simultaneous equations 55, 474
simultaneous transformations 518
sine 63
sine, inverse 72
sine rule 73
single-valued function 37, 427
singular matrix 514
singularity 51
sinh 88
SI units 2, 20
skewness 601